Synthesis of Heteroaromatic Compounds

Synthesis of Heteroaromatic Compounds

Editor

Joseph Sloop

MDPI • Basel • Beijing • Wuhan • Barcelona • Belgrade • Manchester • Tokyo • Cluj • Tianjin

Editor
Joseph Sloop
Chemistry
Georgia Gwinnett College
Lawrenceville
United States

Editorial Office
MDPI
St. Alban-Anlage 66
4052 Basel, Switzerland

This is a reprint of articles from the Special Issue published online in the open access journal *Molecules* (ISSN 1420-3049) (available at: www.mdpi.com/journal/molecules/special_issues/Synthesis_Heteroaromatic_Compounds).

For citation purposes, cite each article independently as indicated on the article page online and as indicated below:

LastName, A.A.; LastName, B.B.; LastName, C.C. Article Title. *Journal Name* **Year**, *Volume Number*, Page Range.

ISBN 978-3-0365-7563-6 (Hbk)
ISBN 978-3-0365-7562-9 (PDF)

© 2023 by the authors. Articles in this book are Open Access and distributed under the Creative Commons Attribution (CC BY) license, which allows users to download, copy and build upon published articles, as long as the author and publisher are properly credited, which ensures maximum dissemination and a wider impact of our publications.

The book as a whole is distributed by MDPI under the terms and conditions of the Creative Commons license CC BY-NC-ND.

Contents

About the Editor vii

Preface to "Synthesis of Heteroaromatic Compounds" ix

Joseph Sloop
Synthesis of Heteroaromatic Compounds
Reprinted from: *Molecules* **2023**, *28*, 3563, doi:10.3390/molecules28083563 1

Luis G. Ardón-Muñoz and Jeanne L. Bolliger
Synthesis of Benzo[4,5]thiazolo[2,3-c][1,2,4]triazole Derivatives via C-H Bond Functionalization of Disulfide Intermediates
Reprinted from: *Molecules* **2022**, *27*, 1464, doi:10.3390/molecules27051464 5

Pavel S. Gribanov, Anna N. Philippova, Maxim A. Topchiy, Lidiya I. Minaeva, Andrey F. Asachenko and Sergey N. Osipov
General Method of Synthesis of 5-(Het)arylamino-1,2,3-triazoles via Buchwald–Hartwig Reaction of 5-Amino- or 5-Halo-1,2,3-triazoles
Reprinted from: *Molecules* **2022**, *27*, 1999, doi:10.3390/molecules27061999 37

Tuyen N. Tran and Maged Henary
Synthesis and Applications of Nitrogen-Containing Heterocycles as Antiviral Agents
Reprinted from: *Molecules* **2022**, *27*, 2700, doi:10.3390/molecules27092700 53

Jing Wu, Bo Feng, Li-Xin Gao, Chun Zhang, Jia Li and Da-Jun Xiang et al.
Synthesis and Biochemical Evaluation of 8H-Indeno[1,2-d]thiazole Derivatives as Novel SARS-CoV-2 3CL Protease Inhibitors
Reprinted from: *Molecules* **2022**, *27*, 3359, doi:10.3390/molecules27103359 79

Anna Maj, Agnieszka Kudelko and Marcin Świątkowski
Synthesis and Luminescent Properties of s-Tetrazine Derivatives Conjugated with the 4H-1,2,4-Triazole Ring
Reprinted from: *Molecules* **2022**, *27*, 3642, doi:10.3390/molecules27113642 91

Gavin R. Hoffman and Allen M. Schoffstall
Syntheses and Applications of 1,2,3-Triazole-Fused Pyrazines and Pyridazines
Reprinted from: *Molecules* **2022**, *27*, 4681, doi:10.3390/molecules27154681 107

Koji Nakano, Ko Takase and Keiichi Noguchi
Furan-Containing Chiral Spiro-Fused Polycyclic Aromatic Compounds: Synthesis and Photophysical Properties
Reprinted from: *Molecules* **2022**, *27*, 5103, doi:10.3390/molecules27165103 135

Amr S. Abouzied, Jehan Y. Al-Humaidi, Abdulrahman S Bazaid, Husam Qanash, Naif K. Binsaleh and Abdulwahab Alamri et al.
Synthesis, Molecular Docking Study, and Cytotoxicity Evaluation of Some Novel 1,3,4-Thiadiazole as Well as 1,3-Thiazole Derivatives Bearing a Pyridine Moiety
Reprinted from: *Molecules* **2022**, *27*, 6368, doi:10.3390/molecules27196368 151

Victor V. Fedotov, Maria I. Valieva, Olga S. Taniya, Semen V. Aminov, Mikhail A. Kharitonov and Alexander S. Novikov et al.
4-(Aryl)-Benzo[4,5]imidazo[1,2-a]pyrimidine-3-Carbonitrile-Based Fluorophores: Povarov Reaction-Based Synthesis, Photophysical Studies, and DFT Calculations
Reprinted from: *Molecules* **2022**, *27*, 8029, doi:10.3390/molecules27228029 171

Marcin Łuczyński, Kornelia Kubiesa and Agnieszka Kudelko
Synthesis of 2,5-Dialkyl-1,3,4-oxadiazoles Bearing Carboxymethylamino Groups
Reprinted from: *Molecules* **2022**, *27*, 7687, doi:10.3390/molecules27227687 . . . . **195**

Hongfei Wu, Xingxing Lu, Jingbo Xu, Xiaoming Zhang, Zhinian Li and Xinling Yang et al.
Design, Synthesis and Fungicidal Activity of N-(thiophen-2-yl) Nicotinamide Derivatives
Reprinted from: *Molecules* **2022**, *27*, 8700, doi:10.3390/molecules27248700 . . . . **207**

Dmitrii L. Obydennov, Diana I. Nigamatova, Alexander S. Shirinkin, Oleg E. Melnikov, Vladislav V. Fedin and Sergey A. Usachev et al.
2-(2-(Dimethylamino)vinyl)-4H-pyran-4-ones as Novel and Convenient Building-Blocks for the Synthesis of Conjugated 4-Pyrone Derivatives
Reprinted from: *Molecules* **2022**, *27*, 8996, doi:10.3390/molecules27248996 . . . . **219**

Tariq Z. Abolibda, Maher Fathalla, Basant Farag, Magdi E. A. Zaki and Sobhi M. Gomha
Synthesis and Molecular Docking of Some Novel 3-Thiazolyl-Coumarins as Inhibitors of VEGFR-2 Kinase
Reprinted from: *Molecules* **2023**, *28*, 689, doi:10.3390/molecules28020689 . . . . **241**

Mariya A. Panova, Konstantin V. Shcherbakov, Ekaterina F. Zhilina, Yanina V. Burgart and Victor I. Saloutin
Synthesis of Mono- and Polyazole Hybrids Based on Polyfluoroflavones
Reprinted from: *Molecules* **2023**, *28*, 869, doi:10.3390/molecules28020869 . . . . **259**

Viktoria V. Viktorova, Elena V. Steparuk, Dmitrii L. Obydennov and Vyacheslav Y. Sosnovskikh
The Construction of Polycyclic Pyridones via Ring-Opening Transformations of 3-hydroxy-3,4-dihydropyrido[2,1-c][1,4]oxazine-1,8-diones
Reprinted from: *Molecules* **2023**, *28*, 1285, doi:10.3390/molecules28031285 . . . . **283**

Yuta Murai and Makoto Hashimoto
Heteroaromatic Diazirines Are Essential Building Blocks for Material and Medicinal Chemistry
Reprinted from: *Molecules* **2023**, *28*, 1408, doi:10.3390/molecules28031408 . . . . **297**

Szymon Rogalski and Cezary Pietraszuk
Application of Olefin Metathesis in the Synthesis of Carbo- and Heteroaromatic Compounds—Recent Advances
Reprinted from: *Molecules* **2023**, *28*, 1680, doi:10.3390/molecules28041680 . . . . **323**

Letizia Crocetti, Gabriella Guerrini, Fabrizio Melani, Claudia Vergelli and Maria Paola Giovannoni
4,5-Dihydro-5-Oxo-Pyrazolo[1,5-a]Thieno[2,3-c]Pyrimidine: A Novel Scaffold Containing Thiophene Ring. Chemical Reactivity and In Silico Studies to Predict the Profile to GABA$_A$ Receptor Subtype
Reprinted from: *Molecules* **2023**, *28*, 3054, doi:10.3390/molecules28073054 . . . . **353**

Andreas S. Kalogirou, Hans J. Oh and Christopher R. M. Asquith
The Synthesis and Biological Applications of the 1,2,3-Dithiazole Scaffold
Reprinted from: *Molecules* **2023**, *28*, 3193, doi:10.3390/molecules28073193 . . . . **373**

Ajay Mallia and Joseph Sloop
Advances in the Synthesis of Heteroaromatic Hybrid Chalcones
Reprinted from: *Molecules* **2023**, *28*, 3201, doi:10.3390/molecules28073201 . . . . **407**

About the Editor

Joseph Sloop

Joseph C. Sloop was born in 1961 in Chapel Hill, North Carolina. In 1983, he received his B.S. with Special Attainments in Chemistry from Davidson College and began service in the U.S. Army. In 1990, he obtained his M.S. at North Carolina State University (NCSU) under the direction of Prof. Carl Bumgardner on the preparation of fluorinated heterocycles. After teaching at the United States Military Academy (USMA) at West Point and continuing his service in the military, he completed doctoral training at NCSU with Prof. David Shultz in the field of magnetochemistry and returned to West Point to teach and conduct research with the USMA Photonics Research Center. In 2009, Dr. Sloop joined the faculty at Georgia Gwinnett College (GGC), where he presently serves as a Professor of Chemistry for the School of Science and Technology. He is the recipient of the Phi Kappa Phi Scholastic Achievement Award (2008), the GGC Award for Scholarship and Creative Activities (2012), a co-recipient of the Blackboard Catalyst Award (2012), and the Chronicle of Higher Education's Top 125 Tech Innovators on The Digital Campus (2013), and he was also named a Governor's Teaching Fellow and a GGC Center for Teaching Excellence Fellow in 2016. He has published more than forty papers, five book chapters, two organic chemistry laboratory manuals, and is the author of *Succeeding in Organic Chemistry: A Systematic Problem-Solving Approach to Structure, Function and Mechanism*. His research interests include selective fluorination methods of ketone and amine derivatives, solventless condensation reactions of indanones, green synthesis of fused-ring, fluorinated heterocyclic systems, as well as QSAR and computational studies of keto-enol tautomerism in diketones and their condensation reactions. Dr. Sloop holds memberships in the American Chemical Society, the International Society of Heterocyclic Chemistry, and the Georgia Academy of Science.

Preface to "Synthesis of Heteroaromatic Compounds"

Dear Colleagues,

The synthesis of heteroaromatic compounds, the examination of their properties, and widespread applications in all areas of science continues to be the subject of intense research. The functionalization of heteroaromatic compounds can productively and unpredictably modulate properties that interest medicinal chemists in drug discovery, theranostics specialists in biomedical imaging, as well as material scientists in terms of their specific optical properties and applications. Many of these topics are examined in this Special Issue.

The use of computational chemistry methods to not only determine optimized molecular structural information and properties but also to model potential biochemical interactions between heteroaromatic molecules and biomolecules has burgeoned over the last two decades. These methods complement and focus on drug discovery, assist in optimizing synthetic research efforts, and lead to more efficacious resource allocation. A number of the articles published in this Special Issue have incorporated computational and molecular modeling methods that show definitive relationships between new heteroaromatic molecules and the desired properties to be imported into the medicinal chemistry and materials science fields of study.

As larger chemical enterprise seeks to improve sustainability in the preparation of heteroaromatic molecules, the application of green chemistry principles to reduce chemical toxicity and minimize waste streams and energy expenditure in chemical synthesis operations has grown dramatically in the last two decades. Several of the articles and reviews in this Special Issue highlight the use of green chemistry methods in the synthesis of heteroaromatic species, often reducing reaction times, milder reaction conditions, and the use of benign reagents and solvents.

While heteroaromatic chemistry has had an enormous global impact on science and society, the exploration of new synthesis methods leading to highly functionalized heteroaromatic molecules continues to flourish. We are delighted that more than ninety international experts in this field have agreed to contribute their research achievements and provide critical reviews to this Special Issue.

Joseph Sloop
Editor

Editorial

Synthesis of Heteroaromatic Compounds

Joseph Sloop

Department of Chemistry, School of Science & Technology, Georgia Gwinnett College, 1000 University Center Lane, Lawrenceville, GA 30043, USA; jsloop@ggc.edu

The synthesis of heteroaromatic compounds has been the subject of intense investigation for well over a century. Studies of the properties exhibited by this broad class of organic molecules have led to countless applications in materials science, agrochemistry and the pharmaceutical industry. This Special Issue, entitled "Synthesis of Heteroaromatic Compounds", is an outstanding collection of fourteen original research papers and six review articles that discuss the advances made in both conventional and green preparatory methods, as well as the properties and applications of heteroaromatic molecules in industrial and medicinal chemistry areas of study [1–20].

In recent years, computational chemistry methods have been brought to bear to determine both the molecular properties and biomolecular interactions that heteroaromatic compounds exhibit. Molecular docking studies of heteroaromatic compounds have identified myriad potential uses in the treatment of a host of illnesses. A number of the articles found in our Special Issue also leverage this important theoretical tool to allow the reader a better understanding of the impact that new molecules bearing heteroaromatic components may have to both the chemical enterprise writ large and the pharmaceutical industry.

This Special Issue brings together contributions from a truly international and diverse array of ninety experts in the fields of heteroaromatic compound synthesis, QSAR studies, computational chemistry and molecular docking, as well as bioactivity studies. Here are some highlights of the original research work and reviews presented in this Special Issue.

Several articles in this Special Issue feature the synthesis of azole derivatives, including 1,2,4-triazoles; imidazoles; pyrazoles; 1,3,4-oxadiazoles; 1,2,3-triazoles; 1,3,4-thiadiazoles; 1,3-thiazoles; benzimidazoles and isoxazoles. Bollinger and Ardón-Muñoz report a new synthetic protocol to prepare a series of benzo[4,5]thiazolo[2,3-c][1,2,4]triazoles derivatives via the C-H bond functionalization of phenyl disulfide intermediates. This method features high functional group tolerance, short reaction times and overall good yields [1]. Kudelko's group prepared a library of 1,2,4,5-tetrazine-4H-1,2,4-triazole conjugates. Compounds of this type may have pharmaceutical uses and possess optoelectronic properties [5]. Saloutin's team functionalized 2-(polyfluorophenyl)-4H-chromen-4-ones via S_NAr conditions with 1,2,4-triazole; imidazole; pyrazole and 1,3,4-oxadiazole. Both monoazole and polyazole products were obtained in modest to good yields. Several products demonstrated luminescent properties with potential application as OLEDs; the fungistatic testing of selected products revealed antifungal activity [14]. Kudelko and coworkers prepared a series of novel carboxymethylamino-substituted 2,5-dialkyl-1,3,4-oxadiazoles via a multistep method employing commercially available acid chlorides, hydrazine hydrate and phosphorus oxychloride in yields as high as 91% [9]. Gribanov's team prepared new examples of 5-(het)arylamino-1,2,3-triazole derivatives in high yields via a palladium-complex-catalyzed Buchwald–Hartwig cross-coupling reaction of 5-amino- or 5-halo-1,2,3-triazoles with (het)aryl halides and amines [2]. Obydennov and coworkers synthesized a new class of conjugated pyrans based on the examination of 2-methyl-4-pyrones with DMF-DMA, including derivatives bearing isoxazole and benzimidazole ring systems. The compounds demonstrated valuable photophysical properties, such as large Stokes shifts and good quantum yield [12]. Sosnovskikh's group synthesized a series of 3-hydroxy-3,4-dihydropyrido[2,1-c][1,4]oxazine-1,8-diones,

Citation: Sloop, J. Synthesis of Heteroaromatic Compounds. *Molecules* **2023**, *28*, 3563. https://doi.org/10.3390/molecules28083563

Received: 16 April 2023
Accepted: 18 April 2023
Published: 19 April 2023

Copyright: © 2023 by the author. Licensee MDPI, Basel, Switzerland. This article is an open access article distributed under the terms and conditions of the Creative Commons Attribution (CC BY) license (https://creativecommons.org/licenses/by/4.0/).

which are novel building blocks for biologically important polycyclic pyridones via an oxazinone ring-opening transformation. The partial aromatization of the heterocycles formed polycyclic benzimidazole-fused pyridines [15]. Fedetov's team prepared a series of novel 4-(aryl)-benzo[4,5]imidazo[1,2-a]pyrimidine-3-carbonitriles via Povarov and oxidation reactions from benzimidazole-2-arylimines. The photophysical properties of the studied compounds included positive emission solvatochromism with large Stokes shifts [10]. Nakano and coworkers designed spiro[indeno[1,2-b][1]benzofuran-10,10'-indeno[1,2-b][1]benzothiophene] as well as the S,S-dioxide derivative and the pyrrole-containing analog. The furan-containing chiral spiro-fused PACs were found to be circularly polarized luminescent materials [7]. Guerrini's team prepared a series of 5-oxo-4,5-dihydropyrazolo[1,5-a]thieno[2,3-e]pyrimidine derivatives as potential GABA$_A$ modulators. These compounds were studied using the 'Proximity Frequency' model with molecular docking and dynamic simulation; all products were determined to have an agonist profile, highlighting the suitability of the nucleus to interact with the receptor protein [18].

Several papers within this collection not only feature the synthesis of new heteroaromatic compounds but also investigate the potential biological applications of those molecules. Xiang's group prepared a series of 8H-indeno[1,2-d]thiazole derivatives and evaluated their inhibitory activities against SARS-CoV-2 3CL^{pro} through a high-throughput screening of their compound collection. One compound was identified as a novel SARS-CoV-2 3CL^{pro} inhibitor and was subjected to molecular docking to predict the binding mode with SARS-CoV-2 3CL^{pro} [4]. Gomha's group synthesized 3-aryl-5-substituted 1,3,4-thiadiazoles, 3-phenyl-4-arylthiazoles and the 4-methyl-3-phenyl-5-substituted thiazoles from 1-(3-cyano-4,6-dimethyl-2-oxopyridin-1(2H)-yl)-3-phenylthiourea and hydrazonoyl halides, α-haloketones, 3-chloropentane-2,4-dione and ethyl 2-chloro-3-oxobutanoate. The new compounds showed anticancer activity against the cell line of human colon carcinoma (HTC-116) as well as hepatocellular carcinoma (HepG-2). Molecular docking studies of the thiadiazol confirmed a binding site with EGFR TK [8]. Another team under Gomha's direction designed and synthesized novel 3-thiazolhydrazinylcoumarins via the reaction of phenylazoacetylcoumarin with various hydrazonoyl halides and α-bromoketones. Molecular docking studies of the resulting 6-(phenyldiazenyl)-2H-chromen-2-one derivatives were assessed against VEGFR-2 and demonstrated comparable activities to that of Sorafenib (an approved medicine). The cytotoxicity of the most active thiazole derivatives was investigated for their efficacy against human breast cancer (MCF-7) cell line and normal cell line LLC-Mk2 using an MTT assay and Sorafenib as the reference drug. Several compounds were found to have higher anticancer activities than Sorafenib [13]. Yang's group prepared a series of new N-(thiophen-2-yl) nicotinamide derivatives via the nucleophilic acyl substitution of nicotinic acid chlorides and aminothiophenes. The in vivo bioassay results of all the compounds against cucumber downy mildew (CDM; *Pseudoperonospora cubensis*) indicated that several compounds exhibited fungicidal activities higher than both diflumetorim and flumorph fungicides [11].

This Special Issue also features reviews of topics that are of ongoing interest to the scientific community in the areas of heteroaromatic synthesis and applications. Henary and Tran reviewed the synthesis and applications of antiviral agents derived from various nitrogen-containing heteroaromatic moieties, such as indole, pyrrole, pyrimidine, pyrazole and quinoline, within the last decade. The synthesized scaffolds target HIV, HCV/HBV, VZV/HSV, SARS-CoV, COVID-19 and influenza [3]. Schoffstall and Hoffman examined the current state of synthesis methodologies used to prepare pyrazines and pyridazines fused to 1,2,3-triazoles. The review also details the use of these heterocycles in medicinal chemistry as c-Met inhibitors or GABA$_A$ modulators, in materials science as fluorescent probes, and as structural units of polymers [6]. Murai and Hashimoto give a comprehensive accounting of both conventional and green synthetic methodologies used to prepare (3-trifluoromethyl)diazirine-substituted heteroaromatics, including pyrimidines, pyridines, benzimidazoles, pyrazoles, benzoxazoles, benzothiophenes and indoles. The authors also highlight medicinal, polymer and materials science applications for these diazirine-substituted heteroaromatics [16]. Rogalski and Pietraszuk explore recent advances in the application of the olefin metathesis reaction, particularly the cyclization of dienes and enynes, in synthesis protocols leading to (hetero)aromatic compounds, including pyrroles, furans, indolizines, benzofurans, pyrimidiniums, indoles, pyridoindoles and carbazoles. Several examples of green preparations of these heteroaromatic compounds are described [17]. Asquith's group provided a thorough review of the synthesis methodologies used to prepare 1,2,3-dithiazoles, their reactivity with other substrates, and their medicinal uses as antifungals, herbicides, antibacterials, anticancer agents, antivirals and antifibrotics, and as melanin and Arabidopsis gibberellin 2-oxidase inhibitors [19]. In their review of heteroaromatic hybrid chalcones, Mallia and Sloop outline the recent advances made in the incorporation of heteroaromatic moieties in the synthesis of hybrid chalcones. Examples of environmentally responsible processes employed in the preparation of this important class of organic compound are also highlighted [20].

Conflicts of Interest: The authors declare no conflict of interest.

References

1. Ardón-Muñoz, L.G.; Bolliger, J.L. Synthesis of Benzo[4,5]thiazolo[2,3-c][1,2,4]triazole Derivatives via C-H Bond Functionalization of Disulfide Intermediates. *Molecules* **2022**, *27*, 1464. [CrossRef]
2. Gribanov, P.S.; Philippova, A.N.; Topchiy, M.A.; Minaeva, L.I.; Asachenko, A.F.; Osipov, S.N. General Method of Synthesis of 5-(Het)arylamino-1,2,3-triazoles via Buchwald–Hartwig Reaction of 5-Amino- or 5-Halo-1,2,3-triazoles. *Molecules* **2022**, *27*, 1999. [CrossRef] [PubMed]
3. Tran, T.N.; Henary, M. Synthesis and Applications of Nitrogen-Containing Heterocycles as Antiviral Agents. *Molecules* **2022**, *27*, 2700. [CrossRef] [PubMed]
4. Wu, J.; Feng, B.; Gao, L.-X.; Zhang, C.; Li, J.; Xiang, D.-J.; Zang, Y.; Wang, W.-L. Synthesis and Biochemical Evaluation of 8H-Indeno[1,2-d]thiazole Derivatives as Novel SARS-CoV-2 3CL Protease Inhibitors. *Molecules* **2022**, *27*, 3359. [CrossRef]
5. Maj, A.; Kudelko, A.; Świątkowski, M. Synthesis and Luminescent Properties of s-Tetrazine Derivatives Conjugated with the 4H-1,2,4-Triazole Ring. *Molecules* **2022**, *27*, 3642. [CrossRef]
6. Hoffman, G.R.; Schoffstall, A.M. Syntheses and Applications of 1,2,3-Triazole-Fused Pyrazines and Pyridazines. *Molecules* **2022**, *27*, 4681. [CrossRef]
7. Nakano, K.; Takase, K.; Noguchi, K. Furan-Containing Chiral Spiro-Fused Polycyclic Aromatic Compounds: Synthesis and Photophysical Properties. *Molecules* **2022**, *27*, 5103. [CrossRef]
8. Abouzied, A.S.; Al-Humaidi, J.Y.; Bazaid, A.S.; Qanash, H.; Binsaleh, N.K.; Alamri, A.; Ibrahim, S.M.; Gomha, S.M. Synthesis, Molecular Docking Study, and Cytotoxicity Evaluation of Some Novel 1,3,4-Thiadiazole as Well as 1,3-Thiazole Derivatives Bearing a Pyridine Moiety. *Molecules* **2022**, *27*, 6368. [CrossRef]
9. Łuczyński, M.; Kubiesa, K.; Kudelko, A. Synthesis of 2,5-Dialkyl-1,3,4-oxadiazoles Bearing Carboxymethylamino Groups. *Molecules* **2022**, *27*, 7687. [CrossRef]
10. Fedotov, V.V.; Valieva, M.I.; Taniya, O.S.; Aminov, S.V.; Kharitonov, M.A.; Novikov, A.S.; Kopchuk, D.S.; Slepukhin, P.A.; Zyryanov, G.V.; Ulomsky, E.N.; et al. 4-(Aryl)-Benzo[4,5]imidazo[1,2-a]pyrimidine-3-Carbonitrile-Based Fluorophores: Povarov Reaction-Based Synthesis, Photophysical Studies, and DFT Calculations. *Molecules* **2022**, *27*, 8029. [CrossRef]
11. Wu, H.; Lu, X.; Xu, J.; Zhang, X.; Li, Z.; Yang, X.; Ling, Y. Design, Synthesis and Fungicidal Activity of N-(thiophen-2-yl) Nicotinamide Derivatives. *Molecules* **2022**, *27*, 8700. [CrossRef]
12. Obydennov, D.L.; Nigamatova, D.I.; Shirinkin, A.S.; Melnikov, O.E.; Fedin, V.V.; Usachev, S.A.; Simbirtseva, A.E.; Kornev, M.Y.; Sosnovskikh, V.Y. 2-(2-(Dimethylamino)vinyl)-4H-pyran-4-ones as Novel and Convenient Building-Blocks for the Synthesis of Conjugated 4-Pyrone Derivatives. *Molecules* **2022**, *27*, 8996. [CrossRef]
13. Abolibda, T.Z.; Fathalla, M.; Farag, B.; Zaki, M.E.A.; Gomha, S.M. Synthesis and Molecular Docking of Some Novel 3-Thiazolyl-Coumarins as Inhibitors of VEGFR-2 Kinase. *Molecules* **2023**, *28*, 689. [CrossRef]
14. Panova, M.A.; Shcherbakov, K.V.; Zhilina, E.F.; Burgart, Y.V.; Saloutin, V.I. Synthesis of Mono-and Polyazole Hybrids Based on Polyfluoroflavones. *Molecules* **2023**, *28*, 869. [CrossRef]
15. Viktorova, V.V.; Steparuk, E.V.; Obydennov, D.L.; Sosnovskikh, V.Y. The Construction of Polycyclic Pyridones via Ring-Opening Transformations of 3-hydroxy-3,4-dihydropyrido[2,1-c][1,4]oxazine-1,8-diones. *Molecules* **2023**, *28*, 1285. [CrossRef]
16. Murai, Y.; Hashimoto, M. Heteroaromatic Diazirines Are Essential Building Blocks for Material and Medicinal Chemistry. *Molecules* **2023**, *28*, 1408. [CrossRef]
17. Rogalski, S.; Pietraszuk, C. Application of Olefin Metathesis in the Synthesis of Carbo-and Heteroaromatic Compounds—Recent Advances. *Molecules* **2023**, *28*, 1680. [CrossRef]

18. Crocetti, L.; Guerrini, G.; Melani, F.; Vergelli, C.; Giovannoni, M.P. 4,5-Dihydro-5-Oxo-Pyrazolo[1,5-a]Thieno[2,3-c]Pyrimidine: A Novel Scaffold Containing Thiophene Ring. Chemical Reactivity and In Silico Studies to Predict the Profile to $GABA_A$ Receptor Subtype. *Molecules* **2023**, *28*, 3054. [CrossRef]
19. Kalogirou, A.S.; Oh, H.J.; Asquith, C.R.M. The Synthesis and Biological Applications of the 1,2,3-Dithiazole Scaffold. *Molecules* **2023**, *28*, 3193. [CrossRef]
20. Mallia, A.; Sloop, J. Advances in the Synthesis of Heteroaromatic Hybrid Chalcones. *Molecules* **2023**, *28*, 3201. [CrossRef]

Disclaimer/Publisher's Note: The statements, opinions and data contained in all publications are solely those of the individual author(s) and contributor(s) and not of MDPI and/or the editor(s). MDPI and/or the editor(s) disclaim responsibility for any injury to people or property resulting from any ideas, methods, instructions or products referred to in the content.

Article

Synthesis of Benzo[4,5]thiazolo[2,3-c][1,2,4]triazole Derivatives via C-H Bond Functionalization of Disulfide Intermediates

Luis G. Ardón-Muñoz and Jeanne L. Bolliger *

Department of Chemistry, Oklahoma State University, 107 Physical Sciences, Stillwater, OK 74078-3071, USA; lardonm@okstate.edu
* Correspondence: jeanne.bolliger@okstate.edu

Abstract: Many nitrogen- and sulfur-containing heterocyclic compounds exhibit biological activity. Among these heterocycles are benzo[4,5]thiazolo[2,3-c][1,2,4]triazoles for which two main synthetic approaches exist. Here we report a new synthetic protocol that allows the preparation of these tricyclic compounds via the oxidation of a mercaptophenyl moiety to its corresponding disulfide. Subsequent C-H bond functionalization is thought to enable an intramolecular ring closure, thus forming the desired benzo[4,5]thiazolo[2,3-c][1,2,4]triazole. This method combines a high functional group tolerance with short reaction times and good to excellent yields.

Keywords: heteroaromatics; C-H bond functionalization; oxidative cyclization

1. Introduction

Sulfur- and nitrogen-containing heterocycles are present in many natural products [1,2], agrochemicals [3–5], commercially available drugs [6,7], and compounds with the potential to become active pharmaceutical ingredients [8–14]. As a result, there is continued interest in developing new methods for the synthesis of biologically active fused heterocycles incorporating the benzothiazole fragment [15–23]. Some of these fused heterocyclic scaffolds with proven biological activities are shown in Figure 1. While the ester-substituted tricyclic benzo[d]imidazo [2,1-b]thiazole, **A**, shows antitumor properties [24], the related phenol derivative, **B**, shows immunosuppressive activity [25]. The bicyclic benzothiazole, **C**, acts as an antibiotic [26], whereas **D** is an antitumor compound with the potential to be applied as a PET imaging agent [27,28]. Among the biologically active benzo[4,5]thiazolo[2,3-c][1,2,4]triazole derivatives, **E–H** are the commercially available fungicide tricyclazole (compound **E**), which is used to treat rice blast, and **F**, which also exhibits antifungal properties [29–32]. **G** displays anti-inflammatory activity [33], and **H** shows promising results as an anticonvulsant agent [34].

Even though benzo[4,5]thiazolo[2,3-c][1,2,4]triazoles are known to exhibit a broad range of biological activities, synthetic methods for obtaining this moiety remain limited and often lack functional group tolerance. The majority of these compounds are prepared by first forming the thiazole ring to obtain a benzothiazole derivative, followed by the construction of the triazole unit (Scheme 1A,B) [15,23,32–34]. Scheme 1A demonstrates this route using tricyclazole (**E**) as an example [35]. The thiourea, **I**, can be obtained by treatment of the appropriate aniline with potassium thiocyanate and is often not isolated before the subsequent oxidative cyclization to the 2-aminobenzothiazole, **J**. An exchange of the amino substituent with a hydrazine results in the 2-hydrazinylbenzothiazole, **K**, which, in the presence of a one-carbon electrophile, such as formic acid, gives the triazole ring of tricyclazole, **E** [35]. As depicted in Scheme 1B, 2-hydrazinylbenzothiazole, **K′**, can alternatively be prepared from 2-mercaptobenzothiazole, **L** [36]. The reaction of **K′** with formamide or formic acid leads to the unsubstituted product, **M**, while the reaction with carbon disulfide has been used to obtain the sulfur derivative, **M′** [37,38]. Substituted triazoles, such as **M″**, can either be obtained directly from **K′** and the appropriate acid chloride or in two steps by using an aldehyde, followed by the addition of an oxidant [32,37,39].

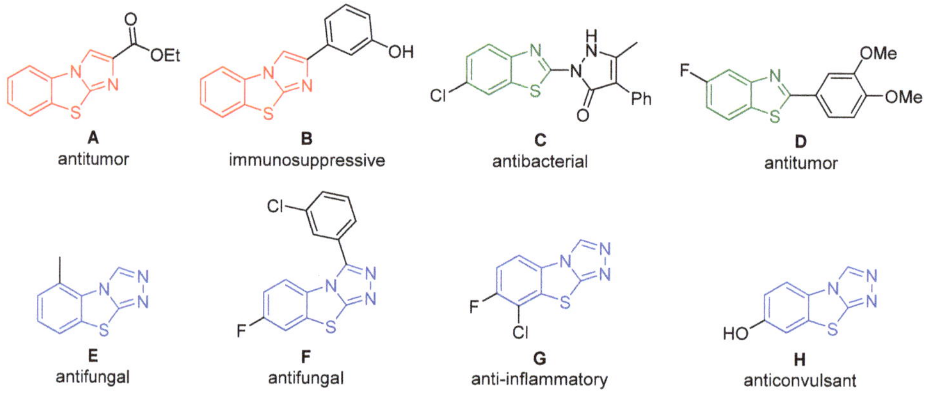

Figure 1. Examples of bioactive benzothiazoles. (**A**) antitumor [24]; (**B**) immunosuppressive [25]; (**C**) antibacterial [26]; (**D**) antitumor [27]; (**E**) antifungal [29–31]; (**F**) antifungal [32]; (**G**) anti–inflammatory [33]; (**H**) anticonvulsant [34].

An alternative approach for the synthesis of benzo[4,5]thiazolo[2,3-c][1,2,4]triazoles starts with a 3-mercaptotriazole derivative that is fused with the benzene ring, thus forming the thiazole ring in the last step (Scheme 1C,F). An example of this synthetic route is shown in Scheme 1C, where 1-chloro-2-isothiocyanatobenzene, **N**, is reacted with a hydrazide derivative to give compound **O** [40]. In the presence of a strong base, such as sodium hydride in boiling DMF, the tricyclic product, **Q**, can be obtained directly from **O**. Insight into the mechanism of this reaction was gained by the use of a weaker base, which allowed the isolation of the reaction intermediate, **P**, upon acidification. Subsequent treatment of this intermediate, **P**, with sodium hydride in DMF under reflux affords the tricyclic compound, **Q**. As shown in Scheme 1D, cyclization of this intermediate, **P′**, can also be achieved photochemically by subjecting it to 254 nm irradiation to yield compound **R** in a moderate yield [41]. Scheme 1E highlights a different route to the triazolo species (**U**) via two sequential oxidation steps. In the first step, a sodium salt (**S**) is converted to its disulfide (**T**), which, in the presence of bromine or iodine, undergoes an oxidative cyclization to the target molecule, **U** [42]. Unfortunately, a very narrow substrate scope, in combination with carbon tetrachloride being used as a solvent, restricts the application of this procedure significantly [43,44]. While the copper-catalyzed sequential diarylation of 2-mercaptotriazole, **W**, with 1-bromo-2-iodobenzene, **V**, in Scheme 1F was also shown to be possible, this reaction gave heterocycle, **M**, in only 14% yield [45].

A third approach leading to the benzo[4,5]thiazolo[2,3-c][1,2,4]triazole ring system is based on forming the thiazole ring in the last reaction step by a bond formation between the sulfur of the 2-mercaptophenyl substituent of a triazole and the unfunctionalized triazole carbon (Scheme 1G,H). The Straub group has observed two of their thiol-containing triazolium salts (**X**) undergo an oxidative cyclization in DMSO (Scheme 1G), thereby forming charged N-substituted heteroaromatic compounds, **Y** [46,47]. As opposed to the acidic triazolium starting materials of the Straub group, we demonstrate in this report that a similar bond formation can be employed to obtain neutral benzo[4,5]thiazolo[2,3-c][1,2,4]triazoles containing a wide variety of functional groups on the benzene ring from non-acidic triazoles (Scheme 1H). We were aiming to include both electron-donating and electron-withdrawing groups, halides, amines, alcohols, carboxylic acids, and other synthetically valuable substituents. As shown in Scheme 1H, we developed a two-step process for the conversion of a 4-(2-((4-methoxybenzyl)thio)phenyl)-4H-1,2,4-triazole species (**Z**) into the target tricyclic compound (**AB**). After the selective removal of the p-methoxybenzyl protecting group in the first step, the resulting free thiol (**AA**) is subsequently oxidized with DMSO to a disulfide intermediate (not shown in this scheme), which, upon deprotonation of the triazole carbon, undergoes an intramolecular ring closure. Herein, we present both the preparation of the starting triazoles (**Z**) from commercially available precursors, as well as their conversion to the benzo[4,5]thiazolo[2,3-c][1,2,4]triazole derivative, **AB**.

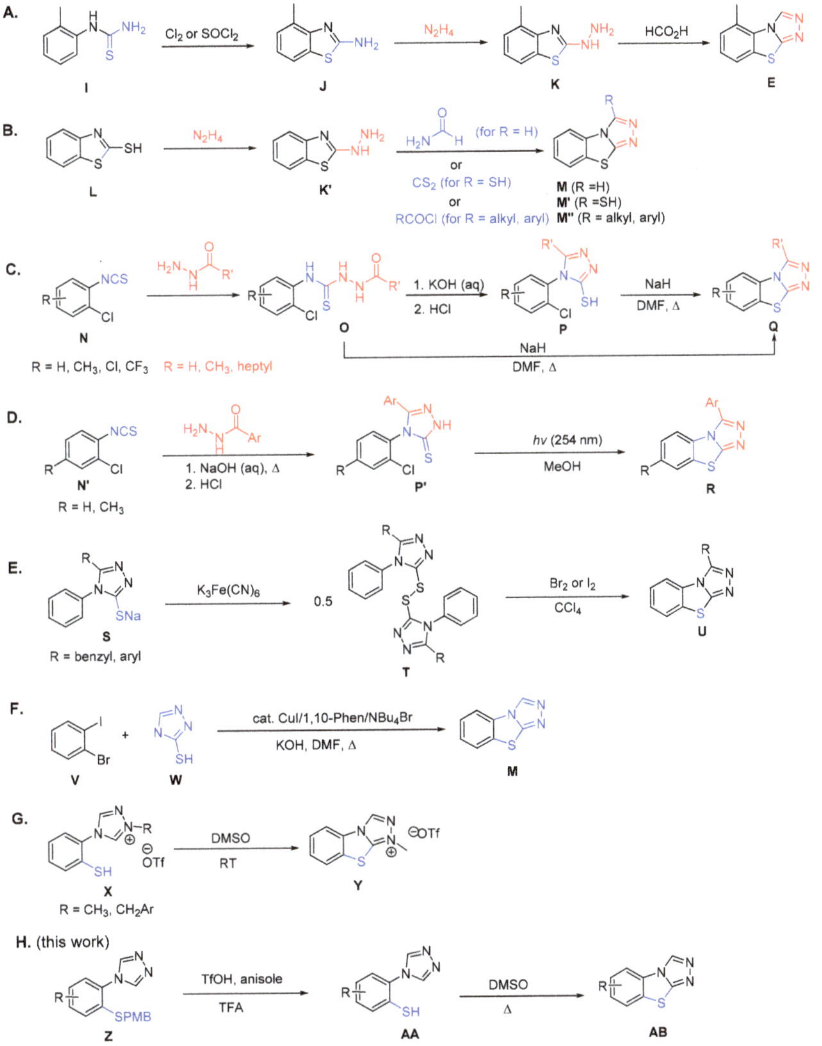

Scheme 1. Synthetic routes leading to benzo[4,5]thiazolo[2,3-c][1,2,4]triazole derivatives. (**A**) Synthesis of the commercially available fungicide tricyclazole via a 2-aminobenzothiazole intermediate [35]. (**B**) A related synthesis starting from a 2-mercaptobenzothiazole derivative [32,36–39]. (**C**) Synthesis of benzothiazolotriazole derivatives from a 1-chloro-2-isothiocyanatobenzene species [40]. (**D**) A photochemical approach by 254 nm irradiation [41]. (**E**) I Oxidative cyclization of sodium salts followed by halogenation [42]. (**F**) Copper catalyzed diarylation of 3-mercaptotriazole [45]. (**G**) Oxidation of benzothiazolotriazolium salts to N-substituted benzothiazolotriazolium derivatives [46,47]. (**H**) This work: Synthesis of benzothiazolotriazole derivatives via C-H bond functionalization.

2. Results and Discussion

2.1. Preparation of Triazole Precursors

Triazole, **3a**, was prepared according to the literature procedure shown in Scheme 2 [46,48]. While the *p*-methoxybenzyl-protected aniline, **2a**, could be obtained in this case by protecting 2-mercaptoaniline, **1a**, under argon atmosphere, the lack of commercial availability of substituted 2-mercaptoanilines required the development of a different synthesis route. An obvious method for introducing a sulfur substituent on an aromatic ring would be a classical nucleophilic aromatic substitution using a 2-halonitrobenzene derivative starting material [49–51]. Subsequent reduction of the nitro group would afford the corresponding aniline derivative.

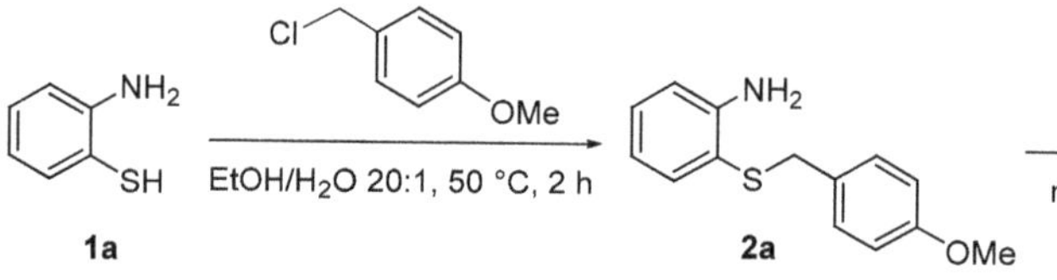

Scheme 2. Synthesis of triazole, **3a**.

This alternative route to the *p*-methoxybenzyl-protected 2-mercaptoanilines, **2b–2r**, starting from 1-fluoro-2-nitrobenzene derivatives, is shown in Scheme 3. By using (4-methoxyphenyl)methanethiol as a reagent in the nucleophilic aromatic substitution, we were able to obtain the *p*-methoxybenzyl-protected 2-mercaptonitrobenzene derivatives, **1b–1r**, in excellent yields (77–99%) in one step. An exception was **1h**, which was prepared from *tert*-butyl(4-fluoro-3-nitrophenoxy)dimethylsilane in 27% yield. The silyl-protecting group was required to prevent self-condensation of the starting material under basic conditions and is cleaved during the nucleophilic aromatic substitution due to the generation of fluoride, thus forming the free phenol **1h**. The ester species **1m′** was obtained quantitatively from the carboxylic acid **1m**. Reduction of the nitrobenzene derivatives **1b–1r** in the presence of excess iron powder and ammonium chloride gave the corresponding amines, **2b–2r**, in excellent yields (79–99%).

Compound **2s** was not available via a nucleophilic substitution followed by a reduction due to the reactivity of the second fluorine substituent. Instead, we prepared **2s** in two steps from 2,2′-disulfanediylbis(4-fluoroaniline) by first reducing it with NaBH$_4$ to free the thiol that was subsequently protected with the *p*-methoxybenzyl group (Scheme 4). The synthesis of 2,2′-disulfanediylbis(4-fluoroaniline) can be found in the experimental procedures (Section 3.6.1).

In analogy to **3a**, which was prepared in 69% yield following a literature procedure [46,49], triazoles **3b–3s** were obtained by heating the aniline derivatives **2b–2s** in the presence of *N,N*-dimethylformamide azine dihydrochloride at 150 °C for 16 h (Scheme 5). While the carboxylic acid **3m** can be formed using the procedure described above, we were unable to separate it from the salts and eventually synthesized it in 81% yield by hydrolysis of the ester **3m′**. The yields of the triazoles obtained using this method were extremely variable (10–72%), as can be seen in Scheme 4, and certainly could use improvement. The low yields depended on the exact structure of the aniline. In many cases, the starting amine could be isolated back. However, leaving the reaction for a longer duration did not increase the amount of product isolated, which suggests that the *N,N*-dimethylformamide azine dihydrochloride may degrade over time. In some cases, the starting amine was not detected at the end of a low-yielding reaction, and we believe that it may undergo decomposition due to the high reaction temperature. Loss of the *p*-methoxybenzyl protecting group was observed in all reactions, generally resulting in the formation of trace amounts of the target benzo[4,5]thiazolo[2,3-*c*][1,2,4]triazoles as byproducts of the triazole formation. However, this reaction was found to be the most dominant transformation in the preparation of

3d, which was obtained in only 17% yield. In addition, dehalogenation further lowered the yields of the bromo- and chloro-derivatives **3n–3q**, whereas deamination was the major cause for the low isolated yields of the pyridine derivatives **3i** and **3j**. While we are currently investigating higher yielding routes to triazoles **3a–3s**, we were able to isolate sufficient material to carry out a thorough investigation of the new oxidative cyclization reaction leading to the benzo[4,5]thiazolo[2,3-c][1,2,4]triazole derivatives.

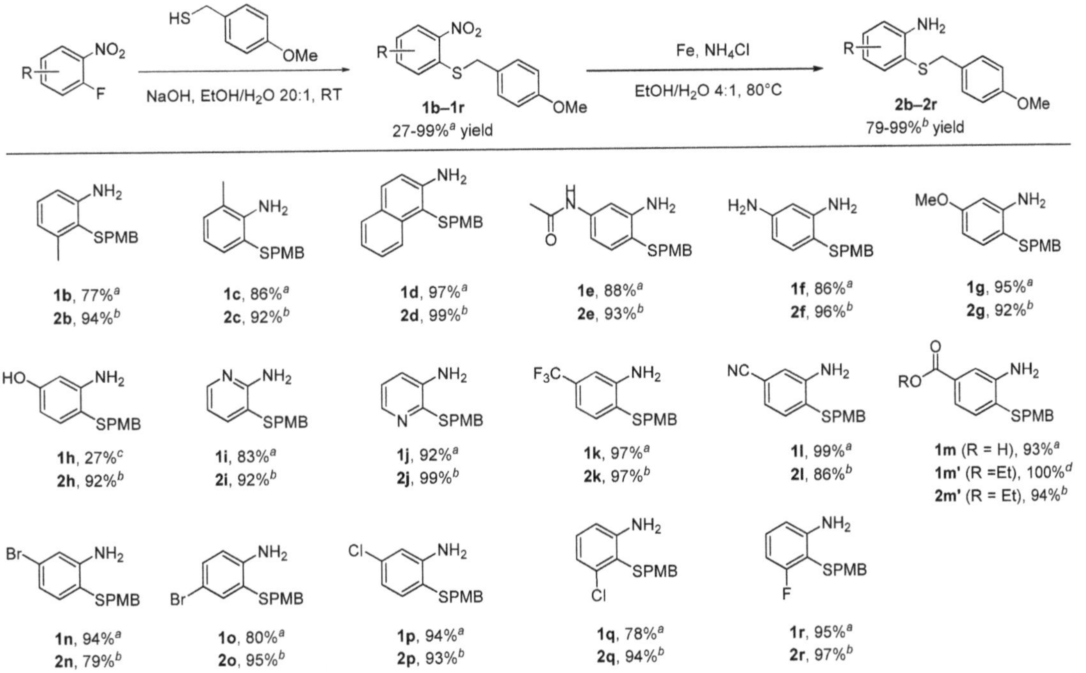

Scheme 3. Preparation of protected 2-mercaptoanilines, **2b–2r**. [a]Isolated yields of p-methoxybenzyl-protected 2-mercaptonitrobenzene derivatives, **1b–1r**. [b]Isolated yields of protected 2-mercaptoanilines, **2b–2r**. [c]Prepared from tert-butyl(4-fluoro-3-nitrophenoxy)dimethylsilane. [d]Prepared from **1m**.

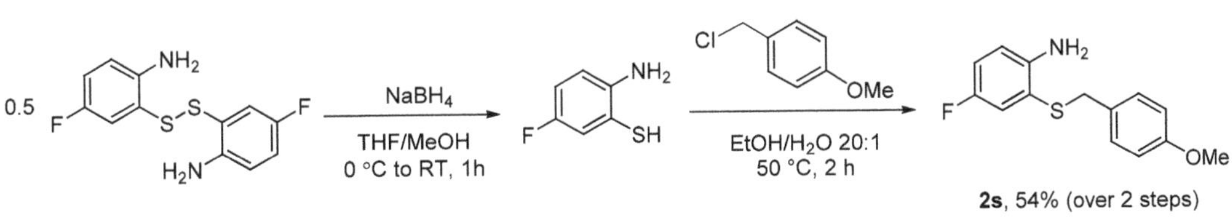

Scheme 4. Synthesis of **2s**.

2.2. Reaction Optimization

Complete deprotection of the thiol group of triazole **3a** was achieved within 1 h by treatment with triflic acid and anisole in TFA under an argon atmosphere at 0 °C, thus forming the free thiol **4a** (Scheme 6).

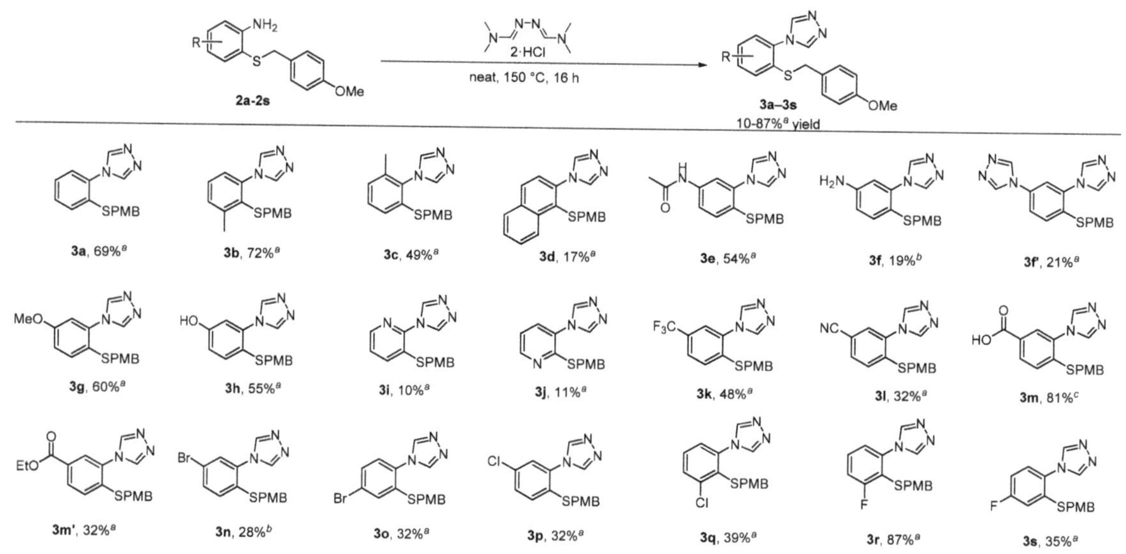

Scheme 5. Preparation of triazole precursors **3a–3s**. [a] All yields shown here are isolated yields. [b] Obtained as a by-product from the synthesis of **3f′**. [c] Synthesized from **3m′**.

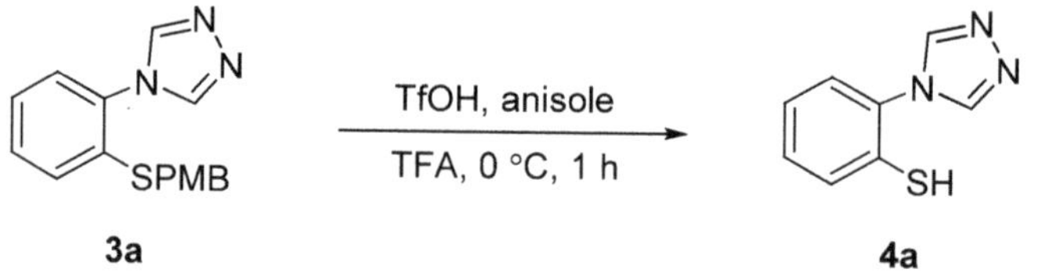

Scheme 6. Synthesis of test compound **4a**.

Compound **4a** served as our model substrate for optimization of the reaction conditions leading to our tricyclic heteroaromatic molecule, **6a** (Table 1). Based on previous studies in our group, we rationalized that an oxidative cyclization should be possible via a symmetrical disulfide intermediate, **5a**. Therefore, standard oxidation conditions for the formation of a disulfide bond were explored for this transformation, and the reaction was followed using LCMS.

The use of stoichiometric hydrogen peroxide in aqueous ethanol, a widely used oxidant for disulfide coupling, led to the disulfide **5a** at room temperature (Entry 1). Increasing the temperature to 80 °C under otherwise identical conditions resulted in complete conversion to the desired product, **6a**, within 16 h, which could be isolated in 78% yield (Entry 2). In an attempt to reduce the reaction time, we used an excess of hydrogen peroxide in ethanol; however, neither the reaction time nor the isolated yield showed any improvements (Entry 3). In addition, hydrogen peroxide was used in 10% aqueous NaOH. In this case, the desired product was not obtained at all since the reaction stopped at the disulfide stage (Entry 4). We decided to change our solvent to DMSO. Using DMSO as both the oxidant and solvent at room temperature, gave only the disulfide **5a**, even after 16 h (Entry 5), and no cyclized product was observed. After increasing the temperature to 100 °C, we observed full conversion to compound **6a** in 4 h and were able to isolate the heterocycle in 83% yield (Entry 6). In an attempt to reduce the reaction temperature, one equivalent of iodine was added to the DMSO solution, but this did not change the outcome of the reaction; at room temperature, the disulfide **5a** was obtained as the sole product (Entry 7), while, at 100 °C, the reaction time for complete conversion to **6a** was still 4 h, with a comparable isolated yield of 82% (Entry 8). With these results in hand, we decided that carrying out the reaction in DMSO at 100 °C (Entry 6) without an additional oxidant were the best reaction conditions to convert thiol **4a** to the tricyclic heteroaromatic species **6a**.

Table 1. Optimization of cyclization conditions [a].

Entry	Oxidant	Solvent	T (°C)	Time (h)	Outcome	Yield (%) [b]
1	1 equiv. H_2O_2	50% aq. Ethanol	RT	16	5a	0
2	1 equiv. H_2O_2	50% aq. Ethanol	80	16	6a	78
3	5 equiv. H_2O_2	Ethanol	80	16	6a	67
4	1 equiv. H_2O_2	10% aq. NaOH	RT	16	5a	0
5	DMSO	DMSO	RT	16	5a	0
6	DMSO	DMSO	100	4	6a	83
7	1 equiv. I_2	DMSO	RT	16	5a	0
8	1 equiv. I_2	DMSO	100	4	6a	82

[a] Reaction conditions: argon atmosphere, thiol **4a** (0.5 mmol), solvent (2 mL), oxidant, solvent, temperature, and time. [b] Isolated product **6a** after column chromatography.

2.3. Oxidative Cyclization Reaction leading to benzo[4,5]thiazolo[2,3-c][1,2,4]triazole Derivatives

We found that, by using the optimized conditions developed above, **3a** could be converted to the tricyclic heteroaromatic compound **6a** in one step without isolating the free thiol **4a**. However, we generally carried out a brief aqueous extraction after the deprotection in order to remove residual acid and characterized the free thiol by LCMS. As shown in Scheme 7, triazoles **3a–3s** were subjected to the deprotection conditions to give the corresponding free thiols **4a–4s**, which were used without purification in the next step after removal of the solvent under high vacuum, followed by an aqueous extraction. During the oxidative cyclization step carried out in DMSO, we observed the free thiols **4a–4s** to rapidly undergo oxidative disulfide coupling, leading to disulfides **5a–5s**. At elevated temperatures, the disulfide intermediates were found to undergo ring closure to give the desired compounds **6a–6s** in 100% conversions by LCMS within less than 4 h. Regardless of whether the substituents on the aryl ring were electron-donating or electron-withdrawing, the tricyclic heteroaromatic compounds **6a–6s** were obtained in good to excellent yields. For example, weakly electron-donating groups, such as methyl (compounds **6b** and **6c**) as well as the naphthalene derivative (**6d**) were obtained in 82–89% yields. Various nitrogen substituents were tolerated (compounds **6e**, **6f**, and **6f′**): While the acetyl-protected amine **6e** was isolated in good yield (82%), a lower yield was observed for the unprotected primary amine **6f** (60%). Compound **6f′**, with its triazole substituent, was, on the other hand, isolated in excellent yield (94%). Since complete conversions were observed by LCMS in all cases and no other byproducts were detected, the lower yield of **6f** is thought to be a result of its water solubility. Both **6e** and **6f′** are not only insoluble in water but also showed poor solubility in organic solvents, such as methylene chloride, methanol, and even DMSO. The two examples of oxygen-containing substituents at position 6 (**6g** and **6h**) also gave good yields (81% and 85%, respectively). Pyridine derivatives (**6i** and **6j**) were found to tolerate our reaction conditions; however, the products were obtained in lower yields (75% and 70% respectively). Strongly electron-withdrawing groups, such as a trifluoromethyl and a carbonitrile substituent in the 6-position, gave the tricyclic compounds **6k** and **6l** in good yields (86% and 80%, respectively). Similarly, the carboxylic acid **6m** and ester

6m′ derivatives were synthesized in 85% and 86% yields, respectively. Additionally, all halogen-containing derivatives 6n–6s were isolated in excellent yields, ranging from 88% to 98%. It is noteworthy that the reaction time for both compounds with halogens at position 7 (6o and 6s) was 30 min shorter than the time required for the cyclization of any of the other compounds under identical conditions.

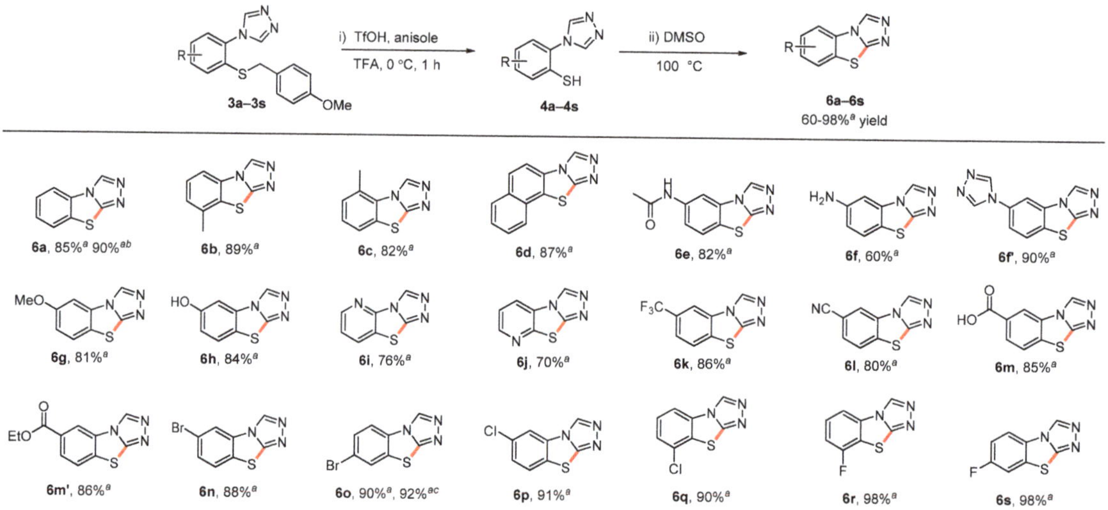

Scheme 7. Formation of benzo[4,5]thiazolo[2,3-c][1,2,4]triazole derivatives 6a–6s in two steps from 3a–3s. [a]Isolated yields for reactions carried out on a 0.25–1 mmol scale, [b]5 mmol scale, and [c]10 mmol scale.

While most of the reactions in Scheme 7 were carried out on a 0.5 mmol or 1 mmol scale, smaller-scale reactions of 0.25 mmol (3g, 3i, and 3j) still allowed the isolation of the products 6g, 6i, and 6j in acceptable yields. However, as demonstrated with 3a and 3o, increasing the reaction scale to 5 mmol and above generally afforded the products 6a and 6o in excellent yields of over 90%. On a 5 mmol scale, heterocycle 6a was obtained in 90% yield compared to 85% on a 1 mmol scale. Likewise, increasing the scale from 1 mmol to 10 mmol for the preparation of 6o led to a small increase in yield from 90% to 92%.

2.4. Basic Mechanistic Investigations

We were interested in following the conversion of thiol 4a to compound 6a by NMR to confirm the disulfide intermediate 5a detected previously by LCMS. Therefore, compound 4a was dissolved in DMSO-d_6 and an ^{1}H NMR was recorded immediately after the addition of the solvent. This NMR showed the presence of a singlet at 8.78 ppm (H_a), which was assigned to the triazole C-H of the free thiol 4a (Figure 2, 5 min RT). The NMR reaction tube was then heated to 100 °C, and a second ^{1}H NMR was measured after 1 h, which showed the presence of two new singlets at 9.64 (H_c) and 8.70 ppm (H_b). The singlet at 9.64 ppm corresponded to H_c of heterocycle 6a, while the singlet at 8.70 ppm was assigned to H_b of the disulfide intermediate 5a. LCMS of the NMR solution confirmed the presence of these two molecules, while no free thiol was detected. After 30 h, the disulfide 5a was consumed completely, and only the compound 6a was observed by both NMR and LCMS.

Interestingly, the reaction carried out in dried deuterated DMSO took longer to reach completion than in reagent grade DMSO, which suggested that water might play a role in the reaction mechanism. While the conversion of each thiol to the heterocycle is expected to generate one equivalent of dimethyl sulfide and one equivalent of water if carried out in DMSO, we were particularly interested to see how the rate of the cyclization step was affected by the amount of water present. To study this effect of water, the disulfide

5a was isolated and heated at 100 °C in DMSO containing known amounts of water (Figure 3). Samples taken at regular intervals were immediately analyzed by LCMS, which, after calibration, allowed the quantification of the species present. Indeed, as expected, we observed that increases in the water content corresponded to increases in the rate of conversion of the disulfide 5a to product 6a. A possible explanation for the rate-enhancing effect of water could be that it is involved in the C-H bond functionalization of the triazole C-H bond. Although it cannot be excluded that some residual acid is present from the deprotection of the thiol despite the aqueous extraction, we hypothesize that the role of water might be twofold. On the one hand, reversible protonation of a triazole nitrogen is expected to significantly increase the acidity of the C-H bond in the resulting triazolium species. However, water is also the strongest base present and is likely to be involved in the deprotonation of this triazolium species, thereby leading to a nucleophilic carbene intermediate, which could attack one of the sulfur atoms in the disulfide bond to afford our heterocyclic product 6a and a thiolate. This thiolate leaving group would immediately undergo oxidative disulfide coupling, thus generating the next active disulfide intermediate.

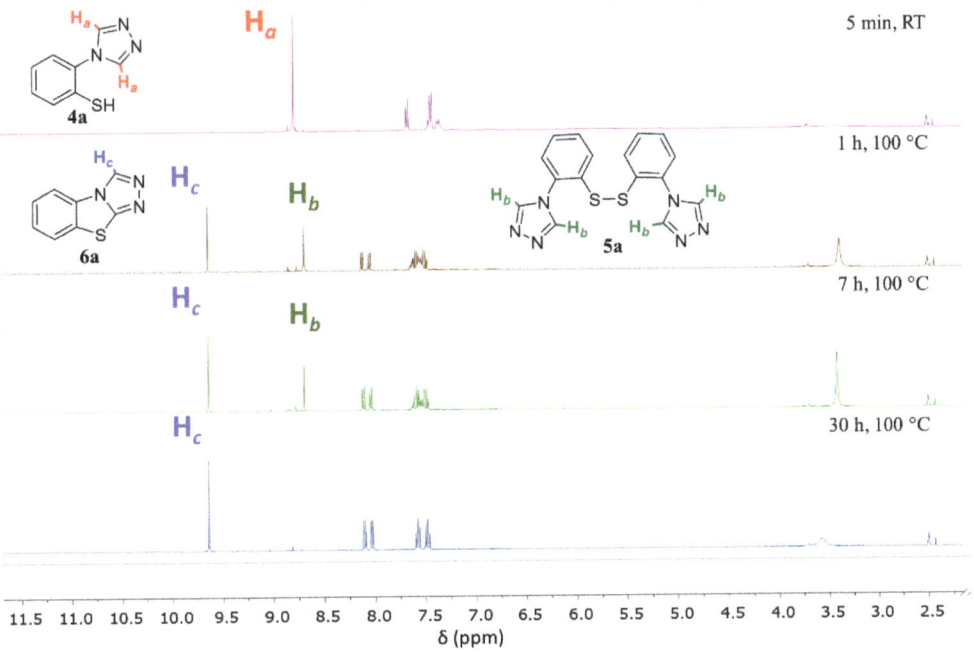

Figure 2. Conversion of thiol 4a in DMSO-d_6 to benzothiazolotriazole 6a via disulfide 5a.

Preliminary NMR studies into substituent effects on the reaction rate show that a very electron-withdrawing group, such as trifluoromethyl, in the *para*-position to the disulfide bond significantly increases the rate of conversion of this disulfide (5k) to the tricyclic heteroaromatic compound 6k (Figure 4). This effect is in agreement with the aforementioned mechanism, as the sulfur atom in the disulfide would become significantly more electrophilic while simultaneously the hydrogen of the C-H bond of the triazole would increase in acidity. Meanwhile, disulfide 5g (containing the electron-donating methoxy group) displayed a similar initial reaction rate as the unsubstituted disulfide 5a. Full conversion to both the unsubstituted heterocycle 6a and the methoxy derivative 6g was only observed after 30 h.

Although our hypothetical mechanism provides both a potential explanation for the role of water and some of the substituent effects observed, we cannot exclude other mechanisms for the formation of the heterocycle in the absence of a more rigorous kinetic investigation.

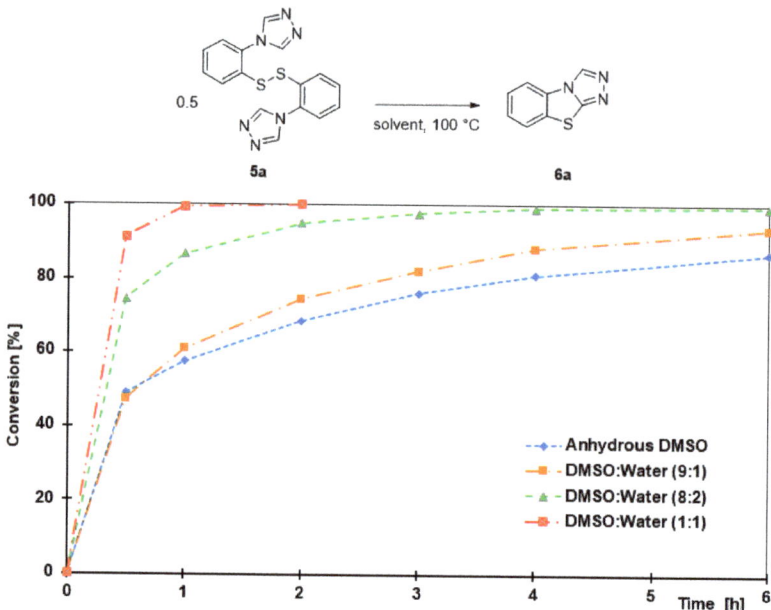

Figure 3. Effect of water on the conversion of disulfide **5a** to heterocycle **6a**.

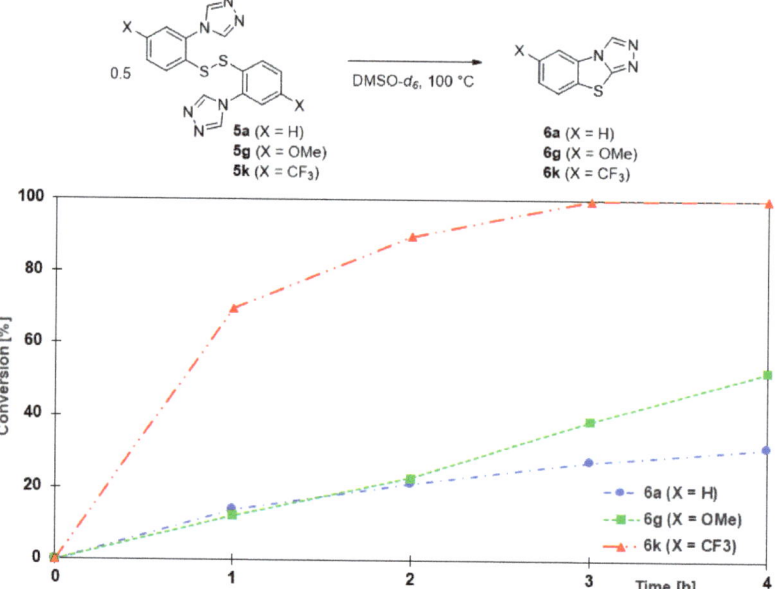

Figure 4. Effect of electron-donating and electron-withdrawing groups on the cyclization step.

3. Materials and Methods

3.1. General Information

Most reagents and solvents were purchased from Fisher Scientific (Waltham, MA, USA), Oakwood Chemical (Estill, SC, USA), TCI America (Portland, OR, USA), and Avantor (Radnor, PA, USA) and were used as supplied unless otherwise noted. Thermo Scientific™ silica gel (for column chromatography, 0.035–0.070 mm, 60Å) from Fisher Scientific (Waltham, MA, USA) was used for chromatographic separations. DMSO-d_6 was dried over molecular sieves.

3.2. Analyses

^{1}H NMR, ^{13}C{^{1}H} NMR spectra, and ^{19}F{^{1}H} NMR spectra were all recorded using a 400 MHz Bruker Avance III spectrometer with a 5 mm liquid-state Smart Probe. Chemical shifts (δ_H and δ_C) are expressed in parts per million (ppm) and reported relative to the resonance of the residual protons of the DMSO-d_6 (δ_H = 2.50 ppm), CD$_3$CN (δ_H = 1.94 ppm), or CDCl$_3$ (δ_H = 7.26 ppm) or in ^{13}C{^{1}H} NMR spectra relative to the resonance of the deuterated solvent DMSO-d_6 (δ_C = 39.52 ppm), CD$_3$CN (δ_C = 1.32 ppm), or CDCl$_3$ (δ_C = 77.16 ppm). Chemical shifts in the ^{19}F{^{1}H} NMR spectra are reported relative to the internal standard fluorobenzene (δ_F = −113.15). The coupling constants (*J*) are given in Hz. All measurements were carried out at 298 K. The abbreviations used in the description of the NMR data are as follows: s, singlet; d, duplet; t, triplet; m, multiplet; and sept., septet. Copies of the ^{1}H NMR and ^{13}C NMR spectra for compounds 1b–1r, 2b–2s, 3b–3s, and 6a–6s are provided in the Supporting Information. High-resolution mass spectrometry (HRMS) data were obtained on an LTQ Orbitrap Fusion in FT orbitrap mode at a resolution of 240,000.

3.3. Preparation of 2-Fluoronitrobenzene Starting Materials for 1e and 1h

3.3.1. N-(4-Fluoro-3-nitrophenyl)acetamide

A 500 mL round-bottomed flask was charged with 4-fluoro-3-nitroaniline (30 mmol, 1 equiv.), acetyl chloride (1.1 equiv.), and THF (150 mL) and stirred overnight at room temperature. The solvent was removed under reduced pressure. Water was added to the residue and the crude product was extracted with ethyl acetate, dried over MgSO$_4$, filtered, and evaporated. Purification of the crude product by column chromatography (silica gel, ethyl acetate/hexanes, 2:1, R_f = 0.50) afforded the product as a beige solid in 80% (4.753 g, 23.98 mmol) yield; m.p. 140–141 °C. ^{1}H NMR (400 MHz, DMSO-d_6, 298 K): δ = 10.37 (s, 1H), 8.49 (dd, *J* = 6.7 Hz, *J* = 2.6 Hz, 1H), 7.84–7.80 (m, 1H), 7.51 (dd, *J* = 11.1 Hz, *J* = 9.2 Hz, 1H), and 2.07 (s, 3H); ^{13}C{^{1}H} NMR (100 MHz, DMSO-d_6, 298 K): δ = 168.9, 150.1 (d, J_{C-F} = 256.2 Hz), 136.5 (d, J_{C-F} = 8.0 Hz), 136.0 (d, J_{C-F} = 3.2 Hz), 126.1 (d, J_{C-F} = 8.1 Hz), 118.7 (d, J_{C-F} = 21.8 Hz), 115.2, and 23.9; ^{19}F{^{1}H} NMR (376 MHz, DMSO-d_6, 298 K, referenced to C$_6$H$_5$F): δ = −125.81.

3.3.2. Tert-Butyl(4-fluoro-3-Nitrophenoxy)dimethylsilane

A 500 mL round-bottomed flask was charged with 4-fluoro-3-nitrophenol (30 mmol, 1 equiv.), *tert*-butylchlorodimethylsilane (1.2 equiv.), imidazole (3 equiv.), and THF (60 mL) and stirred at room temperature for 2 h. The solvent was removed under reduced pressure. Water was added to the residue and the crude product was extracted with diethyl ether, dried over MgSO$_4$, filtered and evaporated. Purification by column chromatography (silica gel, diethyl ether/hexanes 4:1) afforded the product as a red oil in 92% (8.718 g, 28.55 mmol) yield. ^{1}H NMR (400 MHz, CDCl$_3$, 298 K): δ = 7.45 (dd, *J* = 6.1 Hz, *J* = 3.0 Hz, 1H), 7.16–7.05 (m, 2H), 0.97 (s, 9H), and 0.21 (s, 6H); ^{13}C{^{1}H} NMR (100 MHz, CDCl$_3$, 298 K): δ = 151.8 (d, J_{C-F} = 3.0 Hz), 150.4 (d, J_{C-F} = 256.7 Hz), 137.3 (d, J_{C-F} = 7.8 Hz), 127.2 (d, J_{C-F} = 7.6 Hz), 118.9 (d, J_{C-F} = 22.2 Hz), 116.6 (d, J_{C-F} = 5.5 Hz), 25.6, 18.2, and −4.5; ^{19}F{^{1}H} NMR (376 MHz, CDCl$_3$, 298 K, referenced to C$_6$H$_5$F): δ = −127.55.

3.4. General Procedure 1 for the Synthesis of (4-Methoxybenzyl)(2-Nitrophenyl)sulfanes (1b–1r)

The following description is for a 30 mmol scale reaction. The solvent quantities and flask size were adjusted accordingly for smaller-scale reactions.

A 500 mL round-bottomed flask equipped with a stir bar was loaded with the 1-fluoro-2-nitrobenzene derivative (1 equiv.) and 200 mL of ethanol and placed under an atmosphere of argon. (4-methoxyphenyl)methanethiol (1 equiv.) was added with a syringe, followed by a dropwise addition of NaOH (1 equiv.) dissolved in 10 mL of H_2O. The reaction mixture was stirred at room temperature until TLC indicated the completion of the reaction (typically within 2 h). After removing the solvent under reduced pressure, the residue was diluted with 150 mL of H_2O and extracted twice with dichloromethane. The organic phases were combined, dried over $MgSO_4$, filtered, and concentrated. The resulting crude product was purified by recrystallization or column chromatography as described below.

3.4.1. (4-Methoxybenzyl)(2-Methyl-6-Nitrophenyl)sulfane (**1b**)

The title compound was prepared according to general procedure 1 on a 30 mmol scale. Recrystallization from diethyl ether/hexanes (1:1) afforded the product as an off-white powder in 77% (6.701 g, 23.18 mmol) yield; m.p. 63–64 °C. ^{1}H NMR (400 MHz, DMSO-d_6, 298 K): δ = 7.62 (dd, *J* = 7.4 Hz, *J* = 1.0 Hz, 1H), 7.55 (dd, *J* = 7.6 Hz, *J* = 0.8 Hz, 1H), 7.48 (t, *J* = 7.6 Hz, 1H), 7.02 (dt, *J* = 8.6 Hz, *J* = 2.9 Hz, 2H), 6.80 (dt, *J* = 8.4 Hz, *J* = 3.0 Hz, 1H), 3.95 (s, 2H), 3.71 (s, 3H), and 2.39 (s, 3H); ^{13}C{^{1}H} NMR (100 MHz, DMSO-d_6, 298 K): δ = 158.6, 156.2, 145.6, 133.0, 130.0, 130.0, 128.8, 124.2, 120.5, 113.8, 55.0, and 20.6.

3.4.2. (4-Methoxybenzyl)(3-Methyl-2-Nitrophenyl)sulfane (**1c**)

The title compound was prepared according to general procedure 1 on a 5 mmol scale. Recrystallization from diethyl ether afforded the product as a bright yellow powder in 86% (1.238 g, 4.28 mmol) yield; m.p. 90–92 °C. ^{1}H NMR (400 MHz, DMSO-d_6, 298 K): δ = 7.52 (d, *J* = 7.5 Hz, 1H), 7.44 (t, *J* = 7.7 Hz, 1H), 7.30 (d, *J* = 7.5 Hz, 1H), 7.22 (dt, *J* = 8.6 Hz, *J* = 2.8 Hz, 2H), 6.85 (dt, *J* = 8.7 Hz, *J* = 2.9 Hz, 2H), 4.23 (s, 2H), 3.71 (s, 3H), and 3.32 (s, 3H); ^{13}C{^{1}H} NMR (100 MHz, DMSO-d_6, 298 K): δ = 158.5, 151.0, 130.7, 130.1 (2 signals), 129.4, 128.8, 128.5, 128.0, 113.9, 55.0, 37.0, and 17.0.

3.4.3. (4-Methoxybenzyl)(2-Nitronaphthalen-1-yl)sulfane (**1d**)

The title compound was prepared according to general procedure 1 on a 5.21 mmol scale. Recrystallization from acetone/hexanes (1:4) afforded the product as a yellow powder in 97% (1.636 g, 5.03 mmol) yield; m.p. 68–70 °C. ^{1}H NMR (400 MHz, DMSO-d_6, 298 K): δ = 8.57 (d, *J* = 8.2 Hz, 1H), 8.21 (d, *J* = 8.8 Hz, 1H), 8.13 (dd, *J* = 7.8 Hz, *J* = 1.4 Hz, 1H), 7.85 (d, *J* = 8.8 Hz, 1H), 7.81–7.73 (m, 2H), 6.92 (dt, *J* = 8.6 Hz, *J* = 2.6 Hz, 2H), 6.72 (dt, *J* = 8.6 Hz, *J* = 2.6 Hz, 2H), 4.07 (s, 2H), and 3.67 (s, 3H); ^{13}C{^{1}H} NMR (100 MHz, DMSO-d_6, 298 K): δ = 158.6, 153.7, 133.8, 133.6, 131.5, 129.9, 129.0, 128.5, 128.5, 127.3, 123.8, 119.4, 113.8, 55.0, and 40.5.

3.4.4. *N*-(4-((4-Ethoxybenzyl)thio)-3-Nitrophenyl)acetamide (**1e**)

The title compound was prepared according to general procedure 1 from *N*-(4-fluoro-3-nitrophenyl)acetamide (see Section 3.3.1) on a 30 mmol scale. Extraction was carried out with ethyl acetate (instead of dichloromethane) and afforded the pure product as a bright yellow powder in 88% (6.150 g, 19.04 mmol) yield; m.p. 140–141 °C. ^{1}H NMR (400 MHz, DMSO-d_6, 298 K): δ = 10.35 (s, 1H), 8.55 (d, *J* = 2.2 Hz, 1H), 7.80 (dd, *J* = 8.8 Hz, *J* = 2.3 Hz, 1H), 7.67 (d, *J* = 8.8 Hz, 1H), 7.33 (d, *J* = 8.6 Hz, 2H), 6.89 (d, *J* = 8.6 Hz, 2H), 4.44 (s, 2H), 3.73 (s, 3H), and 2.07 (s, 3H); ^{13}C{^{1}H} NMR (100 MHz, DMSO-d_6, 298 K): δ = 168.9, 158.6, 145.4, 136.8, 130.4, 129.6, 128.5, 127.4, 124.4, 114.9, 114.0, 55.0, 35.7, and 24.0.

3.4.5. (2,4-. Dinitrophenyl)(4-Methoxybenzyl)sulfane (**1f**)

The title compound was prepared according to general procedure 1 on a 30 mmol scale. Recrystallization from diethyl ether afforded the product as a light brown powder in 86% (8.718 g, 25.84 mmol) yield; m.p. 113–114 °C. ^{1}H NMR (400 MHz, DMSO-d_6, 298 K): δ = 8.86 (d, J = 2.6 Hz, 2H), 8.44 (dt, J = 9.0 Hz, J = 2.6 Hz, 1H), 7.96 (d, J = 9.1 Hz, 1H), 7.40 (dt, J = 8.7 Hz, J = 2.9 Hz, 2H), 6.92 (dt, J = 8.7 Hz, J = 3.0 Hz, 1H), 4.44 (s, 2H), and 3.74 (s, 3H); ^{13}C{^{1}H} NMR (100 MHz, DMSO-d_6, 298 K): δ = 158.9, 145.7, 144.0, 143.6, 130.6, 128.4, 127.4, 126.1, 121.2, 114.2, 55.1, and 35.8.

3.4.6. (4-. Methoxy-2-Nitrophenyl)(4-Methoxybenzyl)sulfane (**1g**)

The title compound was prepared according to general procedure 1 on a 30 mmol scale. Recrystallization from acetone/hexanes (1:4) afforded the product as a bright orange powder in 95% (8.718 g, 28.55 mmol) yield; m.p. 116–118 °C. ^{1}H NMR (400 MHz, DMSO-d_6, 298 K): δ = 7.65 (d, J = 2.8 Hz, 1H), 7.62 (d, J = 9.0 Hz, 1H), 7.33–7.28 (m, 3 H), 6.87 (dt, J = 8.6 Hz, J = 2.8 Hz, 2H), 4.22 (s, 2 H), 3.83 (s, 3H), and 3.72 (s, 3H); ^{13}C{^{1}H} NMR (100 MHz, DMSO-d_6, 298 K): δ = 158.6, 157.0, 147.1, 130.3, 130.0, 127.6, 126.1, 121.2, 113.9, 109.5, 56.0, 55.0, and 36.2.

3.4.7. 4-((4-Methoxybenzyl)thio)-3-Nitrophenol (**1h**)

The title compound was prepared according to general procedure 1 from *tert*-butyl(4-fluoro-3-nitrophenoxy)dimethylsilane (see 3.3.2) on a 26.5 mmol scale. Purification by column chromatography (silica gel, 1. diethyl ether/hexanes (1:3), and 2. diethyl ether/hexanes (1:2), R_f = 0.21) afforded the product as a bright yellow powder in 27% (8.718 g, 7.06 mmol) yield; m.p. 115–116 °C. ^{1}H NMR (400 MHz, DMSO-d_6, 298 K): δ = 10.35 (s, 1H), 7.53 (d, J = 8.8 Hz, 1H), 7.46 (d, J = 2.7 Hz, 1H), 7.28 (d, J = 8.6 Hz, 2H), 7.13 (dd, J = 8.8 Hz, J = 2.7 Hz, 1H), 8.87 (d, J = 8.6 Hz, 2H), 4.18 (s, 2H), and 3.73 (s, 3H); ^{13}C{^{1}H} NMR (100 MHz, DMSO-d_6, 298 K): δ = 158.5, 155.5, 147.3, 130.4, 130.3, 127.8, 123.8, 122.0, 113.9, 111.2, 55.0, and 36.4.

3.4.8. 3-((4-Methoxybenzyl)thio)-2-Nitropyridine (**1i**)

The title compound was prepared according to general procedure 1 on a 30 mmol scale. Recrystallization from acetone/hexanes (1:4) afforded the product as a bright powder in 83% (6.889 g, 24.75 mmol) yield; m.p. 140–142 °C. ^{1}H NMR (400 MHz, DMSO-d_6, 298 K): δ = 8.39 (d, J = 3.6 Hz, 1H), 8.30 (d, J = 7.8 Hz, 1H), 7.77 (d, J = 4.4 Hz, 1H), 7.34 (d, J = 8.5 Hz, 2H), 6.89 (d, J = 8.5 Hz, 2H), 4.34 (s, 2H), and 3.72 (s, 3H); ^{13}C{^{1}H} NMR (100 MHz, DMSO-d_6, 298 K): δ = 158.7, 153.5, 144.4, 138.8, 131.3, 130.4, 128.7, 126.9, 114.1, 55.1, and 35.4.

3.4.9. 2-((4-Methoxybenzyl)thio)-3-Nitropyridine (**1j**)

The title compound was prepared according to general procedure 1 on a 27.9 mmol scale. Trituration with hexanes afforded the product as a yellow powder in 92% (7.629 g, 27.61 mmol) yield; m.p. 88–90 °C. ^{1}H NMR (400 MHz, DMSO-d_6, 298 K): δ = 8.86 (dd, J = 4.6 Hz, J = 1.5 Hz, 2H), 8.60 (dd, J = 8.3 Hz, J = 1.5 Hz, 1H), 7.46 (dd, J = 8.3 Hz, J = 4.6 Hz, 1H), 7.35 (d, J = 8.6 Hz, 2H), 6.87 (d, J = 8.6 Hz, 2H), 4.42 (s, 2H), and 3.72 (s, 3H); ^{13}C{^{1}H} NMR (100 MHz, DMSO-d_6, 298 K): δ = 158.5, 156.0, 153.8, 141.4, 134.4, 130.5, 128.4, 120.0, 113.9, 55.0, and 33.9.

3.4.10. (4-. Methoxybenzyl)(2-Nitro-4-(Trifluoromethyl)phenyl)sulfane (**1k**)

The title compound was prepared according to general procedure 1 on a 30 mmol scale. Trituration with hexanes afforded the product as a bright yellow powder in 97% (9.967 g, 29.03 mmol) yield; m.p. 111–112 °C. ^{1}H NMR (400 MHz, DMSO-d_6, 298 K): δ = 8.45 (d, J = 1.1 Hz, 1H), 8.04 (dd, J = 8.6 Hz, J = 1.7 Hz, 1H), 7.94 (d, J = 8.6 Hz, 1H), 7.38 (d, J = 8.6 Hz, 2H), 6.91 (d, J = 8.6 Hz, 2H), 4.39 (s, 2H), and 3.74 (s, 3H); ^{13}C{^{1}H} NMR (100 MHz, DMSO-d_6, 298 K): δ = 159.3, 145.2, 142.9, 131.0, 130.4 (q, J_{C-F} = 3.3 Hz), 129.2,

126.9, 125.7 (q, J_{C-F} = 33.6 Hz), 123.6 (q, J_{C-F} = 270.4 Hz), 123.4 (q, J_{C-F} = 4.0 Hz), 114.6, 55.5, and 36.1; ^{19}F{^{1}H} NMR (376 MHz, DMSO-d_6, 298 K, referenced to C_6H_5F): δ = −61.3.

3.4.11. 4-((4-Methoxybenzyl)thio)-3-Nitrobenzonitrile (1l)

The title compound was prepared according to general procedure 1 on a 30 mmol scale. Recrystallization from diethyl ether afforded the product as a bright yellow powder in 99% (8.934 g, 29.74 mmol) yield; m.p. 202–204 °C. ^{1}H NMR (400 MHz, DMSO-d_6, 298 K): δ = 8.70 (d, J = 1.8 Hz, 2H), 8.12 (dt, J = 8.5 Hz, J = 1.8 Hz, 1H), 7.89 (d, J = 8.6 Hz, 1H), 7.38 (dt, J = 8.6 Hz, J = 2.9 Hz, 2H), 6.92 (dt, J = 8.7 Hz, J = 2.9 Hz, 2H), 4.40 (s, 2H), and 3.74 (s, 3H); ^{13}C{^{1}H} NMR (100 MHz, DMSO-d_6, 298 K): δ = 158.8, 144.7, 143.4, 136.2, 130.6, 130.0, 128.4, 126.3, 117.1, 114.2, 107.4, 55.1, and 35.6.

3.4.12. 4-((4-Methoxybenzyl)thio)-3-Nitrobenzoic acid (1m)

The title compound was prepared according to procedure 1 on an 85 mmol scale. Recrystallization from acetone/hexanes (1:4) afforded the product as a bright yellow powder in 93% (25.374 g, 79.46 mmol) yield., m.p. 269–271 °C. ^{1}H NMR (400 MHz, DMSO-d_6, 298 K): δ = 13.55 (s, 1H), 8.61 (d, J = 1.9 Hz, 1H), 8.15 (dd, J = 8.5 Hz, J = 1.9 Hz, 1H), 7.86 (d, J = 8.6 Hz, 1H), 7.39 (d, J = 8.6 Hz, 2H), 6.92 (d, J = 8.7 Hz, 2H), 4.37 (s, 2H), and 3.74 (s, 3H); ^{13}C{^{1}H} NMR (100 MHz, DMSO-d_6, 298 K): δ = 165.4, 158.8, 144.6, 142.5, 133.7, 130.6, 127.8, 127.5, 126.6, 126.4, 114.1, 55.1, and 35.6.

3.4.13. Ethyl 4-((4-Methoxybenzyl)thio)-3-Nitrobenzoate (1m′)

A 250 mL round-bottomed flask was charged with compound 1m (23 mmol, 1 equiv.), oxalyl chloride (3 equiv.), and dichloromethane (50 mL). After adding DMF (3 drops), the reaction mixture was stirred for 1 h at 0 °C in an ice bath. EtOH (1.3 equiv.), triethylamine (1.3 equiv.), and dichloromethane (50 mL) were added at 0 °C, the ice bath was removed, and the reaction flask was fitted with a reflux condenser. The reaction mixture was stirred at 40 °C for 2 h, then cooled to room temperature. After washing the dichloromethane solution with water, the organic phase was dried over MgSO$_4$, filtered, and evaporated. Recrystallization of the crude product from diethyl ether/hexanes (1:2) afforded the product as a bright yellow solid in 99.99% (7.989 g, 22.998 mmol) yield; m.p. 149–150 °C. ^{1}H NMR (400 MHz, DMSO-d_6, 298 K): δ = 8.69 (d, J = 1.9 Hz, 1H), 8.16 (dd, J = 8.5 Hz, J = 1.9 Hz, 1H), 7.89 (d, J = 8.6 Hz, 1H), 7.39 (d, J = 8.6 Hz, 2H), 6.92 (d, J = 8.7 Hz, 2H), 4.39–4.33 (m, 4H), 3.74 (s, 3H), and 1.34 (t, J = 7.1 Hz, 3H); ^{13}C{^{1}H} NMR (100 MHz, DMSO-d_6, 298 K): δ = 164.1, 158.9, 144.7, 143.2, 133.6, 130.8, 128.0, 126.6, 126.5, 126.4, 114.3, 61.7, 55.2, 35.8, and 14.2.

3.4.14. (4-. Bromo-2-Nitrophenyl)(4-Methoxybenzyl)sulfane (1n)

The title compound was prepared according to general procedure 1 on a 30 mmol scale. Recrystallization from diethyl ether/hexanes (1:2) afforded the product as a bright yellow powder in 94% (9.968 g, 28.14 mmol) yield; m.p. 121–124 °C. ^{1}H NMR (400 MHz, DMSO-d_6, 298 K): δ = 8.33 (d, J = 2.2 Hz, 1H), 7.89 (dd, J = 8.7 Hz, J = 2.2 Hz, 1H), 7.67 (d, J = 8.8 Hz, 1H), 7.34 (dt, J = 8.7 Hz, J = 2.8 Hz, 2H), 6.90 (dt, J = 8.7 Hz, J = 3.0 Hz, 2H), 4.31 (s, 2H), and 3.73 (s, 3H); ^{13}C{^{1}H} NMR (100 MHz, DMSO-d_6, 298 K): δ = 158.7, 145.9, 136.6, 136.0, 130.4, 129.6, 128.0, 126.8, 117.0, 114.1, 55.1, and 35.6.

3.4.15. (5-. Bromo-2-nitrophenyl)(4-methoxybenzyl)sulfane (1o)

The title compound was prepared according to general procedure 1 on a 30 mmol scale. Recrystallization from diethyl ether/hexanes (1:5) afforded the product as a bright yellow powder in 80% (8.501 g, 24.00 mmol) yield; m.p. 139–140 °C. ^{1}H NMR (400 MHz, DMSO-d_6, 298 K): δ = 8.10 (d, J = 8.8 Hz, 1H), 7.85 (d, J = 2.0 Hz, 1H), 7.57 (dd, J = 8.8 Hz, J = 2.0 Hz, 1H), 7.35 (dt, J = 8.7 Hz, J = 2.8 Hz, 2H), 7.91 (dt, J = 8.7 Hz, J = 3.0 Hz, 2H), 4.36 (s, 2H), and 3.74 (s, 3H); ^{13}C{^{1}H} NMR (100 MHz, DMSO-d_6, 298 K): δ = 158.7, 144.1, 139.2, 130.5, 129.6, 128.7, 128.1, 127.5, 126.7, 114.1, 55.1, and 35.6.

3.4.16. (4-. Chloro-2-Nitrophenyl)(4-Methoxybenzyl)sulfane (**1p**)

The title compound was prepared according to general procedure 1 on a 30 mmol scale. Recrystallization from diethyl ether/hexanes (1:2) afforded the product as an orange powder in 94% (8.258 g, 28.23 mmol) yield; m.p. 113–114 °C. ^{1}H NMR (400 MHz, DMSO-d_6, 298 K): δ = 8.23 (d, J = 2.2 Hz, 1H), 7.78 (dd, J = 8.7 Hz, J = 2.2 Hz, 1H), 7.74 (d, J = 8.8 Hz, 1H), 7.35 (dt, J = 8.6 Hz, J = 3.0 Hz, 2H), 6.90 (dt, J = 8.7 Hz, J = 3.0 Hz, 2H), 4.31 (s, 2H), and 3.73 (s, 3H); ^{13}C{^{1}H} NMR (100 MHz, DMSO-d_6, 298 K): δ = 158.7, 145.8, 135.6, 133.8, 130.4, 129.4, 126.9, 125.3, 114.1, 55.1, and 35.6.

3.4.17. (2-. Chloro-6-Nitrophenyl)(4-Methoxybenzyl)sulfane (**1q**)

The title compound was prepared according to general procedure 1 on a 30 mmol scale. Recrystallization from diethyl ether/hexanes (1:1) afforded the product as a bright yellow powder in 78% (7.290 g, 23.53 mmol) yield; m.p. 86–88 °C. ^{1}H NMR (400 MHz, DMSO-d_6, 298 K): δ = 7.84 (dd, J = 8.1 Hz, J = 1.2 Hz, 1H), 7.76 (dd, J = 8.0 Hz, J = 1.1 Hz, 1H), 7.58 (t, J = 8.1 Hz, 1H), 7.02 (dt, J = 8.6 Hz, J = 2.7 Hz, 2H), 6.78 (dt, J = 8.6 Hz, J = 2.6 Hz, 2H), 4.09 (s, 2H), and 3.69 (s, 3H); ^{13}C{^{1}H} NMR (100 MHz, DMSO-d_6, 298 K): δ = 158.6, 156.1, 140.5, 132.9, 131.4, 130.0, 128.1, 125.0, 121.9, 113.8, 55.0, and 38.6.

3.4.18. (2-. Fluoro-6-Nitrophenyl)(4-Methoxybenzyl)sulfane (**1r**)

The title compound was prepared according to general procedure 1 on a 15 mmol scale. Recrystallization from acetone/hexanes (1:4) afforded the product as a bright yellow powder in 95% (4.167 g, 14.20 mmol) yield; m.p. 105–107 °C. ^{1}H NMR (400 MHz, DMSO-d_6, 298 K): δ = 7.74–7.71 (m, 1H), 7.63–7.56 (m, 2H), 7.08 (dt, J = 8.6 Hz, J = 2.0 Hz, 2H), 6.78 (dt, J = 8.7 Hz, J = 2.0 Hz, 2H), 4.14 (s, 2H), and 3.69 (s, 3H); ^{13}C{^{1}H} NMR (100 MHz, DMSO-d_6, 298 K): δ = 162.0 (d, J_{C-F} = 246.7 Hz), 158.6, 153.5 (d, J_{C-F} = 3.0 Hz), 130.7 (d, J_{C-F} = 9.4 Hz), 130.0, 128.4, 120.0 (d, J_{C-F} = 5.9 Hz), 119.9 (d, J_{C-F} = 14.8 Hz), 116.6 (d, J_{C-F} = 23.2 Hz), 113.8, 55.0, and 38.0 (d, J_{C-F} = 7.0 Hz); ^{19}F{^{1}H} NMR (376 MHz, DMSO-d_6, 298 K, referenced to C_6H_5F): δ = −102.14

*3.5. General Procedure 2 for the Synthesis of Substituted Anilines (**2b–2r**)*

The following description is for a 20–25 mmol scale reaction. The solvent quantities and flask size were adjusted accordingly for smaller-scale reactions.

A 250 mL round-bottomed flask equipped with a stir bar was loaded with the (4-methoxybenzyl)(2-nitrophenyl)sulfane derivative (**1b–1r**, 1 equiv.), iron power (5.0 equiv.), NH_4Cl (5.0 equiv.), and 150 mL of $EtOH/H_2O$ (4:1). The reaction flask was placed under inert atmosphere (argon), fitted with a reflux condenser, and stirred at 80 °C until TLC indicated a complete reduction (typically between 1 and 4 h). After cooling to room temperature, the reaction mixture was filtered through celite and concentrated under reduced pressure. The residue was then basified with 1M NaOH and extracted twice with dichloromethane, dried over $MgSO_4$, and evaporated. The crystalline solid was washed with hexanes or diethyl ether as described below.

3.5.1. 2-((4-Methoxybenzyl)thio)-3-Methylaniline (**2b**)

The title compound was prepared according to general procedure 2 on a 20.0 mmol scale. The crystalline light brown solid was washed with hexanes, which gave the desired product in 94% (4.897 g, 18.88 mmol) yield; m.p. 74–76 °C. ^{1}H NMR (400 MHz, DMSO-d_6, 298 K): δ = 7.06 (d, J = 8.3 Hz, 2H), 6.92 (t, J = 7.6 Hz, 1H), 7.76 (d, J = 8.3 Hz, 2H), 6.60 (d, J = 7.9 Hz, 1H), 6.41 (d, J = 7.2 Hz, 1H), 5.39 (s, 2H), 3.74 (s, 2H), 3.70 (s, 3H), and 2.13 (s, 3H); ^{13}C{^{1}H} NMR (100 MHz, DMSO-d_6, 298 K): δ = 158.2, 150.5, 143.0, 130.0, 129.9, 129.2, 117.7, 114.8, 113.5, 112.0, 55.0, 36.4, and 21.4. HRMS (ESI) m/z calculated for [M + H]$^+$ = [$C_{15}H_{18}NOS$]$^+$ 260.1104; observed, 260.1105.

3.5.2. 2-((4-Methoxybenzyl)thio)-6-Methylaniline (**2c**)

The title compound was prepared according to general procedure 2 on a 2.00 mmol scale. The crystalline white solid was washed with hexanes, which gave the desired product in 92% (0.474 g, 1.83 mmol) yield; m.p. 92–93 °C. ^{1}H NMR (400 MHz, DMSO-d_6, 298 K): δ = 7.14 (d, J = 8.1 Hz, 2H), 7.03 (d, J = 7.4 Hz, 1H), 6.94 (d, J = 7.0 Hz, 1H), 8.82 (d, J = 7.0 Hz, 2H), 6.44 (t, J = 7.4 Hz, 1H), 5.01 (s, 2H), 3.89 (s, 2H), 3.71 (s, 3H), and 2.11 (s, 3H); ^{13}C{^{1}H} NMR (100 MHz, DMSO-d_6, 298 K): δ = 158.2, 146.9, 132.5, 130.4, 130.0, 129.9, 121.7, 116.2, 116.1, 113.6, 55.0, 37.5, and 18.2. HRMS (ESI) m/z calculated for [M + H]$^+$ = [C$_{15}$H$_{18}$NOS]$^+$ 260.1104 observed, 260.1104.

3.5.3. 1-((4-Methoxybenzyl)thio)naphthalen-2-Amine (**2d**)

The title compound was prepared according to general procedure 2 on a 4.45 mmol scale. The yellow oil was washed with hexanes, which gave the desired product as a light brown solid in 99% (1.300 g, 4.41 mmol) yield; m.p. 58–60 °C. ^{1}H NMR (400 MHz, DMSO-d_6, 298 K): δ = 8.20 (d, J = 8.5 Hz, 2H), 7.66 (t, J = 8.4 Hz, 2H), 7.42–7.38 (m, 1H), 7.17–7.12 (m, 3H), 7.09 (d, J = 8.8 Hz, 1H), 6.78 (dt, J = 8.7 Hz, J = 2.1 Hz, 2H), 5.83 (s, 2H), 3.79 (s, 2H), and 3.69 (s, 3H); ^{13}C{^{1}H} NMR (100 MHz, DMSO-d_6, 298 K): δ = 158.2, 149.5, 136.3, 130.3, 130.0 129.9, 128.3, 127.1, 126.9, 123.3, 121.1, 117.4, 113.6, 104.9, 55.0, and 37.3. HRMS (ESI) m/z calculated for [M + H]$^+$ = [C$_{18}$H$_{18}$NOS]$^+$ 296.1104; observed, 296.1104.

3.5.4. N-(3-Amino-4-((4-Methoxybenzyl)thio)phenyl)acetamide (**2e**)

The title compound was prepared according to general procedure 2 on a 17.05 mmol scale. The crystalline beige powder solid was washed with hexanes, which gave the desired product in 93% (4.807 g, 15.91 mmol) yield; m.p. 111–112 °C. ^{1}H NMR (400 MHz, DMSO-d_6, 298 K): δ = 9.71 (s, 1H), 7.11 (t, J = 8.3 Hz, 3H), 6.97 (d, J = 8.2 Hz, 1H), 6.80 (d, J = 8.3 Hz, 2H), 6.62 (d, J = 7.8 Hz, 1H), 5.33 (s, 2H), 3.81 (s, 2H), 3.70 (s, 3H), and 2.01 (s, 3H); ^{13}C{^{1}H} NMR (100 MHz, DMSO-d_6, 298 K): δ = 168.1, 158.1, 149.9, 140.6, 135.9, 130.2, 130.0, 113.6, 109.7, 107.7, 104.3, 55.0, 37.7, and 24.1. HRMS (ESI) m/z calculated for [M + H]$^+$ = [C$_{16}$H$_{19}$N$_2$O$_2$S]$^+$ 303.1162; observed, 303.1162.

3.5.5. 4-((4-Methoxybenzyl)thio)benzene-1,3-Diamine (**2f**)

The title compound was prepared according to general procedure 2 on a 20.0 mmol scale. The crystalline beige solid was washed with hexanes, which gave the desired product in 96% (4.979 g, 19.12 mmol) yield; m.p. 91–92 °C. ^{1}H NMR (400 MHz, DMSO-d_6, 298 K): δ = 7.08 (d, J = 8.6 Hz, 2H), 6.80 (d, J = 8.6 Hz, 2H), 6.72 (d, J = 8.2 Hz, 1H), 5.93 (d, J = 7.9 Hz, 1H), 5.73 (dd, J = 8.2 Hz, J = 2.3 Hz, 1H), 5.00 (s, 2H), 4.97 (s, 2H), 3.71 (s, 3H), and 3.67 (s, 3H); ^{13}C{^{1}H} NMR (100 MHz, DMSO-d_6, 298 K): δ = 158.0, 150.6, 150.5, 137.4, 130.6, 130.0, 113.5, 104.3, 102.0, 99.0, 55.0, and 38.8. HRMS (ESI) m/z calculated for [M + H]$^+$ = [C$_{14}$H$_{17}$N$_2$OS]$^+$ 261.1056; observed, 261.1057.

3.5.6. 5-Methoxy-2-((4-Methoxybenzyl)thio)aniline (**2g**)

The title compound was prepared according to general procedure 2 on a 27.9 mmol scale. The crystalline dark beige solid was washed with hexanes, which gave the desired product in 92% (7.082 g, 25.72 mmol) yield; m.p. 113–114 °C. ^{1}H NMR (400 MHz, DMSO-d_6, 298 K): δ = 7.10 (dt, J = 8.6 Hz, J = 2.0 Hz, 1H), 6.96 (d, J = 8.4 Hz, 2H), 7.76 (dt, J = 8.6 Hz, J = 2.0 Hz, 2H), 6.30 (d, J = 2.7 Hz, 1H), 6.05 (dd, J = 8.5 Hz, J = 2.7 Hz, 1H), 5.34 (s, 2H), 3.77 (s, 2H), 3.71 (s, 3H), and 3.65 (s, 3H); ^{13}C{^{1}H} NMR (100 MHz, DMSO-d_6, 298 K): δ = 160.9, 158.1, 151.0, 137.2, 130.2, 130.0, 113.6, 107.1, 102.9, 98.9, 55.0, 54.7, and 38.0. HRMS (ESI) m/z calculated for [M + H]$^+$ = [C$_{15}$H$_{18}$NO$_2$S]$^+$ 276.1053; observed, 276.1052.

3.5.7. 3-Amino-4-((4-Methoxybenzyl)thio)phenol (**2h**)

The title compound was prepared according to general procedure 2 on a 2.75 mmol scale. The crystalline light brown powder was washed with diethyl ether/hexanes (1:6), which gave the desired product in 92% (546.4 mg, 2.52 mmol) yield; m.p. 116–118 °C.

^{1}H NMR (400 MHz, CD$_3$CN, 298 K): δ = 7.07 (dt, J = 8.7 Hz, J = 2.0 Hz, 2H), 6.96 (d, J = 8.3 Hz, 1H), 6.86 (s, 2H), 6.79 (dt, J = 8.7 Hz, J = 2.1 Hz, 2H), 6.20 (d, J = 2.6 Hz, 1H), 6.03 (dd, J = 8.3 Hz, J = 2.6 Hz, 1H), 4.62 (s, 2H), 3.76 (s, 2H), and 3.74 (s, 3H); ^{13}C{^{1}H} NMR (100 MHz, CD$_3$CN, 298 K): δ = 159.8, 159.6, 151.9, 139.0, 131.6, 131.0, 114.5, 108.2, 106.2, 101.7, 55.8, and 39.6. HRMS (ESI) m/z calculated for [M + H]$^+$ = [C$_{15}$H$_{15}$NO$_2$S]$^+$ 262.0896; observed, 262.0898.

3.5.8. 3-((4-Methoxybenzyl)thio)pyridin-2-Amine (2i)

The title compound was prepared according to general procedure 2 on a 20.0 mmol scale. The crystalline beige solid was washed with hexanes, which gave the desired product in 92% (4.553 g, 18.48 mmol) yield; m.p. 76–78 °C. ^{1}H NMR (400 MHz, DMSO-d_6, 298 K): δ = 7.87 (dd, J = 4.8 Hz, J = 1.7 Hz, 1H), 7.66 (dd, J = 7.4 Hz, J = 1.6 Hz, 1H), 7.15 (d, J = 8.6 Hz, 2H), 6.82 (d, J = 8.6 Hz, 2H), 6.46 (dd, J = 7.4 Hz, J = 4.9 Hz, 1H), 6.03 (s, 2H), 3.97 (s, 2H), and 3.70 (s, 3H); ^{13}C{^{1}H} NMR (100 MHz, DMSO-d_6, 298 K): δ = 159.1, 158.3, 147.4, 141.6, 130.1, 129.4, 113.7, 112.8, 111.9, 55.0, and 36.4. HRMS (ESI) m/z calculated for [M + H]$^+$ = [C$_{13}$H$_{15}$N$_2$OS]$^+$ 247.0900; observed, 247.0899.

3.5.9. 2-((4-Methoxybenzyl)thio)pyridin-3-Amine (2j)

The title compound was prepared according to general procedure 2 on a 20.0 mmol scale. The crystalline brown powder was washed with hexanes, which gave the desired product in 99% (4.898 g, 19.88 mmol) yield. ^{1}H NMR (400 MHz, DMSO-d_6, 298 K): δ = 7.79 (dd, J = 7.6 Hz, J = 1.4 Hz, 1H), 7.29 (d, J = 7.9 Hz, 1H), 6.90 (t, J = 7.9 Hz, 2H), 6.84 (d, J = 7.6 Hz, 2H), 5.01 (s, 2H), 4.34 (s, 2H), and 3.71 (s, 3H); ^{13}C{^{1}H} NMR (100 MHz, DMSO-d_6, 298 K): δ = 158.2, 141.6, 141.5, 137.2, 130.2, 130.0, 120.5, 119.1, 113.7, 55.0, and 32.7. HRMS (ESI) m/z calculated for [M + H]$^+$ = [C$_{13}$H$_{15}$N$_2$OS]$^+$ 247.0900; observed, 247.0899.

3.5.10. 2-((4-Methoxybenzyl)thio)-5-(Trifluoromethyl)aniline (2k)

The title compound was prepared according to general procedure 2 on a 27.0 mmol scale. The crystalline off-white solid was washed with hexanes, which gave the desired product in 97% (8.167 g, 26.07 mmol) yield; m.p. 79–80 °C. ^{1}H NMR (400 MHz, DMSO-d_6, 298 K): δ = 7.28 (d, J = 8.0 Hz, 1H), 7.21 (d, J = 8.4 Hz, 2H), 7.02 (s, 2H), 6.83 (d, J = 8.4 Hz, 2H), 6.75 (d, J = 7.8 Hz, 1H), 5.66 (s, 2H), 4.04 (s, 3H), and 3.71 (s, 3H); ^{13}C{^{1}H} NMR (100 MHz, DMSO-d_6, 298 K): δ = 158.4, 148.6, 133.1, 130.1, 129.2, 128.7 (q, J_{C-F} = 31.0 Hz), 124.4 (q, J_{C-F} = 270.4 Hz), 121.5, 113.7, 112.0 (q, J_{C-F} = 15.4 Hz), 109.8 (q, J_{C-F} = 3.9 Hz), 55.0, and 36.2; ^{19}F{^{1}H} NMR (376 MHz, DMSO-d_6, 298 K, referenced to C$_6$H$_5$F): δ = −61.45. HRMS (ESI) m/z calculated for [M + H]$^+$ = [C$_{15}$H$_{15}$F$_3$NOS]$^+$ 314.0821; observed, 314.0823.

3.5.11. 3-Amino-4-((4-Methoxybenzyl)thio)benzonitrile (2l)

The title compound was prepared according to general procedure 2 on a 10.0 mmol scale. The crystalline yellow solid was washed with hexanes, which gave the desired product in 86% (2.315 g, 8.56 mmol) yield; m.p. 189–190 °C. ^{1}H NMR (400 MHz, DMSO-d_6, 298 K): δ = 7.28 (d, J = 8.0 Hz, 1H), 7.20 (d, J = 8.6 Hz, 2H), 6.98 (d, J = 1.6 Hz, 1H), 6.84 (d, J = 8.7 Hz, 2H), 6.74 (dd, J = 8.0 Hz, J = 1.4 Hz, 1H), 5.64 (s, 2H), 4.04 (s, 2H), and 3.71 (s, 3H); ^{13}C{^{1}H} NMR (100 MHz, DMSO-d_6, 298 K): δ = 158.8, 144.7, 143.4, 136.2, 130.6, 130.0, 128.4, 126.3, 117.1, 114.2, 107.4, 55.1, and 35.6. HRMS (ESI) m/z calculated for [M + H]$^+$ = [C$_{15}$H$_{15}$N2OS]$^+$ 271.0900; observed, 271.0900.

3.5.12. Ethyl 3-Amino-4-((4-Methoxybenzyl)thio)benzoate (2m′)

The title compound was prepared according to general procedure 2 on a 23.0 mmol scale. The crystalline yellow solid was washed with hexanes, which gave the desired product in 94% (6.894 g, 21.72 mmol) yield; m.p. 89–90 °C. ^{1}H NMR (400 MHz, DMSO-d_6, 298 K): δ = 7.33 (s, 1H), 7.23–7.19 (m, 3H), 7.07 (d, J = 7.9 Hz, 1H), 6.83 (d, J = 8.3 Hz, 2H), 5.44 (s, 2H), 6.88 (q, J = 7.0 Hz, 2H), 4.05 (s, 2H), 3.71 (s, 3H), and 1.29 (t, J = 7.0 Hz, 3H); ^{13}C{^{1}H} NMR (100 MHz, DMSO-d_6, 298 K): δ = 165.9, 158.3, 147.8, 131.8, 130.1, 129.2, 129.2, 123.0,

116.8, 114.3, 113.7, 60.4, 55.0, 55.0, 36.0, and 14.2. HRMS (ESI) m/z calculated for [M + H]$^+$ = [C$_{17}$H$_{20}$NO$_2$S]$^+$ 318.1158; observed, 318.1160.

3.5.13. 5-Bromo-2-((4-Methoxybenzyl)thio)aniline (2n)

The title compound was prepared according to general procedure 2 on a 2.86 mmol scale. The crystalline light brown solid was washed with hexanes, which gave the desired product in 79% (728.1 mg, 2.25 mmol) yield; m.p. 100–102 °C. ^{1}H NMR (400 MHz, DMSO-d_6, 298 K): δ = 7.13 (d, J = 8.5 Hz, 2H), 6.98 (d, J = 8.2 Hz, 1H), 6.90 (d, J = 2.0 Hz, 1H), 6.82 (d, J = 8.6 Hz, 2H), 6.58 (dd, J = 8.2 Hz, J = 2.1 Hz, 1H), 5.56 (s, 2H), 3.89 (s, 2H), and 3.71 (s, 3H); ^{13}C{^{1}H} NMR (100 MHz, DMSO-d_6, 298 K): δ = 158.2, 150.6, 136.3, 130.0, 129.6, 122.2, 118.5, 116.1, 115.2, 113.6, 55.0, and 36.9. HRMS (ESI) m/z calculated for [M + H]$^+$ = [C$_{14}$H$_{15}$BrNOS]$^+$ 324.0052; observed, 324.0052.

3.5.14. 4-Bromo-2-((4-Methoxybenzyl)thio)aniline (2o)

The title compound was prepared according to general procedure 2 on a 30 mmol scale. The crystalline beige solid was washed with hexanes, which gave the desired product in 95% (9.287g, 28.64 mmol) yield; m.p. 86–87 °C. ^{1}H NMR (400 MHz, DMSO-d_6, 298 K): δ = 7.18–7.11 (m, 4H), 6.83 (dt, J = 8.6 Hz, J = 1.9 Hz, 1H), 6.66 (d, J = 8.6 Hz, 1H), 5.44 (s, 2H), 3.95 (s, 2H), and 3.72 (s, 3H); ^{13}C{^{1}H} NMR (100 MHz, DMSO-d_6, 298 K): δ = 158.3, 148.3, 135.6, 131.4, 130,1, 130.0, 118.2, 115.9, 113.6, 105.9, 55.0, and 36.9. HRMS (ESI) m/z calculated for [M + H]$^+$ = [C$_{14}$H$_{15}$BrNOS]$^+$ 324.0052; observed, 324.0050.

3.5.15. 5-Chloro-2-((4-Methoxybenzyl)thio)aniline (2p)

The title compound was prepared according to general procedure 2 on a 2.72 mmol scale. The crystalline pale brown powder was washed with hexanes, which gave the desired product in 93% (7.142 g, 2.55 mmol) yield; m.p. 80–82 °C. ^{1}H NMR (400 MHz, DMSO-d_6, 298 K): δ = 7.13 (d, J = 8.5 Hz, 2H), 7.05 (d, J = 8.2 Hz, 1H), 6.81 (d, J = 2.0 Hz, 1H), 6.75 (d, J = 8.6 Hz, 2H), 6.45 (dd, J = 8.2 Hz, J = 2.1 Hz, 1H), 5.58 (s, 2H), 3.89 (s, 2H), and 3.71 (s, 3H); ^{13}C{^{1}H} NMR (100 MHz, DMSO-d_6, 298 K): δ = 158.2, 150.5, 136.2, 133.6, 130.0, 129.6, 115.6, 114.7, 113.6, 113.2, 55.0, and 37.0. HRMS (ESI) m/z calculated for [M + H]$^+$ = [C$_{14}$H$_{15}$ClNOS]$^+$ 280.0557; observed, 280.0558.

3.5.16. 3-Chloro-2-((4-Methoxybenzyl)thio)aniline (2q)

The title compound was prepared according to general procedure 2 on a 20.0 mmol scale. The crystalline off-white powder was washed with hexanes, which gave the desired product in 94% (5.282 g, 18.88 mmol) yield; m.p. 71–73 °C. ^{1}H NMR (400 MHz, DMSO-d_6, 298 K): δ = 7.14 (d, J = 8.5 Hz, 2H), 7.00 (t, J = 8.0 Hz, 1H), 6.80 (d, J = 8.5 Hz, 2H), 6.67–6.63 (m, 2H), 5.66 (s, 2H), 3.86 (s, 2H), and 3.70 (s, 3H); ^{13}C{^{1}H} NMR (100 MHz, DMSO-d_6, 298 K): δ = 158.3, 152.4, 139.8, 130.5, 130.0, 129.4, 116.5, 113.6, 113.4, 112.8, 55.0, and 37.0. HRMS (ESI) m/z calculated for [M + H]$^+$ = [C$_{14}$H$_{15}$ClNOS]$^+$ 280.0557; observed, 280.0559.

3.5.17. 3-Fluoro-2-((4-Methoxybenzyl)thio)aniline (2r)

The title compound was prepared according to general procedure 2 on a 12 mmol scale. The crystalline off-white powder was washed with hexanes, which gave the desired product in 97% (2.064 g, 11.64 mmol) yield; m.p. 71–72 °C. ^{1}H NMR (400 MHz, DMSO-d_6, 298 K): δ = 7.12 (d, J = 8.6 Hz, 2H), 7.02 (q, J = 6.7 Hz, 1H), 6.79 (d, J = 8.6 Hz, 2H), 6.52 (d, J = 8.2 Hz, 1H), 6.32–6.28 (m, 1H), 5.62 (s, 2H), 3.83 (s, 2H), and 3.70 (s, 3H); ^{13}C{^{1}H} NMR (100 MHz, DMSO-d_6, 298 K): δ = 163.8 (d, J_{C-F} = 239.0 Hz), 158.2, 152.1 (d, J_{C-F} = 3.8 Hz), 130.5 (d, J_{C-F} = 11.0 Hz), 129.9, 129.7, 113.6, 109.8 (d, J_{C-F} = 2.5 Hz), 102.4 (d, J_{C-F} = 21.4 Hz), 101.9 (d, J_{C-F} = 24.0 Hz), 55.0, and 36.9; ^{19}F{^{1}H} NMR (376 MHz, DMSO-d_6, 298 K, referenced to C$_6$H$_5$F): δ = −108.81. HRMS (ESI) m/z calculated for [M + H]$^+$ = [C$_{14}$H$_{15}$FNOS]$^+$ 264.0853; observed, 264.0854.

3.6. Synthesis of 4-Fluoro-2-((4-Methoxybenzyl)thio)aniline (2s)

3.6.1. 2,2'-Disulfanediylbis(4-Fluoroaniline)

A 500 mL round-bottomed flask equipped with a stir bar was charged with 16.802 g (100 mml) of 2-amino-6-fluorobenzothiazole, 100 mL of 10 M NaOH, and 100 mL of 2-methoxyethanol. The flask was fitted with a reflux condenser and the reaction mixture was stirred at 100 °C for 4 days. After cooling to room temperature, the reaction mixture was acidified with HCl to pH 3 and extracted twice with diethyl ether. The combined organic phases were washed with brine, dried over $MgSO_4$, filtered, and evaporated. The product obtained from the extraction was sufficiently pure to be used without further purification. Yield: 69% (9.815 g, 34.52 mmol). ^{1}H NMR (400 MHz, CD_3CN, 298 K): δ = 6.96 (dt, J = 8.7 Hz, J = 3.0 Hz, 2H), 6.85 (dd, J = 8.6 Hz, J = 3.8 Hz, 1H), 6.75 (dd, J = 8.9 Hz, J = 4.9 Hz, 2H), and 4.62 (s, 2H); ^{13}C{^{1}H} NMR (100 MHz, CD_3CN, 298 K): δ = 155.1 (d, J_{C-F} = 234.0 Hz), 147.1 (d, J_{C-F} = 1.7 Hz), 121.8 (d, J_{C-F} = 22.2 Hz), 119.4 (d, J_{C-F} = 22.5 Hz), 118.6 (d, J_{C-F} = 7.3 Hz), and 116.9 (d, J_{C-F} = 7.5 Hz); ^{19}F{^{1}H} NMR (376 MHz, CD_3CN, 298 K, referenced to C_6H_5F): δ = −127.83.

3.6.2. 4-Fluoro-2-((4-Methoxybenzyl)thio)aniline (**2s**)

Reduction of 2,2'-Disulfanediylbis(4-fluoroaniline) to 2-amino-5-fluorobenzenethiol: A 1 L round-bottomed flask equipped with a stir bar was charged with 9.815 g (34.52 mmol, 1 equiv.) of 2,2'-Disulfanediylbis(4-fluoroaniline), 200 mL of THF, and 200 mL of MeOH. While stirring in an ice bath, $NaBH_4$ (5.22 g, 138 mmol, 4 equiv.) was added in small portions over a period of 15 min under a counterflow of argon. The ice bath was then removed, and the reaction mixture was stirred at room temperature for 1h. The reaction mixture was concentrated, water was added, and the product was extracted with diethyl ether. After drying the organic phase over $MgSO_4$, filtration, and removal of the solvent, the crude 2-amino-5-fluorobenzenethiol was used immediately in the next step. Protection of 2-amino-5-fluorobenzenethiol: The crude 2-amino-5-fluorobenzenethiol (max. 69 mmol, 1 equiv.) was dissolved in 100 mL of EtOH and transferred to a 250 mL round-bottomed flask equipped with a stir bar, and 10.41 g (75 mmol, 1.08 equiv.) of 1-(chloromethyl)-4-methoxybenzene was added. After setting the reaction under argon, 3.25 g (81.3 mmol, 1.18 equiv.) of NaOH dissolved in 10 mL of water was added dropwise under a counterflow of argon. Then, the reaction flask was fitted with a reflux condenser and stirred under an argon atmosphere at 50 °C for 4 h. After cooling to room temperature, the reaction mixture was concentrated. Basification with 1 M NaOH, followed by extraction with diethyl ether and drying over $MgSO_4$ gave the crude product as an oil upon filtration and solvent removal. Recrystallization from hot diethyl ether gave the product as a pale yellow powder in 9.747 g (37.01 mmol, 54%) yield. ^{1}H NMR (400 MHz, DMSO-d_6, 298 K): δ = 7.18 (dt, J = 8.6 Hz, J = 2.0 Hz, 2H), 6.92 (dd, J = 9.1 Hz, J = 3.0 Hz, 1H), 6.87–6.81 (m, 3H), 6.69 (dd, J = 8.8 Hz, J = 5.3 Hz, 1H), 5.08 (s, 2H), 3.99 (s, 2H), and 3.71 (s, 3H); ^{13}C{^{1}H} NMR (100 MHz, DMSO-d_6, 298 K): δ = 158.3, 153.6 (d, J_{C-F} = 231.9 Hz), 145.1 (d, J_{C-F} = 1.6 Hz), 130.0, 129.4, 118.8 (d, J_{C-F} = 22.2 Hz), 117.5 (d, J_{C-F} = 7.6 Hz), 115.3 (d, J_{C-F} = 21.8 Hz), 114.8 (d, J_{C-F} = 7.6 Hz), 113.7, 55.0, and 36.5; ^{19}F{^{1}H} NMR (376 MHz, DMSO-d_6, 298 K, referenced to C_6H_5F): δ = −128.70. HRMS (ESI) m/z calculated for [M + H]$^+$ = [$C_{14}H_{15}FNOS$]$^+$ 264.0853; observed, 264.0854.

3.7. General Procedure 3 for the Synthesis of Substituted Triazoles (3b–3s)

A 50 mL round-bottomed flask equipped with a stir bar was loaded the 2-((4-methoxybenzyl)thio)aniline derivative (**2b–2s**, 1 equiv.) and N,N-dimethylformamide azine dihydrochloride (1.1 equiv). After mixing the two compounds thoroughly, the reaction flask was placed under an inert atmosphere (argon) and stirred in an oil bath at 150 °C. The two solid compounds were observed to melt within an hour and turn dark red. After 16 h, the reaction flask was cooled to room temperature. The resulting dark solid was basified with 1 M NaOH and extracted with dichloromethane, dried over $MgSO_4$, filtered, and concentrated. The crude product was purified by column chromatography or recrystallization as described below.

3.7.1. 4-(2-((4-Methoxybenzyl)thio)-3-Methylphenyl)-4H-1,2,4-triazole (3b)

The title compound was prepared according to general procedure 3 on a 3.63 g (14.0 mmol) scale. Purification by column chromatography (silica gel, 1. dichloromethane, and 2. dichloromethane/acetone 4:1; R_f = 0.56) gave the product as an off-white powder in 72% (3.15 g, 10.1 mmol) yield; m.p. 116–118 °C. ^{1}H NMR (400 MHz, CDCl$_3$, 298 K): δ = 7.92 (s, 2H), 7.43 (d, J = 7.6 Hz, 1H), 7.35 (t, J = 7.6 Hz, 1H), 7.04 (t, J = 7.9 Hz, 1H), 6.68 (s, 4H), 3.74 (s, 1H), 3.53 (s, 1H), and 2.61 (s, 1H); ^{13}C{^{1}H} NMR (100 MHz, CDCl$_3$, 298 K): δ = 158.7, 145.5, 143.1, 137.8, 131.6, 129.6, 129.5 (2 signals), 128.7, 124.1, 113.7, 55.1, 38.7, and 21.4. HRMS (ESI) m/z: [M + H]$^+$ calculated for C$_{17}$H$_{17}$N$_3$OS, 312.1165; observed, 312.1165.

3.7.2. 4-(2-((4-Methoxybenzyl)thio)-6-Methylphenyl)-4H-1,2,4-triazole (3c)

The title compound was prepared according to general procedure 3 on a 4.67 g (18.0 mmol) scale. Purification by column chromatography (silica gel, 1. dichloromethane, and 2. dichloromethane/acetone 4:1; R_f = 0.41) gave the product as an off-white powder in 49% (2.76 g, 7.55 mmol) yield; m.p. 191–193 °C. ^{1}H NMR (400 MHz, DMSO-d_6, 298 K): δ = 8.59 (s, 2H), 7.46–7.40 (m, 2H), 7.27 (d, J = 6.7 Hz, 1H), 7.21 (d, J = 8.5 Hz, 1H), 6.86 (d, J = 8.5 Hz, 1H), 4.12 (s, 1H), 3.71 (s, 1H), and 1.96 (s, 1H); ^{13}C{^{1}H} NMR (100 MHz, DMSO-d_6, 298 K): δ = 158.5, 143.2, 135.9, 135.7, 131.3, 130.0 (2 signals), 128.0 (2 signals), 126.2, 113.9, 55.1, 35.7, and 17.2. HRMS (ESI) m/z: [M + H]$^+$ calculated for C$_{17}$H$_{17}$N$_3$OS, 312.1165; observed, 312.1165.

3.7.3. 4-(1-((4-Methoxybenzyl)thio)naphthalen-2-yl)-4H-1,2,4-triazole (3d)

The title compound was prepared according to general procedure 3 on a 975 mg (3.30 mmol) scale. Purification by column chromatography (silica gel, 1. dichloromethane, and 2. dichloromethane/acetone 4:1; R_f = 0.67) gave the product as a pale yellow powder in 17% (195 mg, 0.56 mmol) yield; m.p. 129–130 °C. ^{1}H NMR (400 MHz, DMSO-d_6, 298 K): δ = 8.65 (d, J = 8.4 Hz, 1H), 8.52 (s, 2H), 8.18 (d, J = 8.6 Hz, 1H), 8.13 (d, J = 7.9 Hz, 1H), 7.81–7.77 (m, 1 H), 7.73–7.69 (m, 1 H), 7.57 (d, J = 8.6 Hz, 1H), 6.70 (s, 4H), 3.81 (s, 3H), and 3.68 (s, 3H); ^{13}C{^{1}H} NMR (100 MHz, DMSO-d_6, 298 K): δ = 158.4, 143.6, 136.3, 134.2, 133.3, 131.0, 129.7, 129.0, 128.7, 128.5, 127.7, 127.4, 126.7, 124.4, 113.7, and 55.1. HRMS (ESI) m/z: [M + H]$^+$ calculated C$_{20}$H$_{17}$N$_3$OS, 348.1165; observed, 348.1163.

3.7.4. N-(4-((4-Methoxybenzyl)thio)-3-(4H-1,2,4-Triazol-4-Yl)phenyl)acetamide (3e)

The title compound was prepared according to general procedure 2 on a 3.26 g (10.8 mmol) scale. Purification by column chromatography (silica gel, 1. dichloromethane, and 2. dichloromethane/acetone 4:1; R_f = 0.19) gave the product as a pale brown powder in 54% (2.06 g, 5.80 mmol) yield; m.p. 176–178 °C. ^{1}H NMR (400 MHz, DMSO-d_6, 298 K): δ = 10.26 (s, 1H), 8.58 (s, 1H), 7.69 (d, J = 1.8 Hz, 1H), 7.62 (dd, J = 8.6 Hz, J = 2.0 Hz, 1H), 7.56 (d, J = 8.6 Hz, 1H), 7.06 (d, J = 8.6 Hz, 2H), 6.82 (d, J = 8.6 Hz, 2H), 3.97 (s, 2H), 3.71 (s, 3H), and 2.06 (s, 3H); ^{13}C{^{1}H} NMR (100 MHz, DMSO-d_6, 298 K): δ = 168.9, 158.5, 143.4, 139.1, 134.3, 133.1, 130.0, 128.6, 125.6, 120.1, 117.1, 113.9, 55.1, 37.8, and 24.1. HRMS (ESI) m/z: [M + H]$^+$ calculated C$_{18}$H$_{18}$N$_4$O$_2$S, 355.1223; observed, 355.1225.

3.7.5. 4-((4-Methoxybenzyl)thio)-3-(4H-1,2,4-Triazol-4-Yl)aniline (3f)

The title compound was obtained as a by-product of the synthesis of compound 3f', which was prepared according to general procedure 3 on a 3.64 mg (14.0 mmol) scale; for this reaction, 2.2 equiv. of N,N-dimethylformamide azine dihydrochloride was used. Purification by column chromatography (silica gel, 1. dichloromethane, and 2. dichloromethane/acetone 4:1; R_f = 0.26) gave the product as a pale brown powder in 19% (818 mg, 2.24 mmol) yield; m.p. 147–149 °C. ^{1}H NMR (400 MHz, DMSO-d_6, 298 K): δ = 9.01 (s, 2H), 7.24 (d, J = 8.2 Hz, 1H), 7.18 (dt, J = 8.6 Hz, J = 2.0 Hz, 2H), 6.92 (d, J = 2.4 Hz, 1H),

6.83 (dt, *J* = 8.6 Hz, *J* = 2.0 Hz, 2H), 6.74 (dd, *J* = 8.2 Hz, *J* = 2.4 Hz, 2H), 7.96–7.90 (m, 3 H), 7.86–7.80 (m, 3 H), 7.70 (tt, *J* = 7.4 Hz, *J* = 1.1 Hz, 1H), 5.46 (s, 2H), 3.96 (s, 2H), and 3.71 (s, 3H); ^{13}C{^{1}H} NMR (100 MHz, DMSO-d_6, 298 K): δ = 158.7, 150.7, 141.8, 136.3, 135.0, 130.5, 130.0, 116.5, 114.3, 114.2, 109.3, 106.6, 55.5, and 37.5. HRMS (ESI) *m/z*: [M + H]$^+$ calculated $C_{16}H_{16}N_4OS$, 313.1118; observed, 313.1120.

3.7.6. 4,4'-(4-((4-. Methoxybenzyl)thio)-1,3-Phenylene)bis(4H-1,2,4-Triazole) (3f')

The title compound was prepared according to general procedure 3 on a 3.64 g (14.0 mmol) scale; for this reaction, 2.2 equiv. of *N,N*-dimethylformamide azine dihydrochloride was used. Purification by column chromatography (silica gel, 1. dichloromethane, and 2. dichloromethane/methanol 10:1; R_f = 0.49) gave the product as a pale yellow powder in 21% (1.08 mg, 2.95 mmol) yield; m.p. 250–253 °C. ^{1}H NMR (400 MHz, DMSO-d_6, 298 K): δ = 9.21 (s, 2H), 8.74 (s, 2H), 7.99 (s, 1H), 7.91 (d, *J* = 7.0 Hz, 1H), 7.81 (d, *J* = 7.6 Hz, 1H), 7.20 (d, *J* = 6.2 Hz, 2H), 6.85 (d, *J* = 6.4 Hz, 2H), 4.20 (s, 2H), and 3.71 (s, 3H); ^{13}C{^{1}H} NMR (100 MHz, DMSO-d_6, 298 K): δ = 158.6, 143.3, 141.1, 133.7, 132.8, 132.4, 131.4, 130.0, 127.9, 121.8, 119.7, 113.9, 55.1, and 36.4. HRMS (ESI) *m/z*: [M + H]$^+$ calculated $C_{18}H_{16}N_6OS$, 365.1179; observed, 365.1180.

3.7.7. 4-(5-Methoxy-2-((4-Methoxybenzyl)thio)phenyl)-4H-1,2,4-triazole (3g)

The title compound was prepared according to general procedure 3 on a 4.13 g (15.0 mmol) scale. Purification by column chromatography (silica gel, 1. dichloromethane, and 2. dichloromethane/acetone 4:1; R_f = 0.44) gave the product as a white powder in 60% (2.92 g, 8.93 mmol) yield; m.p. 128–130 °C. ^{1}H NMR (400 MHz, DMSO-d_6, 298 K): δ = 8.56 (s, 2H), 7.57 (dd, *J* = 9.2 Hz, *J* = 2.1 Hz, 1H), 7.11–7.08 (m, 2H), 7.00 (d, *J* = 8.6 Hz, 2H), 6.80 (d, *J* = 8.6 Hz, 2H), 3.89 (s, 2H), 3.80 (s, 3H), and 3.71 (s, 3H); ^{13}C{^{1}H} NMR (100 MHz, DMSO-d_6, 298 K): δ = 159.3, 158.4, 143.3, 136.0, 135.1, 129.9, 128.9, 122.1, 116.0, 113.8, 112.9, 55.8, 55.1, and 36.7 HRMS (ESI) *m/z*: [M + H]$^+$ calculated $C_{17}H_{18}N_3O_2S$, 328.1114; observed 328.1114.

3.7.8. 4-((4-Methoxybenzyl)thio)-3-(4H-1,2,4-Triazol-4-Yl)phenol (3h)

The title compound was prepared according to general procedure 3 on a 900 mg (3.44 mmol) scale. Purification by column chromatography (silica gel, 1. dichloromethane, and 2. dichloromethane/acetone 4:1; R_f = 0.22) gave the product as a white powder in 55% (570 mg, 1.91 mmol) yield; m.p. 212–214 °C. ^{1}H NMR (400 MHz, DMSO-d_6, 298 K): δ = 10.25 (s, 1H), 8.52 (s, 2H), 7.45 (d, *J* = 8.6 Hz, 1H), 6.97 (d, *J* = 8.6 Hz, 2H), 6.91 (dd, *J* = 8.6 Hz, *J* = 2.6 Hz, 1H), 6.80–6.78 (m, 3H) 3.81 (s, 2H), and 3.71 (s, 3H); ^{13}C{^{1}H} NMR (100 MHz, DMSO-d_6, 298 K): δ = 158.3, 158.0, 143.3, 136.4, 136.1, 129.9, 129.0, 119.7, 117.1, 114.1, 113.7, and 55.1. HRMS (ESI) *m/z*: [M + H]$^+$ calculated $C_{16}H_{15}N_3O_2S$, 314.0958; observed, 314.0959.

3.7.9. 3-((4-Methoxybenzyl)thio)-2-(4H-1,2,4-Triazol-4-Yl)pyridine (3i)

The title compound was prepared according to general procedure 3 on a 3.45 g (14.0 mmol) scale. Purification by column chromatography (silica gel, 1. dichloromethane, and 2. dichloromethane/acetone 4:1; R_f = 0.44) gave the product as an orange powder in 10% (421 mg, 1.41 mmol) yield; m.p. 154–156 °C. ^{1}H NMR (400 MHz, DMSO-d_6, 298 K): δ = 8.71 (s, 2H), 8.46 (d, *J* = 6.3 Hz, 1H), 7.77 (d, *J* = 7.2 Hz, 1H), 7.39 (d, *J* = 8.28 Hz, 2H), 6.99–6.92 (m, 3H), 5.83 (s, 2 H), and 3.74 (s, 3 H); ^{13}C{^{1}H} NMR (100 MHz, DMSO-d_6, 298 K): δ = 175.6, 159.0, 143.3, 143.2, 136.1, 133.0, 129.8, 127.5, 114.0, 111.8, 58.2, and 55.1. HRMS (ESI) *m/z*: [M + H]$^+$ calculated $C_{15}H_{14}N_4OS$, 299.0961; observed, 299.0961.

3.7.10. 2-((4-Methoxybenzyl)thio)-3-(4H-1,2,4-Triazol-4-Yl)pyridine (3j)

The title compound was prepared according to general procedure 3 on a 3.70 g (15.0 mmol) scale. Purification by column chromatography (silica gel, 1. dichloromethane, and 2. dichloromethane/acetone 4:1; R_f = 0.48) gave the product as a white powder in 11%

yield (504 mg, 1.70 mmol); m.p. 219–220 °C. ^{1}H NMR (400 MHz, DMSO-d_6, 298 K): δ = 8.85 (s, 2H), 8.42 (dd, J = 4.7 Hz, J = 1.4 Hz, 1H), 8.16 (dd, J = 8.0 Hz, J = 1.4 Hz, 1H), 7.56 (dd, J = 8.0 Hz, J = 4.7 Hz, 1H), 7.18 (d, J = 8.6 Hz, 2H), 6.83 (d, J = 8.6 Hz, 2H), 4.20 (s, 2H), and 3.71 (s, 3H); ^{13}C{^{1}H} NMR (100 MHz, DMSO-d_6, 298 K): δ = 158.6, 146.4, 144.9, 142.4, 140.2, 130.1, 128.5, 127.6, 125.0, 113.9, 55.1, and 36.4. HRMS (ESI) m/z: [M + H]$^+$ calculated $C_{15}H_{14}N_4OS$, 299.0961; observed, 299.0960.

3.7.11. 4-(2-((4-Methoxybenzyl)thio)-5-(Trifluoromethyl)phenyl)-4H-1,2,4-triazole (3k)

The title compound was prepared according to general procedure 3 on an 8.15 g (26.0 mmol) scale. Purification by column chromatography (silica gel, 1. dichloromethane, and 2. dichloromethane/acetone 4:1; R_f = 0.62) gave the product as an off-white powder in 48% (4.54 g, 12.41 mmol) yield; m.p. 126–127 °C. ^{1}H NMR (400 MHz, CD$_3$CN, 298 K): δ = 8.32 (s, 2H), 7.76 (dd, J = 8.5 Hz, J = 1.6 Hz, 1H), 7.73 (d, J = 8.4 Hz, 1H), 7.68 (s, 1H), 7.19 (dt, J = 8.7 Hz, J = 2.0 Hz, 2H), 6.83 (dt, J = 8.7 Hz, J = 2.1 Hz, 2H), 4.16 (s, 2H), and 3.74 (s, 3H); ^{13}C{^{1}H} NMR (100 MHz, CD$_3$CN, 298 K): δ = 160.2, 143.9, 141.2, 133.8, 131.0, 130.7, 128.7 (q, J_{C-F} = 33.2 Hz), 128.4, 127.6 (q, J_{C-F} = 3.7 Hz), 125.3 (q, J_{C-F} = 3.8 Hz), 124.6 (q, J_{C-F} = 269.7 Hz), 115.0, 55.9, and 37.4; ^{19}F{^{1}H} NMR (376 MHz, CD$_3$CN, 298 K, referenced to C_6H_5F): δ = −61.45. HRMS (ESI) m/z: [M + H]$^+$ calculated $C_{17}H_{14}F_3N_3OS$, 366.0883; observed, 366.0882.

3.7.12. 4-((4-Methoxybenzyl)thio)-3-(4H-1,2,4-Triazol-4-yl)benzonitrile (3l)

The title compound was prepared according to general procedure 3 on a 1.94 g (7.16 mmol) scale. Purification by column chromatography (silica gel, 1. dichloromethane, and 2. dichloromethane/acetone 4:1; R_f = 0.53) gave the product as a pale yellow powder in 32% (750 mg, 2.33 mmol) yield; m.p. 209–210 °C. ^{1}H NMR (400 MHz, DMSO-d_6, 298 K): δ = 8.75 (s, 2H), 8.05 (d, J = 1.7 Hz, 2H), 7.97 (dd, J = 8.3 Hz, J = 1.7 Hz, 1H), 7.78 (d, J = 8.4 Hz, 1H), 7.28 (d, J = 8.7 Hz, 2H), 6.87 (d, J = 8.7 Hz, 2H), 4.32 (s, 2H), and 3.72 (s, 3H); ^{13}C{^{1}H} NMR (100 MHz, DMSO-d_6, 298 K): δ = 158.7, 143.2, 142.4, 133.3, 131.7, 130.8, 130.2, 127.9, 126.9, 117.7, 114.0, 108.0, 55.1, and 34.9. HRMS (ESI) m/z: [M + H]$^+$ calculated $C_{17}H_{14}N_4OS$, 323.0961; observed, 323.0962.

3.7.13. 4-((4-Methoxybenzyl)thio)-3-(4H-1,2,4-Triazol-4-Yl)benzoic Acid (3m)

A 50 mL round-bottomed flask was charged with compound 3m' (1.50 mmol, 1 equiv.), KOH (2 equiv.), and MeOH, and then heated to 35 °C for 1 h. After removal of the solvent under reduced pressure, the crude product was purified by column chromatography (silica gel, 1. dichloromethane, and 2. dichloromethane/methanol 10:1, R_f = 0.19), which gave the product as a white powder in 87% (413 mg, 1.21 mmol) yield; m.p. > 360 °C. ^{1}H NMR (400 MHz, DMSO-d_6, 298 K): δ = 13.29 (s, 1H), 8.74 (s, 2H), 8.02 (d, J = 8.0 Hz, 1H), 7.85 (s, 1H), 7.74 (d, J = 8.3 Hz, 1H), 7.27 (d, J = 8.1 Hz, J, 2H), 6.86 (d, J = 8.4 Hz, 2H), 4.27 (s, 2H), and 3.71 (s, 3H); ^{13}C{^{1}H} NMR (100 MHz, DMSO-d_6, 298 K): δ = 166.0, 158.6, 143.3, 140.7, 131.5, 130.4, 130.2, 128.6, 127.8, 127.7, 127.3, 114.0, 7, 55.1, and 35.2. HRMS (ESI) m/z: [M + H]$^+$ calculated $C_{17}H_{15}N_3O_3S$, 342.0907; observed, 342.0908.

3.7.14. Ethyl 4-((4-Methoxybenzyl)thio)-3-(4H-1,2,4-Triazol-4-Yl)benzoate (3m')

The title compound was prepared according to general procedure 3 on a 6.89 g (15.0 mmol) scale. Purification by column chromatography (silica gel, 1. dichloromethane, and 2. dichloromethane/acetone 4:1; R_f = 0.64) gave the product as a pale yellow powder in 32% (1.76 g, 4.76 mmol) yield; m.p. 126–127 °C. ^{1}H NMR (400 MHz, DMSO-d_6, 298 K): δ = 8.74 (s, 2H), 8.03 (dd, J = 8.4 Hz, J = 1.6 Hz, 1H), 7.89 (d, J = 1.5 Hz, 1H), 7.76 (d, J = 8.4 Hz, 1H), 7.28 (d, J = 8.5 Hz, 2H), 6.86 (d, J = 8.6 Hz, 2H), 4.34–4.28 (m, 4H), 3.71 (s, 3H), and 1.31 (t, J = 7.1 Hz, 1H); ^{13}C{^{1}H} NMR (100 MHz, DMSO-d_6, 298 K): δ = 164.5, 158.6, 143.2, 141.5, 131.5, 130.2, 127.7, 127.6, 127.3, 127.1, 114.0, 61.2, 55.1, 35.1, and 14.1. HRMS (ESI) m/z: [M + H]$^+$ calculated $C_{19}H_{19}N_3O_3S$, 370.1220; observed, 370.1222.

3.7.15. 4-(5-Bromo-2-((4-Methoxybenzyl)thio)phenyl)-4H-1,2,4-triazole (**3n**)

The title compound was prepared according to general procedure 3 on a 1.30 g (4.00 mmol) scale. Purification by column chromatography (silica gel, 1. dichloromethane, and 2. dichloromethane/acetone 4:1; R_f = 0.62) gave the product as a pale brown powder in 28% (421 mg, 1.15 mmol) yield; m.p. 163–165 °C. ^{1}H NMR (400 MHz, DMSO-d_6, 298 K): δ = 8.67 (s, 2H), 7.77 (s, 1H), 7.71 (d, J = 8.0 Hz, 1H), 7.57 (d, J = 8.2 Hz, 1H), 7.17 (d, J = 7.3 Hz, 1H), 6.84 (d, J = 7.2 Hz, 1H), 4.13 (s, 2H), and 3.71 (s, 3H); ^{13}C{^{1}H} NMR (100 MHz, DMSO-d_6, 298 K): δ = 158.5, 143.2, 133.9, 133.3, 132.7, 131.5, 130.0 (2 signals), 127.8, 119.0, 113.9, 55.1, and 36.3. HRMS (ESI) m/z: [M + H]$^+$ calculated C$_{16}$H$_{14}$BrN$_3$OS, 376.0114; observed, 376.0113.

3.7.16. 4-(4-Bromo-2-((4-Methoxybenzyl)thio)phenyl)-4H-1,2,4-triazole (**3o**)

The title compound was prepared according to general procedure 3 on a 7.78 g (24.0 mmol) scale. Purification by column chromatography (silica gel, 1. dichloromethane, and 2. dichloromethane/acetone 4:1; R_f = 0.60) gave the product as a pale brown powder in 32% (2.86 g, 7.83 mmol) yield; m.p. 173–175 °C. ^{1}H NMR (400 MHz, DMSO-d_6, 298 K): δ = 8.67 (s, 2H), 7.80 (d, J = 2.1 Hz, 1H), 7.58 (dd, J = 8.6 Hz, J = 2.1 Hz, 1H), 7.40 (d, J = 8.3 Hz, 1H), 7.20 (d, J = 8.6 Hz, 2H), 6.86 (d, J = 8.6 Hz, 2H), 4.22 (s, 2H), and 3.72 (s, 3H); ^{13}C{^{1}H} NMR (100 MHz, DMSO-d_6, 298 K): δ = 158.6, 143.3, 136.6, 131.6, 131.4, 130.2, 129.5, 129.1, 127.5, 123.0, 114.0, 55.1, and 35.9. HRMS (ESI) m/z: [M + H]$^+$ calculated C$_{16}$H$_{14}$BrN$_3$OS, 376.0114; observed, 376.0112.

3.7.17. 4-(5-Chloro-2-((4-Methoxybenzyl)thio)phenyl)-4H-1,2,4-triazole (**3p**)

The title compound was prepared according to general procedure 3 on a 3.36 g (12.0 mmol) scale. Purification by column chromatography (silica gel, 1. dichloromethane, and 2. dichloromethane/acetone 4:1; R_f = 0.60) gave the product as a pale yellow powder in 32% (1.29 g, 3.88 mmol) yield; m.p. 158–159 °C. ^{1}H NMR (400 MHz, CDCl$_3$, 298 K): δ = 8.08 (s, 2H), 7.48 (d, J = 8.5 Hz, 1H), 7.39 (dd, J = 8.5 Hz, J = 2.1 Hz, 1H), 7.19 (d, J = 2.1 Hz, 1H), 6.94 (d, J = 8.6 Hz, 1H), 6.73 (d, J = 8.6 Hz, 1H), 3.85 (s, 1H), and 3.73 (s, 1H); ^{13}C{^{1}H} NMR (100 MHz, CDCl$_3$, 298 K): δ = 159.1, 142.7, 134.9, 133.8, 133.6, 131.7, 130.0 129.8, 127.6, 126.8, 114.1, 55.3, and 39.0. HRMS (ESI) m/z: [M + H]$^+$ calculated C$_{16}$H$_{14}$ClN$_3$OS, 332.0619; observed, 332.0618.

3.7.18. 4-(3-Chloro-2-((4-Methoxybenzyl)thio)phenyl)-4H-1,2,4-triazole (**3q**)

The title compound was prepared according to general procedure 3 on a 4.20 mg (15.0 mmol) scale. Purification by column chromatography (silica gel, 1. dichloromethane, and 2. dichloromethane/acetone 4:1; R_f = 0.43) gave the product as an off-white powder in 39% (1.93 mg, 5.82 mmol) yield; m.p. 121–123 °C. ^{1}H NMR (400 MHz, DMSO-d_6, 298 K): δ = 8.41 (s, 2H), 7.79 (dd, J = 8.1 Hz, J = 1.3 Hz, 1H), 7.57 (t, J = 8.0 Hz, 1H), 7.46 (dd, J = 7.9 Hz, J = 1.3 Hz, 1H), 3.87 (s, 2H), and 3.71 (s, 3H); ^{13}C{^{1}H} NMR (100 MHz, DMSO-d_6, 298 K): δ = 158.5, 143.4, 140.4, 139.1, 131.0 (2 signals), 129.8, 129.8, 128.6, 126.3, 113.8, 55.1, and 38.0. HRMS (ESI) m/z: [M + H]$^+$ calculated C$_{16}$H$_{14}$ClN$_3$OS, 332.0619; observed, 332.0617.

3.7.19. 4-(3-Fluoro-2-((4-Methoxybenzyl)thio)phenyl)-4H-1,2,4-triazole (**3r**)

The title compound was prepared according to general procedure 3 on a 2.11 g (8.00 mmol) scale. Purification by column chromatography (silica gel, 1. dichloromethane, and 2. dichloromethane/acetone 4:1; R_f = 0.41) gave the product as a pale brown powder in 87% (2.20 g, 6.984 mmol) yield; m.p. 218–219 °C. ^{1}H NMR (400 MHz, DMSO-d_6, 298 K): δ = 8.46 (s, 2H), 7.62–7.56 (m, 1H), 7.50 (td, J = 9.7 Hz, J = 1.0 Hz, 1H), 7.35 (d, J = 7.9 Hz, 1H), 6.91 (d, J = 8.6 Hz, 2H), 6.77 (d, J = 8.6 Hz, 2H), 3.88 (s, 2H), and 3.70 (s, 3H); ^{13}C{^{1}H} NMR (100 MHz, DMSO-d_6, 298 K): δ = 163.0 (d, J_{C-F} = 244.5 Hz), 158.5, 143.3, 137.9 (d, J_{C-F} = 3.0 Hz), 131.2 (d, J_{C-F} = 9.9 Hz), 129.8, 128.8, 123.2 (d, J_{C-F} = 3.2 Hz), 118.7 (d, J_{C-F} = 21.0 Hz), 116.9 (d, J_{C-F} = 23.8 Hz), 131.8, 55.1, and 38.0. ^{19}F{^{1}H} NMR (376 MHz,

DMSO-d_6, 298 K, referenced to C$_6$H$_5$F): δ = −102.99. HRMS (ESI) m/z: [M + H]$^+$ calculated C$_{16}$H$_{14}$FN$_3$OS, 316.0915; observed, 316.0913.

3.7.20. 4-(4-Fluoro-2-((4-Methoxybenzyl)thio)phenyl)-4H-1,2,4-triazole (3s)

The title compound was prepared according to general procedure 3 on a 2.788 g (10.58 mmol) scale. Purification by column chromatography (silica gel, 1. dichloromethane, and 2. dichloromethane/acetone 4:1; R$_f$ = 0.38) gave the product as a pale brown powder in 35% (1.148 g, 3.64 mmol) yield; m.p. 127–128 °C. ^{1}H NMR (400 MHz, DMSO-d_6, 298 K): δ = 8.65 (s, 2H), 7.54–7.49 (m, 2H), 7.25–7.20 (m, 3H), 4.23 (s, 2H), and 3.71 (s, 2H); ^{13}C{^{1}H} NMR (100 MHz, DMSO-d_6, 298 K): δ = 162.3 (d, J_{C-F} = 247.1 Hz), 158.6, 143.5, 137.3 (d, J_{C-F} = 9.1 Hz), 130.1, 129.4 (d, J_{C-F} = 9.6 Hz), 128.4 (d, J_{C-F} = 2.8 Hz), 127.5, 115.5 (d, J_{C-F} = 25.4 Hz), 114.0, 113.3 (d, J_{C-F} = 23.0 Hz), 131.8, 55.0, and 36.7. ^{19}F{^{1}H} NMR (376 MHz, DMSO-d_6, 298 K, referenced to C$_6$H$_5$F): δ = -110.77. HRMS (ESI) m/z: [M + H]$^+$ calculated C$_{16}$H$_{14}$FN$_3$OS, 316.0915; observed, 316.0913.

3.8. Synthesis and Isolation of 2-(4H-1,2,4-Triazol-4-Yl)benzenethiol (4a) for Optimization Reactions and NMR Studies

A 25 mL round-bottomed flask equipped with a stir bar was loaded with 594.8 mg of triazole 3a (2 mmol, 1 equiv.) and 7.65 mL of trifluoroacetic acid (100 mmol, 50 equiv.). After flushing with argon, 1.09 mL of anisole (10 mmol, 5 equiv.) and 0.99 mL of trifluoromethanesulfonic acid (10 mmol, 5 equiv.) were added to this solution, and the reaction mixture was stirred at 0 °C for 1 h. The solution was then concentrated under high vacuum. The resulting dark red oil was triturated with hexanes and then diluted with 20 mL of water and extracted twice with dichloromethane (50 mL). The combined organic phases were dried over MgSO$_4$, filtered, and concentrated. The target compound was obtained as a yellow oil (304.8 mg, 86%) and was used without any further purification. ^{1}H NMR (400 MHz, DMSO-d_6, 298 K): δ = 8.78 (s, 2H), 7.66 (dd, J = 8.0 Hz, J = 1.3 Hz, 1H), 7.45–7.41 (m, 2 H), 7.34 (t, J = 7.0 Hz, 1H), and 5.79 (s, 1H); LCMS m/z: [M + H]$^+$ calculated C$_9$H$_7$N$_3$O, 178.0; observed, 178.1.

3.9. General Procedure 4 for the Synthesis of benzo[4,5]thiazolo[2,3-c][1,2,4]triazoles (6a–6s)

A 25 mL round-bottomed flask equipped with a stir bar was loaded with the triazole (3a–3s, 1 equiv.) and trifluoroacetic acid (50 equiv.). After flushing with argon, anisole (5 equiv.) and trifluoromethanesulfonic acid (5 equiv.) were added to this solution, and the reaction mixture was stirred at 0 °C for 1 h. The solution was then concentrated under high vacuum. The resulting dark red oil was diluted with 20 mL of water and extracted twice with dichloromethane (50 mL). Following concentration of the combined organic phases, the crude thiol was dissolved in 2 mL of DMSO and heated to 100 C. After completion of the cyclization reaction, the reaction mixture was diluted with 20 mL of sat. NaHCO$_3$ and extracted twice with dichloromethane (50 mL). The combined organic phases were dried over MgSO$_4$, filtered, and concentrated. The resulting crude product was purified by column chromatography or recrystallization as described below.

3.9.1. benzo[4,5]thiazolo[2,3-c][1,2,4]triazole (6a)

The title compound was prepared twice according to general procedure 4 on a 297 mg (1.0 mmol) and a 1.49 g (5.00 mmol) scale. Purification by column chromatography (silica gel, 1. dichloromethane, and 2. dichloromethane/acetone 4:1, R$_f$ = 0.50) gave the product as a light yellow powder in 85% (149 mg, 0.85 mmol) and 90% (788 mg, 4.5 mmol) yield; m.p. 179–181 °C. ^{1}H NMR (400 MHz, DMSO-d_6, 298 K): δ = 9.64 (s, 1H), 8.11 (dd, J = 8.0 Hz, J = 0.5 Hz, 1H), 8.05 (dd, J = 8.0 Hz, J = 0.4 Hz, 1H), 7.59 (td, J = 8.2 Hz, J = 1.0 Hz, 1H), and 7.49 (td, J = 8.3 Hz, J = 1.1 Hz, 1H); ^{13}C{^{1}H} NMR (100 MHz, DMSO-d_6, 298 K): δ = 154.4, 136.8, 131.5, 129.0, 127.0, 126.6, 125.5, and 114.8. HRMS (ESI) m/z: [M + H] calculated for C$_8$H$_5$N$_3$S, 176.0277; observed, 176.0276.

3.9.2. 8-Methylbenzo[4,5]thiazolo[2,3-c][1,2,4]triazole (6b)

The title compound was prepared according to general procedure 4 on a 156 mg (0.50 mmol) scale. Purification by column chromatography (silica gel, 1. dichloromethane, and 2. dichloromethane/methanol 5:1, R_f = 0.48) gave the product as an off-white powder in 89% (38 mg, 0.445 mmol) yield; m.p. 185–187 °C. ^{1}H NMR (400 MHz, DMSO-d_6, 298 K): δ = 9.62 (s, 1H), 7.92 (d, J = 8.0 Hz, 1H), 7.48 (t, J = 7.7 Hz, 1H), 7.32 (d, J = 7.6 Hz, 1H), and 2.43 (s, 3H); ^{13}C{^{1}H} NMR (100 MHz, DMSO-d_6, 298 K): δ = 153.6, 136.9, 134.2, 131.0, 128.7, 127.1, 127.0, 122.3, and 19.2. HRMS (ESI) m/z: [M + H]$^+$ calculated for $C_9H_7N_3S$, 190.0434; observed, 190.0435.

3.9.3. 5-Methylbenzo[4,5]thiazolo[2,3-c][1,2,4]triazole (6c)

The title compound was prepared according to general procedure 4 on a 311 mg (1.00 mmol) scale. Purification by column chromatography (silica gel, 1. dichloromethane, and 2. dichloromethane/methanol 5:1, R_f = 0.46) gave the product as a light yellow powder in 82% (156 mg, 0.825 mmol) yield; m.p. 189–191 °C. ^{1}H NMR (400 MHz, DMSO-d_6, 298 K): δ = 9.49 (s, 1H), 7.83–7.79 (m, 1H), 7.34 (t, J = 5.3 Hz, 1H), and 2.69 (s, 3H); ^{13}C{^{1}H} NMR (100 MHz, DMSO-d_6, 298 K): δ = 154.4, 138.0, 131.3, 128.6, 128.5, 126.1, 125.6, 122.6, and 18.7. HRMS (ESI) m/z: [M + H]$^+$ calculated for $C_9H_7N_3S$, 190.0434; observed, 190.0433.

3.9.4. Naphtho[2′,1′:4,5]thiazolo[2,3-c][1,2,4]triazole (6d)

The title compound was prepared according to general procedure 4 on a 174 mg (0.50 mmol) scale. Purification by column chromatography (silica gel, 1. dichloromethane, and 2. dichloromethane/methanol 5:1, R_f = 0.46) gave the product as a light yellow powder in 87% (98 mg, 0.433 mmol) yield; m.p. 238–239 °C. ^{1}H NMR (400 MHz, DMSO-d_6, 298 K): δ = 9.72 (s, 1H), 8.25 (d, J = 8.8 Hz, 1H), 8.15 (t, J = 8.6 Hz, 2H), 8.00 (d, J = 8.2 Hz, 1H), 7.74 (td, J = 7.1 Hz, J = 0.9 Hz, 1H), and 7.67 (td, J = 8.1 Hz, J = 1.0 Hz, 1H); ^{13}C{^{1}H} NMR (100 MHz, DMSO-d_6, 298 K): δ = 159.8, 136.8, 131.1, 129.4, 128.5, 128.1, 128.1, 127.2, 126.9, 126.5, 123.2, and 113.9. HRMS (ESI) m/z: [M + H]$^+$ calculated for $C_{12}H_7N_3S$, 226.0434; observed, 266.0433.

3.9.5. N-(benzo[4,5]thiazolo[2,3-c][1,2,4]triazol-6-Yl)acetamide (6e)

The title compound was prepared according to general procedure 4 on a 354 mg (1.00 mmol) scale. Purification by column chromatography (silica gel, 1. dichloromethane, and 2. dichloromethane/methanol 5:1, R_f = 0.05) gave the product as a light yellow powder in 82% (190 mg, 0.82 mmol) yield; m.p. 313–315 °C. ^{1}H NMR (400 MHz, DMSO-d_6, 298 K): δ = 10.43 (s, 1H), 9.69 (s, 1H), 8.56 (d, J = 1.8 Hz, 1H), 7.93 (d, J = 8.8 Hz, 1H), 7.44 (dd, J = 8.8 Hz, J = 2.0 Hz, 1H), and 2.11 (s, 3H); ^{13}C{^{1}H} NMR (100 MHz, DMSO-d_6, 298 K): δ = 168.8, 160.4, 151.4, 138.4, 136.7, 129.1, 125.5, 117.9, 105.3, and 24.0. HRMS (ESI) m/z: [M + H]$^+$ calculated for $C_{10}H_8N_4OS$, 233.0492; observed, 233.0492.

3.9.6. benzo[4,5]thiazolo[2,3-c][1,2,4]triazol-6-Amine (6f)

The title compound was prepared according to general procedure 4 on a 156 mg (0.50 mmol) scale. Purification by column chromatography (silica gel, 1. dichloromethane, and 2. dichloromethane/methanol 5:1, R_f = 0.10) gave the product as a light yellow powder in 60% (114 mg, 0.30 mmol) yield; m.p. 220–222 °C. ^{1}H NMR (400 MHz, DMSO-d_6, 298 K): δ = 9.51 (s, 1H), 7.57 (d, J = 8.7 Hz, 1H), 7.15 (d, J = 2.1 Hz, 1H), 6.71 (dd, J = 8.7 Hz, J = 2.1 Hz, 1H), and 5.68 (s, 2H); ^{13}C{^{1}H} NMR (100 MHz, DMSO-d_6, 298 K): δ = 155.2, 148.7, 136.3, 129.8, 125.5, 115.7, 133.7, and 98.9. HRMS (ESI) m/z: [M + H]$^+$ calculated for $C_8H_6N_4S$, 191.0386; observed, 191.0387.

3.9.7. 6-(4H-1,2,4-Triazol-4-Yl)benzo[4,5]thiazolo[2,3-c][1,2,4]triazole (6f′)

The title compound was prepared according to general procedure 4 on a 182 mg (0.50 mmol) scale. Purification by column chromatography (silica gel, 1. dichloromethane, and 2. dichloromethane/methanol 5:1, R_f = 0.09) gave the product as an off-white powder in 90% (375 mg, 0.471 mmol) yield. ^{1}H NMR (400 MHz, DMSO-d_6, 298 K): δ = 9.53 (s, 1H), 9.17 (s, 2H), 8.60 (d, J = 2.2 Hz, 1H), 8.27 (d, J = 8.8 Hz, 1H), and 7.86 (dd, J = 8.7 Hz, J = 2.2 Hz, 1H); ^{13}C{^{1}H} NMR was not obtained due to poor solubility. HRMS (ESI) m/z: [M + H]$^+$ calculated for $C_{10}H_7N_6S$, 243.0448; observed, 243.0447.

3.9.8. 6-Methoxybenzo[4,5]thiazolo[2,3-c][1,2,4]triazole (6g)

The title compound was prepared according to general procedure 4 on an 82 mg (0.25 mmol) scale. Purification by column chromatography (silica gel, 1. dichloromethane, and 2. dichloromethane/methanol 5:1, R_f = 0.38) gave the product as an off-figurewhite powder in 81% (42 mg, 0.202 mmol) yield; m.p. 223–224 °C. ^{1}H NMR (400 MHz, DMSO-d_6, 298 K): δ = 9.57 (s, 1H), 7.89 (d, J = 8.9 Hz, 1H), 7.82 (d, J = 2.4 Hz, 1H), 7.08 (dd, J = 8.9 Hz, J = 2.5 Hz, 1H), and 3.86 (s, 3H); ^{13}C{^{1}H} NMR (100 MHz, DMSO-d_6, 298 K): δ = 158.8, 155.4, 136.5, 129.7, 126.0, 122.1, 114.1, 100.4, and 55.9. HRMS (ESI) m/z: [M + H]$^+$ calculated for $C_9H_7N_3OS$, 206.0383; observed, 206.0382.

3.9.9. benzo[4,5]thiazolo[2,3-c][1,2,4]triazol-6-Ol (6h)

The title compound was prepared according to general procedure 4 on a 313 mg (1.00 mmol) scale. Purification by column chromatography (silica gel, 1. dichloromethane, and 2. dichloromethane/methanol 10:1, R_f = 0.05) gave the product as an off-white powder in 84% (159.7 mg, 0.84 mmol) yield; m.p. 217–219 °C. ^{1}H NMR (400 MHz, DMSO-d_6, 298 K): δ = 9.57 (s, 1H), 7.78 (d, J = 8.8 Hz, 1H), 7.50 (d, J = 2.3 Hz, 1H), and 6.93 (dd, J = 8.8 Hz, J = 2.3 Hz, 1H); ^{13}C{^{1}H} NMR (100 MHz, DMSO-d_6, 298 K): δ = 157.0, 155.3, 136.6, 129.8, 126.0, 120.2, 115.0, and 102.0. HRMS (ESI) m/z: [M + H]$^+$ calculated for $C_8H_4N_3OS$, 192.0226; observed, 192.0224.

3.9.10. [1,2,4]. triazolo[3′,4′:2,3]thiazolo[4,5-b]pyridine (6i)

The title compound was prepared according to general procedure 4 on a 75 mg (0.25 mmol) scale. Purification by column chromatography (silica gel, 1. dichloromethane, and 2. dichloromethane/methanol 5:1, R_f = 0.55) gave the product as a light orange powder in 76% (34 mg, 0.190 mmol) yield; m.p. 218–220 °C. ^{1}H NMR (400 MHz, DMSO-d_6, 298 K): δ = 9.63 (s, 1H), 8.56–8.52 (m, 2H), and 7.56 (dd, J = 8.1 Hz, J = 4.9 Hz, 1H); ^{13}C{^{1}H} NMR (100 MHz, DMSO-d_6, 298 K): δ = 153.8, 146.2, 141.4, 136.1, 135.1, 127.0, and 122.2. HRMS (ESI) m/z: [M + H]$^+$ calculated for $C_7H_4N_4S$, 177.0230; observed, 177.0230.

3.9.11. [1,2,4]. triazolo[3′,4′:2,3]thiazolo[5,4-b]pyridine (6j)

The title compound was prepared according to general procedure 4 on a 75 mg (0.25 mmol) scale. Purification by column chromatography (silica gel, 1. dichloromethane, and 2. dichloromethane/methanol 5:1, R_f = 0.55) gave the product as a light orange powder in 70% (31 mg, 0.176 mmol) yield; m.p. 220–221 °C. ^{1}H NMR (400 MHz, DMSO-d_6, 298 K): δ = 9.69 (s, 1H), 8.58 (dd, J = 4.8 Hz, J = 1.4 Hz, 1H), 8.51 (dd, J = 8.2 Hz, J = 1.4 Hz, 1H), and 7.68 (dd, J = 8.2 Hz, J = 4.9 Hz, 1H); ^{13}C{^{1}H} NMR (100 MHz, DMSO-d_6, 298 K): δ = 153.9, 151.1, 147.6, 137.7, 125.4, 122.7, and 122.1. HRMS (ESI) m/z: [M + H]$^+$ calculated for $C_7H_4N_4S$, 177.0230; observed, 177.0229.

3.9.12. 6-(Trifluoromethyl)benzo[4,5]thiazolo[2,3-c][1,2,4]triazole (6k)

The title compound was prepared according to general procedure 4 on a 186 mg (0.50 mmol) scale. Purification by column chromatography (silica gel, 1. dichloromethane, and 2. dichloromethane/methanol 5:1, R_f = 0.38) gave the product as a light yellow powder in 86% (105 mg, 0.430 mmol) yield; m.p. 203–204 °C. ^{1}H NMR (400 MHz, DMSO-d_6, 298 K): δ = 9.69 (s, 1H), 8.63 (s, 1H), 8.28 (d, J = 8.5 Hz, 1H), and 7.83 (d, J = 8.4 Hz, 1H); ^{13}C{^{1}H} NMR (100 MHz, DMSO-d_6, 298 K): δ = 154.8, 137.2, 136.6, 129.3, 127.3 (q, J_{C-F} = 32.6 Hz), 126.8, 123.9 (q, J_{C-F} = 270.6 Hz), 122.9 (q, J_{C-F} = 3.7 Hz), and 112. 3 (q, J_{C-F} = 4.1 Hz). ^{19}F{^{1}H} NMR (376 MHz, DMSO-d_6, 298 K, referenced to C_6H_5F): δ = −60.85. HRMS (ESI) m/z: [M + H]$^+$ calculated for $C_9H_4F_3N_3S$, 244.0151; observed, 244.0151.

3.9.13. benzo[4,5]thiazolo[2,3-c][1,2,4]triazole-6-Carbonitrile (6l)

The title compound was prepared according to general procedure 4 on a 322 mg (1.00 mmol) scale. Purification by column chromatography (silica gel, 1. dichloromethane, and 2. dichloromethane/methanol 10:1, R_f = 0.38) gave the product as a light yellow powder in 80% (159 mg, 0.795 mmol) yield; m.p. 267–269 °C. ^{1}H NMR (400 MHz, DMSO-d_6, 298 K): δ = 9.61 (s, 1H), 8.69 (s, 1H), 8.30 (d, J = 8.4 Hz, 1H), and 7.95 (d, J = 8.0 Hz, 1H); ^{13}C{^{1}H} NMR (100 MHz, DMSO-d_6, 298 K): δ = 154.7, 137.8, 137.1, 129.6, 129.3, 126.7, 118.6, 118.1, and 108.9. HRMS (ESI) m/z: [M + H]$^+$ calculated for $C_9H_4N_4S$, 201.0229; observed, 201.0230.

3.9.14. benzo[4,5]thiazolo[2,3-c][1,2,4]triazole-6-Carboxylic acid (6m)

The title compound was prepared according to general procedure 4 on a 341 mg (1.00 mmol) scale. Purification by column chromatography (silica gel, 1. dichloromethane, and 2. dichloromethane/methanol 5:1, R_f = 0.38) gave the product as a white powder in 85% (186 mg, 0.847 mmol) yield; m.p. > 360 °C. ^{1}H NMR (400 MHz, DMSO-d_6, 298 K): δ = 9.78 (s, 1H), 8.74 (d, J = 1.1 Hz, 1H), 8.16 (d, J = 8.4 Hz, 1H), and 8.03 (dd, J = 8.5 Hz, J = 1.6 Hz, 1H); ^{13}C{^{1}H} NMR (100 MHz, DMSO-d_6, 298 K): δ = 167.1, 158.0, 152.9, 138.3, 131.8, 129.1, 125.7, 123.9, and 113.8. HRMS (ESI) m/z: [M + H]$^+$ calculated for $C_9H_5N_3O_2S$, 220.0175; observed, 220.0176.

3.9.15. Ethyl benzo[4,5]thiazolo[2,3-c][1,2,4]triazole-6-Carboxylate (6m')

The title compound was prepared according to general procedure 4 on a 341 mg (1.00 mmol) scale. Purification by column chromatography (silica gel, 1. dichloromethane, and 2. dichloromethane/methanol 5:1, R_f = 0.38) gave the product as a light yellow powder in 86% (173 mg, 0.865 mmol) yield; m.p. 209–210 °C. ^{1}H NMR (400 MHz, DMSO-d_6, 298 K): δ = 9.80 (s, 1H), 8.73 (s, 1H), 8.19 (d, J = 8.5 Hz, 1H), 8.04 (d, J = 8.4 Hz, 1H), 4.39 (q, J = 7.1 Hz, 2H), 3.86 (s, 3H), and 1.37 (t, J = 7.1 Hz, 3H); ^{13}C{^{1}H} NMR (100 MHz, DMSO-d_6, 298 K): δ = 164.8, 137.2, 129.3, 128.5, 126.9, 125.7, 115.5, 61.4, and 14.2. HRMS (ESI) m/z: [M + H]$^+$ calculated for $C_{11}H_9N_3O_2S$, 248.0488; observed, 248.0490.

3.9.16. 6-Bromobenzo[4,5]thiazolo[2,3-c][1,2,4]triazole (6n)

The title compound was prepared according to general procedure 4 on a 118 mg (0.50 mmol) scale. Purification by column chromatography (silica gel, 1. dichloromethane, and 2. dichloromethane/methanol 5:1, R_f = 0.66) gave the product as an off-white powder in 88% (112 mg, 0.442 mmol) yield; m.p. 240–243 °C. ^{1}H NMR (400 MHz, DMSO-d_6, 298 K): δ = 9.59 (s, 1H), 8.47 (s, 1H), 8.01 (d, J = 8.4 Hz, 1H), and 8.67 (d, J = 8.3 Hz, 1H); ^{13}C{^{1}H} NMR (100 MHz, DMSO-d_6, 298 K): δ = 154.8, 136.9, 131.1, 130.0, 129.2, 127.2, 119.2, and 118.0. HRMS (ESI) m/z: [M + H]$^+$ calculated for $C_8H_4BrN_3S$, 253.9382; observed, 253.9381.

3.9.17. 7-Bromobenzo[4,5]thiazolo[2,3-c][1,2,4]triazole (6o)

The title compound was prepared twice according to general procedure 4 on a 376 mg (1.0 mmol) scale and a 3.76 g (10 mmol) scale. The title compound was triturated with acetone to afford an off-white powder in 90% (229 mg, 0.90 mmol) and 92% (2.34 g, 9.2 mmol) yield; m.p. 273–275 °C. ^{1}H NMR (400 MHz, DMSO-d_6, 298 K): δ = 9.65 (s, 1H), 8.35 (d, J = 1.9 Hz, 1H), 8.08 (t, J = 8.6 Hz, 1H), and 7.79 (dd, J = 8.6 Hz, J = 2.0 Hz, 1H); ^{13}C{^{1}H} NMR (100 MHz, DMSO-d_6, 298 K): δ = 154.5, 134.0, 133.8, 129.9, 128.4, 127.8, 118.3, and 116.5. HRMS (ESI) m/z: [M + H]$^+$ calculated for $C_8H_4BrN_3S$, 253.9382; observed, 253.9381.

3.9.18. 6-Chlorobenzo[4,5]thiazolo[2,3-c][1,2,4]triazole (6p)

The title compound was prepared according to general procedure 4 on a 166 mg (0.50 mmol) scale. Purification by column chromatography (silica gel, 1. dichloromethane, and 2. dichloromethane/methanol 5:1, R_f = 0.38) gave the product as an off-white powder in 91% (99 mg, 0.457 mmol) yield; m.p. 229–231 °C. ^{1}H NMR (400 MHz, DMSO-d_6, 298 K): δ = 9.59 (s, 1H), 8.36 (d, J = 1.8 Hz, 1H), 8.08 (d, J = 8.6 Hz, 1H), and 7.56 (dd, J = 8.7 Hz,

J = 2.0 Hz, 1H); ^{13}C{^{1}H} NMR (100 MHz, DMSO-d_6, 298 K): δ = 154.7, 137.4, 131.8, 131.0, 130.3, 127.4, 127.0, and 115.7. HRMS (ESI) m/z: [M + H]$^+$ calculated for C$_8$H$_4$ClN$_3$S, 209.9887; observed, 209.9887.

3.9.19. 8-Chlorobenzo[4,5]thiazolo[2,3-c][1,2,4]triazole (6q)

The title compound was prepared according to general procedure 4 on a 166 mg (1.00 mmol) scale. Purification by column chromatography (silica gel, 1. dichloromethane, and 2. dichloromethane/methanol 5:1, R$_f$ = 0.36) gave the product as an off-white powder in 90% (93.8 mg, 0.448 mmol) yield; m.p. 249–251 °C. ^{1}H NMR (400 MHz, DMSO-d_6, 298 K): δ = 9.68 (s, 1H), 8.14–8.10 (m, 1H), and 7.66–7.64 (m, 2H); ^{13}C{^{1}H} NMR (100 MHz, DMSO-d_6, 298 K): δ = 153.0, 137.4, 130.8, 130.2, 128.6, 127.7, 126.3, and 113.8. HRMS (ESI) m/z: [M + H]$^+$ calculated for C$_8$H$_4$ClN$_3$S, 209.9887; observed, 209.9886.

3.9.20. 8-Fluorobenzo[4,5]thiazolo[2,3-c][1,2,4]triazole (6r)

The title compound was prepared according to general procedure 4 on a 315 mg (1.00 mmol) scale. Purification by column chromatography (silica gel, 1. dichloromethane, and 2. dichloromethane/methanol 5:1, R$_f$ = 0.40) gave the product as a light yellow powder in 98% (188 mg, 0.975 mmol) yield; m.p. 211–212 °C. ^{1}H NMR (400 MHz, DMSO-d_6, 298 K): δ = 9.69 (s, 1H), 7.99 (d, J = 8.1 Hz, 1H), 7.69–7.63 (m, 1H), and 7.46 (t, J = 7.6 Hz, 1H); ^{13}C{^{1}H} NMR (100 MHz, DMSO-d_6, 298 K): δ = 156.5 (d, J_{C-F} = 244.7 Hz), 153.9, 137.3, 130.8 (d, J_{C-F} = 6.8 Hz), 129.0 (d, J_{C-F} = 7.8 Hz), 118.1 (d, J_{C-F} = 22.3 Hz), 112.9 (d, J_{C-F} = 18.1 Hz), and 111.4 (d, J_{C-F} = 3.4 Hz). ^{19}F{^{1}H} NMR (376 MHz, DMSO-d_6, 298 K, referenced to C$_6$H$_5$F): δ = −114.97. HRMS (ESI) m/z: [M + H]$^+$ calculated for C$_8$H$_4$FN$_3$S, 194.0183; observed, 194.0182.

3.9.21. 7-Fluorobenzo[[1,3]4,5]thiazolo[2,3-c][1,2,4]triazole (6s)

The title compound was prepared according to general procedure 4 on a 315 mg (1.00 mmol) scale. Purification by column chromatography (silica gel, 1. dichloromethane, and 2. dichloromethane/methanol 5:1, R$_f$ = 0.38) gave the product as an off-white powder in 98% (188 mg, 0.976 mmol) yield; m.p. 255–258 °C. ^{1}H NMR (400 MHz, DMSO-d_6, 298 K): δ = 9.59 (s, 1H), 8.14 (dd, J = 8.9 Hz, J = 4.6 Hz, 1H), 8.01 (dd, J = 8.8 Hz, J = 2.6 Hz, 1H), and 7.47 (td, J = 9.0 Hz, J = 2.6 Hz, 1H); ^{13}C{^{1}H} NMR (100 MHz, DMSO-d_6, 298 K): δ = 159.7 (d, J_{C-F} = 242.0 Hz), 154.6, 136.8, 133.3 (q, J_{C-F} = 11.0 Hz), 125.9 (q, J_{C-F} = 2.1 Hz), 116.1 (d, J_{C-F} = 9,4 Hz), 114.5 (d, J_{C-F} = 24.7 Hz), and 112.5 (d, J_{C-F} = 28.2 Hz). ^{19}F{^{1}H} NMR (376 MHz, DMSO-d_6, 298 K, referenced to C$_6$H$_5$F): δ = −114.27. HRMS (ESI) m/z: [M + H]$^+$ calculated for C$_8$H$_4$FN$_3$S, 194.0183; observed, 194.0183.

4. Conclusions

In summary, a novel and efficient method for the synthesis of substituted benzo[4,5]thiazolo[2,3-c][1,2,4]triazoles species (6a–6s) from the corresponding p-methoxybenzyl-protected 4-(2-mercaptophenyl)triazoles (3a–3s) has been developed. Following the selective removal of the protecting group, the free thiols (4a–4b) are oxidized to their corresponding disulfides (5a–5s). These disulfides are thought to undergo C-H bond functionalization, thus leading to an intramolecular ring closure, thereby forming the C-S bond of the target heterocycle. Our synthetic approach allows the preparation of benzo[4,5]thiazolo[2,3-c][1,2,4]triazoles (6a–6s) containing synthetically valuable functional groups on their benzene rings. The combination of short reaction times and good to excellent isolated yields, regardless of the nature of the substituents, is a clear advantage of this scalable reaction protocol.

Supplementary Materials: The following supporting information can be downloaded, Copies of the ^{1}H NMR and ^{13}C NMR spectra for compounds 1b–1r, 2b–2s, 3b–3s, and 6a–6s.

Author Contributions: Conceptualization, J.L.B. and L.G.A.-M.; methodology, J.L.B. and L.G.A.-M.; investigation, J.L.B. and L.G.A.-M.; resources, J.L.B.; data curation, J.L.B. and L.G.A.-M.; writing—

original draft preparation, L.G.A.-M.; writing—review and editing, J.L.B.; supervision, J.L.B.; project administration, J.L.B.; funding acquisition, J.L.B. All authors have read and agreed to the published version of the manuscript.

Funding: Acknowledgment is made to the donors of the American Chemical Society Petroleum Research Fund for support of this research (ACS PRF Doctoral New Investigator (DNI) Research Grant; proposal number: 58507 DNI-1).

Institutional Review Board Statement: Not applicable.

Informed Consent Statement: Not applicable.

Data Availability Statement: The data presented in this study are available in supplementary material.

Acknowledgments: We would like to thank Oklahoma State University and the Chemistry Department of Oklahoma State University for their support.

Conflicts of Interest: The authors declare no conflict of interest.

Sample Availability: Samples of the compounds are not available from the authors.

References

1. Bagley, M.C.; Dale, J.W.; Merritt, E.A.; Xiong, X. Thiopeptide Antibiotics. *Chem. Rev.* **2005**, *105*, 685–714. [CrossRef] [PubMed]
2. Hu, D.X.; Withall, D.M.; Challis, G.L.; Thomson, R.J. Structure, Chemical Synthesis, and Biosynthesis of Prodiginine Natural Products. *Chem. Rev.* **2016**, *116*, 7818–7853. [CrossRef] [PubMed]
3. Lamberth, C. Heterocyclic Chemistry in Crop Protection. *Pest Manag. Sci.* **2013**, *69*, 1106–1114. [CrossRef] [PubMed]
4. Maienfisch, P.; Edmunds, A.J.F. Thiazole and Isothiazole Ring–Containing Compounds in Crop Protection. In *Advances in Heterocyclic Chemistry*; Elsevier: Amsterdam, The Netherlands, 2017; Volume 121, pp. 35–88. ISBN 978-0-12-811174-1.
5. Lamberth, C.; Walter, H.; Kessabi, F.M.; Quaranta, L.; Beaudegnies, R.; Trah, S.; Jeanguenat, A.; Cederbaum, F. The Significance of Organosulfur Compounds in Crop Protection: Current Examples from Fungicide Research. *Phosphorus Sulfur Silicon Relat. Elem.* **2015**, *190*, 1225–1235. [CrossRef]
6. Heravi, M.M.; Zadsirjan, V. Prescribed Drugs Containing Nitrogen Heterocycles: An Overview. *RSC Adv.* **2020**, *10*, 44247–44311. [CrossRef]
7. Bhutani, P.; Joshi, G.; Raja, N.; Bachhav, N.; Rajanna, P.K.; Bhutani, H.; Paul, A.T.; Kumar, R.U.S. FDA Approved Drugs from 2015–June 2020: A Perspective. *J. Med. Chem.* **2021**, *64*, 2339–2381. [CrossRef]
8. Pathania, S.; Narang, R.K.; Rawal, R.K. Role of Sulphur-Heterocycles in Medicinal Chemistry: An Update. *Eur. J. Med. Chem.* **2019**, *180*, 486–508. [CrossRef]
9. Vikas, S.; Mohit, G.; Pradeep, K.; Atul, S. A Comprehensive Review on Fused Heterocyclic as DNA Intercalators: Promising Anticancer Agents. *Curr. Pharm. Des.* **2021**, *27*, 15–42.
10. Sbenati, R.M.; Semreen, M.H.; Semreen, A.M.; Shehata, M.K.; Alsaghir, F.M.; El-Gamal, M.I. Evaluation of Imidazo[2,1-b]Thiazole-Based Anticancer Agents in One Decade (2011–2020): Current Status and Future Prospects. *Bioorganic Med. Chem.* **2021**, *29*, 115897. [CrossRef]
11. Slivka, M.V.; Korol, N.I.; Fizer, M.M. Fused Bicyclic 1,2,4-Triazoles with One Extra Sulfur Atom: Synthesis, Properties, and Biological Activity. *J. Heterocycl. Chem.* **2020**, *57*, 3236–3254. [CrossRef]
12. Sharma, P.C.; Bansal, K.K.; Sharma, A.; Sharma, D.; Deep, A. Thiazole-Containing Compounds as Therapeutic Targets for Cancer Therapy. *Eur. J. Med. Chem.* **2020**, *188*, 112016. [CrossRef]
13. Hussain, H.; Al-Harrasi, A.; Al-Rawahi, A.; Green, I.R.; Gibbons, S. Fruitful Decade for Antileishmanial Compounds from 2002 to Late 2011. *Chem. Rev.* **2014**, *114*, 10369–10428. [CrossRef]
14. Rana, A.; Siddiqui, N.; Khan, S.A. Benzothiazoles: A New Profile of Biological Activities. *Indian J. Pharm. Sci.* **2007**, *69*, 10. [CrossRef]
15. Aboelmagd, A.; Ali, I.A.I.; Salem, E.M.S.; Abdel-Razik, M. Synthesis and Antifungal Activity of Some S-Mercaptotriazolo benzothiazolyl Amino Acid Derivatives. *Eur. J. Med. Chem.* **2013**, *60*, 503–511. [CrossRef]
16. Majalakere, K.; Kunhana, S.B.; Rao, S.; Kalal, B.S.; Badiadka, N.; Sanjeev, G.; Holla, B.S. Studies on Imidazo[2,1-b][1,3]Benzothiazole Derivatives as New Radiosensitizers. *SN Appl. Sci.* **2020**, *2*, 1902. [CrossRef]
17. Pushpavalli, S.N.C.V.L.; Janaki Ramaiah, M.; Lavanya, A.; Raksha Ganesh, A.; Kumbhare, R.M.; Bhadra, K.; Bhadra, U.; Pal-Bhadra, M. Imidazo–Benzothiazoles a Potent MicroRNA Modulator Involved in Cell Proliferation. *Bioorganic Med. Chem. Lett.* **2012**, *22*, 6418–6424. [CrossRef]
18. Maddili, S.K.; Yandrati, L.P.; Siddam, S.; Kannekanti, V.K.; Gandham, H. Green Synthesis, Biological and Spectroscopic Study on the Interaction of Multi-Component Mannich Bases of Imidazo[2,1-b]Benzothiazoles with Human Serum Albumin. *J. Photochem. Photobiol. B: Biol.* **2017**, *176*, 9–16. [CrossRef]

19. Shaik, S.P.; Vishnuvardhan, M.V.P.S.; Sultana, F.; Subba Rao, A.V.; Bagul, C.; Bhattacharjee, D.; Kapure, J.S.; Jain, N.; Kamal, A. Design and Synthesis of 1,2,3-Triazolo Linked Benzo[d]Imidazo[2,1-b]Thiazole Conjugates as Tubulin Polymerization Inhibitors. *Bioorganic Med. Chem.* **2017**, *25*, 3285–3297. [CrossRef]
20. Andreani, A.; Granaiola, M.; Leoni, A.; Locatelli, A.; Morigi, R.; Rambaldi, M.; Varoli, L.; Lannigan, D.; Smith, J.; Scudiero, D.; et al. Imidazo[2,1-b]Thiazole Guanylhydrazones as RSK2 Inhibitors [1]. *Eur. J. Med. Chem.* **2011**, *46*, 4311–4323. [CrossRef]
21. Williams, N.S.; Gonzales, S.; Naidoo, J.; Rivera-Cancel, G.; Voruganti, S.; Mallipeddi, P.; Theodoropoulos, P.C.; Geboers, S.; Chen, H.; Ortiz, F.; et al. Tumor-Activated Benzothiazole Inhibitors of Stearoyl-CoA Desaturase. *J. Med. Chem.* **2020**, *63*, 9773–9786. [CrossRef]
22. Chaniyara, R.; Tala, S.; Chen, C.-W.; Lee, P.-C.; Kakadiya, R.; Dong, H.; Marvania, B.; Chen, C.-H.; Chou, T.-C.; Lee, T.-C.; et al. Synthesis and Antitumor Evaluation of Novel Benzo[d]Pyrrolo[2,1-b]Thiazole Derivatives. *Eur. J. Med. Chem.* **2012**, *53*, 28–40. [CrossRef]
23. Abdelazeem, A.H.; Alqahtani, A.M.; Omar, H.A.; Bukhari, S.N.A.; Gouda, A.M. Synthesis, Biological Evaluation and Kinase Profiling of Novel S-benzo[4,5]thiazolo[2,3-c][1,2,4]Triazole Derivatives as Cytotoxic Agents with Apoptosis-Inducing Activity. *J. Mol. Struct.* **2020**, *1219*, 128567. [CrossRef]
24. Trapani, G.; Franco, M.; Latrofa, A.; Reho, A.; Liso, G. Synthesis, in Vitro and in Vivo Cytotoxicity, and Prediction of the Intestinal Absorption of Substituted 2-Ethoxycarbonyl-Imidazo[2,1-b]Benzothiazoles. *Eur. J. Pharm. Sci.* **2001**, *14*, 209–216. [CrossRef]
25. Mase, T.; Arima, H.; Tomioka, K.; Yamada, T.; Murase, K. Imidazo[2,1-b]Benzothiazoles. 2. New Immunosuppressive Agents. *J. Med. Chem.* **1986**, *29*, 386–394. [CrossRef] [PubMed]
26. Stella, A.; Segers, K.; De Jonghe, S.; Vanderhoydonck, B.; Rozenski, J.; Anné, J.; Herdewijn, P. Synthesis and Antibacterial Evaluation of a Novel Series of 2-(1,2-Dihydro-3-Oxo-3H-Pyrazol-2-Yl)Benzothiazoles. *Chem. Biodivers.* **2011**, *8*, 253–265. [CrossRef] [PubMed]
27. Mortimer, C.G.; Wells, G.; Crochard, J.-P.; Stone, E.L.; Bradshaw, T.D.; Stevens, M.F.G.; Westwell, A.D. Antitumor Benzothiazoles. 26. 2-(3,4-Dimethoxyphenyl)-5-Fluorobenzothiazole (GW 610, NSC 721648), a Simple Fluorinated 2-Arylbenzothiazole, Shows Potent and Selective Inhibitory Activity against Lung, Colon, and Breast Cancer Cell Lines. *J. Med. Chem.* **2006**, *49*, 179–185. [CrossRef] [PubMed]
28. Wang, M.; Gao, M.; Mock, B.H.; Miller, K.D.; Sledge, G.W.; Hutchins, G.D.; Zheng, Q.-H. Synthesis of Carbon-11 Labeled Fluorinated 2-Arylbenzothiazoles as Novel Potential PET Cancer Imaging Agents. *Bioorganic Med. Chem.* **2006**, *14*, 8599–8607. [CrossRef] [PubMed]
29. Froyd, J.D. Tricyclazole: A New Systemic Fungicide for Control of Piricularia Oryzae on Rice. *Phytopathology* **1976**, *66*, 1135. [CrossRef]
30. Woloshuk, C.P.; Sisler, H.D.; Tokousbalides, M.C.; Dutky, S.R. Melanin Biosynthesis in Pyricularia Oryzae: Site of Tricyclazole Inhibition and Pathogenicity of Melanin-Deficient Mutants. *Pestic. Biochem. Physiol.* **1980**, *14*, 256–264. [CrossRef]
31. Kumar, M.; Chand, R.; Shah, K. Evidences for Growth-Promoting and Fungicidal Effects of Low Doses of Tricyclazole in Barley. *Plant Physiol. Biochem.* **2016**, *103*, 176–182. [CrossRef]
32. Kukreja, S.; Sidhu, A.; Sharma, V.K. Synthesis of Novel 7-Fluoro-3-Substituted-1,2,4-Triazolo[3,4-b]Benzothiazoles (FTBs) as Potent Antifungal Agents: Molecular Docking and in Silico Evaluation. *Res. Chem Intermed* **2016**, *42*, 8329–8344. [CrossRef]
33. Srinivasa, G.M.; Jayachandran, E.; Shivakumar, B.; Rao, D.S. Inhibition of Albumin Denaturation and Anti-Inflammatory Activity of 8-Fluoro-9-Substituted (1,3)-Benzothiazolo (5,1-B)-3-Substituted 1,2,4-Triazoles. *Orient. J. Chem.* **2004**, *20*, 103–110.
34. Deng, X.-Q.; Song, M.-X.; Wei, C.-X.; Li, F.-N.; Quan, Z.-S. Synthesis and Anticonvulsant Activity of 7-Alkoxy-Triazolo-[3, 4-b] Benzo[d]Thiazoles. *Med. Chem.* **2010**, *6*, 313–320. [CrossRef]
35. Balcer, J.L.; DeAmicis, C.V.; Johnson, P.L.; Klosin, J.; Whiteker, G.T.; Srinivas Rao, C.; Dai, D. Synthesis of New Compounds Related to the Commercial Fungicide Tricyclazole. *Pest Manag. Sci.* **2011**, *67*, 556–559. [CrossRef]
36. Gvozdjakova, A.; Ivanovičová, H. Synthesis and Reactions of Both Tautomers of 2-Hydrazinobenzothiazole. *Chem. Pap.* **1986**, *40*, 797–800.
37. Desmukh, M.B.; Patil, S.S.; Shejwal, R.V. Synthesis of New Triazolo and Pyrazolo Derivatives of Benzothiazole. *Indian J. Heterocycl. Chem.* **2010**, *20*, 163–166.
38. Abdelazeem, A.H.; Alqahtani, A.M.; Arab, H.H.; Gouda, A.M.; Safi El-Din, A.G. Novel benzo[4,5]thiazolo[2,3-c][1,2,4]Triazoles: Design, Synthesis, Anticancer Evaluation, Kinase Profiling and Molecular Docking Study. *J. Mol. Struct.* **2021**, *1246*, 131138. [CrossRef]
39. Demmer, C.S.; Jørgensen, M.; Kehler, J.; Bunch, L. Study of Oxidative Cyclization Using PhI(OAc)2 in the Formation of benzo[4,5]thiazolo[2,3-c][1,2,4]Triazoles and Related Heterocycles–Scope and Limitations. *Synlett* **2014**, *25*, 1279–1282. [CrossRef]
40. Wikel, J.H.; Paget, C.J. Synthesis of S-Triazole[3,4-b]Benzothiazoles. *J. Org. Chem.* **1974**, *39*, 3506–3508. [CrossRef]
41. Jayanthi, G.; Muthusamy, S.; Paramasivam, R.; Ramakrishnan, V.T.; Ramasamy, N.K.; Ramamurthy, P. Photochemical Synthesis of S-Triazolo[3,4-b]Benzothiazole and Mechanistic Studies on Benzothiazole Formation. *J. Org. Chem.* **1997**, *62*, 5766–5770. [CrossRef]
42. Khirpak, S.M.; Staninets, V.I.; Slivka, M.V.; Yu, L. Zborovskii Oxidative Heterocyclization of Sodium Salts of 3-Mercapto-4-phenyl-4H-1,2,4-triazoles. *Ukr. Khimicheskii Zhurnal* **2001**, *67*, 110–113.
43. Recknagel, R.O. Carbon Tetrachloride Hepatotoxicity. *Pharm. Rev.* **1967**, *19*, 145–208.

44. Holbrook, M.T. Carbon Tetrachloride. In *Kirk-Othmer Encyclopedia of Chemical Technology*; John Wiley & Sons, Ltd.: Hoboken, NJ, USA, 2000; ISBN 978-0-471-23896-6.
45. Beresneva, T.; Popelis, J.; Abele, E. Novel Cu-Catalyzed Methods for the Synthesis of Fused Thiazoles Using S,N-Diarylation Reaction. *Chem. Heterocycl. Comp.* **2013**, *49*, 345–347. [CrossRef]
46. Holm, S.C.; Rominger, F.; Straub, B.F. Thiol-Functionalized 1,2,4-Triazolium Salt as Carbene Ligand Precursor. *J. Organomet. Chem.* **2012**, *719*, 54–63. [CrossRef]
47. Seitz, S.C.; Rominger, F.; Straub, B.F. Stepwise Deprotonation of a Thiol-Functionalized Bis(1,2,4-Triazolium) Salt as a Selective Route to Heterometallic NHC Complexes. *Organometallics* **2013**, *32*, 2427–2434. [CrossRef]
48. Hutchinson, S.M.; Ardón-Muñoz, L.G.; Ratliff, M.L.; Bolliger, J.L. Catalytic Preparation of 1-Aryl-Substituted 1,2,4-Triazolium Salts. *ACS Omega* **2019**, *4*, 17923–17933. [CrossRef] [PubMed]
49. Isley, N.A.; Linstadt, R.T.H.; Kelly, S.M.; Gallou, F.; Lipshutz, B.H. Nucleophilic Aromatic Substitution Reactions in Water Enabled by Micellar Catalysis. *Org. Lett.* **2015**, *17*, 4734–4737. [CrossRef] [PubMed]
50. Zhang, X.; Lu, G.; Cai, C. Facile Aromatic Nucleophilic Substitution (SNAr) Reactions in Ionic Liquids: An Electrophile–Nucleophile Dual Activation by [Omim]Br for the Reaction. *Green Chem.* **2016**, *18*, 5580–5585. [CrossRef]
51. Carrot, G.; Hilborn, J.G.; Trollsås, M.; Hedrick, J.L. Two General Methods for the Synthesis of Thiol-Functional Polycaprolactones. *Macromolecules* **1999**, *32*, 5264–5269. [CrossRef]

Article

General Method of Synthesis of 5-(Het)arylamino-1,2,3-triazoles via Buchwald–Hartwig Reaction of 5-Amino- or 5-Halo-1,2,3-triazoles

Pavel S. Gribanov [1,*], Anna N. Philippova [1], Maxim A. Topchiy [2], Lidiya I. Minaeva [2], Andrey F. Asachenko [2] and Sergey N. Osipov [1]

[1] A. N. Nesmeyanov Institute of Organoelement Compounds, Russian Academy of Sciences, 28 Vavilova Str., 119991 Moscow, Russia; anyfil96@gmail.com (A.N.P.); osipov@ineos.ac.ru (S.N.O.)

[2] A.V. Topchiev Institute of Petrochemical Synthesis, Russian Academy of Sciences, Leninskiy Prospect 29, 119991 Moscow, Russia; maxtopchiy@yandex.ru (M.A.T.); minaeva.lidiya@gmail.com (L.I.M.); aasachenko@gmail.com (A.F.A.)

* Correspondence: gribanovps@mail.ru; Tel.: +7-499-135-6212

Abstract: An efficient access to the novel 5-(het)arylamino-1,2,3-triazole derivatives has been developed. The method is based on Buchwald–Hartwig cross-coupling reaction of 5-Amino or 5-Halo-1,2,3-triazoles with (het)aryl halides and amines, respectively. As result, it was found that palladium complex [(THP-Dipp)Pd(cinn)Cl] bearing expanded-ring N-heterocyclic carbene ligand is the most active catalyst for the process to afford the target molecules in high yields.

Keywords: cross-coupling; amination; triazoles; palladium; carbene ligands; heterocycles

---

Citation: Gribanov, P.S.; Philippova, A.N.; Topchiy, M.A.; Minaeva, L.I.; Asachenko, A.F.; Osipov, S.N. General Method of Synthesis of 5-(Het)arylamino-1,2,3-triazoles via Buchwald–Hartwig Reaction of 5-Amino- or 5-Halo-1,2,3-triazoles. *Molecules* **2022**, *27*, 1999. https://doi.org/10.3390/molecules27061999

Academic Editor: Joseph Sloop

Received: 28 February 2022
Accepted: 18 March 2022
Published: 20 March 2022

Publisher's Note: MDPI stays neutral with regard to jurisdictional claims in published maps and institutional affiliations.

Copyright: © 2022 by the authors. Licensee MDPI, Basel, Switzerland. This article is an open access article distributed under the terms and conditions of the Creative Commons Attribution (CC BY) license (https://creativecommons.org/licenses/by/4.0/).

1. Introduction

Nitrogen containing heterocycles, in particular five-membered azole systems, are common structural elements of many natural and synthetic biological active compounds. They serve as universal scaffolds for creating new organic molecules with set properties especially for the needs of biomolecular and medicinal chemistry as well as for materials science [1–6]. In the last few decades fully substituted and variously functionalized 1,2,3-triazoles, whose structure fragment is not found in nature, became one of the most interesting and widely used class of compounds due to their unique physicochemical properties and synthetic accessibility [7,8]. These compounds possess remarkable thermal and metabolic stability, large dipole moment, and capability for H-bond formation making them effective peptide bond isosteres [9–11] that result in a variety of applications in diverse fields of chemistry [12–20]. Among fully substituted 1,2,3-triazoles special attention is focused on 5-amino-1,2,3-triazoles and their 5-arylamino derivatives, which exhibit very promising biological properties such as antiviral, antifungal, antiproliferative and antimetastatic activities. They also serve as activators of potassium channel andchelating agents and have a potential for treating inflammatory kidney diseases (Figure 1) [21–26].

Since the pioneering Dimroth works published in the beginning of the 20th century [27,28], keteniminate-mediated 1,3-dipolar cycloaddition (DCR) of organic azides with nitriles bearing an active methylene group provide one of the most efficient and straightforward methods to access to the 5-amino-1,2,3-triazole synthesis up to date (Scheme 1) [29–32].

Unfortunately this approach is not applicable to 5-amino substituted 1,2,3-triazoles including 5-arylamino derivatives. The scope of the existing methods for the synthesis of these compounds is limited to a few examples and has a number of disadvantages. Thus, previously described methods for the preparation of 5-arylamino-1,2,3-triazoles include: (1) interaction between hard accessible carbodiimides and diazo compounds [21]; (2) three-component amine/enolizible ketone/azide reaction leading to low yields of the target products [33]; (3) high temperature thermolysis of the 5-triazenyl-1,2,3-triazoles to give a large amount of 2H-1,2,3-triazole as a by-product [34]; (4) base-mediated hydrolysis of 1,2,3-triazolo[1,5-a]quinazolin-5(4H)-ones [35] as well as Rh-catalyzed azide-alkyne cycloaddition of internal ynamides to afford N,N-disubstituted amino-1,2,3-triazoles [26].

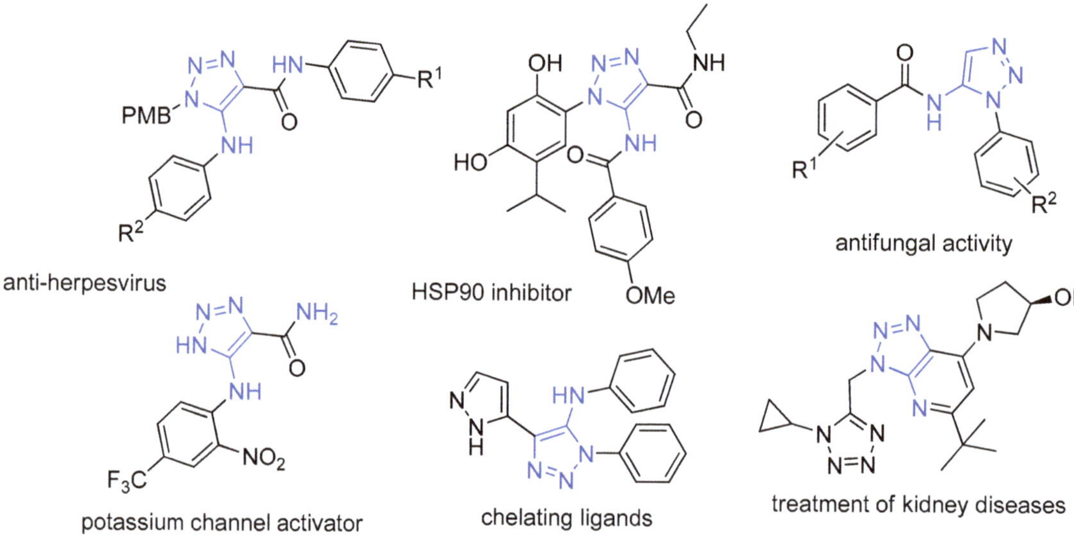

Figure 1. Potential application of 5-amino-1,2,3-triazoles N-substituted derivatives.

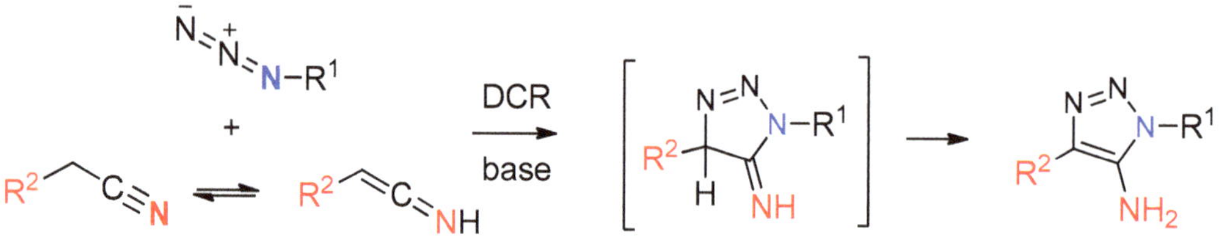

Scheme 1. 1,3-Dipolar cycloaddition reaction (DCR) between aryl azides and monosubstituted acetonitriles.

On the other hand, in the past 30 years, palladium-catalyzed cross-coupling reactions leading to the formation of new C-N bonds have become a widely used tool both in academia and in industry [36,37]. This Buchwald–Hartwig amination is the most popular cross-coupling reaction [38–40] (Figure 2) to access a wide range of N-mono- and N,N-disubstituted arylamines [41]. Despite impressive advances in the field, coupling of heteroaromatic amines with (het)aryl halides still remains problematic, often requiring long reaction times and time-consuming searches for optimal conditions and catalytic systems [3,42–45]. tThere are no examples of Buchwald–Hartwig cross-coupling of 5-halo- and 5-amino-1,2,3-triazoles with (het)aryl amines and halides, respectively, to afford N-aryl amino derivatives except a report on synthesis of related 4-amino-1,2,3-triazoles (with just 3 examples) [46].

Therefore, taking into account the growing popularity of 5-amino-1,2,3-triazole derivatives in medical chemistry, the development of new efficient and robust approaches to their synthesis remains of great interest.

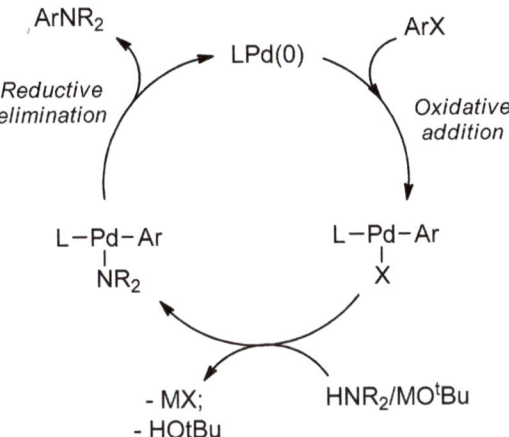

Figure 2. Simplified catalytic cycle for Buchwald–Hartwig amination reaction.

We have recently developed effective methods for obtaining 5-amino- [47] and 5-halo-1,2,3-triazoles [48] via one pot azide-nitrile cycloaddition/Dimroth rearrangement (Scheme 2a) and Cu(I)-catalyzed [3+2] cycloaddition reaction of Cu(I)-acetylide and aryl azides with subsequent Cu-triazolide halogenation (Scheme 2b). Based on our experience in Pd-catalyzed cross-couplings of hetaryl halides [49–52] and halo-1,2,3-triazoles [53,54] we would like to provide details of an efficient route to N-arylamino-1,2,3-triazoles using the Buchwald–Hartwig reaction of 5-amino or 5-halo-1,2,3-triazoles (Scheme 2c).

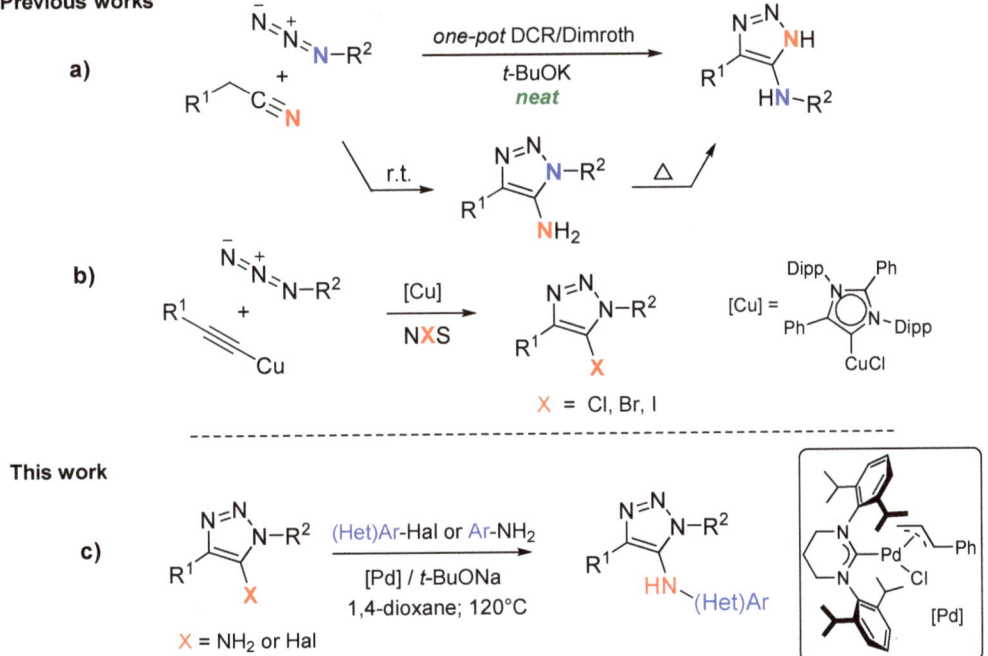

Scheme 2. Synthetic approach to 5-amino-1,2,3-triazoles (**a**), 5-halo-1,2,3-triazoles (**b**) and N-arylamino 1,2,3-triazoles (**c**).

2. Results and Discussion

We commenced our investigation with the reaction between 1-benzyl-4-phenyl-1,2,3-triazole-5-amine and 1-bromo-4-methylbenzene to screen for optimal conditions for the cross-coupling (Table 1). A series of palladium complexes with expanded-ring NHC ligands (Figure 3) were initially tested as they proved to be competent catalysts for Buchwald–Hartwig amination of (het)aryl halides with primary aryl amines [50,51]. We found that the reaction performed in the presence of 1.0 mol% (THP-Dipp)Pd(cinn)Cl and 1.2 equiv. of sodium *tert*-butoxide in 1,4-dioxane at 120 °C for 24 h yielded the desired 5-(*p*-tolyl)amino-1,2,3-triazole **2a** in 53% yield (Table 1, entry 1). The reaction did not reveal the full conversion of the starting materials (TLC and ^{1}H NMR analysis). The prolonged reaction time did not result in a better yield of the product. The increase of the Pd-catalyst loading up to 2 mol% and the base up to 3.0 equiv. almost led to quantitative formation of **2a** (entry 3). Other NHC-Pd complexes with allyl and metallyl ligands exhibited slightly less activity under tested conditions (entries 4, 5). The traditional Pd(OAc)$_2$/phosphine-based catalytic systems [55] were also tested, exhibiting insufficient activity for the process (entries 6–9).

Table 1. Screening of catalytic systems in the BHA reaction [1].

Entry	[Pd] (mol.%)	Base (Equiv.)	Yield, %
1	(THP-Dipp)Pd(cinn)Cl (1)	*t*-BuONa (1.2)	53
2	(THP-Dipp)Pd(cinn)Cl (2)	*t*-BuONa (1.2)	86
3	**(THP-Dipp)Pd(cinn)Cl (2)**	***t*-BuONa (3.0)**	**97**
4	(THP-Dipp)PdAllylCl (2)	*t*-BuONa (1.2)	73
5	(THP-Dipp)PdMetallylCl (2)	*t*-BuONa (1.2)	39
6	Pd(OAc)$_2$ (2 mol %)/RuPhos (4)	*t*-BuONa (1.2)	30
7	Pd(OAc)$_2$ (2 mol %)/SPhos (4)	*t*-BuONa (1.2)	5
8	Pd(OAc)$_2$ (2 mol %)/DavePhos (4)	*t*-BuONa (1.2)	12
9	Pd(OAc)$_2$ (2 mol %)/XPhos (4)	*t*-BuONa (1.2)	6
10	(THP-Dipp)Pd(cinn)Cl (2)	Cs$_2$CO$_3$ (1.2)	16

[1] Reaction conditions: 1-benzyl-4-phenyl-1*H*-1,2,3-triazol-5-amine **1a** (0.5 mmol); 1-bromo-4-methylbenzene (1 equiv.); [Pd], base; 1,4-dioxane (2.5 mL); 120 °C, 24 h.

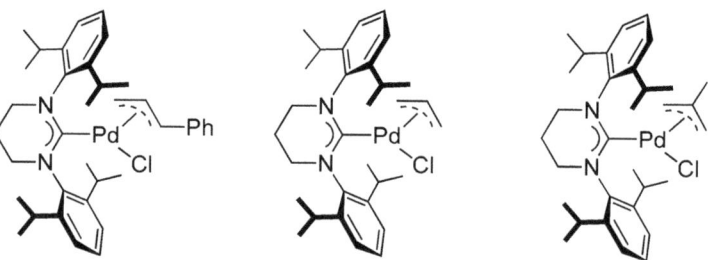

(THP-Dipp)Pd(cinn)Cl   (THP-Dipp)Pd(allyl)Cl   (THP-Dipp)Pd(methallyl)Cl

Figure 3. Structures of (THP-Dipp) Pd complexes.

With these optimized conditions in hand, different 5-amino-1,2,3-triazoles were involved in the Buchwald–Hartwig cross-coupling reactions with a wide range of aromatic and heteroaromatic halides bearing various substituents in their structures. As a result, we found that in all studied cases the nature and location of the substituent in the (het)aryl core of both triazole and halide substrates doesn't not significantly influence the reaction leading to the formation of the corresponding 5-amino-1,2,3-triazoles derivatives **2a–p** including sterically hindered *ortho*-Me aryl derivatives **2b**, **2f**, **2j** in good and excellent yields. It is noteworthy that the reaction works perfectly for both (het)aryl bromides and chlorides (Scheme 3).

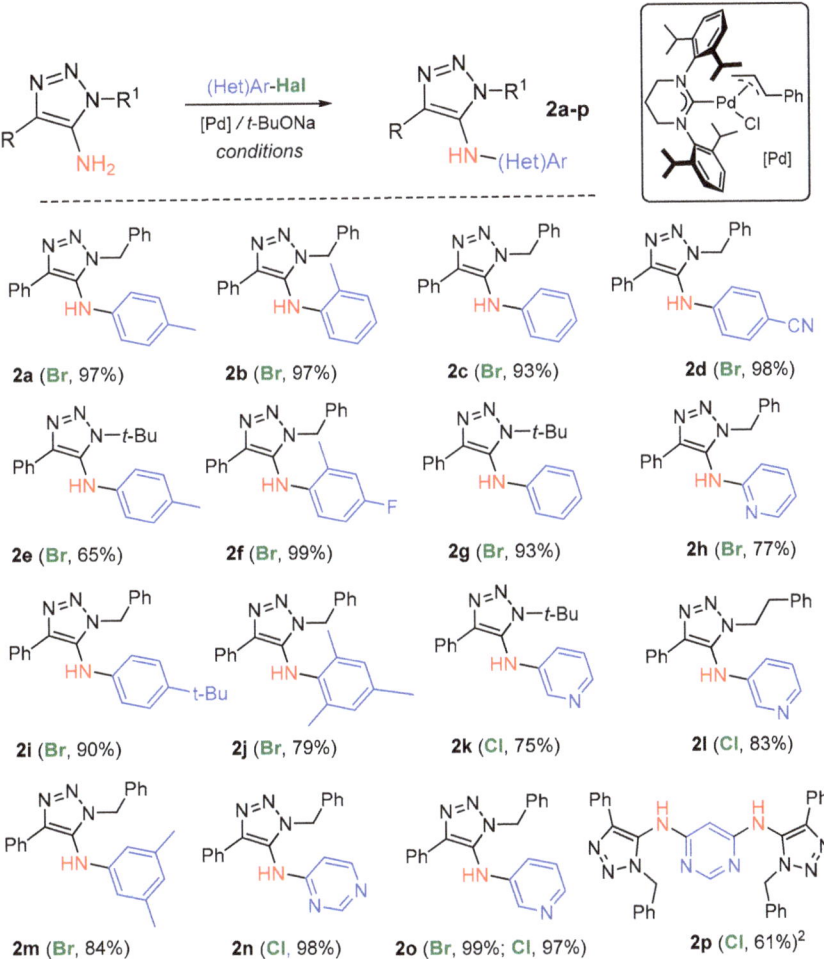

Scheme 3. Buchwald–Hartwig cross-coupling of 5-amino-1,2,3-triazoles [1]. [1] Conditions: 5-amino-1,2,3-triazole (0.5 mmol); (het)aryl-Hal (1 equiv.); (THP-Dipp)Pd(cinn)Cl (2 mol %); *t*-BuONa (3 equiv.); 1,4-dioxane (2.5 mL); 120 °C under argon 24 h; [2] 4,6-Dichloropyrimidine (0.25 mmol); 5-aminotriazole (2.0 equiv.); (THP-Dipp)Pd(cinn)Cl (4 mol %), *t*-BuONa (6 equiv.).

Then, we studied the reversed variant of the Buchwald–Hartwig cross-coupling reaction, namely the interaction of 5-halo-1,2,3-triazoles with aryl amines. Fortunately, we found that the conditions for aminotriazole—aryl halide coupling proved to also be suitable for the combination of halotriazole—aryl amine. Thus, corresponding derivatives of 5-arylamino-1,2,3-triazole such as *N*-(*p*-tolylamino) (**2a**, **2q**) and *N*-(2,4-dimethylamino) (**2r**) triazoles were obtained in good to excellent yields. Arylamines with electron-withdrawing CF$_3$ group(s) in aromatic ring (**2s** and **2t**) can also be successfully used for this reaction.

Example **2u** demonstrates that the method is also applicable for the preparation of 4-(*N*-arylamino)-1,2,3-triazoles from the corresponding 4-halo-1,2,3-triazoles, while their synthesis was previously described via coupling of 4-amino-1,2,3-triazoles [46] (Scheme 4).

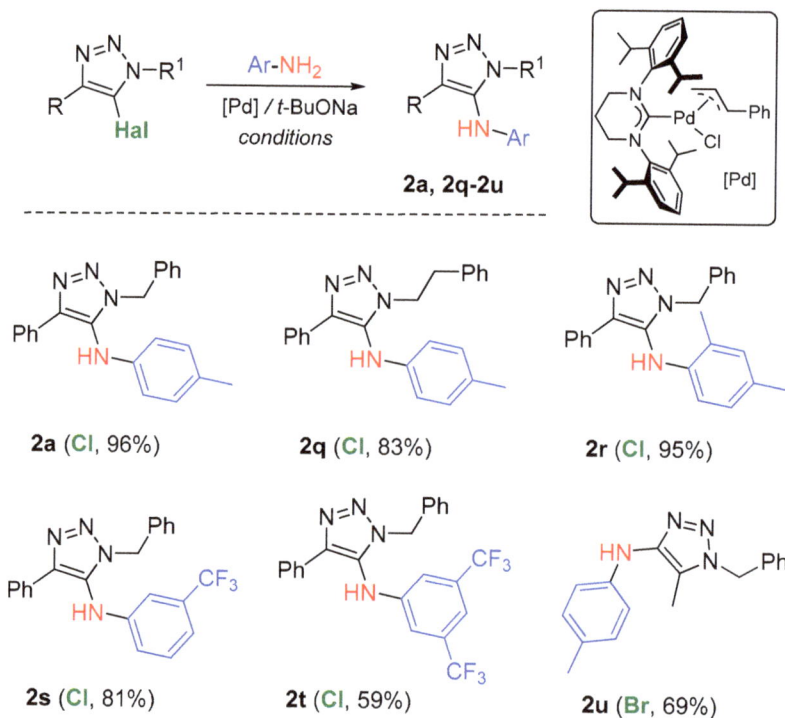

Scheme 4. Buchwald–Hartwig cross-coupling of 4- and 5-halo-1,2,3-triazoles [1]. [1] Reaction conditions: 4- or 5-halo-1,2,3-triazole (0.5 mmol); aryl-NH$_2$ (1 equiv.); (THP-Dipp)Pd(cinn)Cl (2 mol %); *t*-BuONa (3 equiv.); 1,4-dioxane (2.5 mL); 120 °C under argon, 24 h.

3. Materials and Methods

3.1. General Information

All the reactions were carried out under argon atmosphere, and the solvents were distilled from appropriate drying agents prior to use. All reagents were used as purchased from Sigma-Aldrich (Munich, Germany). In the study, 1,4-disubstituted-5-chloro- [48] and 5-amino-1,2,3-triazoles [47] and 1-benzyl-4-bromo-5-methyl-1*H*-1,2,3-triazole [56] were synthesized according to published procedures. (THP-Dipp)Pd(cinn)Cl [57], (THP-Dipp)Pd(allyl)Cl [58] and (THP-Dipp)Pd(metallyl)Cl were synthesized according to published procedure [57] from corresponding NHC-silver (I) complexes. Analytical data was in accordance with the literature data. Analytical TLC was performed with Merck silica gel 60 F 254 plates (Darmstadt, Germany); visualization was accomplished with UV light or iodine vapors. Chromatography was carried out using Merck silica gel (Kieselgel 60, 0.063–0.200 mm, Darmstadt, Germany) and petroleum ether/ethyl acetate as an eluent. The NMR spectra were obtained with Bruker AV-400, Karlsruhe, Germany) (400 MHz ^{1}H, 101 MHz ^{13}C, 376 MHz ^{19}F) using TMS and CCl$_3$F as references for ^{1}H and ^{19}F NMR spectra. respectively. Chemical shifts for ^{1}H and ^{13}C were reported as δ values (ppm).

3.2. General Procedure for Preparation of N-arylamino-1,2,3-triazoles via BHA Reaction of 5-Amino or 4(5)-halo-1,2,3-triazoles

Under argon in a Schlenk tube with magnetic stirring bar, corresponding amino- or halo-1,2,3-triazole (0. 5 mmol), (het)arylhalide or primary amine (1.0 equiv.) were dissolved in dry 1,4-dioxane (2.5 mL) at room temperature. The solution was degassed with three freeze-pump-thaw cycles. Then 6.6 mg (0.01 mmol, 2 mol%) of (THP-Dipp)Pd(cinn)Cl and sodium *tert*-butoxide (3.0 equiv.) were added to the reaction mixture, and the reaction mixture was stirred at 120 °C (oil bath temperature) for 18 h. After cooling to room temperature, the reaction mixture was poured into water and extracted with dichloromethane (3 × 10 mL). The combined organic phases were washed with brine, dried over MgSO$_4$, filtered and concentrated under reduced pressure. Purification by chromatography (eluent—hexane: ethyl acetate 4:1) gave analytically pure corresponding N-arylamino-1,2,3-triazole as a white solid.

3.3. Preparation and Characterization of Novel Compounds

(THP-Dipp)Pd(methallyl)Cl

The title compound was synthesized according to literature procedure [58] from (6-Dipp)AgBr and (2-Methylallyl)palladium(II) chloride dimer as a white powder (88% yield). ^{1}H NMR (400 MHz, Acetone-d_6) δ 7.40–7.11 (m, 6H), 3.85–3.56 (m, 7H), 3.33–3.18 (m, 2H), 2.88 (s, 1H), 2.67–2.59 (m, 2H), 2.53–2.37 (m, 2H), 1.51–1.15 (m, 24H), 1.02 (s, 2H). ^{13}C DEPTQ-135 NMR (Acetone, 101 MHz): δ 214.8, 146.6, 143.8, 130.0, 129.1, 128.2, 125.5, 70.3, 50.3, 49.2, 47.0, 47.0, 29.1, 27.1, 25.2, 24.9, 22.9, 22.1, 21.2, 21.0. HRMS (ESI): calcd for C$_{32}$H$_{47}$N$_2$Pd [(THP-Dipp)Pd(methallyl)]$^+$: 563.2775, 564.2788, 565.2781, 566.2808, 567.2777; found: 563.2776, 564.2795, 565.2788, 566.2810, 567.2780.

1-benzyl-5-(p-tolylamino)-4-phenyl-1H-1,2,3-triazole (**2a**)

From 1-benzyl-4-phenyl-1*H*-1,2,3-triazol-5-amine and 1-bromo-4-methylbenzene (165 mg, 97% yield) or from 1-benzyl-5-chloro-4-phenyl-1*H*-1,2,3-triazole and *p*-toluidine (163 mg, 96% yield), following general procedure, **2a** was obtained as a white solid, m.p. 181–182 °C. ^{1}H NMR (400 MHz, Chloroform-*d*) δ 7.82 (d, *J* = 7.0 Hz, 2H), 7.34–7.26 (m, 6H), 7.22–7.17 (m, 2H), 7.00 (d, *J* = 8.4 Hz, 2H), 6.45 (d, *J* = 8.4 Hz, 2H), 5.36 (s, 2H), 5.05 (s, 1H), 2.27 (s, 3H). ^{13}C{^{1}H} NMR (101 MHz, Chloroform-*d*) δ 141.2, 140.9, 134.8, 132.1, 130.3, 130.0, 129.0, 128.7, 128.4, 128.0, 127.9, 126.0, 114.3, 51.4, 20.6. IR (υ/cm^{-1}): 737.34 (VS), 812 (VS), 1006 (S), 1072 (S), 1177 (S), 1251 (S), 1288 (S), 1325 (S), 1359 (S), 1422 (S), 1441 (S), 1518 (S), 1586 (S), 1610 (S), 1810 (M), 1888 (M), 1955 (M), 2980 (W), 3025 (W), 3249 (M). HRMS (ESI): calcd for C$_{22}$H$_{21}$N$_4$ [M+H]$^+$: 341.1761; found: 341.1769.

1-benzyl-5-(o-tolylamino)-4-phenyl-1H-1,2,3-triazole (**2b**)

From 1-benzyl-4-phenyl-1*H*-1,2,3-triazol-5-amine and 1-bromo-2-methylbenzene, following general procedure, **2b** (165 mg, 97% yield) was obtained as a white solid, m.p. 193–195 °C. ^{1}H NMR (400 MHz, Chloroform-*d*) δ 7.76 (d, *J* = 7.0 Hz, 2H), 7.33–7.25 (m, 6H), 7.18–7.12 (m, 3H), 6.98 (t, *J* = 7.5 Hz, 1H), 6.83 (t, *J* = 7.4 Hz, 1H), 6.25 (d, *J* = 8.1 Hz, 1H), 5.33 (s, 2H), 4.82 (s, 1H), 2.12 (s, 3H). ^{13}C{^{1}H} NMR (101 MHz, Chloroform-*d*) δ 141.5, 141.0, 134.7, 131.9, 131.1, 130.4, 129.0, 128.8, 128.5, 128.1, 127.8, 127.6, 125.9, 123.3, 120.7, 112.7, 51.8,

17.5. IR (υ/cm^{-1}): 3271 (W), 1606 (S), 1586 (S), 1571 (S), 1514 (S), 1496 (S), 1448 (S), 1411 (S), 1362 (S), 1294 (S), 1251 (S), 1159 (S), 1110 (S), 1073 (S), 1006 (S), 769 (VS), 747 (VS), 734 (VS), 717 (VS). HRMS (ESI): calcd for C$_{22}$H$_{21}$N$_4$ [M+H]$^+$: 341.1761; found: 341.1764.

1-benzyl-5-(phenylamino)-4-phenyl-1H-1,2,3-triazole (**2c**)

From 1-benzyl-4-phenyl-1H-1,2,3-triazol-5-amine and bromobenzene, following general procedure, **2c** (151 mg, 93% yield) was obtained as a white solid, m.p. 187–188 °C. ^{1}H NMR (400 MHz, Chloroform-d) δ 7.80 (dd, J = 8.3, 1.4 Hz, 2H), 7.31–7.24 (m, 6H), 7.20–7.15 (m, 4H), 6.87 (t, J = 7.4 Hz, 1H), 6.52 (d, J = 7.6 Hz, 2H), 5.34 (s, 2H), 5.14 (s, 1H). ^{13}C{^{1}H} NMR (101 MHz, Chloroform-d) δ 143.6, 141.1, 134.7, 131.6, 130.2, 129.9, 129.0, 128.8, 128.5, 128.1, 127.9, 126.0, 120.7, 114.3, 51.5. IR (υ/cm^{-1}): 3234 (M), 3180 (W), 2930 (W), 1602 (S), 1582 (S), 1568 (S), 1496 (S), 1445 (S), 1422 (S), 1364 (S), 1325 (S), 1256 (S), 1236 (S), 1176 (S), 1151 (S), 1077 (S), 770 (VS), 752 (VS). HRMS (ESI): calcd for C$_{21}$H$_{19}$N$_4$ [M+H]$^+$: 327.1604; found: 327.1608.

1-benzyl-5-((4-benzonitrile)amino)-4-phenyl-1H-1,2,3-triazole (**2d**)

From 1-benzyl-4-phenyl-1H-1,2,3-triazol-5-amine and 4-bromobenzonitrile, following general procedure, **2d** (172 mg, 98% yield) was obtained as a white solid, m.p. 179–180 °C. ^{1}H NMR (400 MHz, DMSO-d$_6$) δ 9.02 (s, 1H), 7.74 (d, J = 7.9 Hz, 2H), 7.48 (d, J = 8.5 Hz, 2H), 7.37 (t, J = 7.6 Hz, 2H), 7.31–7.24 (m, 4H), 7.19–7.15 (m, 2H), 6.52 (d, J = 8.6 Hz, 2H), 5.45 (s, 2H). ^{13}C{^{1}H} NMR (101 MHz, DMSO-d$_6$) δ 148.5, 139.2, 135.2, 133.8, 130.7, 130.0, 128.8, 128.6, 128.0, 127.9, 127.8, 125.2, 119.6, 113.8, 100.3, 50.2. IR (υ/cm^{-1}): 2962 (M), 2927 (W), 2223 (M), 1604 (S), 1590 (S), 1519 (S), 1456 (S), 1434 (S), 1426 (S), 1358 (S), 1323 (S), 1258 (S), 1173 (S), 1006 (S), 822 (VS), 773 (VS), 734 (VS), 716 (VS), 696 (VS). HRMS (ESI): calcd for C$_{22}$H$_{18}$N$_5$ [M+H]$^+$: 352.1557; found: 352.1556.

1-tert-butyl-5-(p-tolylamino)-4-phenyl-1H-1,2,3-triazole (**2e**)

From 1-*tert*-butyl-4-phenyl-1H-1,2,3-triazol-5-amine and 1-bromo-4-methylbenzene, following general procedure, **2e** (100 mg, 65% yield) was obtained as a white solid, m.p. 241–242 °C. ^{1}H NMR (400 MHz, Chloroform-d) δ 7.80–7.74 (m, 2H), 7.28–7.20 (m, 3H), 6.97 (d, J = 6.5 Hz, 2H), 6.46 (d, J = 6.1 Hz, 2H), 5.25 (s, 1H), 2.22 (s, 3H), 1.68 (s, 9H). ^{13}C{^{1}H} NMR (101 MHz, Chloroform-d) δ 142.6, 142.3, 131.8, 130.6, 130.3, 129.3, 128.6, 127.8, 126.1, 114.2, 61.3, 29.8, 20.6. IR (υ/cm^{-1}): 3233 (M), 2975 (M), 1612 (S), 1593 (S), 1568 (S), 1516 (VS), 1449 (S), 1410 (S), 1371 (S), 1309 (VS), 1235 (S), 1195 (S), 991 (VS), 805 (VS), 762 (VS), 693 (VS). HRMS (ESI): calcd for C$_{19}$H$_{23}$N$_4$ [M+H]$^+$: 307.1917; found: 307.1921.

1-benzyl-5-((4-fluoro-2-methylphenyl)amino)-4-phenyl-1H-1,2,3-triazole (**2f**)

From 1-benzyl-4-phenyl-1H-1,2,3-triazol-5-amine and 1-bromo-4-fluoro-2-methylbenzene, following general procedure, **2f** (177 mg, >99% yield) was obtained as a white solid, m.p. 216–217 °C. ^{1}H NMR (400 MHz, Chloroform-*d*) δ 7.74 (dd, *J* = 8.2, 1.2 Hz, 2H), 7.33 (t, *J* = 7.3 Hz, 2H), 7.31–7.27 (m, 4H), 7.16–7.12 (m, 2H), 6.90 (dd, *J* = 9.1, 2.7 Hz, 1H), 6.70–6.60 (m, 1H), 6.15 (dd, *J* = 8.7, 4.8 Hz, 1H), 5.35 (s, 2H), 4.74 (s, 1H), 2.12 (s, 3H). ^{13}C{^{1}H} NMR (101 MHz, Chloroform-*d*) δ 157.4 (d, *J* = 239.0 Hz), 140.7, 137.5 (d, *J* = 2.0 Hz), 134.6, 132.1, 130.4, 129.0, 128.9, 128.6, 128.2, 127.8, 125.9, 125.4 (d, *J* = 7.6 Hz), 117.8 (d, *J* = 22.8 Hz), 114.1 (d, *J* = 8.2 Hz), 113.6 (d, *J* = 22.3 Hz), 51.8, 17.6. ^{19}F NMR (376 MHz, Chloroform-*d*) δ -123.92. IR (υ/cm^{-1}): 3241 (W), 1610 (S), 1588 (S), 1516 (S), 1498 (S), 1446 (S), 1411 (S), 1362 (S), 1268 (S), 1239 (S), 1199 (S), 1007 (S), 953 (S), 856 (VS), 800 (VS), 771 (VS), 737 (VS), 714 (VS), 697 (VS). HRMS (ESI): calcd for $C_{22}H_{20}FN_4$ [M+H]$^+$: 359.1667; found: 359.1670.

1-tert-butyl-5-(phenylamino)-4-phenyl-1H-1,2,3-triazole (**2g**)

From 1-*tert*-butyl-4-phenyl-1H-1,2,3-triazol-5-amine and bromobenzene, following general procedure, **2g** (136 mg, 93% yield) was obtained as a white solid, m.p. 228–229 °C. ^{1}H NMR (400 MHz, Chloroform-*d*) δ 7.76 (d, *J* = 7.3 Hz, 2H), 7.25–7.19 (m, 3H), 7.16 (t, *J* = 7.1 Hz, 2H), 6.81 (t, *J* = 7.3 Hz, 1H), 6.55 (d, *J* = 7.6 Hz, 2H), 5.58 (s, 1H), 1.68 (s, 9H). ^{13}C{^{1}H} NMR (101 MHz, Chloroform-*d*) δ 144.6, 142.5, 131.6, 130.2, 129.7, 128.6, 127.9, 126.2, 120.1, 114.1, 61.5, 29.8. IR (υ/cm^{-1}): 3346 (W), 3056 (W), 2980 (W), 2931 (W), 1604 (S), 1566 (S), 1498 (S), 1423 (S), 1370 (S), 1309 (S), 1233 (S), 1183 (S), 990 (VS), 768 (VS), 746 (VS), 717 (VS), 690 (VS). HRMS (ESI): calcd for $C_{18}H_{21}N_4$ [M+H]$^+$: 293.1761; found: 293.1766.

1-benzyl-5-((pyridine-2-yl)amino)-4-phenyl-1H-1,2,3-triazole (**2h**)

From 1-benzyl-4-phenyl-1H-1,2,3-triazol-5-amine and 2-bromopyridine, following general procedure, (**2h**) (127 mg, 77% yield) was obtained as a white solid, m.p. 173–174 °C. ^{1}H NMR (400 MHz, DMSO-d_6) δ 8.93 (s, 1H), 7.99–7.96 (m, 1H), 7.77 (d, *J* = 7.0 Hz, 2H), 7.57–7.52 (m, 1H), 7.37 (t, *J* = 7.5 Hz, 2H), 7.32–7.25 (m, 4H), 7.19 (dd, *J* = 7.6, 1.8 Hz, 2H), 6.76–6.72 (m, 1H), 6.62 (d, *J* = 8.5 Hz, 1H), 5.40 (s, 2H). ^{13}C{^{1}H} NMR (101 MHz, Chloroform-*d*) δ 156.1, 148.0, 141.2, 138.6, 134.6, 130.4, 130.2, 128.8, 128.7, 128.4, 128.2, 128.1, 125.9, 115.9, 107.1, 51.6. IR (υ/cm^{-1}): 3140 (W), 3082 (W), 3062 (W), 2914 (M), 2856 (M), 1588 (S), 1522 (S), 1500 (S), 1436 (S), 1361 (S), 1319 (S), 1233 (S), 1213 (S), 1153 (VS), 1101 (S), 1074 (S), 996 (VS), 783 (VS), 772 (VS), 738 (VS). HRMS (ESI): calcd for $C_{20}H_{18}N_5$ [M+H]$^+$: 328.1557; found: 328.1561.

1-benzyl-5-((4-tert-butylphenyl))amino)-4-phenyl-1H-1,2,3-triazole (**2i**)

From 1-benzyl-4-phenyl-1*H*-1,2,3-triazol-5-amine and 1-bromo-4-*tert*-butylbenzene, following general procedure, **2i** (172 mg, 90% yield) was obtained as a white solid, m.p. 169–171 °C. ^{1}H NMR (400 MHz, Chloroform-*d*) δ 7.81 (d, *J* = 7.0 Hz, 2H), 7.31–7.24 (m, 6H), 7.19–7.14 (m, 4H), 6.46 (d, *J* = 8.7 Hz, 2H), 5.33 (s, 2H), 5.07 (s, 1H), 1.27 (s, 9H). ^{13}C{^{1}H} NMR (101 MHz, Chloroform-*d*) δ 143.6, 141.0, 141.0, 134.8, 132.1, 130.4, 128.9, 128.8, 128.4, 128.0, 127.9, 126.6, 126.0, 114.1, 51.4, 34.2, 31.6. IR (υ/cm^{-1}): 3253 (M), 3054 (M), 3034 (M), 2956 (M), 2900 (M), 2857 (M), 1607 (S), 1587 (S), 1568 (S), 1515 (VS), 1400 (S), 1360 (S), 1252 (S), 1190 (S), 922 (S), 814 (S), 770 (VS), 737 (VS), 719 (VS), 695 (VS). HRMS (ESI): calcd for C$_{25}$H$_{27}$N$_4$ [M+H]$^+$: 383.2230; found: 383.2241.

1-benzyl-5-(mesitylamino)-4-phenyl-1H-1,2,3-triazole (**2j**)

From 1-benzyl-4-phenyl-1*H*-1,2,3-triazol-5-amine and 2-bromo- 1,3,5-trimethylbenzene, following general procedure, **2j** (146 mg, 79% yield) was obtained as a white solid, m.p. 132–133 °C. ^{1}H NMR (400 MHz, Chloroform-*d*) δ 7.71 (d, *J* = 7.3 Hz, 2H), 7.31–7.20 (m, 6H), 6.89 (dd, *J* = 7.2, 2.2 Hz, 2H), 6.72 (s, 2H), 5.15 (s, 2H), 4.93 (s, 1H), 2.23 (s, 3H), 1.75 (s, 6H). ^{13}C{^{1}H} NMR (101 MHz, Chloroform-*d*) δ 135.8, 135.1, 134.9, 134.9, 133.6, 131.2, 130.1, 129.8, 128.8, 128.4, 128.2, 127.2, 127.1, 126.2, 51.6, 20.7, 18.2. IR (υ/cm^{-1}): 3339 (M), 3060 (W), 3032 (M), 2913 (M), 2853 (W), 1606 (S), 1586 (S), 1571 (S), 1485 (S), 1445 (S), 1421 (S), 1361 (S), 1317 (S), 1250 (S), 1073 (S), 1029 (S), 994 (S), 840 (S), 769 (VS), 724 (VS), 694 (VS). HRMS (ESI): calcd for C$_{24}$H$_{25}$N$_4$ [M+H]$^+$: 369.2074; found: 369.2074.

1-tert-butyl-5-((pyridine-3-yl)amino)-4-phenyl-1H-1,2,3-triazole (**2k**)

From 1-*tert*-butyl-4-phenyl-1*H*-1,2,3-triazol-5-amine and 3-chloropyridine, following general procedure, **2k** (110 mg, 75% yield) was obtained as a white solid, m.p. 233–234 °C. ^{1}H NMR (400 MHz, DMSO-*d*$_6$) δ 8.27 (s, 1H), 7.94 (s, 1H), 7.90 (d, *J* = 4.7 Hz, 1H), 7.73 (d, *J* = 7.9 Hz, 2H), 7.32 (t, *J* = 7.6 Hz, 2H), 7.24 (t, *J* = 7.6 Hz, 1H), 7.08 (dd, *J* = 8.3, 4.6 Hz, 1H), 6.73 (d, *J* = 8.0 Hz, 1H), 1.65 (s, 9H). ^{13}C{^{1}H} NMR (101 MHz, DMSO-*d*$_6$) δ 141.6, 140.9, 139.9, 136.0, 131.1, 130.3, 128.6, 127.8, 125.4, 124.0, 119.5, 60.7, 29.1. IR (υ/cm^{-1}): 3252 (W), 3002 (W), 2974 (W), 1589 (S), 1580 (S), 1508 (S), 1477 (S), 1449 (S), 1370 (S), 1299 (S), 1239 (S), 990 (VS), 800 (VS), 772 (VS), 709 (VS). HRMS (ESI) calcd for C$_{17}$H$_{20}$N$_5$ [M+H]$^+$: 294.1719; found: 294.1718.

1-phenethyl-5-((pyridine-3-yl)amino)-4-phenyl-1H-1,2,3-triazole (**2l**)

From 1-phenethyl-4-phenyl-1H-1,2,3-triazol-5-amine and 3-chloropyridine, following general procedure, **2l** (142 mg, 83% yield) was obtained as a white solid, m.p. 199–200 °C. ^{1}H NMR (400 MHz, DMSO-d_6) δ 8.46 (s, 1H), 8.01–7.92 (m, 2H), 7.72 (d, J = 7.3 Hz, 2H), 7.35 (t, J = 7.4 Hz, 2H), 7.29–7.16 (m, 4H), 7.12 (d, J = 7.2 Hz, 2H), 7.07 (dd, J = 8.0, 4.6 Hz, 1H), 6.65 (dd, J = 8.5, 1.3 Hz, 1H), 4.44 (t, J = 7.3 Hz, 2H), 3.13 (t, J = 7.6 Hz, 2H). ^{13}C{^{1}H} NMR (101 MHz, DMSO-d_6) δ 140.6, 140.3, 138.3, 137.5, 136.4, 131.5, 130.3, 128.7, 128.6, 128.5, 127.7, 126.6, 125.2, 124.0, 119.7, 47.7, 35.0. IR (υ/cm^{-1}): 3203 (W), 3162 (W), 3083 (W), 3025 (W), 2969 (W), 1582 (S), 1569 (S), 1480 (S), 1455 (S), 1402 (S), 1361 (S), 1312 (S), 1278 (S), 1232 (S), 990 (S), 799 VS, 763 (VS), 743 (VS), 701 (VS). HRMS (ESI): calcd for $C_{21}H_{20}N_5$ [M+H]$^+$: 342.1719; found: 342.1717.

1-benzyl-5-((3,5-dimethylphenyl)amino)-4-phenyl-1H-1,2,3-triazole (**2m**)

From 1-benzyl-4-phenyl-1H-1,2,3-triazol-5-amine and 1-bromo- 3,5-dimethylbenzene, following general procedure, **2m** (149 mg, 84% yield) was obtained as a white solid, m.p. 154–155 °C. ^{1}H NMR (400 MHz, Chloroform-d) δ 7.82 (d, J = 7.0 Hz, 2H), 7.35–7.26 (m, 6H), 7.23–7.19 (m, 2H), 6.54 (s, 1H), 6.15 (s, 2H), 5.34 (s, 2H), 5.06 (s, 1H), 2.18 (s, 6H). ^{13}C{^{1}H} NMR (101 MHz, Chloroform-d) δ 143.7, 141.1, 139.7, 134.8, 131.9, 130.4, 128.9, 128.8, 128.4, 128.1, 128.0, 126.0, 122.7, 112.2, 51.4, 21.5. IR (υ/cm^{-1}): 3266 (W), 2919 (W), 1601 (S), 1585 (S), 1495 (S), 1444 (S), 1353 (S), 1324 (S), 1233 (S), 1170 (S), 1004 (VS), 993 (VS), 837 (VS), 774 (VS), 739 (VS), 727 (VS), 691 (VS). HRMS (ESI): calcd for $C_{23}H_{23}N_4$ [M+H]$^+$: 355.1917; found: 355.1920.

1-benzyl-5-((pyrimidine-4-yl)amino)-4-phenyl-1H-1,2,3-triazole (**2n**)

From 1-benzyl-4-phenyl-1H-1,2,3-triazol-5-amine and 4-chloropyrimidine, following general procedure, **2n** (161 mg, 98% yield) was obtained as a white solid, m.p. 156–157 °C. ^{1}H NMR (400 MHz, DMSO-d_6) δ 9.40 (s, 1H), 8.10 (s, 1H), 7.93 (d, J = 5.6 Hz, 2H), 7.75 (d, J = 7.7 Hz, 2H), 7.38 (t, J = 7.6 Hz, 2H), 7.27 (q, J = 7.7, 6.7 Hz, 4H), 7.18 (d, J = 7.7 Hz, 2H), 5.44 (s, 2H). ^{13}C{^{1}H} NMR (101 MHz, DMSO-d_6) δ 152.5, 141.8, 139.4, 135.3, 135.0, 133.0, 130.5, 130.4, 128.7, 128.5, 127.8, 127.8, 127.8, 125.3, 50.5. IR (υ/cm^{-1}): 3189 (W), 3067 (W), 2953 (W), 1593 (S), 1497 (S), 1472 (S), 1446 (S), 1360 (S), 1318 (S), 1278 (S), 1231 (S), 1150 (S), 996 (S), 825 (VS), 767 (VS), 734 (VS), 694 (VS). HRMS (ESI) calcd for $C_{19}H_{17}N_6$ [M+H]$^+$: 329.1515; found: 329.1514.

1-benzyl-5-((pyridine-3-yl)amino)-4-phenyl-1H-1,2,3-triazole (**2o**)

From 1-benzyl-4-phenyl-1H-1,2,3-triazol-5-amine and 3-chloropyridine (159 mg, 97% yield) or 3-bromopyridine (163 mg, >99% yield), following general procedure, **2o** was obtained as a white solid, m.p. 169–170 °C. ^{1}H NMR (400 MHz, Chloroform-d) δ 8.02–7.97

(m, 2H), 7.73 (dd, *J* = 7.9, 1.6 Hz, 2H), 7.27–7.22 (m, 3H), 7.21–7.17 (m, 3H), 7.15–7.11 (m, 2H), 6.92 (dd, *J* = 8.3, 4.7 Hz, 1H), 6.52 (ddd, *J* = 8.3, 2.7, 1.2 Hz, 1H), 6.30 (s, 1H), 5.36 (s, 2H). ^{13}C{^{1}H} NMR (101 MHz, Chloroform-*d*) δ 141.3, 141.3, 140.3, 136.9, 134.3, 130.6, 129.9, 129.0, 128.8, 128.6, 128.4, 127.9, 125.9, 124.1, 120.3, 51.6. IR (υ/cm^{-1}): 3221 (W), 3173 (M), 3090 (W), 3043 (M), 3027 (M), 2962 (M), 2904 (M), 2780 (M), 1608 (S), 1583 (S), 1570 (S), 1538 (S), 1480 (S), 1427 (S), 1409 (S), 1364 (S), 1321 (S), 1246 (S), 1234 (S), 1048 (S), 1006 (S), 994 (S). HRMS (ESI): calcd for $C_{20}H_{18}N_5$ [M+H]$^+$: 328.1557; found: 328.1561.

N4,N6-bis(1-benzyl-4-phenyl-1H-1,2,3-triazol-5-yl)pyrimidine-4,6-diamine (**2p**)

From 1-benzyl-4-phenyl-1*H*-1,2,3-triazol-5-amine and 4,6-dichloropyrimidine, following general procedure, **2p** (88 mg, 61% yield) was obtained as a white solid, m.p. 263–264 °C. ^{1}H NMR (400 MHz, DMSO-*d*$_6$) δ 9.28 (s, 2H), 7.98 (s, 1H), 7.73 (s, 4H), 7.40 (s, 4H), 7.37–7.29 (m, 3H), 7.28–7.10 (m, 10H), 5.38 (s, 4H). ^{13}C{^{1}H} NMR (101 MHz, DMSO-*d*$_6$) δ 161.3, 158.2, 139.4, 135.2, 130.2, 130.1, 128.7, 128.5, 127.9, 127.8, 125.3, 50.5. IR (υ/cm^{-1}): 3064 (M), 3032 (M), 2927 (M), 1601 (S), 1587 (S), 1496 (S), 1356 (S), 1288 (S), 1237 (S), 1188 (S), 1073 (S), 991 (S), 822 (VS), 769 (VS), 734 (VS), 720 (VS), 692 (VS). HRMS (ESI): calcd for $C_{34}H_{29}N_{10}$ [M+H]$^+$: 577.2571; found: 577.2574.

1-phenethyl-5-(p-tolylamino)-4-phenyl-1H-1,2,3-triazole (**2q**)

From 5-chloro-1-phenethyl-4-phenyl-1*H*-1,2,3-triazole and *p*-toluidine, following general procedure, **2q** (147 mg, 83% yield) was obtained as a white solid, m.p. 152–153 °C. ^{1}H NMR (400 MHz, Chloroform-*d*) δ 7.74 (d, *J* = 7.9 Hz, 2H), 7.30–7.23 (m, 6H), 7.04–6.99 (m, 2H), 6.93 (d, *J* = 8.0 Hz, 2H), 6.29 (d, *J* = 7.5 Hz, 2H), 4.77 (s, 1H), 4.37 (t, *J* = 8.3 Hz, 2H), 3.14 (t, *J* = 8.4 Hz, 2H), 2.21 (s, 3H). ^{13}C{^{1}H} NMR (101 MHz, Chloroform-*d*) δ 140.8, 139.5, 137.6, 132.9, 130.2, 129.9, 129.1, 129.0, 128.7, 128.4, 127.3, 126.2, 125.9, 114.2, 49.2, 36.4, 20.6. IR (υ/cm^{-1}): 3205 (M), 3176 (M), 3085 (M), 3027 (M), 2950 (M), 2931 (M), 1878 (M), 1610 (S), 1585 (S), 1572 (S), 1520 (S), 1498 (S), 1451 (S), 1364 (S), 1258 (S), 1011 (S), 807 (VS), 762 (VS), 748 (VS), 699 (VS). HRMS (ESI): calcd for $C_{23}H_{23}N_4$ [M+H]$^+$: 355.1917; found: 355.1920.

1-benzyl-5-((2,4-dimethylphenyl)amino)-4-phenyl-1H-1,2,3-triazole (**2r**)

From 1-benzyl-5-chloro-4-phenyl-1*H*-1,2,3-triazole and 2,4-dimethylaniline, following general procedure, **2r** (169 mg, 95% yield) was obtained as a white solid, m.p. 194–195 °C. ^{1}H NMR (400 MHz, Chloroform-*d*) δ 7.77 (d, *J* = 7.5 Hz, 2H), 7.33–7.24 (m, 6H), 7.16–7.11 (m, 2H), 6.99 (s, 1H), 6.78 (d, *J* = 8.2 Hz, 1H), 6.16 (d, *J* = 8.1 Hz, 1H), 5.31 (s, 2H), 4.84 (s, 1H), 2.25 (s, 3H), 2.11 (s, 3H). ^{13}C{^{1}H} NMR (101 MHz, Chloroform-*d*) δ 140.5, 139.0, 134.7, 132.5, 131.8, 130.2, 130.1, 129.0, 128.8, 128.5, 128.1, 127.9, 127.9, 126.0, 123.5, 113.0, 51.8, 20.6, 17.5. IR (υ/cm^{-1}): 3260 (W), 2962 (W), 2924 (W), 2857 (W), 1608 (S), 1587 (S), 1571 (S), 1517

(S), 1446 (S), 1360 (S), 1237 (S), 1156 (S), 804 (VS), 766 (VS), 736 (VS), 694 (VS). HRMS (ESI): calcd for $C_{23}H_{23}N_4$ [M+H]$^+$: 355.1917; found: 355.1918.

1-benzyl-5-((3-(trifluoromethyl)phenyl)amino)-4-phenyl-1H-1,2,3-triazole (**2s**)

From 1-benzyl-5-chloro-4-phenyl-1*H*-1,2,3-triazole and 3-(trifluoromethyl)aniline, following general procedure, **2s** (159 mg, 81% yield) was obtained as a white solid, m.p. 115–117 °C. ^{1}H NMR (400 MHz, Chloroform-*d*) δ 7.72 (dd, *J* = 6.4, 2.9 Hz, 2H), 7.23–7.11 (m, 9H), 7.05 (d, *J* = 7.7 Hz, 1H), 6.81 (s, 1H), 6.55 (d, *J* = 8.0 Hz, 1H), 6.41 (m, 1H), 5.32 (s, 2H). ^{13}C{^{1}H} NMR (101 MHz, Chloroform-*d*) δ 144.3, 141.3, 134.2, 131.9 (q, *J* = 32.3 Hz), 131.1, 130.2, 129.8, 128.9, 128.8, 128.5, 128.3, 128.0, 125.9, 124.0 (q, *J* = 272.8 Hz), 116.9, 110.9 (q, *J* = 3.6 Hz), 51.5. ^{19}F NMR (376 MHz, Chloroform-*d*) δ -62.8. IR (υ/cm^{-1}): 3195 (M), 3038 (M), 2927 (M), 1619 (S), 1586 (S), 1571 (S), 1495 (S), 1486 (S), 1444 (S), 1425 (S), 1336 (VS), 1231 (S), 1163 (VS), 1118 (VS), 1099 (S), 1067 (VS), 1006 (S), 996 (S), 916 (S), 871 (S), 791 (S), 769 (VS), 736 (VS), 692 (VS). HRMS (ESI): calcd for $C_{22}H_{18}F_3N_4$ [M+H]$^+$: 395.1478; found: 395.1482.

1-benzyl-5-((3,5-bis(trifluoromethyl)phenyl)amino)-4-phenyl-1H-1,2,3-triazole (**2t**)

From 1-benzyl-5-chloro-4-phenyl-1*H*-1,2,3-triazole and 3,5-bis(trifluoromethyl) aniline, following general procedure, **2t** (136 mg, 59% yield) was obtained as a white solid, m.p. 110–111 °C. ^{1}H NMR (400 MHz, Chloroform-*d*) δ 7.70 (dd, *J* = 7.7, 1.9 Hz, 2H), 7.35–7.26 (m, 4H), 7.24–7.22 (m, 3H), 7.16–7.11 (m, 2H), 6.75 (s, 2H), 5.52 (s, 1H), 5.42 (s, 2H). ^{13}C{^{1}H} NMR (101 MHz, Chloroform-*d*) δ 144.8, 141.9, 133.8, 133.0 (q, *J* = 33.5 Hz), 129.8, 129.5, 129.2, 129.0, 128.8, 128.7, 127.9, 126.0, 125.8, 123.1 (q, *J* = 272.6 Hz), 113.8 (p, *J* = 3.8 Hz), 113.7, 113.6, 52.0. ^{19}F NMR (376 MHz, Chloroform-*d*) δ -63.19. IR (υ/cm^{-1}): 3457 (W), 3204 (W), 3074 (W), 2930 (W), 1616 (S), 1590 (S), 1498 (S), 1471 (S), 1387 (S), 1276 (S), 1182 (S), 1130 (VS), 953 (VS), 873 (VS), 766 (VS), 700 (VS). HRMS (nESI): calcd for $C_{23}H_{17}F_6N_4$ [M+H]$^+$: 463.1357; found: 463.1348.

1-benzyl-4-(p-tolylamino)-5-methyl-1H-1,2,3-triazole (**2u**)

From 1-benzyl-4-bromo-5-methyl-1*H*-1,2,3-triazole and *p*-toluidine, following general procedure, **2u** (97 mg, 69% yield) was obtained as a white solid, m.p. 133–134 °C. ^{1}H NMR (400 MHz, Chloroform-*d*) δ 7.38–7.32 (m, 3H), 7.19 (d, *J* = 6.7 Hz, 2H), 6.98 (d, *J* = 8.3 Hz, 2H), 6.63 (d, *J* = 8.1 Hz, 2H), 5.61 (s, 1H), 5.48 (s, 2H), 2.24 (s, 3H), 2.02 (s, 3H). ^{13}C{^{1}H} NMR (101 MHz, Chloroform-*d*) δ 144.7, 142.4, 134.6, 129.8, 129.2, 129.1, 128.5, 127.3, 125.1, 114.8, 52.9, 29.8, 20.6. IR (υ/cm^{-1}): 3246 (M), 3109 (W), 3034 (M), 2924 (M), 2855 (M), 1884 (M), 1602 (S), 1511 (S), 1455 (S), 1435 (S), 1390 (S), 1345 (S), 1234 (S), 1121 (S), 815 (VS), 725 (VS), 696 (VS). HRMS (ESI): calcd for $C_{17}H_{18}N_5$ [M+H]$^+$: 279.1604; found: 279.1606.

4. Conclusions

In conclusion, we have developed an efficient and robust method for the preparation of a series of new 5-(het)arylamino-1,2,3-triazole derivatives via Buchwald–Hartwig cross-coupling reaction of 5-amino or 5-halo-1,2,3-triazoles with (het)aryl halides and amines respectively. As a result of the careful screening for optimal conditions, a catalytic system based on the palladium complex [(THP-Dipp)Pd(cinn)Cl] with expanded-ring NHC ligand has been revealed as the most active for the process. The reaction functions perfectly in 1,4-dioxane medium at 120 °C in the presence of an excess of t-BuONa to afford a variety of 5-(het)arylamino-1,2,3-triazoles with good to excellent yields. The compounds obtained have major potential to be used in biomolecular chemistry and material science.

Supplementary Materials: The following are available online at https://www.mdpi.com/article/10.3390/molecules27061999/s1, copies of ^{1}H and ^{13}C NMR spectra for all novel compounds.

Author Contributions: Conceptualization, P.S.G., A.F.A., S.N.O.; methodology, P.S.G.; investigation, P.S.G. (synthesis, NMR spectra registering and characterization), A.N.P. (synthesis), A.F.A., L.I.M., M.A.T. (synthesis of (THP-Dipp)Pd(allyl)Cl and (THP-Dipp)PdMetallylCl complexes); writing—original draft preparation, P.S.G., S.N.O.; writing—review and editing, P.S.G., S.N.O.; supervision, P.S.G.; project administration, P.S.G.; funding acquisition, P.S.G. All authors have read and agreed to the published version of the manuscript.

Funding: This work was financially supported by the Russian Science Foundation (grant RSF No. 20-73-00291).

Institutional Review Board Statement: Not applicable.

Informed Consent Statement: Not applicable.

Data Availability Statement: Data are contained within the article and Supplementary Materials.

Acknowledgments: NMR studies and spectral characterization were performed with financial support from the Ministry of Science and Higher Education of the Russian Federation using the equipment of the Center for Molecular Composition Studies of INEOS RAS. Part of this work (synthesis of (THP-Dipp)Pd(allyl)Cl and (THP-Dipp)Pd(methallyl)Cl complexes) was carried out as a part of the State Program of A. V. Topchiev Institute of Petrochemical Synthesis of the Russian Academy of Sciences (TIPS RAS).

Conflicts of Interest: The authors declare no conflict of interest.

Sample Availability: Samples of all of the compounds are available from the authors.

References

1. Thansandote, P.; Lautens, M. Construction of nitrogen-containing heterocycles by C-H bond functionalization. *Chem. Eur. J.* **2009**, *15*, 5874–5883. [CrossRef] [PubMed]
2. Chen, D.; Su, S.-J.; Cao, Y. Nitrogen heterocycle-containing materials for highly efficient phosphorescent OLEDs with low operating voltage. *J. Mater. Chem. C* **2014**, *2*, 9565–9578. [CrossRef]
3. Huang, W.; Buchwald, S.L. Palladium-Catalyzed N-Arylation of Iminodibenzyls and Iminostilbenes with Aryl- and Heteroaryl Halides. *Chem. Eur. J.* **2016**, *22*, 14186–14189. [CrossRef] [PubMed]
4. Bikker, J.A.; Brooijmans, N.; Wissner, A.; Mansour, T.S. Kinase domain mutations in cancer: Implications for small molecule drug design strategies. *J. Med. Chem.* **2009**, *52*, 1493–1509. [CrossRef] [PubMed]
5. Quintas-Cardama, A.; Kantarjian, H.; Cortes, J. Flying under the radar: The new wave of BCR-ABL inhibitors. *Nat. Rev. Drug. Discov.* **2007**, *6*, 834–848. [CrossRef]
6. Topchiy, M.A.; Zharkova, D.A.; Asachenko, A.F.; Muzalevskiy, V.M.; Chertkov, V.A.; Nenajdenko, V.G.; Nechaev, M.S. Mild and Regioselective Synthesis of 3-CF3-Pyrazoles by the AgOTf-Catalysed Reaction of CF3-Ynones with Hydrazines. *Eur. J. Org. Chem.* **2018**, *2018*, 3750–3755. [CrossRef]
7. Sharma, P.; Kumar, A.V.; Upadhyay, S.; Singh, J.; Sahu, V. A novel approach to the synthesis of 1,2,3-triazoles and their SAR studies. *Med. Chem. Res.* **2009**, *19*, 589–602. [CrossRef]
8. Gomes, A.; Martins, P.; Rocha, D.R.; Neves, M.; Ferreira, V.F.; Silva, A.; Cavaleiro, J.; da Silva, F. Consecutive Tandem Cycloaddition between Nitriles and Azides; Synthesis of 5-Amino-1H-[1,2,3]-triazoles. *Synlett* **2013**, *24*, 41–44. [CrossRef]
9. Dheer, D.; Singh, V.; Shankar, R. Medicinal attributes of 1,2,3-triazoles: Current developments. *Bioorg. Chem.* **2017**, *71*, 30–54. [CrossRef]

10. Bonandi, E.; Christodoulou, M.S.; Fumagalli, G.; Perdicchia, D.; Rastelli, G.; Passarella, D. The 1,2,3-triazole ring as a bioisostere in medicinal chemistry. *Drug Discov. Today* **2017**, *22*, 1572–1581. [CrossRef]
11. Lauria, A.; Delisi, R.; Mingoia, F.; Terenzi, A.; Martorana, A.; Barone, G.; Almerico, A.M. 1,2,3-Triazole in Heterocyclic Compounds, Endowed with Biological Activity, through 1,3-Dipolar Cycloadditions. *Eur. J. Org. Chem.* **2014**, *2014*, 3289–3306. [CrossRef]
12. Johansson, J.R.; Beke-Somfai, T.; Said Stalsmeden, A.; Kann, N. Ruthenium-Catalyzed Azide Alkyne Cycloaddition Reaction: Scope, Mechanism, and Applications. *Chem. Rev.* **2016**, *116*, 14726–14768. [CrossRef] [PubMed]
13. Hein, J.E.; Fokin, V.V. Copper-catalyzed azide-alkyne cycloaddition (CuAAC) and beyond: New reactivity of copper(I) acetylides. *Chem. Soc. Rev.* **2010**, *39*, 1302–1315. [CrossRef] [PubMed]
14. Emileh, A.; Duffy, C.; Holmes, A.P.; Rosemary Bastian, A.; Aneja, R.; Tuzer, F.; Rajagopal, S.; Li, H.; Abrams, C.F.; Chaiken, I.M. Covalent conjugation of a peptide triazole to HIV-1 gp120 enables intramolecular binding site occupancy. *Biochemistry* **2014**, *53*, 3403–3414. [CrossRef]
15. Kuntala, N.; Telu, J.R.; Banothu, V.; Nallapati, S.B.; Anireddy, J.S.; Pal, S. Novel benzoxepine-1,2,3-triazole hybrids: Synthesis and pharmacological evaluation as potential antibacterial and anticancer agents. *Med. Chem. Commun.* **2015**, *6*, 1612–1619. [CrossRef]
16. Duan, H.; Sengupta, S.; Petersen, J.L.; Akhmedov, N.G.; Shi, X. Triazole-Au(I) complexes: A new class of catalysts with improved thermal stability and reactivity for intermolecular alkyne hydroamination. *J. Am. Chem. Soc.* **2009**, *131*, 12100–12102. [CrossRef]
17. Nandivada, H.; Jiang, X.; Lahann, J. Click Chemistry: Versatility and Control in the Hands of Materials Scientists. *Adv. Mater.* **2007**, *19*, 2197–2208. [CrossRef]
18. Ashok, D.; Chiranjeevi, P.; Kumar, A.V.; Sarasija, M.; Krishna, V.S.; Sriram, D.; Balasubramanian, S. 1,2,3-Triazole-fused spirochromenes as potential anti-tubercular agents: Synthesis and biological evaluation. *RSC Adv.* **2018**, *8*, 16997–17007. [CrossRef]
19. Azagarsamy, M.A.; Anseth, K.S. Bioorthogonal Click Chemistry: An Indispensable Tool to Create Multifaceted Cell Culture Scaffolds. *ACS Macro Lett.* **2013**, *2*, 5–9. [CrossRef]
20. Johnson, J.A.; Finn, M.G.; Koberstein, J.T.; Turro, N.J. Construction of Linear Polymers, Dendrimers, Networks, and Other Polymeric Architectures by Copper-Catalyzed Azide-Alkyne Cycloaddition "Click" Chemistry. *Macromol. Rapid Commun.* **2008**, *29*, 1052–1072. [CrossRef]
21. Wang, S.; Zhang, Y.; Liu, G.; Xu, H.; Song, L.; Chen, J.; Li, J.; Zhang, Z. Transition-metal-free synthesis of 5-amino-1,2,3-triazoles via nucleophilic addition/cyclization of carbodiimides with diazo compounds. *Org. Chem. Front.* **2021**, *8*, 599–604. [CrossRef]
22. Ferrini, S.; Chandanshive, J.Z.; Lena, S.; Comes Franchini, M.; Giannini, G.; Tafi, A.; Taddei, M. Ruthenium-catalyzed synthesis of 5-amino-1,2,3-triazole-4-carboxylates for triazole-based scaffolds: Beyond the dimroth rearrangement. *J. Org. Chem.* **2015**, *80*, 2562–2572. [CrossRef] [PubMed]
23. Dorokhov, V.A.; Komkov, A.V. Addition of acetylacetone and ethyl acetoacetate to carbodiimides promoted by nickel acetylacetonate. *Russ. Chem. Bull.* **2004**, *53*, 676–680. [CrossRef]
24. Zhao, X.; Lu, B.W.; Lu, J.R.; Xin, C.W.; Li, J.F.; Liu, Y. Design, synthesis and antimicrobial activities of 1,2,3-triazole derivatives. *Chin. Chem. Lett.* **2012**, *23*, 933–935. [CrossRef]
25. Lauria, A.; Abbate, I.; Patella, C.; Martorana, A.; Dattolo, G.; Almerico, A.M. New annelated thieno[2,3-e][1,2,3]triazolo[1,5-a]pyrimidines, with potent anticancer activity, designed through VLAK protocol. *Eur. J. Med. Chem.* **2013**, *62*, 416–424. [CrossRef]
26. Liao, Y.; Lu, Q.; Chen, G.; Yu, Y.; Li, C.; Huang, X. Rhodium-Catalyzed Azide–Alkyne Cycloaddition of Internal Ynamides: Regioselective Assembly of 5-Amino-Triazoles under Mild Conditions. *ACS Catal.* **2017**, *7*, 7529–7534. [CrossRef]
27. Dimroth, O. Ueber eine Synthese von Derivaten des 1,2,3-Triazols. *Eur. J. Inorg. Chem.* **1902**, *35*, 1029–1038.
28. Dimroth, O. Ueber intramolekulare Umlagerungen. Umlagerungen in der Reihe des 1,2,3-Triazols. *Justus Liebigs Ann. Chem.* **1909**, *364*, 183–226. [CrossRef]
29. Krishna, P.M.; Ramachary, D.B.; Peesapati, S. Azide–acetonitrile "click" reaction triggered by Cs2CO3: The atom-economic, high-yielding synthesis of 5-amino-1,2,3-triazoles. *RSC Adv.* **2015**, *5*, 62062–62066. [CrossRef]
30. Lieber, E.; Chao, T.S.; Ramachandra Rao, C.N. Synthesis and Isomerization of Substituted 5-Amino-1,2,3-triazoles1. *J. Org. Chem.* **2002**, *22*, 654–662. [CrossRef]
31. Cottrell, I.F.; Hands, D.; Houghton, P.G.; Humphrey, G.R.; Wright, S.H.B. An improved procedure for the preparation of 1-benzyl-1H-1,2,3-triazoles from benzyl azides. *J. Heterocycl. Chem.* **1991**, *28*, 301–304. [CrossRef]
32. Pokhodylo, N.T.; Matiychuk, V.S.; Obushak, N.B. Synthesis of 1H-1,2,3-triazole derivatives by the cyclization of aryl azides with 2-benzothiazolylacetonone, 1,3-benzo-thiazol-2-ylacetonitrile, and (4-aryl-1,3-thiazol-2-yl)acetonitriles. *Chem. Heterocycl. Compd.* **2009**, *45*, 483–488. [CrossRef]
33. Opsomer, T.; Thomas, J.; Dehaen, W. Chemoselectivity in the Synthesis of 1,2,3-Triazoles from Enolizable Ketones, Primary Alkylamines, and 4-Nitrophenyl Azide. *Synthesis* **2017**, *49*, 4191–4198.
34. Navarro, Y.; Garcia Lopez, J.; Iglesias, M.J.; Lopez Ortiz, F. Chelation-Assisted Interrupted Copper(I)-Catalyzed Azide-Alkyne-Azide Domino Reactions: Synthesis of Fully Substituted 5-Triazenyl-1,2,3-triazoles. *Org. Lett.* **2021**, *23*, 334–339. [CrossRef] [PubMed]
35. Du, W.; Huang, H.; Xiao, T.; Jiang, Y. Metal-Free, Visible-Light Promoted Intramolecular Azole C–H Bond Amination Using Catalytic Amount of I2: A Route to 1,2,3-Triazolo[1,5-a]quinazolin-5(4H)-ones. *Adv. Synth. Catal.* **2020**, *362*, 5124–5129. [CrossRef]
36. Roughley, S.D.; Jordan, A.M. The medicinal chemist's toolbox: An analysis of reactions used in the pursuit of drug candidates. *J. Med. Chem.* **2011**, *54*, 3451–3479. [CrossRef] [PubMed]

37. Flick, A.C.; Ding, H.X.; Leverett, C.A.; Kyne, R.E., Jr.; Liu, K.K.; Fink, S.J.; O'Donnell, C.J. Synthetic Approaches to the New Drugs Approved During 2015. *J. Med. Chem.* **2017**, *60*, 6480–6515. [CrossRef] [PubMed]
38. Corbet, J.P.; Mignani, G. Selected patented cross-coupling reaction technologies. *Chem. Rev.* **2006**, *106*, 2651–2710. [CrossRef]
39. Heravi, M.M.; Kheilkordi, Z.; Zadsirjan, V.; Heydari, M.; Malmir, M. Buchwald-Hartwig reaction: An overview. *J. Organomet. Chem.* **2018**, *861*, 17–104. [CrossRef]
40. Wambua, V.; Hirschi, J.S.; Vetticatt, M.J. Rapid Evaluation of the Mechanism of Buchwald-Hartwig Amination and Aldol Reactions Using Intramolecular (13)C Kinetic Isotope Effects. *ACS Catal.* **2021**, *11*, 60–67. [CrossRef]
41. Liu, Y.-F.; Wang, C.-L.; Bai, Y.-J.; Han, N.; Jiao, J.-P.; Qi, X.-L. A Facile Total Synthesis of Imatinib Base and Its Analogues. *Org. Process Res. Dev.* **2008**, *12*, 490–495. [CrossRef]
42. Tundel, R.E.; Anderson, K.W.; Buchwald, S.L. Expedited palladium-catalyzed amination of aryl nonaflates through the use of microwave-irradiation and soluble organic amine bases. *J. Org. Chem.* **2006**, *71*, 430–433. [CrossRef] [PubMed]
43. Kumar, R.; Ramachandran, U.; Khanna, S.; Bharatam, P.V.; Raichur, S.; Chakrabarti, R. Synthesis, in vitro and in silico evaluation of l-tyrosine containing PPARalpha/gamma dual agonists. *Bioorg. Med. Chem.* **2007**, *15*, 1547–1555. [CrossRef]
44. Reddy, M.; Anusha, G.; Reddy, P. Sterically enriched bulky 1,3-bis(N,N'-aralkyl)benzimidazolium based Pd-PEPPSI complexes for Buchwald–Hartwig amination reactions. *New J. Chem.* **2020**, *44*, 11694–11703. [CrossRef]
45. Deng, Q.; Zhang, Y.; Zhu, H.; Tu, T. Robust Acenaphthoimidazolylidene Palladacycles: Highly Efficient Catalysts for the Amination of N-Heteroaryl Chlorides. *Chem. Asian J.* **2017**, *12*, 2364–2368. [CrossRef] [PubMed]
46. Moss, T.; Addie, M.; Nowak, T.; Waring, M. Room-Temperature Palladium-Catalyzed Coupling of Heteroaryl Amines with Aryl or Heteroaryl Bromides. *Synlett* **2012**, *2012*, 285–289. [CrossRef]
47. Gribanov, P.S.; Atoian, E.M.; Philippova, A.N.; Topchiy, M.A.; Asachenko, A.F.; Osipov, S.N. One-Pot Synthesis of 5-Amino-1,2,3-triazole Derivatives via Dipolar Azide−Nitrile Cycloaddition and Dimroth Rearrangement under Solvent-Free Conditions. *Eur. J. Org. Chem.* **2021**, *2021*, 1378–1384. [CrossRef]
48. Gribanov, P.S.; Topchiy, M.A.; Karsakova, I.V.; Chesnokov, G.A.; Smirnov, A.Y.; Minaeva, L.I.; Asachenko, A.F.; Nechaev, M.S. General Method for the Synthesis of 1,4-Disubstituted 5-Halo-1,2,3-triazoles. *Eur. J. Org. Chem.* **2017**, *2017*, 5225–5230. [CrossRef]
49. Topchiy, M.A.; Asachenko, A.F.; Nechaev, M.S. Solvent-Free Buchwald-Hartwig Reaction of Aryl and Heteroaryl Halides with Secondary Amines. *Eur. J. Org. Chem.* **2014**, *2014*, 3319–3322. [CrossRef]
50. Topchiy, M.A.; Dzhevakov, P.B.; Rubina, M.S.; Morozov, O.S.; Asachenko, A.F.; Nechaev, M.S. Solvent-Free Buchwald-Hartwig (Hetero)arylation of Anilines, Diarylamines, and Dialkylamines Mediated by Expanded-Ring N-Heterocyclic Carbene Palladium Complexes. *Eur. J. Org. Chem.* **2016**, *2016*, 1908–1914. [CrossRef]
51. Chesnokov, G.A.; Gribanov, P.S.; Topchiy, M.A.; Minaeva, L.I.; Asachenko, A.F.; Nechaev, M.S.; Bermesheva, E.V.; Bermeshev, M.V. Solvent-free Buchwald–Hartwig amination with low palladium loadings. *Mendeleev Commun.* **2017**, *27*, 618–620. [CrossRef]
52. Ageshina, A.A.; Sterligov, G.K.; Rzhevskiy, S.A.; Topchiy, M.A.; Chesnokov, G.A.; Gribanov, P.S.; Melnikova, E.K.; Nechaev, M.S.; Asachenko, A.F.; Bermeshev, M.V. Mixed er-NHC/phosphine Pd(ii) complexes and their catalytic activity in the Buchwald-Hartwig reaction under solvent-free conditions. *Dalton. Trans.* **2019**, *48*, 3447–3452. [CrossRef] [PubMed]
53. Gribanov, P.S.; Chesnokov, G.A.; Dzhevakov, P.B.; Kirilenko, N.Y.; Rzhevskiy, S.A.; Ageshina, A.A.; Topchiy, M.A.; Bermeshev, M.V.; Asachenko, A.F.; Nechaev, M.S. Solvent-free Suzuki and Stille cross-coupling reactions of 4- and 5-halo-1,2,3-triazoles. *Mendeleev Commun.* **2019**, *29*, 147–149. [CrossRef]
54. Gribanov, P.S.; Chesnokov, G.A.; Topchiy, M.A.; Asachenko, A.F.; Nechaev, M.S. A general method of Suzuki-Miyaura cross-coupling of 4- and 5-halo-1,2,3-triazoles in water. *Org. Biomol. Chem.* **2017**, *15*, 9575–9578. [CrossRef] [PubMed]
55. Charles, M.D.; Schultz, P.; Buchwald, S.L. Efficient pd-catalyzed amination of heteroaryl halides. *Org. Lett.* **2005**, *7*, 3965–3968. [CrossRef] [PubMed]
56. Afanas'ev, O.I.; Tsyplenkova, O.A.; Seliverstov, M.Y.; Sosonyuk, S.E.; Proskurnina, M.V.; Zefirov, N.S. Homocoupling of bromotriazole derivatives on metal complex catalysts. *Russ. Chem. Bull.* **2016**, *64*, 1470–1472. [CrossRef]
57. Kolychev, E.L.; Asachenko, A.F.; Dzhevakov, P.B.; Bush, A.A.; Shuntikov, V.V.; Khrustalev, V.N.; Nechaev, M.S. Expanded ring diaminocarbene palladium complexes: Synthesis, structure, and Suzuki-Miyaura cross-coupling of heteroaryl chlorides in water. *Dalton. Trans.* **2013**, *42*, 6859–6866. [CrossRef]
58. Farmer, J.L.; Pompeo, M.; Lough, A.J.; Organ, M.G. [(IPent)PdCl2(morpholine)]: A readily activated precatalyst for room-temperature, additive-free carbon-sulfur coupling. *Chem. Eur. J.* **2014**, *20*, 15790–15798. [CrossRef]

Review

Synthesis and Applications of Nitrogen-Containing Heterocycles as Antiviral Agents

Tuyen N. Tran [1] and Maged Henary [1,2,*]

1. Department of Chemistry, Georgia State University, 100 Piedmont Avenue SE, Atlanta, GA 30303, USA; ttran148@student.gsu.edu
2. Center of Diagnostics and Therapeutics, Department of Chemistry, Georgia State University, 100 Piedmont Avenue SE, Atlanta, GA 30303, USA
* Correspondence: mhenary1@gsu.edu; Tel.: +1-404-413-5566; Fax: +1-404-413-5505

Abstract: Viruses have been a long-term source of infectious diseases that can lead to large-scale infections and massive deaths. Especially with the recent highly contagious coronavirus (COVID-19), antiviral drugs were developed nonstop to deal with the emergence of new viruses and subject to drug resistance. Nitrogen-containing heterocycles have compatible structures and properties with exceptional biological activity for the drug design of antiviral agents. They provided a broad spectrum of interference against viral infection at various stages, from blocking early viral entry to disrupting the viral genome replication process by targeting different enzymes and proteins of viruses. This review focused on the synthesis and application of antiviral agents derived from various nitrogen-containing heterocycles, such as indole, pyrrole, pyrimidine, pyrazole, and quinoline, within the last ten years. The synthesized scaffolds target HIV, HCV/HBV, VZV/HSV, SARS-CoV, COVID-19, and influenza viruses.

Keywords: nitrogen-containing heterocycles; synthesis; antiviral agents; viruses; COVID-19; inhibition

1. Introduction

In recent years, outbreaks of infectious viral diseases have been increasing unexpectedly and costing millions of human lives. Despite many developed vaccines and therapeutics for prevention and treatment, viruses have continuously evolved and re-emerged to threaten public health, social relations, and economic stability.

In the early 1980s, the human immunodeficiency virus/acquired immunodeficiency syndrome (HIV/AIDS) infected 79 million people, with 39 million deaths over three decades [1]. The influenza virus series also have a long-term effect on public health. The seasonal influenza viruses infect 2–5 million people and kill 250,000–500,000 people worldwide per year [2]. Other influenza viruses also have caused occasional pandemics throughout history. The H1N1 influenza pandemic of 1918 cost approximately 40 million lives. H2N2 caused another epidemic in 1957, while H3N2 struck in 1968, and H1N1 again in 2009 [2]. During the early 21st century, the severe acute respiratory syndrome coronavirus (SARS-CoV) from a zoonotic source started in Guangdong Province, China. It spread through the global community, causing about 8000 infections and 800 deaths worldwide [3]. Middle East respiratory syndrome coronavirus (MERS-CoV) emerged in 2012 and caused 2029 infections and 704 deaths in 27 countries [4]. The novel coronavirus (SARS-CoV-2) has caused a worldwide pandemic with more than 99 million infections and 2 million deaths within the first twelve months, starting in December 2019 in Wuhan, China. Although some antiviral drugs and vaccines are available for certain viruses, it is necessary to continuously develop new drugs and methods for drug-resistant viruses and viruses newly evolved from mutations.

The most promising antiviral drugs are small organic molecules that can target specific parts of the viruses and interfere with different stages of the viral life cycle [4]. Various strategies have been applied to target different viruses effectively. For example, for SARS-CoV, small molecules were used to target protease activity and inhibit viral replication. For anti-HIV/AIDS agents, viral glycoproteins were targeted to hinder their interaction with the receptors on cell surfaces, which then activate the virus's endocytosis into the cell [4]. Among many chemical scaffolds, nitrogen-containing heterocyclic small compounds have been exploited extensively due to their broad range of applications in biological and pharmacological activities. The nitrogen-containing heterocyclic bases have high versatility in synthesis with different moieties. Those heterocyclic backbones have rigid aromatic structures that can be incorporated into the binding pockets and provide various molecular interactions, such as ionic bonding, hydrogen bonding, hydrophobic interaction, non-covalent bonding, etc., for ligand binding with receptor proteins [5,6]. The interaction can mimic specific properties that can effectively inhibit the activities of various biological enzymes and components that play vital roles in the development of viral infections [7,8]. Previous reviews focused on single types of nitrogen-containing heterocycles such as indole and/or imidazothiazole derivatives in designing antiviral agents [7,9]. Other reviews covered broader details on various biological applications such as antimicrobial, anti-inflammatory, anti-tubercular, anti-depressant, and anti-cancer activities [10–12]. Our current review focuses specifically on the common nitrogen-containing heterocycles including indoles, pyrroles, pyrimidines, pyrazoles, and quinolines that have been applied in drug design for antiviral purposes during the past ten years. These heterocycles are significant backbones of pharmaceutical products with exceptional biological activity to interfere with various viral infections [8]. While the indole cores are known for their wide existence in natural products with biological activity [7], the properties of pyrrole can be expanded for chemical design and are suitable for biological systems [13]. Pyrimidine derivatives have widespread therapeutic applications, as they are essential building blocks of nucleic acids in DNA and RNA [14]. Pyrazoles can be fused with other heterocycles to extend their active biological potential [15]. Quinoline derivatives have versatile chemical properties for synthesis and biological activities [16]. Our review discusses the synthesis and biological characteristics of nitrogen-containing heterocycle derivatives in detail. Further modifications and functions of nitrogen heterocycles are introduced along with various antiviral purposes.

2. Indole

Indole derivatives are one of the well-known scaffolds in drug discovery that can inhibit a wide diversity of enzymes by binding with potential ligands [7]. The first indole was prepared using Fischer indole synthesis, which was reported in 1866. Nowadays, most indole synthesis pathways start with incorporating a benzene ring with additional factors that stabilize the formation with the fused pyrrole ring. Indole is an aromatic ring with ten π electrons and a lone pair of nitrogens. The lone pair is not available for protonation but is involved in the delocalization of the indole conjugated system. The protonation and other electrophilic substitutions mostly occur at the C-3 position, which has the highest electron density and is most thermodynamically stable for modifying antiviral agents [17].

Hassam et al. designed new scaffolds of HIV non-nucleoside reverse transcriptase (RT) inhibitor with the addition of a cyclopropyl group at the C-3 of the indole base [18]. The RT enzyme plays a crucial role in the reverse transcription of viral RNA to single-stranded DNA in host cells. The binding of specific cyclopropyl chemical moiety to the RT pocket can significantly inhibit the enzyme's function by improving electrostatic interaction with the hydrophobic pocket [18]. As shown in Scheme 1, the ester-substituted indole **1** reacted with benzoyl chloride for the electrophilic substitution at the C-3 position. Tert-butyloxycarbonyl (Boc) was added to intermediate **2** to prevent reactions at the amine before further modifying cyclopropyl moiety. Intermediate **3** underwent a Wittig reaction with methyl triphenylphosphonium ylide reagent to convert ketone to alkene **4**, which then reacted with diiodomethane and zinc alloy for cyclopropyl addition to furnish product 5 [18].

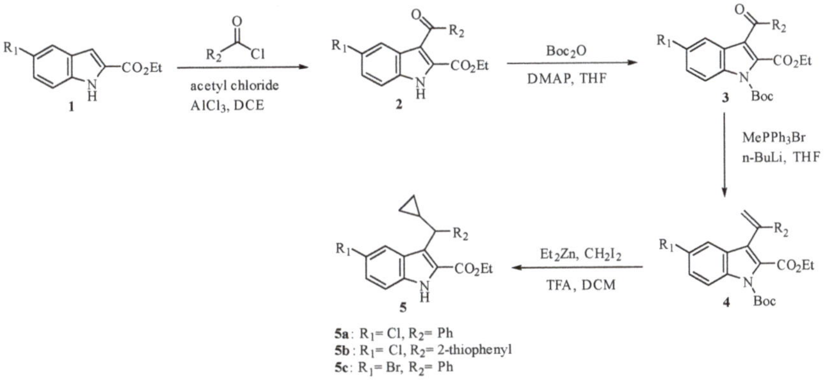

Scheme 1. The synthesis of non-nucleoside scaffolds as HIV reverse transcriptase inhibitors.

As shown in Figure 1, in the in vitro phenotypic assay, compounds **5a–c** showed high inhibitory activity (IC_{50}), with low cytotoxicity (CC_{50}) in **5a** and **5c**. In the modeling study, **5a** was better at accommodating the binding site. The indole NH and ester moiety was incorporated for hydrogen bonding in the reverse transcriptase's allosteric site at Lys 101. The ester group associates well in the binding site with the ethyl directed out of the site. Modification of the ethyl group decreased the potency of the compound. The cyclopropyl moiety was implemented to bind with Val 179 binding pocket, a small, hydrophobic cleft located near the catalytic site. The aromatic ring of R_2 was favorable for the interaction with Tyr 188 and Trp 299. The modification at R_2 from phenyl to 2-thiophenyl resulted in a slightly better inhibitory value but increased toxicity in cells. Increasing the size of the halogen at R_3 also enhanced IC_{50} and CC_{50} values [18].

Compound	IC_{50} (μM)	CC_{50} (μM)
5a	0.085	30.3
5b	0.065	67.1
5c	0.066	24.5

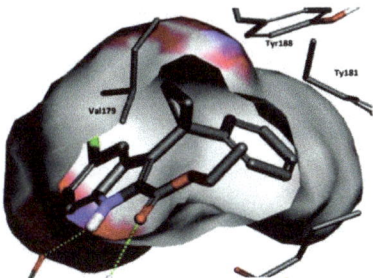

Figure 1. In vitro phenotypic assay values and corresponding cytotoxicity values against HIV. The molecular docking study of **5a** in the binding site of HIV reverse transcriptase [18].

Another study of antiviral agents with substitution at C-3 of indole was utilized to target the hepatitis C virus (HCV). HCV infection can lead to acute or chronic liver diseases, such as hepatocellular carcinoma and liver cirrhosis. Specific treatments have been approved for clinical applications. However, those treatments possessed low effectiveness in the patient population with additional side effects [19]. Han et al. [20] introduced the small molecules of N-protected indole scaffolds (NINS) to inhibit HCV. As shown in Scheme 2, scaffold **6** reacted with the racemic epibromohydrin to form N-protection derivative **7** with two enantiomers. The substitution of NH was further extended by ring-opening of epoxide with the nucleophilic amine chain to afford **8**.

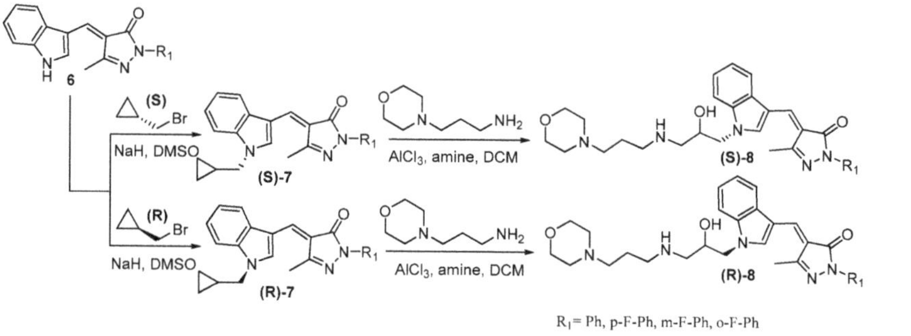

Scheme 2. The synthesis of NINS for anti-HVC activity.

In the structure–activity relationship (SAR) study, multiple modifications of the phenyl ring at R_1 were analyzed. As shown in Table 1, p-F-Ph and m-F-Ph improved inhibitory potency, but o-F-Ph reduced anti-HCV activity. The m-F-Ph exhibited lower cytotoxicity compared to p-F-Ph. The (R)-enantiomer can possess better anti-HCV potency and less cytotoxicity than its corresponding (S)-enantiomer of both m-F-Ph and p-F-Ph. The results indicated that the position of substitution and chirality has a high impact on their inhibitory effect. The scaffold was also proved to inhibit viral entry into the viral cycle rather than interfere in any viral RNA replication in host cells with a mechanism of action (MoA) study.

Table 1. The in vitro SAR study of compounds against HCV [20].

R_1	EC_{50} (μM)	CC_{50} (μM)	SI
Ph	2.04 ± 0.11	35.83 ± 0.25	17.56
p-F-Ph	1.02 ± 0.10	46.47 ± 0.24	45.56
(R) p-F-Ph	0.72 ± 0.09	>50	>69.44
(S) p-F-Ph	7.12 ± 0.21	34.57 ± 0.53	4.86
m-F-Ph	0.92 ± 0.06	21.46 ± 0.34	23.33
(R) m-F-Ph	0.74 ± 0.11	31.06 ± 0.37	41.97
(S) p-F-Ph	5.87 ± 0.18	23.31 ± 0.30	3.97
o-F-Ph	2.79 ± 0.17	34.21 ± 0.43	12.26

The varicella-zoster virus (VZV) infection can cause acute varicella and herpes zoster, leading to various disease complications in the central nervous system from latent viruses [21]. Most approved drugs for VZV-associated treatments were nucleosides that competitively inhibit the viral DNA polymerase and interfere with DNA replication. However, those nucleosides implied multiple drug resistance and produced low efficacy in the anti-VZV virus. Mussela et al. [22] synthesized a family of indole derivatives as non-nucleoside antivirals to mimic deoxythymidine. The compounds can inhibit thymidine kinase (TK) by interfering with the series phosphorylation of thymidine triphosphates, which can be used as a complementary base in the replication process [23]. As shown in Scheme 3, the 3-(2-bromoethyl)-1H-indole **9** was reacted with methyl iodide for N-alkylation. The nucleophilic displacement of bromine in **10** was conducted with chain-linked amine and used palladium as a catalyst under microwave conditions to obtain **11**, which was then treated with acetyl chloride under the basic condition to yield the final compound **12**.

Scheme 3. Synthesis of tryptamine derivative for targeting TK against VZV.

Compound **12** has low cytotoxicity in cells and good inhibitory activity at low EC_{50} due to the alkylation of NH and acylation of the amine group in C-3 substitution [22]. This tryptamine derivative has lower potency than reference drugs Acyclovir and Brivudine but displayed similar inhibiting activity against TK in anti-VZV mechanisms. Additionally, the derivative **12** selectively targeted VZV strains (OKA, 07-1, and YS-R) only when tested with other members in the same family of Herpesviridae (HMCV, HSV-1, and HSV-2) and various RNA virus strains such as HIV and influenza, as shown in Table 2. Compound **12** can be a leading compound for further investigation in inhibiting VZV specifically.

Table 2. Activity of compound **12** against various viral strains [22].

Virus Strain	OKA (TK+)	07-1 (TK−)	YS-R (TK−)	AD 169 (HMCV)	Davis (HMCV)	KOS (HSV-1)	TK− KOS ACV$^{r\,1}$ (HSV-1)	HSV-2
EC_{50} (μM)	2.5	1.7	2.1	>4	>4	>100	>100	>100[2]

[1] TK− KOS ACV^r: HSV-1; KOS and thymidine kinase-deficient acyclovir-resistant.

Other synthesized potential antiviral agents have substitution at C-2 of indole, the second most reactive site for electrophiles. For targeting SARS-CoV viruses, Thanigaimalai et al. [24] discovered compounds to target the chymotrypsin-like protease (CL^{pro}), which plays a vital role in cleaving polyproteins to produce functional proteins directly involved in viral replication and transcription [25]. The analogs were peptidomimetic covalent inhibitors that can mimic the substrate of CoV 3CL^{pro}. As shown in Scheme 4, to synthesize compound **17**, the peptidomimetic chain was added to the commercially available indole derivative **13**, through peptide coupling with the carboxyl group at C2 to form the peptide **14**. For the other intermediate, γ-lactam acid **15** coupled with N, O-dimethylhydroxylamine to form substituted amide from the carboxylic end using Weinreb–Nahm synthesis. The amide continuously reacted with benzothiazole to create **16**, which then deprotected and coupled with the peptide **14** in the presence of peptide coupling reagent to furnish compound **17** [24].

The synthetic inhibitor **17** has four main features that are suitable for different pockets in the active site of CoV CL^{pro}. The compounds included ketone to target cysteine residue's thiol (Cys 145) in the S1″ pocket and (S)-γ-lactam ring for the S1 pocket. The hydrophobic leucine was used for the S2 position [24,26]. In another study, the leucine was replaced by the π conjugated system and hydrophobic interaction of the aryl or cyclohexyl group to enhance access of the S2 pocket to target SARS-CoV-2 [27]. Based on its SAR study, compound **17** exhibited great inhibitory (Ki = 0.006 μM or IC_{50} = 0.74 μM) activity with the methoxy at C-4 of the indole. Indole possessed the best inhibitory effect among other heterocycles due to its NH, which can provide hydrogen bonding interaction with Gln 166, as shown in Figure 2 from the molecular docking study.

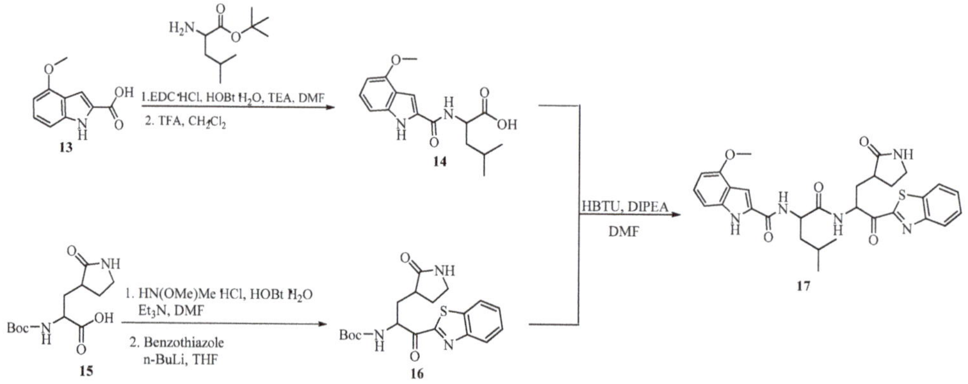

Scheme 4. The synthesis of dipeptide inhibitor against SAR-CoV 3CL[pro] [24].

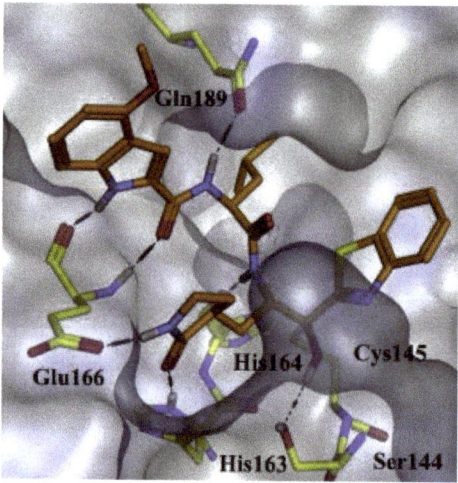

Figure 2. Molecular docking poses and binding interaction of **17** bound to SARS-CoV 3CL[pro] [24].

3. Pyrrole

Pyrrole is a heterocyclic aromatic five-membered ring that was first observed in coal tar and bone oil in 1834. Many investigations reported pyrrole as an integral part of different natural compounds. The delocalization of the lone pair of electrons from the nitrogen atom provided additional stabilization of the ring. In total, six π electrons delocalized over the five-membered ring formed the isoelectronic system and allowed electrophilic attack in different reactions [28].

Curreli et al. [29] have developed drugs to block HIV-1 envelope glycoprotein (gp120) from binding the receptor CD4 of the host cells and prevent the entry of viral RNA [30]. The scaffolds were designed to mimic receptor CD4 and act as HIV-1 entry antagonists. Intermediates **18** and **19** were synthesized and coupled to yield compound **20**, as shown in Scheme 5 [28,31]. The deprotonation of amine followed to afford product **21**. Intermediate **19** can be synthesized from R- and S-isomer imines that derived isomers in compounds **20** and **21** [29].

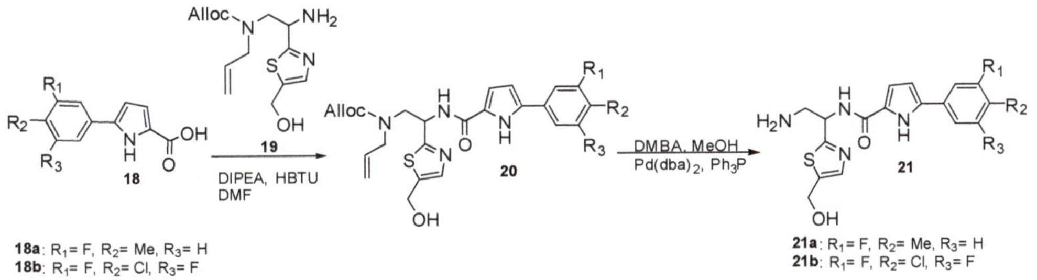

Scheme 5. The synthesis of pyrrole derivatives as glycoprotein inhibitors of HIV-1.

Based on X-ray crystal structure analysis, pyrrole allowed a potential hydrogen bond of NH with the residue Asn425 of gp120. Methylation of NH lost an H-bond donor atom and increased steric hindrance, which interferes with binding capability. The scaffold loses its antiviral activity when replacing the pyrrole ring with imidazole. According to Table 3, the compounds exhibited high antiviral potency with low cytotoxicity and good selectivity for HIV-1 gp120. The methyl substitution at R_2 of **21a** enhanced antiviral potency, while the addition of fluoro at R_3 of **21b** showed similar activity but improved metabolic stability. Because **(R) 21a** (NBD-14088) and **(S) 21b** (NBD-14107) have better selectivity, they were used to measure HIV-1 entry antagonist properties with cell-to-cell fusion inhibition assay and infectivity in cells, as shown in Figure 3. Even though **(R) 21a** and **(S) 21b** required higher IC_{50} compared to NBD-556 (HIV-1 entry agonist) and NBD-11021 (HIV-1 entry antagonist), negative and positive controls, respectively, two synthesized compounds still exhibited antagonist properties to reduce infections. The compounds also can inhibit HIV-1 reverse transcriptase from converting viral RNA into complementary DNA in hosts [29].

Table 3. The SAR study of the compounds against HIV-1 in TZM-bl and MT-2 cells [29].

Compound	TZM-bl Cells		MT-2-Cells	
	IC_{50} (µM)	CC_{50} (µM)	IC_{50} (µM)	CC_{50} (µM)
(S) 21a	0.85 ± 0.06	39.2 ± 0.8	1.6 ± 0.08	35.2 ± 0.8
(R) 21a	0.45 ± 0.05	38.8 ± 0.06	0.76 ± 0.3	38 ± 1
(S) 21b	0.64 ± 0.06	39.5 ± 2.3	0.96 ± 0.1	37 ± 1.5
(R) 21b	0.48 ± 0.1	20.6 ± 0.2	0.97 ± 0.2	17.4 ± 0.4

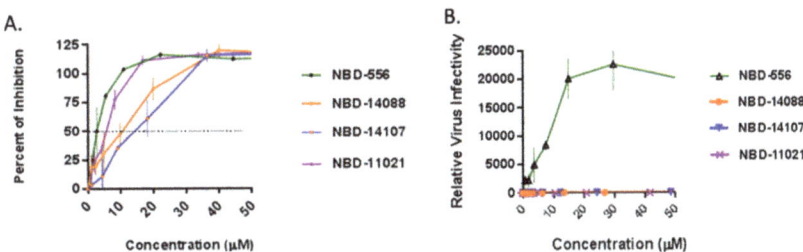

Figure 3. (**A**) Inhibition assay of tested compounds **(R) 21a** and **(S) 21b** with the controls NBD-556 and NBD-11021. (**B**) Infectivity assay of the compounds and controls with CD4-dependent HIV-1$_{ADA}$ [29].

Herpes simplex virus (HSV) is another virus in the family of α-herpesviruses with VZV that can cause common, self-resolving diseases of skin or mucosa, such as herpes labialis and other infectious diseases. However, HSV and VZV possess differences in the route of infection, spread pathway, and range of hosts [32]. Due to the viral strains highly resistant to the nucleoside treatments, non-nucleoside compounds were synthesized to inhibit thymidine kinase (TK) from early DNA replication. According to Hilmy et al. [33], the pyrrole analogs **23a–d** were synthesized using different substituted 2-amino-3-cyano-1,5-diarylpyrroles to react with various aryl aldehydes under the basic condition with phosphorous pentoxide, as shown in Scheme 6.

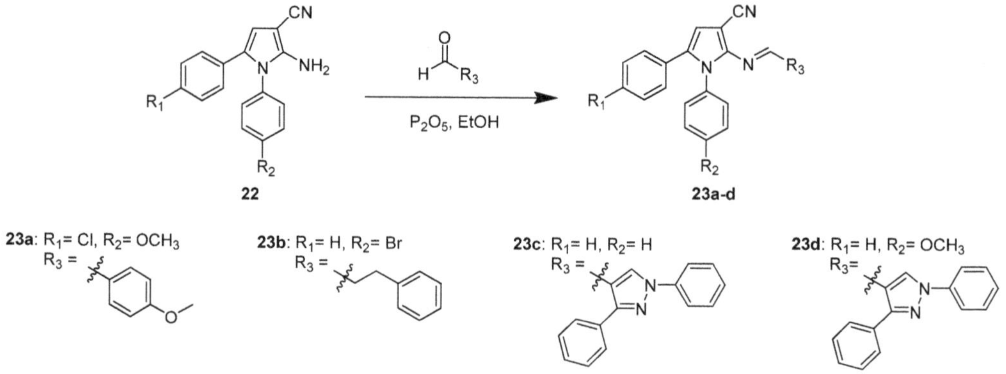

Scheme 6. Synthesis of thymidine kinase inhibitors to target HSV.

The analogs were compared with Acyclovir (ACV), a standard drug used to treat HSV, for anti-HSV activity and cytotoxicity. As shown in Table 4, all new compounds exhibited a high percentage of reduction (94–99%) in the number of virus plaques. Compounds **23a** (99%) and **23d** (97%) even had better results than ACV due to their similarity at substituted N of pyrrole with 4-methoxyphenyl. Compound **23a** had the highest activity with the 4-methoxybenzylideneamino at position C-2 of pyrrole. The synthesized compounds had better docking scores than ACV, indicating that they had better ligand–receptor interaction in the TK active site. Comparing compounds **23a** and **23d** in Figure 4, they had different interactions in the binding pocket. The two oxygens of the methoxy group in **23a** had hydrogen bonding with Lys62 and Tyr132. Another hydrogen bond was formed between the cyano nitrogen and Arg222. However, for **23d**, the methoxy group formed two hydrogen bonds with Arg176 and Tyr101. Other hydrogen bonds were also established from the nitrogens of the compounds. Both also formed hydrophobic and van der Waals interactions with other amino acids. However, **23d** interacted with crucial amino acids that actively contribute to the function of TK; hence, **23d** showed better inhibitory effect compared to **23a**. Additionally, compounds **27b** and **27c** might have different anti-HSV activity mechanisms rather than targeting TK [33].

Table 4. Anti-HSV-1 activity and docking score of analogs **27a–d** and Acyclovir [33].

Compound	Cytotoxicity (μM)	% Reduction in Cytopathic Effect of HSV-1	Docking Score
ACV	10^{-2}	96	−12.74
23a	10^{-2}	99	−11.30
23b	10^{-2}	96	−11.44
23c	10^{-2}	94	-8.98
23d	10^{-2}	97	−7.32

Lin et al. [34] introduced a class of anti-influenza agents targeting the viral nucleoprotein (NP), a binding protein that contributes to the transcription and packaging processes. The synthesized pyrimido-pyrrolo-quinoxalinedione analogs were aimed to inhibit the synthesis of NP and interrupt viral replication [35]. As shown in Scheme 7, substituted pyrimidinedione **24** was methylated at the two atoms of nitrogenusing dimethyl sulfate, followed by Friedel–Crafts acylation with benzyl chloride to furnish compound **26**. The synthesis was continued with the addition of bromine and the formation of fused pyrrole **28** from the reaction of **27** with 2-amino-2-methylpropan-1-ol. The annulation of intermediate **28** with *F*- substituted aryl aldehyde produced the final product, compound **29**.

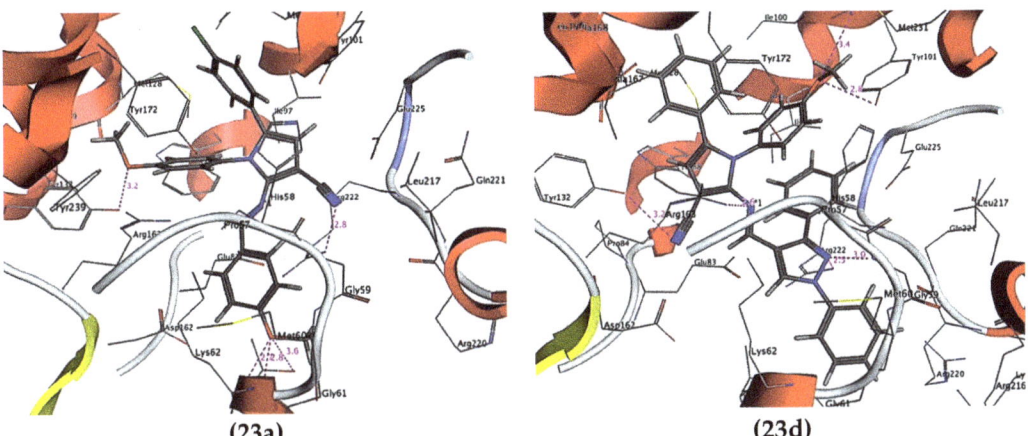

Figure 4. Docking study of **23a** and **23d** in the binding site of HSV-1 thymidine kinase [33].

Scheme 7. The synthesis of pyrimido-pyrrolo-quinoxalinedione for inhibiting nucleoprotein of influenza A H1N1 virus.

The inhibiting effect of compound **29** (PPQ-581) was compared with that of the nucleozin 3061, a potent antagonist of NP. Both have a similar trend of effectively inhibiting nucleoprotein (NP) synthesis shortly after infection. Treatment with **29** and nucleozin from 3–8 h post-infection partially inhibits NP synthesis. The docking study of compound **29** in the influenza A nucleoprotein is shown in Figure 5. The oxygen of ketone formed hydrogen bonding with S377, which was a crucial binding area for compound **29**. The mutation of the S377 sidechain can result in the loss of the anti-influenza activity of compound **29**. Compound **29** also inhibited the influenza RNA-dependent RNA polymerase (RdRP) activity of nucleozin-resistant influenza strains.

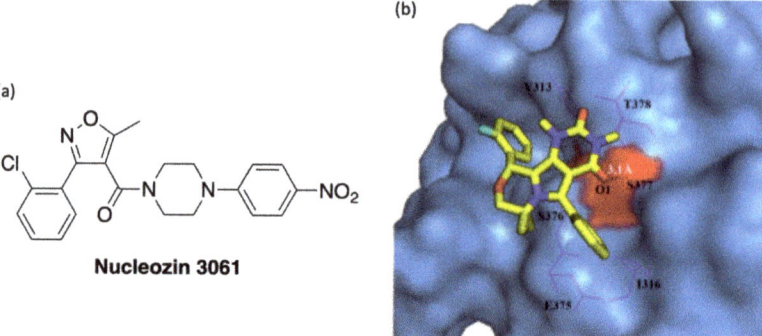

Figure 5. (a) Chemical structure of nucleozin 3061. (b) The molecular docking study of compound **29** in the influenza A NP [34].

4. Pyrimidine

Pyrimidine is a heterocycle composed of a six-membered ring with two nitrogens at positions 1 and 3. The first pyrimidine derivative was synthesized in 1818 by Gaspare Brugnatelli. Pyrimidines were continuously developed for applications in biological systems, as their nitrogen bases were associated with DNA and RNA. The two nitrogens in pyrimidines are electron withdrawers, leaving specific carbons with electron deficiency [36]. Pyrimidine has also been fused with other heterocycles to improve its applicable spectrum of biological activity [37].

Currently, HIV/AIDS is still causing a major global health issue. The treatments for HIV/AIDS require a long-term commitment to target different stages of viral DNA replication. Malancona et al. [38] introduced the 5,6-dihydroxypyrimidine scaffolds to target the HIV nucleocapsid (NC), which is a multifunctional protein that plays crucial roles in binding amino acids for their sequence-specific binding in reverse transcription. NC is also a nucleic acid chaperone that can manipulate nucleic acid structures for their thermodynamically stable conformations [39]. As shown in Scheme 8, the central dihydroxypyrimidine cores were obtained by reacting substituted 2-pyridinecarbonitrile with hydroxylamine hydrochloride to form intermediate **31**. The synthesis was continued with the reaction of compound **31** with dimethyl acetylenedicarboxylate (DMAD) for thermal cyclization to produce **32**, which coupled with (R)-1-cyclohexylethanamine to furnish **33a–33d**.

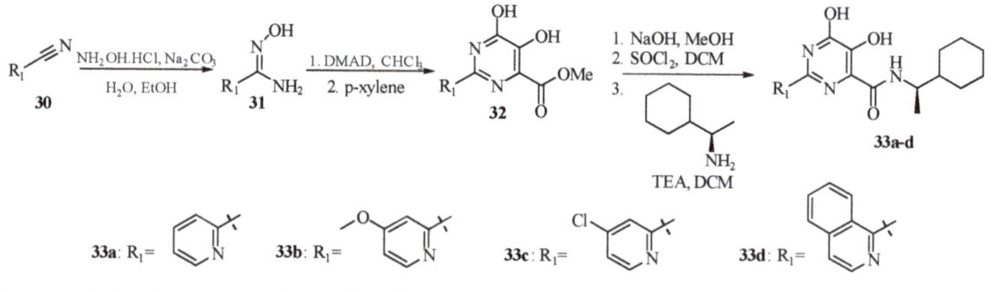

Scheme 8. Synthesis of nucleocapsid inhibitors as anti-HIV/AIDS activity.

The 2-(pyridin-2-yl) substituent at R1 was essential for the inhibitory activity of the analogs. The amide moiety (N-cyclohexylmethyl) targeted the hydrophobic binding site of the protein. As shown in Table 5, further modification with 5-methoxy, 5-chlorine at the pyridine group, and quinoline resulted in **33b–33d** with 40–80 folds of improved potency and higher SI values (25–75) than without any modification of 2-(pyridin-2-yl) substituent. As shown in Figure 6, the molecular docking of compound **33c** in the NC demonstrated that 2-(pyridin-2-yl) substituent has π-stacking interaction with Trp37, while the dihydroxypyrimidine formed H-bonds with Gly35, Met46, and Gln45. The amide moiety interacted with the hydrophobic pocket, including Phe16, Ala25, Trp37, and Met46. Other in vitro and in vivo tests against multiple drug-resistant HIV-1 strains were also analyzed in the study. The analogs were also active in different HIV-1 resistant strains, with high oral bioavailability and excellent in vitro metabolic stability in rat and human samples and half-life in the in vivo study [38].

Table 5. The inhibiting activity of analogs 33 against HIV nucleocapsid and cells [38].

ID	NC Inhibition IC_{50} (µM)	BiCycle Wild Type Strain NL4-3 EC_{50} (µM)	HeLa CC_{50} (µM)	SI	PMBC CC_{50} (µM)	SI
33a	167 ± 7	1.2 ± 0.7	3.6 ± 0.7	3	2.6 ± 1.7	2
33b	22 ± 7	0.04 ± 0.03	3.0 ± 0.1	75	2.7 ± 0.1	75
33c	2 ± 1	0.1 ± 0.03	3.0 ± 0.5	30	2.5 ± 0.2	25
33d	4 ± 1	0.1 ± 0.05	2.1 ± 0.3	21	5 ± 2	50

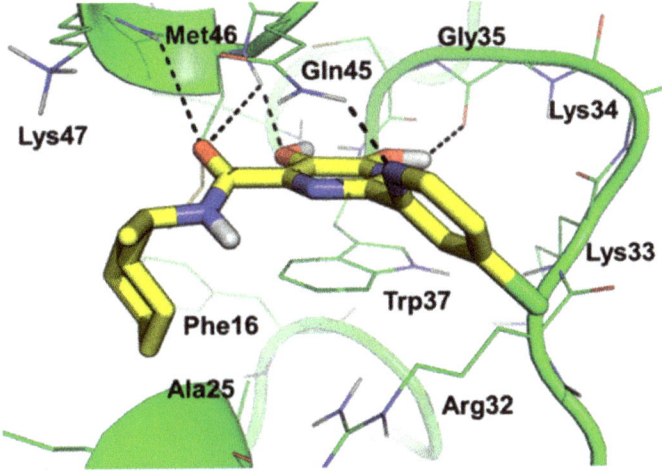

Figure 6. The molecular interaction of 33c in the HIV nucleocapsid [38].

Mohamed et al. [40] reported the synthesis of non-nucleoside antiviral agents for hepatitis C virus (HCV), which has the highest infection rate in Egypt and causes 350,000 deaths worldwide annually. The HCV NS5B polymerase functioned as RNA-dependent RNA polymerase (RdRp) and enabled the catalyzation of viral genome synthesis [41]. The synthesis of pyrrolopyrimidine derivative 39 as a non-nucleoside purine scaffold to inhibit HCC NS5B polymerase is summarized in Scheme 9. The benzoin 34 was condensed with aryl amine to form intermediate 35, which then reacted with malononitrile to produce pyrrole derivative 36 via Dakin West reaction. The 2-aminopyrrole-3-carbonitrile 36 coupled with formic acid to furnish associated pyrrolopyrimidine derivative 37. The carbonyl was replaced by chlorine at C-4 pyrimidine, resulting in intermediate 38 for further modification at this position. Aryl amine was introduced to produce 4-aryl amino derivative 39.

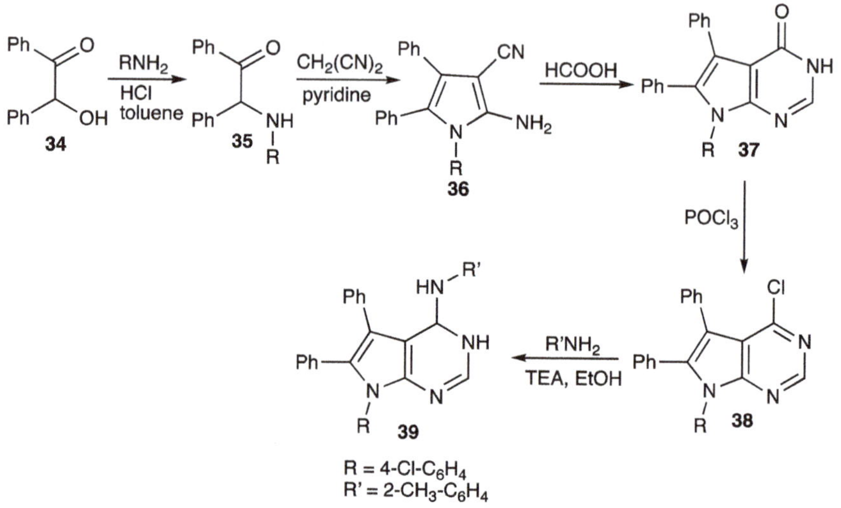

Scheme 9. Synthesis of anti-HCV agents targeting HCV polymerase (NS5B).

Compound **39** exhibited the highest antiviral activity in the study. The arylamino group ('') enhanced the toxicity of the scaffold in HCV genotype 4a cells. In the molecular docking, it was found that **39** bound strongly with Mg^{2+} in the docked site. Other measurements in the docking system proved that **39** was stable in the binding pocket and improved binding affinity. As shown in Figure 7, the carbonyl C=O of Gln446 in the HCV RdRp catalytic site can form a hydrogen bond with aniline NH. At the same time, the NH_2 interacted with nitrogen in the pyrimidine ring to enhance the binding of compound **39** to the inhibition site.

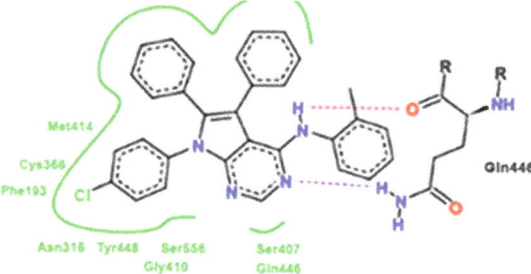

Figure 7. Molecular interaction of compound **39** in the binding site of HCV polymerase [40].

Pyrimidine derivatives were also employed in the study of anti-HSV activity. Currently, Acylclovir (Figure 8) is one of the drugs used to treat the herpes simplex virus (HSV). However, in some cases, the viruses can resist the drug; hence, new agents are continuously developed and tested for antiviral activity. Mohamed et al. [42] reported the synthesis of pyrimidine derivatives as anti-HSV agents. As shown in Scheme 10, acetophenone **40** was coupled with 3,4-dimethoxybenzaldehyde via aldol condensation to afford chalcone **41**, which was reacted with hydrogen peroxide to form epoxide **42**. The condensation of compound **42** with thiourea resulted in the cyclic formation of compound **43**, which could either react with 3-chloroacetylacetone to produce compound **44** or be condensed with 2-bromopropionic acid to form methylthiazolo compound **45**.

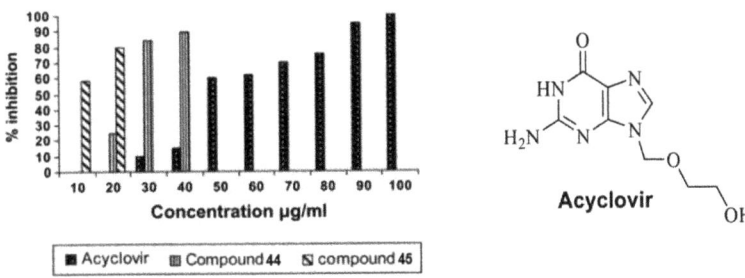

Figure 8. Comparison of effective percentages of inhibition against HSV infection among compounds **44**, **45**, and Acyclovir [42].

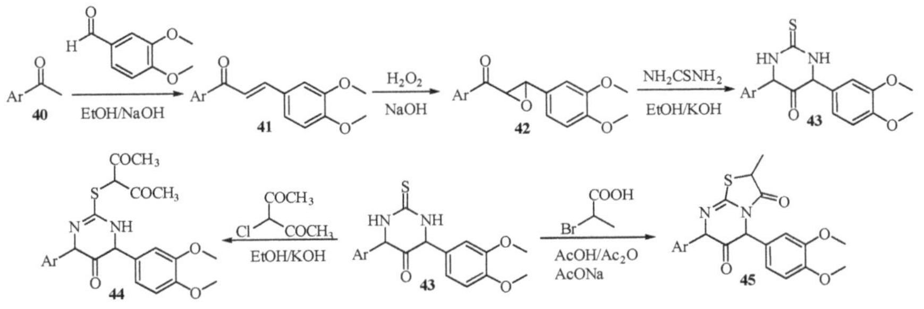

Scheme 10. Synthesis of pyrimidine derivatives as anti-HSV agents.

For the antiviral activity of compounds **44** and **45**, a plaque reduction assay was conducted using concentrations of 2 and 5 µg/mL, and the viral count was recorded after adding the compounds. As shown in Table 6, compounds **44** and **45** showed a slightly higher percentage of inhibition with the 5 µg/mL concentration. Compound **44** was more effective than compound **45**, achieving 100% inhibition with 5 µg/mL in the test sample. As shown in Figure 8, both compounds showed a higher percentage of inhibition at much lower concentrations compared to Acyclovir®. For the antiviral mechanism, **44** and **45** showed viricidal activity that possibly altered viral epitopes to inhibit binding to cells. Besides viricidal activity, compound **44** also has the potential to interfere with the replication processes of HSV. The two compounds are highly promising as potential new anti-HSV agents.

Table 6. The calculated percentage of inhibition in the plaque reduction assay of compounds **44** and **45** [42].

ID	Concentration (µg/mL)	Viral Count (Control)	Viral Count (Extract)	% Inhibition
44	2	0.84×10^7	0.08×10^7	90
	5	0.84×10^7	0	100
45	2	0.84×10^7	0.08×10^7	90
	5	0.84×10^7	0.04×10^7	95

During the current COVID-19 pandemic, Remdesivir® (Figure 9) was approved for use in the treatment of patients with confirmed SARS-CoV-2 infection. Remdesivir® is an adenosine nucleoside prodrug that exhibits a broad antiviral spectrum. Due to the rapid emergence of coronavirus infections that threaten millions of people in the global community, a safe, approved drug like Remdesivir® was temporarily employed for treatment. It showed in vitro antiviral activity with animal and human coronaviruses, including

SARS-CoV-2. As shown in Figure 10, Remdesivir® can effectively inhibit viral infection at a low concentration (EC$_{50}$ = 0.77 µM) with a high SI value (129.87 µM). The antiviral activity of Remdesivir® was confirmed with an immunofluorescence assay. The results indicated complete viral reduction, as the viral nucleoproteins could not be observed with a 3.70 µM concentration at 48 h post-infection [43].

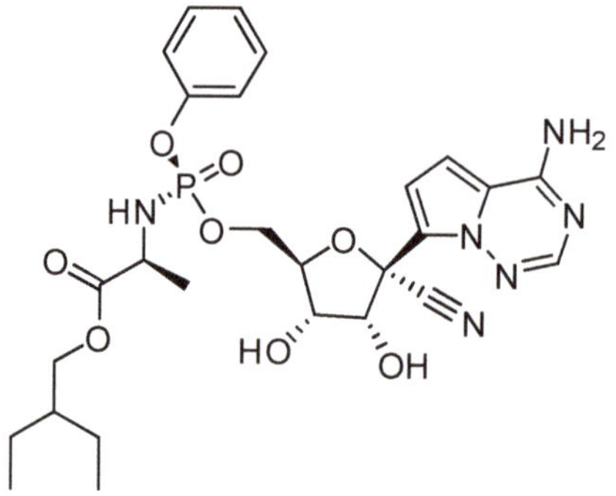

Figure 9. The molecular structure of Remdesivir®.

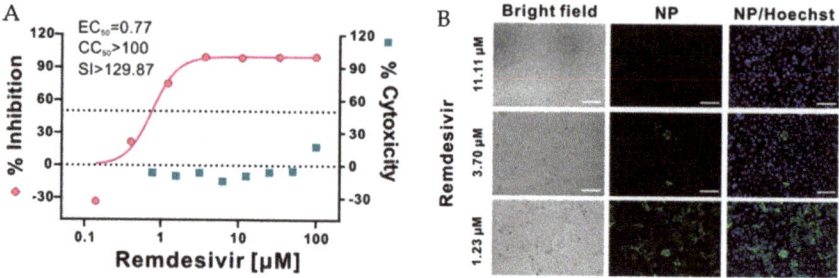

Figure 10. (**A**) Percentage of inhibition measured by qRT-PCR and cytotoxicity of Remdesivir® when used to treat SARS-CoV-2-infected Vero E6 cells. (**B**) Immunofluorescence assay of viral infection with Remdesivir® treatment [43].

Remdesivir® was originally targeted to inhibit RNA-dependent RNA polymerase (RdRp), which has adenosine triphosphate (ATP) as the main substrate. Zhang et al. [44] reported the homology modeling of NSP12 (the RdRp complex with multiple nonstructural protein units) of SARS-CoV-2 RdRp (SARS-CoV-2 NSP12). When accumulated in cells, Remdesivir® was hydrolyzed and coupled with triphosphate (RemTP) to compete with ATP actively. As shown in Figure 11, the triphosphate of RemTP has other interactions with crucial residues compared to ATP interactions to inhibit NSP12. The adenosine of ATP and the Remdesivir® core also interacted differently from each other in the binding pocket. Additionally, RemTP affected D618, which is an essential residue for the function of SARS-CoV-2 RdRp by forming a hydrogen bond with K798. The binding free energies of ATP and RemTP were represented by FEP calculations, which showed that RemTP binds more strongly than ATP in the pocket of SARS-CoV-2 RdRp with approximately 800 folds in K_d value to effectively inhibit SARS-CoV-2 RdRp activity in RNA reproduction [44].

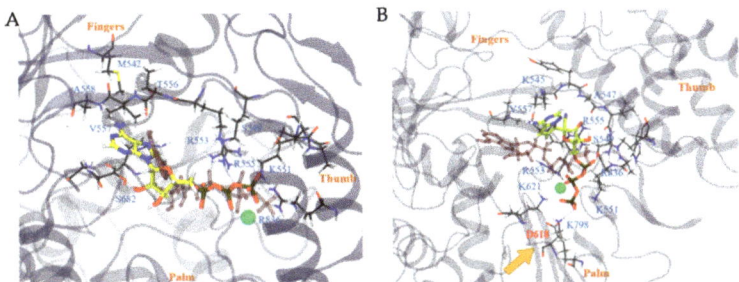

Figure 11. (**A**) The binding interaction of ATP with COVID-19 NSP12. (**B**) The binding interaction of RemTP with COVID-19 NSP12. The green sphere represents the Mg^{2+} ion [44].

In addition, Grein et al. [45] further investigated Remdesivir® clinical application conducted within a cohort of hospitalized patients diagnosed with COVID-19. The data were collected from 53 patients across three continents. The majority of patients (34 out of 53) had severe symptoms and were under invasive oxygen support. The treatment plan was designed for a 10-day course with 200 mg administered intravenously on day 1 and 100 mg daily for the next nine days. Only 40 patients completed ten days of treatments, while others discontinued treatment due to serious adverse effects. After 18 days of day one treatment, 12 patients (100%) previously with no oxygen or low-flow oxygen support were discharged. Five out of seven patients (71%) previously under non-invasive ventilation were also discharged. Nineteen out of thirty-four showed improvement within the oxygen-support group, as they were discharged without oxygen aid and under non-invasive oxygen support. The other patients in this group either showed no improvement (9 out of 34) or died (6 out of 34). About 83% experienced various common to serious adverse effects. Overall, Remdesivir in this study showed improvement in the treatment of COVID-19 based on the oxygen support scale. Among the patients, 68% showed improvement in oxygen supporting status or were fully discharged, with a 13% mortality rate after treatment.

5. Pyrazole

Pyrazole is an aromatic heterocycle composed of a five-membered ring with two nitrogens. Pyrazole core structures have been employed in wide applications from natural components to pharmaceutical activities [46]. With high electron densities of nitrogen at positions 1 and 2, pyrazole is considered as a π-excessive heterocycle. The ring is available for various modifications with multiple reagents in synthetic reactions. Substituted pyrazole can form chelation with metal ions to inhibit enzymes [47].

Currently, HIV-1 is still considered one of the most threatening infectious diseases to people with a high rate of infection. The crucial step of viral infection is the integration of viral RNA into host DNA by reverse transcriptase (RT). RT is composed of a DNA polymerase domain and a ribonuclease H domain (RNase H) [48]. Messore et al. [49] reported the synthesis of specific inhibitory analogs to target RNase H. As shown in Scheme 11, the synthesis started with the aldol condensation of substituted benzaldehyde **46** with acetone, then its reaction with TosMIC to furnish pyrrole derivative **47**. Compound **47** coupled with substituted aryl or alkyl chloride for the *N*-substitution of pyrrole ring to form compound **48**. Diethyl oxalate was coupled with intermediate **48** for aldol condensation product **49**. Then, hydrazine reacted with compound **49** for the formation of the pyrazole ring of product **50**. The ester of product **50** was hydrolyzed under a basic condition to yield carboxylic acid for a separate analysis of antiviral activity.

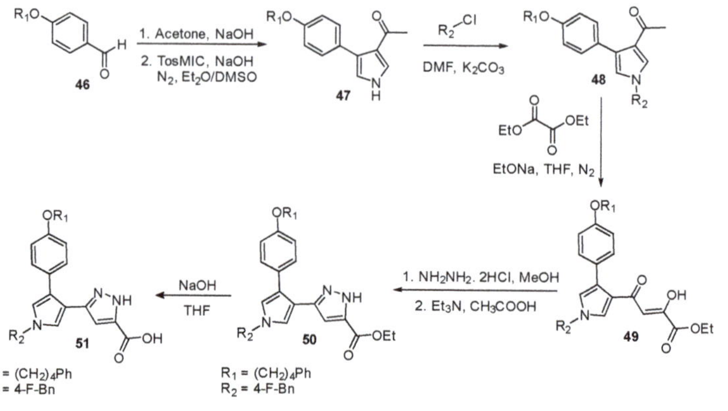

Scheme 11. Synthesis of inhibitors against HIV-1 RNase H.

In in vitro screening of RNase H inhibitory activity, compound **50** showed better inhibitory activity of 0.27 ± 0.05 μM compared to its ester version **50** (17.8 ± 1.2). However, compound **50** had a better effective concentration (EC_{50}), higher cytotoxicity, and selectivity index; hence, compound **50** was chosen for analysis in the docking study, as shown in Figure 12. The nitrogen of pyrazole ring and ester moiety formed ligands with two magnesium ions in the core of RNase H ligands. The ester moiety also had hydrogen interaction with H539, while the substituted benzyl group was possibly linked with W535 and K540 by π–cation interaction. The phenyl moiety was linked with Y501, S499, Q475, E478, N474, and Q475. The scaffold exhibited high potency with the HIV-1 RT pocket and could be optimized for drug development.

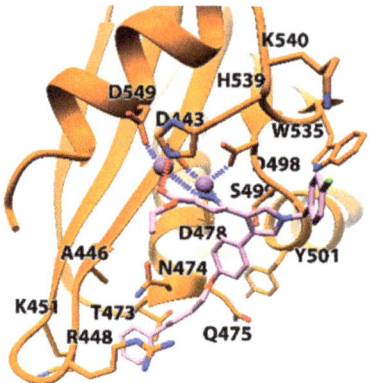

Figure 12. The molecular docking study of compound **50** in the HIV-1 RNase H [49].

Annually, the influenza virus causes respiratory infection, known as flu, in millions of people in the United States and worldwide. A vaccine is available to prevent the massive outbreak of the disease, but it does not efficiently protect children and elders from the infection. Hence, inhibitors were continuously developed for a long-term battle against influenza viruses. Among many molecular drug targets of influenza viruses, only neuraminidase (NA) inhibitors were approved for treatment [50,51]. Meng et al. [52] synthesized pyrazole derivatives as neuraminidase inhibitors. As shown in Scheme 12, 4-methoxy substituted benzaldehyde **52**, phenylhydrazine **53**, and ethyl acetoacetate **54** were allowed to react in the presence of cerium (IV) ammonium nitrate (CAN) and PEG 400 in a one-pot synthesis. The reaction time and yield were optimized with CAN compared to other catalysts to yield pyrazole derivative **55**.

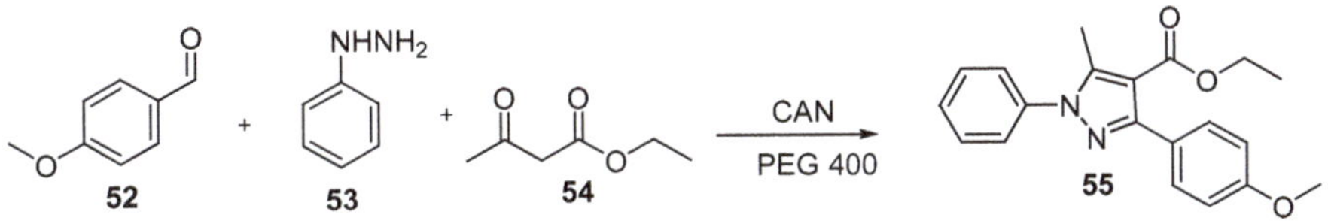

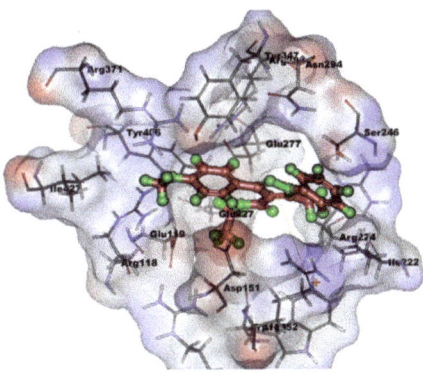

Scheme 12. One-pot synthesis of influenza neuraminidase inhibitor.

As shown in Table 7, compound **55** has a high potency to inhibit the replication of viruses with an IC_{50} of 5.4 ± 0.34 µM and a high selectivity index of 102.20 µM in the assay with the influenza A (H1N1) virus. Additionally, the analog also showed great potency (1.32 ± 0.06 µM) against neuraminidase. As shown in Figure 13, the molecular docking of compound **55** in the 3D structure of neuraminidase (NA) was observed. Compound **55** can fit into the binding pocket of neuraminidase and form two hydrogen bonds with residues Arg371 and Arg118 from the methoxy of the compound. The aromatic ring also stabilized with Arg152 through π–cation interaction. With all of the interacted factors in the binding site of NA, compound **55** efficiently exerted inhibitory activity against the targeted protein.

Table 7. The inhibitory activity of compound **55** against influenza H1N1 virus and neuraminidase.

Compound	CC_{50} (µM)	Influenza H1N1 Virus		NA Inhibitor IC_{50} (µM)
		IC_{50} (µM)	SI (CC_{50}/IC_{50}) (µM)	
55	552.14 ± 187.34	5.4 ± 0.34	102.20	1.32 ± 0.06

Figure 13. Molecular interaction of compound **55** in the binding site of H1N1 neuraminidase [52].

Hepatitis B is another viral infection that is caused by a member of the Hepadnaviridae family. Hepatitis B virus (HBV) has infected millions of people worldwide and the deaths of more than 700,000 people per year. Its replication process is similar to that of RNA retrovirus using the reverse transcription of RNA progenome. Jia et al. [53] developed non-nucleoside HBV inhibitors to combat HBV. As shown in Scheme 13, diethyl carbonate **56** coupled with 4-methylpentan-2-one to produce diketone **57**. Compound **57** reacted with 1,1-dimethoxy-N,N-dimethylmethanamine for α-substitution to furnish compound **58**. The substituted hydrazine condensed with compound **58** for pyrazole ring formation of compound **59**. Pyrazole derivative **60** was produced via base hydrolysis of compound **59**. The carboxylic acid of **60** continued to react with an aryl-substituted amine under the coupling condition to form amide **61**.

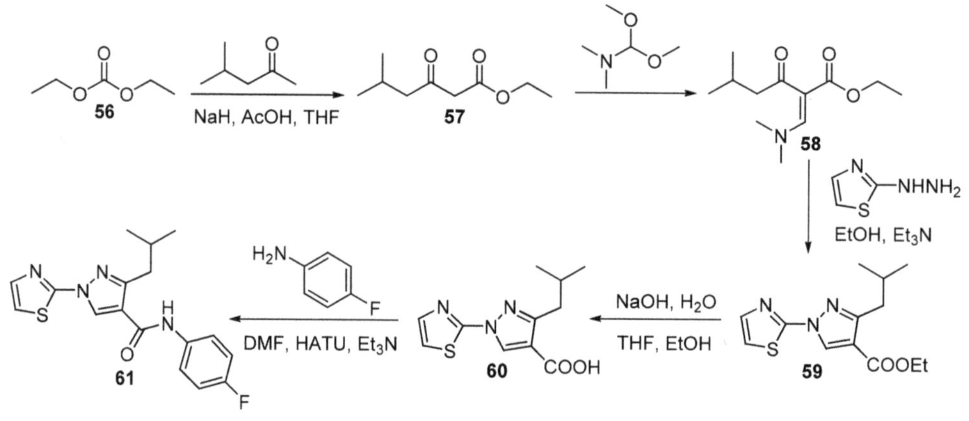

Scheme 13. Synthesis of pyrazole derivatives as anti-HBV agents.

As shown in Table 8, compound **61** demonstrated excellent cytotoxicity in the SAR study. Comparing the inhibitory activity, compound **61** seemed to inhibit viral proteins (HBeAg) of HBV more effectively with a lower IC_{50} (2.22 μM) and corresponding selectivity index (37.69 μM) than the surface antigen (HBsAg). According to the study, compound **61** showed moderate inhibitory activity of approximately 50% against HBV DNA replication. As a result, compound **61** needs further investigation to optimize its inhibitory properties in targeting viral proteins and inhibiting the fusing of viral DNA into host cells.

Table 8. The SAR study of compound **61** against HBV.

Compound	CC_{50} (μM)	HBsAg		HBeAg	
		IC_{50} (μM)	SI	IC_{50} (μM)	SI
61	83.67	24.33	3.44	2.22	37.69

6. Quinoline

Quinoline is a heterocycle made of benzene and a fused pyridine ring found in shale oil, coal, and petroleum. This heterocycle attracted attention in organic chemistry for its feasible synthesis and availability for substitution modifications. Quinoline derivatives have been employed in various applications, including those in the pharmaceutical, bioorganic, and industrial chemistry fields [54]. Quinoline skeletons exhibited a broad range of antiviral activities and were submitted for potential clinical applications.

With the progress in developing inhibitors at the RNase H of HIV RT, quinoline scaffolds also joined the race to combat HIV-1. While HIV RT synthesizes DNA from viral RNA, RNase H plays a crucial role in degrading RNA to form double-stranded DNA. Overacker et al. [55] synthesized quinoline derivatives to target HIV RNase H. As shown in Scheme 14, the methanolysis of carbamate moiety in compound **62** took place to afford quinol derivative **63**, which then underwent Williamson ether synthesis, converting to isopropyl ether to furnish product **64**.

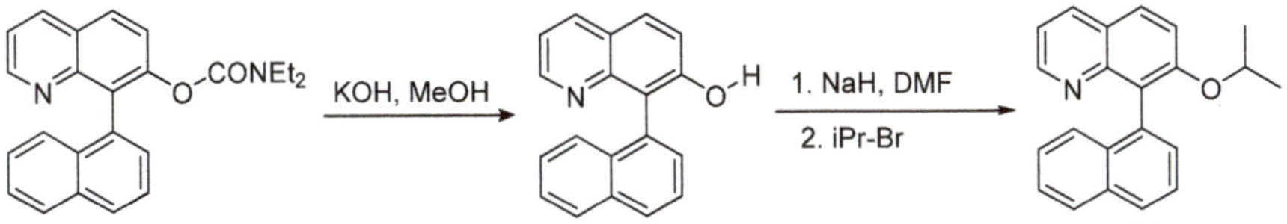

![Scheme 14 structures]

Scheme 14. Synthesis of quinoline derivative **64** as HIV RNase H inhibitor.

As shown in Table 9, quinoline derivative **64** required a low concentration for antiviral activity and a high concentration for cytotoxicity in the in vitro infectivity assay with pseudoviruses. The compound also had a high selectivity index to inhibit the growth of HIV. For the mode of action, the compound showed better inhibition against the HIV-1 RNase H enzyme compared to weak inhibition (>100 μM) in HIV-1 integrase and HIV-1 RT. As shown in Figure 14, no changes could be detected from the UV spectra, indicating that compound **64** did not chelate with Mg^{2+} ion in RNase H's core despite varying concentrations. The structure possibly acted differently in a cell environment or worked as a non-competitive inhibitor of RNase H. This compound was a lead compound for further modification to target RNase H.

Table 9. The inhibitory activity of compound **64** against different HIV strains and HIV-1 RNase H.

Compound	HIV-IC_{50} (μM)			CC_{50} (μM) TZM-bl	SI (CC_{50}/IC_{50})	IC_{50} (μM) HIV-1 RNase H
	HXB2	YU2	89.6			
64	6.7 ± 0.9	8.9 ± 0.6	4.7 ± 1.6	68.5 ± 17.1	14.6	46.8 ± 7.3

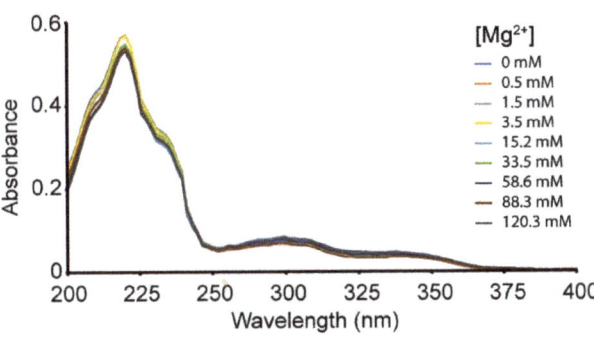

Figure 14. UV absorbance spectra of compound **64** with different concentrations of Mg^{2+} ion [55].

Due to the rapid emergence of hepatitis C virus (HCV) drug resistance, Shah et al. [56] reported the synthesis of quinoline derivatives as HCV drug candidates to target the NS3/4a protease, which plays crucial roles in processing viral protein and replicating viral RNA. It also inhibits the production of interferons that can enhance the immune system against viral infection [57]. The new scaffold was developed based on previous clinical candidate **MK-5172** (Figure 15), with a quinoline moiety instead of quinoxaline. As shown in Scheme 15, compound **65** was synthesized by linking the amine derivative with quinoline via five steps. The hydroxyl group was protected using *N*-Boc-4-hydroxypiperidine before the methyl ester was hydrolyzed and coupled with an amine-acyl sulfonamide chain. The *N*-Boc-piperidine was then deprotected for the addition of bicyclic piperidine to produce product **66**.

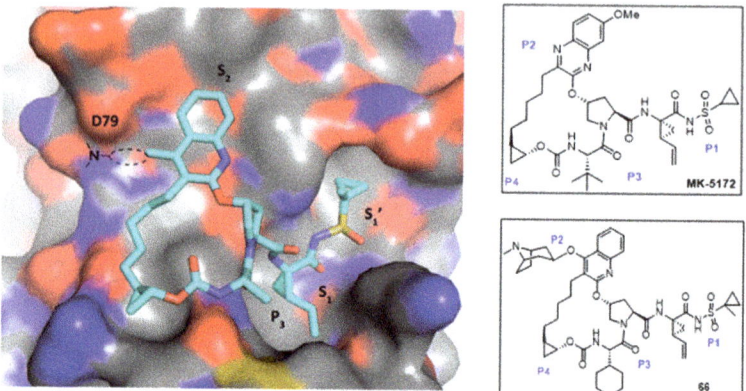

Figure 15. Visualization of **66** in the HCV NS3 gt-1b protease active site for interaction with subsites [56].

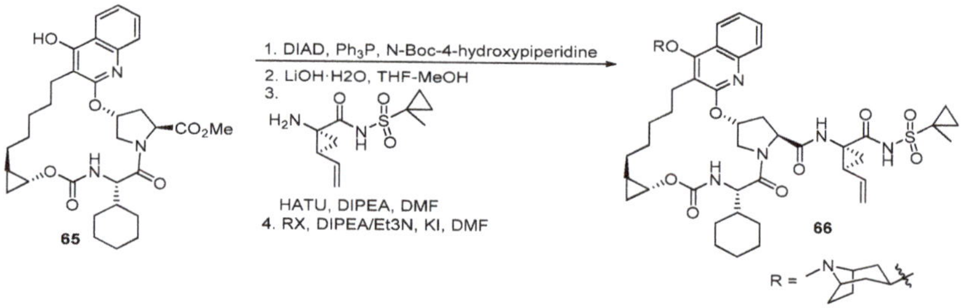

Scheme 15. Synthesis of the inhibitor to target HCV NS3/4a protease.

Compound **66** exhibited improved potency against HCV genotype 3a and A165 mutants compared to reference MK-5172 and had a moderate pharmacokinetic profile in rats in the SAR study. Compared to MK-5172, the replacement of quinoxaline for quinoline in compound **66** enhanced its interaction with the binding pocket HCV NS3 protease. As shown in Figure 15, the substitution at C-4 of quinoline mainly improved the interaction with D79. The replacement of t-butyl with cyclohexane near P3 or modifying cyclopropyl group seemed to not impact the activity of the scaffold in the HCV NS3 protease binding pocket.

Wang et al. [58] reported the synthesis of indoloquinoline or quindoline derivatives as potential anti-influenza A agents to counter drug resistance of viral strains. As shown in Scheme 16, the 11-chloroquindoline **67** reacted with benzene-1,2-diamine to produce compound **68**. In the study, benzene-1,2-diamine could be changed to 1,3 and 1,4-diamine, which then further affected the position of carboxyphenylboronic acid for different analogs of the scaffold. The final product, compound **69**, was produced by the coupling reaction of compound **68** and 2-carboxyphenylboronic acid.

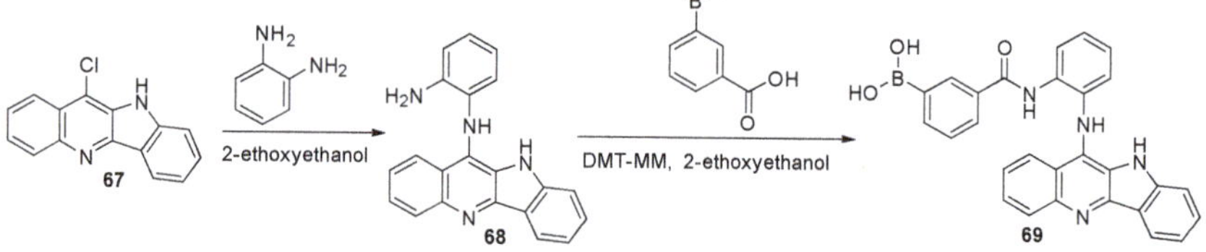

Scheme 16. Synthesis of quindoline derivative **69** as anti-influenza A agent.

Compound **69** exhibited superior properties compared to other derivatives in this scaffold with low cytotoxicity. As shown in Figure 16, the plaque reduction assay showed that plaque development was effectively inhibited, indicating that the compound could inhibit the replication of influenza H1N1 and H3N1. In the confocal imaging, compound **69** was detected in the cytoplasm to target viral neuraminidase and prevent the import of viruses. Compound **69** could interfere with cellular signaling pathways that are essential for viral replication and improve the survival rate of the mouse model in a histopathology study. The quindoline with boronic acid possessed broad anti-influenza A activity for future references and applications.

Figure 16. (**A**) Plaque reduction assay at different concentrations of **69** with PR8, Cal09, Minnesota influenza strains in MDCK cells. (**B**) Corresponding plaque number from the assay with three viruses at different concentrations [58].

In the early study of COVID-19, besides Remdesivir®, Chloroquine®, an anti-malarial drug (Figure 17), exhibited antiviral activity against 2019-n-CoVs infection in vitro. As shown in Figure 18, Chloroquine® could effectively inhibit viral infection at a low concentration and with good cytotoxicity compared to other FDA-approved antiviral agents. Additionally, Chloroquine® also effectively reduced viral copies with 10 μM after 48 h post-infection in the nucleoprotein through immunofluorescence assay [43].

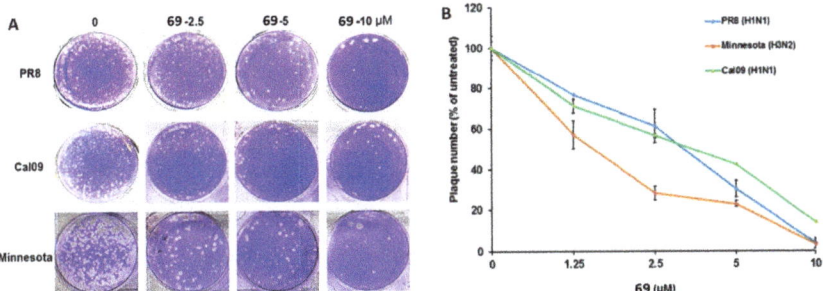

Figure 17. The molecular structure of Chloroquine® [43].

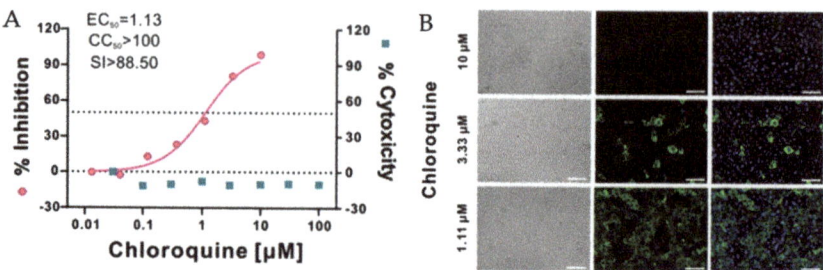

Figure 18. (**A**) Percentage of inhibition measured by qRT-PCR and cytotoxicity of Chloroquine® when treating 2019 n-CoV infected Vero E6 cells. (**B**) Immunofluorescence assay of viral infection with Chloroquine® treatment [43].

As shown in Figure 18, Chloroquine showed its effectiveness in blocking virus infection at EC_{50} = 1.13 µM concentration with a high selectivity index (SI). The immunofluorescence assay also confirmed the efficiency of Chloroquine in inhibiting viruses, as the viral nucleoproteins were eliminated completely with 10 µM. Together with Chloroquine®, Hydroxychloroquine® was also tested for its anti-2019-nCoV activity. Hydroxychloroquine possessed less toxicity in cells compared to Chloroquine® with anti-inflammatory properties [59]. However, both Chloroquine® and Hydroxychloroquine® have side effects, interfering with lysosomal activity and causing cardiac and skeletal muscle problems [60].

7. Conclusions

Nitrogen-containing heterocycle derivatives have a long history against a broad spectrum of viral agents. Their structures are similar to the biological components that are vital for the infection mechanism and replication of viruses. Incorporating nitrogen-containing heterocycles enhances the binding affinity of the scaffolds by increasing interaction with residues and matching with the pocket shape to deactivate the function of targeted enzymes. This article summarized the history, synthesis, and antiviral applications of various nitrogen-containing heterocycle scaffolds, such as indoles, pyrroles, pyrimidines, pyrazoles, and quinolines. The scaffolds were applied to a variety of viruses, ranging from HIV and HCV/HBV to VZV/HSV, SARS-CoV, and influenza viruses. Some approved scaffolds exhibit multiple functions in targeting different viruses, similar to applying Remdesivir®, Chloroquine®, and Hydrochloroquine® for the recent SARS-Co-V 2 viruses. Most scaffolds exhibited potential antiviral activity that could lead to further optimization and in vivo studies for drug development. For example, for HCV treatment, FDA-approved drugs such as Grazoprevir, Voxilaprevir, and Glecaprevir were modified from scaffold **66**. Viruses have high rates of emergence that demand continual drug discovery of new antiviral agents. Research on nitrogen heterocycles will progressively expand for future drug development.

Funding: This work was funded in part by a Georgia State University training grant (NIH T34GM131939) to support the undergraduate student TNT.

Institutional Review Board Statement: Not applicable.

Informed Consent Statement: Not applicable.

Data Availability Statement: The data presented in this study are available on request from the corresponding author.

Acknowledgments: M.H. expresses gratitude for the support received through the Georgia State University Brains, Behavior Seed Grant, Atlanta Clinical and Translational Science Institute Healthcare Innovation Seed Grant, and Georgia Research Alliance Ventures Phase 1 Grant for the support. M.H. and T.N.T. are also thankful for GSU Chemistry Department support.

Conflicts of Interest: The authors have no conflict of interest.

References

1. Ghosh, A.K.; Osswald, H.L.; Prato, G. Recent Progress in the Development of HIV-1 Protease Inhibitors for the Treatment of HIV/AIDS. *J. Med. Chem.* **2016**, *59*, 5172–5208. [CrossRef] [PubMed]
2. Krammer, F.; Palese, P. Advances in the development of influenza virus vaccines. *Nat. Rev. Drug Discov.* **2015**, *14*, 167–182. [CrossRef] [PubMed]
3. Bolles, M.; Donaldson, E.; Baric, R. SARS-CoV and emergent coronaviruses: Viral determinants of interspecies transmission. *Curr. Opin. Virol.* **2011**, *1*, 624–634. [CrossRef] [PubMed]
4. Wang, X.; Zou, P.; Wu, F.; Lu, L.; Jiang, S. Development of small-molecule viral inhibitors targeting various stages of the life cycle of emerging and re-emerging viruses. *Front. Med.* **2017**, *11*, 449–461. [CrossRef]
5. Korkmaz, A.; Bursal, E. An in vitro and in silico study on the synthesis and characterization of novel bis(sulfonate) derivatives as tyrosinase and pancreatic lipase inhibitors. *J. Mol. Struct.* **2022**, *1259*, 132734. [CrossRef]
6. Cetin, A.; Bursal, E.; Türkan, F. 2-methylindole analogs as cholinesterases and glutathione S-transferase inhibitors: Synthesis, biological evaluation, molecular docking, and pharmacokinetic studies. *Arab. J. Chem.* **2021**, *14*, 103449. [CrossRef]
7. Zhang, M.Z.; Chen, Q.; Yang, G.F. A review on recent developments of indole-containing antiviral agents. *Eur. J. Med. Chem.* **2015**, *89*, 421–441. [CrossRef]
8. Mohana Roopan, S.; Sompalle, R. Synthetic chemistry of pyrimidines and fused pyrimidines: A review. *Synth. Commun.* **2016**, *46*, 645–672. [CrossRef]
9. Fascio, M.L.; Errea, M.I.; D'Accorso, N.B. Imidazothiazole and related heterocyclic systems. Synthesis, chemical and biological properties. *Eur. J. Med. Chem.* **2015**, *90*, 666–683. [CrossRef]
10. Shalini, K.; Sharma, P.K.; Kumar, N. Imidazole and its biological activities: A review. *Der Chem. Sin.* **2010**, *1*, 36–47.
11. Singh, N.; Pandurangan, A.; Rana, K.; Anand, P.; Ahamad, A.; Tiwari, A.K. Benzimidazole: A short review of their antimicrobial activities. *Int. Curr. Pharm. J.* **2012**, *1*, 110–118. [CrossRef]
12. Gupta, V.; Kant, V. A review on biological activity of imidazole and thiazole moieties and their derivatives. *Sci. Int.* **2013**, *1*, 253–260. [CrossRef]
13. Bhardwaj, V.; Gumber, D.; Abbot, V.; Dhiman, S.; Sharma, P. Pyrrole: A resourceful small molecule in key medicinal hetero-aromatics. *RSC Adv.* **2015**, *5*, 15233–15266. [CrossRef]
14. Sharma, V.; Chitranshi, N.; Agarwal, A.K. Significance and biological importance of pyrimidine in the microbial world. *Int. J. Med. Chem.* **2014**, *2014*, 1–31. [CrossRef]
15. Raffa, D.; Maggio, B.; Raimondi, M.V.; Cascioferro, S.; Plescia, F.; Cancemi, G.; Daidone, G. Recent advanced in bioactive systems containing pyrazole fused with a five membered heterocycle. *Eur. J. Med. Chem.* **2015**, *97*, 732–746. [CrossRef]
16. Marella, A.; Tanwar, O.P.; Saha, R.; Ali, M.R.; Srivastava, S.; Akhter, M.; Shaquiquzzaman, M.; Alam, M.M. Quinoline: A versatile heterocyclic. *Saudi Pharm. J.* **2013**, *21*, 1–12. [CrossRef]
17. Sundberg, R.J. *Indoles*; Academic Press: London, UK; San Diego, CA, USA, 1996.
18. Hassam, M.; Basson, A.E.; Liotta, D.C.; Morris, L.; van Otterlo, W.A.L.; Pelly, S.C. Novel Cyclopropyl-Indole Derivatives as HIV Non-Nucleoside Reverse Transcriptase Inhibitors. *ACS Med. Chem. Lett.* **2012**, *3*, 470–475. [CrossRef]
19. Lavanchy, D. Evolving epidemiology of hepatitis C virus. *Clin. Microbiol. Infect.* **2011**, *17*, 107–115. [CrossRef]
20. Han, Z.; Liang, X.; Wang, Y.; Qing, J.; Cao, L.; Shang, L.; Yin, Z. The discovery of indole derivatives as novel hepatitis C virus inhibitors. *Eur. J. Med. Chem.* **2016**, *116*, 147–155. [CrossRef]
21. Chen, J.J.; Gershon, A.A.; Li, Z.; Cowles, R.A.; Gershon, M.D. Varicella zoster virus (VZV) infects and establishes latency in enteric neurons. *J. Neurovirol.* **2011**, *17*, 578–589. [CrossRef]
22. Musella, S.; di Sarno, V.; Ciaglia, T.; Sala, M.; Spensiero, A.; Scala, M.C.; Ostacolo, C.; Andrei, G.; Balzarini, J.; Snoeck, R.; et al. Identification of an indol-based derivative as potent and selective varicella zoster virus (VZV) inhibitor. *Eur. J. Med. Chem.* **2016**, *124*, 773–781. [CrossRef] [PubMed]
23. Topalis, D.; Gillemot, S.; Snoeck, R.; Andrei, G. Thymidine kinase and protein kinase in drug-resistant herpesviruses: Heads of a Lernaean Hydra. *Drug Resist. Updat.* **2018**, *37*, 1–16. [CrossRef] [PubMed]
24. Thanigaimalai, P.; Konno, S.; Yamamoto, T.; Koiwai, Y.; Taguchi, A.; Takayama, K.; Yakushiji, F.; Akaji, K.; Chen, S.E.; Naser-Tavakolian, A.; et al. Development of potent dipeptide-type SARS-CoV 3CL protease inhibitors with novel P3 scaffolds: Design, synthesis, biological evaluation, and docking studies. *Eur. J. Med. Chem.* **2013**, *68*, 372–384. [CrossRef] [PubMed]
25. Khan, R.J.; Jha, R.K.; Amera, G.M.; Jain, M.; Singh, E.; Pathak, A.; Singh, R.P.; Muthukumaran, J.; Singh, A.K. Targeting SARS-CoV-2: A systematic drug repurposing approach to identify promising inhibitors against 3C-like proteinase and 2′-O-ribose methyltransferase. *J. Biomol. Struct. Dyn.* **2021**, *39*, 2679–2692. [CrossRef]
26. Thanigaimalai, P.; Konno, S.; Yamamoto, T.; Koiwai, Y.; Taguchi, A.; Takayama, K.; Yakushiji, F.; Akaji, K.; Kiso, Y.; Kawasaki, Y.; et al. Design, synthesis, and biological evaluation of novel dipeptide-type SARS-CoV 3CL protease inhibitors: Structure-activity relationship study. *Eur. J. Med. Chem.* **2013**, *65*, 436–447. [CrossRef]
27. Dai, W.; Zhang, B.; Jiang, X.-M.; Su, H.; Li, J.; Zhao, Y.; Xie, X.; Jin, Z.; Peng, J.; Liu, F.; et al. Structure-based design of antiviral drug candidates targeting the SARS-CoV-2 main protease. *Science* **2020**, *368*, 1331–1335. [CrossRef]
28. Jones, R.; Bean, G. *The Chemistry of Pyrroles*; Academic Press: London, UK, 1997.

29. Curreli, F.; Kwon, Y.D.; Belov, D.S.; Ramesh, R.R.; Kurkin, A.V.; Altieri, A.; Kwong, P.D.; Debnath, A.K. Synthesis, Antiviral Potency, in Vitro ADMET, and X-ray Structure of Potent CD4 Mimics as Entry Inhibitors That Target the Phe43 Cavity of HIV-1 gp120. *J. Med. Chem.* **2017**, *60*, 3124–3153. [CrossRef]
30. Checkley, M.A.; Luttge, B.G.; Freed, E.O. HIV-1 envelope glycoprotein biosynthesis, trafficking, and incorporation. *J. Mol. Biol.* **2011**, *410*, 582–608. [CrossRef]
31. Ezquerra, J.; Pedregal, C.; Rubio, A.; Valenciano, J.; Navio, J.L.G.; Alvarez-Builla, J.; Vaquero, J.J. General method for the synthesis of 5-arylpyrrole-2-carboxylic acids. *Tetrahedron Lett.* **1993**, *34*, 6317–6320. [CrossRef]
32. Kinchington, P.R.; Leger, A.J.; Guedon, J.M.; Hendricks, R.L. Herpes simplex virus and varicella zoster virus, the house guests who never leave. *Herpesviridae* **2012**, *3*, 5. [CrossRef]
33. Hilmy, K.M.; Soliman, D.H.; Shahin, E.B.; Abd Alhameed, R. Synthesis and molecular modeling study of novel pyrrole Schiff Bases as anti-HSV-1 agents. *Life Sci. J.* **2012**, *9*, 736–745.
34. Lin, M.I.; Su, B.H.; Lee, C.H.; Wang, S.T.; Wu, W.C.; Dangate, P.; Wang, S.Y.; Huang, W.I.; Cheng, T.J.; Lin, O.A.; et al. Synthesis and inhibitory effects of novel pyrimido-pyrrolo-quinoxalinedione analogues targeting nucleoproteins of influenza A virus H1N1. *Eur. J. Med. Chem.* **2015**, *102*, 477–486. [CrossRef] [PubMed]
35. Turrell, L.; Lyall, J.W.; Tiley, L.S.; Fodor, E.; Vreede, F.T. The role and assembly mechanism of nucleoprotein in influenza A virus ribonucleoprotein complexes. *Nat. Commun.* **2013**, *4*, 1591. [CrossRef] [PubMed]
36. Brown, D.J. *The Pyrimidines*; John Wiley & Sons: New York, NY, USA, 1994; pp. 1–8.
37. Dinakaran, V.S.; Bomma, B.; Srinivasan, K.K. Fused pyrimidines: The heterocycle of diverse biological and pharmacological significance. *Der Pharma Chem.* **2012**, *4*, 255–265.
38. Malancona, S.; Mori, M.; Fezzardi, P.; Santoriello, M.; Basta, A.; Nibbio, M.; Kovalenko, L.; Speziale, R.; Battista, M.R.; Cellucci, A.; et al. 5,6-Dihydroxypyrimidine Scaffold to Target HIV-1 Nucleocapsid Protein. *ACS Med. Chem. Lett.* **2020**, *11*, 766–772. [CrossRef]
39. Levin, J.G.; Mitra, M.; Mascarenhas, A.; Musier-Forsyth, K. Role of HIV-1 nucleocapsid protein in HIV-1 reverse transcription. *RNA Biol.* **2010**, *7*, 754–774. [CrossRef]
40. Mohamed, M.S.; Sayed, A.I.; Khedr, M.A.; Soror, S.H. Design, synthesis, assessment, and molecular docking of novel pyrrolopyrimidine (7-deazapurine) derivatives as non-nucleoside hepatitis C virus NS5B polymerase inhibitors. *Bioorg. Med. Chem.* **2016**, *24*, 2146–2157. [CrossRef]
41. Zhao, C.; Wang, Y.; Ma, S. Recent advances on the synthesis of hepatitis C virus NS5B RNA-dependent RNA-polymerase inhibitors. *Eur. J. Med. Chem.* **2015**, *102*, 188–214. [CrossRef]
42. Mohamed, S.F.; Flefel, E.M.; Amr Ael, G.; Abd El-Shafy, D.N. Anti-HSV-1 activity and mechanism of action of some new synthesized substituted pyrimidine, thiopyrimidine and thiazolopyrimidine derivatives. *Eur. J. Med. Chem.* **2010**, *45*, 1494–1501. [CrossRef]
43. Wang, M.; Cao, R.; Zhang, L.; Yang, X.; Liu, J.; Xu, M.; Shi, Z.; Hu, Z.; Zhong, W.; Xiao, G. Remdesivir and chloroquine effectively inhibit the recently emerged novel coronavirus (2019-nCoV) in vitro. *Cell Res.* **2020**, *30*, 269–271. [CrossRef]
44. Zhang, L.; Zhou, R. Structural Basis of the Potential Binding Mechanism of Remdesivir to SARS-CoV-2 RNA-Dependent RNA Polymerase. *J. Phys. Chem. B* **2020**, *124*, 6955–6962. [CrossRef] [PubMed]
45. Grein, J.; Ohmagari, N.; Shin, D.; Diaz, G.; Asperges, E.; Castagna, A.; Feldt, T.; Green, G.; Green, M.L.; Lescure, F.X.; et al. Compassionate Use of Remdesivir for Patients with Severe COVID-19. *N. Engl. J. Med.* **2020**, *382*, 2327–2336. [CrossRef] [PubMed]
46. Ansari, A.; Ali, A.; Asif, M.; Shamsuzzaman. Review: Biologically active pyrazole derivatives. *New J. Chem.* **2017**, *41*, 16–41. [CrossRef]
47. Eicher, T.; Hauptmann, S.; Speicher, A. *The Chemistry of Heterocycles: Structure, Reactions, Syntheses, and Applications*; Wiley-VCH: Weinheim, Germany, 2003.
48. Christen, M.T.; Menon, L.; Myshakina, N.S.; Ahn, J.; Parniak, M.A.; Ishima, R. Structural basis of the allosteric inhibitor interaction on the HIV-1 reverse transcriptase RNase H domain. *Chem. Biol. Drug Des.* **2012**, *80*, 706–716. [CrossRef]
49. Messore, A.; Corona, A.; Madia, V.N.; Saccoliti, F.; Tudino, V.; De Leo, A.; Scipione, L.; De Vita, D.; Amendola, G.; Di Maro, S.; et al. Pyrrolyl Pyrazoles as Non-Diketo Acid Inhibitors of the HIV-1 Ribonuclease H Function of Reverse Transcriptase. *ACS Med. Chem. Lett.* **2020**, *11*, 798–805. [CrossRef]
50. Blyth, C.C.; Jacoby, P.; Effler, P.V.; Kelly, H.; Smith, D.W.; Robins, C.; Willis, G.A.; Levy, A.; Keil, A.D.; Richmond, P.C. Effectiveness of trivalent flu vaccine in healthy young children. *Pediatrics* **2014**, *133*, e1218–e1225. [CrossRef]
51. David, J.K.; Amber, F. Extremely low vaccine effectiveness against influenza H3N2 in the elderly during the 2012/2013 flu season. *J. Infect. Dev. Ctries* **2013**, *7*, 299–301. [CrossRef]
52. Meng, F.J.; Sun, T.; Dong, W.Z.; Li, M.H.; Tuo, Z.Z. Discovery of Novel Pyrazole Derivatives as Potent Neuraminidase Inhibitors against Influenza H1N1 Virus. *Arch. Pharm.* **2016**, *349*, 168–174. [CrossRef]
53. Jia, H.; Bai, F.; Liu, N.; Liang, X.; Zhan, P.; Ma, C.; Jiang, X.; Liu, X. Design, synthesis and evaluation of pyrazole derivatives as non-nucleoside hepatitis B virus inhibitors. *Eur. J. Med. Chem.* **2016**, *123*, 202–210. [CrossRef]
54. Prajapati, S.M.; Patel, K.D.; Vekariya, R.H.; Panchal, S.N.; Patel, H.D. Recent advances in the synthesis of quinolines: A review. *RSC Adv.* **2014**, *4*, 24463–24476. [CrossRef]
55. Overacker, R.D.; Banerjee, S.; Neuhaus, G.F.; Milicevic Sephton, S.; Herrmann, A.; Strother, J.A.; Brack-Werner, R.; Blakemore, P.R.; Loesgen, S. Biological evaluation of molecules of the azaBINOL class as antiviral agents: Inhibition of HIV-1 RNase H activity by 7-isopropoxy-8-(naphth-1-yl)quinoline. *Bioorg. Med. Chem.* **2019**, *27*, 3595–3604. [CrossRef] [PubMed]

56. Shah, U.; Jayne, C.; Chackalamannil, S.; Velázquez, F.; Guo, Z.; Buevich, A.; Howe, J.A.; Chase, R.; Soriano, A.; Agrawal, S.; et al. Novel Quinoline-Based P2–P4 Macrocyclic Derivatives As Pan-Genotypic HCV NS3/4a Protease Inhibitors. *ACS Med. Chem. Lett.* **2014**, *5*, 264–269. [CrossRef] [PubMed]
57. Schiering, N.; D'Arcy, A.; Villard, F.; Simic, O.; Kamke, M.; Monnet, G.; Hassiepen, U.; Svergun, D.I.; Pulfer, R.; Eder, J.; et al. A macrocyclic HCV NS3/4A protease inhibitor interacts with protease and helicase residues in the complex with its full-length target. *Proc. Natl. Acad. Sci. USA* **2011**, *108*, 21052–21056. [CrossRef] [PubMed]
58. Wang, W.; Yin, R.; Zhang, M.; Yu, R.; Hao, C.; Zhang, L.; Jiang, T. Boronic Acid Modifications Enhance the Anti-Influenza A Virus Activities of Novel Quindoline Derivatives. *J. Med. Chem.* **2017**, *60*, 2840–2852. [CrossRef]
59. Liu, J.; Cao, R.; Xu, M.; Wang, X.; Zhang, H.; Hu, H.; Li, Y.; Hu, Z.; Zhong, W.; Wang, M. Hydroxychloroquine, a less toxic derivative of chloroquine, is effective in inhibiting SARS-CoV-2 infection in vitro. *Cell Discov.* **2020**, *6*, 16. [CrossRef]
60. Pereira, B.B. Challenges and cares to promote rational use of chloroquine and hydroxychloroquine in the management of coronavirus disease 2019 (COVID-19) pandemic: A timely review. *J. Toxicol. Environ. Health B Crit. Rev.* **2020**, *23*, 177–181. [CrossRef]

Article

Synthesis and Biochemical Evaluation of 8H-Indeno[1,2-d]thiazole Derivatives as Novel SARS-CoV-2 3CL Protease Inhibitors

Jing Wu [1,†], Bo Feng [2,3,†], Li-Xin Gao [1,3], Chun Zhang [1], Jia Li [2,3,4], Da-Jun Xiang [5,*], Yi Zang [3,*] and Wen-Long Wang [1,*]

1. School of Life Sciences and Health Engineering, Jiangnan University, Wuxi 214122, China; 6191504014@stu.jiangnan.edu.cn (J.W.); lxgao@simm.ac.cn (L.-X.G.); zhangchun@jiangnan.edu.cn (C.Z.)
2. School of Life Science and Biopharmaceutics, Shenyang Pharmaceutical University, Shenyang 110016, China; fengboo@live.com (B.F.); jli@simm.ac.cn (J.L.)
3. State Key Laboratory of Drug Research, Shanghai Institute of Materia Medica, Chinese Academy of Sciences, Shanghai 201203, China
4. Zhongshan Institute for Drug Discovery, Shanghai Institute of Materia Medica, Chinese Academy of Sciences, Zhongshan Tsuihang New District, Zhongshan 528400, China
5. Xishan People's Hospital of Wuxi City, Wuxi 214105, China
* Correspondence: xiangdjxshospital@yeah.net (D.-J.X.); yzang@simm.ac.cn (Y.Z.); wenlongwang@jiangnan.edu.cn (W.-L.W.)
† These authors contributed equally to this work.

Abstract: The COVID-19 pandemic caused by SARS-CoV-2 is a global burden on human health and economy. The 3-Chymotrypsin-like cysteine protease (3CL^{pro}) becomes an attractive target for SARS-CoV-2 due to its important role in viral replication. We synthesized a series of 8H-indeno[1,2-d]thiazole derivatives and evaluated their biochemical activities against SARS-CoV-2 3CL^{pro}. Among them, the representative compound **7a** displayed inhibitory activity with an IC$_{50}$ of 1.28 ± 0.17 µM against SARS-CoV-2 3CL^{pro}. Molecular docking of **7a** against 3CL^{pro} was performed and the binding mode was rationalized. These preliminary results provide a unique prototype for the development of novel inhibitors against SARS-CoV-2 3CL^{pro}.

Keywords: COVID-19; M^{pro} inhibitors; drug design and synthesis; structure-activity relationships (SAR)

1. Introduction

The global pandemic of coronavirus disease (COVID-19) caused by the severe acute respiratory syndrome coronavirus 2 (SARS-CoV-2) has posted major challenges to public health systems and the economy worldwide [1–5]. There have been 434 million confirmed cases of COVID-19 worldwide as of the end of February 2022, and almost 6 million deaths have been reported [6]. Although multiple effective vaccines against COVID-19 are available, reinfections and breakthrough infections are frequently reported [7,8]. In addition, the virus is continuing to evolve, and a new variant named Omicron enables the virus to evade the immune protective barrier due to a large number of mutations in the receptor binding sites [9–11]. Therefore, it is urgent to develop effective drugs and specific treatments for people who are infected by COVID-19 with severe symptoms.

3CL^{pro} (also called M^{pro}) plays an essential role during replication and transcription of SARS-CoV-2 and has been regarded as an attractive target for treating COVID-19 and other coronavirus-caused diseases [12–14]. The development of 3CL^{pro} inhibitors has attracted much attention from medicinal chemists and the pharmaceutical industry. The collective efforts culminated in the recent approval of Paxlovid (nirmatrelvir) by FDA for the treatment of SARS-CoV-2 [15]. As shown in Figure 1, Most known 3CL^{pro} inhibitors are peptidomimetic inhibitors containing a warhead of Michael acceptor, such as nirmatrelvir with nitrile [16], YH-53 with benzothiazolyl ketone [17], compound **1** with α-ketoamide [18], and compound **2** with aldehyde [19]. Others are nonpeptidic inhibitors including covalent and noncovalent inhibitors. Covalent inhibitors, such as Carmofur, Shikonin [20], and **3** [21], are identified by high-throughput screening. Noncovalent inhibitor CCF0058981 [22] and flavonoid analogs (baicalin, baicalein, and 4′-*O*-Methylscutellarein) [23,24] were obtained through structure-based optimization and from traditional Chinese medicines, respectively.

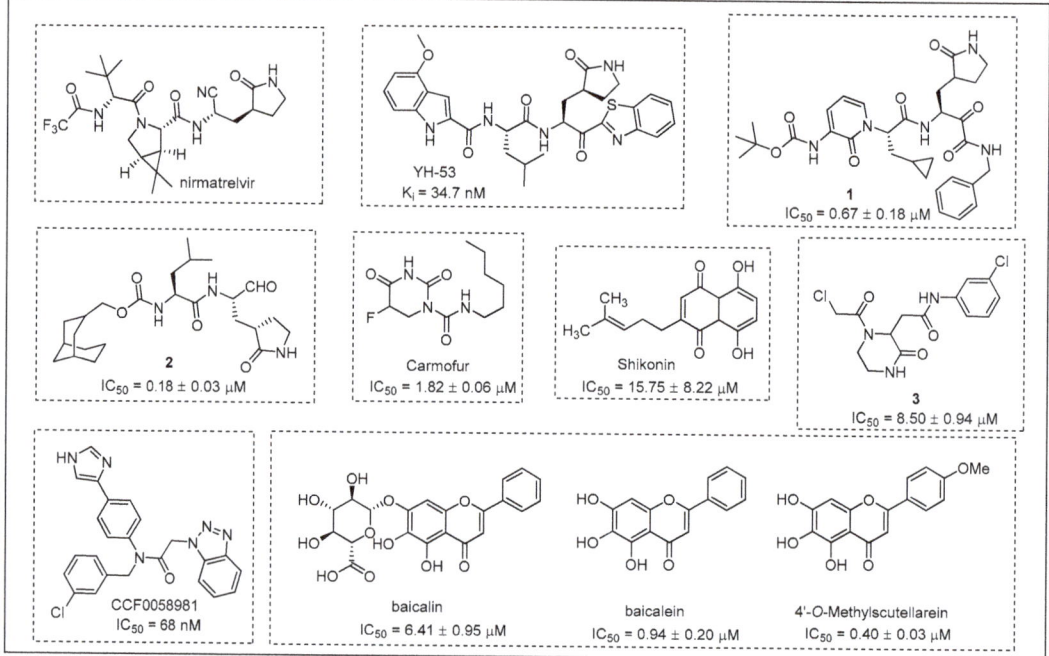

Figure 1. SARS-CoV-2 3CL^{pro} inhibitors.

In pursuit of novel 3CL^{pro} inhibitors, we identified 8*H*-indeno[1,2-*d*]thiazole derivative **4** as a novel SARS-CoV-2 3CL^{pro} inhibitor (IC$_{50}$ = 6.42 ± 0.90 μM) through high-throughput screening of our compound collection (Figure 2). This result provided us with an opportunity to explore novel small molecule inhibitors against SARS-CoV-2 3CL^{pro}. Herein, we designed and synthesized a series of 8*H*-indeno[1,2-*d*]thiazole derivatives, evaluated their inhibitory activities against SARS-CoV-2 3CL^{pro}, and elucidated the SARs. Selected compound **7a** was subjected to molecular docking to predict the binding mode with SARS-CoV-2 3CL^{pro}.

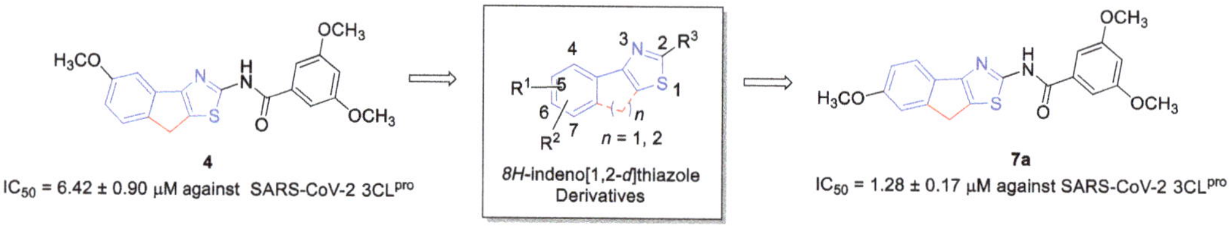

Figure 2. Structure of 8*H*-indeno[1,2-*d*]thiazole derivatives.

2. Results and Discussion

2.1. Design and Synthesis of 8H-Indeno[1,2-d]thiazole Derivatives

Based on the structure of compound **4**, 14 new 8H-indeno[1,2-d]thiazole derivatives (compounds **7a–7l**, and **10a–10b**) (shown in Schemes 1 and 2) were designed and synthesized through a two-step synthesis from the appropriate ketone and thiourea [25–28]. Adjusting the methoxy group of compound **4** from position 5 to position 6 afforded compound **7a**. Considering the effects of steric hindrance and electron withdrawing, compounds **7b–7e** were synthesized by substitution of the methoxy group for the butoxy, isobutoxy, and methyl groups and for the chlorine atom. After replacing the 3,5-dimethoxybenzamido moiety in compound **7a** with 3,4,5-trimethoxybenzamido, 3,5-diacetoxybenzamido, 3-methoxybenzamido, 3-fluorobenzamido, thiophene-2-carboxamido, and 4-chlorobenzamido, compounds **7f–7k** were obtained. To evaluate the effect of ring expansion, compound **7l** was synthesized. Finally, ring opening analogues **10a** and **10b** were synthesized to elucidate the effect of the central ring on the inhibition of 3CL^{pro}.

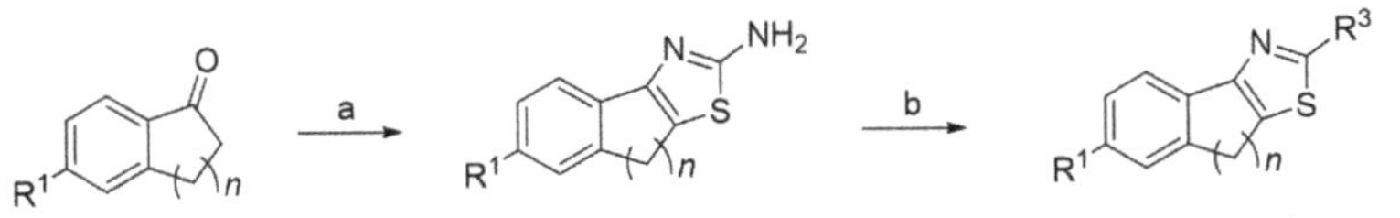

5a - 5f

6a: R^1 = methoxy, n = 1
6b: R^1 = butoxy, n = 1
6c: R^1 = isobutoxy, n = 1
6d: R^1 = methyl, n = 1
6e: R^1 = chloro, n = 1
6f: R^1 = methoxy, n = 2

7a: R^1 = methoxy, R^3 = 3,5-dimethoxybenzamido, n = 1
7b: R^1 = butoxy, R^3 = 3,5-dimethoxybenzamido, n = 1
7c: R^1 = isobutoxy, R^3 = 3,5-dimethoxybenzamido, n = 1
7d: R^1 = methyl, R^3 = 3,5-dimethoxybenzamido, n = 1
7e: R^1 = chloro, R^3 = 3,5-dimethoxybenzamido, n = 1
7f: R^1 = methoxy, R^3 = 3,4,5-trimethoxybenzamido, n = 1
7g: R^1 = methoxy, R^3 = 3,5-diacetoxybenzamido, n = 1
7h: R^1 = methoxy, R^3 = 3-methoxybenzamido, n = 1
7i: R^1 = methoxy, R^3 = 3-fluorobenzamido, n = 1
7j: R^1 = methoxy, R^3 = thiophene-2-carboxamido, n = 1
7k: R^1 = methoxy, R^3 = 4-chlorobenzamido, n = 1
7l: R^1 = methoxy, R^3 = 3,5-dimethoxybenzamido, n = 2

Scheme 1. (a) thiourea, bromine, ethanol, 100 °C, 5–6 h; (b) aromatic acid, HATU, DIPEA, DMF, r t, 2–3 h, 25–50%.

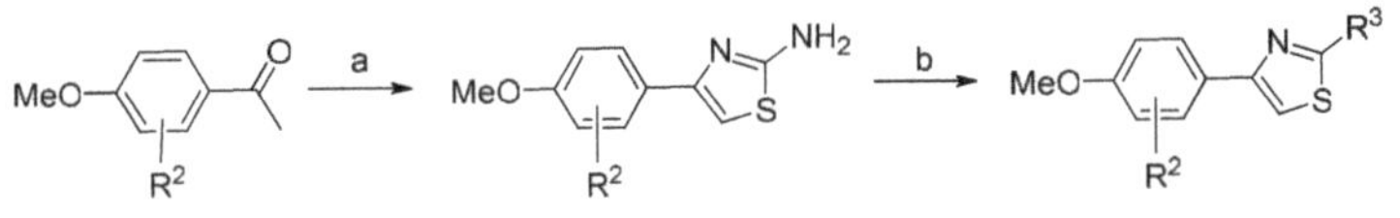

8a: R^2 = 2-methyl
8b: R^2 = 3-methyl

9a: R^2 = 2-methyl
9b: R^2 = 3-methyl

10a: R^2 = 2-methyl, R^3 = 3,5-dimethoxybenzamido
10b: R^2 = 3-methyl, R^3 = 3,5-dimethoxybenzamido

Scheme 2. (a) thiourea, iodine, 110 °C, 10 h; (b) aromatic acid, HATU, DIPEA, DMF, r t, 2–3 h, 35–40%.

2.2. SARS-CoV-2 3CL^{pro} Inhibitory Activities and Structure-Activity Relationships

All synthesized compounds were evaluated for inhibitory activity against SARS-CoV-2 3CL^{pro} using PF-07321332 as positive control [29–31], and the results were detailed in Table 1. We initially prepared **7a** from the commercially available compound **5a** by the route outlined in Scheme 1. We noticed that compound **7a** with 6-methoxy group on the phenyl ring exhibited inhibitory activity against SARS-CoV-2 3CL^{pro} with 1.28 ± 0.17 μM, about five times more potent than compound **4** with 5-methoxy group on the phenyl ring. The result indicated that the position of the methoxy group on the phenyl ring significantly affected inhibitory activities against SARS-CoV-2 3CL^{pro}. To explore the SAR of this seemingly important position, methoxy group on compound **7a** was replaced by butoxy (**7b**), isobutoxy (**7c**), methyl groups (**7d**), and chlorine atom (**7e**); the inhibitory activities of the corresponding compounds **7b–7e** were completely abolished. These results demonstrated that the effect of steric hindrance at this position was detrimental to inhibitory activities. The SAR of **R³** was explored next. Replacement of the 3,5-dimethoxybenzamido moiety with 3,4,5-trimethoxybenzamido moiety, 3,5-diacetoxybenzamido moiety, 3-methoxybenzamido moiety, 3-fluorobenzamido moiety, thiophene-2-carboxamido moiety, and 4-chlorobenzamido moiety led to compounds **7f, 7g, 7h, 7i, 7j,** and **7k**, respectively. The inhibitory activity of compounds **7f** and **7g** dropped significantly, while compound **7h** almost maintained its inhibitory activities. These results indicated that the extra steric hindrance had negative impact on the inhibitory activities. Compared to compound **7h**, the inhibitory activities of compounds **7i–7k** diminished; these results indicated that introduction of an electron-withdrawing group or heterocyclic ring on the scaffold of 8H-indeno[1,2-d]thiazole took negative roles for inhibitory activities. Expanding the five-membered ring on compound **7a** to a six-membered ring led to compound **7l**, which unfortunately did not show any inhibitory activity against SARS-CoV-2 3CL^{pro}. Opening the five-membered ring on compound **7a** resulted in compounds **10a** and **10b**, which also lost inhibitory activities. These results indicated that the five-membered ring on compound **7a** is important for the inhibitory activity against SARS-CoV-2 3CL^{pro}.

Table 1. Inhibitory activities of target compounds against SARS-CoV-2 3CL^{pro}.

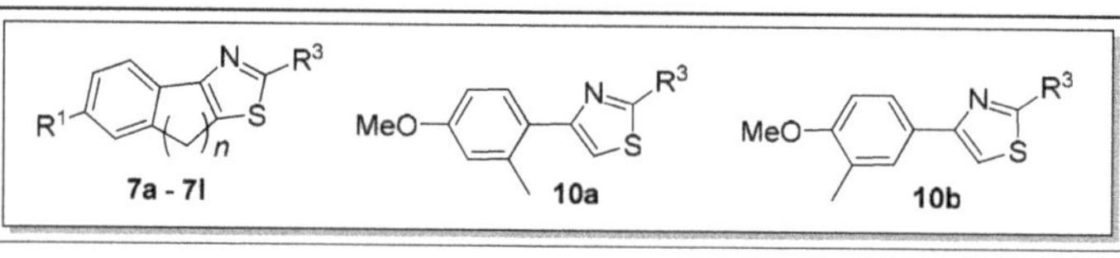

Compd.	R¹	R³	n	SARS-CoV-2 3CL^{pro}	
				Inhibition (%) at 20 μM	IC$_{50}$ (μM)
7a	methoxy	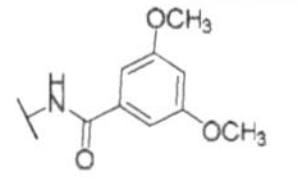 3,5-dimethoxybenzamido	1	89.5 ± 2.0	1.28 ± 0.17
7b	butoxy	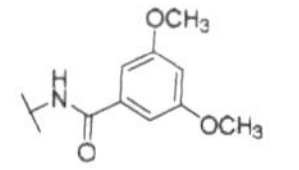 3,5-dimethoxybenzamido	1	0.5 ± 4.9	>20
7c	isobutoxy	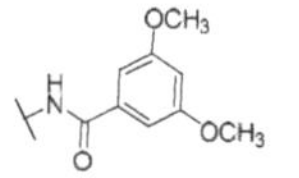 3,5-dimethoxybenzamido	1	−3.1 ± 1.7	>20
7d	methyl	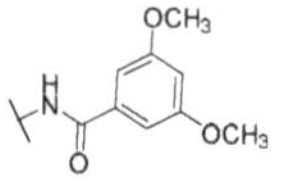 3,5-dimethoxybenzamido	1	21.7 ± 2.2	>20
7e	chloro	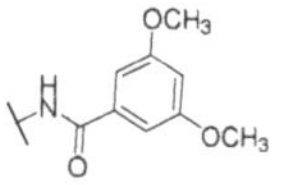 3,5-dimethoxybenzamido	1	27.2 ± 5.3	>20
7f	methoxy	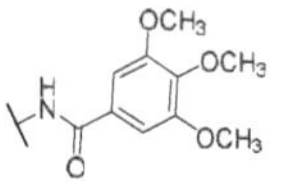 3,4,5-trimethoxybenzamido	1	5.0 ± 5.6	>20
7g	methoxy	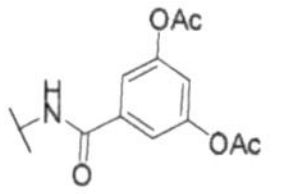 3,5-diacetoxybenzamido	1	32.6 ± 6.8	>20

Table 1. Cont.

Compd.	R¹	R³	n	SARS-CoV-2 3CL^{pro} Inhibition (%) at 20 μM	IC$_{50}$ (μM)
7h	methoxy	(3-methoxybenzamido)	1	72.5 ± 6.1	2.86 ± 0.11
7i	methoxy	(3-fluorobenzamido)	1	20.3 ± 4.7	>20
7j	methoxy	(thiophene-2-carboxamido)	1	31.9 ± 18.2	>20
7k	methoxy	(4-chlorobenzamido)	1	1.5 ± 4.5	>20
7l	methoxy	(3,5-dimethoxybenzamido)	2	−13.1 ± 1.7	>20
10a	-	(3,5-dimethoxybenzamido)	-	1.9 ± 2.1	>20
10b	-	(3,5-dimethoxybenzamido)	-	1.8 ± 3.5	>20
PF-07321332 (nirmatrelvir)				99.5 ± 0.1	0.012 ± 0.001

2.3. Predicting Binding Mode of 7a with 3CL^{pro}

To explore the interaction mode between small molecule **7a** and 3CL^{pro} (PDB code: 6M2N) [23], we carried out molecular docking by applying AutoDock 4.2 program [31–34]. Figure 3a showed that **7a** docked well into the binding pockets S1 and S2 of 3CL^{pro}, in which the S1, S2 sites play a key role in substrate recognition [35]. As illustrated in Figure 3b, the indene moiety of compound **7a** buried deeply into the hydrophobic S2 subsite with π-electrons with Arg188 and hydrophobic interaction with Met165; the 3,5-dimethoxybenzamido moiety of compound **7a** formed strong H-bonds with Asn142, Glu166 on S1 subsite, while compounds **4** and **7h** escaped from S1 subsite, as shown in Supplementary Materials Figures S1 and S2.

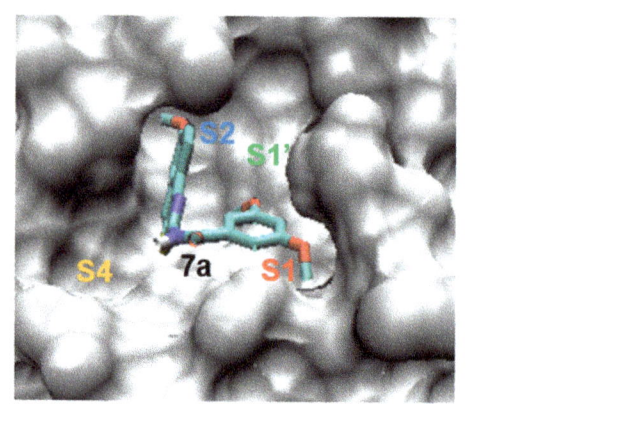 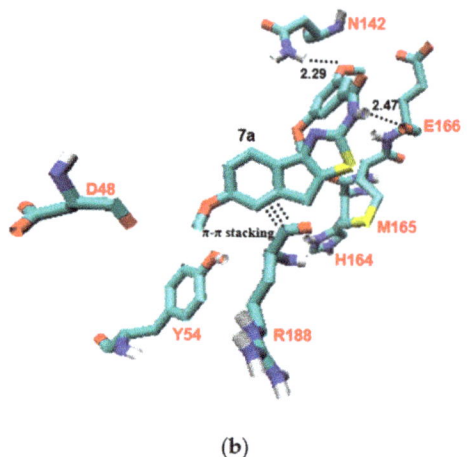

(a) (b)

Figure 3. (a) surf representation of the compound **7a** (bonds representation) in the 3CL^{pro} S1 (red), S2 (blue), S1′ (green), S4 (orange) binding pocket; (b) the docking results of **7a** and 3CL^{pro} (PDB code: 6M2N, active residues in 3.0 Å range around **7a**).

3. Materials and Methods

3.1. Chemistry

All chemical reagents are reagent grade and used as purchased. ^{1}H NMR (400 MHz) spectra were recorded on a Bruker AVIII 400 MHz spectrometer (Bruker, Billerica, MA, USA). The chemical shifts were reported in parts per million (ppm) using the 2.50 signal of DMSO (^{1}H NMR) and the 39.52 signal of DMSO (^{13}C NMR) as internal standards. ESI Mass spectra (MS) were obtained on a SHIMADZU 2020 Liquid Chromatograph Mass Spectrometer (SHIMADZU, Kyoto, Japan).

3.1.1. General Procedure for the Synthesis of Compounds **7a**–**7k** (Exemplified by **7a**)

To a solution of **5a** (6.2 mmol, 1.0 equiv) in dry ethanol (25 mL) were added thiourea (12.4 mmol, 2.0 equiv) and bromine (6.8 mmol, 1.1 equiv) at room temperature. The reaction solution was stirred at 100 °C for 5–6 h, At the end of the reaction, the solvent was evaporated, and aqueous ammonium hydroxide (25%) was added to the residue. The precipitated solid was collected without purification for the next step. The mixture of **6a** (2.2 mmol, 1.1 equiv), aromatic acid (2.0 mmol, 1.0 equiv), HATU (2.0 mmol, 1.0 equiv), and DIPEA (6.0 mmol, 3.0 equiv) in DMF (15 mL) was stirred at room temperature for 2 h. The reaction mixture was quenched with water. The aqueous layer was extracted with EtOAc (30 mL × 2). The combined organic layers were dried over Na$_2$SO$_4$. The residue was purified by column chromatography on silica gel (eluting with DCM) to afford compound **7a** as a yellow solid (280.0 mg, yield 37%). ^{1}H NMR (400 MHz, DMSO-d_6) δ 12.81 (s, 1H), 7.46 (d, J = 8.4 Hz, 1H), 7.33 (d, J = 2.0 Hz, 2H), 7.22 (d, J = 2.0 Hz, 1H), 6.94 (dd, J = 8.0, 2.4 Hz, 1H), 6.74 (t, J = 2.0 Hz, 1H), 3.87 (s, 2H), 3.84 (s, 6H), 3.80 (s, 3H) ppm. ^{13}C NMR (100 MHz, DMSO-d_6) δ 164.22, 162.12, 160.52, 157.76, 155.03, 147.98, 133.82, 130.05, 128.39, 118.28, 112.37, 111.83, 105.74, 105.08, 55.60, 55.36, 32.43 ppm. MS (ESI): m/z calcd for C$_{20}$H$_{19}$N$_2$O$_4$S [M + H]$^+$ 383.11, found 383.20.

N-(6-butoxy-8H-indeno[1,2-d]thiazol-2-yl)-3,5-dimethoxybenzamide (**7b**), eluting with DCM, yield = 32%; ^{1}H NMR (400 MHz, DMSO-d_6) δ 12.81 (s, 1H), 7.43 (d, J = 8.0 Hz, 1H), 7.33 (d, J = 2.4 Hz, 2H), 7.18 (s, 1H), 6.91 (dd, J = 8.4, 2.4 Hz, 1H), 6.73 (d, J = 2.4 Hz, 1H), 3.99 (t, J = 6.4 Hz, 2H), 3.84 (s, 2H), 3.83 (s, 6H), 1.74–1.67 (m, 2H), 1.49–1.40 (m, 2H), 0.95–0.91 (m, 3H) ppm. ^{13}C NMR (100 MHz, DMSO-d_6) δ 164.25, 162.15, 160.54, 157.20, 155.04, 147.96, 133.85, 129.95, 128.32, 118.29, 112.97, 112.39, 105.76, 105.10, 67.48, 55.62, 32.43, 30.89, 18.84, 13.76 ppm. MS (ESI): m/z calcd for C$_{23}$H$_{25}$N$_2$O$_4$S [M + H]$^+$ 425.15, found 425.10.

N-(6-isobutoxy-8H-indeno[1,2-d]thiazol-2-yl)-3,5-dimethoxybenzamide (**7c**), eluting with DCM, yield = 40%; ^{1}H NMR (400 MHz, DMSO-d_6) δ 8.27 (s, 1H), 7.43 (dd, *J* = 8.4, 3.2 Hz, 1H), 7.36–7.32 (m, 2H), 7.19 (d, *J* = 2.8 Hz, 1H), 6.92 (dd, *J* = 8.4, 3.2 Hz, 1H), 6.73 (t, *J* = 2.4 Hz, 1H), 3.85 (s, 2H), 3.83 (s, 6H), 3.78 (d, *J* = 6.4 Hz, 2H), 2.06–2.00 (m, 1H), 0.99 (d, *J* = 7.2 Hz, 6H) ppm. ^{13}C NMR (100 MHz, DMSO-d_6) δ 164.25, 162.17, 160.50, 157.26, 154.99, 147.94, 133.86, 129.95, 128.31, 118.26, 112.99, 112.45, 105.74, 105.06, 74.11, 55.59, 32.41, 27.78, 19.11 ppm. MS (ESI): *m/z* calcd for $C_{23}H_{25}N_2O_4S$ [M + H]$^+$ 425.15, found 425.20.

3,5-dimethoxy-N-(6-methyl-8H-indeno[1,2-d]thiazol-2-yl)benzamide (**7d**), eluting with DCM, yield = 47%; ^{1}H NMR (400 MHz, DMSO-d_6) δ 12.81 (s, 1H), 7.45 (d, *J* = 7.6 Hz, 1H), 7.39 (s, 1H), 7.34 (d, *J* = 2.4 Hz, 2H), 7.18 (d, *J* = 7.6 Hz, 1H), 6.74 (t, *J* = 2.0 Hz, 1H), 3.87 (s, 2H), 3.84 (s, 6H), 2.38 (s, 3H) ppm; ^{13}C NMR (100 MHz, DMSO-d_6) δ 164.26, 162.14, 160.50, 155.22, 146.31, 134.42, 134.30, 133.78, 129.84, 127.40, 126.00, 117.56, 105.73, 105.09, 55.58, 32.16, 21.13 ppm. MS (ESI): *m/z* calcd for $C_{20}H_{19}N_2O_3S$ [M + H]$^+$ 367.11, found 366.95.

N-(6-chloro-8H-indeno[1,2-d]thiazol-2-yl)-3,5-dimethoxybenzamide (**7e**), eluting with DCM, yield = 39%; ^{1}H NMR (400 MHz, DMSO-d_6) δ 12.85 (s, 1H), 7.64 (d, *J* = 1.6 Hz, 1H), 7.54 (d, *J* = 8.0 Hz, 1H), 7.42 (dd, *J* = 8.0, 2.0 Hz, 1H), 7.33 (d, *J* = 2.0 Hz, 2H), 6.74 (t, *J* = 2.0 Hz, 1H), 3.94 (s, 2H), 3.84 (s, 6H) ppm. ^{13}C NMR (100 MHz, DMSO-d6) δ 164.36, 162.60, 160.48, 154.14, 148.13, 135.82, 133.64, 131.60, 129.72, 126.87, 125.43, 118.83, 105.75, 105.13, 55.57, 32.49 ppm. MS (ESI): *m/z* calcd for $C_{19}H_{16}ClN_2O_3S$ [M + H]$^+$ 387.06, found 387.15.

3,5-dimethoxy-N-(5-methoxy-8H-indeno[1,2-d]thiazol-2-yl)benzamide (**4**), eluting with DCM, yield = 50%; ^{1}H NMR (400 MHz, DMSO-d_6) δ 12.80 (s, 1H), 7.44 (d, *J* = 8.4 Hz, 1H), 7.34 (d, *J* = 2.4 Hz, 2H), 7.07 (d, *J* = 2.4 Hz, 1H), 6.81 (dd, *J* = 8.4, 2.4 Hz, 1H), 6.73 (t, *J* = 2.4 Hz, 1H), 3.83 (s, 6H), 3.82 (s, 2H), 3.81 (s, 3H) ppm. ^{13}C NMR (100 MHz, DMSO-d_6) δ 164.32, 162.22, 160.50, 158.81, 138.17, 137.75, 133.77, 132.04, 125.68, 110.29, 105.74, 105.10, 104.65, 103.93, 55.58, 55.20, 31.64 ppm. MS (ESI): *m/z* calcd for $C_{20}H_{19}N_2O_4S$ [M + H]$^+$ 383.11, found 383.15.

3,4,5-trimethoxy-N-(6-methoxy-8H-indeno[1,2-d]thiazol-2-yl)benzamide (**7f**), eluting with DCM, yield = 44%; ^{1}H NMR (400 MHz, DMSO-d_6) δ 12.76 (s, 1H), 7.52 (s, 2H), 7.45 (d, *J* = 8.4 Hz, 1H), 7.21 (d, *J* = 2.4 Hz, 1H), 6.93 (dd, *J* = 8.4, 2.4 Hz, 1H), 3.89 (s, 6H), 3.86 (s, 2H), 3.80 (s, 3H), 3.75 (s, 3H) ppm. ^{13}C NMR (100 MHz, DMSO-d_6) δ 163.94, 162.29, 157.73, 154.98, 152.79, 147.95, 141.00, 130.07, 128.23, 126.73, 118.20, 112.35, 111.81, 105.61, 60.14, 56.11, 55.35, 32.41 ppm. MS (ESI): *m/z* calcd for $C_{21}H_{21}N_2O_5S$ [M + H]$^+$ 413.12, found 413.15.

5-((6-methoxy-8H-indeno[1,2-d]thiazol-2-yl)carbamoyl)-1,3-phenylene diacetate (**7g**), eluting with DCM, yield = 25%; ^{1}H NMR (400 MHz, DMSO-d_6) δ 12.91 (s, 1H), 7.86 (d, *J* = 2.0 Hz, 2H), 7.46 (d, *J* = 8.4 Hz, 1H), 7.33 (t, *J* = 2.0 Hz, 1H), 7.22 (d, *J* = 2.0 Hz, 1H), 6.94 (dd, *J* = 8.4, 2.0 Hz, 1H), 3.88 (s, 2H), 3.80 (s, 3H), 2.33 (s, 6H) ppm. ^{13}C NMR (100 MHz, DMSO-d_6) δ 172.06, 169.05, 164.01, 158.37, 157.74, 151.38, 147.94, 134.01, 130.00, 128.36, 119.23, 118.29, 113.15, 112.35, 111.83, 55.35, 32.42, 20.86 ppm. MS (ESI): *m/z* calcd for $C_{22}H_{19}N_2O_6S$ [M + H]$^+$ 439.10, found 439.05

3-methoxy-N-(6-methoxy-8H-indeno[1,2-d]thiazol-2-yl)benzamide (**7h**), eluting with DCM, yield = 47%; ^{1}H NMR (400 MHz, DMSO-d_6) δ 12.81 (s, 1H), 7.72–7.70 (m, 2H), 7.48–7.44 (m, 2H), 7.22–7.18 (m, 2H), 6.94 (dd, *J* = 8.0, 2.4 Hz, 1H), 3.88 (s, 2H), 3.86 (s, 3H), 3.80 (s, 3H) ppm. ^{13}C NMR (100 MHz, DMSO-d_6) δ 164.43, 162.14, 159.31, 157.73, 155.00, 147.95, 133.25, 130.05, 129.78, 128.32, 120.45, 119.02, 118.26, 112.58, 112.34, 111.80, 55.41, 55.34, 32.40 ppm. MS (ESI): *m/z* calcd for $C_{19}H_{17}N_2O_3S$ [M + H]$^+$ 353.10, found 353.15.

3-fluoro-N-(6-methoxy-8H-indeno[1,2-d]thiazol-2-yl)benzamide (**7i**), eluting with DCM, yield = 34%; ^{1}H NMR (400 MHz, DMSO-d_6) δ 12.92 (s, 1H), 8.00–7.94 (m, 2H), 7.65–7.59 (m, 1H), 7.52 (dd, *J* = 8.4, 2.4 Hz, 1H), 7.47 (d, *J* = 8.0 Hz, 1H), 7.22 (d, *J* = 2.4 Hz, 1H), 6.94 (dd, *J* = 8.0, 2.4 Hz, 1H), 3.88 (s, 2H), 3.80 (s, 3H) ppm. ^{13}C NMR (100 MHz, DMSO-d_6) δ 163.43,163.21, 161.90, 160.77, 157.77, 147.93, 134.26, 130.84 (d, *J* = 8.0 Hz), 129.95, 128.48, 124.36 (d, *J* = 3.0 Hz), 119.52 (d, *J* = 21.0 Hz), 118.30, 114.91 (d, *J* = 23.0 Hz), 112.36, 111.80, 55.34, 32.42 ppm. MS (ESI): *m/z* calcd for $C_{18}H_{14}FN_2O_2S$ [M + H]$^+$ 341.08, found 341.05.

N-(6-methoxy-8H-indeno[1,2-d]thiazol-2-yl)thiophene-2-carboxamide (**7j**), eluting with DCM, yield = 30%; ^{1}H NMR (400 MHz, DMSO-d_6) δ 12.92 (s, 1H), 8.28 (d, *J* = 8.0 Hz, 1H), 7.98 (d,

J = 4.8 Hz, 1H), 7.45 (d, *J* = 8.0 Hz, 1H), 7.27 (t, *J* = 4.8 Hz, 1H), 7.22 (d, *J* = 2.4 Hz, 1H), 6.94 (dd, *J* = 8.0, 2.4 Hz, 1H), 3.87 (s, 2H), 3.80 (s, 3H) ppm. ^{13}C NMR (100 MHz, DMSO-d_6) δ 161.76, 159.35, 157.74, 154.97, 147.91, 137.30, 133.61, 130.69, 129.96, 128.64, 128.29, 118.24, 112.34, 111.78, 55.33, 32.43 ppm. MS (ESI): *m/z* calcd for $C_{16}H_{13}N_2O_2S_2$ [M + H]$^+$ 329.04, found 329.10.

4-chloro-N-(6-methoxy-8H-indeno[1,2-d]thiazol-2-yl)benzamide (**7k**), eluting with DCM, yield = 38%; ^{1}H NMR (400 MHz, DMSO-d_6) δ 12.90 (s, 1H), 8.14 (dt, *J* = 8.8, 2.0 Hz, 2H), 7.63 (dt, *J* = 8.4, 2.0 Hz, 2H), 7.47 (d, *J* = 8.4 Hz, 1H), 7.22 (d, *J* = 2.0 Hz, 1H), 6.94 (dd, *J* = 8.4, 2.4 Hz, 1H), 3.88 (s, 2H), 3.80 (s, 3H) ppm. ^{13}C NMR (100 MHz, DMSO-d_6) δ 163.77, 162.03, 157.74, 154.90, 147.93, 137.48, 130.82, 130.03, 129.97, 128.73, 128.38, 118.28, 112.34, 111.79, 55.34, 32.41 ppm. MS (ESI): *m/z* calcd for $C_{18}H_{14}ClN_2O_2S$ [M + H]$^+$ 357.05, found 356.90.

3.1.2. Procedure for the Synthesis of Compound **7l**

To a solution of **5f** (528.2 mg, 3.0 mmol) in dry ethanol (10 mL) were added thiourea (456.7 mg, 6.0 mmol) and bromine (0.2 mL, 3.3 mmol) at room temperature. The reaction solution was stirred at 100 °C for 5–6 h. At the end of the reaction, the solvent was evaporated and aqueous ammonium hydroxide (25%) was added to the residue. The precipitated solid **6f** was collected without purification for the next step. The mixture of 6f (255.2 mg, 1.1 mmol), 3,5-dimethoxybenzoic acid (182.1 mg, 1.0 mmol), HATU (380.2 mg, 1.0 mmol), and DIPEA (0.5 mL 3.0 mmol) in DMF (6 mL) was stirred at room temperature for 2 h. The reaction mixture was quenched with water. The aqueous layer was extracted with EtOAc (20 mL × 2). The combined organic layers were dried over Na_2SO_4. The residue was purified by column chromatography on silica gel (eluting with DCM) to afford compound **7l** (103.0 mg, yield 26%) as a white solid.

^{1}H NMR (400 MHz, DMSO-d_6) δ 12.66 (s, 1H), 7.66 (dd, *J* = 8.4, 2.0 Hz, 1H), 7.32 (t, *J* = 2.0 Hz, 2H), 6.88 (s, 1H), 6.85 (dd, *J* = 8.4, 2.4 Hz, 1H), 6.73 (d, *J* = 2.4 Hz, 1H), 3.83 (s, 6H), 3.77 (s, 3H), 3.00–2.91 (m, 4H) ppm. ^{13}C NMR (100 MHz, DMSO-d_6) δ 164.31, 160.45, 158.42, 156.44, 143.66, 136.68, 133.90, 124.23, 123.33, 121.55, 114.08, 111.82, 105.75, 105.06, 55.59, 55.09, 28.65, 20.74 ppm. MS (ESI): *m/z* calcd for $C_{21}H_{21}N_2O_4S$ [M + H]$^+$ 397.12, found 396.95.

3.1.3. General Procedure of Synthesis of **10a–10b** (Exemplified by **10a**)

A mixture of **8a** (10.0 mmol, 1.0 equiv), thiourea (20.0 mmol, 2.0 equiv), and iodine (10.0 mmol, 1.0 equiv) was stirred at 110 °C for 10 h. After the reaction was completed, the residue was triturated with MTBE and adjusted to pH 9–10 with 25% ammonium hydroxide. The precipitated solid was collected and washed with EtOAc (30 mL × 2) and $NaHCO_3$ (15 mL × 2) aqueous solution. The separated organic layer dried over Na_2SO_4 and evaporated to dryness to afford crude product **9a**. The mixture of **9a** (3.3 mmol, 1.1 equiv), aromatic acid (3.0 mmol, 1.0 equiv), HATU (3.0 mmol, 1.0 equiv), and DIPEA (9.0 mmol, 3.0 equiv) in DMF (20 mL) was stirred at room temperature for 2 h. Then the reaction mixture was quenched with water. The aqueous layer was extracted with EtOAc (30 mL × 2). The combined organic layers were dried over Na_2SO_4. The residue was purified by column chromatography on silica gel (eluting with DCM) to afford compound **10a** as a white solid (406.7 mg, yield 35%). ^{1}H NMR (400 MHz, DMSO-d_6) δ 12.67 (s, 1H), 7.57 (d, *J* = 8.4 Hz, 1H), 7.32 (d, *J* = 2.0 Hz, 2H), 7.21 (s, 1H), 6.86 (d, *J* = 2.4 Hz, 1H), 6.83 (dd, *J* =8.4, 2.8 Hz, 1H), 6.74 (t, *J* = 2.4 Hz, 1H), 3.83 (s, 6H), 3.77 (s, 3H), 2.43 (s, 3H) ppm. ^{13}C NMR (100 MHz, DMSO-d_6) δ 164.74, 160.45, 158.60, 157.76, 149.01, 136.97, 134.22, 130.74, 127.31, 115.96, 111.22, 110.14, 105.80, 104.90, 55.58, 55.04, 21.26 ppm. MS (ESI): *m/z* calcd for $C_{20}H_{21}N_2O_4S$ [M + H]$^+$ 385.12, found 385.20.

3,5-dimethoxy-N-(4-(4-methoxy-3-methylphenyl)thiazol-2-yl)benzamide (**10b**), eluting with DCM, yield = 40%; ^{1}H NMR (400 MHz, DMSO-d_6) δ 12.70 (s, 1H), 7.77 (d, *J* = 2.4 Hz, 1H), 7.75 (s, 1H), 7.49 (s, 1H), 7.33 (d, *J* = 2.4 Hz, 2H), 6.99 (d, *J* = 8.8 Hz, 1H), 6.74 (t, *J* = 2.4 Hz, 1H), 3.84 (s, 6H), 3.82 (s, 3H), 2.20 (s, 3H) ppm. ^{13}C NMR (100 MHz, DMSO-d_6) δ 164.59, 160.46, 158.26, 157.15, 149.27, 133.86, 128.07, 126.72, 125.68, 124.62, 110.36, 106.38, 105.79,

105.09, 55.59, 55.31, 16.21 ppm. MS (ESI): m/z calcd for $C_{20}H_{21}N_2O_4S$ [M + H]$^+$ 385.12, found 385.25.

3.2. Molecule Docking

The protease structure, SARS-CoV-2 3CL^{pro} enzyme (PDB code: 6M2N) with 2.2 Å, was obtained from the the Protein Data Bank at the RCSB site (http://www.rcsb.org (accessed on 6 March 2022)). The molecule docking used the Lamarckian genetic algorithm local search method and the semiempirical free energy calculation method in the AutoDock 4.2 program. Additionally, the charge was added by Kollman in AutoDock 4.2, The docking method was employed on rigid receptor conformation, all the rotatable torsional bonds of compound **7a** were set free, the size of grid box was set at to 10.4 nm × 12.6 nm × 11.0 nm points with a 0.0375 nm spacing and grid center (−33.798 −46.566 39.065), and the other parameters were maintained at their default settings.

3.3. Enzymatic Activity and Inhibition Assays

The enzyme activity and inhibition assays of SARS-CoV-2 3CL^{pro} have been described previously [20,36]. Briefly, the recombinant SARS-CoV-2 3CL^{pro} (40 nM at a final concentration) was mixed with each compound in 50 µL of assay buffer (20 mM Tris, pH 7.3, 150 mM NaCl, 1% Glycerol, 0.01% Tween-20) and incubated for 10 min. The reaction was initiated by adding the fluorogenic substrate MCA-AVLQSGFRK (DNP) K (GL Biochem, Shanghai, China), with a final concentration of 40 µM. After that, the fluorescence signal at 320 nm (excitation)/405 nm (emission) was immediately measured by continuous 10 points for 5 min with an EnVision multimode plate reader (Perkin Elmer, Waltham, MA, USA). The initial velocity was measured when the protease reaction was proceeding in a linear fashion; plots of fluorescence units versus time were fitted with linear regression to determine initial velocity. Plots of initial velocity versus inhibitor concentration were fitted using a four-parameter concentration–response model in GraphPad Prism 8 to calculate the IC$_{50}$ values. All data are shown as mean ± SD, n = 3 biological replicates.

4. Conclusions

In summary, we synthesized a series of 8*H*-Indeno[1,2-*d*]thiazole derivatives and evaluated their biochemical activities against SARS-CoV-2 3CL^{pro}. Among them, the representative compound **7a** displayed inhibitory activity with an IC$_{50}$ of 1.28 ± 0.17 µM against SARS-CoV-2 3CL^{pro}. Molecular docking elucidated that **7a** was well-docked into the binding pockets S1 and S2 of 3CL^{pro}. These preliminary results could provide a possible opportunity for the development of novel inhibitors against SARS-CoV-2 3CL^{pro} with optimal potency and improved pharmacological properties.

Supplementary Materials: The following supporting information can be downloaded at: https://www.mdpi.com/article/10.3390/molecules27103359/s1, copies of the ^{1}H NMR and ^{13}C NMR spectra for compounds **4, 7a–7l, 10a–10b** and Figure S1. surf representation of the compound **4** (bonds representation) in the 3CL^{pro} S1 (red), S2 (blue), S1' (green), S4 (orange) binding pockets, Figure S2. surf representation of the compound **7h** (bonds representation) in the 3CL^{pro} S1 (red), S2 (blue), S1' (green), S4 (orange) binding pockets.

Author Contributions: Investigation, J.W. (synthesis); B.F. and L.-X.G. (bioassay); C.Z. (molecule docking); Conceptualization, J.L., D.-J.X., Y.Z. and W.-L.W.; writing—original draft preparation, J.W., B.F. and D.-J.X.; writing—review and editing, Y.Z. and W.-L.W.; supervision, Y.Z. and W.-L.W.; project administration, D.-J.X. and W.-L.W.; funding acquisition. All authors have read and agreed to the published version of the manuscript.

Funding: This work was financially supported by the science and technology development foundation of Wuxi (N2020X016) and the Natural Science Foundation of Jiangsu Province (BK20190608).

Institutional Review Board Statement: Not applicable.

Informed Consent Statement: Not applicable.

Data Availability Statement: Data are available upon request to the Corresponding Authors.

Acknowledgments: The authors express their gratitude to the BioDuro-Sundia in Wuxi for NMR spectral data and mass spectral data.

Conflicts of Interest: The authors declare no conflict of interest.

Sample Availability: Some of the compounds may be available in mg quantities upon request from the corresponding authors.

References

1. Zhu, N.; Zhang, D.; Wang, W.; Li, X.; Yang, B.; Song, J.; Zhao, X.; Huang, B.; Shi, W.; Lu, R.; et al. A Novel Coronavirus from Patients with Pneumonia in China, 2019. *N. Engl. J. Med.* **2020**, *382*, 727–733. [CrossRef] [PubMed]
2. Zhou, P.; Yang, X.L.; Wang, X.G.; Hu, B.; Zhang, L.; Zhang, W.; Si, H.R.; Zhu, Y.; Li, B.; Huang, C.L.; et al. A pneumonia outbreak associated with a new coronavirus of probable bat origin. *Nature* **2020**, *579*, 270–273. [CrossRef] [PubMed]
3. Wang, C.; Horby, P.W.; Hayden, F.G.; Gao, G.F. A novel coronavirus outbreak of global health concern. *Lancet* **2020**, *395*, 470–473. [CrossRef]
4. Wu, F.; Zhao, S.; Yu, B.; Chen, Y.M.; Wang, W.; Song, Z.G.; Hu, Y.; Tao, Z.W.; Tian, J.H.; Pei, Y.Y.; et al. A new coronavirus associated with human respiratory disease in China. *Nature* **2020**, *579*, 1–8. [CrossRef] [PubMed]
5. Coronaviridae Study Group of the International Committee on Taxonomy of V. The species Severe acute respiratory syndrome-related coronavirus: Classifying 2019-nCoV and naming it SARS-CoV-2. *Nat. Microbiol.* **2020**, *5*, 536–544. [CrossRef] [PubMed]
6. WHO. COVID-19 Dashboard with Vaccination Data. Available online: https://covid19.who.int/ (accessed on 28 February 2022).
7. Abu-Raddad, L.J.; Chemaitelly, H.; Ayoub, H.H.; Tang, P.; Coyle, P.; Hasan, M.R.; Yassine, H.M.; Benslimane, F.M.; Al-Khatib, H.A.; Al-Kanaani, Z.; et al. Relative infectiousness of SARS-CoV-2 vaccine breakthrough infections, reinfections, and primary infections. *Nat. Commun.* **2022**, *13*, 532. [CrossRef]
8. CDC COVID-19 Vaccine Breakthrough Case Investigations Team. COVID-19 Vaccine Breakthrough Infections Reported to CDC—United States, January 1–April 30, 2021. *MMWR Morb. Mortal. Wkly. Rep.* **2021**, *70*, 792–793. [CrossRef]
9. Chen, J.; Wang, R.; Gilby, N.B.; Wei, G.W. Omicron Variant (B.1.1.529): Infectivity, Vaccine Breakthrough, and Antibody Resistance. *J. Chem. Inf. Model* **2022**, *62*, 412–422. [CrossRef]
10. Zhang, L.; Li, Q.; Liang, Z.; Li, T.; Liu, S.; Cui, Q.; Nie, J.; Wu, Q.; Qu, X.; Huang, W.; et al. The significant immune escape of pseudotyped SARS-CoV-2 variant Omicron. *Emerg. Microbes. Infect.* **2022**, *11*, 1–5. [CrossRef]
11. Hoffmann, M.; Kleine-Weber, H.; Schroeder, S.; Kruger, N.; Herrler, T.; Erichsen, S.; Schiergens, T.S.; Herrler, G.; Wu, N.H.; Nitsche, A.; et al. SARS-CoV-2 Cell Entry Depends on ACE2 and TMPRSS2 and Is Blocked by a Clinically Proven Protease Inhibitor. *Cell* **2020**, *181*, 271–280. [CrossRef]
12. Mody, V.; Ho, J.; Wills, S.; Mawri, A.; Lawson, L.; Ebert, M.; Fortin, G.M.; Rayalam, S.; Taval, S. Identification of 3-chymotrypsin like protease (3CLPro) inhibitors as potential anti-SARS-CoV-2 agents. *Commun. Biol.* **2021**, *4*, 93. [CrossRef] [PubMed]
13. V'Kovski, P.; Kratzel, A.; Steiner, S.; Stalder, H.; Thiel, V. Coronavirus biology and replication: Implications for SARS-CoV-2. *Nat. Rev. Microbiol.* **2021**, *19*, 155–170. [CrossRef] [PubMed]
14. Pillaiyar, T.; Meenakshisundaram, S.; Manickam, M. Recent discovery and development of inhibitors targeting coronaviruses. *Drug Discov. Today* **2020**, *25*, 668–688. [CrossRef]
15. Wen, W.; Chen, C.; Tang, J.; Wang, C.; Zhou, M.; Cheng, Y.; Zhou, X.; Wu, Q.; Zhang, X.; Feng, Z.; et al. Efficacy and safety of three new oral antiviral treatment (molnupiravir, fluvoxamine and Paxlovid) for COVID-19a meta-analysis. *Ann. Med.* **2022**, *54*, 516–523. [CrossRef] [PubMed]
16. Chia, C.S.B. Novel Nitrile Peptidomimetics for Treating COVID-19. *ACS Med. Chem. Lett.* **2022**, *13*, 330–331. [CrossRef]
17. Konno, S.; Kobayashi, K.; Senda, M.; Funai, Y.; Seki, Y.; Tamai, I.; Schäkel, L.; Sakata, K.; Pillaiyar, T.; Taguchi, A.; et al. 3CL Protease Inhibitors with an Electrophilic Arylketone Moiety as Anti-SARS-CoV-2 Agents. *J. Med. Chem.* **2022**, *65*, 2926–2939. [CrossRef]
18. Zhang, L.L.; Lin, D.Z.; Sun, X.Y.Y.; Curth, U.; Drosten, C.; Sauerhering, L.; Becker, S.; Rox, K.; Hilgenfeld, R. Crystal structure of SARS-CoV-2 main protease provides a basis for design of improved α-ketoamide inhibitors. *Science* **2020**, *368*, 409–412. [CrossRef]
19. Dampalla, C.S.; Kim, Y.; Bickmeier, N.; Rathnayake, A.D.; Nguyen, H.N.; Zheng, J.; Kashipathy, M.M.; Baird, M.A.; Battaile, K.P.; Lovell, S. Structure-Guided Design of Conformationally Constrained Cyclohexane Inhibitors of Severe Acute Respiratory Syndrome Coronavirus-2 3CL Protease. *J. Med. Chem.* **2021**, *64*, 10047–10058. [CrossRef]
20. Jin, Z.; Du, X.; Xu, Y.; Deng, Y.; Liu, M.; Zhao, Y.; Zhang, B.; Li, X.; Zhang, L.; Peng, C.; et al. Structure of M(pro) from SARS-CoV-2 and discovery of its inhibitors. *Nature* **2020**, *582*, 289–293. [CrossRef]
21. Xiong, M.; Nie, T.; Shao, Q.; Li, M.; Su, H.; Xu, Y. In silico screening-based discovery of novel covalent inhibitors of the SARS-CoV-2 3CL protease. *Eur. J. Med. Chem.* **2022**, *231*, 114130. [CrossRef]
22. Han, S.H.; Goins, C.M.; Arya, T.; Shin, W.J.; Maw, J.; Hooper, A.; Sonawane, D.P.; Porter, M.R.; Bannister, B.E.; Crouch, R.D. Structure-Based Optimization of ML300-Derived, Noncovalent Inhibitors Targeting the Severe Acute Respiratory Syndrome Coronavirus 3CL Protease (SARS-CoV-2 3CLpro). *J. Med. Chem.* **2022**, *65*, 2880–2904. [CrossRef] [PubMed]

23. Su, H.X.; Yao, S.; Zhao, W.F.; Li, M.J.; Liu, J.; Shang, W.J.; Xie, H.; Ke, C.Q.; Hu, H.C.; Gao, M.N.; et al. Anti-SARS-CoV-2 activities in vitro of Shuanghuanglian preparations and bioactive ingredients. *Acta Pharmacol. Sin.* **2020**, *41*, 1167–1177. [CrossRef] [PubMed]
24. Wu, Q.; Yan, S.; Wang, Y.; Li, M.; Xiao, Y.; Li, Y. Discovery of 4′-O-methylscutellarein as a potent SARS-CoV-2 main protease inhibitor. *Biochem. Biophys. Res. Commun.* **2022**, *604*, 76–82. [CrossRef] [PubMed]
25. Goblyos, A.; Santiago, S.N.; Pietra, D.; Mulder-Krieger, T.; von Frijtag Drabbe Kunzel, J.; Brussee, J.; Ijzerman, A.P. Synthesis and biological evaluation of 2-aminothiazoles and their amide derivatives on human adenosine receptors. Lack of effect of 2-aminothiazoles as allosteric enhancers. *Bioorg. Med. Chem.* **2005**, *13*, 2079–2087. [CrossRef]
26. Kocyigit, U.M.; Aslan, O.N.; Gulcin, I.; Temel, Y.; Ceylan, M. Synthesis and Carbonic Anhydrase Inhibition of Novel 2-(4-(Aryl)thiazole-2-yl)-3a,4,7,7a-tetrahydro-1H-4,7-methanoisoindole-1,3(2H)-di one Derivatives. *Arch. Pharm.* **2016**, *349*, 955–963. [CrossRef]
27. Chordia, M.D.; Murphree, L.J.; Macdonald, T.L.; Linden, J.; Olsson, R.A. 2-Aminothiazoles: A new class of agonist allosteric enhancers of A1 adenosine receptor. *Bioorg. Med. Chem. Lett.* **2002**, *12*, 1563–1566. [CrossRef]
28. Chordia, M.D.; Zigler, M.; Murphree, L.J.; Figler, H.; Macdonald, T.L.; Olsson, R.A.; Linden, J. 6-Aryl-8H-indeno[1,2-d]thiazol-2-ylamines: A1 Adenosine Receptor Agonist Allosteric Enhancers Having Improved Potency. *J. Med. Chem.* **2005**, *48*, 5131–5139. [CrossRef]
29. Catlin, N.R.; Bowman, C.J.; Campion, S.N.; Cheung, J.R.; Nowland, W.S.; Sathish, J.G.; Stethem, C.M.; Updyke, L.; Cappon, G.D. Reproductive and developmental safety of nirmatrelvir (PF-07321332), an oral SARS-CoV-2 M(pro) inhibitor in animal models. *Reprod. Toxicol.* **2022**, *108*, 56–61. [CrossRef]
30. Macchiagodena, M.; Pagliai, M.; Procacci, P. Characterization of the non-covalent interaction between the PF-07321332 inhibitor and the SARS-CoV-2 main protease. *J. Mol. Graph. Model.* **2022**, *110*, 108042. [CrossRef]
31. Zhao, Y.; Fang, C.; Zhang, Q.; Zhang, R.; Zhao, X.; Duan, Y.; Wang, H.; Zhu, Y.; Feng, L.; Zhao, J.; et al. Crystal structure of SARS-CoV-2 main protease in complex with protease inhibitor PF-07321332. *Protein Cell.* **2021**, 1–5. [CrossRef]
32. Meng, X.D.; Gao, L.X.; Wang, Z.J.; Feng, B.; Zhang, C.; Satheeshkumar, R.; Li, J.; Zhu, Y.L.; Zhou, Y.B.; Wang, W.L. Synthesis and biological evaluation of 2,5-diaryl-1,3,4-oxadiazole derivatives as novel Src homology 2 domain-containing protein tyrosine phosphatase 2 (SHP2) inhibitors. *Bioorg. Chem.* **2021**, *116*, 105384. [CrossRef] [PubMed]
33. Morris, G.M.; Huey, R.; Lindstrom, W.; Sanner, M.F.; Belew, R.K.; Goodsell, D.S.; Olson, A.J. AutoDock4 and AutoDockTools4: Automated docking with selective receptor flexibility. *J. Comput. Chem.* **2009**, *30*, 2785–2791. [CrossRef] [PubMed]
34. Peralta, J.; Ogliaro, F.; Bearpark, M.; Heyd, J.; Brothers, E.; Kudin, K.; Staroverov, V.; Kobayashi, R.; Normand, J.; Raghavachari, K. *Gaussian 09, Revision. D. 01*; Gaussian, Inc.: Wallingford, CT, USA, 2013.
35. Yang, H.; Xie, W.; Xue, X.; Yang, K.; Ma, J.; Liang, W.; Zhao, Q.; Zhou, Z.; Pei, D.; Ziebuhr, J.; et al. Design of wide-spectrum inhibitors targeting coronavirus main proteases. *PLoS Biol.* **2005**, *3*, e324. [CrossRef]
36. Dai, W.H.; Zhang, B.; Jiang, X.M.; Su, H.X.; Li, J.; Zhao, Y.; Xie, X.; Jin, Z.M.; Peng, J.J.; Liu, F.J. Structure-based design of antiviral drug candidates targeting the SARS-CoV-2 main protease. *Science* **2020**, *368*, eabb4489. [CrossRef]

Article

Synthesis and Luminescent Properties of *s*-Tetrazine Derivatives Conjugated with the 4*H*-1,2,4-Triazole Ring

Anna Maj [1], Agnieszka Kudelko [1,*] and Marcin Świątkowski [2]

[1] Department of Chemical Organic Technology and Petrochemistry, The Silesian University of Technology, Krzywoustego 4, PL-44100 Gliwice, Poland; anna.maj@polsl.pl
[2] Institute of General and Ecological Chemistry, Lodz University of Technology, Zeromskiego 116, PL-90924 Lodz, Poland; marcin.swiatkowski@p.lodz.pl
* Correspondence: agnieszka.kudelko@polsl.pl; Tel.: +48-32-237-1729

Abstract: New derivatives obtained by the combination of unique 1,2,4,5-tetrazine and 4*H*-1,2,4-triazole rings have great application potential in many fields. Therefore, two synthetic few-step methodologies, which make use of commercially available 4-cyanobenzoic acid (method A) and ethyl diazoacetate (method B), were applied to produce two groups of the aforementioned heterocyclic conjugates. In both cases, the target compounds were obtained in various combinations, by introducing electron-donating or electron-withdrawing substituents into the terminal rings, together with aromatic or aliphatic substituents on the triazole nitrogen atom. Synthesis of such designed systems made it possible to analyze the influence of individual elements of the structure on the reaction course, as well as the absorption and emission properties. The structure of all products was confirmed by conventional spectroscopic methods, and their luminescent properties were also determined.

Keywords: *s*-tetrazine; 4*H*-1,2,4-triazole; pinner reaction

1. Introduction

Over the years, scientists from around the world have been keen to study heterocyclic organic compounds, and nitrogen-rich systems have proven to be particularly valuable. One of the most interesting areas of this research is the synthesis and properties of 1,2,4,5-tetrazine derivatives (*s*-tetrazine). This unique ring contains four nitrogen atoms, which is the maximum content in a stable six-membered system. This specific structure has attracted scientists' attention as an important candidate for high energy density materials (HEDMs, A, Scheme 1), as its thermal decomposition leads to ring opening and the release of a nitrogen molecule [1–3]. The high nitrogen content has also encouraged research into its biological activity (B, Scheme 1), which has resulted in compounds that have anti-tubercular, anti-cancer, or anti-malarial effects [4–6]. Moreover, its high reactivity in Diels–Alder reactions with inverse electron demand determines its application potential in bioorthogonal chemistry (C, Scheme 1) [7–10]. Important features of the *s*-tetrazine ring are its low-energy n→π electronic transitions, which are especially valuable from the point of view of optoelectronics (Scheme 1). It can be used in the production of organic light-emitting diodes (OLEDs), organic field-effect transistors (OFETs), and solar cells. Due to the high electronegativity of nitrogen, the ring in question is also characterized by a high electron deficit, and thus a high electron affinity. Consequently, it is also a promising building block in ambipolar and n-type materials [11,12].

The five-membered compound, which, like *s*-tetrazine, shows high nitrogen content, is 4*H*-1,2,4-triazole. In this case, too, the presence of nitrogen is associated with a high affinity toward biological macromolecules, which results in biological activity, such as the possession of antiviral, anti-migraine, antifungal, anti-cancer, or psychotropic properties, and various commercially available products incorporate 4*H*-1,2,4-triazole rings (E, Scheme 1) [13–16]. Another consequence of the nitrogen atoms is the aforementioned change in the electron density distribution, and the associated ability to transport electrons, making it an acceptor unit (F, Scheme 1). Therefore, 4H-1,2,4-triazole derivatives are often used in the production of blue OLEDs [17–20].

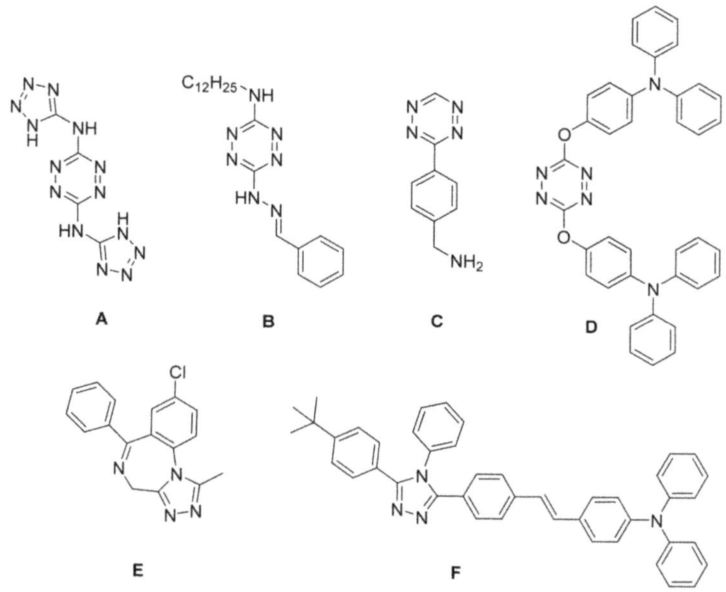

Scheme 1. Derivatives of the title heterocycles with great application potential [3,4,7,12,13,17].

Many synthetic methods can be found in the literature for both s-tetrazine and 4H-1,2,4-triazole derivatives. The five-membered heterocycle is usually obtained from acyclic compounds such as N,N'-diacylhydrazines, N-cyanoguanidine, isothiocyanates, hydrazides, aminoethylidenehydrazones, aldehydes, and semicarbazides [21]. For the six-membered s-tetrazine system, the Pinner method is the most popular: cyclization, supported by an activating agent, occurs as a result of the reaction of carbonitriles with hydrazine hydrate. The product of this transformation is the corresponding dihydro derivative that requires oxidation to give the desired ring [22,23]. This approach is distinguished by a wide range of substrates, but also the ability to synthesize both symmetrical and unsymmetrical products. Our research to date proves that, among its other uses, it is perfect for the preparation of complex conjugated systems that contain additional five-membered rings. In recent years, we have successfully synthesized s-tetrazine conjugated via a 1,4-phenylene linker with a range of 1,3,4-oxadiazoles, 1,3,4-thiadiazoles and 4H-1,2,4-triazoles; however, in the latter case, we have so far only obtained symmetrical systems [24–26]. In a continuation of our research, we decided to use the Pinner method to prepare unsymmetrical ones. Moreover, encouraged by the improvement in the luminescent properties after the introduction of the 4H-1,2,4-triazole ring, we found that the directly connected heterocycles could be the basis of very promising products. Therefore, we focused on modifying the methodology used to prepare analogous compounds containing 1,3,4-oxadiazole and 1,3,4-thiadiazole, so as to introduce the triazole ring instead [27]. This study was planned to make it possible not only to obtain new, unknown compounds, but also to analyze the influence of their structure on their absorption and emission properties.

2. Results

2.1. Synthesis

As already mentioned in previous studies, we obtained a series of symmetric s-tetrazine derivatives conjugated via a 1,4-phenylene linker with a 4H-1,2,4-triazole ring. For this purpose, it was necessary to prepare appropriate precursors for the Pinner reaction, i.e., carbonitriles containing a five-membered ring (Scheme 2, **6a–h**). Initially, from the commercially available 4-cyanobenzoic acid (**1**), we obtained the hydrazide (**2**) in a two-step reaction sequence. The original assumption was to treat it with acid chlorides (**3a–d**) in order to obtain diacyl derivatives (**4a–d**), and then convert them into the corresponding imidoyl chlorides (**5a–d**), which, under the influence of amines, would be cyclized to the assumed products (**6a–h**). This approach, however, turned out to be very troublesome due to the formation of the undesirable products **7a–d**. This prompted us to change the reaction path by synthesizing other imidoyl chlorides (**8a–h**) from the corresponding amides. These intermediates were treated with hydrazide (**3**), resulting in the target precursors (**6a–h**) in satisfactory yields [26].

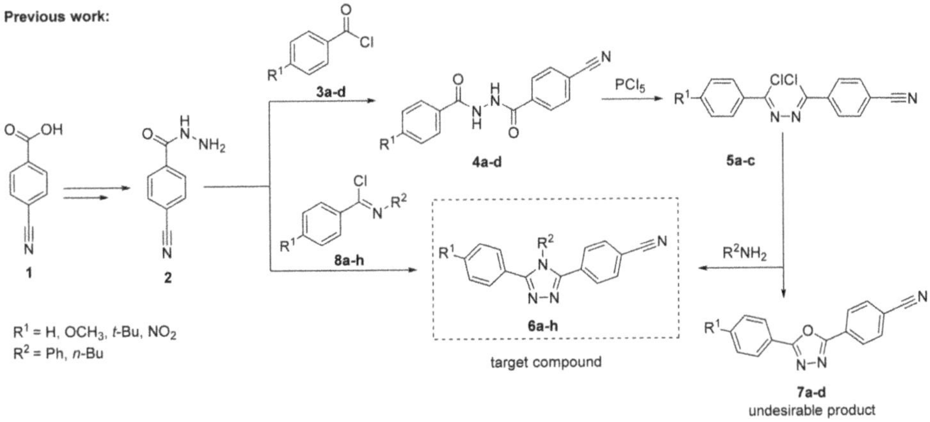

Scheme 2. Synthesis of precursors containing a 4H-1,2,4-triazole ring (**6a–h**) [26].

The presence of the carbonitrile moiety allows the formation of a second heterocycle, which is s-tetrazine. Under the conditions of the Pinner method, the treatment of the precursors **6a–h** with hydrazine hydrate, in the presence of an activating agent, leads to the formation of unoxidized derivatives of the assumed products **9a–l**. One of the popular activating agents is sulfur, with the help of which we have successfully obtained symmetrical s-tetrazine derivatives connected via a 1,4-phenylene linker with a 4H-1,2,4-triazole ring, and extended systems containing 1,3,4-oxadiazole and 1,3,4-thiadiazole cores [24–26]. Therefore, we also began to research the synthesis of unsymmetrical compounds with the use of this methodology, which allowed us to obtain the product **10a** with a yield of 42% (Entry 1, Table 1). In connection with literature reports on the possibility of improving this yield with the use of zinc catalysts [28,29], we attempted to repeat the described transformation with its participation and, as a result, the yield increased to 56% (Entry 2, Table 1). An analogous test was performed for derivatives containing an aliphatic chain attached to the triazole nitrogen atom, instead of an aromatic ring (**10g**). Again, the yield improved from 35% to 50% (Entries 8 and 9, Table 1). These results were an important reason to modify the previously used procedure. Such a modified approach resulted in obtaining a series of unsymmetrical systems containing both electron-donating and electron-withdrawing substituents in the terminal ring. Traces of two symmetrical products were also detected. As in the previous studies, the oxidation was carried out with hydrogen peroxide (Scheme 3).

Table 1. The yield of the reaction for the preparation of s-tetrazine derivatives conjugated via a 1,4-phenylene linker with a 4H-1,2,4-triazole ring (**10a–l**).

Entry	Product	R^1	R^3	R^2	Activating Agent	Yield [%]
1	10a	H	OCH$_3$	Ph	S	42
2					Zn(CF$_3$SO$_3$)$_2$	56
3	10b	H	t-Bu	Ph	Zn(CF$_3$SO$_3$)$_2$	52
4	10c	H	NO$_2$	Ph	Zn(CF$_3$SO$_3$)$_2$	49
5	10d	OCH$_3$	t-Bu	Ph	Zn(CF$_3$SO$_3$)$_2$	56
6	10e	OCH$_3$	NO$_2$	Ph	Zn(CF$_3$SO$_3$)$_2$	54
7	10f	t-Bu	NO$_2$	Ph	Zn(CF$_3$SO$_3$)$_2$	52
8	10g	H	OCH$_3$	n-Bu	S	35
9					Zn(CF$_3$SO$_3$)$_2$	50
10	10h	H	t-Bu	n-Bu	Zn(CF$_3$SO$_3$)$_2$	47
11	10i	H	NO$_2$	n-Bu	Zn(CF$_3$SO$_3$)$_2$	45
12	10j	OCH$_3$	t-Bu	n-Bu	Zn(CF$_3$SO$_3$)$_2$	51
13	10k	OCH$_3$	NO$_2$	n-Bu	Zn(CF$_3$SO$_3$)$_2$	48
14	10l	t-Bu	NO$_2$	n-Bu	Zn(CF$_3$SO$_3$)$_2$	47

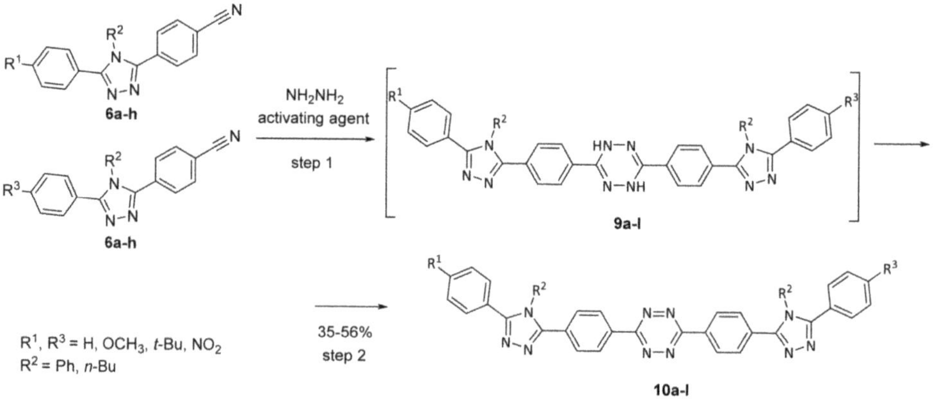

Scheme 3. Synthesis of s-tetrazine derivatives conjugated via a 1,4-phenylene linker with a 4H-1,2,4-triazole ring (**10a–l**). Reaction conditions: step 1: two precursors (**6a–h**, 0.5 mmol of each compound), activating agent (zinc trifluoromethanesulfonate (0.009 g, 5 mol%) or sulfur (0.02 g, 125 mol%), ethanol (25 mL), hydrazine hydrate (hydrazine 64%, 0.1 mL), reflux 12 h; step 2: methanol (10 mL), hydrogen peroxide (solution 34.5–36.5%, 11 mL), rt, 24 h.

The next step was the synthesis of products in which s-tetrazine is directly linked to the 4H-1,2,4-triazole ring. As part of our previous research, we had already obtained similar compounds containing 1,3,4-oxadiazole and 1,3,4-thiadiazole, but their synthesis required the use of microwave irradiation [27]. The methodology was based on the use of commercially available ethyl diazoacetate (**11**), which was transformed into a dihydrazide (**12**) in a sequence of several transformations (Scheme 4). The product was then treated with acid chlorides to prepare bisdiacyl derivatives (**13**). In this case, too, we intended to convert these compounds into imidoyl chlorides (**14**), which could then be cyclized to triazoles (**15a**) under the influence of amines. However, the high reactivity of such derivatives again caused serious difficulties. Despite the maximum shortening of the reaction times, which had a beneficial effect in previous studies, the observed undesirable derivatives of 1,3,4-oxadiazole (**16**) were predominantly formed. Additionally, isolation of the desired product from the reaction mixture was extremely problematic and, as a result, only traces of the target compound were obtained.

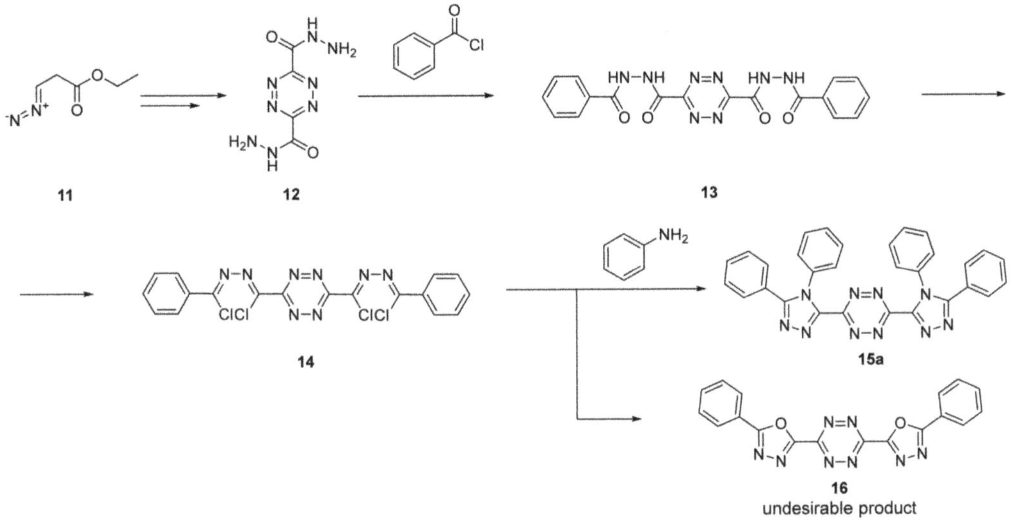

Scheme 4. An attempt to synthesize s-tetrazine derivatives directly conjugated to the 4H-1,2,4-triazole ring.

Based on the experience of obtaining triazole precursors for the Pinner reaction, where we encountered a similar problem, we decided to use an alternative methodology. For this purpose, the dihydrazide **12** was reacted with a range of imidoyl chlorides (**8a–h**) previously obtained from amides (Scheme 5). This approach was effective for both systems containing an aromatic ring (**15a–d**) and an aliphatic chain (**15e–h**) on the triazole nitrogen atom. In addition, derivatives containing both electron-donating and electron-withdrawing moieties attached to a terminal aromatic ring were obtained. Compared to the unsubstituted products, the electron-withdrawing nitro group showed a decreased yield (Entries 4 and 8, Table 2), while for the electron-donating groups (methoxy and *tert*-butyl) the yield was increased (Entries 2, 3, 6, 7, Table 2). The presence of an aliphatic chain also had a beneficial effect on the reaction yield (Entries 5–8, Table 2).

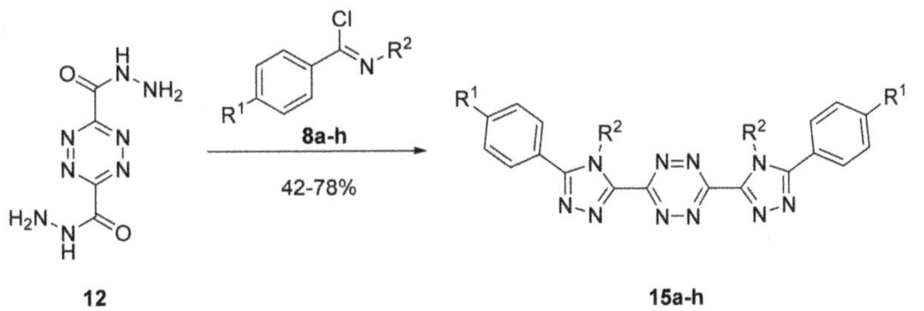

R^1 = H, OCH$_3$, *t*-Bu, NO$_2$
R^2 = Ph, *n*-Bu

Scheme 5. Synthesis of s-tetrazine derivatives directly conjugated to the 4H-1,2,4-triazole ring (**15a–h**). Reaction conditions: 1,2,4,5-tetrazine-3,6-dicarbohydrazide (**12**, 0.50 g, 2.5 mmol), imidoyl chloride (**8a–h**, 5.5 mmol), chloroform (20 mL), reflux, 24 h.

Table 2. The yield of the reaction for the preparation of s-tetrazine derivatives directly conjugated to the 4H-1,2,4-triazole ring (15a–h).

Entry	Product	R¹	R²	Yield [%]
1	15a	H	Ph	45
2	15b	OCH₃	Ph	68
3	15c	t-Bu	Ph	59
4	15d	NO₂	Ph	40
5	15e	H	n-Bu	56
6	15f	OCH₃	n-Bu	78
7	15g	t-Bu	n-Bu	73
8	15h	NO₂	n-Bu	42

The structure of all the obtained intermediates and final products was confirmed by ^{1}H- and ^{13}C-NMR spectroscopy. Both in the case of systems containing a 1,4-phenylene linker, and with directly conjugated heterocycles, the ^{13}C-NMR spectra were the most characteristic. The presence of the 4H-1,2,4-triazole ring was confirmed by signals above 140 ppm, and the presence of the s-tetrazine ring by signals above 160 ppm. The introduction of individual groups to the terminal aromatic ring conditioned the appearance of specific signals for the benzene carbon attached to them: above 160 ppm for the methoxy group, above 150 ppm for the *tert*-butyl group, and above 140 ppm for the nitro group. The lowest shifts corresponded to the carbon atoms of the aliphatic chain (13–45 ppm), the methoxy group (about 55 ppm), and the *tert*-butyl group (30–35 ppm). The ^{1}H-NMR spectra mainly included aromatic signals. Additionally, the protons of the aliphatic chain (butyl) gave a series of signals in the range of 0.6–4.5 ppm, the methoxy group a peak around 3.8 ppm, and the *tert*-butyl group a peak around 1.3 ppm.

2.2. Luminescent Properties

UV-Vis and 3D fluorescence spectra were registered for compounds 10a–l and 15a–h (Figures S40–S64, Supplementary Materials). The fluorescence was completely quenched in the case of 15d and 15h, due to the presence of two NO₂ groups in their structure. The rest of the compounds exhibited a maximum of one emission. The range of emission wavelengths is 375–412 nm for the 10a–l series (Entries 1–12, Table 3) and 353–375 nm for the 15a–h series (Entries 13–20, Table 3). It shows that the separation of fluorophore moieties by phenyl ring leads to a bathochromic shift of fluorescence. In the tetrazine and triazole derivatives, the n→π* transitions are a source of fluorescence [30–33]. The location of emission maximum (excitation wavelength—λ_{ex} and emission wavelength—λ_{em}) is dependent on substituents R¹, R², and R³, which indicates that both tetrazine and triazole rings are involved in the orbitals from which the excitation occurs. The influence of substituents on λ_{ex} and λ_{em} is the same as in previously reported symmetrically substituted analogs of the 10a–l series [26]. The R² affects the λ_{ex}, whereas R¹ and R³ affect the λ_{em}. The Ph substituent as R² induces the bathochromic shift of λ_{ex} (Entries 1–6 and 13–16, Table 3) in comparison to n-Bu (Entries 7–12 and 17–20, Table 3, red color vs. blue color in Figure S65). In the case of the 15a–h series, which consists of the symmetrically substituted compounds, the λ_{em} increases together with the rising electron-donating strength of R¹ (H < t-Bu < OCH₃), which is typical for tetrazine derivatives [34,35]. A partially similar relationship is observed in the unsymmetrically substituted 10a–l series. Taking into account compounds with the same substituent as one of R¹/R³, e.g., NO₂, the λ_{em} shifts bathochromically in line with the electron-donating properties of the second R¹/R³ substituent, i.e., H < t-Bu < OCH₃. However, there are some exceptions to that rule in this series because, compared to compounds containing OCH₃/ t-Bu and OCH₃/NO₂ substituents (10d vs. 10e and 10j vs. 10k, Entries 4, 5, 10 and 11, Table 3), those with NO₂ (which is an electron-withdrawing group) unexpectedly possess a larger λ_{em}. This shows that the changes in the electron density distribution induced by different substituents in unsymmetrically substituted compounds are difficult to predict, thus inferring their absorption-emission properties based only on a molecular structure can be misleading. The quantum yield (Φ) is directly related to the fluorescence intensity for the studied compounds (Figure S66). Generally, the compounds with Ph as R^2 exhibit higher Φs than those with *n*-Bu, which is in agreement with previous findings [26]. However, most of the studied compounds are not efficient fluorescent materials, because their Φs do not exceed 0.3 (Table 3). The relatively favorable conjugation occurs only for three compounds, i.e., **10a**, **10b**, and **10d**. It shows that the direct coupling of tetrazine and triazole rings, as well as *n*-Bu as R^2 and NO_2 as R^1/R^3, decreases the population of fluorescent transitions.

Table 3. Spectroscopic data for the studied *s*-tetrazine derivatives. λ_{abs}—wavelength of absorption maximum directly preceding λ_{em}. λ_{ex} and λ_{em}—excitation and emission wavelength at global fluorescence maximum. Stokes shift was calculated as $\lambda_{em} - \lambda_{abs}$. UV-Vis absorption and 3D fluorescence spectra were registered in dichloromethane solutions (c = 5×10^{-6} mol/dm^3). The quantum yields Φ were determined according to the method described by Brouwer [36] by comparison with two standards: quinine sulphate (qn-SO_4^{2-}) [37] and *trans,trans*-1,4-diphenyl-1,3-butadiene (dpb) [38].

Entry	Compound	λ_{abs} (nm)	ε (mol^{-1} dm^3 cm^{-1})	λ_{ex} (nm)	λ_{em} (nm)	Stokes Shift (nm)	Φ qn-SO_4^{2-}	Φ dpb
1	10a	283	43,774	295	386	103	0.50	0.49
2	10b	284	43,560	300	382	98	0.70	0.69
3	10c	293	50,920	294	375	82	0.24	0.24
4	10d	287	41,880	302	391	104	0.67	0.66
5	10e	303	48,280	303	409	106	0.14	0.14
6	10f	292	44,760	299	384	92	0.29	0.28
7	10g	242	32,860	288	399	157	0.19	0.19
8	10h	232	32,680	291	378	146	0.20	0.20
9	10i	236	21,760	291	375	139	0.04	0.04
10	10j	253	38,120	288	396	143	0.05	0.05
11	10k	257	32,260	304	412	155	0.03	0.03
12	10l	239	36,940	296	386	147	0.22	0.21
13	15a	278	30,180	297	354	76	0.26	0.26
14	15b	256	36,160	309	375	119	0.26	0.25
15	15c	276	32,600	298	362	86	0.30	0.29
16	15d	298	31,300	-	-	-	-	-
17	15e	257	27,900	270	353	96	0.07	0.07
18	15f	253	37,100	284	373	120	0.21	0.20
19	15g	259	15,100	283	361	102	0.11	0.10
20	15h	269	12,800	-	-	-	-	-

Summarizing the current and previous research on *s*-tetrazine derivatives in terms of their Φs, it can be stated that they are moderately efficient fluorescent materials. Most of the investigated tetrazine derivatives exhibit Φ no higher than 0.60, but there are some examples, which achieve Φ close to 1, which shows their great potential to use as functional materials, e.g., in optoelectronic applications. In the case of *s*-tetrazines conjugated via phenylene linkers with different 5-membered rings (Scheme 6, Table 4), the Φ changes approximately according to the following order, Triazole (R^2 = *n*-Bu) < Oxadiazole ≤ Thiadiazole < Triazole (R^2 = Ph). On the other hand, the analogical order for *s*-tetrazines directly conjugated with the same 5-membered rings is as follows, Triazole (R^2 = *n*-Bu) < Triazole (R^2 = Ph) < Oxadiazole < Thiadiazole (Scheme 7, Table 5). The greatest similarities are between oxadiazoles and thiadiazoles bearing *s*-tetrazine, due to small structural changes resulting from the replacement of oxygen with sulfur (atoms with similar electronic properties). Notably, the separation of tetrazine rings and triazole rings via phenylene linkers is more favorable for the fluorescence efficiency than the direct conjugation of them. This is in agreement with the study on the nature of the absorption–emission properties of tetrazine derivatives, which revealed that fluorescence is dependent on the character of HOMO and HOMO-1 orbitals [34]. Fluorescence occurs when the orbital involved in the excitation has a nonbonding n character, but if it is π orbital, the fluorescence is quenched. In this research, it was found that tetrazine derivatives directly conjugated with heteroatomic rings did not exhibit fluorescence, while diphenyl s-tetrazine was reported to be weakly fluorescent [34,39]. It showed that the conjugation with phenyl rings allows for the retention of the nonbonding n character of the excited orbitals, whereas the direct conjugation with heteroatomic rings changes its character to the π one.

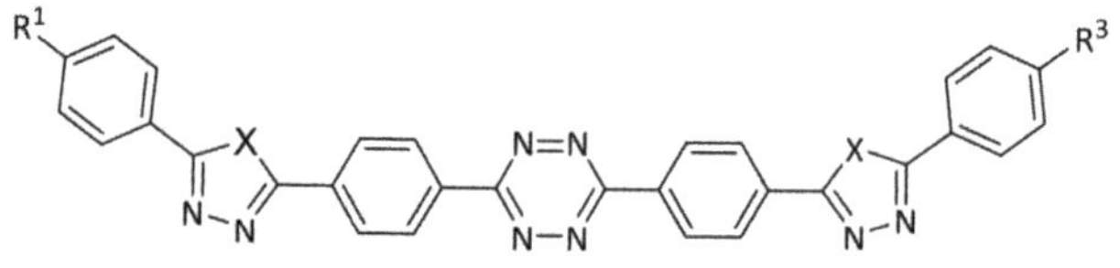

R^1, R^3 = H, OCH$_3$, t-Bu, NO$_2$
X = O, S, N-R^2
R^2 = Ph, n-Bu

Scheme 6. Structure of s-tetrazine derivatives conjugated via phenylene linkers with oxadiazole, thiadiazole, and triazole rings.

Table 4. Comparison of the quantum yields of s-tetrazine derivatives conjugated via phenylene linkers with oxadiazole [24], thiadiazole [25], and triazole rings (symmetrically substituted from [26], and unsymmetrically substituted from current work).

Entry	R^1	R^3	Oxadiazole	Thiadiazole	Triazole R^2 = Ph	Triazole R^2 = n-Bu
1	H	H	0.09	0.46	0.69	0.59
2	OCH$_3$	OCH$_3$	0.39	0.60	>0.98	0.49
3	t-Bu	t-Bu	0.43	0.58	0.33	0.51
4	NO$_2$	NO$_2$	0.09	0.14	0.02	0.02
5	H	OCH$_3$	0.41	0.44	0.50	0.19
6	H	t-Bu	0.51	0.40	0.70	0.20
7	H	NO$_2$	0.57	0.26	0.24	0.04
8	OCH$_3$	t-Bu	0.54	0.53	0.67	0.05
9	OCH$_3$	NO$_2$	0.39	0.38	0.14	0.03
10	t-Bu	NO$_2$	0.05	0.26	0.29	0.22

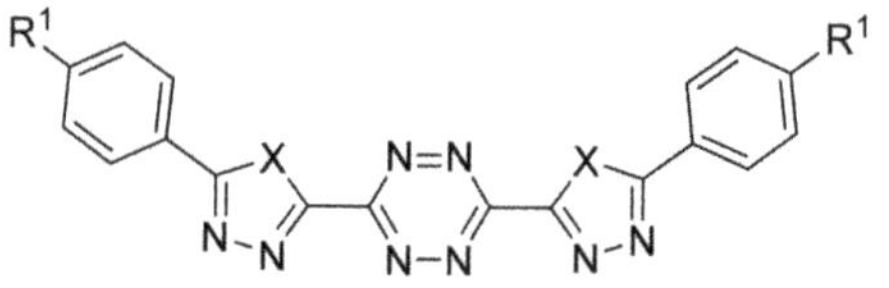

R^1 = H, OCH$_3$, t-Bu, NO$_2$
X = O, S, N-R^2
R^2 = Ph, n-Bu

Scheme 7. Structure of s-tetrazine derivatives directly conjugated with oxadiazole, thiadiazole, and triazole rings.

Table 5. Comparison of the quantum yields of *s*-tetrazine derivatives directly conjugated with oxadiazole [27], thiadiazole [27], and triazole rings (current work).

Entry	R1	Oxadiazole	Thiadiazole	Triazole R^2 = Ph	Triazole R^2 = *n*-Bu
1	H	0.10	0.74	0.26	0.07
2	OCH$_3$	>0.98	>0.98	0.26	0.21
3	*t*-Bu	*	*	0.30	0.11
4	NO$_2$	0.08	0.50	-	-

* compound was not synthesized.

3. Experimental Section

3.1. General Information

All reagents were purchased from commercial sources and used without further purification. Melting points were measured on a Stuart SMP3 melting point apparatus (Staffordshire, UK). NMR spectra were recorded at 25 °C on an Agilent 400-NMR spectrometer (Agilent Technologies, Waldbronn, Germany) at 400 MHz for ^{1}H and 100 MHz for ^{13}C, using CDCl$_3$ or DMSO as the solvent and TMS as the internal standard. UV-Vis absorption and 3D fluorescence spectra were registered in dichloromethane solutions (c = 5 × 10^{-6} mol/dm^3) with Jasco V-660 (Jasco Corporation, Tokyo, Japan) and Jasco F-6300 (Jasco Corporation, Tokyo, Japan) spectrometers, respectively. FT-IR spectra were measured between 4000 and 650 cm^{-1} on an FT-IR Nicolet 6700 apparatus (Thermo Fischer Scientific, Wesel, Germany) with a Smart iTR accessory. Elemental analyses were performed with a VarioELanalyser (Elementar UK Ltd., Stockport, UK). High-resolution mass spectra were obtained by means of a Waters ACQUITY UPLC/Xevo G2QT instrument (Waters Corporation, Milford, MA, USA). Thin-layer chromatography was performed on silica gel 60 F254 (Merck, Merck KGaA, Darmstadt, Germany) thin-layer chromatography plates using chloroform, chloroform/ethyl acetate (1:1 *v*/*v*), or chloroform/ethyl acetate (5:1 *v*/*v*) as the mobile phases.

3.2. Synthesis and Characterization

Compounds **6**, **8** and **12** were synthesized according to the literature [26,27].

3.2.1. Synthesis of *s*-Tetrazine Derivatives Coupled via a 1,4-Phenylene Linkage with a 4*H*-1,2,4-Triazole Ring (**10a–l**)

Two of substrates (**6a–h**, 0.5 mmol of each compound) and zinc trifluoromethanesulfonate (0.009 g, 5 mol%) were suspended in ethanol (25 mL) and hydrazine hydrate (hydrazine 64%, 0.1 mL) was added dropwise. It was heated under reflux for 12 h, then filtered and evaporated on a rotary evaporator. The obtained crude intermediate (**9a–l**) was dissolved in methanol (10 mL), hydrogen peroxide was added (hydrogen peroxide solution 34.5–36.5%, 11 mL), and it was stirred at room temperature for 24 h. The resulting mixture was filtered and concentrated on a rotary evaporator. The crude product (**10a–l**) was purified by column chromatography using chloroform/ethyl acetate (1:1 *v*/*v*) as the mobile phases.

3-(4-(4,5-Diphenyl-4*H*-1,2,4-triazol-3-yl)phenyl)-6-(4-(5-(4-methoxyphenyl)-4-phenyl-4*H*-1,2,4-triazol-3-yl)phenyl)-1,2,4,5-tetrazine (**10a**)

The product was obtained as yellow powder (0.20 g, 56%); m.p. 187–188 °C. UV (CH$_2$Cl$_2$) λ_{max} (log ε) 257 (4.76), 283 (4.64) nm; IR (ATR) ν_{max} 3064, 2947, 2232, 2187, 2141, 2129, 2098, 1696, 1683, 1609, 1565, 1533, 1494, 1472, 1445, 1256, 1179, 1077, 1019, 991, 972, 932, 848, 790, 772, 751, 730, 713, 699, 678 cm^{-1}; ^{1}H-NMR (400 MHz, CDCl$_3$): δ 3.79 (s, 3H, OCH$_3$), 6.81 (d, 2H, *J* = 8.0 Hz, Ar), 7.17–7.21 (m, 2H, Ar), 7.30–7.38 (m, 7H, Ar), 7.52–7.58 (m, 12H, Ar), 7.76 (d, 2H, *J* = 12.0 Hz, Ar), 8.11 (d, 2H, *J* = 8.0 Hz, Ar); ^{13}C-NMR (100 MHz, CDCl$_3$): δ 55.3, 113.4, 114.1, 117.6, 117.7, 118.1 126.2, 127.7, 127.7, 128.6, 128.8, 129.0, 129.9, 130.1,

130.2, 130.3, 130.3, 130.4, 130.4, 131.1, 132.2, 123.5, 134.2, 134.7, 152.7, 153.0, 155.3, 155.5, 161.0, 169.2, 171.1. Anal. calc. for $C_{43}H_{30}N_{10}O$: C, 73.49; H, 4.30; N, 19.93. Found: C, 73. 46; H, 4.32; N, 19.91; HRMS (ESI): m/z calcd for $C_{43}H_{30}N_{10}O + H^+$: 703.2682; found: 703.2684.

3-(4-(5-(4-(*tert*-Butyl)phenyl)-4-phenyl-4*H*-1,2,4-triazol-3-yl)phenyl)-6-(4-(4,5-diphenyl-4*H*-1,2,4-triazol-3-yl)phenyl)-1,2,4,5-tetrazine (**10b**)

The product was obtained as pink powder (0.19 g, 52%); m.p. 174–175 °C. UV (CH_2Cl_2) λ_{max} (log ε) 284 (4.64) nm; IR (ATR) ν_{max} 3062, 2964, 2868, 2232, 2167, 2155, 2028, 2007, 1966, 1695, 1610, 1527, 1494, 1473, 1435, 1362, 1305, 1269, 1201, 1181, 1156, 1108, 1078, 1019, 973, 932, 850, 837, 790, 773, 749, 730, 699 cm^{-1}; ^{1}H-NMR (400 MHz, CDCl$_3$): δ 1.28 (s, 9H, C(CH$_3$)$_3$), 7.18 (t, 2H, J = 8.0 Hz, Ar), 7.29–7.39 (m, 9H, Ar), 7.48–7.57 (m, 14H, Ar), 7.78 (d, 2H, J = 8.0 Hz, Ar); ^{13}C-NMR (100 MHz, CDCl$_3$): δ 31.1, 34.8, 113.2, 113.3, 118.1, 118.1, 123.4, 125.5, 126.2, 127.6, 127.7, 128.3, 128.5, 128.8, 128.9, 129.0, 129.0, 130.0, 130.2, 130.2, 131.3, 132.2, 132.4, 134.7, 134.9, 152.9, 153.0, 153.3, 155.5, 155.5, 166.7, 167.6. Anal. calc. for $C_{46}H_{36}N_{10}$: C, 75.80; H, 4.98; N, 19.22. Found: C, 75.81; H, 4.99; N, 19.20; HRMS (ESI): m/z calcd for $C_{46}H_{36}N_{10} + H^+$: 729.3203; found: 729.3202.

3-(4-(4,5-Diphenyl-4*H*-1,2,4-triazol-3-yl)phenyl)-6-(4-(5-(4-nitrophenyl)-4-phenyl-4*H*-1,2,4-triazol-3-yl)phenyl)-1,2,4,5-tetrazine (**10c**)

The product was obtained as yellow powder (0.18 g, 49%); m.p. 199–200 °C. UV (CH_2Cl_2) λ_{max} (log ε) 293 (4.71) nm; IR (ATR) ν_{max} 3053, 2232, 2172, 2142, 2129, 2003, 1965, 1698, 1608, 1550, 1515, 1494, 1468, 1446, 1428, 1406, 1337, 1317, 1277, 1202, 1181, 1152, 1108, 1078, 1018, 1002, 973, 933, 848, 790, 773, 760, 739, 713, 698, 685 cm^{-1}; ^{1}H-NMR (400 MHz, CDCl$_3$): δ 7.22 (d, 2H, J = 8.0 Hz, Ar), 7.54–7.63 (m, 21H, Ar), 7.88 (d, 2H, J = 8.0 Hz, Ar), 8.16 (d, 2H, J= 8.0 Hz, Ar); ^{13}C-NMR (100 MHz, CDCl$_3$): δ 113.3, 113.8, 117.9, 118.1, 123.8, 126.4, 127.5, 127.7, 128.1, 128.5, 128.8, 129.0 129.0, 129.4, 130.1, 130.2, 130.4, 130.8, 131.3, 132.2, 132.3, 134.2, 134.8, 148.4, 153.0, 153.5, 153.8, 155.5, 163.6, 164.0. Anal. calc. for $C_{42}H_{27}N_{11}O_2$: C, 70.28; H, 3.79; N, 21.47. Found: C, 70.25; H, 3.77; N, 21.45; HRMS (ESI): m/z calcd for $C_{42}H_{27}N_{11}O_2 + H^+$: 718.2427; found: 718.2425.

3-(4-(5-(4-(*tert*-Butyl)phenyl)-4-phenyl-4*H*-1,2,4-triazol-3-yl)phenyl)-6-(4-(5-(4-methoxyphenyl)-4-phenyl-4*H*-1,2,4-triazol-3-yl)phenyl)-1,2,4,5-tetrazine (**10d**)

The product was obtained as pink powder (0.21 g, 56%); m.p. 159–160 °C. UV (CH_2Cl_2) λ_{max} (log ε) 238 (4.56), 287 (4,62) nm; IR (ATR) ν_{max} 3060, 2966, 2268, 2232, 2172, 2140, 2032, 2003, 1972, 1948, 1911, 1690, 1609, 1565,1531, 1496, 1475, 1459, 1434, 1362, 1305, 1254, 1200, 1175, 1156, 1099, 1076, 1020, 992, 972, 920, 851, 837, 789, 774, 749, 737, 714, 699 cm^{-1}; ^{1}H-NMR (400 MHz, CDCl$_3$): δ 1.28 (s, 9H, C(CH$_3$)$_3$), 3.79 (s, 3H, OCH$_3$), 6.80 (d, 2H, J = 8.0 Hz, Ar), 7.17 (m, 4H, Ar), 7.29–735 (m, 6H, Ar), 7.49–7.58 (m, 14H, Ar); ^{13}C-NMR (100 MHz, CDCl$_3$): δ 31.1, 34.8, 55.3, 113.2, 114.0, 118.1, 118.6, 120.2, 123.3, 125.5, 125.7, 127.7, 127.7, 128.3, 129.0, 129.9, 130.2, 130.2, 130.4, 131.3, 132.2, 132.4, 132.5, 132.9, 134.9, 152.9, 153.3, 154.6, 155.4, 155.5, 160.9, 164.0, 154.8. Anal. calc. for $C_{47}H_{38}N_{10}O$: C, 74.39; H, 5.05; N, 18.46. Found: C, 74.38; H, 5.07; N, 18.44; HRMS (ESI): m/z calcd for $C_{47}H_{38}N_{10}O + H^+$: 759.3308; found: 759.3309.

3-(4-(5-(4-Methoxyphenyl)-4-phenyl-4*H*-1,2,4-triazol-3-yl)phenyl)-6-(4-(5-(4-nitrophenyl)-4-phenyl-4*H*-1,2,4-triazol-3-yl)phenyl)-1,2,4,5-tetrazine (**10e**)

The product was obtained as orange powder (0.20 g, 54%); m.p. 189–190 °C. UV (CH_2Cl_2) λ_{max} (log ε) 303 (4.68) nm; IR (ATR) ν_{max} 3073, 2957, 2228, 2175, 2138, 2030, 2014, 1978, 1960, 1697, 1684, 1607, 1577, 1515, 1493, 1472, 1434, 1407, 1337, 1316, 1288, 1253, 1178, 1108, 1068, 1021, 992, 972, 848, 834, 784, 771, 752, 741, 698 cm^{-1}; ^{1}H-NMR (400 MHz, DMSO-d$_6$): δ 3.75 (s, 3H, OCH$_3$), 6.92 (d, 2H, J = 8.0 Hz, Ar), 7.33 (d, 2H, J = 12.0 Hz, Ar), 7.47–7.59 (m, 10H, Ar), 7.67 (d, 4H, J = 8.0 Hz, Ar), 7.88 (d, 4H, J = 8.0 Hz, Ar), 8.23 (d,

4H, J = 8.0 Hz, Ar); ^{13}C-NMR (100 MHz, DMSO-d$_6$): δ 55.2, 112.4, 113.95, 116.5, 118.1, 118.8, 123.7, 127.6, 128.1, 128.9, 129.1, 129.5, 129.7, 129.9, 130.0, 130.2, 130.4, 131.0, 131.5, 132.4, 132.5, 133.9, 134.6, 147.9, 152.6, 153.2, 153.6, 154.7, 160.3, 161.2, 163.3. Anal. calc. for C$_{43}$H$_{29}$N$_{11}$O$_3$: C, 69.07; H, 3.91; N, 20.60. Found: C, 69.09; H, 3.94; N, 20.58; HRMS (ESI): m/z calcd for C$_{43}$H$_{29}$N$_{11}$O$_3$ + H$^+$: 748.2533; found: 748.2531.

3-(4-(5-(4-(*tert*-Butyl)phenyl)-4-phenyl-4*H*-1,2,4-triazol-3-yl)phenyl)-6-(4-(5-(4-nitrophenyl)-4-phenyl-4*H*-1,2,4-triazol-3-yl)phenyl)-1,2,4,5-tetrazine (**10f**)

The product was obtained as orange powder (0.20 g, 52%); m.p. 124–125 °C. UV (CH$_2$Cl$_2$) λ$_{max}$ (log ε) 292 (4.65) nm; IR (ATR) ν$_{max}$ 3062, 2962, 2229, 2159, 2136, 2127, 2099, 2028, 1989, 1974, 1966, 1700, 1608, 1523, 1498, 1476, 1433, 1407, 1338, 1268, 1200, 1156, 1109, 1075, 1019, 972, 842, 787, 771, 752, 739, 729, 698 cm^{-1}; ^{1}H-NMR (400 MHz, CDCl$_3$): δ 1.28 (s, 9H, C(CH$_3$)$_3$), 7.23 (d, 4H, J = 8.0 Hz, Ar), 7.32 (d, 2H, J = 8.0 Hz, Ar), 7.37 (d, 2H, J = 8.0 Hz, Ar), 7.48–7.64 (m, 16H, Ar), 8.16 (d, 2H, J = 8.0 Hz, Ar); ^{13}C-NMR (100 MHz, CDCl$_3$): δ 31.1, 34.9, 113.6, 113.8, 117.9, 118.0, 122.2, 123.8, 125.7, 127.5, 127.8, 128.6, 129.1, 129.2, 129.4, 130.5, 130.6, 130.8, 130.9, 132.2, 132.3, 134.2, 134.3, 148.5, 152.9, 153.5, 153.8, 154.0, 155.2, 164.6, 166.0. Anal. calc. for C$_{46}$H$_{35}$N$_{11}$O$_2$: C, 71.40; H, 4.56; N, 19.91. Found: C, 71.42; H, 4.54; N, 19.90; HRMS (ESI): m/z calcd for C$_{46}$H$_{35}$N$_{11}$O$_2$ + H$^+$: 774.3054; found: 774.3056.

3-(4-(4-Butyl-5-(4-methoxyphenyl)-4*H*-1,2,4-triazol-3-yl)phenyl)-6-(4-(4-butyl-5-phenyl-4*H*-1,2,4-triazol-3-yl)phenyl)-1,2,4,5-tetrazine (**10g**)

The product was obtained as orange powder (0.16 g, 52%); m.p. 173–174 °C. UV (CH$_2$Cl$_2$) λ$_{max}$ (log ε) 242 (4.52) nm; IR (ATR) ν$_{max}$ 3103, 3075, 3053, 2329, 2231, 2175, 2138, 1945, 1695, 1682, 1607, 1566, 1504, 1403, 1317, 1294, 1243, 1176, 1130, 1112, 1052, 1024, 990, 869, 856, 844, 769, 751, 676 cm^{-1}; ^{1}H-NMR (400 MHz, CDCl$_3$): δ 0.63–0.67 (m, 6H, CH$_3$), 0.91–0.93 (m, 4H CH$_2$), 1.36–1.43 (m, 4H, CH$_2$), 3.87 (s, 3H, OCH$_3$), 3.93–3.96 (m, 4H, CH$_2$), 7.13 (d, 2H, J = 8.0 Hz, Ar), 7.73–7.78 (m, 9H, Ar), 7.90 (d, 2H, J = 8.0 Hz, Ar), 7.97–8.01 (m, 4H, Ar); ^{13}C-NMR (100 MHz, CDCl$_3$): δ 14.2, 14.3, 22.7, 22.8, 29.9, 30.4, 47.6, 47.6, 55.7, 113.7, 114.1, 117.7, 117.8, 128.2, 129.5, 129.7, 130.0, 130.2, 130.4, 131.7, 131.9, 132.3, 133.1, 134.3, 151.9, 152.1, 154.6, 155.4, 162.2, 164.1, 165.1. Anal. calc. for C$_{39}$H$_{38}$N$_{10}$O: C, 70.67; H, 5.78; N, 21.13. Found: C, 70.69; H, 5.75; N, 21.11; HRMS (ESI): m/z calcd for C$_{39}$H$_{38}$N$_{10}$O + H$^+$: 663.3308; found: 663.3309.

3-(4-(4-Butyl-5-(4-(*tert*-butyl)phenyl)-4*H*-1,2,4-triazol-3-yl)phenyl)-6-(4-(4-butyl-5-phenyl-4*H*-1,2,4-triazol-3-yl)phenyl)-1,2,4,5-tetrazine (**10h**)

The product was obtained as pink powder (0.16 g, 47%); m.p. 90–91 °C. UV (CH$_2$Cl$_2$) λ$_{max}$ (log ε) 232 (4.51) nm; IR (ATR) ν$_{max}$ 3316, 3067, 2958, 2932, 2867, 2231, 2193, 2170, 2134, 2034, 1978, 1959, 1721, 1637, 1578, 1541, 1490, 1465, 1395, 1364, 1308, 1275, 1249, 1221, 1178, 1154, 1109, 1074, 1018, 993, 946, 845, 803, 772, 694 cm^{-1}; ^{1}H-NMR (400 MHz, CDCl$_3$): δ 0.63–0.67 (m, 6H, CH$_3$), 0.92–0.95 (m, 4H CH$_2$), 1.36–1.41 (m, 13H, CH$_2$, C(CH$_3$)$_3$), 3.42–3.46 (m, 4H, CH$_2$), 7.45–7.50 (m, 3H, Ar), 7.53 (d, 2H, J = 8.0 Hz, Ar), 7.60 (d, 2H, J = 8.0 Hz, Ar), 7.68–7.71 (m, 2H, Ar), 7.74–7.77 (m, 4H, Ar), 7.84 (d, 4H, J = 4.0 Hz, Ar); ^{13}C-NMR (100 MHz, CDCl$_3$): δ 13.1, 13.8, 19.3, 20.2, 31.2, 31.2, 21.7, 34.9, 44.8, 44.9, 113.9, 114.7, 118.0, 118.1, 124.0, 126.9, 127.8, 128.5, 129.4, 129.5, 129.9, 130.1, 131.3, 132.3, 132.5, 153.7, 153.8, 154.8, 155.3, 156.4, 167.5, 167.6. Anal. calc. for C$_{42}$H$_{44}$N$_{10}$: C, 73.23; H, 6.44; N, 20.33. Found: C, 73.21; H, 6.46; N, 20.32; HRMS (ESI): m/z calcd for C$_{42}$H$_{44}$N$_{10}$ + H$^+$: 689.3829; found: 689.3827.

3-(4-(4-Butyl-5-(4-nitrophenyl)-4*H*-1,2,4-triazol-3-yl)phenyl)-6-(4-(4-butyl-5-phenyl-4*H*-1,2,4-triazol-3-yl)phenyl)-1,2,4,5-tetrazine (**10i**)

The product was obtained as orange powder (0.15 g, 45%); m.p. 183–184 °C. UV (CH$_2$Cl$_2$) λ$_{max}$ (log ε) 236 (4.34) nm; IR (ATR) ν$_{max}$ 3307, 3067, 2958, 2928, 2872, 2231,

2173, 2136, 1697, 1637, 1602, 1578, 1526, 1490, 1466, 1346, 1307, 1248, 1178, 1108, 1074, 1016, 995, 853, 803, 771, 753, 694 cm^{-1}; ^{1}H-NMR (400 MHz, CDCl$_3$): δ 0.94–0.97 (m, 6H, CH$_3$), 1.36–1.46 (m, 4H CH$_2$), 1.57–1.64 (m, 4H, CH$_2$), 3.43–3.48 (m, 4H, CH$_2$), 7.42 (t, 3H, J = 8.0 Hz, Ar), 7.47–7.49 (m, 2H, Ar), 7.74–7.76, m, 8H, Ar), 7.94 (d, 2H, J = 12.0 Hz, Ar), 8.24 (d, 2H, J = 8.0 Hz, Ar); ^{13}C-NMR (100 MHz, CDCl$_3$): δ 13.1, 13.8, 19.3, 20.2, 31.6, 31.7, 39.9, 40.2, 114.4, 114.8, 117.9, 119.0, 123.7, 126.8, 127.7, 128.2, 128.6, 129.5, 129.9, 130.4, 131.4, 132.5, 132.9, 149.5, 151.8, 152.0, 154.8, 155.1, 165.6, 167.7. Anal. calc. for C$_{38}$H$_{35}$N$_{11}$O$_2$: C, 67.34; H, 5.21; N, 22.73. Found: C, 67.33; H, 5.23; N, 22.74; HRMS (ESI): m/z calcd for C$_{38}$H$_{35}$N$_{11}$O$_2$ + H$^+$: 678.3054; found: 678.3053.

3-(4-(4-Butyl-5-(4-(*tert*-butyl)phenyl)-4*H*-1,2,4-triazol-3-yl)phenyl)-6-(4-(4-butyl-5-(4-methoxyphenyl)-4*H*-1,2,4-triazol-3-yl)phenyl)-1,2,4,5-tetrazine (**10j**)

The product was obtained as pink powder (0.18 g, 51%); m.p. 95–96 °C. UV (CH$_2$Cl$_2$) λ$_{max}$ (log ε) 253 (4.58) nm; IR (ATR) ν$_{max}$ 3265, 2957, 2871, 2229, 1632, 1607, 1544, 1504, 1464, 1396, 1365, 1307, 1253, 1222, 1176, 1113, 1031, 978, 918, 841, 772 cm^{-1}; ^{1}H-NMR (400 MHz, CDCl$_3$): δ 0.65–0.67 (m, 6H, CH$_3$), 0.92–0.96 (m, 4H CH$_2$), 1.37–1.41 (m, 13H, CH$_2$, C(CH$_3$)$_3$), 3.84 (s, 3H, OCH$_3$), 3.86–3.89 (m, 4H, CH$_2$), 7.04 (d, 2H, J = 8.0 Hz, Ar), 7.54 (d, 2H, J = 8.0 Hz, Ar), 7.61 (d, 2H, J = 4.0 Hz, Ar), 7.66–7.74 (m, 8H, Ar), 7.92 (d, 2H, J = 8.0 Hz, Ar); ^{13}C-NMR (100 MHz, CDCl$_3$): δ 13.2, 13.5, 19.2, 19.7, 31.2, 31.6, 31.8, 34.9, 44.8, 44.9, 55.4, 113.7, 114.6, 117.4, 118.2, 125.4, 126.0, 127.8, 128.6, 128.7, 129.4, 130.5, 131.5, 132.3, 133.0, 153.8, 154.7, 154.8, 155.2, 155.8, 162.0, 167.1, 167.5. Anal. calc. for C$_{43}$H$_{46}$N$_{10}$O: C, 71.84; H, 6.45; N, 19.48. Found: C, 71.86; H, 6.44; N, 19.45; HRMS (ESI): m/z calcd for C$_{43}$H$_{46}$N$_{10}$O + H$^+$: 719.3934; found: 719.3935.

3-(4-(4-Butyl-5-(4-methoxyphenyl)-4*H*-1,2,4-triazol-3-yl)phenyl)-6-(4-(4-butyl-5-(4-nitrophenyl)-4*H*-1,2,4-triazol-3-yl)phenyl)-1,2,4,5-tetrazine (**10k**)

The product was obtained as orange powder (0.17 g, 48%); m.p. 194–195 °C. UV (CH$_2$Cl$_2$) λ$_{max}$ (log ε) 257 (4.51) nm; IR (ATR) ν$_{max}$ 2964, 2842, 2228, 2128, 1601, 1578, 1519, 1437, 1308, 1256, 1171, 1105, 1050, 1033, 1021, 919, 837, 762, 747, 727, 686 cm^{-1}; ^{1}H-NMR (400 MHz, CDCl$_3$): δ 0.65–0.69 (m, 6H, CH$_3$), 0.94–0.99 (m, 4H CH$_2$), 1.38–1.46 (m, 4H, CH$_2$), 3.88–3.95 (m, 7H, CH$_2$, OCH$_3$), 6.92 (d, 2H, J = 8.0 Hz, Ar), 7.60 (d, 2H, J = 12.0 Hz, Ar), 7.72 (d, 2H, J = 8.0 Hz, Ar), 7.81–7.84 (m, 4H, Ar), 7.90–7.93 (m, 4H, Ar), 8.27–8.29 (m, 2H, Ar); ^{13}C-NMR (100 MHz, CDCl$_3$): δ 13.8, 13.8, 20.1, 20.2, 31.8, 31.9, 39.8, 40.2, 55.5, 113.7, 114.7, 117.5, 118.2, 124.1, 128.0, 128.6, 129.1, 129.0, 129.4, 129.6, 130.3, 130.4, 130.7, 133.0, 148.0, 150.8, 150.8, 155.4, 156.2, 160.6, 165.1, 165.8. Anal. calc. for C$_{39}$H$_{37}$N$_{11}$O$_3$: C, 66.18; H, 5.27; N, 21.77. Found: C, 66.15; H, 5.29; N, 21.76; HRMS (ESI): m/z calcd for C$_{39}$H$_{37}$N$_{11}$O$_3$ + H$^+$: 708.3159; found: 707.3157.

3-(4-(4-Butyl-5-(4-(*tert*-butyl)phenyl)-4*H*-1,2,4-triazol-3-yl)phenyl)-6-(4-(4-butyl-5-(4-nitrophenyl)-4*H*-1,2,4-triazol-3-yl)phenyl)-1,2,4,5-tetrazine (**10l**)

The product was obtained as orange powder (0.17 g, 47%); m.p. 99–100 °C. UV (CH$_2$Cl$_2$) λ$_{max}$ (log ε) 239 (4.57) nm; IR (ATR) ν$_{max}$ 3265, 3067, 2958, 2933, 2867, 2230, 2149, 2132, 1636, 1611, 1526, 1501, 1477, 1464, 1395, 1364, 1346, 1304, 1286, 1269, 1200, 1154, 1111, 1016, 977, 841, 773, 751, 710, 693 cm^{-1}; ^{1}H-NMR (400 MHz, CDCl$_3$): δ 0.64–0.69 (m, 6H, CH$_3$), 1.02–1.05 (m, 4H CH$_2$), 1.38–1.44 (m, 13H, CH$_2$, C(CH$_3$)$_3$), 4.12–4.20 (m, 4H, CH$_2$), 7.55 (d, 2H, J = 8.0 Hz, Ar), 7.61 (d, 2H, J = 8.0 Hz, Ar), 7.69–7.71 (m, 4H, Ar), 7.84–7.90 (m, 4H, Ar), 7.96 (d, 2H, J = 12.0 Hz, Ar), 8.25 (d, 2H, J = 12.0 Hz, Ar); ^{13}C-NMR (100 MHz, CDCl$_3$): δ 13.1, 13.8, 19.3, 20.2, 31.2, 31.6, 31.8, 35.0, 39.7, 40.2, 114.8, 114.9, 117.9, 118.1, 123.7, 125.5, 126.1, 126.6, 127.7, 128.2, 128.7, 129.3, 132.0, 132.4, 149.4, 151.2, 151.4, 154.8, 155.2, 155.7, 165.7, 167.5. Anal. calc. for C$_{42}$H$_{43}$N$_{11}$O$_2$: C, 68.74; H, 5.91; N, 20.99. Found: C, 68.75; H, 5.94; N, 20.97; HRMS (ESI): m/z calcd for C$_{42}$H$_{43}$N$_{11}$O$_2$ + H$^+$: 734.3679; found: 734.3678.

3.2.2. Synthesis of s-Tetrazine Derivatives Directly Conjugated with a 4H-1,2,4-Triazole Ring (15a–h)

The crude imidoyl chloride (**8a–h**, 5.5 mmol) and 1,2,4,5-tetrazine-3,6-dicarbohydrazide (**12**, 0.50 g, 2.5 mmol) were dissolved in chloroform (20 mL) and heated under reflux for 24 h. The mixture was then cooled to room temperature, filtered, and evaporated on a rotary evaporator. For systems containing an aromatic ring attached to a triazole nitrogen atom (**15a–d**) and compound **15h**, residue was washed with a small amount of cold ethanol to produce a pure product. For systems with an aliphatic chain, except compound **15h** (**15e–g**), a small amount of ethanol (5 mL) was added, filtered, and the filtrate was evaporated again to give the product as an oil.

3,6-Bis(4,5-diphenyl-4H-1,2,4-triazol-3-yl)-1,2,4,5-tetrazine (**15a**)

The product was obtained as brown powder (0.59 g, 45%); m.p. 208–209 °C. UV (CH$_2$Cl$_2$) λ_{max} (log ε) 278 (4.48) nm; IR (ATR) ν_{max} 3352, 3061, 1967, 1685, 1596, 1541, 1497, 1466, 1444, 1385, 1317, 1261, 1188, 1074, 1017, 1000, 973, 931, 803, 781, 769, 730, 715, 692 cm^{-1}; ^{1}H-NMR (400 MHz, DMSO-d$_6$): δ 7.39–7.45 (m, 12H, Ar), 7.50–7.56 (m, 8H, Ar); ^{13}C-NMR (100 MHz, DMSO-d$_6$): δ 120.3, 123.6, 127.6, 128.3, 128.4, 128.5, 128.6, 134.2, 154.5, 155.3, 164.1. Anal. calc. for C$_{30}$H$_{20}$N$_{10}$: C, 69.22; H, 3.87; N, 26.91. Found: C, 69.25; H, 3.89; N, 26.90; HRMS (ESI): m/z calcd for C$_{30}$H$_{20}$N$_{10}$ + H$^+$: 521.1951; found: 521.1952.

3,6-Bis(5-(4-methoxyphenyl)-4-phenyl-4H-1,2,4-triazol-3-yl)-1,2,4,5-tetrazine (**15b**)

The product was obtained as yellow powder (0.99 g, 68%); m.p. 217–218 °C. UV (CH$_2$Cl$_2$) λ_{max} (log ε) 256 (4.56) nm; IR (ATR) ν_{max} 3308, 3212, 2938, 2840, 2038, 1712, 1697, 1686, 1604, 1578, 1551, 1535, 1512, 1458, 1432, 1363, 1318, 1307, 1276, 1252, 1172, 1105, 1073, 1020, 916, 887, 851, 832, 795, 771, 741, 697 cm^{-1}; ^{1}H-NMR (400 MHz, DMSO-d$_6$): δ 3.83 (s, 6H, OCH$_3$), 7.04 (d, 4H, J = 8.0 Hz, Ar), 7.27–7.51 (m, 10H, Ar), 7.92 (d, 4H, J = 12.0 Hz, Ar); ^{13}C-NMR (100 MHz, DMSO-d$_6$): δ 55.5, 113.9, 120.3, 122.1, 127.7, 129.3, 130.1, 131.1, 153.8, 155.4, 163.0, 165.4. Anal. calc. for C$_{32}$H$_{24}$N$_{10}$O$_2$: C, 66.20; H, 4.17; N, 24.12. Found: C, 66.21; H, 4.19; N, 24.11; HRMS (ESI): m/z calcd for C$_{32}$H$_{24}$N$_{10}$O$_2$ + H$^+$: 581.2162; found: 581.2160.

3,6-Bis(5-(4-(*tert*-butyl)phenyl)-4-phenyl-4H-1,2,4-triazol-3-yl)-1,2,4,5-tetrazine (**15c**)

The product was obtained as orange powder (0.93 g, 59%); m.p. 198–199 °C. UV (CH$_2$Cl$_2$) λ_{max} (log ε) 276 (4.51) nm; IR (ATR) ν_{max} 3058, 2957, 2866, 2238, 2184, 2174, 2019, 1982, 1958, 1697, 1596, 1541, 1495, 1466, 1439, 1394, 1363, 1316, 1269, 1201, 1112, 1076, 1017, 963, 915, 841, 751, 731, 711, 692 cm^{-1}; ^{1}H-NMR (400 MHz, DMSO-d$_6$): δ 1.22 (s, 18H, C(CH$_3$)$_3$), 7.14–7.19 (m, 4H, Ar), 7.24 (d, 4H, J = 8.0 Hz, Ar), 7.47–7.53 (m, 10H, Ar); ^{13}C-NMR (100 MHz, DMSO-d$_6$): δ 29.8, 33.5, 119.2, 124.4, 126.8, 127.0, 127.2, 127.8, 132.9, 151.8, 153.0, 153.3, 163.1. Anal. calc. for C$_{38}$H$_{36}$N$_{10}$: C, 72.13; H, 5.73; N, 22.14. Found: C, 72.11; H, 5.76; N, 22.12; HRMS (ESI): m/z calcd for C$_{38}$H$_{36}$N$_{10}$ + H$^+$: 633.3203; found: 633.3204.

3,6-Bis(5-(4-nitrophenyl)-4-phenyl-4H-1,2,4-triazol-3-yl)-1,2,4,5-tetrazine (**15d**)

The product was obtained as orange powder (0.61 g, 40%); m.p. 172–173 °C. UV (CH$_2$Cl$_2$) λ_{max} (log ε) 298 (4.50) nm; IR (ATR) ν_{max} 3064, 2851, 2206, 2166, 2030, 1983, 1948, 1698, 1653, 1598, 1576, 1520, 1494, 1441, 1343, 1205, 1178, 1108, 1075, 1014, 965, 919, 853, 756, 708, 692 cm^{-1}; ^{1}H-NMR (400 MHz, DMSO-d$_6$): δ 7.61–7.69 (m, 6H, Ar), 7.80 (d, 4H, J = 8.0 Hz, Ar), 8.21 (d, 4H, J = 8.0 Hz, Ar), 8.37 (d, 4H, J = 8.0 Hz, Ar); ^{13}C-NMR (100 MHz, DMSO-d$_6$): δ 120.5, 123.5, 123.7, 124.1, 128.7, 129.2, 138.7, 149.1, 153.2, 153.4, 163.8. Anal. calc. for C$_{30}$H$_{18}$N$_{12}$O$_4$: C, 59.02; H, 2.97; N, 27.53. Found: C, 59.03; H, 2.99; N, 27.51; HRMS (ESI): m/z calcd for C$_{30}$H$_{18}$N$_{12}$O$_4$ + H$^+$: 611.1652; found: 611.1650.

3,6-Bis(4-butyl-5-phenyl-4H-1,2,4-triazol-3-yl)-1,2,4,5-tetrazine (**15e**)

The product was obtained as brown oil (0.67 g, 56%). UV (CH$_2$Cl$_2$) λ_{max} (log ε) 257 (4.45) nm; IR (ATR) ν_{max} 3264, 2957, 2932, 2873, 2212, 2165, 1636, 1541, 1491, 1449, 1378, 1308, 1220, 1157, 1113, 1074, 1026, 930, 802, 772, 694 cm^{-1}; ^{1}H-NMR (400 MHz, CDCl$_3$): δ

0.92 (t, 6H, J = 8.0 Hz, CH$_3$), 1.37 (sextet, 4H, J = 8.0 Hz, CH$_2$), 1.63 (quintet, 4H, J = 8.0 Hz, CH$_2$), 3.48 (t, 4H, J = 8.0 Hz, CH$_2$), 7.39 (t, 4H, J = 8.0 Hz, Ar), 7.51 (t, 2H, J = 8.0 Hz, Ar), 7.85 (d, 4H, J = 8.0 Hz, Ar); ^{13}C-NMR (100 MHz, CDCl$_3$): δ 13.7, 20.1, 31.0, 41.3, 128.1, 128.6, 130.9, 132.8, 149.3, 153.8, 169.8. Anal. calc. for C$_{26}$H$_{28}$N$_{10}$: C, 64.98; H, 5.87; N, 29.15. Found: C, 64.96; H, 5.88; N, 29.17; HRMS (ESI): m/z calcd for C$_{26}$H$_{28}$N$_{10}$ + H$^+$: 481.2577; found: 481.2578.

3,6-Bis(4-butyl-5-(4-methoxyphenyl)-4H-1,2,4-triazol-3-yl)-1,2,4,5-tetrazine (**15f**)

The product was obtained as brown oil (1.05 g, 78%). UV (CH$_2$Cl$_2$) λ$_{max}$ (log ε) 253 (4.57) nm; IR (ATR) ν$_{max}$ 3299, 2957, 2932, 2872, 2213, 2151, 1697, 1608, 1577, 1541, 1506, 1464, 1440, 1365, 1295, 1251, 1176, 1112, 1027, 971, 837, 801, 770 cm^{-1}; ^{1}H-NMR (400 MHz, CDCl$_3$): δ 0.95 (t, 6H, J = 8.0 Hz, CH$_3$), 1.40 (sextet, 4H, J = 8.0 Hz, CH$_2$), 1.59 (quintet, 4H, J = 8.0 Hz, CH$_2$), 4.44 (t, 4H, J = 8.0 Hz, CH$_2$), 3.84 (s, 6H, OCH$_3$), 6.90 (d, 4H, J = 8.0 Hz, Ar), 7.74 (d, 4H, J = 12.0 Hz, Ar); ^{13}C-NMR (100 MHz, CDCl$_3$): δ 13.9, 20.3, 31.9, 39.8, 55.5, 113.8, 114.6, 128.7, 144.6, 155.0, 161.3, 167.1. Anal. calc. for C$_{28}$H$_{32}$N$_{10}$O$_2$: C, 62.21; H, 5.97; N, 25.91. Found: C, 62.24; H, 5.99; N, 25.90; HRMS (ESI): m/z calcd for C$_{28}$H$_{32}$N$_{10}$O$_2$ + H$^+$: 541.2788; found: 541.2789.

3,6-Bis(4-butyl-5-(4-(*tert*-butyl)phenyl)-4H-1,2,4-triazol-3-yl)-1,2,4,5-tetrazine (**15g**)

The product was obtained as brown oil (1.08 g, 73%). UV (CH$_2$Cl$_2$) λ$_{max}$ (log ε) 259 (4.18) nm; IR (ATR) ν$_{max}$ 3265, 2958, 2867, 2240, 2212, 2170, 2049, 1978, 1958, 1698, 1612, 1541, 1504, 1464, 1363, 1302, 1254, 1219, 1177, 1114, 1024, 924, 839, 771, 751 cm^{-1}; ^{1}H-NMR (400 MHz, CDCl$_3$): δ 0.94 (t, 6H, J = 8.0 Hz, CH$_3$), 1,35–1,38 (m, 22H, CH$_2$, C(CH$_3$)$_3$), 1.58 (quintet, 4H, J = 8.0 Hz, CH$_2$), 3.72 (t, 4H, J = 8.0 Hz, CH$_2$), 7.42 (d, 4H, J = 8.0 Hz, Ar), 7.69 (d, 4H, J = 8.0 Hz, Ar); ^{13}C-NMR (100 MHz, CDCl$_3$): δ 13.9, 20.3, 31.3, 31.9, 35.1, 39.8, 125.5, 126.2, 126.8, 145.4, 154,8, 155.0, 167.5. Anal. calc. for C$_{34}$H$_{44}$N$_{10}$: C, 68.89; H, 7.48; N, 23.63. Found: C, 68.87; H, 7.49; N, 23.65; HRMS (ESI): m/z calcd for C$_{34}$H$_{44}$N$_{10}$ + H$^+$: 593.3829; found: 593.3827.

3,6-Bis(4-butyl-5-(4-nitrophenyl)-4H-1,2,4-triazol-3-yl)-1,2,4,5-tetrazine (**15h**)

The product was obtained as yellow powder (0.60 g, 42%); m.p. 103–104 °C. UV (CH$_2$Cl$_2$) λ$_{max}$ (log ε) 269 (4.11) nm; IR (ATR) ν$_{max}$ 3303, 3110, 2938, 2864, 2167, 2142, 2038, 2029, 2004, 1949, 1635, 1599, 1518, 1481, 1466, 1422, 1343, 1317, 1294, 1255, 1181, 1153, 1132, 1108, 1011, 973, 938, 868, 855, 841, 762, 723, 710, 691 cm^{-1}; ^{1}H-NMR (400 MHz, CDCl$_3$): δ 0.97 (t, 6H, J = 8.0 Hz, CH$_3$), 1.43 (sextet, 4H, J = 8.0 Hz, CH$_2$), 1.63 (quintet, 4H, J = 8.0 Hz, CH$_2$), 3.49 (t, 4H, J = 8.0 Hz, CH$_2$), 7.93 (d, 4H, J = 12.0 Hz, Ar), 8.28 (d, 4H, J = 12.0 Hz, Ar); ^{13}C-NMR (100 MHz, CDCl$_3$): δ 13.9, 20.3, 31.7, 40.3, 123.9, 128.2, 130.2, 140.6, 145.1, 149.6, 165.6. Anal. calc. for C$_{26}$H$_{26}$N$_{12}$O$_4$: C, 54.73; H, 4.59; N, 29.46. Found: C, 54.71; H, 4.58; N, 29.47; HRMS (ESI): m/z calcd for C$_{26}$H$_{26}$N$_{12}$O$_4$ + H$^+$: 571.2278; found: 571.2277.

4. Conclusions

Two effective methodologies for the synthesis of extended systems containing 1,2,4,5-tetrazine and 4H-1,2,4-triazole have been presented. The first methodology, comprising the Pinner reaction of carbonitriles bearing a 4H-1,2,4-triazole scaffold, is useful for obtaining unsymmetrical derivatives with heterocycles connected via a 1,4-phenylene linker. The second procedure, which makes use of imidoyl chloride and s-tetrazine-3,6-dicarbohydrazide, has proven to be successful for symmetrical systems with directly conjugated rings. In both cases, the approach leads to the desired products in satisfactory yields, regardless of the nature of the substituents attached to the terminal rings, as well as the type of groups on the triazole nitrogen atom. The obtained compounds exhibit mainly violet luminescence in CH$_2$Cl$_2$ solution. Their absorption–emission properties are directly related to the compound structure. The spectroscopic investigation revealed the dependency between the electron-donating strength of substituents and the emission wavelength, as well as the relationship between the quantum yield and the separation or direct conjunction of fluorophore moieties (tetrazine and triazole rings).

Supplementary Materials: The following are available online at https://www.mdpi.com/article/10.3390/molecules27113642/s1. Copies of the ^{1}H-NMR, ^{13}C-NMR, UV-Vis and fluorescent spectra of the title compounds are available in the online Supplementary Materials.

Author Contributions: A.M. and A.K. conceived and designed the experiments, performed the experiments and analyzed the data. M.Ś. performed emission measurements. A.M. and A.K. wrote the manuscript with the help of M.Ś. All authors have read and agreed to the published version of the manuscript.

Funding: The synthetic part of the project was financially supported by The Silesian University of Technology (Gliwice, Poland) Grant No. 04/050/RGP20/0115 and 04/050/BKM22/0147, BKM-608/RCh5/2022.

Institutional Review Board Statement: Not applicable.

Informed Consent Statement: Not applicable.

Data Availability Statement: Not applicable.

Conflicts of Interest: The authors declare no conflict of interest.

Sample Availability: Samples of the compounds **10a–l**, **15a–h** are available from the authors.

References

1. Wang, T.; Zheng, C.; Yang, J.; Zhang, X.; Gong, X.; Xia, M. Theoretical studies on a new high energy density compound 6-amino-7-nitropyrazino[2,3-e][1,2,3,4]tetrazine-1,3,5-trioxide. *J. Mol. Model.* **2014**, *20*, 2261–2271. [CrossRef]
2. Saracoglu, N. Recent advances and applications in 1,2,4,5-tetrazine chemistry. *Tetrahedron* **2007**, *63*, 4199–4235. [CrossRef]
3. Sinditskii, V.; Egorshev, V.Y.; Rudakov, G.F.; Filatov, S.A.; Burzhava, A.V. High-Nitrogen Energetic Materials of 1,2,4,5-Tetrazine Family: Thermal and Combustion Behaviors. In *Chemical Rocket Propulsion a Comprehensive Survey of Energetic Materials*; de Luca, L.T., Shimada, T., Sinditskii, V.P., Calabro, M., Eds.; Springer International Publishing: Cham, Switzerland, 2017; Volume 45, pp. 89–125. [CrossRef]
4. Ishmetova, R.I.; Ignatenko, N.K.; Ganebnykh, I.N.; Tolschina, S.G.; Korotina, A.V.; Kravchenko, M.A.; Skornyakov, S.N.; Rusinov, G.L. Synthesis and tuberculostatic activity of amine-substituted 1,2,4,5-tetrazines and pyridazines. *Russ. Chem. Bull.* **2014**, *63*, 1423–1430. [CrossRef]
5. Hu, W.X.; Rao, G.W.; Sun, Y.Q. Synthesis and antitumor activity of s-tetrazine derivatives. *Bioorg. Med. Chem. Lett.* **2004**, *14*, 1177–1181. [CrossRef] [PubMed]
6. Werbel, L.M.; Mc Namara, D.J.; Colbry, N.L.; Johnson, J.L.; Degan, M.J.; Whitney, B. Synthesis and antimalarial effects of N,N-dialkyl-6-(substituted phenyl)-1,2,4,5-tetrazin-3-amines. *J. Heterocycl. Chem.* **1979**, *16*, 881–894. [CrossRef]
7. Devaraj, N.K.; Upadhyay, R.; Haun, J.B.; Hilderbrand, S.A.; Weissleder, R. Fast and sensitive pretargeted labeling of cancer cells through a tetrazine/trans-cyclooctene cycloaddition. *Angew. Chem. Int. Ed.* **2009**, *48*, 7013–7016. [CrossRef] [PubMed]
8. Wang, M.; Svatunek, D.; Rohlfing, K.; Liu, Y.; Wang, H.; Giglio, B.; Yuan, H.; Wu, Z.; Li, Z.; Fox, J. Conformationally strained trans-cyclooctene (sTCO) enables the rapid construction of ^{18}F-PET probes via tetrazine ligation. *Theranostics* **2016**, *6*, 887–895. [CrossRef]
9. Brown, S.P.; Smith, A.B. Peptide/protein stapling and unstapling: Introduction of s-tetrazine, photochemical release, and regeneration of the peptide/protein. *J. Am. Chem. Soc.* **2015**, *137*, 4034–4037. [CrossRef] [PubMed]
10. Li, J.; Jia, S.; Chen, P.R. Diels-Alder reaction–triggered bioorthogonal protein decaging in living cells. *Nat. Chem. Biol.* **2014**, *10*, 1003–1005. [CrossRef]
11. Moral, M.; Garzon, A.; Olivier, Y.; Muccioli, L.; Sancho-Garcia, J.C.; Granadino-Roldan, J.M.; Fernandez-Gomez, M. Bis(aryleneethynylene)-s-tetrazines: A promising family of n-type organic semiconductors? *J. Phys. Chem. C* **2015**, *119*, 18945–18955. [CrossRef]
12. Pluczyk, S.; Zassowski, P.; Quinton, C.; Audebert, P.; Alain-Rizzo, V.; Łapkowski, M. Unusual electrochemical properties of the electropolymerized thin layer based on a s-tetrazine-triphenylamine monomer. *J. Phys. Chem. C* **2016**, *120*, 4382–4391. [CrossRef]
13. Gaber, M.; Fathalla, S.K.; El-Ghamry, H.A. 2,4-Dihydroxy-5-[(5-mercapto-1H-1,2,4-triazole-3-yl)diazenyl]benzaldehyde acetato, chloro and nitrato Cu(II) complexes: Synthesis, structural characterization, DNA binding and anticancer and antimicrobial activity. *Appl. Organomet. Chem.* **2019**, *33*, e4707. [CrossRef]
14. Shcherbyna, R.O.; Danilchenko, D.M.; Khromykh, N.O. The study of 2-((3-R-4-R1-4H-1,2,4-triazole-5-yl)thio) acetic acid salts as growth stimulators of winter wheat sprouts. *Visn. Farm.* **2017**, *89*, 61–65. [CrossRef]

15. Kaproń, B.; Łuszczki, J.J.; Płazińska, A.; Siwek, A.; Karcz, T.; Gryboś, A.; Nowak, G.; Makuch-Kocka, A.; Walczak, K.; Langner, E.; et al. Development of the 1, 2, 4-triazole-based anticonvulsant drug candidates acting on the voltage-gated sodium channels. Insights from in-vivo, in-vitro, and in-silico studies. *Eur. J. Pharm. Sci.* **2019**, *129*, 42–57. [CrossRef]
16. Jalihal, P.C.; Kashaw, V. Synthesis, antimicrobial and anti-inflammatory activity of some bioactive 1,2,4-triazoles analogues. *Int. J. Pharm. Biol. Sci.* **2018**, *8*, 94–104.
17. Maindron, T.; Wang, Y.; Dodelet, J.P.; Miyatake, K.; Hlil, A.R.; Hay, A.S.; Tao, Y.; D'Iorio, M. Highly electroluminescent devices made with a conveniently synthesized triazole-triphenylamine derivative. *Thin Solid Film.* **2004**, *466*, 209–216. [CrossRef]
18. Dutta, R.; Kalita, D.J. Charge injection and hopping transport in bridged-dithiophene-triazole-bridged-dithiophene (DT-Tr-DT) conducting oligomers: A DFT approach. *Comput. Theor. Chem.* **2018**, *1132*, 42–49. [CrossRef]
19. Tang, Y.; Zhuang, J.; Xie, L.; Chen, X.; Zhang, D.; Hao, J.; Su, W.; Cui, Z. Thermally cross-linkable host materials for solution-processed OLEDs: Synthesis, characterization, and optoelectronic properties. *Eur. J. Org. Chem.* **2016**, *22*, 3737–3747. [CrossRef]
20. Tsai, L.R.; Yun, C. Hyperbranched luminescent polyfluorenes containing aromatic triazole branching units. *J. Polym. Sci. A Polym. Chem.* **2007**, *45*, 4465–4476. [CrossRef]
21. Curtis, N.J.; Jennings, N. 1,2,4-Triazoles. In *Comprehensive Heterocyclic Chemistry*, 3rd ed.; Katritzky, A.R., Ramsden, C.A., Scriven, E.F.V., Taylor, R.J.K., Eds.; Elsevier: Amsterdam, The Netherlands, 2009; Volume 5, pp. 159–209. [CrossRef]
22. Clavier, G.; Audebert, P. s-Tetrazines as building blocks for new functional molecules and molecular materials. *Chem. Rev.* **2010**, *110*, 3299–3314. [CrossRef]
23. Savastano, M.; García-Gallarín, C.; Dolores López de la Torre, M.; Bazzicalupi, C.; Bianchi, A.; Melguizo, M. Anion-π and lone pair-π interactions with s-tetrazine-based ligands. *Coord. Chem. Rev.* **2019**, *397*, 112–137. [CrossRef]
24. Kędzia, A.; Kudelko, A.; Świątkowski, M.; Kruszyński, R. Highly fluorescent 1,2,4,5-tetrazine derivatives containing 1,3,4-oxadiazolering conjugated via a 1,4-phenylene linker. *Dyes Pigm.* **2020**, *183*, 108715–108723. [CrossRef]
25. Maj, A.; Kudelko, A.; Świątkowski, M. 1,3,4-Thiadiazol-2-ylphenyl-1,2,4,5-tetrazines: Efficient synthesis via Pinner reaction and their luminescent properties. *Arkivoc* **2021**, *8*, 167–178. [CrossRef]
26. Maj, A.; Kudelko, A.; Świątkowski, M. Novel conjugated s-tetrazinederivatives bearing a 4H-1,2,4-triazole scaffold: Synthesis and luminescent properties. *Molecules* **2022**, *27*, 459. [CrossRef] [PubMed]
27. Kędzia, A.; Kudelko, A.; Świątkowski, M.; Kruszyński, R. Microwave-promoted synthesis of highly luminescent s-tetrazine-1,3,4-oxadiazole and s-tetrazine-1,3,4-thiadiazole hybrids. *Dyes Pigm.* **2020**, *172*, 107865–107872. [CrossRef]
28. Fan, X.; Ge, Y.; Lin, F.; Yang, Y.; Zhang, G.; Ngai, W.S.C.; Lin, Z.; Zheng, S.; Wang, J.; Zhao, J.; et al. Optimized tetrazine derivatives for rapid bioorthogonal decaging in living cells. *Angew. Chem. Int. Ed.* **2016**, *55*, 14046–14050. [CrossRef]
29. Yang, J.; Karver, M.R.; Li, W.; Sahu, S.; Devaraj, N.K. Metal-catalyzed one-pot synthesis of tetrazines directly from aliphatic nitriles and hydrazine. *Angew. Chem. Int. Ed.* **2012**, *51*, 5222–5225. [CrossRef]
30. Chowdhury, M.; Goodman, L. Fluorescence of s-tetrazine. *J. Chem. Phys.* **1962**, *36*, 548–549. [CrossRef]
31. Chowdhury, M.; Goodman, L. Nature of s-tetrazine emission spectra. *J. Chem. Phys.* **1963**, *38*, 2979–2985. [CrossRef]
32. Choi, S.-K.; Kim, J.; Kim, E. Overview of syntheses and molecular-design strategies for tetrazine-based fluorogenic probes. *Molecules* **2021**, *26*, 1868. [CrossRef]
33. Liu, K.; Shi, W.; Cheng, P. The coordination chemistry of Zn(II), Cd(II) and Hg(II) complexes with 1,2,4-triazole derivatives. *Dalton Trans.* **2011**, *40*, 8475–8490. [CrossRef] [PubMed]
34. Li, C.; Ge, H.; Yin, B.; She, M.; Liu, P.; Lia, X.; Li, J. Novel 3,6-unsymmetrically disubstituted-1,2,4,5-tetrazines: S-induced one-pot synthesis, properties and theoretical study. *RSC Adv.* **2015**, *5*, 12277–12886. [CrossRef]
35. Gong, Y.-H.; Miomandre, F.; Méallet-Renault, R.; Badré, S.; Galmiche, L.; Tang, J.; Audebert, P.; Clavier, G. Synthesis and physical chemistry of s-tetrazines: Which ones are fluorescent and why? *Eur. J. Org. Chem.* **2009**, 6121–6128. [CrossRef]
36. Brouwer, A.M. Standards for photoluminescence quantum yield measurements in solution (IUPAC technical report). *Pure Appl. Chem.* **2011**, *83*, 2213–2228. [CrossRef]
37. Melhuish, W.H. Quantum efficiencies of fluorescence of organic substances: Effect of solvent and concentration of the fluorescent solute. *J. Phys. Chem.* **1961**, *65*, 229–235. [CrossRef]
38. Birks, J.B.; Dyson, D.J. The relations between the fluorescence and absorption properties of organic molecules. *Proc. R. Soc. Lond. Ser. A Math. Phys. Sci.* **1963**, *275*, 135–148. [CrossRef]
39. Ghosh, S.; Chowdhury, M. S1 (n,π*), T1 (n,π*) and S2 (n,π*) emissions in 3,6-diphenyl-s-tetrazine. *Chem. Phys. Lett.* **1982**, *85*, 233–238. [CrossRef]

Review

Syntheses and Applications of 1,2,3-Triazole-Fused Pyrazines and Pyridazines

Gavin R. Hoffman and Allen M. Schoffstall *

Department of Chemistry and Biochemistry, University of Colorado Colorado Springs, Colorado Springs, CO 80918, USA; ghoffman@uccs.edu
* Correspondence: aschoffs@uccs.edu; Tel.: +1-719-255-3479

Abstract: Pyrazines and pyridazines fused to 1,2,3-triazoles comprise a set of heterocycles obtained through a variety of synthetic routes. Two typical modes of constructing these heterocyclic ring systems are cyclizing a heterocyclic diamine with a nitrite or reacting hydrazine hydrate with dicarbonyl 1,2,3-triazoles. Several unique methods are known, particularly for the synthesis of 1,2,3-triazolo[1,5-*a*]pyrazines and their benzo-fused quinoxaline and quinoxalinone-containing analogs. Recent applications detail the use of these heterocycles in medicinal chemistry (c-Met inhibition or GABA$_A$ modulating activity) as fluorescent probes and as structural units of polymers.

Keywords: synthesis; 1,2,3-triazole; fused 1,2,3-triazole; 1,2,3-triazolo[4,5-*b*]pyrazine; 1,2,3-triazolo[4,5-*c*]pyridazine; 1,2,3-triazolo[4,5-*d*]pyridazine; 1,2,3-triazolo[1,5-*a*]pyrazine; 1,2,3-triazolo[1,5-*b*]pyridazine; triazolopyrazine; triazolopyridazine; practical applications

1. Introduction

Within the 1,2,3-triazole-fused pyrazines and pyridazines, a series of congeners exists depending on whether a nitrogen atom occupies a position at the ring fusion (Figure 1).

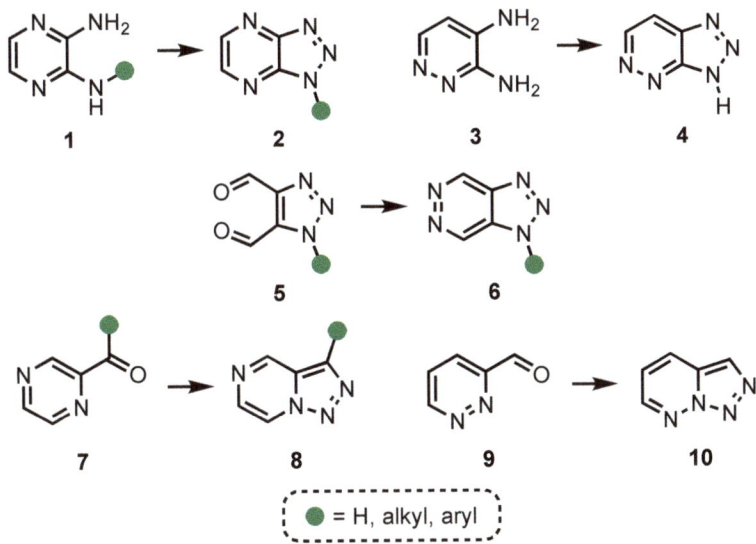

Figure 1. Structures of 1,2,3-triazole-fused pyrazines and pyridazines: 1*H*-1,2,3-triazolo[4,5-*b*]pyrazine (**2**); 1*H*-1,2,3-triazolo[4,5-*c*]pyridazine (**4**); 1*H*-1,2,3-triazolo[4,5-*d*]pyridazine (**6**); 1,2,3-triazolo[1,5-*a*]pyrazine (**8**); 1,2,3-triazolo[1,5-*b*]pyridazine (**10**); common precursors **1, 3, 5, 7, 9**.

We became interested in structures containing heterocyclic nuclei **2**, **4**, **6**, **8** and **10** following reports detailing potent mesenchymal–epithelial transition factor (c-Met) protein kinase inhibition, such as the current clinical candidate Savolitinib [1] (Figure 2, Structure A) and specifically those containing substructures **2** and **8** [1,2]. In addition to c-Met inhibition, structures containing these heterocyclic nuclei have shown GABA$_A$ allosteric modulating activity [3] (Figure 2, Structure B), have been incorporated into polymers for use in solar cells [4,5] (Figure 2, Structure C), and have demonstrated β-secretase 1 (BACE-1) inhibition [6] (Figure 2, Structure D). Their piperazine derivatives have demonstrated potent PDP-IV inhibition [7].

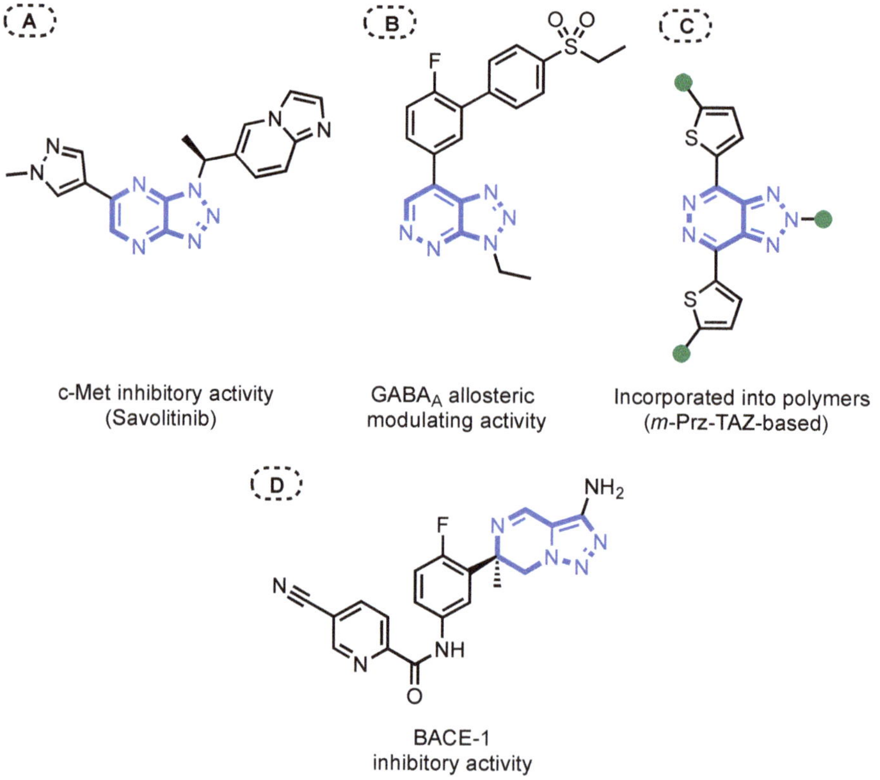

Figure 2. Examples of useful structures containing 1,2,3-triazole-fused pyrazines and pyridazines: (**A**) c-Met inhibitor Savolitinib [1], containing a 1,2,3-triazolo[4,5-*b*]pyrazine, (**B**) a compound with GAGA$_A$ allosteric modulating activity containing a 1,2,3-triazolo[4,5-*c*]pyridazine [3], (**C**) a 1,2,3-triazolo[4,5-*d*]pyridazine derivative used in polymers for solar cells [4,5], and (**D**), a 1,2,3-triazolo[1,5-*a*]pyrazine derivative with BACE-1 inhibitory activity [6].

Emphasis in this review is placed on the more common derivatives of **2** and **8**. In comparison to the heterocyclic scaffolds outlined in Figure 2, derivatives of **4**, **6** and **10** are less common in the literature. Among fused heterocycles containing the more well-known fused 1,2,4-triazoles, both 1,2,4-triazolo[1,5-*a*]pyrimidines [8] and 1,2,4-triazolo[4,3-*a*]pyrazines [9] have been recently reviewed. Kumar and coworkers [10] surveyed 1,2,3-triazoles fused to various rings, both aromatic and non-aromatic. In the present review, we address approaches to the synthesis of 1,2,3-triazole-fused pyrazines and pyridazines and their related congeners, while setting two limitations:

1. This review covers synthetic methods of preparing structures containing fused heterocycles **2**, **4**, **6**, **8**, **10** (Figure 1). Tricyclic and tetracyclic congeners containing these heterocycles are included.
2. 1,2,3-Triazolopyrimidines do not appear in this review. They have received attention in the literature on purine chemistry [11–13].

1,2,3-Triazolopyrimidines, which form the core structure of 8-azapurines, 8-azaadenines, and 8-azaguanines, have been well-studied and reviewed [13–15] owing to their similarity to the respective nucleobases. With both scope and limitations in place, this review addresses synthetic approaches to the 1,2,3-triazolodiazine family: 1,2,3-triazolo[4,5-b]pyrazine, 1,2,3-triazolo[4,5-c]pyridazine, 1,2,3-triazolo[4,5-d]pyridazine, 1,2,3-triazolo[1,5-a]pyrazine, and 1,2,3-triazolo[1,5-b]pyridazine. The literature covered includes articles published since the most recent review of each type of compound, or earlier if no review exists. Reports are covered until the spring of 2022 and exclude tetrahydro-derivatives.

2. Synthetic Approaches

This overview of synthetic methods is organized according to the type of heterocycle. In the case of 1H-1,2,3-triazolo[1,5-a]pyrazines, methods are subdivided into pyrazines and benzopyrazines. Reaction times are included along with solvents, catalysts, and other reagents in most examples. Commercial availability of precursors is emphasized where applicable.

2.1. Syntheses of 1H-1,2,3-Triazolo[4,5-b]pyrazines

One of the first reported preparations of a 1H-1,2,3-triazolo[4,5-b]pyrazine came from Lovelette and coworkers [16], who utilized condensation of a 4,5-diamino-1,2,3-triazole, **14**, and a 1,2-dicarbonyl compound **15** (Scheme 1) to give the desired triazolopyrazines **16** in yields in the range 30–35%. A useful precursor, 4,5-diamino-1,2,3-1H-triazole **14**, was prepared by reacting carbamate **13** with a strong base. This carbamate was readily prepared from the carbonyl azide by refluxing in ethanol. The carbonyl azide can be prepared from benzyl azide **11**, ethyl cyanoacetate **12**, and sodium ethoxide, all commercially available starting materials.

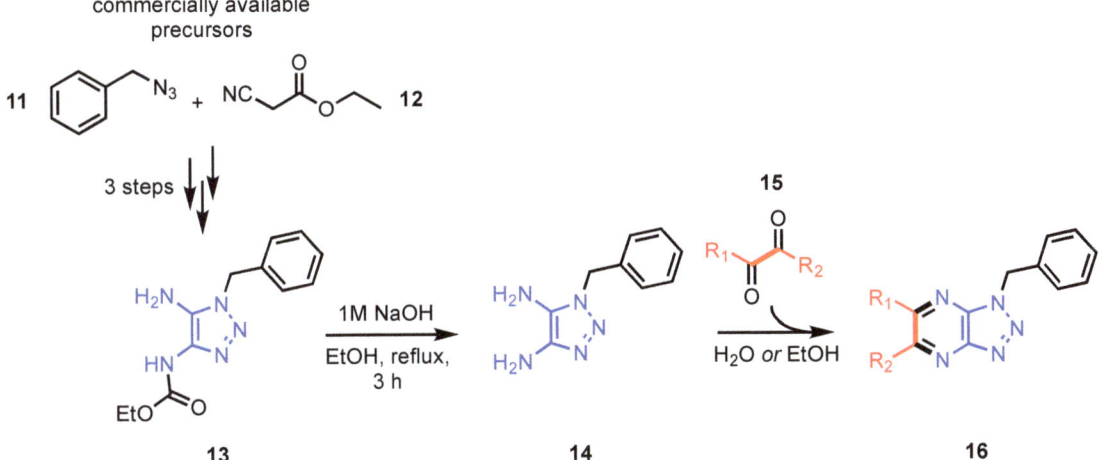

Scheme 1. One of the first reported syntheses of 1H-1,2,3-triazolo[4,5-b]pyrazines, **16**. (R_1, R_2 = alkyl, aryl, H.)

Dicarbonyl compounds included glyoxal ($R_1 = R_2 = H$), benzil ($R_1 = R_2 = Ph$), and others. This was one of the first reports of 1,2,3-triazole-fused pyrazines, highlighted within a study of fused 1,2,3-triazoles. This method offers three-point diversity, one from the triazole substituent, and the other two from the respective dicarbonyl substituents. Despite this, a potential drawback lay in the restriction to a symmetrically substituted 1,2-dicarbonyl species to avoid mixtures of isomers. Indeed, the authors noted the two condensation products using an asymmetrically substituted diketone, where $R_1 = CH_3$ and $R_2 = H$, as being indistinguishable.

Monge and coworkers [17] prepared benzo-fused 1H-1,2,3-triazolo[4,5-b]pyrazines through the acid-catalyzed cyclization of 2-azido-3-cyanoquinoxaline, **18**, obtained from 2-chloro-3-cyanoquinoxaline **17**, yielding 1-hydroxy-1H-1,2,3-triazolo[4,5-b]quinoxaline **19** (Scheme 2) in 52% yield. Though uncommon, acid-catalyzed cyclization of *ortho*-substituted azidocyanoaryl species may represent an underutilized method of obtaining structures with the 1,2,3-triazolo[4,5-b]pyrazine core. Despite this, the use of costly starting materials hinders wider applicability.

Scheme 2. Conversion of 2-chloro-3-cyanoquinoxaline **17** to 2-azido-3-cyanoquinoxaline **18** and benzo-fused 1H-1,2,3-triazolo[4,5-b]pyrazine **19**.

Unexpectedly, Starchenkov and coworkers [18] determined that, upon treatment of diamine **20** with trifluoroacetic anhydride (TFAA) and HNO_3 and proceeding via intermediate **21**, triazolopyrazine N-oxide **22** was formed (Scheme 3). This was one of the first reports of the preparation of a fused 1,2,3-triazole 2-N-oxide, namely [1,2,5]oxadiazolo[3,4-b][1,2,3]triazolo[4,5-e]pyrazine-6-oxide **22**, formed in 92% yield.

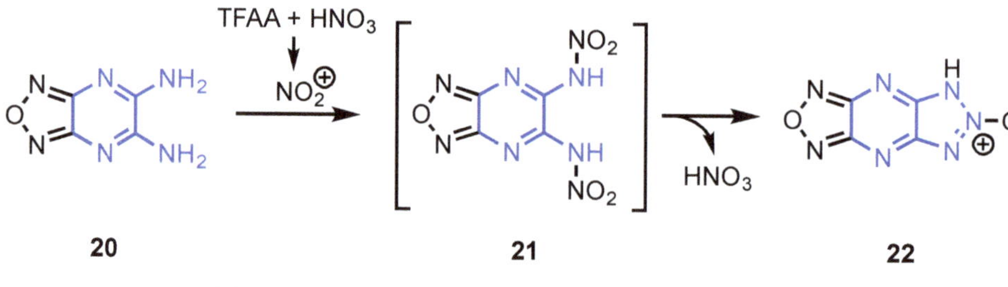

Scheme 3. Formation of triazolopyrazine N-oxide **22** from diaminopyrazine **20**, proceeding via intermediate **21**.

Forming a mesoionic ring system while studying luminescence, Slepukhin and coworkers [19] obtained the 1H-1,2,3-triazolo[4,5-b]pyrazine core within the azapentalene inner salt **27** in 50% yield after intramolecular cyclization of 8-(benzotriazole-1-yl)tetrazolo[1,5-a]pyrazine **25** in refluxing DMF, causing loss of nitrogen via intermediate **26** and formation of 5H-pyrazino[2′,3′:4,5][1,2,3]triazol[1,2-a]benzotriazol-6-ium, inner salt **27** (Scheme 4). Pyrazine **25** was prepared in 48% yield by nucleophilic aromatic substitution of chloride by the benzotriazolyl ion after deprotonation of 1H-1,2,3-benzotriazole **24** by carbonate.

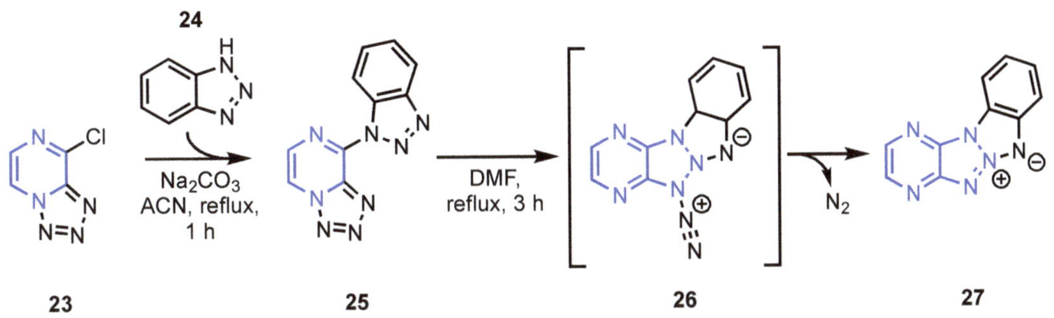

Scheme 4. Synthesis of an azapentalene, 5H-pyrazino[2′,3′:4,5][1,2,3]triazolo[1,2-a]benzotriazol-6-ium, inner salt **27**, after intramolecular cyclization of **25** and loss of nitrogen via intermediate **26**.

Azapentalenes, containing the 1H-1,2,3-triazolo[4,5-b]pyrazine nucleus, have gained attention for their useful properties, such as in luminescence and complexation [19]. Compounds of this type have demonstrated low toxicity, high solubility, and other properties desirable as potential fluorescence probes [20]. This intramolecular approach has remained popular in obtaining various substituted azapentalenes, another example being that of Nyfenegger and coworkers [21]. Here, the azapentalene, 5H-pyrazolo[1′,2′:1,2][1,2,3]triazolo[4,5-b]pyrazin-6-ium, inner salt, **31**, was obtained in yields up to 85% via cyclization with loss of nitrogen after amination of 2-azido-3-chloropyrazine, **28**, with either pyrazole **29** or 1,2,4-triazole affording 2-azido-3-(1H-pyrazol-1-yl)pyrazine **30** (Scheme 5). Other derivatives using nitro-substituted pyrazoles were formed in yields in the range 63–97%. This method offers convenience in that a precursor to **28**, 2,3-dichloropyrazine, is commercially available.

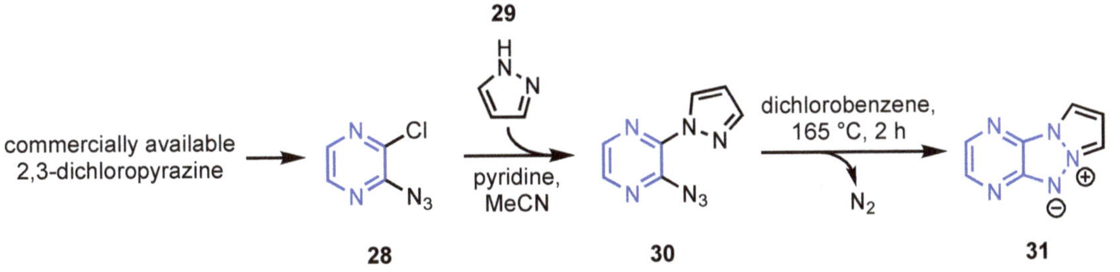

Scheme 5. Synthesis of the azapentalene **31**, from pyrazine **30**, derived from 2-azido-3-chloropyrazine **28**.

Notably, in addition to having an azido group substituted *ortho* to the pyrazole of **30** [20,21], reports have also made use of the respective amine via ring closure by displacement of an N-iodonium intermediate by an adjacent nitrogen atom of the attached pyrazole to form azapentalenes [22,23]. Compounds of this type have been thoroughly characterized via NMR spectroscopy [24]. A Pfizer patent [25] filed in 2007 detailed the use of either isoamyl nitrite in DMF or $NaNO_2$ in aqueous acetic acid, after first aminating commercially available 2-amino-3,5-dibromopyrazine **32** in the presence of a sterically hindered base, N,N-diisopropylethylamine (DIPEA), then treating diaminopyrazine **33** with nitrite to form 3,5-disubstituted 1H-1,2,3-triazolo[4,5-b]pyrazine **34** (Scheme 6).

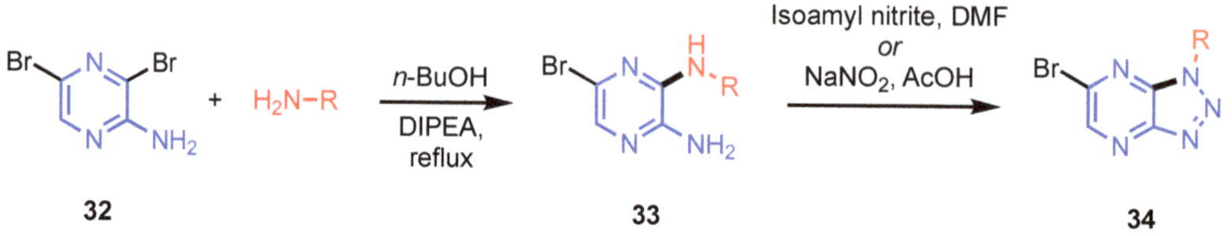

Scheme 6. Amination of 2-amino-3,5-dibromopyrazine **32** to form diaminopyrazine **33**, followed by cyclization to form triazolo[4,5-*b*]pyrazine **34**. (R = alkyl or aryl.)

The use of nitrite for triazole cyclization, via the nitrosonium ion, has also been reported by Ye and coworkers [26,27], Cui and coworkers [2] (who were cited in the original patent [25]), and others. Thottempudi and coworkers [28] used a combination of TFAA/HNO$_3$ as an in situ nitronium source, giving a triazole 2-N-oxide, while Jia and coworkers [1], and others [29,30] used nitrosonium generated from nitrite. Both syntheses offer straightforward introduction of the triazole based on the amine chosen during amination. They also have the advantage of short reaction times and little or no required purification. Likely owing to these benefits, cyclization using nitrite to generate nitrosonium ion, such as in **33** to **34** (Scheme 6), continues to dominate reports in the literature. Indeed, the reaction of various diazinyl diamines with nitrite represents a central theme throughout the discussion of syntheses of 1*H*-1,2,3-triazole-fused pyrazines and pyridazines.

2.2. Syntheses of 1,2,3-Triazolo[1,5-*a*]pyrazines

More well-known than 1,2,3-triazolo[4,5-*b*]pyrazines are the fused [1,5-*a*]pyrazine derivatives. While benzo[*b*]pyrazines (i.e., quinoxalines) are not commonly encountered as part of 1,2,3-triazolo[4,5-*b*]pyrazines, they are widespread in the literature in compounds containing the 1,2,3-triazolo[1,5-*a*]pyrazine nucleus. Therefore, this section is organized into the syntheses of benzo-fused structures (e.g., 1,2,3-triazolo[1,5-*a*]pyrazines containing quinoxaline or quinoxalinone), and those that are bicyclic 1,2,3-triazolo[1,5-*a*]pyrazines. A recent brief review of 4,5,6,7-tetrahydro[1–3]triazolo[1,5-*a*]pyrazines has been published [31]. An earlier review detailed aspects of the chemistry of 1,2,3-triazolo[1,5-*a*]pyrazines [32]. A review on the synthesis of triazoloquinazolines also appeared in 2016 [33].

2.2.1. Syntheses of Bicyclic 1,2,3-Triazolo[1,5-*a*]pyrazines

The first method of synthesizing a 1,2,3-triazolo[1,5-*a*]pyrazine by Wentrup [34] was, at the time, the synthesis of a novel purine isomer. Wentrup utilized the thermolysis of 5-(2-pyrazinyl)tetrazole **36** (400 °C, 10^{-5} Torr), affording **38**, 1,2,3-triazolo[1,5-*a*]pyrazine in 20% yield proceeding via diazo intermediate **37**. The precursor 2-(2*H*-tetrazol-5-yl)pyrazine **36** was readily prepared from 2-cyanopyrazine, **35**, upon treatment with hydrazoic acid generated in situ from ammonium chloride and sodium azide (Scheme 7). This method, while suffering from harsh reaction conditions and poor yields, was the first utilizing intramolecular cyclization of diazo intermediates in the formation of 1,2,3-triazolo[1,5-*a*]pyrazines. Lead tetraacetate oxidation of the hydrazone of pyrazine-2-carbaldehyde similarly gave **38** in 75% yield [35].

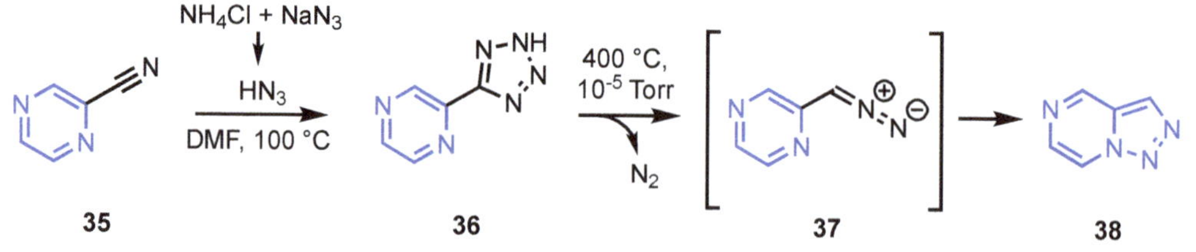

Scheme 7. Conversion of 2-cyanopyrazine **35** to 1,2,3-triazolo[1,5-*a*]pyrazine **38** via tetrazolylpyrazine **36**.

In addition to syntheses of neutral compounds of this type, several reports have appeared for the preparation of fused pyrazinium salts. A method by Beres and coworkers [36] afforded 1-(4-bromophenyl)-3-methyl-1,2,3-triazolo[1,5-*a*]pyrazinium tetrafluoroborates **31** in 55% yield (when R_1 = *p*-chlorophenyl) and 81% yield (when R_1 = CH_3) after reaction of 4-bromophenylhydrazones **39** (prepared from the respective 2-pyrazinyl ketone) with tribromophenol bromine (TBB) and NH_4BF_4 (Scheme 8). When R_1 = CH_3, the yield of **40** was 81%. Interestingly, after treatment of **40** with pyrrolidine in methanol, the ring-opened 2-aza-1,3-butadienes can be valuable starting materials for other conversions. For example, a ring-opened triazolyl-2-aza-1,3-butadiene was converted to a fused pyridine after treatment with N-phenylmaleinimide, or an imidazoline when treated with tosyl azide [36].

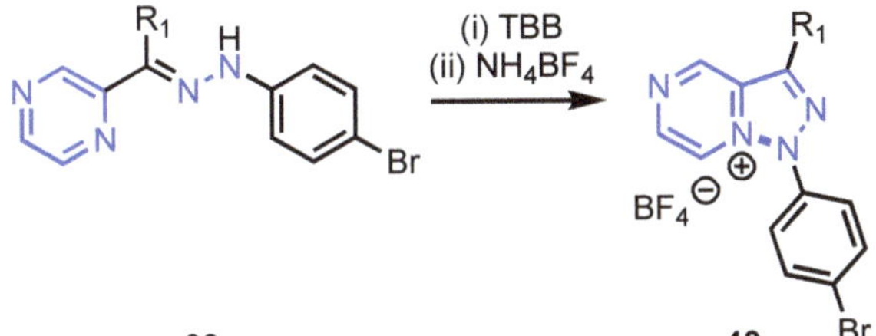

Scheme 8. Intramolecular cyclization of a 4-bromophenylhydrazone **39** forming triazolopyrazinium tetrafluoroborate **40**. (R_1 = alkyl or aryl.)

Methods have been reported for the preparation of 1,2,3-triazol[1,5-*a*]pyrazinones. In work by Nein and coworkers [37,38], the reaction of 5-hydroxy-N-diphenyl-1H-1,2,3-triazole-4-carboxamide **41** with chloroacetonitrile in DMF and base gave the alkylated product **42**, which, after refluxing in sodium ethoxide, gave 6-amino-4-oxo-2,5-diphenyl-4,5-dihydro-2H-1,2,3-triazolo[1,5-*a*]pyrazinium-5-olate **43** in 80% yield (Scheme 9). They proposed the geometry of 3-phenacyl- and 3-cyanomethyl derivatives of triazolium-5-olates indicated interaction of the carboxamide nitrogen at position 4 of the triazole with cyano groups, which was then confirmed experimentally after obtaining the desired mesoionic **42** [37].

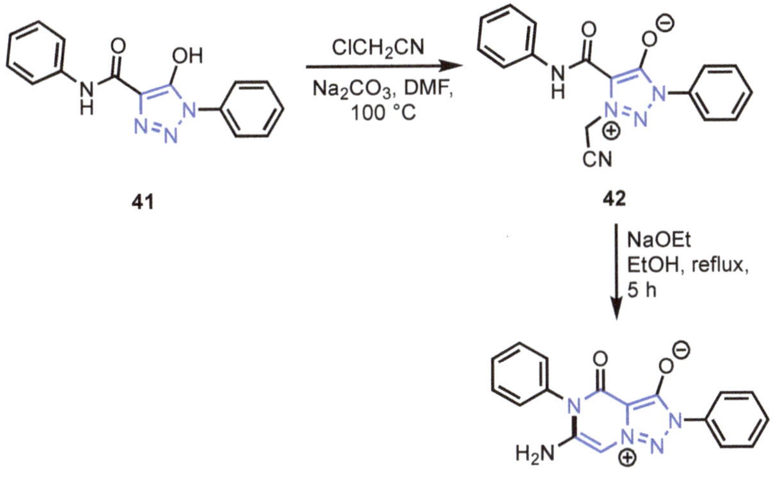

Scheme 9. Intramolecular cyclization of a triazolium salt **42**, forming zwitterionic **43**.

Similarly involving reaction of triazolium olates such as **42**, during a synthesis of 1,2,5-triazepines by Savel'eva and coworkers [39], [1,5-*a*]triazolopyrazines were formed as byproducts (5–7%) from the intramolecular cyclization of 1-amino-3-(*p*-phenacyl)-4-{[2-(1-methylethylidene)hydrazino]carbonyl}-[1,2,3]-triazolium-5-olates. Jug and coworkers [40] took a novel approach for the reaction of 4-(ethoxymethylene)-2-phenyloxazol-5(4*H*)-one **44** with commercially available diaminomaleonitrile **45**, forming adduct **46** which, after conversion to triazole **47** with nitrite, afforded the substituted 1,2,3-triazolo[1,5-*a*]pyrazine **48** (Scheme 10). Later, derivatives of **48** such as ethyl 4-amino-3-cyano-1,2,3-triazolo[1,5-*a*]pyrazine-6-carboxylate were further reacted by Trcek and coworkers [41] to form 1,2,3-triazolo[1,5-*a*]-1,2,4-triazolo[5,1-*c*]pyrazines in 55–65% yield.

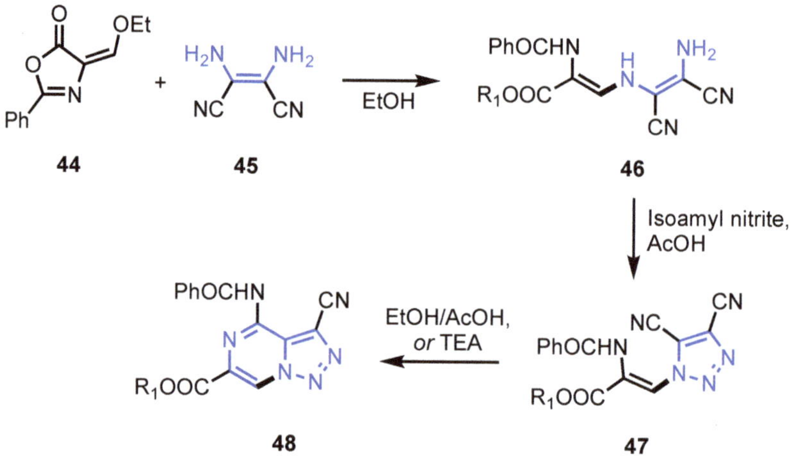

Scheme 10. Reaction of 4-(ethoxymethylene)-2-phenyloxazol-5(4*H*)-one **44** and diaminomaleonitrile **45** forming an adduct **46**, which led to triazole **47** and pyrazine **48**. (R_1 = alkyl.)

Raghavendra and coworkers [42] reported a triazolopyrazine synthesis employing solid-phase polystyrene *p*-toluenesulfonyl hydrazide, a common carbonyl scavenging resin. After reaction of the polystyrene *p*-toluenesulfonyl hydrazide **49** with an acetylpyrazine **50** in the presence of 5% $TiCl_4$ in MeOH, hydrazone **51** was obtained. Reaction of **51** with morpholine gave the desired 1,2,3-triazolo[1,5-*a*]pyrazines, **52** (Scheme 11), in yields ranging from 33–62%. This regiospecific, traceless protocol represented the first solid-phase assisted synthesis of a triazolopyrazine and was also used for the synthesis of several non-fused 1,2,3-triazoles in the same report in yields up to 60%.

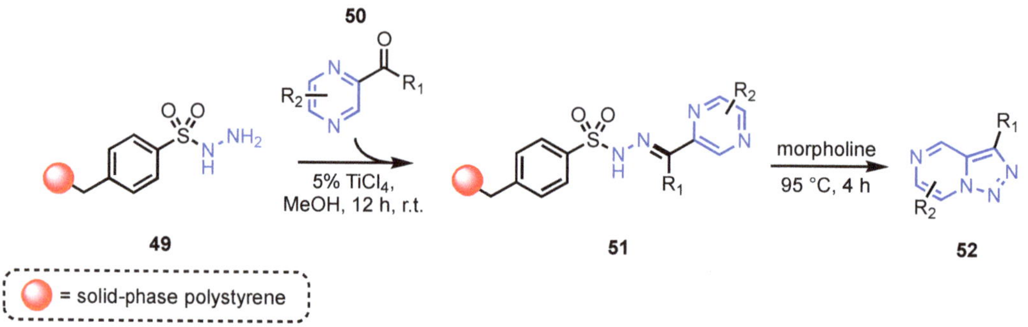

Scheme 11. Intramolecular cyclization of hydrazone **51**, derived from tosylhydrazide **49**, in the formation of disubstituted 1,2,3-triazolo[1,5-*a*]pyrazines **52**. (R_1, R_2 = alkyl.)

Copper-catalyzed [3 + 2] cycloaddition of propiolamide **53**, followed by halide displacement to form a fused product, was utilized in the synthesis of saturated derivatives of 1,2,3-triazolo[1,5-a]pyrazine (i.e., triazolopiperazines) **54** in 80% yield [43] (Scheme 12). Koguchi and coworkers used ynones and β-amino azides to afford 6,7-dihydro-1,2,3-triazolo[1,5-a]pyrazines. These authors verified that the one-pot reaction gave cycloaddition of the alkyne and azide first, followed by reaction of the amine with the ketone [44].

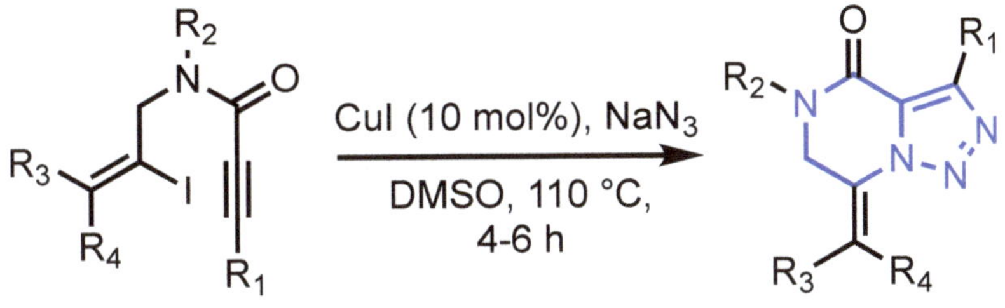

Scheme 12. Cycloaddition of propiolamide **53** and displacement of iodide to form triazolopiperazine **54**. (R_1, R_2, R_3, R_4 = alkyl or aryl.)

2.2.2. Syntheses of Benzo-Fused 1,2,3-Triazolo[1,5-a]pyrazines

One of the first reported preparations of a 1,2,3-triazolo[1,5-a]pyrazine by Kauer and coworkers [45] started with dimethyl l-(o-nitrophenyl)-1H-triazole-4,5-dicarboxylate **59**, and upon treatment with tributyl phosphine in refluxing toluene, afforded methyl 4-methoxy-1,2,3-triazolo[3,4-a]quinoxaline-3-carboxylate **60** (Scheme 13). Triazole **59** was readily prepared from o-azidonitrobenzene **57** (which in turn was prepared from o-chloronitrobenzene **55** or o-aminonitrobenzene **56**) and dimethyl acetylenedicarboxylate **58** in $CHCl_3$.

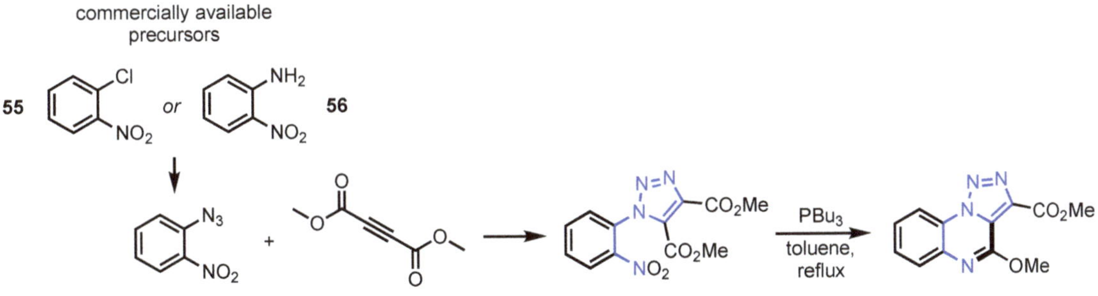

Scheme 13. Treatment of triazolyl-o-nitrobenzene **59** with tributyl phosphine, PBu_3, resulting in 1,2,3-triazolo[3,4-a]quinoxaline **60**.

Through a different approach, Cue and coworkers [46] accessed 1,2,3-triazolo[l,5-a]quinoxaline N-oxides **62** in yields ranging from 52–70% by cyclization of quinoxaline-3-carboxaldehyde-1-oxide-p-toluenesulfonylhydrazone **61** (Scheme 14). The starting sulfonylhydrazone **61** was prepared by reaction of a 3-substituted quinoxaline N-oxide with p-toluenesulfonylhydrazide [45].

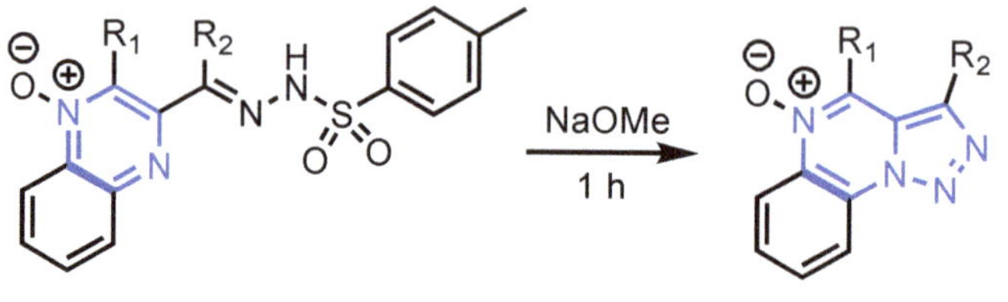

Scheme 14. Synthesis of substituted 1,2,3-triazolo[1,5-a]quinoxaline N-oxides, **62**, via intramolecular cyclization of sulfonylhydrazone **61**. (R_1, R_2 = H, methyl.)

For the intramolecular cyclization of *ortho*-substituted amines to prepare 1,2,3-triazoles using nitrite, as is commonly reported for non-fused derivatives [2,25], Ager and coworkers [47] illustrated that the amines used in cyclization do not need to be primary. Through reaction of a secondary amine within a ring and a primary amine **63** with isoamyl nitrite in chloroacetic acid, they obtained 1,2,3-triazoles fused to both lactones and lactams, **64**, in yields in the range 54–76% (Scheme 15). In the case of lactams, 1,2,3-triazoloquinoxalinones were formed.

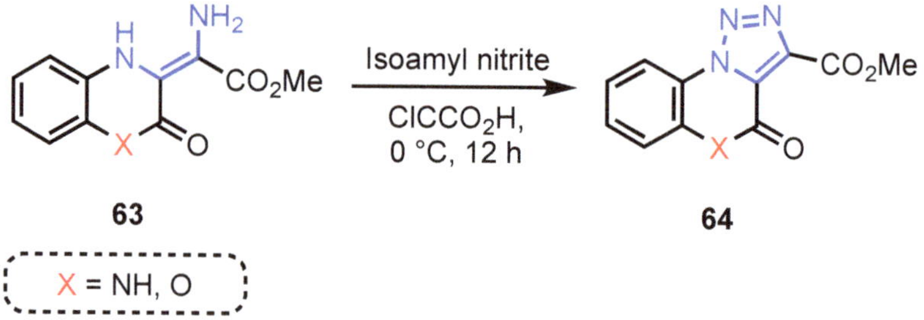

Scheme 15. Synthesis of 1,2,3-triazole-fused quinoxalinones, using a diamine in which one amine is secondary.

Synthesizing compounds of the same type, Bertelli, and coworkers [48] first formed a triazole diester on a ring *ortho* to a nitro group, **65**, which was intramolecularly cyclized to form ethyl 4,5-dihydro-4-oxo-[1,2,3]triazolo[1,5-a]quinoxaline-3-carboxylate, **66** (Scheme 16). This reaction was conducted by hydrogenation with a 10% Pd/C catalyst or by reaction with $FeCl_3$ and Fe powder. Biagi and coworkers [49] cyclized the triazole diester into 1,2,3-triazoloquinoxalinone **66** with 10% Pd/C in ethanol in an excellent 98% yield. Shen and coworkers further modified the ester group of **66** to prepare a derivative suitable for biological testing [50].

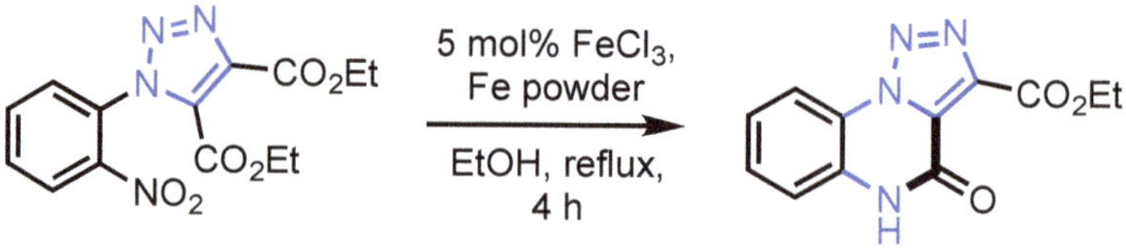

Scheme 16. Synthesis of 1,2,3-triazole-fused quinoxalinone **66**, using a pre-formed 1,2,3-triazole **65** in the presence of iron catalysts.

Abbott and coworkers [51] prepared 1,2,3-triazoloquinoxalines in an analogous manner, but opted for use of an amide instead of a nitro group, giving mesoionic 1,2,3-triazoles **68**, which were derived from the lithium salt of [2-(acetylamino)phenyl]amino acetic acid **67** (Scheme 17). A series of 1,2,3-triazoloquinoxalines, **69**, was synthesized after cyclization with *p*-toluenesulfonic acid (*p*-TSA) in refluxing toluene in yields in the range 16–59%.

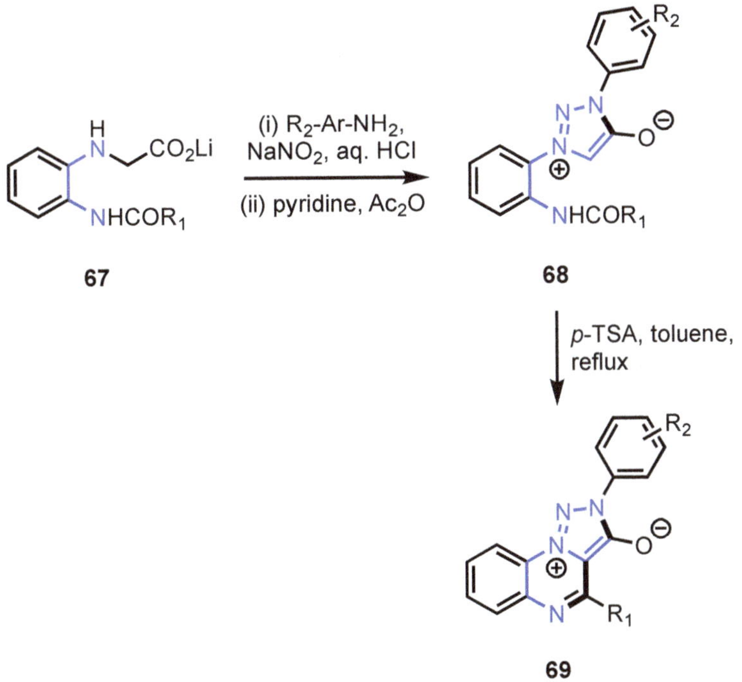

Scheme 17. Synthesis of mesoionic 1,2,3-triazoloquinoxalines **69** from *ortho*-substituted 1,2,3-triazolobenzamides **68**. (R_1, R_2 = alkyl or aryl.)

Saha and coworkers [52] used the intramolecular cyclization of *ortho*-substituted anilines with tethered 1,2,3-triazoles, **72**, a Pictet–Spengler reaction, to form 1,2,3-triazoloquinoxalines **73** in yields in the range 61–70% (Scheme 18). This sequence offers two-point diversity: one from **72**, and the other from an aryl aldehyde **73**. The prerequisite triazole **72** was conveniently prepared from readily available starting materials, including *o*-fluoronitrobenzene **70**, phenylacetylene **71**, and sodium azide.

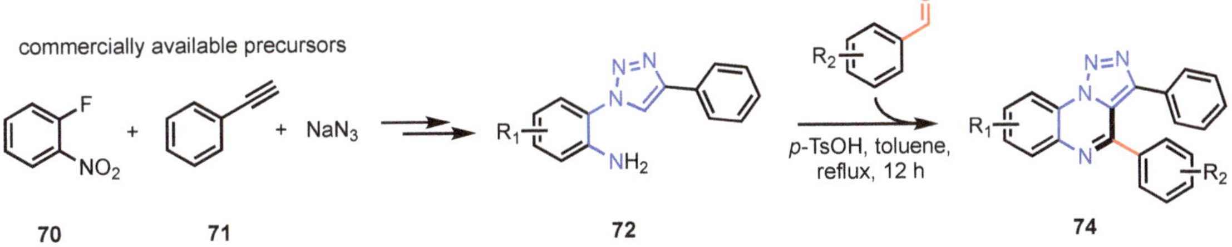

Scheme 18. 3-Phenyl-4-*p*-tolyl-1,2,3-triazolo[1,5-*a*]quinoxalines **74** synthesized from *ortho*-substituted 2-(4-phenyl-[1,2,3]triazol-1-yl)anilines **72**. (R_1, R_2 = alkyl or aryl.)

Chen and coworkers [53] used a novel approach for the synthesis of 4-(trifluoromethyl)-1,2,3-triazolo[1,5-*a*]quinoxaline **76** via cascade reactions of N-(*o*-haloaryl)alkynylimine **75** with sodium azide in the presence of copper iodide and L-proline (Scheme 19). Among a series of amine-containing catalysts, L-proline resulted in a 98% isolated yield, while tetramethylethylenediamine and N,N′-dimethylethylenediamine gave lower yields, and higher percentages of the uncyclized imine product.

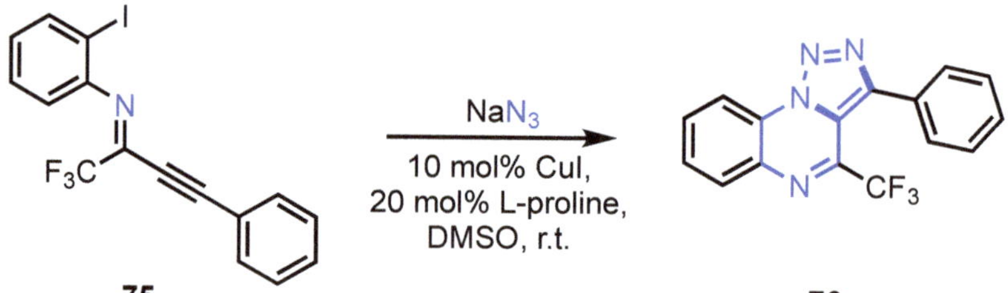

Scheme 19. Synthesis of 3-phenyl-4-(trifluoromethyl)-1,2,3-triazolo[1,5-*a*]quinoxaline **76** from N-(*o*-haloaryl)alkynylimine **75**.

Using photoredox catalysis, He and coworkers [54] used [*fac*-Ir(ppy)$_3$] as a photocatalyst to afford the corresponding 1,2,3-triazoloquinoxaline **78** from isonitrile **77** in 60% yield (Scheme 20). Due to poor solubility of the catalyst, ACN resulted in decreased yields compared to DMF. This work is a rare example of free-radical generation of 1,2,3-triazole-fused ring systems, as cyclohexyl radicals are proposed to have formed from phenyliodine(III)dicarboxylate. The radicals yield isonitrile carbon radicals, followed by reaction with carbon 5 of the triazole. Various fused rings were synthesized in addition to 1,2,3-triazoles including tetrazoles, pyrazoles, and imidazoles in yields as high as 80%.

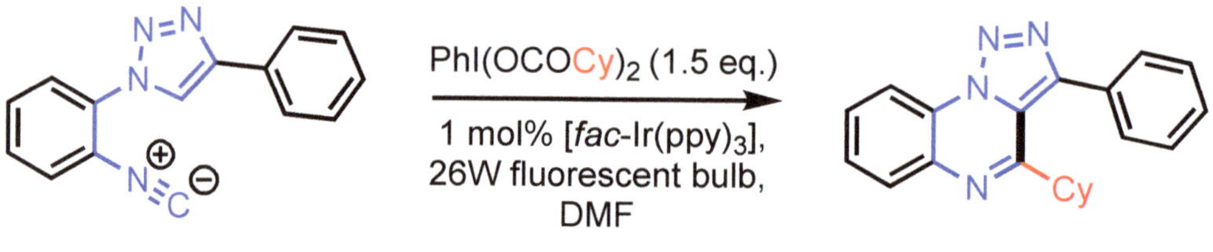

Scheme 20. Photoredox approach for the synthesis of a 1,2,3-triazoloquinoxaline **78**. (Cy = cyclohexyl, fac = facial, ppy = 2-phenylpyridine).

In the presence of Cu(OAc)$_2$ and base in DMSO/THF, Li and coworkers [55] reported an efficient one-pot synthesis of 1,2,3-triazolo[1,5-*a*]quinoxalines **81** from 1-azido-2-isocyanoarenes **79** in yields in the range 40–84% (Scheme 21). They outlined the option of using terminal acetylenes **80** or substituted acetaldehydes **82**, the former being cyclized into **81** in one step (in yields ranging from 40–83%), and the latter forming uncyclized triazole **83**, which was annulated using Togni's reagent II and tetra-*n*-butylammonium iodide (TBAI), forming **84**, or phenylboronic acid, forming **85**. Derivatives of **84** were prepared in yields in the range 26–78%, and one synthesis of **85** yielded 86%.

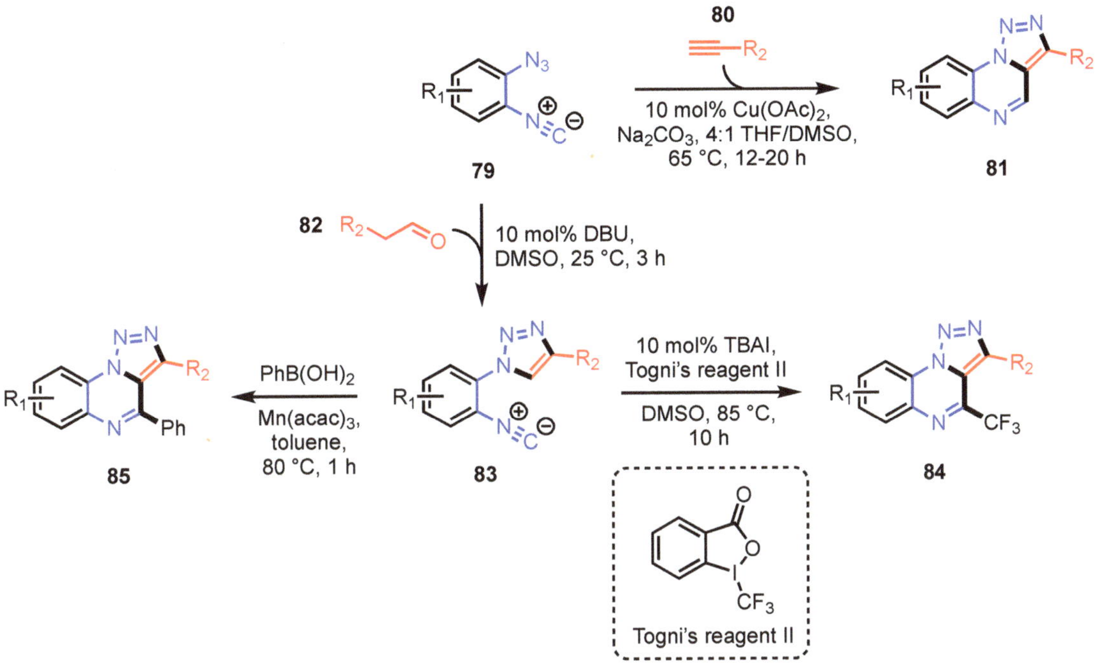

Scheme 21. Copper-catalyzed synthesis of 1,2,3-triazolo[1,5-a]quinoxalines **81** from 1-azido-2-isocyanoarenes **79**. Annulation of intermediate **83** with 1) Togni's reagent II and catalytic TBAI forming **84**, or 2) phenylboronic acid and a manganese catalyst forming **85**. (R_1, R_2 = alkyl or aryl.)

Owing to the versatility of intermediate **83**, many functionalized 1,2,3-triazoloquinoxalines were prepared, and indeed, Li and coworkers reported several compounds containing the 1,2,3-triazolo[1,5-a]quinoxaline core with a variety of functionalities. Additionally, in this report, the fused products were further reacted into diversified quinoxaline derivatives via Rh(II)-catalyzed carbenoid insertion reactions [55].

Employing a Pd-catalyzed intramolecular cyclization of triazole **86**, Kotovshchikov and coworkers [56] synthesized 3-butyl-[1,2,3]triazolo[1,5-a]quinoxalin-4(5H)-one **87** in 77% yield. As this reaction was conducted under CO (1 atm), the carbonyl carbon of the quinoxalinone was introduced by Pd-catalyzed insertion of CO (Scheme 22).

Scheme 22. Pd-catalyzed synthesis of quinoxalinone **87** from *ortho*-substituted aniline **86**.

Xiao and coworkers [43] and Chen and coworkers [57] used in situ conversion of N-propargyl-N-(2-iodoaryl)amides **88** to azides, which underwent 1,3-dipolar cycloaddition with the adjacent alkyne to form substituted 1,2,3-triazolo[1,5-a]quinoxalines **89** (Scheme 23) in yields in the range 58–91%. Chen and coworkers suggested that cycloaddition might occur first. The sequence was conducted in the presence of DIPEA and 1,2-dimethylethylenediamine (DMEDA).

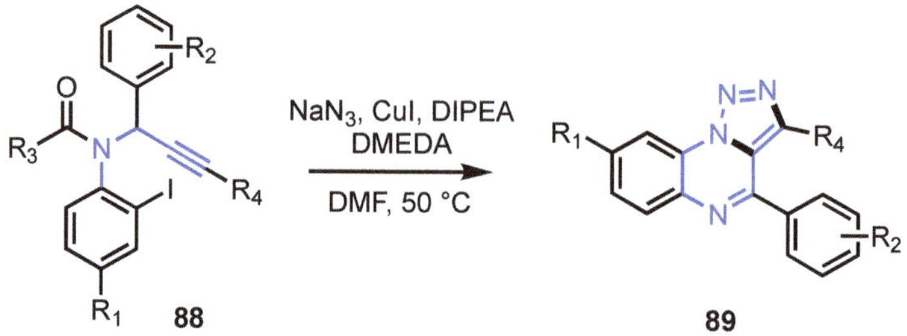

Scheme 23. Intramolecular cyclization of N-propargyl-N-(2-iodoaryl)amides **88**, yielding 1,2,3-triazoloquinoxalines **89** after in situ conversion of **88** to the azide. (R_1, R_2, R_3, R_4 = alkyl or aryl.)

Preparative thermolysis of tetrazoloquinoxaline **90** proceeded by loss of nitrogen through diazo intermediate **91** and then to 1,2,3-triazolo[1,5-*a*]quinoxaline **92** in 67% yield (Scheme 24) [58]. Using a ring-closure method similar to that used by both Raghavendra and coworkers [42] and Cue and coworkers [46], Vogel and Lippmann [59] developed a route to derivatives of **92** in 47–89% yield via conversion from tosylhydrazones **93** using base (Bamford-Stevens conditions) or, in certain cases, heat (Scheme 24).

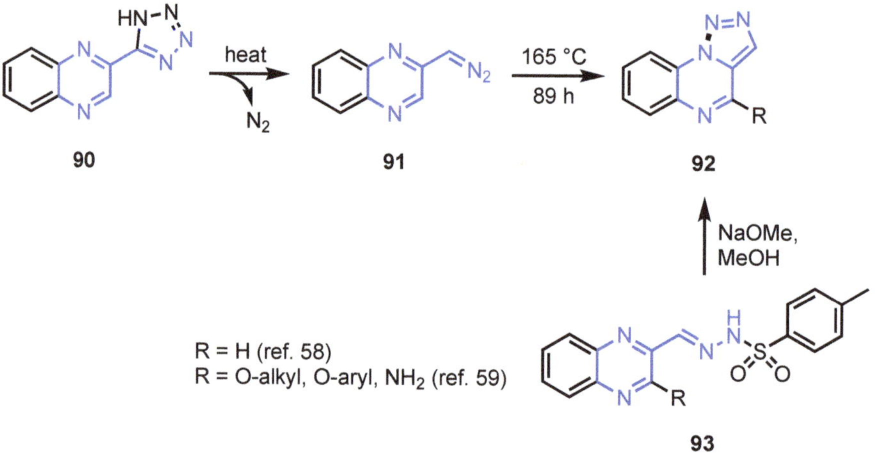

Scheme 24. Cyclization methods for preparing 1,2,3-triazolo[1,5-*a*]quinoxalines.

Overall, there exist diverse methods for the synthesis of both bicyclic 1,2,3-triazolo[1,5-*a*]pyrazines and 1,2,3-triazolo[1,5-*a*]quinoxalines.

2.3. Syntheses of 1H-1,2,3-Triazolo[4,5-d]pyridazines

Livi and coworkers [60] reviewed syntheses of this heterocyclic system covering reports prior to 1996. Another review on condensed 1,2,3-triazoles appeared in 2008, which includes synthesis of 1*H*-1,2,3-triazolo[4,5-*d*]pyridazines [32]. Here, we summarize both older and newer reports. A common theme in the literature regarding the synthesis of 1*H*-1,2,3-triazolo[4,5-*d*]pyridazines is the reaction of 1,2,3-triazole dicarbonyl species with hydrazine hydrate. This yields a diacylhydrazide, which can be cyclized with either high heat or acid. One of the first examples (Scheme 25) is from Fournier and Miller [61], who used 2-(4,5-dibenzoyl-1*H*-1,2,3-triazol-1-ylmethyl)-3,4,6-trimethylhydroquinone diacetate and hydrazine hydrate in ethanol to form 4,5-diphenyl-1*H*-1,2,3-triazolo[4,5-*d*]pyridazine. In a comparable manner, Erichomovitch [62] used triazole diesters **94** to obtain diacylhydrazides **95**, which were heated to form 1*H*-1,2,3-triazolo[4,5-*d*]pyridazines **96** in 80% yield with loss of hydrazine.

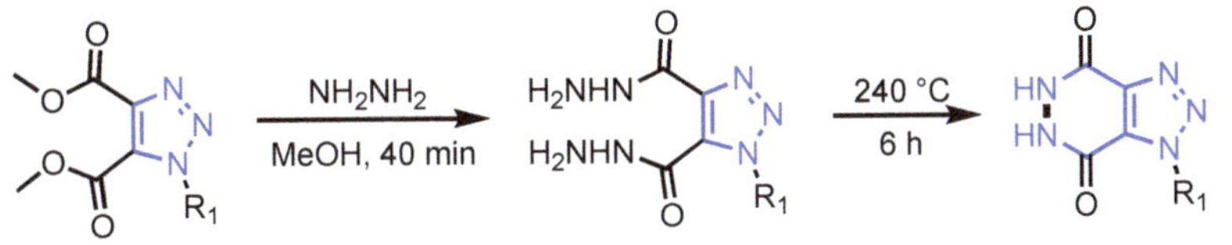

Scheme 25. Intramolecular cyclization of diacylhydrazide **95**, forming 1,2,3-triazolo[4,5-*d*]pyridazine **96** upon high heat with loss of hydrazine. (R_1 = alkyl.)

Janietz and coworkers [63] developed a scheme that proceeded through dichlorotriazole **97**, which, after conversion to a dinitrone and subsequent treatment with acid, afforded the dialdehyde **98**, which cyclized to form the desired 1*H*-1,2,3-triazolo[4,5-*d*]pyridazine **99** after treatment with hydrazine (Scheme 26).

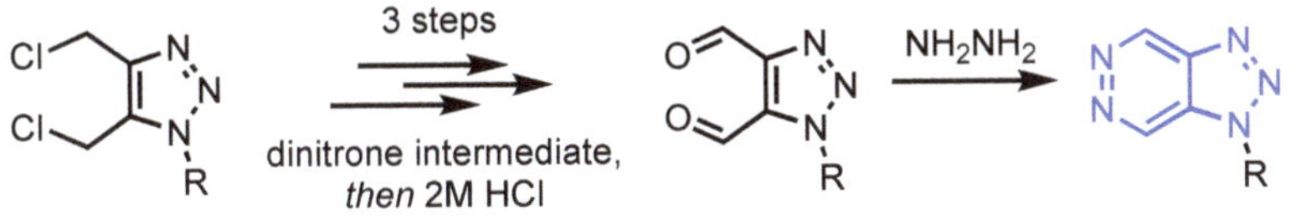

Scheme 26. Synthesis of substituted 1,2,3-triazolo[4,5-*d*]pyridazines **99** from 4,5-dichloromethyltriazoles **97**, proceeding through dialdehyde **98**. (R = aryl.)

Reports of forming 1,2,3-triazolo[4,5-*d*]pyridazones or pyridazines using this method include those of Gilchrist [64,65], Milhelcic [66], Ramesh [67], Theocharis [68], Bussolari [69], Biagi [70–72], Abu-Orabi [73], Ramanaiah [74], Bankowska [75], and others [5,76–78].

Martin and Castle [79] used ring closure by nitrosonium ion in their treatment of a 4,5-diamino-6-pyridazinone **101** in forming 3,5-dihydro-4*H*-1,2,3-triazolo[4,5-*d*]pyridazin-4-one **102** in 91% yield (Scheme 27). Commercially available 4,5-dichloro-3(2*H*)-pyridazinone **100** was converted to **101** in three steps. Similar methods of reacting substituted diaminopyridazines with nitrite have been conducted by Yanai [80] (conversion of **103** to **104** in Scheme 27), Chen [81], Draper [82], and Mataka [83].

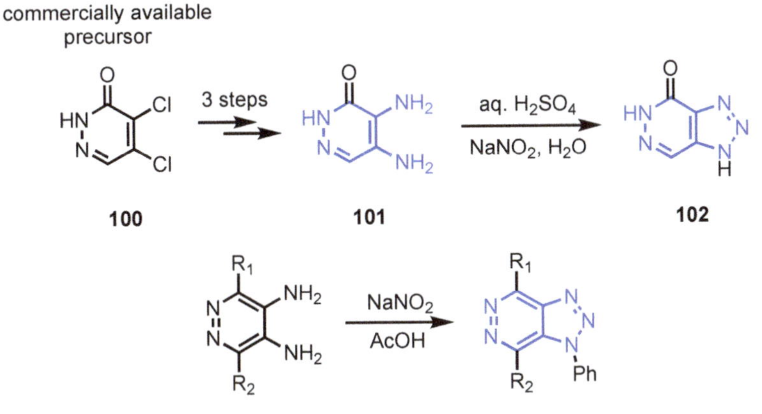

Scheme 27. Formation of 3,5-dihydro-4*H*-1,2,3-triazolo[4,5-*d*]pyridaz-4-one **102** upon treatment of 4,5-diamino-6-pyridazone **101** with nitrite, and a similar reaction of diaminopyridazine **103** cyclizing to **104** with nitrite. (R_1 = H, O-alkyl, SH, SCH_3, OH, NH_2, R_2 = H, CH_3, O-alkyl, OH, Cl.)

Smolyar and coworkers [84] reported a novel synthesis of a 1H-1,2,3-triazolo[4,5-d]pyridaz-4-one, **106** by a ring-opening/ring-closing "cyclotransformation" involving treatment of 1H-1,2,3-triazole-fused 5-nitropyridin-2(1H)-ones **105** with a large excess of hydrazine hydrate (Scheme 28). They reported that after heating for 3–4 h, at 140 °C, the desired pyrazinone was obtained in 86% yield with no chromatography required. 5-Nitropyridin-2(1H)-ones fused with benzene and pyridine were also studied in this report.

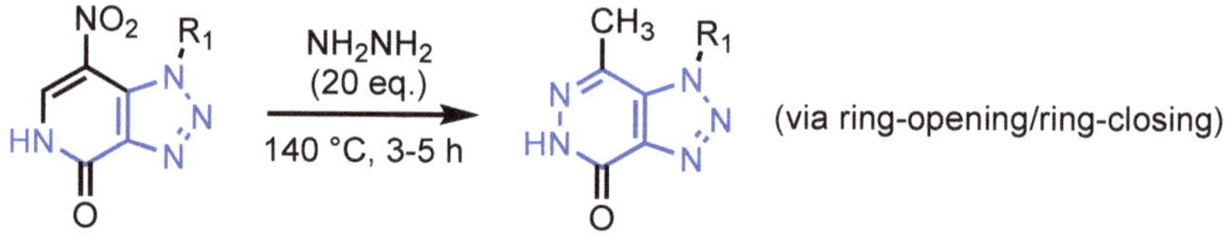

Scheme 28. Cyclotransformation of 1,2,3-triazole-fused lactams **105** to 1,5-dihydro-1,7-dimethyl-1,5-4H-1,2,3-triazolo[4,5-d]pyridazin-4-ones **106** in the presence of excess hydrazine hydrate and high heat. (R_1 = methyl, ethyl, butyl, cyclohexyl, and $(CH_2)_3NMe_2$.)

A number of methods exist for the preparation of molecules containing the 1,2,3-triazolo[4,5-d]pyridazine core, the majority of which involve the treatment of 1,2,3-triazole dicarbonyl species with hydrazine hydrate followed by acid or heat-promoted cyclization, or the cyclization of a diaminopyridazine with nitrite.

2.4. Syntheses of 1,2,3-Triazolo[1,5-b]pyridazines

Despite being reported as early as 1949 by Schofield and coworkers [85] in their study of cinnolines, 1,2,3-triazolo[1,5-b]pyridazines remain rare in the literature, in part owing to few methods available for their synthesis. While synthesizing azepinones, Evans and coworkers [86] instead serendipitously obtained 3,6-diphenyl-1,2,3-triazolo[1,5-b]pyridazine **108**. This was obtained from the intramolecular cyclization of diketo-oxime **107** (Scheme 29) after refluxing in HCl. This gave up to 22% of a pyrazinylhydrazone byproduct. A similar method in the same report used HOAc, but this resulted in poor yields (about 15%) and up to three products.

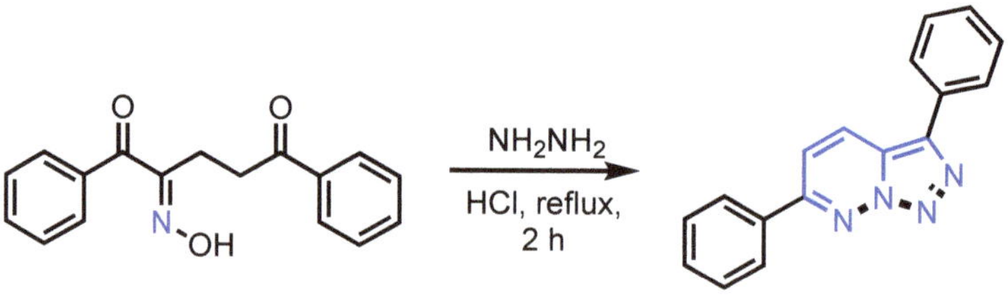

Scheme 29. Formation of a 1,2,3-triazolo[1,5-b]pyridazine **108** after the intramolecular cyclization of oxime **107**.

A fluoroborate salt was prepared by Riedl and coworkers [87] in a manner similar to that of Beres and coworkers [36]. The acyl-substituted pyridazine, **111**, after treatment with p-bromophenyl hydrazine hydrochloride **112** gave the hydrazone **113**. Tribromophenol bromine (TBP) in DCM afforded the desired ring-closed product **114** in 67% yield (Scheme 30). The initial bromide salt was converted to the fluoroborate salt with 40% fluoroboric acid in ACN. Ketone **111** was prepared by the same group via reaction of a commercially available 3-cyanopyridizine **109** with p-chlorophenylmagnesium bromide **110**, also synthesized from commercially available p-chlorobromobenzene and Mg. This was followed by acidic workup to afford the desired ketone. Compounds of this type were also prepared by Vasko and coworkers [88] using a similar method, which gave a 27% yield. A third method for the synthesis of 1,2,3-triazolo[1,5-*b*]pyridazines consisted of intramolecular oxidative ring closure of a hydrazone derived from **111** to afford the neutral 1,2,3-triazolo[1,5-*b*]pyridazine **115** [89]. Kvaskoff and coworkers employed MnO$_2$ as an oxidant using a similar procedure [35,89,90], where purification by sublimation afforded the desired product **115** (where R$_1$ = R$_2$ = H) in 71% yield.

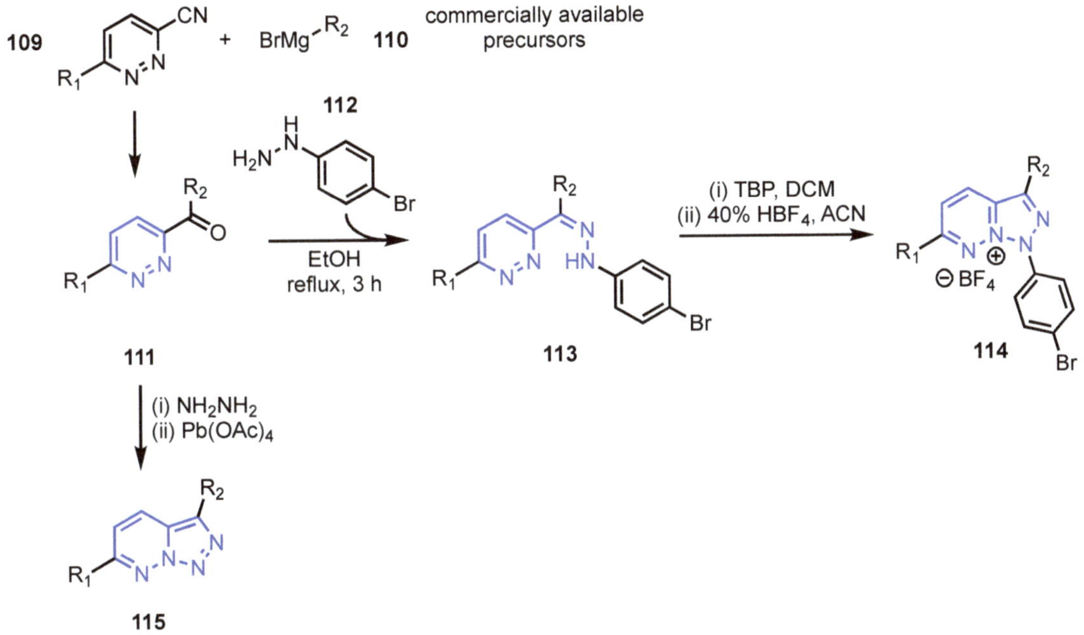

Scheme 30. Cyclization of hydrazone **113**, derived from acylpyridazine **111**, to afford the 1,2,3-triazolo[1,5-*b*]pyridazinium salt **114**, or 1,2,3-triazolo[1,5-*b*]pyridazine **115**. (R$_1$, R$_2$ = alkyl or aryl.)

2.5. Syntheses of 1,2,3-Triazolo[4,5-c]pyridazines

More prevalent in the literature than 1,2,3-triazolo[1,5-*b*]pyridazines but still uncommon are the 1,2,3-triazolo[4,5-*c*]pyridazines. One of the first reports of such a compound came from Gerhardt and coworkers [91], whereas in previous reports, nitrite was used to cyclize 5-chloro-3,4-diaminopyridazine **116** to afford 7-chloro-3*H*-1,2,3-triazolo[4,5-*c*]pyridazine **117** (Scheme 31) in 83% yield. Nitrite in the presence of an acid catalyst has been used for the synthesis of this heterocyclic ring system from the respective diaminopyridazines in other reports by Murakami [92], Lunt [93], Ramanaiah [74], and Owen [3].

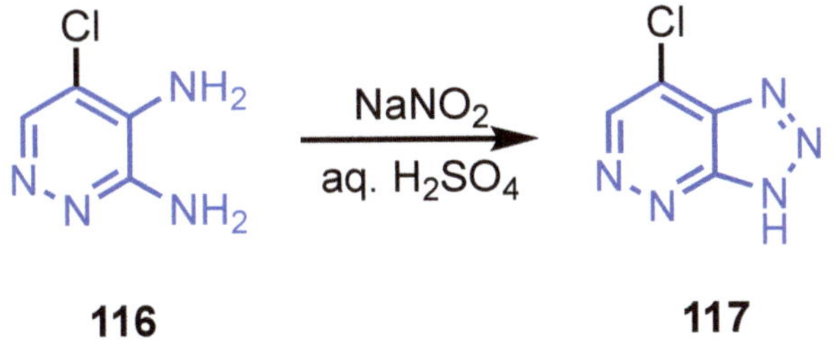

Scheme 31. Cyclization of diaminopyridazine **116** to give the triazolo[4,5-*c*]pyridazine **117**.

In a report by Pokhodylo and coworkers [94], nitrite was used in the synthesis of a substituted 1,2,3-triazolo[4,5-c]pyridazine despite only having one amine group present (as opposed to other cyclizations, which have two amine groups present). For example, 4-(3,4-dimethoxyphenyl)-1-phenyl-1H-1,2,3-triazol-5-amine **118** was reacted with sodium nitrite and glacial acetic acid to give the desired 3-(4-chlorophenyl)-7,8-dimethoxy-3H-[1,2,3]triazolo[4,5-c]cinnoline **119** in 35% yield (Scheme 32). Yields may have been low compared to other nitrite cyclizations due to the formation of a C-N bond directly with a carbon of an aromatic ring.

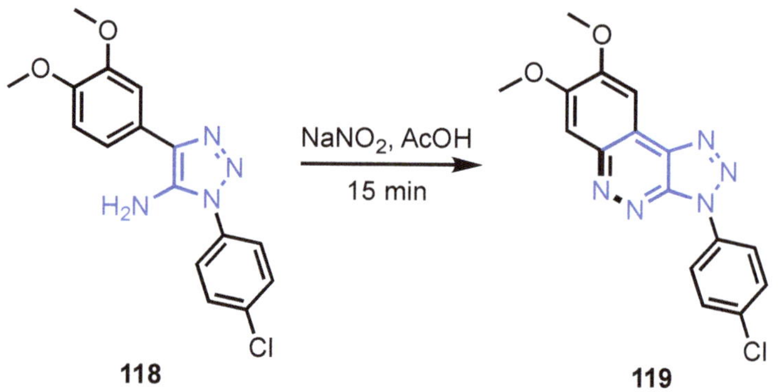

Scheme 32. Cyclization of aminotriazole **118** to give triazolo[4,5-c]cinnoline **119**.

Daniel and coworkers [22] formed tricyclic ylides **121** in 65% yield by oxidative cyclization of the respective *ortho*-substituted amino pyridazine **120** (Scheme 33). Unfortunately, compounds containing the 1,2,3-triazolo[4,5-c]pyridazine nucleus remain rare in the literature, and little is known of their biological or pharmacological properties.

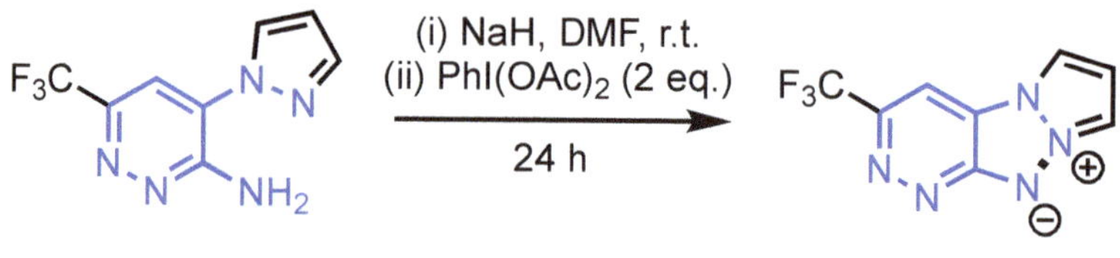

Scheme 33. Cyclization of aminotriazole **120** into 1,2,3-triazolo[4,5-c]pyridazinium ylide **121**.

3. Applications

Recent applications of the aforementioned heterocyclic systems, covering both medicinal and non-medicinal topics, are discussed in the following section.

3.1. Applications of 1H-1,2,3-Triazolo[4,5-b]pyrazines

In the last decade, 1H-1,2,3-triazolo[4,5-b]pyrazines have garnered an interest within the field of medicinal chemistry for serving as the scaffold of selective c-Met inhibitors. Medicinal studies of 1H-1,2,3-triazolo[4,5-b]pyrazines have extended well into the patent literature, with one patent even exploring antiviral efficacy against SARS-CoV-2 [95]. The first notable report of physiological activity came from Cui and coworkers [2], who reported the discovery of PF-04217903, a 1,2,3-triazolo[4,5-b]pyrazine that demonstrated potent (IC_{50} = 0.005 µM) and selective inhibition of over 200 c-Met kinases [2]. This heterocyclic scaffold in general gave rise to derivatives (altering substituents at the 2 and 6 ring positions) with potent inhibition, of which PF-04217903 was the best. This compound was selected as a preclinical candidate for the treatment of cancer [96].

Later, using PF-04217903 as a reference, Jia, and coworkers [1] reported the discovery of a compound now known as Savolitinib (Figure 3). This compound, also an exquisite c-Met inhibitor with an equal IC$_{50}$ of 0.005 µM, demonstrated favorable pharmacokinetic properties in mice [1]. Savolitinib possessed equal potency. Having recently passed phase II clinical trials for the treatment of metastatic non-small cell lung cancer, papillary and clear cell renal cell carcinoma, gastric cancer, and colorectal cancer, Savolitinib has been granted conditional approval for use in China at the time of this review [97]. A review of c-Met inhibitors in non-small cell lung cancer has recently appeared [98].

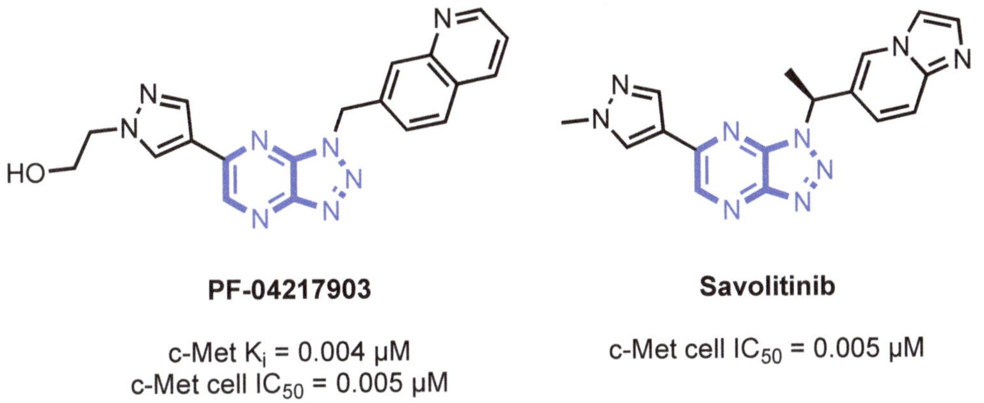

PF-04217903

c-Met K$_i$ = 0.004 µM
c-Met cell IC$_{50}$ = 0.005 µM

Savolitinib

c-Met cell IC$_{50}$ = 0.005 µM

Figure 3. Two potent and selective c-Met inhibitors containing the 1,2,3-triazolo[4,5-*b*]pyrazine core: PF-04217903 and Savolitinib.

Sirbu and coworkers [20] recently reported a novel class of small molecules containing the 1,2,3-triazolo[4,5-*b*]pyrazine scaffold with excellent properties for use as versatile fluorescent probes in optical imaging (Figure 4). Specifically, a phenyl ester derivative was used to dye HeLa cells in epifluorescence microscopy. Compared to commercially available LysoTracker Green DND-26, the tested triazolopyrazine derivative demonstrated comparable properties. In addition, it showed low cytotoxicity when evaluated in Alamar Blue assay (>95% cell viability up to 170 µM) and showed high solubility with a variety of desirable characteristics. A phenyl ester derivative, when evaluated as a dye in HeLa cells, showed high photostability and low cytotoxicity [20].

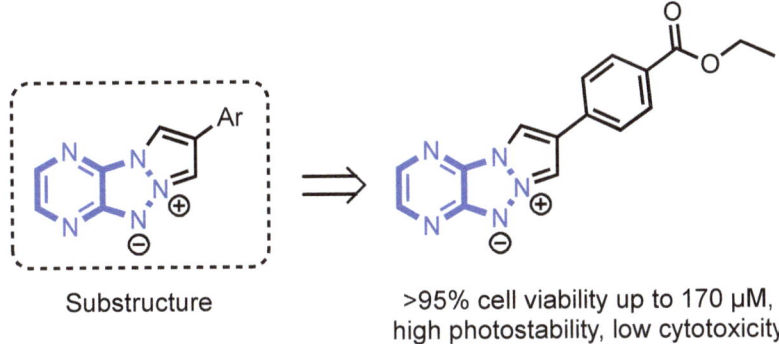

Substructure

>95% cell viability up to 170 µM, high photostability, low cytotoxicity

Figure 4. A novel class of small molecule fluorescent probes developed by Sirbu and coworkers [20] for use in optical and/or cellular imaging.

Intriguingly, another application lay in the monitoring of hypoxic regions within tumor cells. This was explored by Janczy-Cempa and coworkers [23], who looked at the fluorescent products produced after reduction of nitrotriazolopyrazine probes by nitroreductases (enzymes often overexpressed in tumor regions). Both probes studied (Figure 5) had very weak fluorescence in normoxic regions, but their reduction by nitroreductases led to a 15-fold increase in intensity in hypoxic regions. This was evaluated using the human melanoma cell line A2058. In contrast to the fluorescence probes developed by Sirbu and coworkers [20], probes in this study had substitutions on the pyrazine ring as opposed to the triazole-fused pyrazole. While additional work is still to be done, this report demonstrates the potential for these highly conjugated compounds to be useful in biomedical monitoring. Legentil and coworkers [99] obtained compounds similar to the structure on the right in Figure 5 in yields as high as 79%, which were used to develop a luminescence layered double-hydroxide filter. This material was dispersed into a polymer for use as a dye.

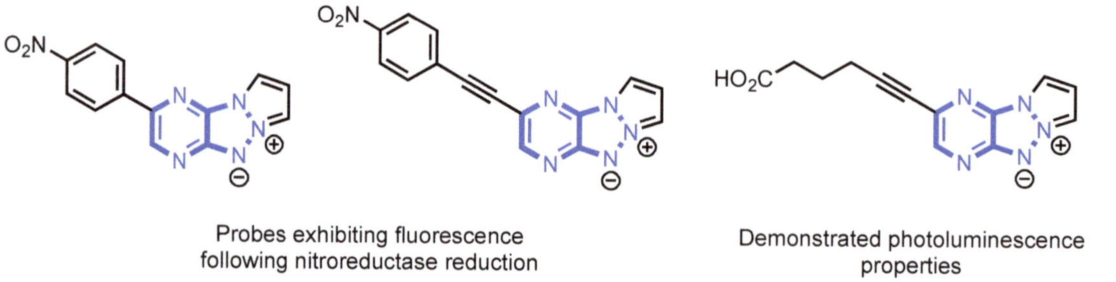

Probes exhibiting fluorescence following nitroreductase reduction

Demonstrated photoluminescence properties

Figure 5. Compounds containing 1,2,3-triazolo[4,5-*b*]pyrazines based on the nitro-pyrazinotriazapentalene scaffold [20].

Overall, applications of compounds containing 1,2,3-triazolo[4,5-*b*]pyrazines in the current literature are focused on c-Met inhibition (i.e., the treatment of distinct types of cancers), and optical and/or cellular imaging, with triazapentalene-type molecules demonstrating a wide range of favorable characteristics as fluorescent probes.

3.2. Applications of 1H-1,2,3-Triazolo[4,5-c]pyridazines

After being initially evaluated by Gerhardt and coworkers [91] as potential purine antagonists, 1H-1,2,3-triazolo[4,5-*c*]pyridazines have since found broader interest within medicinal chemistry. In a report by Owen and coworkers [3], a 1H-1,2,3-triazolo[4,5-*c*]pyridazine was found to have $GABA_A$ modulating activity during a structure–activity relationship study of the respective imidazolopyridazine. Compounds containing the 1,2,3-triazolo[4,5-*c*]pyridazine scaffold have been investigated in the patent literature for the treatment of Huntington's disease [100] and as modulators of Janus-family kinase-related diseases [101].

Other recent patents have been filed regarding fused pyridazines with herbicidal activity, of which 1,2,3-triazolo[4,5-*c*]pyridazine is included [102]. In another recent patent, compounds of this type were implicated in controlling unwanted plant growth [103].

Reports of compounds containing the 1,2,3-triazolo[4,5-*c*]pyridazine scaffold are uncommon in the current literature beyond synthetic reports and patents. Undoubtedly, there is still work to be done in exploring the potential applications of this unique heterocyclic system.

3.3. Applications of 1H-1,2,3-Triazolo[4,5-d]pyridazines

In a recent development, Li, and coworkers [4] outlined a series of triazole-based structures for the construction of conjugated polymers for solar cells. In addition to demonstrating desirable properties as units incorporated into polymers (Figure 6), their reported synthetic route uses affordable, commercially available starting materials and produces units compatible with other monomers. Structures containing 1,2,3-triazolo[4,5-d]pyridazine components offer a privileged, conjugated unit for the construction of polymers owing in part to the convenient para substitution of the pyridazine ring and perpendicular N2 substitution of the triazole ring.

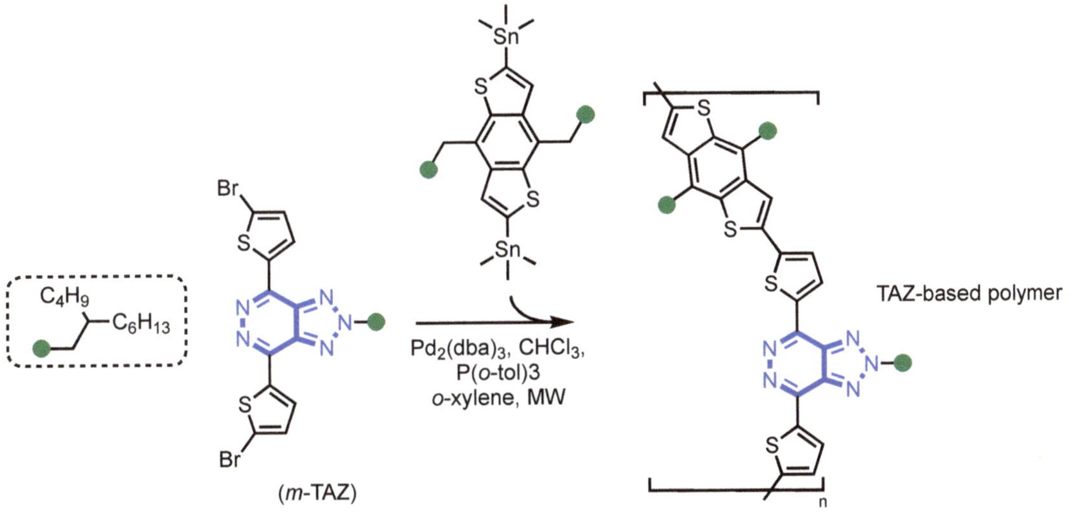

Figure 6. Structures containing the 1,2,3-triazolo[4,5-d]pyridazine-based monomer, *m*-TAZ, used to construct highly conjugated TAZ-based polymers.

Another notable outcome of the study of 1,2,3-triazolo[4,5-d]pyridazines was that from Biagi and coworkers [104], who reported compounds of this type with high selectivity for the A_1 receptor subtype in radioligand binding assays at bovine brain adenosine A_1 and A_{2A} receptors. The most potent compound contained a 4-amino-substituted 7-hydroxy-1,2,3-triazolo[4,5-d]pyridazine, and after substitution of the hydroxyl group for a chlorine, affinity decreased and suggested a hydrogen-bond donating substituent at position 7 was critical for binding affinity.

3.4. Applications of 1,2,3-Triazolo[1,5-a]pyrazines

Among applications of compounds containing the 1,2,3-triazolo[1,5-a]pyrazine unit are those of benzo-fused 1,2,3-triazoloquinoxalines and saturated 1,2,3-triazole-fused piperidines. In a recent report by Pérez Morales and coworkers [105], a 1,2,3-triazoloquinoxalinone (Structure A, Figure 7) was identified via high-throughput screening as inducing expression of Rgg2/3-regulated genes in the presence of short hydrophobic pheromones at low concentrations. This work stemmed from interest in the Rgg2/3 quorum sensing circuit of the pathogen *Streptococcus pyogenes*, with the objective of manipulating and inhibiting the bacteria. After analyzing its mode of action, it was determined this compound directly uncompetitively inhibited recombinant PepO in vitro, and induced quorum sensing signaling by stabilizing short hydrophobic pheromones.

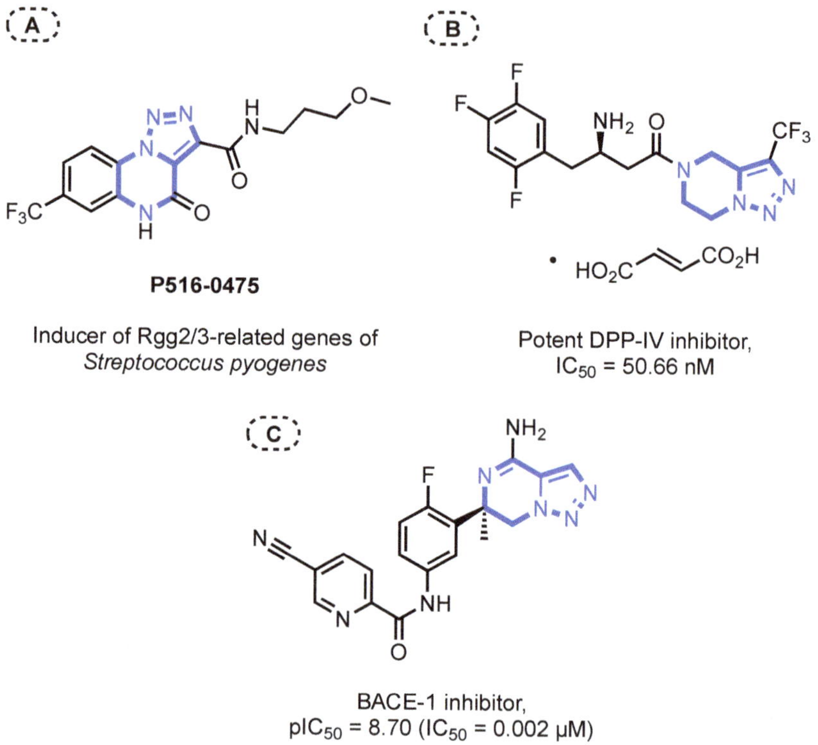

Figure 7. Compounds containing congeners of the 1,2,3-triazolo[1,5-*a*]pyrazine core with a diverse set of biological activities: (**A**) an inducer of Rgg2/3-related genes of the human pathogen *Streptococcus pyogenes* [105], (**B**) a potent DPP-IV inhibitor evaluated for the treatment of type II diabetes [7], and (**C**), an identified BACE-1 inhibitor [6].

Based on the antidiabetic 1,2,4-triazolopiperazine-containing drug Sitagliptin (brand name Januvia), Shan and coworkers [7] identified a dipeptidyl peptidase (DPP) IV inhibitor containing a 1,2,3-triazolopiperazine (Structure B, Figure 7) for use in the treatment of type II diabetes.

Partially saturated 1,2,3-triazolo[1,5-*a*]pyrazines have demonstrated BACE-1 inhibition, an enzyme implicated in the formation of amyloid beta in Alzheimer's disease. Oehlrich and coworkers [6] identified one such candidate, (*R*)-*N*-(3-(4-amino-6-methyl-6,7-dihydro-[1,2,3]triazolo[1,5-*a*]pyrazin-6-yl)-4-fluorophenyl)-5-cyanopicolinamide, (Structure C, Figure 7). This demonstrated an inhibition of the BACE-1 enzyme of pIC$_{50}$ = 8.70.

These reports, while not exhaustive, demonstrate recent applications of compounds containing the 1,2,3-triazolo[1,5-*a*]pyrazine scaffold or congeners thereof. Particularly prominent in the literature are benzo-fused and piperazine-containing analogs.

3.5. Applications of 1,2,3-Triazolo[1,5-b]pyridazines

There are no reported applications of compounds containing the 1,2,3-triazolo[1,5-*b*]pyridazine ring system, and little regarding its physiological and/or pharmacological effects are known. Aside from one recent patent [106] regarding immunoregulatory functions, additional applications remain scarce at the time of this review.

4. Conclusions

In reviewing synthetic approaches to and reported applications of members of the 1,2,3-triazolodiazine family of fused bicyclic heterocycles, the following conclusions can be drawn regarding the most common synthetic methods and applications in the present literature:

(a) *1,2,3-Triazolo[4,5-b]pyrazines*: The most common synthetic method is cyclization of an *ortho*-substituted diaminopyrazine [2], in which one of the amines does not need to be primary [18,25]. Given the current commercial availability and affordability of 2-amino-3,5-dibromopyrazine, this serves as a convenient starting material. Other methods include condensation of a dicarbonyl species with a 4,5-diamino-1,2,3-triazole [16], cyclization of a 2-azido-3-cyanoquinoxaline [17], or formation of azapentalenes from tetrazolopyrazines [19] or pyrazolopyrazines [21] with loss of nitrogen. *Primary Applications*: Primarily c-Met inhibition [1,2,25,96] and use as fluorescent probes in optical and/or cellular imaging [20,23,99].

(b) *1,2,3-Triazolo[1,5-a]pyrazines*: For non-fused derivatives, the most common methods are: intramolecular cyclization of pyrazinyl hydrazones [36,42], formation of 1,2,3-triazolo[1,5-a]pyrazinium-5-olates from cyano and amide groups [37,40], or reaction of iodopropiolamides to form triazolopiperazine [43]. For benzo-fused derivatives (i.e., those containing quinoxaline or quinoxalinone), the most common methods are: cyclization of a ring-bound 1,2,3-triazole with an *ortho*-substituted amine [52] or nitro [45] group (if a nitro group, either PBu$_3$ to give a quinoxaline [45] or FeCl$_3$ [48] to give a quinoxalinone), cyclization of 1-azido-2-isocyanoarenes or 1-triazolyl-2-isocyanoarenes [54,55], or intramolecular cyclization of alkynes [53,57]. *Primary Applications*: Primarily GABA$_A$ modulating activity [3], and patents detailing use as Janus-family kinase modulators [101] or for the treatment of Huntington's disease [100]. There also exist recent patents describing use as herbicides [102] and plant growth attenuators [103].

(c) *1,2,3-Triazolo[4,5-d]pyridazines*: The most common synthetic method is reaction of a 4,5-dicarbonyl-1,2,3-triazole species with hydrazine to form the hydrazone, followed by acid or heat promoted cyclization [5,64–70,73–78,104]. The second most common method is treatment of the respective diaminopyridazine with nitrite [80–83]. Ring-opening/ring-closing of lactams has also been reported [84]. *Primary Applications*: Use as highly conjugated linkers in triazole-based polymers [4] for the evaluation of solar cell materials is the main reported application.

(d) *1,2,3-Triazolo[1,5-b]pyridazines*: The most common synthetic method is treatment of a keto-substituted pyridazine with *p*-bromophenyl hydrazine hydrochloride forming the hydrazone, then treatment with TBP in DCM [87,88]. A report of intramolecular cyclization of a diketo-oxime has been reported [86]. *Primary Applications:* Benzo-fused or saturated piperazine-containing analogs are common. Notable reports include identification of a 1,2,3-triazole-fused quinoxalinone as inducing Rgg2/3-related gene expression in the human pathogen *Streptococcus pyogenes*, as a potent DPP IV inhibitor [7], and as a BACE-1 inhibitor [6].

(e) *1,2,3-Triazolo[4,5-c]pyridazines*: The most common method is cyclization of the respective diaminopyridazine with nitrite [3,74,91–93]. The intramolecular cyclization to form a pyridazine [94] or a tricyclic ylide have also been reported [22]. *Primary Applications*: Outside of a patent [106] detailing immunoregulatory functions, no other applications exist in the literature at the time of this review.

The potential for new synthetic contributions is considerable for the triazole-fused pyrazines and pyridazines. Given the diversity of synthetic methods summarized in this review, new contributions that could be most beneficial are new routes to some of the precursors of the fused systems. In many of the reports cited, the starting materials are either not available commercially or are very expensive. For example, some diamino pyrazines are available as unsubstituted compounds or as halogenated derivatives, but all are USD 500–1000 per gram. Future studies of methods employing additional intramolecular cycloadditions leading to 1,2,3-triazolo[1,5-*a*]pyrazine derivatives would appear to have potential. Work on synthesis of the 1H-1,2,3-triazolo[4,5-c]pyridazines and the 1,2,3-triazolo[1,5-b]pyridazines would be welcome for these less frequently studied areas.

Overall, diverse methods exist for the preparation of 1,2,3-triazole-fused diazines, spanning the last seven decades with numerous reports in the last five years. Currently, drugs containing these ring systems remain scarce with only a handful of exceptions, particularly containing either the 1,2,3-triazolo[4,5-b]pyrazine or 1,2,3-triazolo[4,5-c]pyridazine scaffold. Applications of the aforementioned types of compounds span from medicinal chemistry into the development of dyes, probes, and inhibitors of enzymes implicated in various diseases. Despite this, there lies underrealized and exciting potential for employing triazolopyrazines and triazolopyridazines as diverse substrates in the generation of novel molecules with a wide array of applications.

Author Contributions: Conceptualization: G.R.H., A.M.S. (these authors contributed equally); writing—original draft preparation: G.R.H.; preparation of graphics: G.R.H.; validation: A.M.S.; writing—review and editing: G.R.H., A.M.S. (these authors contributed equally). All authors have read and agreed to the published version of the manuscript.

Funding: This research received no external funding.

Institutional Review Board Statement: Not applicable.

Informed Consent Statement: Not applicable.

Data Availability Statement: Not applicable.

Acknowledgments: G.R.H. would like to acknowledge his grandparents, Bob and Shirley Hoffman, for their support and encouragement throughout the research and writing process.

Conflicts of Interest: The authors declare no conflict of interest.

Sample Availability: Samples of compounds are not available from the authors.

References

1. Jia, H.; Dai, G.; Weng, J.; Zhang, Z.; Wang, Q.; Zhou, F.; Jiao, L.; Cui, Y.; Ren, Y.; Fan, S.; et al. Discovery of (S)-1-(1-(imidazo[1,2-a]pyridin-6-yl)ethyl)-6-(1-methyl-1H-pyrazol-4-yl)-1H-[1,2,3]triazolo[4,5-b]pyrazine (Volitinib) as a highly potent and selective mesenchymal-epithelial transition factor (c-Met) inhibitor in clinical development for treatment of cancer. *J. Med. Chem.* **2014**, *57*, 7577–7589. [CrossRef] [PubMed]
2. Cui, J.J.; McTigue, M.; Nambu, M.; Tran-Dubé, M.; Pairish, M.; Shen, H.; Jia, L.; Cheng, H.; Hoffman, J.; Le, P.; et al. Discovery of a novel class of exquisitely selective mesenchymal-epithelial transition factor (c-MET) protein kinase inhibitors and identification of the clinical candidate 2-(4-(1-(quinolin-6-ylmethyl)-1H-[1,2,3]triazolo[4,5-b]pyrazin-6-yl)-1H-pyrazol-1-yl)ethanol (PF-04217903) for the treatment of cancer. *J. Med. Chem.* **2012**, *55*, 8091–8109. [CrossRef] [PubMed]
3. Owen, R.M.; Blakemore, D.; Cao, L.; Flanagan, N.; Fish, R.; Gibson, K.R.; Gurrell, R.; Huh, C.W.; Kammonen, J.; Mortimer-Cassen, E.; et al. Design and identification of a novel, functionally subtype selective GABAA positive allosteric modulator (PF-06372865). *J. Med. Chem.* **2019**, *62*, 5773–5796. [CrossRef] [PubMed]
4. Li, W.; Yan, L.; Zhou, H.; You, W. A general approach toward electron deficient triazole units to construct conjugated polymers for solar cells. *Chem. Mater.* **2015**, *27*, 6470–6476. [CrossRef]
5. Swarup, H.A.; Kemparajegowda; Mantelingu, K.; Rangappa, K.S. Effective and transition-metal-free construction of disubstituted, trisubstituted 1,2,3-NH-triazoles and triazolo pyridazine via intermolecular 1,3-dipolar cycloaddition reaction. *ChemistrySelect* **2018**, *3*, 703–708. [CrossRef]
6. Oehlrich, D.; Peschiulli, A.; Tresadern, G.; van Gool, M.; Vega, J.A.; de Lucas, A.I.; de Diego, S.A.A.; Prokopcova, H.; Austin, N.; van Brandt, S.; et al. Evaluation of a series of β-Secretase 1 inhibitors containing novel heteroaryl-fused-piperazine amidine warheads. *ACS Med. Chem. Lett.* **2019**, *10*, 1159–1165. [CrossRef]
7. Shan, Z.; Peng, M.; Fan, H.; Lu, Q.; Lu, P.; Zhao, C.; Chen, Y. Discovery of potent dipeptidyl peptidase IV inhibitors derived from β-aminoamides bearing substituted [1,2,3]-triazolopiperidines for the treatment of type 2 diabetes. *Bioorg. Med. Chem. Lett.* **2011**, *21*, 1731–1735. [CrossRef]
8. Fischer, G. Recent advances in 1,2,4-triazolo[1,5-a]pyrimidine chemistry. In *Advances in Heterocyclic Chemistry*; Academic Press Inc.: Cambridge, MA, USA, 2019; Volume 128, pp. 1–101.
9. Bhavsar, Z.A.; Patel, H.D. New dimensions in triazolo[4,3-a]pyrazine derivatives: The land of opportunity in organic and medicinal chemistry. *Add abbrev. Arab. J. Chem.* **2020**, *13*, 8532–8591.
10. Kumar, H.; Dhameja, M.; Rizvi, M.; Gupta, P. Progress in the synthesis of fused 1,2,3-triazoles. *ChemistrySelect* **2021**, *6*, 4889–4947. [CrossRef]

11. Wierzchowski, J.; Antosiewicz, J.M.; Shugar, D. 8-Azapurines as Isosteric Purine Fluorescent Probes for Nucleic Acid and Enzymatic Research. *Mol. BioSyst.* **2014**, *10*, 2756–2774. [CrossRef]
12. Wierzchowski, J. Excited-State Proton Transfer and Phototautomerism in Nucleobase and Nucleoside Analogs: A Mini-Review. *Nucleosides Nucleotides Nucleic Acids* **2014**, *33*, 626–644. [CrossRef] [PubMed]
13. Giorgi, I.; Scartoni, V. 8-Azapurine Nucleus: A versatile scaffold for different targets. *Mini. Rev. Med. Chem.* **2009**, *9*, 1367–1378. [CrossRef] [PubMed]
14. Vorbrueggen, H.; Ruh-Pohlenz, C. Synthesis of nucleosides. In *Organic Reactions*; John Wiley & Sons: Hoboken, NJ, USA, 2000; p. 55. [CrossRef]
15. Albert, A. Chemistry of 8-azapurines (1,2,3-triazolo[4,5-d]pyrimidines). *Adv. Heterocycl. Chem.* **1986**, *39*, 117–180. [CrossRef]
16. Lovelette, C.A.; Long, L., Jr. Nonbridgehead fused nitrogen heterocycles. Fused 1,2,3-triazoles. *J. Org. Chem.* **1972**, *37*, 4124–4128.16. [CrossRef]
17. Monge, A.; Palop, J.A.; Piñol, A.; Martinex-Crespo, F.J.; Narro, S.; Gonzalez, M.; Sáinz, Y.; de Cerain, A.L. 3-amino-2-quinoxalinecarbonitrile. New fused quinoxalines with potential cytotoxic activity. *J. Heterocycl. Chem.* **1994**, *31*, 1135–1139. [CrossRef]
18. Starchenkov, B.; Andrianov, V.G.; Mishnev, A.F. Chemistry of Furazano[3,4-b]pyrazine. 5. 1,2,3-triazolo[4,5-e]furazano[3,4-b]pyrazine 6-oxides. *Chem. Heterocycl. Compd. Transl. Khimiya Geterotsiklicheskikh Soedin.* **1997**, *33*, 1355–1359. [CrossRef]
19. Slepukhin, P.A.; Rusinov, G.L.; Dedeneva, S.S.; Charushin, V.N. Transformations of 8-substituted tetrazolo[1,5-a]pyrazines. *Russ. Chem. Bull.* **2007**, *56*, 345–350. [CrossRef]
20. Sirbu, D.; Diharce, J.; Martinić, I.; Chopin, N.; Eliseeva, S.V.; Guillaumet, G.; Petoud, S.; Bonnet, P.; Suzenet, F. An original class of small sized molecules as versatile fluorescent probes for cellular imaging. *Chem. Commun.* **2019**, *55*, 7776–7779. [CrossRef]
21. Nyffenegger, C.; Pasquinet, E.; Suzenet, F.; Poullain, D.; Guillaumet, G. Synthesis of nitro-functionalized polynitrogen tricycles bearing a central 1,2,3-triazolium ylide. *Synlett* **2009**, *8*, 1318–1320. [CrossRef]
22. Daniel, M.; Hiebel, M.A.; Guillaumet, G.; Pasquinet, E.; Suzenet, F. Intramolecular metal-free N−N bond formation with heteroaromatic amines: Mild access to fused-triazapentalene derivatives. *Chem. Eur. J.* **2020**, *26*, 1525–1529. [CrossRef]
23. Janczy-Cempa, E.; Mazuryk, O.; Sirbu, D.; Chopin, N.; Żarnik, M.; Zastawna, M.; Colas, C.; Hiebel, M.A.; Suzenet, F.; Brindell, M. Nitro-pyrazinotriazapentalene scaffolds–nitroreductase quantification and in vitro fluorescence imaging of hypoxia. *Sens. Actuators B Chem.* **2021**, *346*, 130504. [CrossRef]
24. Palmas, P.; Nyffenegger, C.; Pasquinet, E.; Guillaumetb, G. ^{1}H, ^{13}C and ^{15}N NMR Spectral Assignments for New Triazapentalene Derivatives. *Magn. Reson. Chem.* **2009**, *47*, 752–756. [CrossRef] [PubMed]
25. Cheng, H.; Cui, J.J.; Hoffman, J.E.; Jia, L.; Johnson, M.C.; Kania, R.S.; Le, P.T.Q.; Nambu, M.D.; Pairish, M.A.; Shen, H.; et al. Triazolopyrazine Derivatives Useful as Anti-Cancer Agents. WO2007138472A2, 6 December 2007.
26. Ye, L.; Tian, Y.; Li, Z.; Zhang, J.; Wu, S. Design and synthesis of some novel 2,3,4,5-tetrahydro-1H-pyrido[4,3-b]indoles as potential c-Met inhibitors. *Helv. Chim. Acta* **2012**, *95*, 320–326. [CrossRef]
27. Ye, L.; Tian, Y.; Li, Z.; Jin, H.; Zhu, Z.; Wan, S.; Zhang, J.; Yu, P.; Zhang, J.; Wu, S. Design, synthesis and molecular docking studies of some novel spiro[indoline-3, 4′-piperidine]-2-ones as potential c-Met inhibitors. *Eur. J. Med. Chem.* **2012**, *50*, 370–375. [CrossRef] [PubMed]
28. Thottempudi, V.; Yin, P.; Zhang, J.; Parrish, D.A.; Shreeve, J.M. 1,2,3-Triazolo[4,5-e]furazano[3,4,-b]pyrazine 6-oxide-a fused heterocycle with a roving hydrogen forms a new class of insensitive energetic materials. *Chem. Eur. J.* **2014**, *20*, 542–548. [CrossRef]
29. Zhao, F.; Zhang, L.D.; Hao, Y.; Chen, N.; Bai, R.; Wang, Y.J.; Zhang, C.C.; Li, G.S.; Hao, L.J.; Shi, C.; et al. Identification of 3-substituted-6-(1-(1H-[1,2,3]triazolo[4,5-b]pyrazin-1-yl)-ethylquinoline derivatives as highly potent and selective mesenchymal-epithelial transition factor (c-Met) inhibitors via metabolite profiling-based structural optimization. *Eur. J. Med. Chem.* **2017**, *134*, 147–158. [CrossRef]
30. Adlington, N.K.; Agnew, L.R.; Campbell, A.D.; Cox, R.J.; Dobson, A.; Barrat, C.F.; Gall, M.A.Y.; Hicks, W.; Howell, G.P.; Jawor-Baczynska, A.; et al. Process design and optimization in the pharmaceutical industry: A Suzuki-Miyaura procedure for the synthesis of savolitinib. *J. Org. Chem.* **2019**, *84*, 4735–4747. [CrossRef]
31. Tupychak, M.A.; Obushak, M.D. New methods for the synthesis of substituted 4,5,6,7-tetrahydro[1,2,3]triazolo[1,5-a]pyrazines (microreview). *Chem. Heterocycl. Compd.* **2021**, *57*, 1164–1166. [CrossRef]
32. Shafran, E.A.; Bakulev, V.A.; Rozin, Y.A.; Shafran, Y.M. Condensed 1,2,3-triazoles. *Chem. Heterocycl. Compd.* **2008**, *44*, 1040–1069. [CrossRef]
33. Baashen, M.A.; Abdel-Wahab, B.F.; El-Hitl, G.L. Syntheses of triazoloquinoxalines. *Heterocycles* **2016**, *92*, 1931–1952. [CrossRef]
34. Wentrup, C. [1,2,3]Triazoloazine/(Diazomethyl)Azine Valence Tautomers from 5-Azinyltetrazoles. *Helv. Chim. Acta* **1978**, *61*, 1755–1761. [CrossRef]
35. Maury, G.; Meziane, D.; Srairi, D.; Paugan, J.P.; Paugam, R. 1,2-Triazolo[1,5]azines and other nitrogen heterocycles derived from azinecarboxaldehydes. *Bull. Soc. Chim. Belges.* **1982**, *91*, 153–161. [CrossRef]
36. Beres, M.; Hajos, G.; Riedl, Z.; Timdri, G.; Messmer, A.; Holly, S.; Schantl, J.G. Ring opening of 1,2,3-triazolo[1,5-a]pyrazinium salts: Synthesis and some transformations of a novel type of 2-aza-1,3-butadienes. *Tetrahedron* **1997**, *53*, 9393–9400. [CrossRef]
37. Nein, Y.I.; Morzherin, Y.Y.; Rozin, Y.A.; Bakulev, V.A. Synthesis of [1,2,3]triazolo-[1,5-a]pyrazinium-3-oleate. *Khimiya Geterotsiklicheskikh Soedin.* **2002**, *9*, 1302–1303. [CrossRef]
38. Nein, Y.I.; Savel'eva, E.A.; Rozin, Y.A.; Bakulev, V.A.; Morzherin, Y.Y. Synthesis of condensed mesoionic heterocycles. Intramolecular cyclization of 3-acetonyl(phenacyl)-1,2,3-triazolium-5-oleates. *Chem. Heterocycl. Compd.* **2006**, *42*, 412–413. [CrossRef]

39. Savel'eva, E.A.; Rozin, Y.A.; Kodess, M.I.; van Meervelt, L.; Dehaen, W.; Morzherin, Y.Y.; Bakulev, V.A. Synthesis of mesoionic[1,2,3]triazolo[5,1-d][1,2,5]triazepines. *Tetrahedron* **2004**, *60*, 5367–5372. [CrossRef]
40. Jug, T.; Polak, M.; Trcek, T.; Vercek, B. A novel approach to [1,2,3]triazolo[1,5-a]pyrazines. *Heterocycles* **2002**, *56*, 353. [CrossRef]
41. Trcek, T.; Vercek, B. Synthesis of [1,2,3]triazolo[1,5-a][1,2,4]triazolo[5,1-c]pyrazines. *Arkivoc* **2003**, *14*, 246–252. [CrossRef]
42. Raghavendra, M.S.; Lam, Y. Regiospecific solid-phase synthesis of substituted 1,2,3-triazoles. *Tetrahedron Lett.* **2004**, *45*, 6129–6132. [CrossRef]
43. Xiao, G.; Wu, K.; Zhou, W.; Cai, Q. Access to triazolopiperidine derivatives via copper(I)-catalyzed [3+2] cycloaddition/alkenyl C−N coupling tandem reactions. *Adv. Synth. Catal.* **2021**, *363*, 4988–4991. [CrossRef]
44. Koguchi, S.; Sakurai, A.; Niwa, K. One-pot synthesis of [1,2,3]triazolo[1,5-a]pyrazine derivatives from ynones and amino azide. *Heterocycles* **2015**, *91*, 41–48. [CrossRef]
45. Kauer, J.C.; Carboni, R.A. Aromatic azapentalenes. III. 1,3a,6,6a-tetraazapentalenes. *J. Am. Chem. Soc.* **1967**, *89*, 2633–2637. [CrossRef]
46. Cue, B.W., Jr.; Czuba, L.J.; Dirlam, J.P. Azoloquinoxaline N-oxides. *J. Org. Chem.* **1978**, *43*, 4125–4128. [CrossRef]
47. Ager, I.R.; Barnes, A.C.; Danswan, G.W.; Hairsine, P.W.; Kay, D.P.; Kennewell, P.D.; Matharu, S.S.; Miller, P.; Robson, P.; Rowlands, D.A.; et al. Synthesis and oral antiallergic activity of carboxylic acids derived from imidazo[2,1-c][1,4]benzoxazines, imidazo[1,2-a]quinolines, imidazo[1,2-a]quinoxalines, imidazo[1,2-a]quinoxalinones, pyrrolo[1,2-a]quinoxalinones, pyrrolo[2,3-a]quinoxalinones, and imidazo[2,1-b]benzothiazoles. *J. Med. Chem.* **1988**, *31*, 1098–1115. [CrossRef] [PubMed]
48. Bertelli, L.; Biagi, G.; Giorgi, I.; Manera, C.; Livi, O.; Scartoni, V.; Betti, L.; Giannaccini, G.; Trincavelli, L.; Barili, P.L. 1,2,3-triazolo[1,5-a]quinoxalines: Synthesis and binding to benzodiazepine and adenosine receptors. *Eur. J. Med. Chem.* **1998**, *33*, 113–122. [CrossRef]
49. Biagi, G.; Giorgi, I.; Livi, O.; Scartoni, V.; Betti, L.; Giannaccini, G.; Trincavelli, M.L. New 1,2,3-triazolo[1,5-a]quinoxalines: Synthesis and binding to benzodiazepine and adenosine receptors. II. *Eur. J. Med. Chem.* **2002**, *37*, 565–571. [CrossRef]
50. Shen, H.C.; Ding, F.-X.; Deng, Q.; Wilsie, L.C.; Krsmanovic, M.L.; Taggart, A.K.; Carballo-Jane, E.; Ren, N.; Cai, T.-Q.; Wu, T.-J.; et al. Discovery of Novel Tricyclic Full Agonists for the G Protein Coupled Niacin Receptor 109A with Minimized Flushing in Rats. *J. Med. Chem.* **2009**, *52*, 2587–2602. [CrossRef]
51. Abbott, P.A.; Bobbert, R.V.; Caffrey, M.V.; Cage, P.A.; Cooke, A.J.; Donald, D.K.; Furber, M.; Hill, S.; Withnall, J. Fused mesoionic heterocycles: Synthesis of [1,2,3]triazolo[1,5-a]quinoline, [1,2,3]triazolo[1,5-a]quinazoline, [1,2,3]triazolo[1,5-a]quinoxaline and [1,2,3]triazolo[5,1-c]benzotriazine derivatives. *Tetrahedron* **2002**, *58*, 3185–3198. [CrossRef]
52. Saha, B.; Sharma, S.; Sawant, D.; Kundu, B. Application of the Pictet-Spengler Reaction to aryl amine substrates linked to deactivated aromatic heterosystems. *Tetrahedron* **2008**, *64*, 8676–8684. [CrossRef]
53. Chen, Z.; Zhu, J.; Xie, H.; Li, S.; Wu, Y.; Gong, Y. Copper(I)-catalyzed synthesis of novel 4-(trifluoromethyl)-[1,2,3]triazolo[1,5-a]quinoxalines via cascade reactions of N-(o-haloaryl)alkynylimine with sodium azide. *Adv. Syn. Cat.* **2010**, *352*, 1296–1300. [CrossRef]
54. He, Z.; Bae, M.; Wu, J.; Jamison, T.F. Synthesis of highly functionalized polycyclic quinoxaline derivatives using visible-light photoredox catalysis. *Angew. Chem. Int. Ed.* **2014**, *53*, 14451–14455. [CrossRef]
55. Li, D.; Mao, T.; Huang, J.; Zhu, Q. A one-pot synthesis of [1,2,3]triazolo[1,5-a]quinoxalines from 1-azido-2-isocyanoarenes with high bond forming efficiency. *Chem. Commun.* **2017**, *53*, 1305–1308. [CrossRef]
56. Kotovshchikov, Y.N.; Latyshev, G.V.; Beletskaya, I.P.; Lukashev, N.V. Regioselective approach to 5-carboxy-1,2,3-triazoles based on palladium-catalyzed carbonylation. *Synthesis* **2018**, *50*, 1926–1934. [CrossRef]
57. Chen, W.; Tu, X.; Xu, M.; Chu, Y.; Zhu, Y. Copper(I) iodide catalyzed tandem reactions of N-propargyl-N-(2-iodoaryl)amides with sodium azide: An efficient synthesis of [1,2,3]triazolo[1,5-a]quinoxalines. *Synlett* **2021**, *32*, 805–809. [CrossRef]
58. Addicott, C.; Luerssen, H.; Kuzaj, M.; Kvaskoff, D.; Wentrup, C. 4-Quinolylnitrene and 2-quinoxalinylcarbene. *J. Phys. Org. Chem.* **2011**, *24*, 999–1008. [CrossRef]
59. Vogel, M.; Lippmann, E. Synthesis of 4-substituted 1,2,3-triazolo[1,5-a]quinoxalines. *J. Prakt. Chem.* **1987**, *329*, 101–107. [CrossRef]
60. Livi, O.; Scartoni, V. 1,2,3-triazolo[4,5-d]pyridazines. *Il Farm.* **1997**, *52*, 205–211.
61. Jane, O.F.; Miller, J.B. Preparation of 4,5-disubstituted 1,2,3-triazoles. *J. Heterocycl. Chem.* **1965**, *2*, 488–490. [CrossRef]
62. Erichomovitch, L.; Chubb, F.L. La synthése de quelques nouvelles pyridazines bicycliques. *Can. J. Chem.* **1966**, *44*, 2095–2100. [CrossRef]
63. Janietz, D.; Rudorf, W.-D. Simple synthesis of 1-aryl-1,2,3-triazole-4,5-dialdehydes. *Z. Chem.* **1988**, *28*, 211–212. [CrossRef]
64. Gilchrist, T.L.; Gymer, G.E.; Rees, C.W. Benzonitrile N-(phthalimido)imide, a functionalized 1,3-dipole. Preparation of 4,5,8-triphenylpyridazino[4,5-d]triazine and generation of 3,6-diphenyl-4,5-didehydropyridazine. *J. Chem. Soc. Perkin Trans. 1 Org. Bio-Org. Chem.* **1975**, *18*, 1747–1750. [CrossRef]
65. Gilchrist, T.L.; Gymer, G.E.; Rees, C.W. Fragmentation of 3,6-diphenyl-4,5-dehydropyridazine to diphenylbutadiyne. *Chem. Commun.* **1973**, *21*, 819–820. [CrossRef]
66. Mihelcic, B.; Simonic, S.; Stanovni, B.; Tisler, M. Pyridazines. LXXI. Addition of azidoazolopyridazines to triple bonds. *Croat. Chem. Acta* **1974**, *46*, 275–278.
67. Ramesh, K.; Panzica, R.P. A convenient synthesis of 1-(β-D-ribofuranosyl)imidazo[4,5-d]pyridazin-4(5H)-one (2-aza-3-deazainosine) and its 2′-deoxy counterpart by ring closure of imidazole nucleosides. *J. Chem. Soc. Perkin Trans. 1* **1989**, *10*, 1769–1774. [CrossRef]
68. Theocharis, A.B.; Alexandrou, N.E. Generation and dienophilic properties of 1-benzyl-1H-1,2,3-triazolo[4,5-d]pyridazine-4,7-dione. *J. Heterocycl. Chem.* **1990**, *27*, 1741–1744. [CrossRef]

69. Bussolari, J.C.; Ramesh, K.; Stoeckler, J.D.; Chen, S.F.; Panzica, R.P. Synthesis and biological evaluation of N4-substituted imidazo- and v-triazolo[4,5-d]pyridazine nucleosides. *J. Med. Chem.* **1993**, *36*, 4113–4120. [CrossRef]
70. Biagi, G.; Ferretti, M.; Giorgi, I.; Livi, O.; Scartoni, V. 1,2,3-Triazole[4,5-d]pyridazines I. Analogues of Prostaglandin synthesis inhibitors. *Il Farm.* **1993**, *48*, 1159–1165.
71. Biagi, G.; Ciambrone, F.; Giorgi, I.; Livi, O.; Scartoni, V.; Barili, P.L. New 1,2,3-triazolo[4,5-d]-1,2,4-triazolo[3,4-b]pyridazine derivatives II. *J. Heterocycl. Chem.* **2002**, *39*, 889–893. [CrossRef]
72. Biagi, G.; Giorgi, I.; Livi, O.; Manera, C.; Scartoni, V. 1,2,3-Triazolo[4,5-d]-1,2,4-triazolo[4,3-b]pyridazines. *J. Heterocycl. Chem.* **1997**, *34*, 65–69. [CrossRef]
73. Abu-Orabi, S.T.; Al-Hamdany, R.; Shahateet, S.; Abu-Shandi, K. Synthesis of substituted benzyl-1H-1,2,3-triazolo[4,5-d]pyridazidine-4,7-diones. *Heterocycl. Commun.* **2000**, *6*, 443–449. [CrossRef]
74. Ramanaiah, K.V.C.; Stevens, E.D.; Trudell, M.L. Synthesis of 1-substituted [1,2,3]triazolo[4,5-d]pyridazines as precursors for novel tetraazapentalene derivatives. *J. Heterocycl. Chem.* **2000**, *37*, 1597–1602. [CrossRef]
75. Bankowska, E.; Balzarini, J.; Głowacka, I.E.; Wróblewski, A.E. Design, synthesis, antiviral and cytotoxic evaluation of novel acyclic phosphonate nucleotide analogues with a 5,6-dihydro-1H-[1,2,3]triazolo[4,5-d] pyridazine-4,7-dione system. *Monatsh. Chem.* **2014**, *145*, 663–673. [CrossRef]
76. Pokhodylo, N.T.; Shyyka, O.Y.; Obushak, M.D. Convenient synthetic path to ethyl 1-aryl-5-formyl-1H-1,2,3-triazole-4-carboxylates and 1-aryl-1,5-dihydro-4H-[1,2,3]triazolo[4,5-d]pyridazin-4-ones. *Chem. Heterocycl. Compd.* **2018**, *54*, 773–779. [CrossRef]
77. Cui, X.; Zhang, X.; Wang, W.; Zhong, X.; Tan, Y.; Wang, Y.; Zhang, J.; Li, Y.; Wang, X. Diazo transfer reaction for the synthesis of 1,4,5-trisubstituted 1,2,3-triazoles and subsequent regiospecific construction of 1,4-disubstituted 1,2,3-triazoles via C-C bond cleavage. *J. Org. Chem.* **2021**, *86*, 4071–4080. [CrossRef]
78. Cao, W. 4,5-dicyano-1,2,3-triazole—A aromising precursor for a new family of energetic compounds and its nitrogen-rich derivatives: Synthesis and srystal structures. *Molecules* **2021**, *26*, 6735. [CrossRef]
79. Martin, S.F.; Castle, R.N. The synthesis of imidazo[4,5-d] pyridazines. VI. v-triazolo[4,5-d] pyridazines, pyrazino[2,3-d] pyridazines and 7H-imidazo[4,5-d] tetrazolo[1,5-b]pyridazine. *J. Heterocycl. Chem.* **1969**, *6*, 93–98. [CrossRef]
80. Yanai, M.; Kinoshita, T.; Takeda, S.; Mori, M.; Sadaki, H.; Watanabe, H. Studies on the synthesis of pyridazine derivatives. XIII. Synthesis of imidazolo-[4,5-d]- and v-triazolo[4,5-d]pyridazine derivatives and reaction of 4-alkoxy-v-triazolo[4,5-d]pyridazines with various amines. *Chem. Pharm. Bull.* **1970**, *18*, 1685–1692. [CrossRef]
81. Chen, S.-F.; Panzica, R.P. Synthesis and biological evaluation of certain 4-alkylaine and 4-arylalkylamino derivatives of the imidazolo[4,5-d]pyridazine and v-triazolo[4,5-d]pyridazine ring systems (1a). *J. Heterocycl. Chem.* **1982**, *19*, 285–288. [CrossRef]
82. Draper, R.E.; Castle, R.N. The synthesis of 3,5-dimethylpyrazol-1-yl-v-triazolo[4,5-d]pyridazines and substituted 3,5-dimethylpyrazol-1-ylimidazo[4,5-d]pyridazines. *J. Heterocycl. Chem.* **1983**, *20*, 193–197. [CrossRef]
83. Mataka, S.; Misumi, O.; Lin, W.H.; Tashiro, M. The effect of fused heterocycles on the liquid crystalline properties of di(alkoxy-substituted-phenyl)pyridazines and di(alkoxy-substittued-phenyl)pyridines. *J. Heterocycl. Chem.* **1992**, *29*, 87–92. [CrossRef]
84. Smolyar, N.N.; Yutilov, Y.M. Cyclotransformation in the series of fused 5-nitropyridin-2(1H)-ones. *Russ. J. Org. Chem.* **2008**, *44*, 274–281. [CrossRef]
85. Schofield, K.; Theobald, R.S. Cinnolines. XXIII. Some derivatives of 5-, 7-, and 8-nitro-4-hydroxycinnoline. 4-hydroxy-7-acetylcinnoline. *J. Chem. Soc.* **1949**, 2404–2408. [CrossRef]
86. Evans, N.A.; Johns, R.B.; Markham, K.R. II Syntheses of substituted pyridazines. *Aust. J. Chem.* **1967**, *20*, 713–722. [CrossRef]
87. Reidl, Z.; Hajós, G.; Messmer, A. Synthesis and selective reactions of v-triazolo-[1,5-b]pyridazinium salts with nucleophiles. A facile access to functionalized ethenyl-1,2,3-triazoles. *J. Heterocycl. Chem.* **1993**, *30*, 819–823. [CrossRef]
88. Vasko, G.A.; Riedl, Z.; Egyed, O.; Hajos, G. Synthesis of a new tricyclic ring system: [1,2,3]triazolo[1,5-b]cinnolinium salt. *Arkivoc* **2008**, *3*, 25–32. [CrossRef]
89. Kosary, J. Synthesis of some derivatives of 1,2,3-triazolo[1,5-b]pyridazine. Studies in the field of pyridazine compounds. *Magy. Kem. Foly.* **1980**, *86*, 564–566. [CrossRef]
90. Kvaskoff, D.; Bednarek, P.; Wentrup, C. 2-Pyridylnitrene and 3-pyridazylcarbene and their relationship via ring-expansion, ring-opening, ring-contraction, and fragmentation. *J. Org. Chem.* **2010**, *75*, 1600–1611. [CrossRef]
91. Gerhardt, G.A.; Castle, R.N. The synthesis of v-triazolo[4,5-c]pyridazines, a new heterocyclic ring system as potential purine antagonists. *J. Heterocycl. Chem.* **1964**, *1*, 247–250. [CrossRef]
92. Murakami, H.; Castle, R.N. Synthesis of imidazo[4,5-c]- and v-triazolo-[4,5-c]pyridazines. *J. Heterocycl. Chem.* **1967**, *4*, 555–563. [CrossRef]
93. Lunt, E.; Washbourn, K.; Wragg, W.R. A new cinnoline synthesis. Part IV: Nitration and sulphonation of chloro-(5,6,7,8)-4-hydroxy-cinnolines. *J. Chem. Soc. C.* **1968**, 687–695. [CrossRef]
94. Pokhodylo, N.T.; Shyyka, O.Y. New cascade reaction of azides with malononitrile dimer to polyfunctional [1,2,3]triazolo[4,5-b]pyridine. *Synth. Commun.* **2017**, *47*, 1096–1101. [CrossRef]
95. Betz, U.; Otto, G.P. Pharmaceutical Compositions Comprising CMET Kinase Inhibitors for the Treatment of Viral Infections Such as Coronavirus and COVID-19 and Methods of Preparation Thereof. WO 2022063869A2, 31 March 2022.
96. Cui, J.J.; Shen, H.; Tran-Dubé, M.; Nambu, M.; McTigue, M.; Grodsky, N.; Ryan, K.; Yamazaki, S.; Aguirre, S.; Parker, M.; et al. Lessons from (S)-6-(1-(6-(1-methyl-1H-pyrazol-4-yl)-[1,2,4]triazolo[4,3-b]pyridazin-3-yl)ethyl)quinoline (PF-04254644), an inhibitor of receptor tyrosine kinase c-Met with high protein kinase selectivity but broad phosphodiesterase family inhibition leading to myocardial degeneration in rats. *J. Med. Chem.* **2013**, *56*, 6651–6665. [CrossRef]

97. Markham, A. Savolitinib: First approval. *Drugs* **2021**, *81*, 1665–1670. [CrossRef]
98. Santarpia, M.; Massafra, M.; Gebbia, V.; D'Aquino, A.; Garipoli, C.; Altavilla, G.; Rosell, R. A narrative review of MET inhibitors in non-small cell lung cancer with MET exon 14 skipping mutations. *Transl. Lung Cancer Res.* **2021**, *10*, 1536–1556. [CrossRef]
99. Legentil, P.; Chadeyron, G.; Therias, S.; Chopin, N.; Sirbu, D.; Suzenet, F.; Leroux, F. Luminescent N-heterocycles based molecular backbone interleaved within LDH host structure and dispersed into polymer. *Appl. Clay Sci.* **2020**, *189*, 105561. [CrossRef]
100. Nadiya, S.; Rauful, A.M.; Michael, A.A.; Suresh, B.; Anuradha, B.; Guangming, C.; Aleksey, I.G.; Mitchell, K.G.; Andrew, J.K.; Anthony, R.M.; et al. Preparation of Heterocyclic and Heteroaryl Compounds for Treating Huntington's Disease. WO2020005873, 2 January 2020.
101. Barawkar, D.; Bandyopadhyay, A.; Zahler, R.; Sarangthem, R.; Waman, Y.; Bonagiri, R.; Jadhav, D.; Mukhopadhyay, P. Preparation of Tricyclic Compounds as JAK Modulators for Treatment of JAK-Associated Diseases. WO 2014045305A1, 27 March 2014.
102. Scutt, J.N.; Willetts, N.J.; Sonawane, R.; Kandukuri, S.R. Preparation of Herbicidal Fused Pyridazine Compounds. WO 2020161209A1, 13 August 2020.
103. Scutt, J.N.; Willetts, N.J. Preparation of Pyridazinium Compounds for Use in Controlling Unwanted Plant Growth. WO 2020164973A1, 20 August 2020.
104. Biagi, G.; Giorgio, I.; Livi, O.; Scartoni, V.; Lucacchini, A. Synthesis of 4,6-disubstituted- and 4,5,6-trisubstituted-2-phenyl-pyrimidines and their affinity towards A1 adenosine receptors. *Farmaco* **1997**, *52*, 61–65.
105. Morales, T.G.P.; Ratia, K.; Wang, D.-S.; Gogos, A.; Driver, T.G.; Federle, M.J. A novel chemical inducer of streptococcus quorum sensing acts by inhibiting the pheromone-degrading endopeptidase PepO. *J. Biol. Chem.* **2018**, *293*, 931–940. [CrossRef]
106. Zhang, Y.; Deng, J.; Jiang, L.; Lu, X.; Shang, K.; Shou, J.; Wang, B.; Xu, X.; Xu, Y. Preparation and Application of Class of N-Containing Heterocyclic Compounds Having Immunoregulatory Function. CN 110790758A, 14 February 2020.

Article

Furan-Containing Chiral Spiro-Fused Polycyclic Aromatic Compounds: Synthesis and Photophysical Properties

Koji Nakano [1,*], Ko Takase [1] and Keiichi Noguchi [2]

[1] Department of Applied Chemistry, Graduate School of Engineering, Tokyo University of Agriculture and Technology, 2-24-16 Naka-cho, Koganei, Tokyo 184-8588, Japan
[2] Instrumentation Analysis Center, Tokyo University of Agriculture and Technology, 2-24-16 Naka-cho, Koganei, Tokyo 184-8588, Japan
* Correspondence: k_nakano@cc.tuat.ac.jp; Tel.: +81-42-388-7162

Abstract: Spiro-fused polycyclic aromatic compounds (PACs) have received growing interest as rigid chiral scaffolds. However, furan-containing spiro-fused PACs have been quite limited. Here, we design spiro[indeno[1,2-*b*][1]benzofuran-10,10′-indeno[1,2-*b*][1]benzothiophene] as a new family of spiro-fused PACs that contains a furan unit. The compound was successfully synthesized in enantiopure form and also transformed to its *S*,*S*-dioxide derivative and the pyrrole-containing analog via aromatic metamorphosis. The absorption and emission properties of the obtained furan-containing chiral spiro-fused PACs are apparently different from those of their thiophene analogs that have been reported, owing to the increased electron-richness of furan compared to thiophene. All of the furan-containing chiral spiro-fused PACs were found to be circularly polarized luminescent materials.

Keywords: spiro π-conjugated compound; chiral compound; thiophene; furan; pyrrole; nucleophilic aromatic substitution; circular dichroism; circularly polarized luminescence

Citation: Nakano, K.; Takase, K.; Noguchi, K. Furan-Containing Chiral Spiro-Fused Polycyclic Aromatic Compounds: Synthesis and Photophysical Properties. *Molecules* **2022**, *27*, 5103. https://doi.org/10.3390/molecules27165103

Academic Editor: Joseph Sloop

Received: 22 July 2022
Accepted: 9 August 2022
Published: 11 August 2022

Publisher's Note: MDPI stays neutral with regard to jurisdictional claims in published maps and institutional affiliations.

Copyright: © 2022 by the authors. Licensee MDPI, Basel, Switzerland. This article is an open access article distributed under the terms and conditions of the Creative Commons Attribution (CC BY) license (https://creativecommons.org/licenses/by/4.0/).

1. Introduction

Spiro-fused polycyclic aromatic compounds (PACs), in which two planarized biaryls are connected perpendicularly by a tetragonal spiro atom, have been intensively studied in the last few decades, owing to their unique three-dimensional structure, high thermal stability, superior processability, and promising photophysical and electronic properties [1–3]. These intriguing characteristics offer advantages in (opto)electronic applications, such as organic field effect transistors [4,5], organic light emitting diodes [6–8], and organic solar cells [9–12]. The prototypical motif of spiro-fused PACs is 9,9′-spirobi[fluorene] (SBF). A variety of SBF-based compounds with functional substituent(s) and/or a π-extended structure have been developed for the purpose of tuning the photophysical and electronic properties [1,2]. Incorporation of heterocycle unit(s) into π-conjugated systems or a heteroatom as a spiro atom has also been a promising strategy to achieve the desired functions of spiro-fused PACs [3,13–16].

When two dissymmetric biaryls are linked by a spiro atom, the resulting spiro compound is chiral. Two different dissymmetric biaryls give a C_1-symmetric spiro compound, while two identical dissymmetric biaryls afford a C_2-symmetric one. Although chiral spiro-fused PACs have attracted less attention until recently, there have been several reports on their applications to molecular recognition [17,18], diastereoselective self-assembly [19,20], and asymmetric catalysts [21,22]. In 2016, Kuninobu, Takai, and co-workers reported the first example of a chiral spiro-fused PAC [23] that exhibited circularly polarized luminescence (CPL), which has received significant interest, owing to its potential applications [24–28]. This report has triggered many studies on spiro-fused PACs that exhibit CPL [29–37]. We have also developed a series of chiral spiro-fused PACs, such as **spiro-SS**, **-SS(O)$_2$**, and **-SN** (Figure 1) [16,38–42]. These spiro-fused PACs exhibit CPL and their optical properties were found to depend on the substituents and the incorporated heterocycle units. The recent studies clearly demonstrate the excellent potential of spiro-fused PACs as chiral materials, and further investigations are still required for elucidating the structure–property relationship, and thus developing spiro-fused PACs with tailored properties.

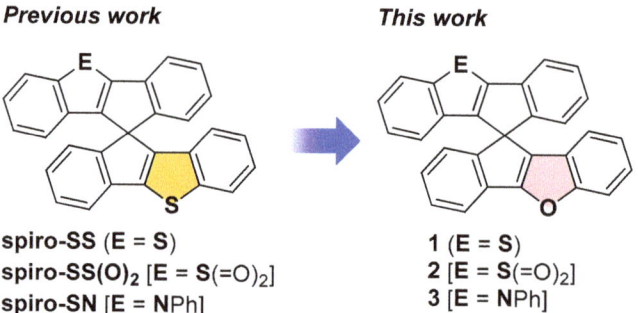

Figure 1. Molecular structures of previously reported chiral spiro-fused PACs and **1–3** in this study.

In the course of our studies, we have focused on incorporating a furan unit in a chiral spiro-fused PAC skeleton. Furan, the oxygen analog of thiophene, has reduced aromaticity and is more electron-rich compared to thiophene [43]. Therefore, the electronic perturbation of a furan unit to a π-conjugated system would be different from that of a thiophene unit, inducing unique electronic and photophysical properties. To date, a wide range of furan-containing π-conjugated compounds have been reported as organic functional materials [44–48]. However, the synthetic attempt of spiro-fused PACs containing furan unit(s) have been quite limited. In 2010, Ohe and co-workers reported the first examples of furan-containing spiro-fused PACs [49,50]. These compounds were found to be highly emissive. More recently, the chiral spiro-fused PAC with a furan unit has been reported by Nakamura and co-workers, demonstrating CPL properties [35]. Herein, we report the synthesis and photophysical properties of furan-containing chiral spiro-fused PACs **1–3**, which are oxygen analogs of **spiro-SS**, **-SS(O)₂**, and **-SN** (Figure 1). The S,S-dioxide derivative **2** and the pyrrole-containing compound **3** can be synthesized from **1** through aromatic metamorphosis. Incorporation of a furan unit in place of a thiophene unit was found to have great impact on photophysical properties.

2. Results

Our first attempt to prepare racemic **1** (*rac*-**1**) is illustrated in Scheme 1. 3-Bromo-2-(bromophenyl)-1-benzothiophene (**4**) was dilithiated with BuLi, and then treated with the ester **5**. The subsequent Friedel–Crafts cyclization of the resulting tertiary alcohol **6** successfully gave the desired compound *rac*-**1** (37% yield from **4**). The structure was confirmed by X-ray crystallographic analysis (Figure 2). Both enantiomers were found to be contained in the unit cell. The two spiro-linked π-conjugated planes are almost completely perpendicular (87.3°). Screening of optical resolution conditions with HPLC on a chiral stationary phase (chiral HPLC) showed that baseline separation of enantiomers cannot be achieved.

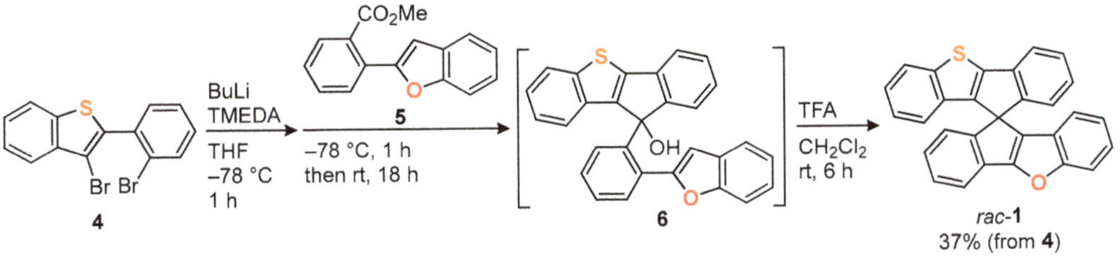

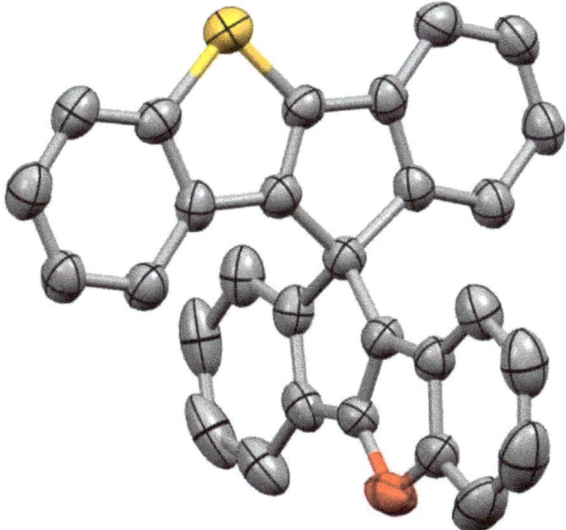

Scheme 1. Synthesis of *rac*-1.

Figure 2. ORTEP drawing of *rac*-1 with thermal ellipsoids at 50% probability. All hydrogen atoms are omitted for clarity.

In order to prepare each enantiomer of **1**, we designed the hydroxy-substituted spiro compound **11**, which would be converted into **1** via a two-step reaction (Scheme 2). Previously, Lützen and co-workers have reported the efficient optical resolution of 9,9′-spirobi[fluorene]-2,2′-diol and its derivatives with chiral HPLC [51]. We have also demonstrated that the dihydroxylated derivative of **spiro-SS** can be separated into enantiomers with chiral HPLC more efficiently than its parent compound **spiro-SS** [39]. Furthermore, Nakamura and co-workers reported the optical resolution of the furan-containing chiral spiro-fused PAC with one hydroxy group by using chiral HPLC [35]. Accordingly, we envisaged that incorporation of a hydroxy group on **1** would make an efficient optical resolution with chiral HPLC possible.

The synthesis of the hydroxy-substituted compound **11** and its transformation to **1** are illustrated in Scheme 2. First, 2-(2-bromo-4-methoxyphenyl)-1-benzothiophene (**7**) was lithiated with BuLi, and then treated with 10*H*-indeno[1,2-*b*][1]benzofuran-10-one (**8**). The resulting tertiary alcohol **9** was converted into the methoxy-substituted spiro compound *rac*-**10** via acid-promoted Friedel–Crafts cyclization (66% yield from **8**). Finally, demethylation of *rac*-**10** with BBr$_3$ gave the hydroxy-substituted compound **11** (58% yield) in a racemic form. As expected, the optical resolution of *rac*-**11** was achieved by HPLC with a CHIRALPAK® IA column with hexane/CHCl$_3$ (50/50) as an eluent (Figure S18). In addition, the hydroxy group of **11** was found to be cleaved off via the transformation to the triflate **12** (85% yield with *rac*-**11**; 87% yield with (+)-**11**) and the following palladium-catalyzed reduction with formic acid (96% yield with *rac*-**12**; 89% yield with (+)-**12**) [39]. This two-step reaction should not affect the chiral center. Therefore, the transformation of enantiopure **11** can give the corresponding enantiomer of **1** without racemization.

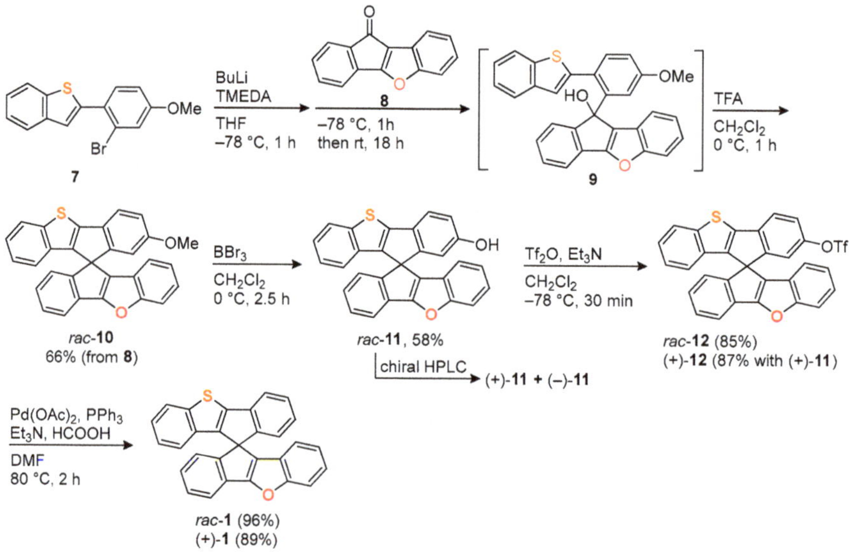

Scheme 2. Synthesis of the hydroxy-substituted spiro compound **11** and its transformation to **1**.

A benzothiophene skeleton can be transformed to an indole one by aromatic metamorphosis, which includes oxidation of a thiophene unit and the inter/intra molecular S_NAr reaction of the resulting *S,S*-dioxide unit with a primary amine [52]. Recently, we applied this transformation to **spiro-SS** [40]. The resulting spiro-fused compounds with one [**spiro-SS(O)$_2$**] or two *S,S*-dioxide units or with one (**spiro-SN**) or two pyrrole units showed photophysical properties that were quite different from the parent compound **spiro-SS**. In this context, we investigated the transformation of **1** to the *S,S*-dioxide derivative **2** and the pyrrole-containing compound **3** (Scheme 3). The oxidation of *rac*-**1** was readily achieved by using an excess amount of 3-chloroperbenzoic acid (*m*CPBA) as an oxidant, affording *rac*-**2** in high yield (96%). Furthermore, the reaction of *rac*-**2** with aniline in the presence of KHMDS (potassium hexamethyldisilazide) successfully gave *rac*-**3** in a moderate yield (64%). Enantiomers of **2** and **3** were also prepared from enantiopure **1**. The oxidation and the inter/intra molecular S_NAr reaction should not affect the chiral center. Therefore, these transformations could proceed without racemization to afford enantiopure **2** and **3**.

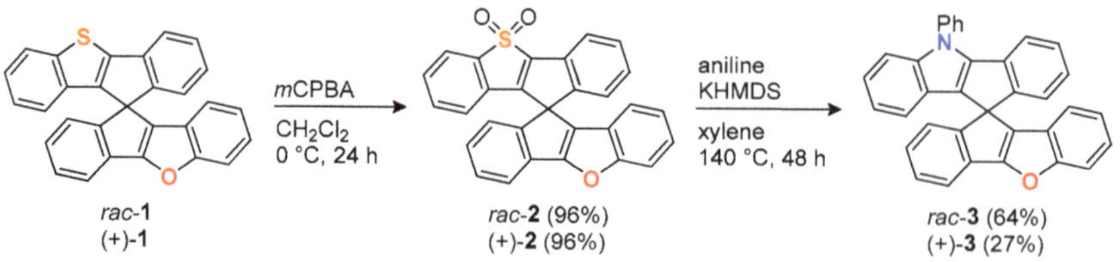

Scheme 3. Synthesis of the *S,S*-dioxide derivative **2** and the pyrrole-containing compound **3**.

The UV–vis absorption and photoluminescence (PL) spectra of spiro-fused PACs **1**–**3** are shown in Figure 3. The photophysical data are summarized in Table 1, together with those of the previously reported **spiro-SS**, **spiro-SS(O)$_2$**, and **spiro-SN** for comparison.

We also performed theoretical calculations by density functional theory (DFT) and time-dependent (TD) DFT methods at the B3LYP/6-31G(d) level of theory to understand the experimental photophysical properties.

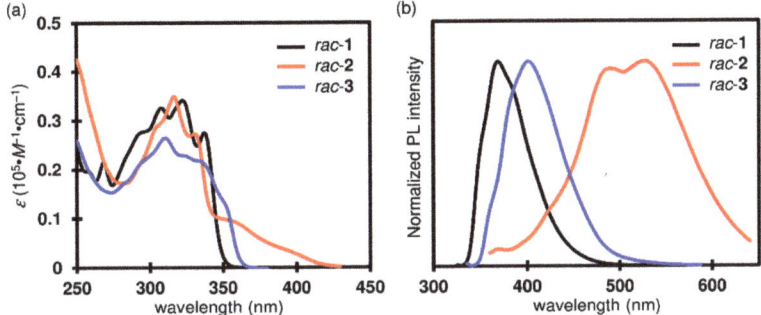

Figure 3. (a) UV–vis absorption and (b) PL spectra of *rac*-**1–3** in CH_2Cl_2.

Table 1. Photophysical properties of spiro-fused PACs **1–3**.

	λ_{abs} (nm) [a]	λ_{em} (nm) [b]	Φ [c]
rac-**1**	337	369 (320)	13
rac-**2**	332, 355 (sh)	480, 519 (330)	<1
rac-**3**	335, 353 (sh)	402 (330)	16
rac-**SS** [d]	340	368 (330)	6
rac-**SS(O)$_2$** [d]	336, 356 (sh)	459 (380)	1
rac-**SN** [e]	334, 350 (sh)	414 (330)	2

[a] The longest absorption maximum in CH_2Cl_2. sh: shoulder. [b] Emission maximum in CH_2Cl_2. Excitation wavelength in parenthesis. [c] Absolute quantum yield in CH_2Cl_2. [d] Reference [38]. [e] Reference [40].

Compound *rac*-**1** gave a well-resolved absorption spectrum, with the longest absorption maximum (λ_{abs}) at 337 nm (Figure 3a). In contrast, the absorption spectrum of the *S*,*S*-dioxide derivative *rac*-**2** exhibited well-resolved absorption bands in the <340 nm range and broad absorption bands in the longer wavelength range. Such a difference in the absorption properties between *rac*-**1** and *rac*-**2** is the same as that observed for **spiro-SS** and **spiro-SS(O)$_2$** [38]. By analogy with the discussion on **spiro-SS** and **spiro-SS(O)$_2$** in our previous report, the well-resolved absorption bands and the broader absorption bands were derived from the indeno[1,2-*b*][1]benzofuran subunit and the indeno[1,2-*b*][1]benzothiophene *S*,*S*-dioxide subunit, respectively. Each of these two subunits works as an almost-independent chromophore, since their perpendicular arrangement through a spiro carbon atom allows a limited orbital interaction between them in the ground state. The absorption spectrum of *rac*-**3** is slightly broader and red-shifted in comparison to that of *rac*-**1**, exhibiting a shoulder peak at 353 nm. The absorption spectra of *rac*-**1–3** are almost independent of solvent polarity (Figure S19a–c). The TD-DFT calculations demonstrated that the spiro-fused PACs *rac*-**1–3** exhibit the calculated longest absorption bands at 352 nm, 428 nm, and 361 nm, respectively, all of which are assigned to the transitions dominated by the HOMO→LUMO transition (Table S5). The obtained calculation results are qualitatively coincident with their experimental absorption spectra. The absorption spectra of *rac*-**1–3** are very similar to those of their thiophene analogs **spiro-SS**, **-SS(O)$_2$**, and **-SN**, respectively [38,40]. Therefore, the replacement of a thiophene unit with a furan unit was found to have little impact on absorption properties.

In the PL spectra, *rac*-**1** exhibited emission maximum (λ_{em}) at 369 nm in CH_2Cl_2 (Figure 3b). A slight red-shift was observed with an increase in solvent polarity (λ_{em}: 361 nm (hexane), 366 nm (toluene), 366 nm (THF), 369 nm (CH_2Cl_2), and 381 nm (acetonitrile) Figure S19d). The *S*,*S*-dioxide derivative *rac*-**2** exhibited emission maximum at

434 nm in non-polar hexane, which is largely red-shifted compared to that of *rac*-**1**. With the increase in solvent polarity, a significant positive solvatochromic shift was observed (Figure 4a). In addition, the second emission band clearly appeared at the longer wavelength (519 nm) in CH_2Cl_2 and became predominant in the most polar acetonitrile (542 nm). According to our previous report [38], the shorter-wavelength emission band reflects the feature of the indeno[1,2-*b*][1]benzothiophene *S,S*-dioxide subunit. On the other hand, the longer-wavelength one could be ascribed to the photo-induced intramolecular charge transfer (ICT), in which the indeno[1,2-*b*][1]benzofuran and the indeno[1,2-*b*][1]benzothiophene *S,S*-dioxide subunits work as electron-donating and electron-accepting units, respectively. Indeed, the HOMO and LUMO estimated by DFT calculation are mainly localized in indeno[1,2-*b*][1]benzofuran and the indeno[1,2-*b*][1]benzothiophene *S,S*-dioxide subunits, respectively, by reflecting the donor–acceptor-type structure (Figure 5). The emission band of the pyrrole-containing compound *rac*-**3** is also red-shifted compared to that of *rac*-**1**. In addition, the PL spectrum of *rac*-**3** exhibited a stronger dependence on solvent polarity than that of *rac*-**1** (Figure 4b). The HOMO of *rac*-**3** is delocalized both on indeno[1,2-*b*][1]benzofuran and indeno[1,2-*b*]indole subunits, but the apparently larger distribution is demonstrated on the latter subunit. In contrast, the LUMO is mainly located on the indeno[1,2-*b*][1]benzofuran subunit. Therefore, indeno[1,2-*b*][1]benzofuran and indeno[1,2-*b*]indole subunits act as electron-accepting and electron-donating units, respectively, inducing an ICT character of *rac*-**3** in its emissive state and a clear positive solvatochromic shift in the PL spectrum.

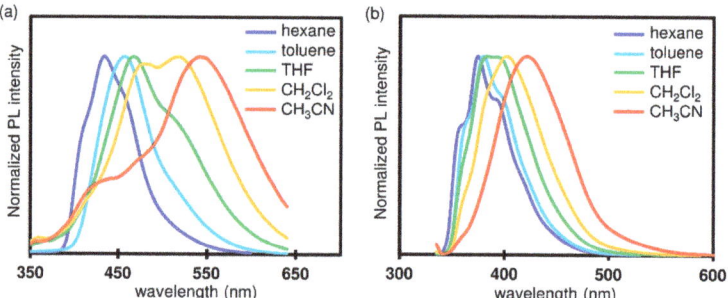

Figure 4. Solvent effect on PL spectra of (**a**) *rac*-**2** and (**b**) *rac*-**3**.

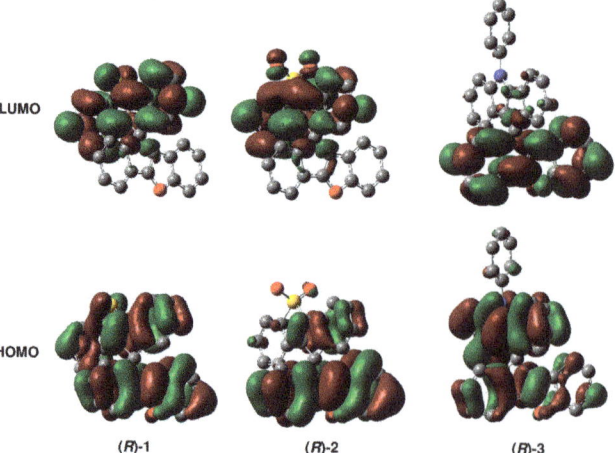

Figure 5. Frontier molecular orbitals of (*R*)-**1**–**3**.

Next, the effect of the furan unit on the emission properties of *rac*-**1**–**3** was evaluated through a comparison with those of the thiophene analogs **spiro-SS**, **SS(O)₂**, and **-SN**. As described above, the PL spectrum of *rac*-**2** in CH₂Cl₂ clearly shows the longer-wavelength emission band, owing to a photo-induced ICT, and it is predominant in acetonitrile (Figure 4a). On the other hand, such an emission band is less clear and observed only as a shoulder in the case of **spiro-SS(O)₂** [38]. This difference could be attributed to the more electron-rich furan unit, which works as a stronger electron-donating unit and allows more efficient photo-induced ICT. The degree of the solvatochromic shift of *rac*-**3** (λ_{em}: 374 nm (hexane) and 421 (acetonitrile)) (Figure 4b) is slightly smaller than that of the thiophene analog **spiro-SN** (λ_{em}: 375 nm (hexane) and 430 (acetonitrile)) [40]. The indeno[1,2-*b*]indole unit is the electron-donating unit both in *rac*-**3** and **spiro-SN**, and the indeno[1,2-*b*][1]benzofuran and the indeno[1,2-*b*][1]benzothiophene subunits work as electron-accepting units in *rac*-**3** and **spiro-SN**, respectively. Therefore, the smaller solvatochromic shift of *rac*-**3** could be due to the less efficient electron-accepting character of the more electron-rich indeno[1,2-*b*][1]benzofuran unit. The absolute quantum yields of *rac*-**1** and *rac*-**3** are 0.13 and 0.16, respectively (Table 1), which are much higher than those of **spiro-SS** (0.06) and **spiro-SN** (0.02) [38,40]. The *S,S*-dioxide derivative *rac*-**2** is merely emissive (<0.01), similar to **spiro-SS(O)₂** [38].

Chiroptical properties of **1**–**3** were investigated by CD and CPL spectroscopies (Figure 6 and Table 2). The CD spectrum of (+)-**1** exhibited two large positive Cotton effects at 340 nm and 328 nm and several negative and positive ones around 305 nm and 260 nm, respectively, which are quite similar to those of (+)-(*S*)-**spiro-SS** [38,39]. Therefore, the absolute configuration of (+)-**1** is considered to be *S*. The TD-DFT calculation results also support this assignment. CD spectra of (+)-**2** and (+)-**3** are slightly different from those of the thiophene analogs **spiro-SS(O)₂** and **spiro-SN**, respectively, but the sign of the first Cotton effect of (+)-**2** and (+)-**3** is identical to that of (+)-(*S*)-isomers of **spiro-SS(O)₂** and **spiro-SN** [38,40]. The dissymmetry factors in absorption, g_{abs} ($\Delta\varepsilon/\varepsilon = (\varepsilon_L - \varepsilon_R)/[(\varepsilon_L + \varepsilon_R)/2]$), were estimated to be approximately 1.2×10^{-3} for (+)-**1**, 1.4×10^{-3} for (+)-**2**, and $+1.4 \times 10^{-3}$ for (+)-**3**. In the CPL spectra, (+)- and (−)-**1**–**3** gave mirror-image CPL spectra with each other. The dissymmetry factors in luminescence, g_{lum} [$2\Delta I/I = 2(I_L - I_R)/(I_L + I_R)$, I_L and I_R: luminescence intensities of left and right circularly polarized light, respectively] [24], of (+)-**1**–**3** were estimated to be $+0.90 \times 10^{-3}$, $+2.8 \times 10^{-3}$, and $+0.97 \times 10^{-3}$, respectively, which are comparable to those of **spiro-SS**, **SS(O)₂**, and **-SN** [38,40] and the previously reported chiral small organic molecules with significant CPL activity. The solvent effect on CPL was also investigated for (+)-**2**, since its PL spectrum exhibits significant positive solvatochromism. As expected, with the increase in solvent polarity, the CPL maximum shifted to the longer wavelength range (Figure 6b).

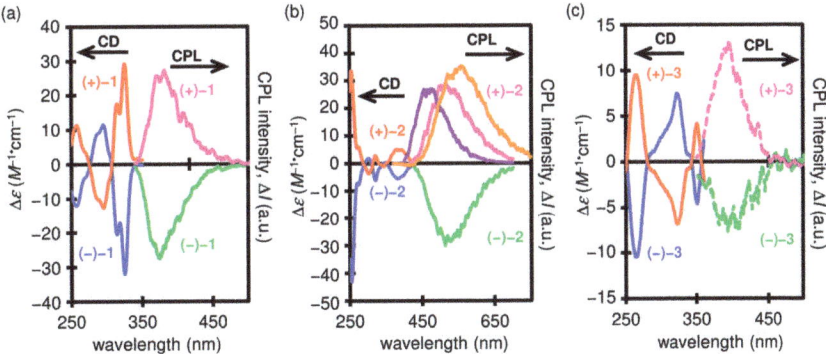

Figure 6. CD and CPL spectra of (**a**) **1**, (**b**) **2**, and (**c**) **3** in CH₂Cl₂ (red and blue lines for CD; pink and green lines for CPL; purple and orange lines in (**b**) for CPL in toluene and acetonitrile, respectively).

Table 2. Chiroptical date of spiro-fused PACs 1–3 in CH_2Cl_2.

	g_{abs} ($\times 10^{-3}$) [a]	g_{lum} ($\times 10^{-3}$) [b]	$|[\alpha]_D^{25}|$ [c]
(+)-1	+1.2 (341 nm)	+0.90 (381 nm)	154 (c 0.298)
(+)-2	+1.4 (378 nm)	+2.8 (517 nm)	151 (c 0.150)
(+)-3	+0.31 (351 nm)	+0.97 (394 nm)	175 (c 0.094)

[a] Calculated at the longest CD maximum. Wavelength for the calculated g_{abs} in parentheses. [b] Calculated at the CPL maximum. Wavelength for the calculated g_{lum} in parentheses. [c] In CH_2Cl_2. Concentration in parentheses.

3. Materials and Methods

3.1. General Procedures

All manipulations that involved air- and/or moisture-sensitive compounds were carried out with the standard Schlenk technique under argon. Analytical thin-layer chromatography was performed on glass plates coated with 0.25-mm 230–400 mesh silica gel that contained a fluorescent indicator. Column chromatography was performed by using silica gel (spherical neutral, particle size 63–210 μm). The recycling preparative HPLC was performed with YMC–GPC T–2000 and T–4000 columns (chloroform as an eluent). Most of the reagents were purchased from commercial suppliers, such as Sigma-Aldrich Co. LLC (St. Louis, MO, USA), Tokyo Chemical Industry Co., Ltd. (Tokyo, Japan), and Kanto Chemical Co., Inc. (Tokyo, Japan), and used without further purification, unless otherwise specified. Commercially available anhydrous solvents were used for air- and/or moisture sensitive reactions. Compound 4 was prepared according to the literature [53].

NMR spectra were recorded in $CDCl_3$ on a JEOL-ECX400 spectrometer (JEOL Ltd., Tokyo, Japan) (^{1}H 400 MHz; ^{13}C 101 MHz; ^{19}F 376 MHz). Chemical shifts were reported in ppm relative to the internal standard signal (0 ppm for Me_4Si in $CDCl_3$ and acetone-d_6) for ^{1}H and the deuterated solvent signal (77.16 ppm for $CDCl_3$ and 29.84 ppm for acetone-d_6) for ^{13}C. Data are presented as follows: chemical shift, multiplicity (s = singlet, brs = broad singlet, d = doublet, t = triplet, m = multiplet and/or multiple resonances), coupling constant in hertz (Hz), and signal area integration in natural numbers. Melting points were determined on SRS OptiMelt melting point apparatus (Stanford Research Systems, Sunnyvale, CA, USA). High resolution mass spectra were taken with a Bruker Daltonics micrOTOF–QII mass spectrometer (Bruker Corporation, Billerica, MA, USA) by the atmospheric pressure chemical ionization-time-of-flight (APCI–TOF) method. UV–vis absorption spectra were recorded on a JASCO V-650 spectrophotometer (JASCO Corporation, Tokyo, Japan). Photoluminescence spectra were recorded on a JASCO FP-6500 spectrofluorometer (JASCO Corporation, Tokyo, Japan). Absolute quantum yields were determined by an absolute quantum yield measurement system with a JASCO ILF–533 integrating sphere (JASCO Corporation, Tokyo, Japan). HPLC analyses and optical resolution were carried out using a DAICEL CHIRALPAK® IA-3 column (4.6 mm × 250 mm) and a DAICEL CHIRALPAK® IA column (20 mm × 250 mm) (Daicel Corporation, Tokyo, Japan), respectively. Circular dichroism (CD) spectra were recorded on a JASCO J–725 spectrometer (JASCO Corporation, Tokyo, Japan). CPL spectra were measured by using a JASCO CPL–300 spectrometer (JASCO Corporation, Tokyo, Japan). Optical rotations were measured on a JASCO P–2200 polarimeter (JASCO Corporation, Tokyo, Japan) using a 50-mm cell.

3.2. Synthesis

3.2.1. Methyl 2-(1-Benzofuran-2-yl)benzoate (5)

A mixture of methyl 2-bromobenzoate (0.97 mL, 6.9 mmol), (1-benzofuran-2-yl)boronic acid (1.23 g, 7.6 mmol), $Pd_2(dba)_3$ (0.13 g, 0.14 mmol), SPhos (0.22 g, 0.53 mmol), and K_2CO_3 (2.86 g, 21 mmol) in anhydrous MeCN (30 mL) and deionized water (3 mL) was placed in a 100-mL Schlenk tube and degassed by three freeze–pump–thaw cycles. After stirring at 80 °C for 24 h under argon, the reaction mixture was cooled to room temperature. To the reaction mixture, 1 M aqueous HCl was slowly added, and the resulting mixture was extracted with CH_2Cl_2 (10 mL × 3). The combined organic layers were dried over anhydrous Na$_2$SO$_4$, filtered, and concentrated under reduced pressure. The resulting crude residue was purified by silica-gel column chromatography with hexane/CH$_2$Cl$_2$ (3/2, R_f = 0.48) as an eluent to give **5** as a colorless oil (1.55 g, 89% yield). The ^{1}H and ^{13}C NMR data were identical to those reported in the literature [54].

3.2.2. rac-Spiro[indeno[1,2-b][1]benzofuran-10,10′-indeno[1,2-b][1]benzothiophene] (rac-1)

A mixture of **4** (0.74 g, 2.0 mmol) and N,N,N′,N′-tetramethylethylenediamine (1.2 mL, 8.0 mmol) in anhydrous THF (15 mL) was placed in a 50-mL Schlenk tube and cooled to −78 °C. To the mixture was added BuLi (2.3 M in hexane, 1.8 mL, 4.2 mmol) dropwise, and the resulting mixture was stirred at −78 °C for 1 h. Compound **5** (0.56 g, 2.2 mmol) in anhydrous THF (15 mL) was slowly added to the reaction mixture at −78 °C, and the resulting mixture was stirred at −78 °C for 1 h, warmed to ambient temperature, and then stirred for 18 h, before being quenched with water. The resulting mixture was extracted with EtOAc (5 mL × 3), and the combined organic layers were dried over anhydrous Na$_2$SO$_4$, filtered, and concentrated under reduced pressure. The resulting crude residue containing the tertiary alcohol **6** was used in the following step without purification.

The obtained crude residue was dissolved in CH$_2$Cl$_2$ (7 mL). To the solution, trifluoroacetic acid (1.5 mL) was added at ambient temperature, and the reaction mixture was stirred at ambient temperature for 6 h, before being quenched with aqueous saturated NaHCO$_3$. The resulting mixture was extracted with CH$_2$Cl$_2$ (10 mL × 4), and the combined organic layers were dried over anhydrous Na$_2$SO$_4$, filtered, and concentrated under reduced pressure. The resulting crude residue was purified by silica-gel column chromatography with hexane (R_f = 0.38) as an eluent to give rac-**1** as a colorless solid (0.30 g, 37% yield); mp 245.5–246.6 °C; ^{1}H NMR (400 MHz, CDCl$_3$) δ 7.82 (d, J = 8.2 Hz, 1H), 7.74 (d, J = 7.3 Hz, 1H), 7.66 (d, J = 7.8 Hz, 1H), 7.58 (d, J = 8.2 Hz, 1H), 7.39–7.35 (m, 2H), 7.18–7.12 (m, 2H), 7.07 (td, J = 7.3, 0.9 Hz, H), 7.04 (td, J = 7.8, 0.9 Hz, H), 7.01–6.93 (m, 2H), 6.83 (d, J = 7.3 Hz, 1H), 6.77–6.73 (m, 3H); ^{13}C NMR (101 MHz, CDCl$_3$) δ 162.4, 160.2, 149.8, 148.7, 144.4, 144.0, 141.7, 139.2, 133.4, 128.3, 128.2, 127.2, 126.9, 125.5, 124.9, 124.4, 124.2, 124.0, 123.9, 123.8, 123.5, 123.3, 121.1, 120.1, 119.3, 118.2, 112.4, 57.4 (one missing signal for an aromatic carbon is presumed to overlap one of the signals that was observed.); HRMS–APCI$^+$ (m/z) calcd for C$_{29}$H$_{17}$OS$^+$ ([M + H]$^+$), 413.0995, found 413.0994.

3.2.3. 2-(2-Bromo-4-methoxyphenyl)-1-benzothiophene (7)

A mixture of (1-benzothiophen-2-yl)boronic acid (99 mg, 0.55 mmol), 2-bromo-1-iodo-4-methoxybenzene (0.15 g, 0.49 mmol), Na$_2$CO$_3$ (0.11 g, 1.1 mmol), and Pd(PPh$_3$)$_4$ (29 mg, 25 μmol) in anhydrous 1,4-dioxane (2.1 mL) and deionized water (0.30 mL) was placed in a 30-mL Schlenk tube and degassed by three freeze–pump–thaw cycles. After stirring at 90 °C for 48 h under argon, the reaction mixture was cooled to room temperature. To the reaction mixture, 1 M aqueous HCl was slowly added, and the resulting mixture was extracted with CH$_2$Cl$_2$ (5 mL × 3). The combined organic layers were dried over anhydrous Na$_2$SO$_4$, filtered, and concentrated under reduced pressure. The resulting crude residue was purified by silica-gel column chromatography with hexane/EtOAc (9/1, R_f = 0.52) as an eluent and preparative HPLC to give **7** as a colorless solid (0.11 g, 67% yield): mp

82.3–84.1 °C; ^{1}H NMR (400 MHz, CDCl$_3$) δ 7.84 (dd, J = 8.2, 0.9 Hz, 1H), 7.80 (dd, J = 7.3, 1.4 Hz, 1H), 7.46 (d, J = 8.2 Hz, 1H), 7.42 (s, 1H), 7.37 (td, J = 6.9, 1.4 Hz, 1H), 7.32 (td, J = 7.3, 1.4 Hz, 1H), 7.25 (d, J = 2.7 Hz, 1H), 6.92 (dd, J = 8.7, 2.7 Hz, 1H), 3.85 (s, 3H); ^{13}C NMR (101 MHz, CDCl$_3$) δ 160.0, 142.1, 140.2, 140.0, 132.9, 127.8, 124.5, 124.4, 124.2, 123.8, 123.6, 122.2, 118.9, 113.8, 55.8; HRMS–APCI$^+$ (m/z) calcd for C$_{15}$H$_{12}$BrOS$^+$ ([M + H]$^+$), 318.9787, found 318.9791.

3.2.4. 2-(1-Benzofuran-2-yl)benzoic Acid (13)

To a solution of **5** (1.75 g, 6.9 mmol) in methanol (20 mL), a mixture of NaOH (1.68 g, 42 mmol) and water (20 mL) was added at room temperature. After stirring at 60 °C for 4.5 h, the reaction mixture was cooled to room temperature and acidified with 1 M aqueous HCl. The resulting mixture was extracted with Et$_2$O (10 mL × 4). The combined organic layers were dried over anhydrous Na$_2$SO$_4$, filtered, and concentrated under reduced pressure to give **5** as a colorless solid (1.60 g, 97% yield); mp 132.8–135.3 °C; ^{1}H NMR (400 MHz, acetone-d_6) δ 11.30 (br, 1H), 7.84 (d, J = 7.3 Hz, 1H), 7.83 (d, J = 7.3 Hz, 1H), 7.67–7.62 (m, 2H), 7.55 (td, J = 7.8, 1.4 Hz, 1H), 7.51 (ddd, J = 8.2, 1.8, 0.9 Hz, 1H), 7.3 (ddd, J = 8.7, 7.3, 1.4 Hz, 1H), 7.25 (td, J = 7.6, 1.4 Hz, 1H), 7.12 (d, J = 0.9 Hz, 1H); ^{13}C NMR (101 MHz, acetone-d_6) δ 169.6, 155.93, 155.88, 132.8, 131.9, 130.4, 130.2, 130.1, 130.0, 129.7, 125.3, 123.8, 122.1, 111.8, 105.3; HRMS–APCI$^+$ (m/z) calcd for C$_{15}$H$_{11}$O$_3^+$ ([M + H]$^+$), 239.0703, found 239.0702.

3.2.5. 10H-Indeno[1,2-b][1]benzofuran-10-one (8)

A mixture of **13** (3.83 g, 16 mmol) and thionyl chloride (1.4 mL, 19 mmol) in anhydrous CH$_2$Cl$_2$ (160 mL) was placed in a 300-mL three-neck flask with a condenser at ambient temperature. To the mixture, anhydrous DMF (100 µL) was added, and the resulting mixture was stirred for 1.5 h at 40 °C. The volatiles were removed under reduced pressure to afford the acid chloride intermediate.

A mixture of AlCl$_3$ (3.26 g, 24 mmol) and anhydrous CH$_2$Cl$_2$ (180 mL) was charged in a 500-mL three-neck flask with a condenser and a dropping funnel and cooled to 0 °C. To the mixture, a solution of the acid chloride intermediate in anhydrous CH$_2$Cl$_2$ (160 mL) was slowly added via the dropping funnel over 2 h at 0 °C under argon atmosphere. The reaction mixture was stirred for 18 h at 40 °C, cooled to ambient temperature, and then poured into ice-cooled 1 M aqueous HCl. The resulting mixture was extracted with CH$_2$Cl$_2$ (100 mL × 3), and the combined organic layers were dried over anhydrous Na$_2$SO$_4$, filtered, and concentrated under reduced pressure. The resulting crude residue was purified by silica-gel column chromatography with hexane/EtOAc (9/1, R_f = 0.49) as an eluent to give **8** as an orange solid (1.44 g, 41% yield). The ^{1}H and ^{13}C NMR data were identical to those reported in the literature [55].

3.2.6. rac-2′-Methoxyspiro[indeno[1,2-b][1]benzofuran-10,10′-indeno[1,2-b][1] benzothiophene] (rac-**10**)

A mixture of **7** (0.48 g, 1.5 mmol) and $N,N,N′,N′$-tetramethylethylenediamine (0.45 mL, 3.0 mmol) in anhydrous THF (12 mL) was placed in a 100-mL Schlenk flask and cooled to −78 °C. To the mixture, BuLi (2.6 M in hexane, 0.59 mL, 1.5 mmol) was added dropwise, and the resulting mixture was stirred at −78 °C for 1 h. Compound **8** (0.27 g, 1.3 mmol) in anhydrous THF (12 mL) was slowly added to the reaction mixture at −78 °C, and the resulting mixture was stirred at −78 °C for 1 h, warmed to ambient temperature, and then stirred for 18 h, before being quenched with water. The resulting mixture was extracted with CH$_2$Cl$_2$ (10 mL × 3), and the combined organic layers were dried over anhydrous Na$_2$SO$_4$, filtered, and concentrated under reduced pressure. The resulting crude residue was roughly purified by silica-gel column chromatography with hexane/EtOAc (9/1) as an eluent. The fractions (R_f = 0.16) that contained the tertiary alcohol **9** and some impurities were concentrated under reduced pressure.

The resulting residue was dissolved in CH$_2$Cl$_2$ (10 mL) and cooled to 0 °C. To the solution, trifluoroacetic acid (0.4 mL) was added at 0 °C, and the reaction mixture was stirred at 0 °C for 1 h, before being quenched with aqueous saturated NaHCO$_3$. The resulting mixture was extracted with CH$_2$Cl$_2$ (10 mL × 3), and the combined organic layers were dried over anhydrous Na$_2$SO$_4$, filtered, and concentrated under reduced pressure. The resulting crude residue was purified by silica-gel column chromatography with hexane/EtOAc (9/1, R_f = 0.49) as an eluent to give *rac*-**10** as a colorless solid (0.37 g, 66% yield from **8**); mp 216.4–221.6 °C; ^{1}H NMR (400 MHz, CDCl$_3$) δ 7.78 (ddd, *J* = 8.2, 1.8, 0.9 Hz, 1H), 7.73 (ddd, *J* = 7.3, 1.8, 0.9 Hz, 1H), 7.57 (d, *J* = 8.2, 1H), 7.55 (d, *J* = 8.2, 1H), 7.36 (td, *J* = 7.6, 0.9 Hz, 1H), 7.15 (ddd, *J* = 8.2, 6.9, 1.4, 1H), 7.10 (ddd, *J* = 8.2, 6.9, 1.4, 1H), 7.04 (td, *J* = 7.6, 1.4 Hz, 1H), 6.98–6.94 (m, 2H), 6.89 (dd, *J* = 8.2, 2.3 Hz, 1H), 6.79–6.77 (m, 2H), 6.69 (ddd, *J* = 7.8, 1.4, 0.9 Hz, 1H), 6.39 (d, *J* = 2.3 Hz, 1H), 3.63 (s, 3H); ^{13}C NMR (101 MHz, CDCl$_3$) δ 162.3, 160.2, 159.4, 150.9, 150.1, 144.4, 143.5, 140.1, 133.7, 133.4, 132.2, 128.2, 127.2, 125.9, 124.8, 124.4, 124.1, 123.9, 123.7, 123.6, 123.5, 120.7, 120.6, 119.4, 118.1, 113.4, 112.4, 109.9, 57.4, 55.6; HRMS–APCI$^+$ (*m/z*) calcd for C$_{30}$H$_{19}$O$_2$S$^+$ ([M + H]$^+$), 443.1100, found 443.1103.

3.2.7. *rac*-Spiro[indeno[1,2-*b*][1]benzofuran-10,10′-indeno[1,2-*b*][1]benzothiophene]-2′-ol (*rac*-**11**)

To a mixture of *rac*-**10** (1.04 g, 2.4 mmol,) and anhydrous CH$_2$Cl$_2$ (24 mL), BBr$_3$ (1 M in CH$_2$Cl$_2$, 4.7 mL, 4.7 mmol) was added dropwise at 0 °C. After stirring at 0 °C for 2.5 h, the reaction was quenched with water. The resulting mixture was extracted with CH$_2$Cl$_2$ (15 mL × 3) and the combined organic layers were dried over anhydrous Na$_2$SO$_4$, filtered, and concentrated under reduced pressure. The resulting crude residue was purified by silica-gel column chromatography with hexane/EtOAc (3/1, R_f = 0.46) as an eluent to give *rac*-**11** as a colorless solid (0.59 g, 58% yield). *rac*-**11** can be separated into enantiopure (−)-**11** and (+)-**11** by HPLC equipped with an analytical DAICEL CHIRALPAK® IA-3 column (Φ4.6 mm × 250 mm) (t_R = 12.7 min for (−)-**11** and 14.5 min for (+)-**11** (flow rate: 1.0 mL/min; eluent: hexane/CHCl$_3$ = 50/50)): [α]$_D^{25}$ for (+)-**11** +175 (*c* 0.112, CH$_2$Cl$_2$); mp 261.2–265.7 °C; ^{1}H NMR (500 MHz, CDCl$_3$) δ 7.80 (d, *J* = 8.0 Hz, 1H), 7.74 (d, *J* = 7.4 Hz, 1H), 7.58 (d, *J* = 8.6 Hz, 1H), 7.52 (d, *J* = 8.0 Hz, 1H), 7.38 (t, *J* = 7.2 Hz, 1H), 7.18 (td, *J* = 8.0, 1.2 Hz, 1H), 7.12 (td, *J* = 8.0, 1.2 Hz, 1H), 7.06 (td, *J* = 7.5, 1.2 Hz, 1H), 7.00–6.96 (m, 2H), 6.84 (dd, *J* = 8.0, 2.3 Hz, 1H), 6.80–6.79 (m, 2H), 6.70 (d, *J* = 8.0 Hz, 1H), 6.32 (d, *J* = 2.6 Hz, 1H), 4.54 (brs, 1H); ^{13}C NMR (101 MHz, CDCl$_3$) δ 162.3, 160.2, 155.1, 151.1, 150.0, 144.4, 143.5, 140.0, 133.6, 133.3, 132.3, 128.2, 127.2, 125.7, 124.8, 124.4, 124.1, 124.0, 123.75, 123.69, 123.5, 120.9, 120.7, 119.3, 118.2, 115.0, 112.4, 111.2, 57.3; HRMS–APCI$^+$ (*m/z*) calcd for C$_{29}$H$_{17}$O$_2$S$^+$ ([M + H]$^+$), 429.0944, found 429.0931.

3.2.8. *rac*-Spiro[indeno[1,2-*b*][1]benzofuran-10,10′-indeno[1,2-*b*][1]benzothiophene]-2′-yl Trifluoromethanesulfonate (*rac*-**12**)

A mixture of *rac*-**11** (82 mg, 0.19 mmol) and anhydrous triethylamine (0.14 mL, 1.0 mmol) in anhydrous CH$_2$Cl$_2$ (2 mL) was placed in a 20-mL Schlenk tube and cooled to −78 °C. To the mixture, trifluoromethanesulfonic anhydride (45 μL, 0.27 mmol) was added dropwise, and the resulting mixture was stirred at −78 °C for 30 min. The reaction was quenched with water, and then the resulting mixture was extracted with CH$_2$Cl$_2$ (5 mL × 3). The combined organic layers were dried over anhydrous Na$_2$SO$_4$, filtered, and concentrated under reduced pressure. The resulting crude residue was purified by silica-gel column chromatography with hexane/EtOAc (9/1, R_f = 0.43) as an eluent to give *rac*-**12** as a colorless solid (92 mg, 85% yield); ^{1}H NMR (400 MHz, CDCl$_3$) δ 7.84 (d, *J* = 7.8 Hz, 1H), 7.77 (d, *J* = 7.3 Hz, 1H), 7.70 (d, *J* = 8.2 Hz, 1H), 7.60 (d, *J* = 8.2 Hz, 1H), 7.42 (dd, *J* = 7.6, 6.6 Hz,

1H), 7.32 (dd, J = 8.2, 2.3 Hz, 1H), 7.22–7.18 (m, 2H), 7.09 (td, J = 7.6, 0.9 Hz, 1H), 7.02 (td, J = 7.3, 0.9 Hz, 1H), 6.99 (t, J = 7.8 Hz, 1H), 6.77–6.72 (m, 4H); ^{13}C NMR (101 MHz, CDCl$_3$) δ 162.6, 160.2, 151.5, 148.3, 144.4, 143.7, 142.4, 139.5, 133.3, 133.1, 128.7, 127.5, 125.2, 124.9, 124.3, 124.2, 124.0, 123.92, 123.89, 123.7, 121.5, 121.4, 120.8, 119.0, 118.70 (q, J_{CF} = 321 Hz), 118.5, 116.9, 112.6, 57.5 (one missing signal for an aromatic carbon is presumed to overlap one of the signals that was observed.); ^{19}F NMR (376 MHz, CDCl$_3$) δ − 72.8; HRMS–APCI$^+$ (m/z) calcd for C$_{30}$H$_{16}$F$_3$O$_4$S$_2{}^+$ ([M + H]$^+$), 561.0437, found 561.0443.

The crude product of (+)-**12** was also obtained by using (+)-**11** (62 mg, 0.14 mmol), anhydrous triethylamine (0.10 mL, 0.72 mmol), and trifluoromethanesulfonic anhydride (32 μL, 0.20 mmol), according to the procedure for *rac*-**12**. Purification by silica-gel column chromatography with hexane/EtOAc (9/1, R_f = 0.43) as an eluent gave (+)-**12** as a colorless solid (70 mg, 87% yield): [α]$_D^{25}$ +128 (*c* 0.158, CH$_2$Cl$_2$). The ^{1}H and ^{13}C NMR spectra were identical to those of *rac*-**12**.

3.2.9. *rac*-Spiro[indeno[1,2-*b*][1]benzofuran-10,10′-indeno[1,2-*b*][1]benzothiophene] (*rac*-**1**)

A mixture of *rac*-**12** (56 mg, 0.10 mmol), Pd(OAc)$_2$ (1.2 mg, 5.3 μmol), PPh$_3$ (2.4 mg, 9.2 μmol), anhydrous triethylamine (85 μL, 0.61 mmol), formic acid (15 μL, 0.39 mmol), and anhydrous DMF (1.0 mL) was placed in a 20-mL Schlenk tube and degassed by three freeze–pump–thaw cycles. After stirring at 80 °C for 2 h under argon, the reaction mixture was cooled to room temperature. The reaction was quenched by adding 1 M aqueous HCl slowly, and the resulting mixture was extracted with CH$_2$Cl$_2$ (5.0 mL × 3). The combined organic layers were dried over anhydrous Na$_2$SO$_4$, filtered, and concentrated under reduced pressure. The resulting crude residue was purified by silica-gel column chromatography with hexane/EtOAc (9/1, R_f = 0.70) as an eluent to give *rac*-**1** as a colorless solid (40 mg, 96% yield). The ^{1}H and ^{13}C NMR spectra were identical to those of *rac*-**1** described above.

The crude product of (+)-**1** was also obtained by using (+)-**12** (79 mg, 0.12 mmol), Pd(Oac)$_2$ (1.4 mg, 6.2 μmol), PPh$_3$ (3.1 mg, 12 μmol), anhydrous trimethylamine (0.12 mL, 0.83 mmol), formic acid (22 μL, 0.53 mmol), and DMF (1.4 mL), according to the procedure for *rac*-**1**. Purification by silica-gel column chromatography with hexane/EtOAc (9/1, R_f = 0.70) as an eluent gave (+)-**1** as a colorless solid (52 mg, 89% yield): [α]$_D^{25}$ +154 (*c* 0.298, CH$_2$Cl$_2$). The ^{1}H and ^{13}C NMR spectra were identical to those of *rac*-**1**.

3.2.10. *rac*-Spiro[indeno[1,2-*b*][1]benzofuran-10,10′-indeno[1,2-*b*][1]benzothiophene] 5′,5′-Dioxide (*rac*-**2**)

To a mixture of *rac*-**1** (0.12 g, 0.30 mmol) and CH$_2$Cl$_2$ (2.0 mL), *m*-chloroperoxybenzoic acid (contains ca. 30% water, 0.19 g, 1.1 mmol) was added at 0 °C. The resulting mixture was stirred at 0 °C for 24 h. The reaction was quenched with a mixture of an aqueous saturated Na$_2$S$_2$O$_3$ and NaHCO$_3$, and then the resulting mixture was extracted with CH$_2$Cl$_2$ (5.0 mL × 3). The combined organic layers were dried over anhydrous Na$_2$SO$_4$, filtered, and concentrated under reduced pressure. The resulting crude residue was purified by silica-gel column chromatography with CHCl$_3$ (R_f = 0.81) as an eluent to give *rac*-**2** as a colorless solid (0.13 mg, 96% yield); ^{1}H NMR (400 MHz, CDCl$_3$) δ 7.83 (d, J = 7.8 Hz, 1H), 7.77 (d, J = 7.3 Hz, 1H), 7.66 (d, J = 7.3 Hz, 1H), 7.59 (d, J = 8.2 Hz, 1H), 7.46–7.41 (m, 2H), 7.28 (t, J = 7.6 Hz, 1H), 7.24–7.20 (m, 1H), 7.18–7.09 (m, 3H), 7.06 (t, J = 7.6 Hz, 1H), 6.94 (m, 2H), 6.84 (d, J = 7.8 Hz, 1H), 6.35 (d, J = 7.3 Hz, 1H); ^{13}C NMR (101 MHz, CDCl$_3$) δ 162.8, 160.2, 151.4, 147.7, 146.3, 144.5, 142.9, 134.0, 133.5, 133.3, 130.0, 129.2, 128.9, 128.5, 128.0, 127.8, 124.6, 124.3, 124.0, 123.8, 123.7, 122.9, 121.8, 121.6, 119.2, 118.8, 112.6, 57.7 (one missing signal for an aromatic carbon is presumed to overlap one of the signals that was observed.); HRMS–APCI$^+$ (m/z) calcd for C$_{29}$H$_{17}$O$_3$S$^+$ ([M + H]$^+$), 445.0893, found 445.0891.

The crude product of (+)-**2** was obtained by using (+)-**1** (27 mg, 65 μmol), *m*-chloroperoxybenzoic acid (41 mg, 0.17 mmol), and CH$_2$Cl$_2$ (0.7 mL), according to the procedure for *rac*-**2**. Purification by silica-gel column chromatography with CHCl$_3$ (R_f = 0.81) as an eluent gave (+)-**2** as a colorless solid (28 mg, 96% yield): $[\alpha]_D^{25}$ +151 (c 0.15, CH$_2$Cl$_2$). The ^{1}H and ^{13}C NMR spectra were identical to those of *rac*-**2**.

3.2.11. *rac*-5′-Phenyl-5′H-spiro[indeno[1,2-b][1]benzofuran-10,10′-indeno[1,2-b]indole] (*rac*-**3**)

A solution of KHMDS (0.5 M in toluene, 0.19 mL, 93 µmol) was charged in a 20-mL Schlenk tube, and toluene was removed under reduced pressure. The resulting residue was dissolved with anhydrous xylene (0.3 mL) under argon. The resulting xylene solution of KHMDS was added to a mixture of *rac*-**2** (14 mg, 31 µmol) and aniline (5.7 µL, 62 µmol) in another 20-mL Schlenk tube under argon at ambient temperature. The resulting solution was stirred at 140 °C for 48 h. The reaction mixture was cooled to an ambient temperature, and 1 M aqueous HCl was added slowly to the reaction mixture. The resulting mixture was extracted with CH$_2$Cl$_2$ (5 mL × 4), and the combined organic layers were dried over Na$_2$SO$_4$, filtered, and concentrated under reduced pressure. The crude residue was purified by silica-gel column chromatography with hexane/EtOAc (9/1, R_f = 0.48) as an eluent to give *rac*-**3** as a colorless solid (9.4 mg, 64% yield); mp 255.1–259.2 °C; ^{1}H NMR (400 MHz, CDCl$_3$) δ 7.75–7.66 (m, 5H), 7.60–7.54 (m, 2H), 7.40–7.34 (m, 2H), 7.18–7.13 (m, 3H), 7.08–7.03 (m, 2H), 7.00–6.94 (m, 2H), 6.89–6.78 (m, 5H); ^{13}C NMR (101 MHz, CDCl$_3$) δ 162.3, 160.2, 151.3, 150.6, 144.6, 142.5, 138.1, 135.4, 133.5, 129.8, 128.0, 127.8, 127.5, 127.1, 127.0, 126.4, 126.2, 124.7, 123.8, 123.69, 123.67, 123.34, 123.29, 122.8, 122.3, 120.9, 119.4, 119.0, 118.9, 117.9, 112.3, 111.1, 54.3; HRMS–APCI$^+$ (m/z) calcd for C$_{35}$H$_{22}$NO$^+$ ([M + H]$^+$), 472.1696, found 472.1687.

The crude product of (+)-**3** was obtained by using (+)-**2** (16 mg, 36 µmol), aniline (7 µL, 77 µmol), KHMDS (0.5 M in toluene, 0.22 mL), and xylene (0.36 mL), according to the procedure for *rac*-**3**. Purification by silica-gel column chromatography with hexane/EtOAc (9/1, R_f = 0.48) as an eluent gave (+)-**3** as a colorless solid (4.7 mg, 27% yield): $[\alpha]_D^{25}$ +175 (c 0.094, CH$_2$Cl$_2$). The ^{1}H and ^{13}C NMR spectra were identical to those of *rac*-**3**.

3.3. Computational Studies

The DFT and TD-DFT calculations were performed by using the Gaussian 09 program [56] at the B3LYP/6-31G(d) level of theory in the gas phase. The starting molecular models for DFT geometry optimizations were built and optimized with MMFF molecular mechanics by using the Spartan '08 package (Wavefunction, Inc., Irvine, CA, USA). Six singlet states were calculated in the TD-DFT calculations. The visualization of the molecular orbitals was performed using GaussView 5.

3.4. X-ray Crystallography

For X-ray crystallographic analysis, a suitable single crystal was selected under ambient conditions, mounted using a nylon loop filled with paraffin oil, and transferred to the goniometer of a RIGAKU R-AXIS RAPID diffractometer (Rigaku Corporation, Tokyo, Japan), with graphite-monochromated Cu–Kα irradiation (λ = 1.54187 Å). The structure was solved by a direct method (SIR 2008 [57]) and refined by full-matrix least-squares techniques against F^2 (SHELXL-2014 [58,59]). The intensities were corrected for Lorentz and polarization effects. All non-hydrogen atoms were refined anisotropically. Hydrogen atoms were placed using AFIX instructions.

Crystal data for *rac*-**1**: formula: C$_{29}$H$_{16}$OS (M = 412.48 g/mol): monoclinic, space group $P2_1/n$ (No. 14), a = 14.2753(3) Å, b = 9.8401(2) Å, c = 15.7643(3) Å, β = 114.4630(10)°, V = 2015.63(7) Å^3, Z = 4, T = 193(2) K, μ(CuKα) = 1.566 mm^{-1}, D_{calc} = 1.359 g/cm^3, 34,993 reflections measured (3.519° ≤ θ ≤ 68.226°), 3684 unique (R_{int} = 0.0448; R_{sigma} = 0.0213), which were used in all calculations. The final R_1 was 0.0688 ($I > 2\sigma(I)$) and wR_2 was 0.2028 (all data). CCDC 2,191,229 contains the supplementary crystallographic data for this paper. The data can be obtained free of charge from the Cambridge Crystallographic Data Centre via www.ccdc.cam.ac.uk/structures, accessed on 8 August 2022.

4. Conclusions

In summary, we have synthesized furan-containing chiral spiro-fused PACs **1–3**. Aromatic metamorphosis strategy was applied to **1**, affording the pyrrole-containing **3**, as well as the *S,S*-dioxide derivative **2**. The hydroxylated derivatives of **1** were found to be separated into enantiopure isomers by chiral HPLC, which allowed us to prepare enantiomers of **1–3**. Introduction of a furan unit has proven to have a great impact on their photophysical properties. The observed impact could be attributed to the increased electron-richness of furan compared to thiophene. The synthesized furan-containing chiral spiro-fused PACs demonstrated CPL with a g_{lum} value of <2.3 × 10^{-3}, which is comparable to those of the reported chiral small organic molecules.

Supplementary Materials: The following supporting information can be downloaded at: https://www.mdpi.com/article/10.3390/molecules27165103/s1, Figures S1–S17: ^{1}H, ^{13}C, and ^{19}F NMR spectra for **1–3**, **7**, and **10–13**; Table S1: Crystallographic data for *rac*-**1**; Figure S18: HPLC chart for optical resolution of **11**; Figure S19: UV–vis (**1–3**) and PL (**1**) spectra in various solvents; Figures S20–S22: Molecular orbitals for (*R*)-**1–3**; Tables S2–S4: Coordinates and absolute energy of the optimized structures for (*R*)-**1–3**; Table S5: The selected absorption of (*R*)-**1–3** calculated by TD-DFT method.

Author Contributions: Conceptualization, K.T. and K.N. (Koji Nakano); validation, K.T. and K.N. (Koji Nakano); formal analysis, K.T. and K.N. (Koji Nakano); investigation, K.T., K.N. (Keiichi Noguchi) and K.N. (Koji Nakano); data curation, K.T., K.N. (Keiichi Noguchi) and K.N. (Koji Nakano); writing—original draft preparation, K.N. (Koji Nakano); writing—review and editing, K.T. and K.N. (Koji Nakano); supervision, K.N. (Koji Nakano); project administration, K.N. (Koji Nakano); funding acquisition, K.N. (Koji Nakano). All authors have read and agreed to the published version of the manuscript.

Funding: This work was partially supported by MEXT KAKENHI, grant number 16H00824.

Institutional Review Board Statement: Not applicable.

Informed Consent Statement: Not applicable.

Data Availability Statement: The data presented in this study are available in the Supplementary Materials.

Acknowledgments: The authors are grateful to Kyoko Nozaki (The University of Tokyo) for chiral HPLC separation. The computations were performed using the Research Center for Computational Science, Okazaki, Japan.

Conflicts of Interest: The authors declare no conflict of interest.

Sample Availability: Samples of the compounds are available from the authors.

References

1. Pudzich, R.; Fuhrmann-Lieker, T.; Salbeck, J. Spiro compounds for organic electroluminescence and related applications. *Adv. Polym. Sci.* **2006**, *199*, 83–142. [CrossRef]
2. Saragi, T.P.I.; Spehr, T.; Siebert, A.; Fuhrmann-Lieker, T.; Salbeck, J. Spiro Compounds for Organic Optoelectronics. *Chem. Rev.* **2007**, *107*, 1011–1065. [CrossRef] [PubMed]
3. Liu, S.; Xia, D.; Baumgarten, M. Rigidly Fused Spiro-Conjugated π-Systems. *ChemPlusChem* **2021**, *86*, 36–48. [CrossRef] [PubMed]
4. Ohshita, J.; Lee, K.-H.; Hamamoto, D.; Kunugi, Y.; Ikadai, J.; Kwak, Y.-W.; Kunai, A. Synthesis of Novel Spiro-condensed Dithienosiloles and the Application to Organic FET. *Chem. Lett.* **2004**, *33*, 892–893. [CrossRef]
5. Saragi, T.P.I.; Fuhrmann-Lieker, T.; Salbeck, J. Comparison of Charge-Carrier Transport in Thin Films of Spiro-Linked Compounds and Their Corresponding Parent Compounds. *Adv. Funct. Mater.* **2006**, *16*, 966–974. [CrossRef]
6. Poriel, C.; Rault-Berthelot, J. Structure–property relationship of 4-substituted-spirobifluorenes as hosts for phosphorescent organic light emitting diodes: An overview. *J. Mater. Chem. C* **2017**, *5*, 3869–3897. [CrossRef]
7. Poriel, C.; Sicard, L.; Rault-Berthelot, J. New generations of spirobifluorene regioisomers for organic electronics: Tuning electronic properties with the substitution pattern. *Chem. Commun.* **2019**, *55*, 14238–14254. [CrossRef] [PubMed]
8. Qu, Y.-K.; Zheng, Q.; Fan, J.; Liao, L.-S.; Jiang, Z.-Q. Spiro Compounds for Organic Light-Emitting Diodes. *Acc. Mater. Res.* **2021**, *2*, 1261–1271. [CrossRef]
9. Ma, S.; Fu, Y.; Ni, D.; Mao, J.; Xie, Z.; Tu, G. Spiro-fluorene based 3D donor towards efficient organic photovoltaics. *Chem. Commun.* **2012**, *48*, 11847–11849. [CrossRef]

10. Yan, Q.; Zhou, Y.; Zheng, Y.-Q.; Pei, J.; Zhao, D. Towards rational design of organic electron acceptors for photovoltaics: A study based on perylenediimide derivatives. *Chem. Sci.* **2013**, *4*, 4389. [CrossRef]
11. Wu, X.-F.; Fu, W.-F.; Xu, Z.; Shi, M.; Liu, F.; Chen, H.-Z.; Wan, J.-H.; Russell, T.P. Spiro Linkage as an Alternative Strategy for Promising Nonfullerene Acceptors in Organic Solar Cells. *Adv. Funct. Mater.* **2015**, *25*, 5954–5966. [CrossRef]
12. Yi, J.; Wang, Y.; Luo, Q.; Lin, Y.; Tan, H.; Wang, H.; Ma, C.Q. A 9,9′-spirobi[9H-fluorene]-cored perylenediimide derivative and its application in organic solar cells as a non-fullerene acceptor. *Chem. Commun.* **2016**, *52*, 1649–1652. [CrossRef] [PubMed]
13. Murakami, K.; Ooyama, Y.; Higashimura, H.; Ohshita, J. Synthesis, Properties, and Polymerization of Spiro[(dipyridinogermole)(dithienogermole)]. *Organometallics* **2015**, *35*, 20–26. [CrossRef]
14. Ohshita, J.; Hayashi, Y.; Adachi, Y.; Enoki, T.; Yamaji, K.; Ooyama, Y. Optical and Photosensitizing Properties of Spiro(dipyridinogermole)(dithienogermole)s with Eletron-Donating Amino and Electron-Withdrawing Pyridinothiadiazole Substituents. *ChemistrySelect* **2018**, *3*, 8604–8609. [CrossRef]
15. Ohshita, J.; Kondo, K.; Adachi, Y.; Song, M.; Jin, S.-H. Synthesis of spirodithienogermole with triphenylamine units as a dopant-free hole-transporting material for perovskite solar cells. *J. Mater. Chem. C* **2021**, *9*, 2001–2007. [CrossRef]
16. Terada, N.; Uematsu, K.; Higuchi, R.; Tokimaru, Y.; Sato, Y.; Nakano, K.; Nozaki, K. Synthesis and Properties of Spiro-double Sila[7]helicene: The LUMO Spiro-conjugation. *Chem. Eur. J.* **2021**, *27*, 9342–9349. [CrossRef] [PubMed]
17. Dobler, M.; Miljnko, D.; Martin, E.; Vladmir, P.; Dumić, M.; Egli, M.; Prelog, V. Chiral Poly(9,9′-spirobifluorene) Crown Ethers. *Angew. Chem. Int. Ed.* **1985**, *24*, 792–794. [CrossRef]
18. Alcazar, V.; Diederich, F. Enantioselective Complexation of Chiral Dicarboxylic Acids in Clefts of Functionalized 9,9′-Spirobifluorenes. *Angew. Chem. Int. Ed.* **1992**, *31*, 1521–1523. [CrossRef]
19. Hovorka, R.; Meyer-Eppler, G.; Piehler, T.; Hytteballe, S.; Engeser, M.; Topic, F.; Rissanen, K.; Lützen, A. Unexpected Self-Assembly of a Homochiral Metallosupramolecular M_4L_4 Catenane. *Chem. Eur. J.* **2014**, *20*, 13253–13258. [CrossRef]
20. Hovorka, R.; Hytteballe, S.; Piehler, T.; Meyer-Eppler, G.; Topic, F.; Rissanen, K.; Engeser, M.; Lützen, A. Self-assembly of metallosupramolecular rhombi from chiral concave 9,9′-spirobifluorene-derived bis(pyridine) ligands. *Beilstein J. Org. Chem.* **2014**, *10*, 432–441. [CrossRef]
21. Cheng, X.; Xie, J.H.; Li, S.; Zhou, Q.L. Asymmetric hydrogenation of α,β-unsaturated carboxylic acids catalyzed by ruthenium(II) complexes of spirobifluorene diphosphine (SFDP) ligands. *Adv. Synth. Catal.* **2006**, *348*, 1271–1276. [CrossRef]
22. Cheng, X.; Zhang, Q.; Xie, J.-H.; Wang, L.-X.; Zhou, Q.-L. Highly Rigid Diphosphane Ligands with a Large Dihedral Angle Based on a Chiral Spirobifluorene Backbone. *Angew. Chem. Int. Ed.* **2005**, *44*, 1118–1121. [CrossRef] [PubMed]
23. Murai, M.; Takeuchi, Y.; Yamauchi, K.; Kuninobu, Y.; Takai, K. Rhodium-Catalyzed Synthesis of Chiral Spiro-9-silabifluorenes by Dehydrogenative Silylation: Mechanistic Insights into the Construction of Tetraorganosilicon Stereocenters. *Chem. Eur. J.* **2016**, *22*, 6048–6058. [CrossRef] [PubMed]
24. Tanaka, H.; Inoue, Y.; Mori, T. Circularly Polarized Luminescence and Circular Dichroisms in Small Organic Molecules: Correlation between Excitation and Emission Dissymmetry Factors. *ChemPhotoChem* **2018**, *2*, 386–402. [CrossRef]
25. Chen, N.; Yan, B. Recent Theoretical and Experimental Progress in Circularly Polarized Luminescence of Small Organic Molecules. *Molecules* **2018**, *23*, 3376. [CrossRef] [PubMed]
26. Ma, J.L.; Peng, Q.; Zhao, C.H. Circularly Polarized Luminescence Switching in Small Organic Molecules. *Chem. Eur. J.* **2019**, *25*, 15441–15454. [CrossRef]
27. Li, X.; Xie, Y.; Li, Z. The Progress of Circularly Polarized Luminescence in Chiral Purely Organic Materials. *Adv. Photonics Res.* **2021**, *2*, 2000136. [CrossRef]
28. Wan, S.-P.; Lu, H.-Y.; Li, M.; Chen, C.-F. Advances in circularly polarized luminescent materials based on axially chiral compounds. *J. Photochem. Photobiol. C Photochem. Rev.* **2022**, *50*, 100500. [CrossRef]
29. Shintani, R.; Misawa, N.; Takano, R.; Nozaki, K. Rhodium-Catalyzed Synthesis and Optical Properties of Silicon-Bridged Arylpyridines. *Chem. Eur. J.* **2017**, *23*, 2660–2665. [CrossRef]
30. Miki, K.; Noda, T.; Gon, M.; Tanaka, K.; Chujo, Y.; Mizuhata, Y.; Tokitoh, N.; Ohe, K. Near-Infrared Circularly Polarized Luminescence through Intramolecular Excimer Formation of Oligo(p-phenyleneethynylene)-Based Double Helicates. *Chem. Eur. J.* **2019**, *25*, 9211–9216. [CrossRef]
31. Feng, J.; Fu, L.; Geng, H.; Jiang, W.; Wang, Z. Designing a near-infrared circularly polarized luminescent dye by dissymmetric spiro-fusion. *Chem. Commun.* **2019**, *56*, 912–915. [CrossRef] [PubMed]
32. Oniki, J.; Moriuchi, T.; Kamochi, K.; Tobisu, M.; Amaya, T. Linear [3]Spirobifluorenylene: An S-Shaped Molecular Geometry of p-Oligophenyls. *J. Am. Chem. Soc.* **2019**, *141*, 18238–18245. [CrossRef] [PubMed]
33. Zhu, K.; Kamochi, K.; Kodama, T.; Tobisu, M.; Amaya, T. Chiral cyclic [n]spirobifluorenylenes: Carbon nanorings consisting of helically arranged quaterphenyl rods illustrating partial units of woven patterns. *Chem. Sci.* **2020**, *11*, 9604–9610. [CrossRef] [PubMed]
34. Yang, S.Y.; Wang, Y.K.; Peng, C.C.; Wu, Z.G.; Yuan, S.; Yu, Y.J.; Li, H.; Wang, T.T.; Li, H.C.; Zheng, Y.X.; et al. Circularly Polarized Thermally Activated Delayed Fluorescence Emitters in Through-Space Charge Transfer on Asymmetric Spiro Skeletons. *J. Am. Chem. Soc.* **2020**, *142*, 17756–17765. [CrossRef] [PubMed]
35. Hamada, H.; Nakamuro, T.; Yamashita, K.; Yanagisawa, H.; Nureki, O.; Kikkawa, M.; Harano, K.; Shang, R.; Nakamura, E. Spiro-Conjugated Carbon/Heteroatom-Bridgedp-Phenylenevinylenes: Synthesis, Properties, and Microcrystal Electron Crystallographic Analysis of Racemic Solid Solutions. *Bull. Chem. Soc. Jpn.* **2020**, *93*, 776–782. [CrossRef]

36. Hamada, H.; Itabashi, Y.; Shang, R.; Nakamura, E. Axially Chiral Spiro-Conjugated Carbon-Bridged *p*-Phenylenevinylene Congeners: Synthetic Design and Materials Properties. *J. Am. Chem. Soc.* **2020**, *142*, 2059–2067. [CrossRef]
37. Yang, S.Y.; Feng, Z.Q.; Fu, Z.; Zhang, K.; Chen, S.; Yu, Y.J.; Zou, B.; Wang, K.; Liao, L.S.; Jiang, Z.Q. Highly Efficient Sky-Blue π-Stacked Thermally Activated Delayed Fluorescence Emitter with Multi-Stimulus Response Properties. *Angew. Chem. Int. Ed.* **2022**, *61*, in press. [CrossRef]
38. Takase, K.; Noguchi, K.; Nakano, K. Circularly Polarized Luminescence from Chiral Spiro Molecules: Synthesis and Optical Properties of 10,10′-Spirobi(indeno[1,2-*b*][1]benzothiophene) Derivatives. *Org. Lett.* **2017**, *19*, 5082–5085. [CrossRef]
39. Takase, K.; Noguchi, K.; Nakano, K. [1]Benzothiophene-Fused Chiral Spiro Polycyclic Aromatic Compounds: Optical Resolution, Functionalization, and Optical Properties. *J. Org. Chem.* **2018**, *83*, 15057–15065. [CrossRef]
40. Takase, K.; Noguchi, K.; Nakano, K. Synthesis of Pyrrole-Containing Chiral Spiro Molecules and Their Optical and Chiroptical Properties. *Bull. Chem. Soc. Jpn.* **2019**, *92*, 1008–1017. [CrossRef]
41. Kubo, M.; Takase, K.; Noguchi, K.; Nakano, K. Solvent-sensitive circularly polarized luminescent compounds bearing a 9,9′-spirobi[fluorene] skeleton. *Org. Biomol. Chem.* **2020**, *18*, 2866–2876. [CrossRef] [PubMed]
42. Kubo, M.; Noguchi, K.; Nakano, K. Chiral Benzo[*b*]silole-Fused 9,9′-Spirobi[fluorene]: Synthesis, Chiroptical Properties, and Transformation to pi-Extended Polycyclic Arene. *ChemPlusChem* **2021**, *86*, 171–175. [CrossRef]
43. Bulumulla, C.; Gunawardhana, R.; Gamage, P.L.; Miller, J.T.; Kularatne, R.N.; Biewer, M.C.; Stefan, M.C. Pyrrole-Containing Semiconducting Materials: Synthesis and Applications in Organic Photovoltaics and Organic Field-Effect Transistors. *ACS Appl. Mater. Interfaces* **2020**, *12*, 32209–32232. [CrossRef] [PubMed]
44. Qiu, B.; Yuan, J.; Zou, Y.; He, D.; Peng, H.; Li, Y.; Zhang, Z. An asymmetric small molecule based on thieno[2,3-*f*]benzofuran for efficient organic solar cells. *Org. Electron.* **2016**, *35*, 87–94. [CrossRef]
45. Tsuji, H.; Nakamura, E. Design and Functions of Semiconducting Fused Polycyclic Furans for Optoelectronic Applications. *Acc. Chem. Res.* **2017**, *50*, 396–406. [CrossRef]
46. Peng, H.; Luan, X.; Qiu, L.; Li, H.; Liu, Y.; Zou, Y. New naphtho[1,2-*h*:5,6-*h*′]difuran based two-dimensional conjugated small molecules for photovoltaic application. *Opt. Mater.* **2017**, *72*, 147–155. [CrossRef]
47. Cao, H.; Rupar, P.A. Recent Advances in Conjugated Furans. *Chem. Eur. J.* **2017**, *23*, 14670–14675. [CrossRef]
48. Zheng, B.; Huo, L. Recent Advances of Furan and Its Derivatives Based Semiconductor Materials for Organic Photovoltaics. *Small Methods* **2021**, *5*, e2100493. [CrossRef]
49. Kowada, T.; Ohe, K. Synthesis and Characterization of Highly Fluorescent and Thermally Stable π-Conjugates involving Spiro[fluorene-9,4′-[4*H*]indeno[1,2-*b*]furan]. *Bull. Korean Chem. Soc.* **2010**, *31*, 577–581. [CrossRef]
50. Kowada, T.; Kuwabara, T.; Ohe, K. Synthesis, structures, and optical properties of heteroarene-fused dispiro compounds. *J. Org. Chem.* **2010**, *75*, 906–913. [CrossRef]
51. Stobe, C.; Seto, R.; Schneider, A.; Lützen, A. Synthesis, Chiral Resolution, and Absolute Configuration of C_2-Symmetric, Chiral 9,9′-Spirobifluorenes. *Eur. J. Org. Chem.* **2014**, *2014*, 6513–6518. [CrossRef]
52. Bhanuchandra, M.; Murakami, K.; Vasu, D.; Yorimitsu, H.; Osuka, A. Transition-Metal-Free Synthesis of Carbazoles and Indoles by an S_NAr-Based "Aromatic Metamorphosis" of Thiaarenes. *Angew. Chem. Int. Ed.* **2015**, *54*, 10234–10238. [CrossRef]
53. Hung, T.Q.; Dang, T.T.; Villinger, A.; Sung, T.V.; Langer, P. Efficient synthesis of thieno[3,2-*b*:4,5-*b*′]diindoles and benzothieno[3,2-*b*]indoles by Pd-catalyzed site-selective C-C and C-N coupling reactions. *Org. Biomol. Chem.* **2012**, *10*, 9041–9044. [CrossRef] [PubMed]
54. Bates, C.G.; Saejueng, P.; Murphy, J.M.; Venkataraman, D. Synthesis of 2-Arylbenzo[*b*]furans via Copper(I)-Catalyzed Coupling of o-Iodophenols and Aryl Acetylenes. *Org. Lett.* **2002**, *4*, 4727–4729. [CrossRef]
55. Wei, B.; Zhang, D.; Chen, Y.H.; Lei, A.; Knochel, P. Preparation of Polyfunctional Biaryl Derivatives by Cyclolanthanation of 2-Bromobiaryls and Heterocyclic Analogues Using $nBu_2LaCl_4 \cdot LiCl$. *Angew. Chem. Int. Ed.* **2019**, *58*, 15631–15635. [CrossRef]
56. Frisch, M.J.; Trucks, G.W.; Schlegel, H.B.; Scuseria, G.E.; Robb, M.A.; Cheeseman, J.R.; Scalmani, G.; Barone, V.; Mennucci, B.; Petersson, G.A.; et al. *Gaussian 09 Rev. E.01*; Gaussian, Inc.: Wallingford, CT, USA, 2016.
57. Burla, M.C.; Caliandro, R.; Camalli, M.; Carrozzini, B.; Cascarano, G.L.; De Caro, L.; Giacovazzo, C.; Polidori, G.; Siliqi, D.; Spagna, R. IL MILIONE: A suite of computer programs for crystal structure solution of proteins. *J. Appl. Crystallogr.* **2007**, *40*, 609–613. [CrossRef]
58. Sheldrick, G.M. A short history of SHELX. *Acta Crystallogr. A* **2008**, *64*, 112–122. [CrossRef]
59. Sheldrick, G.M. Crystal structure refinement with SHELXL. *Acta Crystallogr. Sect. C Struct. Chem.* **2015**, *71*, 3–8. [CrossRef]

Article

Synthesis, Molecular Docking Study, and Cytotoxicity Evaluation of Some Novel 1,3,4-Thiadiazole as Well as 1,3-Thiazole Derivatives Bearing a Pyridine Moiety

Amr S. Abouzied [1,2], Jehan Y. Al-Humaidi [3], Abdulrahman S Bazaid [4], Husam Qanash [4,5], Naif K. Binsaleh [4], Abdulwahab Alamri [6], Sheikh Muhammad Ibrahim [7] and Sobhi M. Gomha [8,*]

1. Department of Pharmaceutical Chemistry, College of Pharmacy, University of Hail, Hail 81442, Saudi Arabia
2. Department of Pharmaceutical Chemistry, National Organization for Drug Control and Research (NODCAR), Giza 12311, Egypt
3. Department of Chemistry, College of Science, Princess Nourah Bint Abdulrahman University, Riyadh 11671, Saudi Arabia
4. Department of Medical Laboratory Science, College of Applied Medical Sciences, University of Hail, Hail 55476, Saudi Arabia
5. Molecular Diagnostics and Personalized Therapeutics Unit, University of Hail, Hail 55476, Saudi Arabia
6. Department of Pharmacology and Toxicology, College of Pharmacy, University of Hail, Hail 55211, Saudi Arabia
7. Chemistry Department, Faculty of Science, Islamic University of Madinah, Madinah 42351, Saudi Arabia
8. Chemistry Department, Faculty of Science, Cairo University, Giza 12613, Egypt
* Correspondence: s.m.gomha@cu.edu.eg

Citation: Abouzied, A.S.; Al-Humaidi, J.Y.; Bazaid, A.S.; Qanash, H.; Binsaleh, N.K.; Alamri, A.; Ibrahim, S.M.; Gomha, S.M. Synthesis, Molecular Docking Study, and Cytotoxicity Evaluation of Some Novel 1,3,4-Thiadiazole as Well as 1,3-Thiazole Derivatives Bearing a Pyridine Moiety. *Molecules* 2022, 27, 6368. https://doi.org/10.3390/molecules27196368

Academic Editors: Joseph Sloop and Alessandro Pedretti

Received: 31 August 2022
Accepted: 20 September 2022
Published: 27 September 2022

Publisher's Note: MDPI stays neutral with regard to jurisdictional claims in published maps and institutional affiliations.

Copyright: © 2022 by the authors. Licensee MDPI, Basel, Switzerland. This article is an open access article distributed under the terms and conditions of the Creative Commons Attribution (CC BY) license (https://creativecommons.org/licenses/by/4.0/).

Abstract: Pyridine, 1,3,4-thiadiazole, and 1,3-thiazole derivatives have various biological activities, such as antimicrobial, analgesic, anticonvulsant, and antitubercular, as well as other anticipated biological properties, including anticancer activity. The starting 1-(3-cyano-4,6-dimethyl-2-oxopyridin-1(2H)-yl)-3-phenylthiourea (**2**) was prepared and reacted with various hydrazonoyl halides **3a–h**, α-haloketones **5a–d**, 3-chloropentane-2,4-dione **7a** and ethyl 2-chloro-3-oxobutanoate **7b**, which afforded the 3-aryl-5-substituted 1,3,4-thiadiazoles **4a–h**, 3-phenyl-4-arylthiazoles **6a–d** and the 4-methyl-3- phenyl-5-substituted thiazoles **8a,b**, respectively. The structures of the synthesized products were confirmed by spectral data. All of the compounds also showed remarkable anticancer activity against the cell line of human colon carcinoma (HTC-116) as well as hepatocellular carcinoma (HepG-2) compared with the Harmine as a reference under in vitro condition. 1,3,4-Thiadiazole **4h** was found to be most promising and an excellent performer against both cancer cell lines (IC_{50} = 2.03 ± 0.72 and 2.17 ± 0.83 µM, respectively), better than the reference drug (IC_{50} = 2.40 ± 0.12 and 2.54 ± 0.82 µM, respectively). In order to check the binding modes of the above thiadiazole derivatives, molecular docking studies were performed that established a binding site with EGFR TK.

Keywords: pyridines; 1,3,4-thiadiazoles; 1,3-thiazoles; hydrazonoyl halides; molecular docking; anticancer activity

1. Introduction

Designing new, effective, selective, highly potent, although more tolerant, anticancer drugs through the identification of novel structures remains a considerable challenge for the researchers in the field of medicinal chemistry. Hybrid drug design has emerged during the past few years as a leading technique for the creation of innovative anticancer medicines that, in theory, can address many of the pharmacokinetic drawbacks of conventional anticancer medications [1]. Medical researchers have focused on pyridines, 1,3,4-thiadiazole and 1,3-thiazole systems that have led to somewhat more effective and promising results in recent years. To name just a few, the following pyridine-based small compounds have received approval as anticancer medications: Vismodegib III, Crizotinib IV, Regorafenib II, and Sorafenib I (Figure 1) [2–4]. Different pyridine derivatives were studied for a variety of human cancer cell lines as a tool for novel anticancer drugs through the topoisomerase inhibitory activity. These results show various reports regarding different derivatives, such as bioisosteres of α-terthiophene as a potent for protein kinase C inhibitor [5], promising topoisomerase I and/or II inhibitory activity, as well as cytotoxicity against a variety of human cancer cell lines [6–10].

Among those, 1,3,4-thiadiazoles gained substantial interest due to their widespread biological activity, including antimicrobial, anti-inflammatory, antithrombotic, antihypertensive, antituberculosis, anesthetic, anticonvulsant and antiulcer activities [11–16]. Furthermore, different researchers also particularly report 1, 3, 4-thiadiazole derivatives for their excellent anticancer activity, which are confirmed by desirable IG_{50} and IC_{50} values in inhibitory effect, such as Filanesib and compounds I–III (Figure 1) [16–20].

1,3-Thiazoles, which derived from thiosemicarbazone derivatives, is also known for its various pharmacological applications, as its scaffold is useful for several natural, non-natural and semi-synthetic drugs, including anti-inflammatory, anti-parasitic and antineoplastic properties [13,21–31]. Numerous studies suggested that medications such as Tiazofurin, Dasatinib, and Dabrafenib that contain thiazoles may have anticancer properties against different cancer types (Figure 1) [32–34].

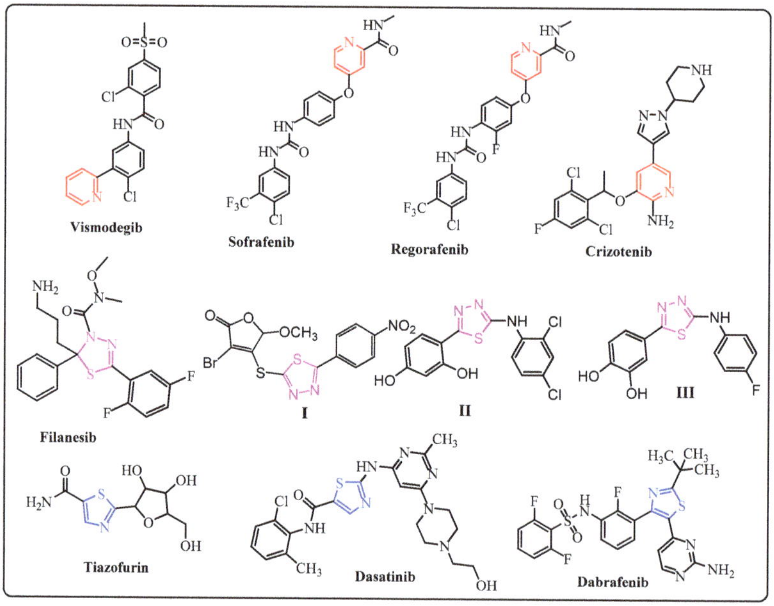

Figure 1. Examples of anticancer drugs bearing pyridine, thiazole, and thiadiazole moieties.

Under the influence of the findings mentioned above, continuous efforts were performed to synthesize innovative anticancer compounds [13,35–46]. The present report aims to elaborate on the new series of thiadiazole-pyridines as well as thiazole-pyridines that might have cytotoxic effects via the inhibition of protein Epidermal Growth Factor Tyrosine Kinase receptor (EGFR TK), which plays an essential mediating role in cell proliferation, angiogenesis, apoptosis, and metastatic spread compared with reference drugs.

2. Results and Discussion

1-(3-Cyano-4,6-dimethyl-2-oxopyridin-1(2H)-yl)-3-phenylthiourea (**2**) [47] was synthesized through the reaction of phenyl isothiocyanate with 1-amino-4,6-dimethyl-2-oxo-

1,2-dihydropyridine-3-carbonitrile (**1**) under the influence of a catalytic amount of KOH in absolute ethanol as a solvent, as shown in Scheme 1.

The conduct of compound **2** towards various hydrazonoyl chlorides was explored to synthesize a novel series of 1,3,4-thiadiazole derivatives. Therefore, when compound **2** was treated with hydrazonoyl chlorides **3a–h** in ethanol as a solvent in the presence of a TEA as a catalyst, it afforded a series of single product, recognized as 4,6-dimethyl-2-oxo-1-((3-aryl-5-substituted-1,3,4-thiadiazol-2(3*H*)-ylidene)amino)-1,2-dihydropyridine-3-carbonitriles **4a–h** (Scheme 1). The 1,3,4-thiadiazole **4** was formed through the alkylation of the thiol group present in thiosemicarbazone moiety, an intramolecluar cyclization and finally elimination of aniline molecule.

All the structures of products **4a–h** were confirmed by elemental analyses followed by spectral data. In general, the ^{1}H-NMR spectra of **4a** (see Supplementary Materials), taken as an example, showed a singlet (1*H*) at δ 6.39 ppm corresponding to the pyridine proton, three singlets at δ 2.01, 2.33 and 2.46 ppm corresponding to the three CH$_3$ groups and a multiplet δ 7.09–7.46 ppm corresponding to the five aromatic protons. The ^{13}C-NMR spectrum of **4a** revealed two signals at δ = 25.6, 194.7 ppm, which is characteristic for the acetyl group (CH$_3$C=O). The disappearance of the two NH absorption bands was also observed in the IR spectra because of the elimination of amine groups from the starting material **2**. Moreover, the mass spectrum of the products **4a–h** revealed a molecular ion peaks at the expected *m/z* values.

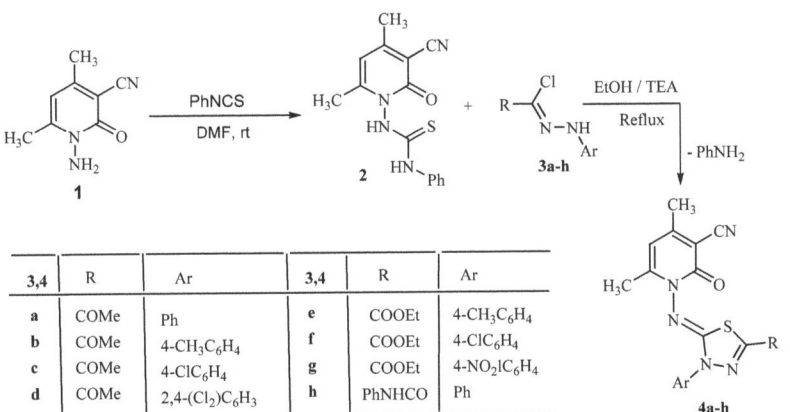

Scheme 1. Synthesis of thiadiazoles **4a–h**.

In addition to this, the chemical reactivity of compound **2** with several of α-haloketones was also investigated to synthesize a series of novel thiazole derivatives. Accordingly, compound **2** reacted with α-haloketones **5a–d** in the presence of TEA as a catalyst under refluxing condition using EtOH as a solvent that resulted a corresponding thiazoles **6a–d** series, as shown in Scheme 2 (see Experimental).

All the structures of the series of products **6a–e** were also confirmed through the analytical followed by spectral data analysis (see Experimental). Compound **6c** showed a typical singlet signal that appeared at δ 3.82, 6.55, and 6.80 ppm due to the OCH$_3$ group, pyridine-H5, and thiazole-H5, respectively, in the ^{1}H-NMR spectra. In addition to this, a multiplet is observed in the region: 7.01–7.47 ppm assignable to the nine aromatic hydrogens. On the other hand, the ^{13}C-NMR spectrum of **6c** showed four signals at δ = 17.8, 21.2, 56.7 and 162.4 ppm characteristic for 2 Ar-CH$_3$, Ar-OCH$_3$ and C=O groups, respectively, in addition to sixteen aromatic carbon signals in the range of 107.3–149.4 ppm.

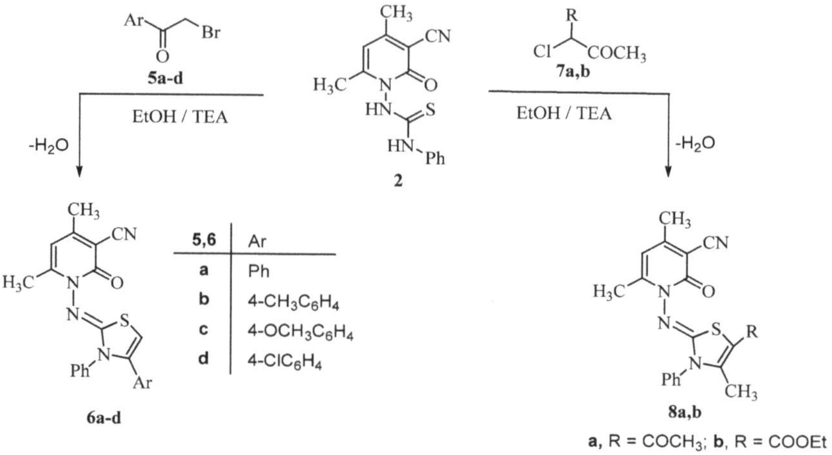

Scheme 2. Synthesis of thiazoles **6a–d** and **8a,b**.

Finally, compound **2** reacted with α-chloro compounds **7a,b** under refluxing condition in the presence of EtOH as a solvent and TEA as a catalyst that resulted a single product, identified as 4,6-dimethyl-1-((4-methyl-3-phenyl-5-substitutedthiazol-2(3H)-ylidene) amino)4yridine-2(1H)-ones **8a,b**, as outlined in Scheme 2.

The structure of the isolated product **8** was inferred from its IR and ^{1}H-NMR spectral data and elemental analysis (see Experimental).

2.1. Anti-Cancer Activity

The series of prepared compounds **4a–h** and **6a–d** were investigated against human colon carcinoma (HCT-116) followed by the hepatocellular carcinoma (HepG2) cell lines to obtain pharmacological activities using Harmine as a reference drug through colorimetric MTT assay under in vitro conditions. The survival curve was obtained by plotting the relation between the concentrations of the drugs against the surviving cells, resulting in the 50% inhibitory concentration (IC_{50}). The anti-proliferative activity is also achieved through the expression of the mean IC_{50} by three independent experiments (μM) ± standard deviation calculated from three replicates.

Table 1 and Figure 2 summarize the structure- and concentration-dependent anticancer activities of the series of compounds against HTC-116 cell lines. An in vitro inhibition activity shows a positive trend along with all tested compounds. Compounds such as **4c**, **4d**, **4f** and **4g** show comparable activity to that of Harmine (IC_{50} = 2.40 ± 0.12 μM) as a reference, whereas compound **4h** demonstrates even better results compared with the same reference. A similar trend of results was also observed for the hepatocellular carcinoma (HepG2) cell line assay, where **4c**, **4d**, **4f**, **4g** and **4h** show either comparable or improved inhibitory activity, with **6h** (IC_{50} = 2.17 ± 0.83 μM) showing the maximum effect in comparison with the reference Harmine (IC_{50} = 2.54 ± 0.82 μM).

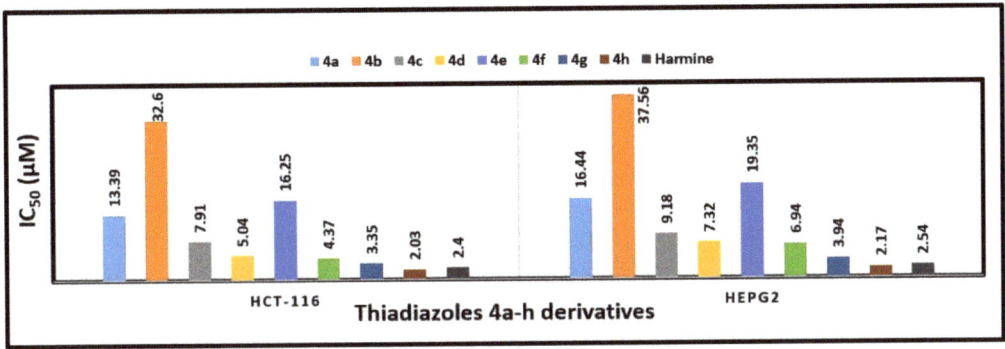

Figure 2. The anticancer activity of series of compounds **4a–h** against HCT-116 and HepG2 cell lines.

Table 1. The anticancer activity of the series of compounds **4a–h**, and **6a–d** towards human colon carcinoma (HCT-116) and hepatocellular carcinoma (HepG2) cell lines expressed as IC$_{50}$ values (µM) ± standard deviation from three replicates.

Tested Compounds	R	X (or Y)	IC$_{50}$ (µM)	
			HCT-116	HepG2
4a	COCH$_3$	H	13.39 ± 1.04	16.44 ± 1.06
4b	COCH$_3$	CH$_3$	32.57 ± 2.37	37.56 ± 1.24
4c	COCH$_3$	Cl	7.91 ± 0.83	9.18 ± 0.91
4d	COCH$_3$	2,4-diCl	5.04 ± 0.59	7.32 ± 0.75
4e	COOEt	CH$_3$	16.25 ± 1.05	19.35 ± 1.30
4f	COOEt	Cl	4.37 ± 0.28	6.94 ± 0.69
4g	COOEt	NO$_2$	3.35 ± 0.46	3.94 ± 0.80
4h	CONHPh	H	2.03 ± 0.72	2.17 ± 0.83
6a	–	H	15.57 ± 1.30	19.12 ± 1.36
6b	–	CH$_3$	36.29 ± 1.32	25.90 ± 0.70
6c	–	OCH$_3$	21.00 ± 1.28	19.37 ± 1.29
6d	–	Cl	9.61 ± 0.88	7.36 ± 0.85
Harmine	–	–	2.40 ± 0.12	2.54 ± 0.82

For thiadiazoles **4a–h**: **4h** (amidophenyl, has a phenyl ring along with electron withdrawing amido group resulting strongest activity) > **4g** (strong electron withdrawing nitro group, increases activity) > **4f** (with ester group along with one electron withdrawing Cl atom) > **4d** (acetyl group with electron withdrawing 2 Cl atom) > **4c** (acetyl group with mild electron withdrawing one Cl atom) > **4a** (with acetyl group with un-substituted phenyl group) > **4e** (ester with methyl group) > **4b** (acetyl group with methyl group, electron donating group decreases activity). Overall, electron releasing groups decrease the activity, whereas strong electron withdrawing groups increase the activity. A selective high activity is observed, particularly with **4h**, possibly due to the fact that **4h** possess one extra phenyl ring connected with the pyridine group in the amido side, which significantly enhances its aromatic π-π interaction with the Phe, Tyr and Trp residues. This is in coherence with the harmine where one phenyl and one pyridine is fused with an indole group, resulting in the same kind of interaction with amino acid residues containing an aromatic group. In addition to this, a noteworthy mention would be, in **4h**, the nitrogen in the amido group is engaged in a tautomeric structure, thereby restricting the electron releasing power of nitrogen in the phenyl ring. Such interactions are absent for the rest of thiadiazole derivatives **4a–g**, where only electron withdrawing group is present in the lone available phenyl ring.

This is in analogous with the thiazoles derivatives **6a–d**, where only electron releasing groups are present with the single phenyl group available, resulting in less activity compared to thiadiazoles **4a–h** with the only exception of **6d** (an electron withdrawing Cl atom is present), where moderate anticancer activity is detected (Figure 3).

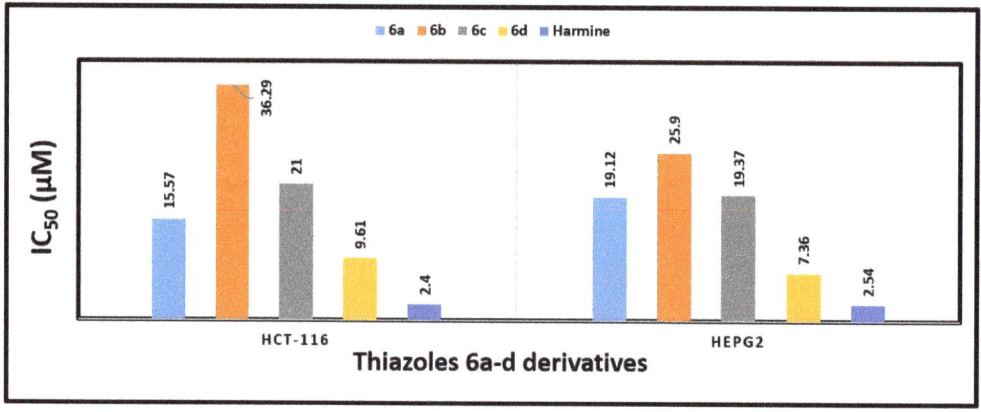

Figure 3. The anticancer activity of series of compounds **6a–d** against HCT-116 and HepG2 cell lines.

2.2. Docking Study for Cytotoxicity

The protein Epidermal Growth Factor Receptor Tyrosine Kinase Domain (EGFR TK) was selected for this study, where the ability to inhibit this receptor ultimately leads to the blockade of the growth pathway, giving a promising anti-cancer agent [48]. The lower binding energy resulting from the association of the compound with the targeted protein is an indication of a higher binding efficiency. The results of the docking protocol were validated by the re-docking of the co-crystallized ligand (W19) inside the active site of EGFR TK (Figures 4 and 5). Harmine was used in this study as an EGFR TK inhibitor. By comparing the binding affinity of different screened synthesized compounds with Harmine (ΔG of −7.1), it was found that compound **4h** showed the best binding affinity with ΔG of −10.8, and **4b, 4c, 4d, 4e, 4f, 6a, 6b**, and **6c** showed binding activity ΔG −8.1 to −9.2. The screened compounds showed a possible interaction with EGFR TK active sites as depicted in Table 2 and Figures 6–13.

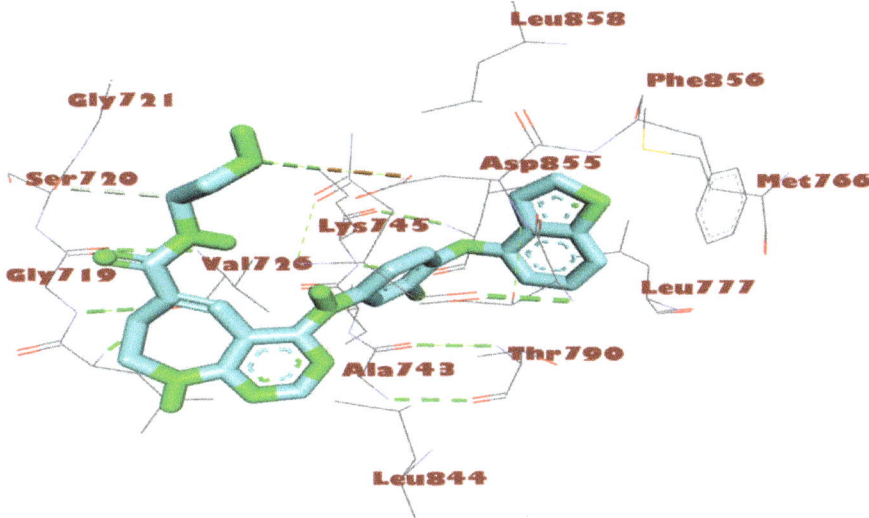

Figure 4. 3D molecular interactions of re-docked co-crystalized ligand (W19) with EGFR TK residues. The hydrogen bonds are represented as green dashed lines; the pi interactions are represented as orange lines.

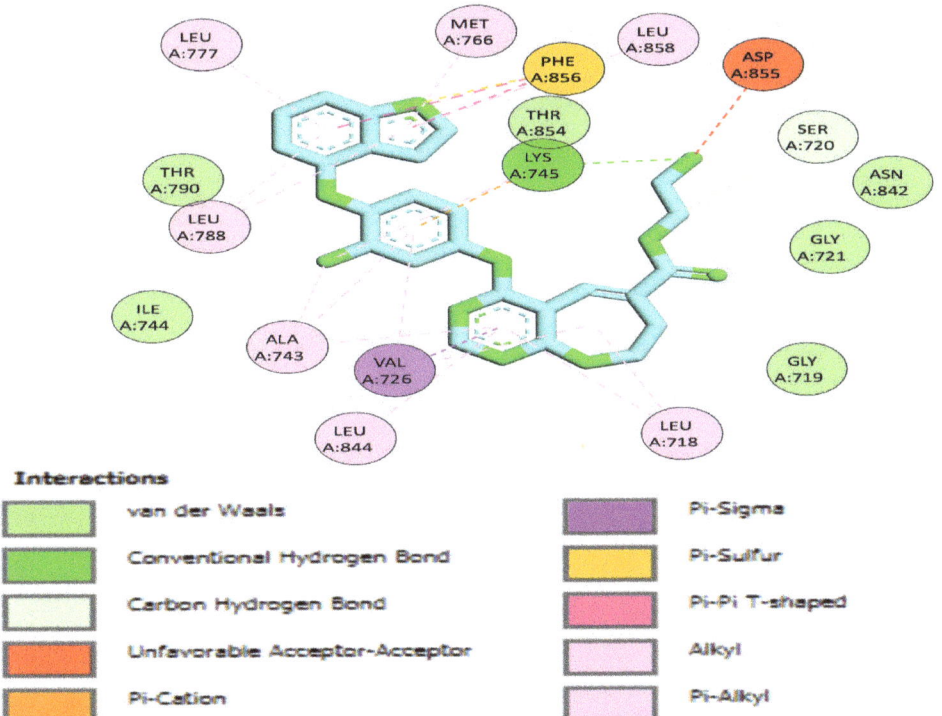

Figure 5. Two-dimensional molecular interactions of re-docked co-crystalized ligand (W19) with EGFR TK residues.

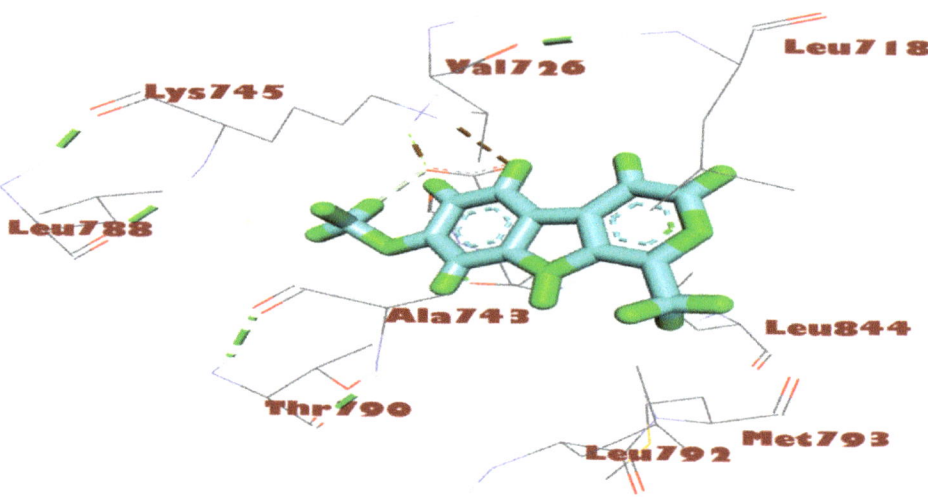

Figure 6. 3D molecular interactions of Harmine reference drug with EGFR TK residues. The hydrogen bonds are represented as green dashed lines; the pi interactions are represented as orange lines.

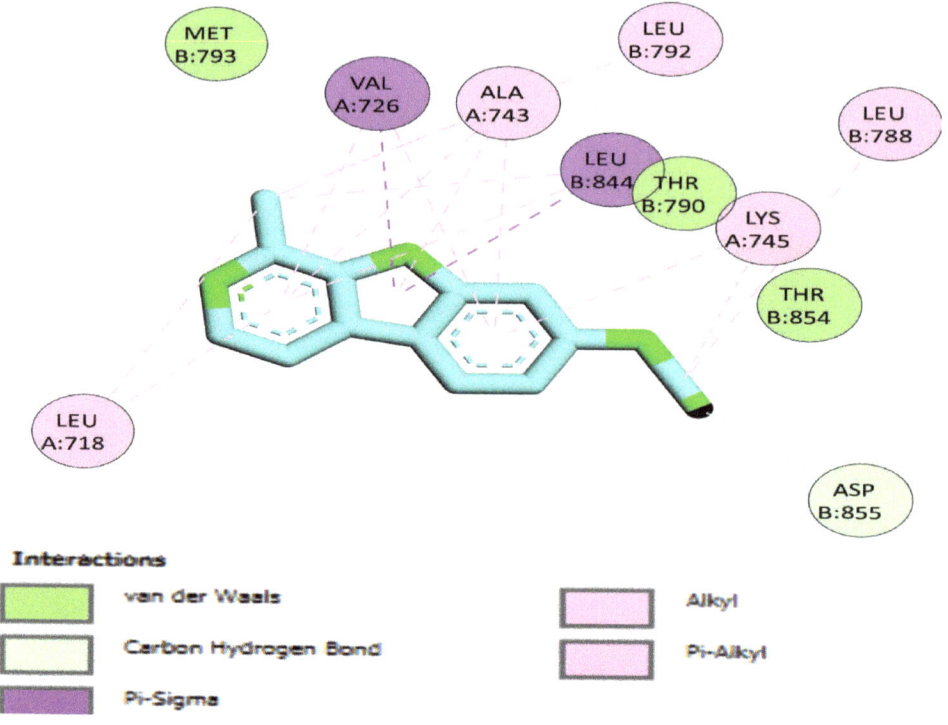

Figure 7. Two-dimensional molecular interactions of Harmine reference drug with EGFR TK residues.

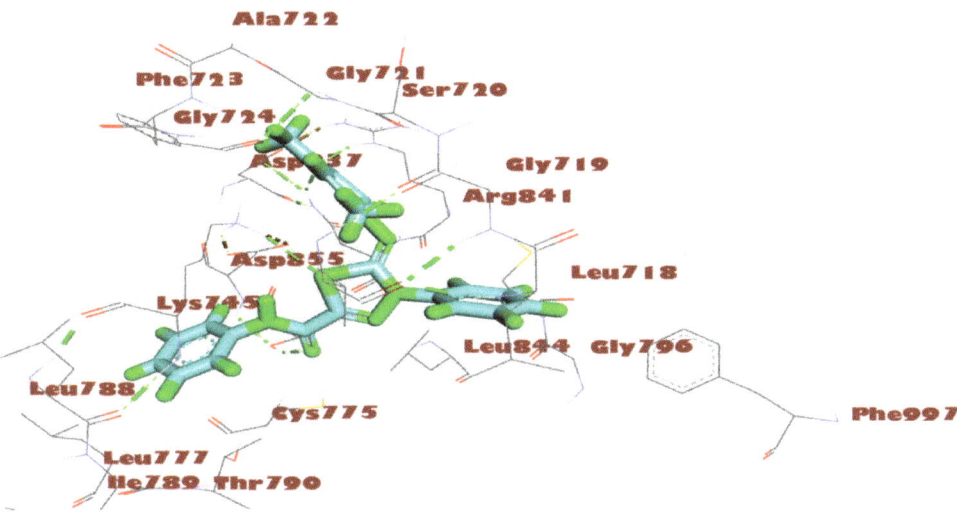

Figure 8. 3D molecular interactions of compound **4h** with EGFR TK residues. The hydrogen bonds are represented as green dashed lines; the pi interactions are represented as orange lines.

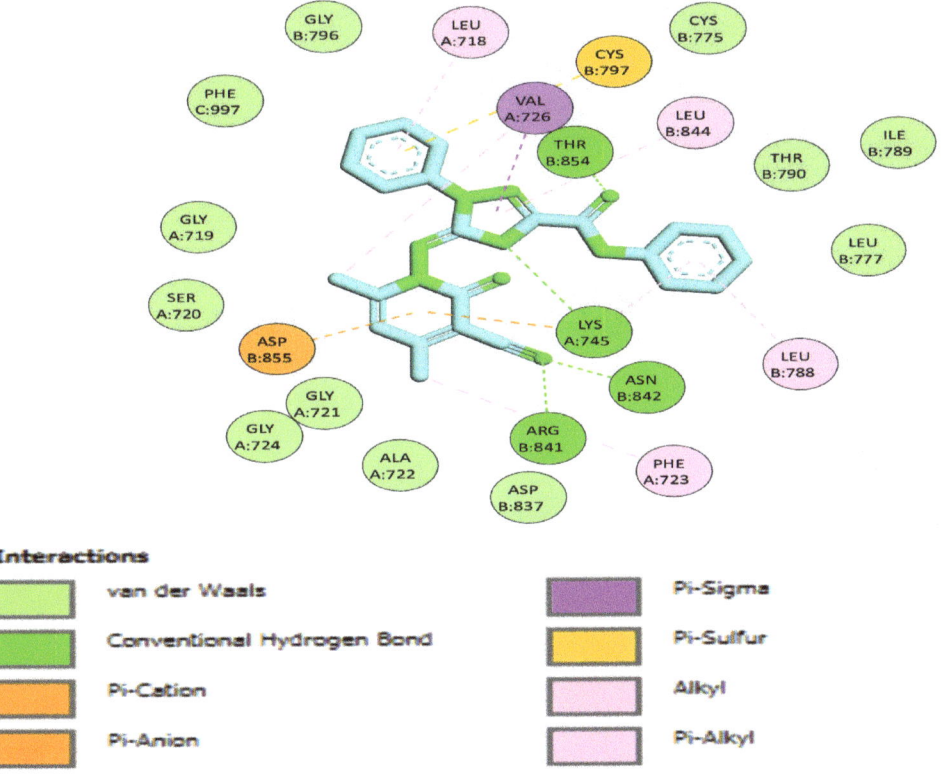

Figure 9. Two-dimensional molecular interactions of compound **4h** with EGFR TK residues.

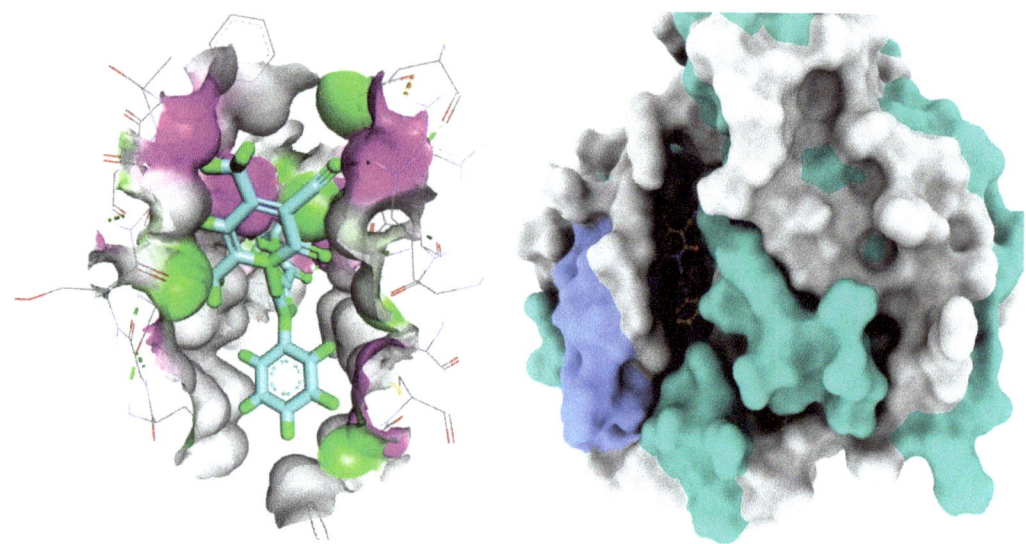

Figure 10. Mapping surface showing compound **4h** occupying the active pocket of EGFR TK.

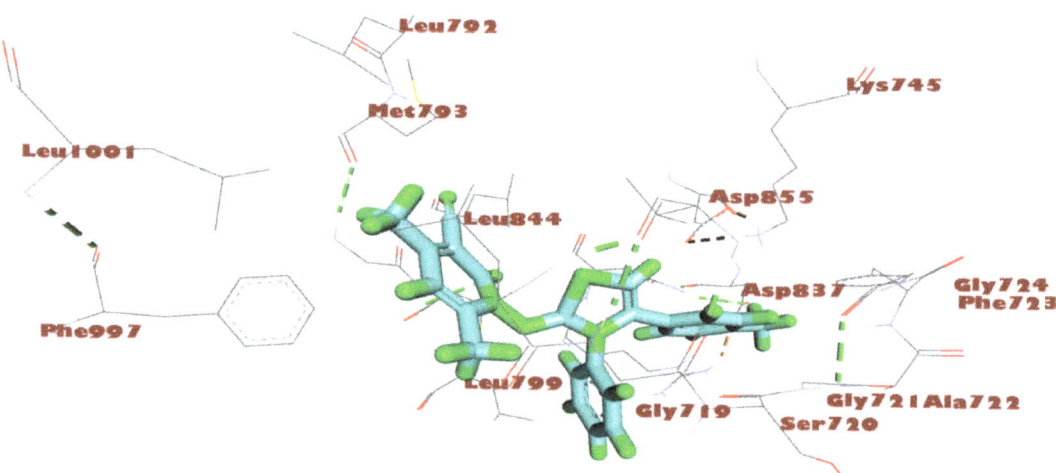

Figure 11. 3D molecular interactions of compound **6b** with EGFR TK residues. The hydrogen bonds are represented as green dashed lines; the pi interactions are represented as orange lines.

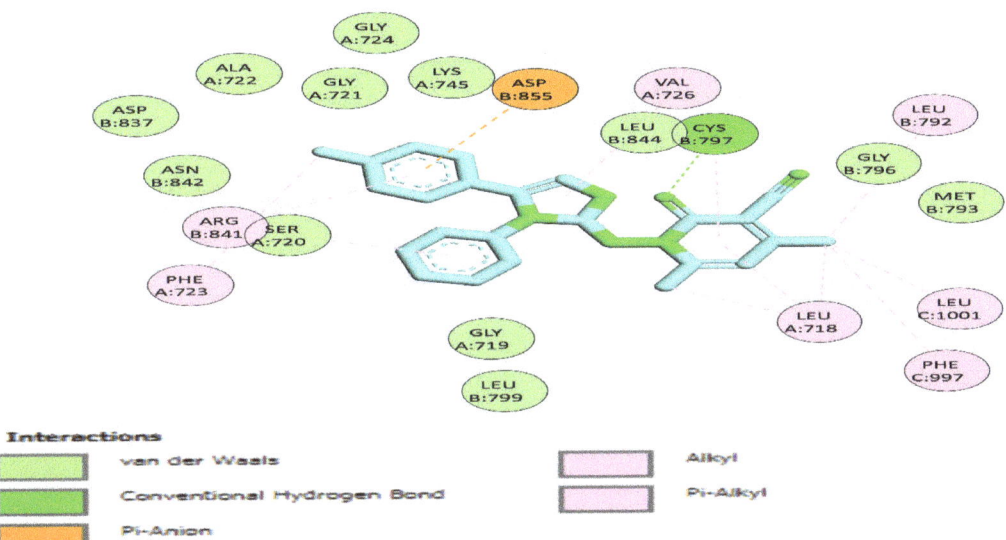

Figure 12. Two-dimensional molecular interactions of compound **6b** with EGFR TK residues.

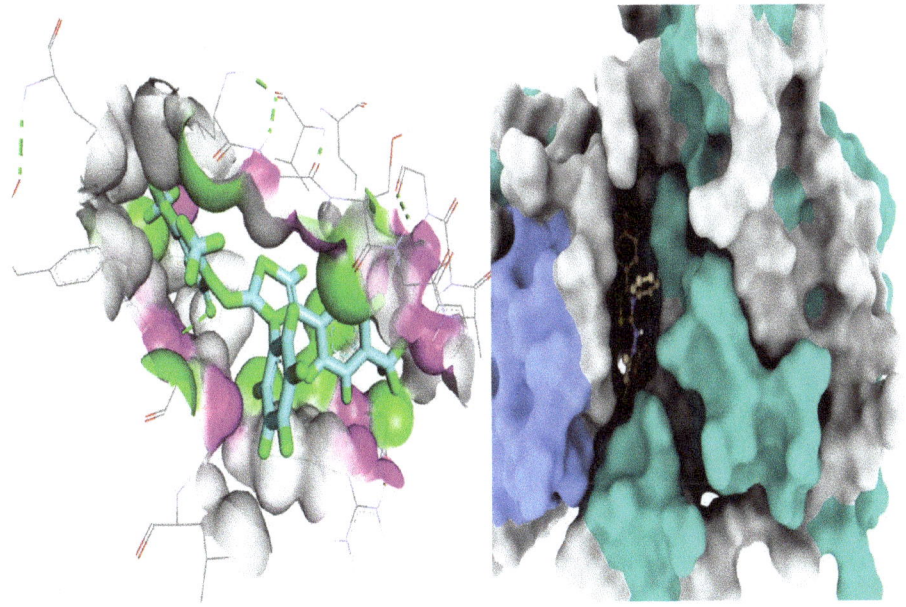

Figure 13. Mapping surface showing compound **6b** occupying the active pocket of EGFR TK.

Table 2. The binding scores and interactions of the examined compounds and the Harmine inhibitor inside the binding pocket of receptor of (3W33) for EGFR TK.

Compounds	Binding Scores (kcal/mol)	Hydrogen Bond Interactions	Distance (Å)	Hydrophobic Interactions	Distance (Å)
4a	−8.5	MET769	2.08	LEU694 VAL702 LYS721 LEU820 THR830	3.78 3.75 3.46, 3.93 3.24, 3.90 3.77
4b	−9.2	MET769	1.96	LEU694 VAL702 ALA719 LYS721 LEU764 LEU820	3.74 3.69 3.89 3.48, 3.71 3.77 3.39
4c	−8.9	MET769	2.07	LEU694 VAL702 ALA719 LYS721 LEU820 THR830	3.78 3.69 3.87 3.47 3.21, 3.94 3.75
4d	−9.2	GLU738 THR830 ASP831 PHE832	2.71 2.64 2.20 3.17	LEU694 PHE699 VAL702 ALA719 LYS721 LEU764 THR766 LEU820	3.55 3.66 3.04 3.56 3.95 3.59 3.50 3.58
4e	−8.7	GLU738 THR830 ASP831 PHE832	2.71 2.64 2.20 3.17	LEU694 PHE699 VAL702 ALA719 LYS721 LEU764 THR766 LEU820	3.55 3.66 3.04 3.56 3.95 3.59 3.50 3.58
4f	−8.8	MET769	2.10	LEU694 VAL702 ALA719 LYS721 LEU820 THR830	3.88 3.58 3.87 3.42 3.33, 3.86 3.92
4g	−9.1	MET769	2.56	LEU694 VAL702 ALA719 LYS721 LEU820	3.74 3.38 3.68 3.59 3.31, 3.81
4h	−10.8	ARG841 ASN842 LYS745 THR854	2.08 2.27 3.48 3.45	LEU718 VAL726 ALA743 LYS745 LEU788 THR790	3.74, 3.50 3.61, 3.30 3.67 3.36, 3.89 3.66 3.80

Table 2. Cont.

Compounds	Binding Scores (kcal/mol)	Hydrogen Bond Interactions	Distance (Å)	Hydrophobic Interactions	Distance (Å)
6a	−8.8			LEU694	3.69
				VAL702	3.13, 3.91
				ALA719	3.61
				LYS721	3.73, 3.87
				LEU820	3.41, 3.64
6b	−8.9	CYS797	2.29	LEU718	3.90, 3.39, 3.47
				VAL726	3.64
				ASP855	3.67
				LEU777	3.08
				LEU788	3.48
				THR790	3.48
				LEU844	3.53
				PHE997	3.29
6c	−7.9	THR766	2.38	LEU694	3.39
				VAL702	3.32, 3.51
				LYS721	3.69, 3.73
				LEU820	3.39
6d	−8.8	THR766	2.58	PHE699	3.57
				VAL702	3.95
				ALA719	3.50
				MET769	3.93
				ARG817	3.78
				LEU820	3.91, 3.48
Harmine	−7.1			LEU718	3.61
				VAL726	3.58, 3.64
				LYS745	3.76
				THR790	3.63
				LEU792	3.72
W19	−10.8	LYS745	2.36	LEU718	3.67
				VAL726	3.65, 3.94
				LYS745	3.85
				LEU777	3.79
				LEU788	3.97
				THR790	3.73
				THR845	3.79

3. Experimental

Elementar vario LIII CHNS analyzer (Elementar Analysensysteme GmbH, Langenselbold, Germany) is used to measure all elemental analysis. Electrothermal IA 9000 series Digital Melting Point Apparatus (Shanghai Jiahang Instruments Co., Jiading District, Shanghai, China) was used to obtain melting points data. Shimadzu FTIR 8101 PC infrared spectrophotometers (Shimadzu Co., Kyoto, Japan) were used to record IR spectra data in KBr discs on Pye Unicam SP 3300. Varian Mercury VX-300 NMR spectrometer (Bruker Biospin, Karlsruhe, Germany) was used with the operating frequency of 300 MHz (^{1}H-NMR) in deuterated dimethylsulfoxide (DMSO-$d6$) solvent to record NMR spectra, where chemical shifts were related to the solvent used. Shimadzu GCeMS-QP1000 EX mass spectrometer (Shimadzu Co., Kyoto, Japan) was used to record mass spectra at 70 eV. The cytotoxicity of the prepared compounds was measured by the Regional Center for Mycology and Biotechnology in Al-Azhar University, Cairo, Egypt.

Synthesis of 1-(3-cyano-4,6-dimethyl-2-oxopyridin-1(2H)-yl)-3-phenylthiourea (2)

A mixture of 1-amino-4,6-dimethyl-2-oxo-1,2-dihydropyridine-3-carbonitrile (1) (1.63 g, 10 mmol), KOH (0.56 g, 10 mmol) in DMF (30 mL), was stirred for 10 min. Then, PhNCS (1.35 g, 10 mmol) is added under stirring condition and continued for the next 6 h. Afterwards, the solution was diluted with 30 mL of distilled water, followed by neutralization by adding aqueous AcOH dropwise, resulting in a solid recrystallized from dioxin to obtain yellowish brown crystals (71%) as a pure product of compound 2; mp = 209–211 °C (Lit mp = 205–207 °C [47]); ^{1}H-NMR (DMSO-d_6): δ 2.22 (s, 3H, CH_3), 2.30 (s, 3H, CH_3), 6.34 (s, 1H, Pyridine-H5), 7.32–7.69 (m, 5H, Ar-H), 8.55 (s, br, 1H, NH), 8.95 (s, br, 1H, NH) ppm; IR (KBr): v 3372, 3241 (2NH), 3033, 2951 (CH), 2218 (CN), 1675 (C=O), 1599 (C=N), 1335 (C=S) cm^{-1}; MS m/z (%): 298 (M$^+$, 85). Anal Calcd for $C_{15}H_{14}N_4OS$ (298.36): C, 60.38; H, 4.73; N, 18.78. Found: C, 60.24; H, 4.59; N, 18.58%.

Synthesis of 4,6-dimethyl-2-oxo-1-((3-aryl-5-substituted-1,3,4-thiadiazol-2(3H)-ylidene)amino)-1,2-dihydropyridine-3-carbonitriles (4a–h).

A mixture of compound 2 (0.298 g, 1 mmol) and appropriate hydrazonoyl halides 3a–h (1 mmol) in DMF (20 mL) containing Et_3N (0.1 g, 1 mmol) was heated under reflux for 3–6 h. The resultant solid product was recrystallized by appropriate solvent to give thiadiazoles 4a–h. Below is a list of the spectrum information and physical characteristics of the products 4a–h.

1-((5-Acetyl-3-phenyl-1,3,4-thiadiazol-2(3H)-ylidene)amino)-4,6-dimethyl-2-oxo-1,2-dihydropyridine-3-carbonitrile (4a).

Yellow solid (79%); m.p. 233–235 °C (DMF); ^{1}H-NMR (DMSO-d_6): δ 2.01 (s, 3H, CH_3), 2.33 (s, 3H, CH_3), 2.46 (s, 3H, CH_3), 6.39 (s, 1H, Pyridine-H5), 7.09–7.46 (m, 5H, Ar-H) ppm; ^{13}C-NMR (DMSO-d_6): δ 19.6, 21.2, 25.6 (3CH_3), 107.8, 115.9, 116.4, 118.6, 122.7, 123.1, 124.0, 125.8, 138.1, 142.3, 152.3 (Ar-C and C=N), 163.2, 194.7 (2 C=O) ppm; IR (KBr): v 3047, 2933 (CH), 2220 (CN),1704, 1651 (C=O), 1599 (C=N) cm^{-1}; MS m/z (%): 365 (M$^+$, 49). Anal. Calcd. for $C_{18}H_{15}N_5O_2S$ (365.41): C, 59.17; H, 4.14; N, 19.17. Found C, 59.30; H, 4.04; N, 19.11%.

1-((5-Acetyl-3-(p-tolyl)-1,3,4-thiadiazol-2(3H)-ylidene)amino)-4,6-dimethyl-2-oxo-1,2-dihydropyridine-3-carbonitrile (4b).

Yellow solid (80%); m.p. 243–245 °C (DMF); ^{1}H-NMR (DMSO-d_6): δ 2.01 (s, 3H, CH_3), 2.32 (s, 3H, CH_3), 2.40 (s, 3H, CH_3), 2.45 (s, 3H, CH_3), 6.39 (s, 1H, Pyridine-H5), 7.02–7.80 (m, 4H, Ar-H) ppm; ^{13}C-NMR (DMSO-d_6): δ 19.1, 20.7, 21.2, 25.4 (4CH_3), 107.3, 116.5, 122.9, 127.5, 130.1, 130.6, 133.8, 138.2, 145.2, 154.2 (Ar-C and C=N), 162.1, 194.2 (2 C=O) ppm; IR (KBr): v 3028, 2940 (CH), 2217 (CN),1703, 1667 (C=O), 1597 (C=N) cm^{-1}; MS m/z (%): 379 (M$^+$, 38). Anal. Calcd. for $C_{19}H_{17}N_5O_2S$ (379.44): C, 60.14; H, 4.52; N, 18.46. Found C, 60.05; H, 4.42; N, 18.29%.

1-((5-Acetyl-3-(4-chlorophenyl)-1,3,4-thiadiazol-2(3H)-ylidene)amino)-4,6-dimethyl-2-oxo-1,2-dihydropyridine-3-carbonitrile (4c).

Yellow solid (78%); m.p. 228–230 °C (dioxane); ^{1}H-NMR (DMSO-d_6): δ 2.03 (s, 3H, CH_3), 2.25 (s, 3H, CH_3), 2.41 (s, 3H, CH_3), 6.39 (s, 1H, Pyridine-H5), 7.11–7.85 (m, 4H, Ar-H) ppm; IR (KBr): v 3052, 2944 (CH), 2218 (CN), 1709, 1663 (C=O), 1598 (C=N) cm^{-1}; MS m/z (%): 401 (M$^+$+ 2, 31), 399 (M$^+$, 100). Anal. Calcd. for $C_{18}H_{14}ClN_5O_2S$ (399.85): C, 54.07; H, 3.53; N, 17.52. Found C, 54.01; H, 3.36; N, 17.48%.

1-((5-Acetyl-3-(2,4-dichlorophenyl)-1,3,4-thiadiazol-2(3H)-ylidene)amino)-4,6-dimethyl-2-oxo-1,2-dihydropyridine-3-carbonitrile (4d).

Brown solid (79%); m.p. 262–264 °C (DMF); ^{1}H-NMR (DMSO-d_6): δ 2.04 (s, 3H, CH_3), 2.31 (s, 3H, CH_3), 2.43 (s, 3H, CH_3), 6.39 (s, 1H, Pyridine-H5), 7.26–7.86 (m, 3H, Ar-H) ppm; IR (KBr): v 3047, 2937 (CH), 2222 (CN), 1711, 1672 (C=O), 1601 (C=N) cm^{-1}; MS m/z (%): 434 (M$^+$, 81). Anal. Calcd. for $C_{18}H_{13}Cl_2N_5O_2S$ (434.30): C, 49.78; H, 3.02; N, 16.13. Found C, 49.83; H, 3.00; N, 16.04%.

Ethyl 5-((3-cyano-4,6-dimethyl-2-oxopyridin-1(2H)-yl)imino)-4-(p-tolyl)-4,5-dihydro-1,3,4-thiadiazole-2-carboxylate (4e).

Yellow solid (77%); m.p. 189–191 °C (EtOH\DMF); ^{1}H-NMR (DMSO-d_6): δ 1.17–1.21 (t, 3H, CH$_3$), 2.01 (s, 3H, CH$_3$), 2.32 (s, 3H, CH$_3$), 2.42 (s, 3H, CH$_3$), 4.13–4.17 (q, 2H, CH$_2$), 6.45 (s, 1H, Pyridine-H5), 7.12–7.59 (m, 4H, Ar-H) ppm; ^{13}C-NMR (DMSO-d_6): δ 12.9, 17.5, 21.1, 21.7 (4CH$_3$), 117.3, 119.0, 121.2, 121.7, 122.4, 124.1, 125.0, 130.1, 139.4, 142.5, 151.2 (Ar-C and C=N), 161.2, 163.5 (2 C=O) ppm; IR (KBr): v 3049, 2930 (CH), 2219 (CN), 1723, 1669 (C=O), 1600 (C=N) cm^{-1}; MS m/z (%): 409 (M$^+$, 14). Anal. Calcd. for C$_{20}$H$_{19}$N$_5$O$_3$S (409.46): C, 58.67; H, 4.68; N, 17.10. Found C, 58.52; H, 4.55; N, 17.02%.

Ethyl 4-(4-chlorophenyl)-5-((3-cyano-4,6-dimethyl-2-oxopyridin-1(2H)-yl)imino)-4,5-dihydro-1,3,4-thiadiazole-2-carboxylate (4f).

Yellow solid (77%); m.p. 185–187 °C (EtOH); ^{1}H-NMR (DMSO-d_6): δ 1.20–1.27 (t, 3H, CH$_3$), 2.03 (s, 3H, CH$_3$), 2.38 (s, 3H, CH$_3$), 4.11–4.17 (q, 2H, CH$_2$), 6.46 (s, 1H, Pyridine-H5), 7.23–7.60 (m, 4H, Ar-H) ppm; IR (KBr): v 3046, 2937 (CH), 2217 (CN),1722, 1660 (C=O), 1601 (C=N) cm^{-1}; MS m/z (%): 431 (M$^+$$_+$ 2, 20), 429 (M$^+$, 63). Anal. Calcd. for C$_{19}$H$_{16}$ClN$_5$O$_3$S (429.88): C, 53.09; H, 3.75; N, 16.29. Found C, 53.18; H, 3.58; N, 16.14%.

Ethyl 5-((3-cyano-4,6-dimethyl-2-oxopyridin-1(2H)-yl)imino)-4-(4-nitrophenyl)-4,5-dihydro-1,3,4-thiadiazole-2-carboxylate (4g).

Yellow solid (73%); mp = 217–219 °C (EtOH\DMF); ^{1}H-NMR (DMSO-d_6): δ 1.20–1.27 (t, 3H, CH$_3$), 2.29 (s, 3H, CH$_3$), 2.42 (s, 3H, CH$_3$), 4.18–4.26 (q, 2H, CH$_2$), 6.84 (s, 1H, Pyridine-H5), 7.27–8.34 (m, 4H, Ar-H) ppm; IR (KBr): v 3039, 2943 (CH), 2218 (CN),1719, 1665 (C=O), 1600 (C=N) cm^{-1}; MS m/z (%): 440 (M$^+$, 70). Anal. Calcd. for C$_{19}$H$_{16}$N$_6$O$_5$S (440.43): C, 51.81; H, 3.66; N, 19.08. Found C, 51.66; H, 3.50; N, 19.03%.

5-((3-Cyano-4,6-dimethyl-2-oxopyridin-1(2H)-yl)imino)-N,4-diphenyl-4,5-dihydro-1,3,4-thiadiazole-2-carboxamide (4h).

Yellow solid (86%); mp = 277–279 °C (DMF); ^{1}H-NMR (DMSO-d_6): δ 2.16 (s, 3H, CH$_3$), 2.38 (s, 3H, CH$_3$), 6.42 (s, 1H, Pyridine-H5), 7.03–7.79 (m, 10H, Ar-H), 10.21 (s, 1H, NH) ppm; ^{13}C-NMR (DMSO-d_6): δ 17.9, 21.4 (2CH$_3$), 112.4, 115.9, 119.3, 120.5, 121.5, 122.1, 124.3, 124.7, 125.2, 128.6, 132.4, 134.6, 139.4, 141.5, 152.4 (Ar-C and C=N), 161.2, 162.7 (2 C=O) ppm; IR (KBr): v 3278 (NH), 3061, 2947 (CH), 2219 (CN), 1678, 1663 (C=O), 1597 (C=N) cm^{-1}; MS m/z (%): 442 (M$^+$, 18). Anal. Calcd. for C$_{23}$H$_{18}$N$_6$O$_2$S (442.50): C, 62.43; H, 4.10; N, 18.99. Found C, 62.52; H, 4.04; N, 18.75%.

Synthesis of thiazoles 6a–d or 8a,b.

A mixture of **2** (0.298 g, 1 mmol) and α-haloketones **5a–d** or 3-chloropentane-2,4-dione (**7a**) or ethyl 2-chloro-3-oxobutanoate (**7b**) (1 mmol) in DMF (15 mL) was refluxed for 4–6 h and was continuously monitored in TLC. The separation of the product was clearly observed during the course of the reaction. The resultant solid product was then filtered, washed several times with water, dried and recrystallized in the proper solvent to give the corresponding thiazoles **6a–d** or **8a,b**, respectively.

1-((3,4-Diphenylthiazol-2(3H)-ylidene)amino)-4,6-dimethyl-2-oxo-1,2-dihydropyridine-3-carbonitrile (6a).

Yellow solid (74%); mp = 230–232 °C (DMF); ^{1}H-NMR (DMSO-d_6): δ 2.23 (s, 3H, CH$_3$), 2.34 (s, 3H, CH$_3$), 6.37 (s, 1H, Pyridine-H5), 6.67 (s, 1H, Thiazole-H5), 7.19–8.05 (m, 10H, Ar-H) ppm; IR (KBr): v 3047, 2934 (CH), 2217 (CN), 1667 (C=O), 1599 (C=N) cm^{-1}; MS m/z (%): 398 (M$^+$, 37). Anal. Calcd. for C$_{23}$H$_{18}$N$_4$OS (398.48): C, 69.33; H, 4.55; N, 14.06. Found C, 69.17; H, 4.42; N, 14.04%.

4,6-Dimethyl-2-oxo-1-((3-phenyl-4-(p-tolyl)thiazol-2(3H)-ylidene)amino)-1,2-dihydropyridine-3-carbonitrile (6b).

Yellow solid (77%); mp = 213–215 °C (EtOH); ^{1}H-NMR (DMSO-d_6): δ 2.23 (s, 3H, CH$_3$), 2.24 (s, 3H, CH$_3$), 2.34 (s, 3H, CH$_3$), 6.37 (s, 1H, Pyridine-H5), 7.07 (s, 1H, Thiazole-H5),

7.33–7.44 (m, 9H, Ar-H) ppm; IR (KBr): v 3042, 2950 (CH), 2219 (CN), 1671 (C=O), 1602 (C=N) cm^{-1}; MS m/z (%): 412 (M$^+$, 18). Anal. Calcd. for $C_{24}H_{20}N_4OS$ (412.51): C, 69.88; H, 4.89; N, 13.58. Found C, 69.70; H, 4.83; N, 13.37%.

1-((4-(4-Methoxyphenyl)-3-phenylthiazol-2(3H)-ylidene)amino)-4,6-dimethyl-2-oxo-1,2-dihydropyridine-3-carbonitrile (6c).

Yellow solid (80%); mp = 221–223 °C (EtOH\DMF); ^{1}H-NMR (DMSO-d_6): δ 2.23 (s, 3H, CH$_3$), 2.32 (s, 3H, CH$_3$), 3.82 (s, 3H, OCH$_3$), 6.55 (s, 1H, Pyridine-H5), 6.80 (s, 1H, Thiazole-H5), 7.01–7.47 (m, 9H, Ar-H) ppm; ^{13}C-NMR (DMSO-d_6): δ 17.8, 21.2 (2CH$_3$), 56.7 (OCH$_3$), 107.8, 113.9, 115.0, 118.3, 121.9, 122.5, 123.6, 130.0, 131.8, 135.6, 138.9, 140.8, 144.0, 152.0, 152.5, 154.3 (Ar-C and C=N), 162.4 (C=O) ppm; IR (KBr): v 3074, 2928 (CH), 2218 (CN), 1683 (C=O), 1603 (C=N) cm^{-1}; MS m/z (%): 403 (M$^+$, 23). Anal. Calcd. for $C_{24}H_{20}N_4O_2S$ (428.51): C, 67.27; H, 4.70; N, 13.08. Found C, 67.36; H, 4.61; N, 13.02%.

1-((4-(4-Chlorophenyl)-3-phenylthiazol-2(3H)-ylidene)amino)-4,6-dimethyl-2-oxo-1,2-dihydropyridine-3-carbonitrile (6d).

Yellow solid (79%); mp = 217–219 °C (DMF); ^{1}H-NMR (DMSO-d_6): δ 2.23 (s, 3H, CH$_3$), 2.34 (s, 3H, CH$_3$), 6.76 (s, 1H, Pyridine-H5), 7.01 (s, 1H, Thiazole-H5), 7.12–7.77 (m, 9H, Ar-H) ppm;IR (KBr): v 3048, 2927 (CH), 2218 (CN), 1669 (C=O), 1602 (C=N) cm^{-1}; MS m/z (%): 409 (M$^+$+ 2, 13), 407 (M$^+$, 41). Anal. Calcd. for $C_{23}H_{17}ClN_4OS$ (432.93): C, 63.81; H, 3.96; N, 12.94. Found C, 63.62; H, 3.77; N, 12.73%.

1-((5-Acetyl-4-methyl-3-phenylthiazol-2(3H)-ylidene)amino)-4,6-dimethyl-2-oxo-1,2-dihydropyridine-3-carbonitrile (8a).

Yellow solid (83%); mp = 195–197 °C (EtOH); ^{1}H-NMR (DMSO-d_6): δ 2.18 (s, 3H, CH$_3$), 2.25 (s, 3H, CH$_3$), 2.33 (s, 3H, CH$_3$), 2.43 (s, 3H, CH$_3$), 6.36 (s, 1H, Pyridine-H5), 7.55–7.66 (m, 5H, Ar-H) ppm; IR (KBr): v 3048, 2935 (CH), 2219 (CN),1709, 1671 (C=O), 1601 (C=N) cm^{-1}; MS m/z (%): 353 (M$^+$, 35). Anal. Calcd. for $C_{20}H_{18}N_4O_2S$ (378.45): C, 63.47; H, 4.79; N, 14.80. Found C, 63.35; H, 4.62; N, 14.69%.

Ethyl 2-((3-cyano-4,6-dimethyl-2-oxopyridin-1(2H)-yl)imino)-4-methyl-3-phenyl-2,3-dihydrothiazole-5-carboxylate (8b).

Yellow solid (85); mp = 190–192 °C (EtOH); ^{1}H-NMR (DMSO-d_6): δ 1.21–1.25 (t, 3H, CH$_3$), 2.19 (s, 3H, CH$_3$), 2.22 (s, 3H, CH$_3$), 2.38 (s, 3H, CH$_3$), 4.21–4.25 (q, 2H, CH$_2$), 6.39 (s, 1H, Pyridine-H5), 7.45–7.62 (m, 5H, Ar-H) ppm; ^{13}C-NMR (DMSO-d_6): δ 14.5, 17.1, 21.3 (3CH$_3$), 62.5 (CH$_2$), 106.4, 113.3, 115.3, 119.8, 123.7, 124.5, 127.6, 131.5, 133.0, 140.2, 144.1, 148.7 (Ar-C and C=N), 162.5, 164.1 (2 C=O) ppm; IR (KBr): v 3037, 2943 (CH), 2218 (CN), 1724, 1668 (C=O), 1600 (C=N) cm^{-1}; MS m/z (%): 408 (M$^+$, 74). Anal. Calcd. for $C_{21}H_{20}N_4O_3S$ (408.48): C, 61.75; H, 4.94; N, 13.72. Found C, 61.63; H, 4.81; N, 13.53%.

3.1. Cytotoxic Activity

The cytotoxicity of the newly synthesized series of compounds was studied against HCT-116 and HepG2 cells using the MTT assay through an incubation period of 24 h [49,50].

Mammalian cell line: HCT-116 and HepG2 cells were collected from VACSERA Tissue Culture Unit, Cairo, Egypt.

3.2. Docking Method

The MOE 2019.012 suite (Chemical Computing Group ULC, Montreal, Canada) [51] was applied in order to carry all docking studies for the newly synthesized compounds to suggest their plausible mechanism of action as the protein Epidermal Growth Factor Receptor Tyrosine Kinase Domain (EGFR TK) inhibitors by evaluating their binding grooves and modes to compare with Harmine as a reference standard.

The newly prepared derivatives were placed into the MOE window, where they were treated to partial charge addition and energy minimization [52,53]. The produced compounds also were placed into a single database with the Harmine and saved as an MDB file, which was then uploaded to the ligand icon during the docking process. The Protein Data Bank was used to generate an X-ray of the targeted Epidermal Growth Factor Receptor Tyrosine Kinase Domain (EGFR TK) 3W33. Available online: https://www.rcsb.org/structure/3W33 (accessed on 17 July 2022). [48]. Furthermore, it was readied for the docking process by following the previously detailed stages [54,55]. Additionally, the downloaded protein was error-corrected, 3D hydrogen-loaded and energy-minimized [56,57].

In a general docking procedure, the newly prepared derivatives were substituted for the ligand site. After modifying the default program requirements previously stated, the co-crystallized ligand site was selected as the docking site, and the docking process was started [58]. In a nutshell, the docking site was chosen using the dummy atoms method [59]. The placement and scoring procedures, respectively, Triangle matcher and London dG, were chosen. The stiff receptor was used as the scoring method, and the GBVI/WSA dG was used as the refining method, respectively, to select the top 10 poses out of a total of 100 poses for each docked molecule [60,61]. For further research, the optimal pose for each ligand with the highest score, binding mode and RMSD value was chosen. It is important to note that the applied MOE program underwent the first step of program verification by docking Harmine to its ligand binding of the prepared Target [62,63]. By obtaining a low RMSD value (1.43) between the newly created compounds with docked Harmine, a valid performance was demonstrated.

4. Conclusions

In this paper, two new series of aryl substituted novel pyridine-1,3,4-thiadiazoles, and pyridine-thiazoles were synthesized starting with 1-(3-cyano-4,6-dimethyl-2-oxopyridin-1(2H)-yl)-3-phenylthiourea and several available reagents. All the structures were confirmed through elemental and spectral analysis, where the plausible mechanistic approach for their formation was discussed. All the prepared derivatives showed effectiveness towards the inhibition of human colon carcinoma (HCT-116) as well as hepatocellular carcinoma (HepG2) cell lines through in vitro evaluation and an in silico docking study. From the obtained results, compound **4h** (amidophenyl has a phenyl ring along with the electron withdrawing amido group resulting in the strongest activity) was found to be the strongest and most effective with 2.03 ± 0.72 µM against HCT-116, contributing its activity through a variety of interactions, such as hydrophobic interaction, hydrogen bonding in addition to aromatic stacking interactions with selected target (EGFR TK) pockets, compared with Harmine as a reference drug. A detailed analysis of all the series confirmed the electron withdrawing group present in the aryl substitution, resulting in the enhancement of anticancer activity, which could be promising for the future generation of new efficient anticancer drugs based on 1,3,5-thiadiazole and 1,3-thiazole derivatives.

Supplementary Materials: The following supporting information can be downloaded at: https://www.mdpi.com/article/10.3390/molecules27196368/s1, Samples of ^{1}H-NMR and ^{13}C-NMR spectra.

Author Contributions: Supervision, Investigation, Methodology, Resources, Formal analysis, Data curation, Funding acquisition, Writing—original draft, Writing—review & editing: S.M.G., A.S.A., J.Y.A.-H., A.S.B., H.Q., N.K.B., A.A. and S.M.I. All authors have read and agreed to the published version of the manuscript.

Institutional Review Board Statement: Not applicable.

Informed Consent Statement: Not applicable.

Data Availability Statement: The data presented in this study are available on request from corresponding author.

Acknowledgments: This research was funded by the Scientific Research Deanship at University of Ha'il—Saudi Arabia via a project number (RG-21081).

Conflicts of Interest: The authors declare no conflict of interest.

Sample Availability: Samples of all compounds are available from the authors.

References

1. Singh, A.K.; Kumar, A.; Singh, H.; Sonawane, P.; Paliwal, H.; Thareja, S.; Pathak, P.; Grishina, M.; Jaremko, M.; Emwas, A.-H.; et al. Concept of hybrid drugs and recent advancements in anticancer hybrids. *Pharmaceuticals* **2022**, *15*, 1071. [CrossRef]
2. Wilhelm, S.; Carter, C.; Lynch, M.; Lowinger, T.; Dumas, J.; Smith, R.A.; Schwartz, B.; Simantov, R.; Kelley, S. Discovery and development of sorafenib: A multikinase inhibitor for treating cancer. *Nat. Rev. Drug Discov.* **2006**, *5*, 835–844. [CrossRef]
3. DiGiulio, S. FDA Approves Stivarga for Advanced GIST. *Oncol. Times* **2013**, *35*, 12.
4. Cui, J.J.; Tran-Dubé, M.; Shen, H.; Nambu, M.; Kung, P.P.; Pairish, M.; Jia, L.; Meng, J.; Funk, L.; Botrous, I.; et al. Structure Based Drug Design of Crizotinib (PF-02341066), a Potent and Selective Dual Inhibitor of Mesenchymal-Epithelial Transition Factor (c-MET) Kinase and Anaplastic Lymphoma Kinase (ALK). *J. Med. Chem.* **2011**, *54*, 6342–6363. [CrossRef] [PubMed]
5. Zhao, L.X.; Moon, Y.S.; Basnet, A.; Kim, E.K.; Jahng, Y.; Park, J.G.; Jeong, T.C.; Cho, W.J.; Choi, S.U.; Lee, C.O.; et al. The discovery and synthesis of novel adenosine receptor (A2A) antagonists. *Bioorg. Med. Chem. Lett.* **2004**, *14*, 1333–1336. [CrossRef]
6. Zhao, L.X.; Sherchan, J.; Park, J.K.; Jahng, Y.; Jeong, B.S.; Jeong, T.C.; Lee, C.S.; Lee, E.S. Synthesis, cytotoxicity and structure-activity relationship study of terpyridines. *Arch. Pharm. Res.* **2006**, *29*, 1091–1095. [CrossRef] [PubMed]
7. Son, J.K.; Zhao, L.X.; Basnet, A.; Thapa, P.; Karki, R.; Na, Y.; Jahng, Y.; Jeong, T.C.; Jeong, B.S.; Lee, C.S.; et al. Synthesis of 2,6-diaryl-substituted pyridines and their antitumor activities. *Eur. J. Med. Chem.* **2008**, *43*, 675–682. [CrossRef] [PubMed]
8. Thapa, P.; Karki, R.; Thapa, U.; Jahng, Y.; Jung, M.J.; Nam, J.M.; Na, Y.; Kwon, Y.; Lee, E.S. 2-Thienyl-4-furyl-6-aryl pyridine derivatives: Synthesis, topoisomerase I and II inhibitory activity, cytotoxicity, and structure–activity relationship study. *Bioorg. Med. Chem.* **2010**, *18*, 377–386. [CrossRef]
9. Thapa, P.; Karki, R.; Choi, H.Y.; Choi, J.H.; Yun, M.; Jeong, B.S.; Jung, M.J.; Nam, J.M.; Na, Y.; Cho, W.J.; et al. Synthesis of 2-(thienyl-2-yl or -3-yl)-4-furyl-6-aryl pyridine derivatives and evaluation of their topoisomerase I and II inhibitory activity, cytotoxicity, and structure–activity relationship. *Bioorg. Med. Chem.* **2010**, *18*, 2245–2254. [CrossRef] [PubMed]
10. Jeong, B.S.; Choi, H.Y.; Thapa, P.; Karki, R.; Lee, E.; Nam, J.M.; Na, Y.; Ha, E.-M.; Kwon, Y.; Lee, E.-S. Synthesis, topoisomerase I and II inhibitory activity, cytotoxicity, and structure-activity relationship study of rigid analogues of 2,4,6-trisubstituted pyridine containing 5,6-dihydrobenzo[h]quinoline moiety. *Bull. Korean Chem. Soc.* **2011**, *32*, 303–306. [CrossRef]
11. Omar, A.Z.; Alshaye, N.A.; Mosa, T.M.; El-Sadany, S.K.; Hamed, E.A.; El-Atawy, M.A. Synthesis and antimicrobial activity screening of piperazines bearing n,n'-bis(1,3,4-thiadiazole) moiety as probable enoyl-ACP reductase inhibitors. *Molecules* **2022**, *27*, 3698. [CrossRef]
12. Gomha, M.S.; Kheder, N.A.; Abdelaziz, M.R.; Mabkhot, Y.N.; Alhajoj, A.M. A facile synthesis and anticancer activity of some novel thiazoles carrying 1,3,4-thiadiazole moiety. *Chem. Cent. J.* **2017**, *11*, 25. [CrossRef]
13. Gomha, S.M.; Abdelaziz, M.R.; Kheder, N.A.; Abdel-aziz, H.M.; Alterary, S.; Mabkhot, Y.N. A Facile access and evaluation of some novel thiazole and 1,3,4-thiadiazole derivatives incorporating thiazole moiety as potent anticancer agents. *Chem. Cent. J.* **2017**, *11*, 105. [CrossRef]
14. Janowska, S.; Khylyuk, D.; Andrzejczuk, S.; Wujec, M. Design, synthesis, antibacterial evaluations and in silico studies of novel thiosemicarbazides and 1,3,4-thiadiazoles. *Molecules* **2022**, *27*, 3161. [CrossRef] [PubMed]
15. Ikhlass, M.A.; Sobhi, G.; Elaasser, M.M.; Bauomi, M.A. Synthesis and biological evaluation of new pyridines containing imidazole moiety as antimicrobial and anticancer agents. *Turk. J. Chem.* **2015**, *39*, 334–346.
16. Zhang, Z.-P.; Zhong, Y.; Han, Z.-B.; Zhou, L.; Su, H.-S.; Wang, J.; Liu, Y.; Cheng, M.-S. Synthesis, molecular docking analysis and biological evaluations of saccharide-modified thiadiazole sulfonamide derivatives. *Int. J. Mol. Sci.* **2021**, *22*, 5482. [CrossRef] [PubMed]
17. Oleson, J.J.; Slobada, A.; Troy, W.P.; Halliday, S.L.; Landes, M.J.; Angier, R.B.; Semb, J.; Cyr, K.; Williams, J.H. The Carcinostatic activity of some 2-amino-1,3,4-thiadiazole. *J. Am. Chem. Soc.* **1955**, *77*, 6713–6714. [CrossRef]
18. Matysiak, J.; Opolski, A. Synthesis and antiproliferative activity of N-substituted 2-amino-5-(2,4-dihydroxyphenyl)-1,3,4-thiadiazoles. *Bioorg. Med. Chem.* **2006**, *14*, 4483–4489. [CrossRef]
19. Kumar, D.; Kumar, N.M.; Chang, K.; Shah, K. Synthesis and anticancer activity of 5-(3-indolyl)-1,3,4-thiadiazoles. *Eur. J. Med. Chem.* **2010**, *45*, 4664–4668. [CrossRef] [PubMed]
20. Bhole, R.P.; Bhusari, K.P. 3-Benzhydryl-1,3,4-thiadiazole-2(3H)-thione. *Med. Chem. Res.* **2010**, *20*, 695. [CrossRef]
21. Sujatha, K.; Vedula, R.R. Novel one-pot expeditious synthesis of 2, 4-disubstituted thiazoles through a three-component reaction under solvent free conditions. *Synth. Commun.* **2018**, *48*, 302–308. [CrossRef]
22. Abdalla, M.A.; Gomha, S.M.; Abdelaziz, M.; Serag, N. Synthesis and antiviral evaluation of some novel thiazoles and 1,3-thiazines substituted with pyrazole moiety against rabies virus. *Turk. J. Chem.* **2016**, *40*, 441–453. [CrossRef]
23. Gomha, S.M.; Khalil, K.D. A convenient ultrasound-promoted synthesis and cytotoxic activity of some new thiazole derivatives bearing a coumarin nucleus. *Molecules* **2012**, *17*, 9335–9347. [CrossRef] [PubMed]
24. Nayak, S.; Gaonkar, S.L. A Review on recent synthetic strategies and pharmacological importance of 1, 3-thiazole derivatives. *Mini Rev. Med. Chem.* **2019**, *19*, 215–238. [CrossRef]
25. Kumar, S.; Aggarwal, R. Thiazole: A Privileged motif in marine natural products. *Mini Rev. Org. Chem.* **2019**, *16*, 26–34. [CrossRef]
26. Sharma, S.; Devgun, M.; Narang, R.; Lal, S.; Rana, A.C. Thiazoles: A retrospective study on synthesis, structure-activity relationship and therapeutic significance. *Indian J. Pharm. Educ. Res.* **2022**, *56*, 646–666. [CrossRef]
27. Hassan, S.; Abdullah, M.; Aziz, D. An efficient one-pot three-component synthesis, molecular docking, ADME and DFT predictions of new series thiazolidin-4-one derivatives bearing a sulfonamide moiety as potential antimicrobial and antioxidant agents. *Egypt. J. Chem.* **2022**, *65*, 133–146. [CrossRef]

28. Nastasă, C.; Tamaian, R.; Oniga, O.; Tiperciuc, B. 5-Arylidene(chromenyl-methylene)-thiazolidinediones: Potential new agents against mutant oncoproteins K-Ras, N-Ras and B-Raf in colorectal cancer and melanoma. *Medicina* **2019**, *55*, 85. [CrossRef]
29. Donarska, B.; Świtalska, M.; Wietrzyk, J.; Płaziński, W.; Mizerska-Kowalska, M.; Zdzisińska, B.; Łączkowski, K.Z. Discovery of new 3,3-diethylazetidine-2,4-dione based thiazoles as nanomolar human neutrophil elastase inhibitors with broad-spectrum antiproliferative activity. *Int. J. Mol. Sci.* **2022**, *23*, 7566. [CrossRef]
30. Nastasă, C.; Vodnar, D.C.; Ionuţ, I.; Stana, A.; Benedec, D.; Tamaian, R.; Oniga, O.; Tiperciuc, B. antibacterial evaluation and virtual screening of new thiazolyl-triazole schiff bases as potential dna-gyrase inhibitors. *Int. J. Mol. Sci.* **2018**, *19*, 222. [CrossRef]
31. Stana, A.; Vodnar, D.C.; Marc, G.; Benedec, D.; Tiperciuc, B.; Tamaian, R.; Oniga, O. Antioxidant activity and antibacterial evaluation of new thiazolin-4-one derivatives as potential tryptophanyl-tRNA synthetase inhibitors. *J. Enzym. Inhibit. Med. Chem.* **2019**, *34*, 898–908. [CrossRef] [PubMed]
32. Gomha, S.M.; Edrees, M.; Altalbawy, F. Synthesis and characterization of some new bis-pyrazolyl-thiazoles incorporating the thiophene moiety as potent anti-tumor agents. *Int. J. Mol. Sci.* **2016**, *17*, 1499. [CrossRef] [PubMed]
33. Gomha, S.M.; Salah, T.A.; Abdelhamid, A.O. Synthesis, characterization, and pharmacological evaluation of some novel thiadiazoles and thiazoles incorporating pyrazole moiety as anticancer agents. *Monatsh. Chem.* **2015**, *146*, 149–158. [CrossRef]
34. dos Santos Silva, T.D.; Bomfim, L.M.; da Cruz Rodrigues, A.C.B.; Dias, R.B.; Sales, C.B.S.; Rocha, C.A.G.; Soares, M.B.P.; Bezerra, D.P.; de Oliveira Cardoso, M.V.; Leite, A.C.L.; et al. Anti-liver cancer activity in vitro and in vivo induced by 2-pyridyl 2, 3-thiazole derivatives. *Toxicol. Appl. Pharmacol.* **2017**, *329*, 212–223. [CrossRef] [PubMed]
35. Said, M.A.; Riyadh, S.M.; Al-Kaff, N.S.; Nayl, A.A.; Khalil, K.D.; Bräse, S.; Gomha, S.M. Synthesis and greener pastures biological study of bis-thiadiazoles as potential covid-19 drug candidates. *Arab. J. Chem.* **2022**, *15*, 104101. [CrossRef]
36. Gomha, S.M.; Muhammad, Z.A.; Abdel-aziz, M.R.; Abdel-aziz, H.M.; Gaber, H.M.; Elaasser, M.M. One Pot Synthesis of new thiadiazolyl-pyridines as anticancer and antioxidant agents. *J. Heterocycl. Chem.* **2018**, *55*, 530–536. [CrossRef]
37. Gomha, S.M.; Abdelhady, H.A.; Hassain, D.Z.H.; Abdelmonsef, A.H.; El-Naggar, M.; Elaasser, M.M.; Mahmoud, H.K. Thiazole based thiosemicarbazones: Synthesis, cytotoxicity evaluation and molecular docking study. *Drug Des. Dev. Ther.* **2021**, *15*, 659–677. [CrossRef]
38. Gomha, S.M.; Edrees, M.M.; Muhammad, Z.A.; El-Reedy, A.A.M. 5-(Thiophen-2-yl)-1,3,4-thiadiazole derivatives: Synthesis, molecular docking and in-vitro cytotoxicity evaluation as potential anticancer agents. *Drug Des. Dev. Ther.* **2018**, *12*, 1511–1523. [CrossRef]
39. Gomha, S.M.; Riyadh, S.M.; Huwaimel, B.; Zayed, M.E.M.; Abdellattif, M.H. Synthesis, molecular docking study and cytotoxic activity on mcf cells of some new thiazole clubbed thiophene scaffolds. *Molecules* **2022**, *27*, 4639. [CrossRef]
40. Gomha, S.M.; Abdel-aziz, H.M.; El-Reedy, A.A.M. Facile synthesis of pyrazolo[3,4-c]pyrazoles bearing coumarine ring as anticancer agents. *J. Heterocycl. Chem.* **2018**, *55*, 1960–1965. [CrossRef]
41. Edrees, M.M.; Abu-Melha, S.; Saad, A.M.; Kheder, N.A.; Gomha, S.M.; Muhammad, Z.A. Eco-friendly synthesis, characterization and biological evaluation of some new pyrazolines containing thiazole moiety as potential anticancer and antimicrobial agents. *Molecules* **2018**, *23*, 2970. [CrossRef] [PubMed]
42. Abu-Melha, S.; Edrees, M.M.; Salem, H.H.; Kheder, N.A.; Gomha, S.M.; Abdelaziz, M.R. Synthesis and biological evaluation of some novel thiazole-based heterocycles as potential anticancer and antimicrobial agents. *Molecules* **2019**, *24*, 539. [CrossRef] [PubMed]
43. Gomha, S.M.; Muhammad, Z.A.; Abdel-aziz, H.M.; Matar, I.K.; El-Sayed, A.A. Green synthesis, molecular docking and anticancer activity of novel 1,4-dihydropyridine-3,5-dicarbohydrazones under grind-stone chemistry. *Green Chem. Lett. Rev.* **2020**, *13*, 6–17. [CrossRef]
44. Sayed, A.R.; Abd El-lateef, H.M.; Gomha, S.M.; Abolibda, T.Z. L-Proline catalyzed green synthesis and anticancer evaluation of novel bioactive benzil bis-hydrazones under grinding technique. *Green Chem. Lett. Rev.* **2021**, *14*, 179–188. [CrossRef]
45. Alshabanah, L.A.; Al-Mutabagani, L.A.; Gomha, S.M.; Ahmed, H.A. Three-component synthesis of some new coumarin derivatives as anti-cancer agents. *Front. Chem.* **2022**, *9*, 762248. [CrossRef] [PubMed]
46. Nayl, A.A.; Arafa, W.A.A.; Ahmed, M.; Abd-Elhamid, A.I.; El-Fakharany, E.M.; Abdelgawad, M.A.; Gomha, S.M.; Ibrahim, H.M.; Aly, A.A.; Bräse, S.; et al. Novel pyridinium based ionic liquid promoter for aqueous knoevenagel condensation: Green and efficient synthesis of new derivatives with their anticancer evaluation. *Molecules* **2022**, *27*, 2940. [CrossRef]
47. Khidre, R.E.; Abu-Hashem, A.A.; El-Shazly, M. Synthesis and anti-microbial activity of some 1-substituted amino-4,6-dimethyl-2-oxo-pyridine-3-carbonitrile derivatives. *Eur. J. Med. Chem.* **2011**, *46*, 5057–5064. [CrossRef]
48. Kawakita, Y.; Seto, M.; Ohashi, T.; Tamura, T.; Yusa, T.; Miki, H.; Iwata, H.; Kamiguchi, H.; Tanaka, T.; Sogabe, S.; et al. Design and synthesis of novel pyrimido[4,5-b]azepine derivatives as HER2/EGFR dual inhibitors. *Bioorg. Med. Chem.* **2013**, *21*, 2250–2261. [CrossRef]
49. Sayed, A.R.; Gomha, S.M.; Abdelrazek, F.M.; Farghaly, M.S.; Hassan, S.A.; Metz, P. Design, efficient synthesis and molecular docking of some novel thiazolyl-pyrazole derivatives as anticancer agents. *BMC Chem.* **2019**, *13*, 116. [CrossRef]
50. Gomha, S.M.; Riyadh, S.M.; Mahmmoud, E.A.; Elaasser, M.M. Synthesis and anticancer activity of arylazothiazoles and 1,3,4-thiadiazoles using chitosan-grafted-poly(4-vinylpyridine) as a novel copolymer basic catalyst. *Chem. Heterocycl. Compd.* **2015**, *51*, 1030–1038. [CrossRef]
51. Chemical Computing Group Inc. *Molecular Operating Environment (MOE)*; Chemical Computing Group Inc.: Montreal, QC, Canada, 2016.

52. Ma, C.; Taghour, M.S.; Belal, A.; Mehany, A.B.M.; Mostafa, N.; Nabeeh, A.; Eissa, I.H.; Al-Karmalawy, A.A. Synthesis of new quinoxaline derivatives as potential histone deacetylase inhibitors targeting hepatocellular carcinoma: In Silico, In Vitro, and SAR studies. *Front. Chem.* **2021**, *9*, 725135. [CrossRef]
53. El-Shershaby, M.H.; Ghiaty, A.; Bayoumi, A.H.; Al-Karmalawy, A.A.; Husseiny, E.M.; El-Zoghbi, M.S.; Abulkhair, H.S. The antimicrobial potential and pharmacokinetic profiles of novel quinoline-based scaffolds: Synthesis and in silico mechanistic studies as dual DNA gyrase and DHFR inhibitors. *New J. Chem.* **2021**, *45*, 13986–14004. [CrossRef]
54. Ghanem, A.; Emara, H.A.; Muawia, S.; Abd El Maksoud, A.I.; Al-Karmalawy, A.A.; Elshal, M.F. Tanshinone IIA synergistically enhances the antitumor activity of doxorubicin by interfering with the PI3K/AKT/mTOR pathway and inhibition of topoisomeraseII: In Vitro and molecular docking studies. *New J. Chem.* **2020**, *44*, 17374–17381. [CrossRef]
55. Khattab, M.; Al-Karmalawy, A.A. Revisiting Activity of Some Nocodazole Analogues as a Potential Anticancer Drugs Using Molecular Docking and DFT Calculations. *Front. Chem.* **2021**, *9*, 628398. [CrossRef] [PubMed]
56. Soltan, M.A.; Eldeen, M.A.; Elbassiouny, N.; Mohamed, I.; El-Damasy, D.A.; Fayad, E.; Abu Ali, O.A.; Raafat, N.; Eid, R.A.; Al-Karmalawy, A.A. Proteome based approach defines candidates for designing a multitope vaccine against the nipah virus. *Int. J. Mol. Sci.* **2021**, *22*, 9330. [CrossRef]
57. Alesawy, M.S.; Al-Karmalawy, A.A.; Elkaeed, E.B.; Alswah, M.; Belal, A.; Taghour, M.S.; Eissa, I.H. Design and discovery of new 1,2,4-triazolo[4,3-c]quinazolines as potential DNA intercalators and topoisomerase II inhibitors. *Arch. Der Pharm.* **2020**, *354*, e2000237. [CrossRef] [PubMed]
58. Eliaa, S.G.; Al-Karmalawy, A.A.; Saleh, R.M.; Elshal, M.F. Empagliflozin and Doxorubicin Synergistically Inhibit the Survival of Triple-Negative Breast Cancer Cells via Interfering with the mTOR Pathway and Inhibition of Calmodulin: In Vitro and Molecular Docking Studies. *ACS Pharmacol. Transl. Sci.* **2020**, *3*, 1330–1338. [CrossRef]
59. Grima, J.N.; Gatt, R.; Bray, T.G.C.; Alderson, A.; Evans, K.E. Empirical modelling using dummy atoms (EMUDA): An alternative approach for studying "auxetic" structures. *Mol. Simul.* **2005**, *31*, 915–924. [CrossRef]
60. El-Shershaby, M.H.; Ghiaty, A.; Bayoumi, A.H.; Al-Karmalawy, A.A.; Husseiny, E.M.; El-Zoghbi, M.S.; Abulkhair, H.S. From triazolophthalazines to triazoloquinazolines: A bioisosterism-guided approach toward the identification of novel PCAF inhibitors with potential anticancer activity. *Bioorg. Med. Chem.* **2021**, *42*, 116266. [CrossRef]
61. Soltan, M.A.; Elbassiouny, N.; Gamal, H.; Elkaeed, E.B.; Eid, R.A.; Eldeen, M.A.; Al-Karmalawy, A.A. In Silico Prediction of a Multitope Vaccine against Moraxella catarrhalis: Reverse Vaccinology and Immunoinformatics. *Vaccines* **2021**, *9*, 669. [CrossRef]
62. McConkey, B.J.; Sobolev, V.; Edelman, M. The Performance of Current Methods in Ligand–Protein Dockin. *Curr. Sci.* **2002**, *83*, 845–856.
63. Abdallah, A.E.; Alesawy, M.S.; Eissa, S.I.; El-Fakharany, E.M.; Kalaba, M.H.; Sharaf, M.H.; Shama, N.M.A.; Mahmoud, S.H.; Mostafa, A.; Al-Karmalawy, A.A.; et al. Design and synthesis of new 4-(2-nitrophenoxy)benzamide derivatives as potential antiviral agents: Molecular modeling and in vitro antiviral screening. *New J. Chem.* **2021**, *36*, 16557–16571. [CrossRef]

Article

4-(Aryl)-Benzo[4,5]imidazo[1,2-a]pyrimidine-3-Carbonitrile-Based Fluorophores: Povarov Reaction-Based Synthesis, Photophysical Studies, and DFT Calculations

Victor V. Fedotov [1,*], Maria I. Valieva [1], Olga S. Taniya [1,*], Semen V. Aminov [1], Mikhail A. Kharitonov [1], Alexander S. Novikov [2], Dmitry S. Kopchuk [1], Pavel A. Slepukhin [1], Grigory V. Zyryanov [1], Evgeny N. Ulomsky [1], Vladimir L. Rusinov [1] and Valery N. Charushin [1]

[1] Chemical Engineering Institute, Ural Federal University, 19 Mira St., 620002 Yekaterinburg, Russia
[2] Institute of Chemistry, Saint Petersburg State University, 7/9 Universitetskaya Nab., 199034 Saint Petersburg, Russia
* Correspondence: viktor.fedotov@urfu.ru (V.V.F.); olga.tania@urfu.ru (O.S.T.)

Abstract: A series of novel 4-(aryl)-benzo[4,5]imidazo[1,2-a]pyrimidine-3-carbonitriles were obtained through the Povarov (aza-Diels–Alder) and oxidation reactions, starting from benzimidazole-2-arylimines. Based on the literature data and X-ray diffraction analysis, it was discovered that during the Povarov reaction, [1,3] sigmatropic rearrangement leading to dihydrobenzimidazo[1,2-a]pyrimidines took place. The structures of all the obtained compounds were confirmed based on the data from ^{1}H- and ^{13}C-NMR spectroscopy, IR spectroscopy, and elemental analysis. For all the obtained compounds, their photophysical properties were studied. In all the cases, a positive emission solvatochromism with Stokes shifts from 120 to 180 nm was recorded. Aggregation-Induced Emission (AIE) has been illustrated for compound **6c** using different water fractions (fw) in THF. The compounds **6c** and **6f** demonstrated changes in emission maxima or/and intensities after mechanical stimulation.

Keywords: pyrimidine; benzimidazole; aza-Diels–Alder reaction; Povarov reaction; oxidation; fluorescence; aggregation-induced emission; mechanochromic properties

Citation: Fedotov, V.V.; Valieva, M.I.; Taniya, O.S.; Aminov, S.V.; Kharitonov, M.A.; Novikov, A.S.; Kopchuk, D.S.; Slepukhin, P.A.; Zyryanov, G.V.; Ulomsky, E.N.; et al. 4-(Aryl)-Benzo[4,5]imidazo[1,2-a]pyrimidine-3-Carbonitrile-Based Fluorophores: Povarov Reaction-Based Synthesis, Photophysical Studies, and DFT Calculations. *Molecules* **2022**, *27*, 8029. https://doi.org/10.3390/molecules27228029

Academic Editor: Joseph Sloop

Received: 31 October 2022
Accepted: 15 November 2022
Published: 19 November 2022

Publisher's Note: MDPI stays neutral with regard to jurisdictional claims in published maps and institutional affiliations.

Copyright: © 2022 by the authors. Licensee MDPI, Basel, Switzerland. This article is an open access article distributed under the terms and conditions of the Creative Commons Attribution (CC BY) license (https://creativecommons.org/licenses/by/4.0/).

1. Introduction

Azolopyrimidines are ubiquitous heterocyclic systems, particularly important in living organisms as a core of purine bases, and these heterocycles are widely present among biologically active compounds, including those with antiviral [1–4], anticancer [5–7], antibacterial [8,9], and antidiabetic activity [10,11]. In addition to a wide range of biological activities, azolopyrimidines are considered promising candidates for important fluorescence applications [12–15]. Furthermore, strongly electron-withdrawing pyrimidine derivatives have found applications for the synthesis of push-pull molecules and the construction of functionalized π-conjugated materials such as dye-sensitized solar cells [16], non-doped OLED and laser dyes [17], and nonlinear optical materials [18]. Among the methods for the structural modification of azolopyrimidines, the approaches based on the creation of polycyclic fused analogs of azolopyrimidines such as benzo[4,5]imidazo[1,2-a]pyrimidines are of growing interest and significance [19–21]. Since polycyclic fused systems with a conjugated planar structure exhibit relevant photophysical properties, they have found applications as phosphors in optoelectronics or as fluorescent dyes for textile and polymer materials [22].

Among the methods for constructing heterocyclic systems is the aza-Diels–Alder [4 + 2] cycloaddition reaction between various dienophiles and N-aryl-substituted imines, which yields a wide range of azaheterocycles. This reaction, also known as the Povarov reaction [23–26], is a convenient tool for the construction of six-membered rings with high molecular complexity via the direct construction of carbon–carbon and carbon–heteroatom bonds [27]. In addition, the Povarov reaction is considered an important and efficient approach for creating large libraries of bioactive compounds in drug discovery programs [28]. From this point of view, the use of such a powerful synthetic methodology can be useful for the creation of new derivatives of azolopyrimidines, in particular benzo[4,5]imidazo[1,2-a]pyrimidines.

The use of molecules with aggregation-induced emission (AIE) properties, including those with reversible mechanochromism properties, is of great research interest due to their potential applications in biomedical imaging, sensors, and organic light-emitting diodes [29]. Additionally, fluorophores based on acceptor azaheteroarene domains, such as triazoles, oxadiazoles, thiadiazoles, benzothiazoles, quinoxalines, s- or as-triazines, and pyrimidines, are of particular interest [30–34]. Apart from these acceptors, imidazole-based units have been reported as electron acceptors for blue emission acquisition due to their low LUMO energy level [35]. However, the imidazole unit has been less studied for the development of efficient fluorescent materials due to its weak electron-accepting ability [36,37]. Wang et al. reported the synthesis of TPE-substituted phenanthroimidazole derivatives [38]. These compounds exhibited AIE properties as well as an intriguing mechanofluorochromism: after a short-time grinding, the blue emitting in a solid-state fluorophores (with maxima around 438 nm) changed their emission color to sky blue with a maxima near 450 nm. The functionalization of the imidazole-containing domain with a strongly electron-withdrawing cyano-group and a reduced singlet-triplet energy gap, on the other hand, has received special attention as a universal and appealing strategy for creating AIE-active fluorophores, including those with thermally activated delayed fluorescence (TADF) [39]. For instance, the authors of [40] recently developed TADF materials with C3-functionalized cyano-group 2-phenylimidazopyrazine as an acceptor unit linked to either acridine or phenoxazine donor units, and for these fluorophores an EQE of about 12.7% was achieved. In addition, the use of 2-phenylimidazo[1,2-a]pyridine containing cyano-group as an acceptor has been reported as a tool for designing dark blue emitters with a relatively high fluorescence quantum yield [36,41].

We recently reported the synthesis of asymmetric donor-acceptor azoloazine fluorophors based on 4-heteroaryl-substituted 2-phenyl-2H-benzo[4,5]imidazo[1,2-a][1,2,3]triazolo[4,5-e]pyrimidine via the reaction of nucleophilic aromatic hydrogen substitution (S_NH) and studied their microenvironmental sensitivity in the PLICT process (Scheme 1) [42].

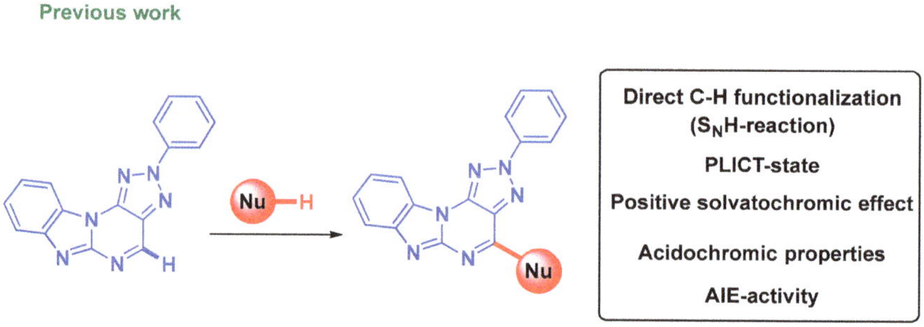

Previous work

Scheme 1. Nucleophilic substitution of hydrogen (S_NH) in 2-phenyl-2H-benzo[4,5]imidazo[1,2-a][1,2,3]triazolo[4,5-e]pyrimidine [42].

Herein, we wish to report a synthetic design of novel benzo[4,5]imidazo[1,2-a]pyrimidines bearing cyano-group (instead of a 1,2,3-triazole fragment) via the combination of the Povarov reaction and oxidative aromatization of the resulting dihydro derivatives, as well as studies of their aggregation-induced fluorescence behavior and mechanofluorochromic properties, as well as structure-property correlation studies involving DFT methods.

2. Results

2.1. Synthesis

Arylimines (the diene component) and various dienophiles are the classical substrates used for the Povarov reaction (the aza-Diels–Alder reaction). For the preparation of arylimines, Brønsted acid catalysis [43–45] and Lewis acid catalysis [46,47] are traditionally used, as are various modifications, including those involving microwave radiation [48–50]. Within the frame of current research, we have proposed a new catalyst-free and solvent-free method for obtaining benzimidazole-2-arylimine **3a–f** by heating 2-aminobenzimidazoles **1a,b** and aromatic aldehydes **2a–c** at 130 °C for 3 h. This method afforded desired diene substrates **3a–f** in good to excellent yields (83–90%) (Scheme 2).

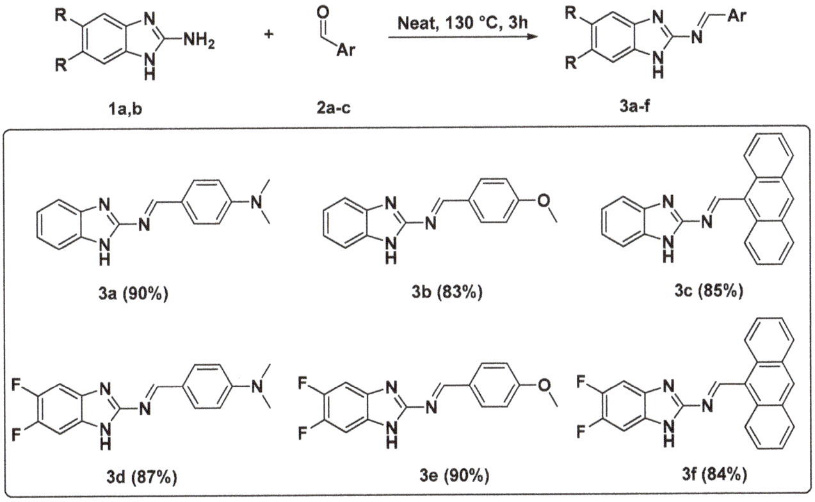

Scheme 2. Scope of benzimidazole-2-arylimine **3a–f**.

The structure of all intermediates **3a–f** was confirmed by means of the data from ^{1}H NMR spectroscopy, as well as ^{13}C NMR spectroscopy, IR spectroscopy, and elemental analysis. These data were also considered for the identification of previously undescribed benzimidazole-2-arylimines **3a**, **3c–f** (Figures S4–S8 and S21–S23, Supplementary Materials).

It is worth mentioning that Chen et al. previously reported an unprecedented in situ [1,3] sigmatropic rearrangement that resulted in 4,10-dihydropyrimido[1,2-*a*]benzimidazoles [49]. Additionally, the same rearrangement was observed by us in the case of using *N*-2-substituted benzimidazoles (Scheme 3).

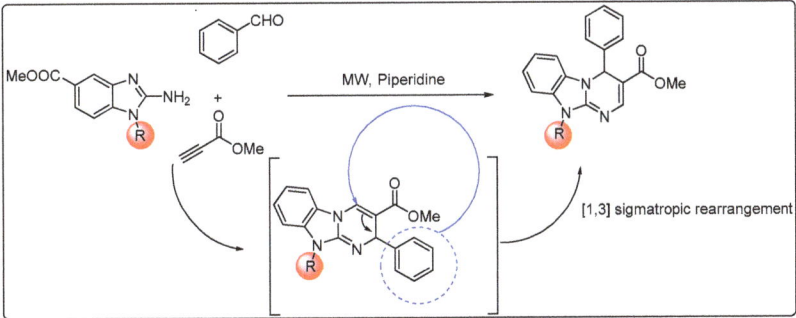

Scheme 3. Povarov reaction and rearrangement [49].

Inspired by this fact, we decided to investigate the possibility of rearrangement in the case of unsubstituted benzimidazole-2-arylimine. To test this possibility, derivatives **3a–f** were used as diene substrates in the Povarov reaction, and 3-morpholinoacrylonitrile **4** was chosen as an EWG-dienophile (Table 1). A careful literature survey revealed that the most commonly used catalysts for this type of reaction are Brønsted acids [51,52] and Lewis acids [27,53]. However, there are examples of using basic catalysts [49] as well as electrochemical methods [23]. To optimize the synthetic procedure for the reaction between benzimidazole-2-arylimine **3a** and 3-morpholinoacrylonitrile **4**, leading to the target, 4-(4-(dimethylamino)phenyl)-1,4-dihydrobenzo[4,5]imidazo[1,2-a]pyrimidine-3-carbonitrile **5a** was chosen. Next, the influence of the nature of the solvents and activating agents, their amounts, as well as the reaction time, on the yields of the target product was assessed (Table 1). The obtained results clearly demonstrated that $BF_3 \cdot Et_2O$ was the best activating agent when used at an amount of 1.5 equivalents in *n*-BuOH for 5 h (Table 1, entry 9).

Table 1. Optimization of the reaction conditions for dihydropyrimidin **5a** [1].

No.	Solvent [2]	Activating Agent (Catalysts)	X, Equiv	Reaction Condition [3]	Yield, % [4]
entry 1	EtOH	$BF_3 \cdot Et_2O$	0.5	reflux, 5 h	35
entry 2	*i*-PrOH	$BF_3 \cdot Et_2O$	0.5	reflux, 5 h	46
entry 3	*n*-BuOH	$BF_3 \cdot Et_2O$	0.5	reflux, 5 h	50
entry 4	Toluene	$BF_3 \cdot Et_2O$	0.5	reflux, 5 h	-
entry 5	*n*-BuOH	$BF_3 \cdot Et_2O$	0.5	reflux, 6 h	51
entry 6	AcOH	-	-	reflux, 5 h	-
entry 7	*n*-BuOH	Et_3N	0.5	reflux, 5 h	-
entry 8	*n*-BuOH	$BF_3 \cdot Et_2O$	1.0	reflux, 5 h	63
entry 9	*n*-BuOH	$BF_3 \cdot Et_2O$	1.5	reflux, 5 h	74
entry 10	*n*-BuOH	$BF_3 \cdot Et_2O$	2.0	reflux, 5 h	76

[1] Reaction conditions: **3a** (0.10 mmol) and **4** (0.10 mmol); [2] amount of solvent—5 mL; [3] conventional heating with an oil bath; and [4] isolated yield.

As a next step, by using the optimized reaction conditions, we have prepared a series of annulated dihydropyrimidines **5a–f** in moderate to good yields (59–74%) (Scheme 4).

The structures of the obtained dihydropyrimidines **5a–f** were confirmed by means of IR-, ^{1}H-, and ^{13}C-NMR spectroscopy as well as elemental analysis data. Due to the very low solubility of derivatives **5a–f** a mixture of $CDCl_3 - CF_3COOD$ (*v*/*v* = 10:1) was used as a solvent for NMR measurements. All the prepared compounds provided satisfactory analytical data. The signals H-4 are the characteristic ones for the products **5a–f** in the corresponding ^{1}H NMR spectra. It should be noted that in compounds **5a,b** and **5d,e** the H-4 signals are located at δ 6.16–6.42 ppm, whereas for the derivatives **5c** and **5f**, bearing an anthracene fragment, the H-4 proton shifts downfield to the region of δ 7.60–7.63 ppm. Apparently, it occurs due to the deshielding effect of the H-4 proton because of the presence of the anthracene substituent. In the IR spectra, for all the series of dihydropyrimidines **5a–f** the characteristic stretching vibrations of (-C≡N) bonds are observed at ν 2202–2215 cm^{-1} (see Supplementary Materials).

The Povarov reaction is a versatile and efficient method to access the tetrahydroquinoline scaffolds [26], and, as a rule, the research on this reaction is limited only by the availability of such systems. At the same time, the oxidative aromatization products of the

Povarov reaction may be of interest from the point of view of studying their properties, in particular their photophysical ones. Therefore, as a next step, the aromatization of these novel dihydropyrimidine systems **5a–f** was carried out.

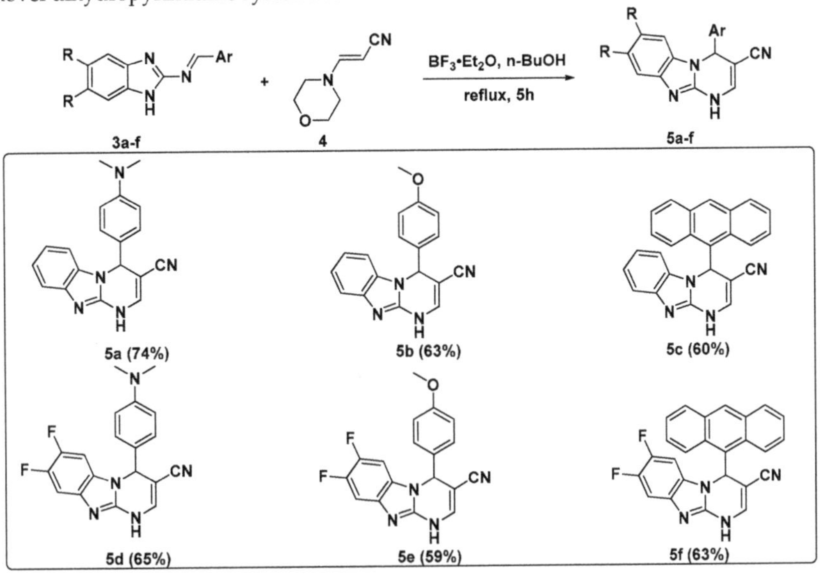

Scheme 4. Substrate scope of dihydropyrimidines **5a–f**.

By using compound **5a** as a key heterocyclic substrate, the most suitable solvent for the oxidation reaction was selected (Table 2). Thus, DMF seems to be the most suitable solvent for the reaction since substrate **5a** has good solubility in this solvent. Moreover, the boiling point of DMF makes it possible to carry out the reaction at high temperatures. As a first step, the blank experiments without oxidation agents (Table 2 entries 1–4) were carried out. It was found that heating the substrate **5a** in DMF resulted in the formation of the oxidation product **6a** in some amounts (according to TLC data), possibly, due to the oxidation in the ambient air. However, even after the prolonged heating (12 h) at the evaluated 140 °C temperature, the complete conversion of compound **5a** to the target product **6a** was not observed. The use of mild oxidizing agents at 120 °C, such as MnO_2, reduced the reaction time to 6 h (Table 2 entries 5-8). Subsequently, the increase in the amount of MnO_2 to four equivalents resulted in the complete conversion of **5a** within 1 h (Table 2, entry 8).

This newly developed methodology was then used to synthesize a series of new 4-(aryl)benz[4,5]imidazo[1,2-a]pyrimidine-3-carbonitriles **6b–f** with yields in the range of 80–90% (Scheme 5).

All derivatives **6a–f** were obtained with comparable yields, which indicates an insignificant influence of the nature of the substituents on the oxidation process. All the synthesized compounds were fully characterized by ^{1}H-NMR, ^{13}C-NMR, IR-spectroscopy, and elemental analysis (Supplementary Materials). In particular, in the ^{1}H-NMR spectra, the aromatic proton signals were observed at δ 5.00–9.41 ppm, whereas the aliphatic proton signals were observed at δ 3.11–3.96 ppm. In the ^{13}C-NMR spectra, (hetero)aryl carbon nuclei are located at δ 94.6–162.2 ppm, while signals corresponding to aliphatic carbon were observed at δ 39.6–55.6 ppm. It should be emphasized that for the difluoro derivatives **6d–f** in both the ^{1}H- and ^{13}C-NMR spectra, a characteristic multiplicity was observed, due to the spin–spin interaction of the H-F and C-F nuclei. It is also interesting that in the IR spectra of compounds **6a–f**, the characteristic stretching vibrations of (-C≡N) bonds at ν 2227–2230 cm^{-1} were observed.

Table 2. Optimization of the oxidation reactions for dihydropyrimidin 5a [1].

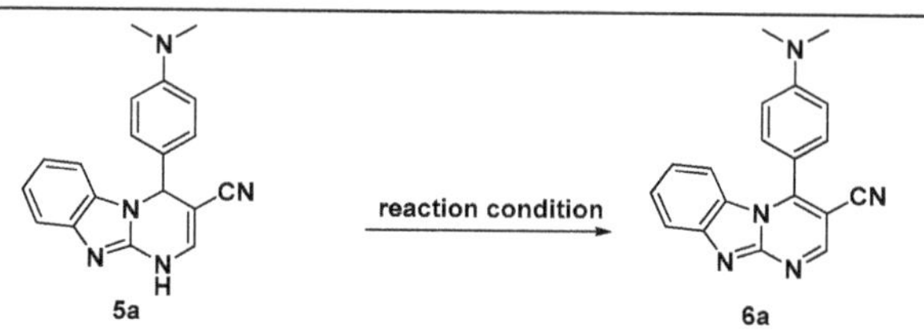

No.	Solvent [2]	Oxidant [3]	X, Equiv [3]	Reaction Condition [4]	Conversion, % [5]	Yield, % [6]
entry 1	DMF	-	-	Heating 100 °C, 1 h	10	5
entry 2	DMF	-	-	Heating 140 °C, 1 h	15	6
entry 3	DMF	-	-	Heating 140 °C, 4 h	30	17
entry 4	DMF	-	-	Heating 140 °C, 12 h	40	30
entry 5	DMF	MnO_2	1.0	Heating 120 °C, 6 h	100	83
entry 6	DMF	MnO_2	2.0	Heating 120 °C, 4 h	100	84
entry 7	DMF	MnO_2	3.0	Heating 120 °C, 1.5 h	100	86
entry 8	DMF	MnO_2	4.0	Heating 120 °C, 1.0 h	100	85

[1] Reaction conditions: **5a** (0.10 mmol); [2] amount of solvent—5 mL; [3] X equivalent of oxidant; [4] conventional heating with an oil bath; [5] in accordance with TLC; and [6] isolated yield.

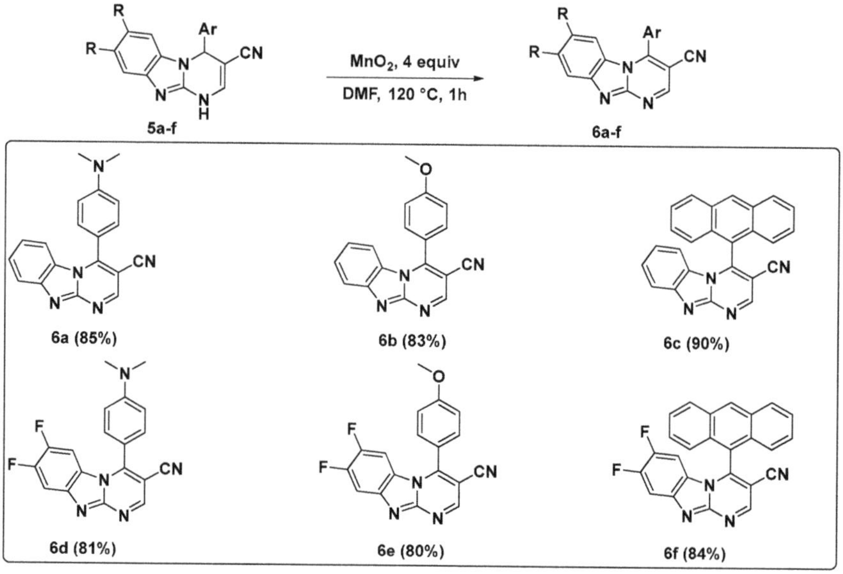

Scheme 5. Scope of the 4-(aryl)benzo[4,5]imidazo[1,2-*a*]pyrimidine-3-carbonitriles **6a–f**.

As previously stated, an unprecedented in situ [1,3] sigmatropic rearrangement was reported for the related *N*-10 substituted systems. However, the spectral data obtained for compounds **5a–f** and **6a–f** do not allow one to determine the position of the Ar substituent in the dihydropyrimidine system with certainty. Single crystal X-ray diffraction analysis was performed on compound **6c** to confirm the structure of the obtained compound and to prove the hypotheses about the possibility of rearrangement in the case of unsubstituted benzimidazole-2-arylimine (Figure 1).

According to the XRD data, in compound **6c**, the (Ar) substituent is located in the position of C4 of the pyrimidine ring, which indicates the possibility of the rearrangement in the herein reported systems.

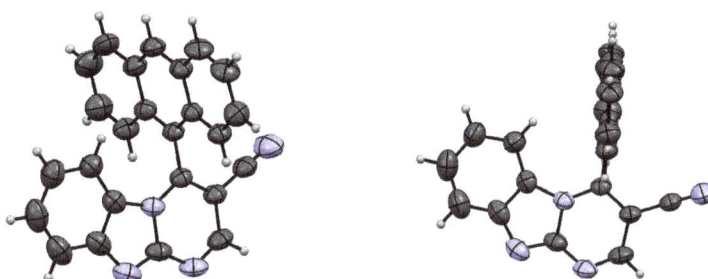

Figure 1. Molecular structure of **6c**.

The proposed mechanism of the interaction between benzimidazole-2-arylimines **3a–f** and 3-morpholinoacrylonitrile (**4**), based on the reactivity of these substrates and literature data [23,26,49], is shown in Scheme 6.

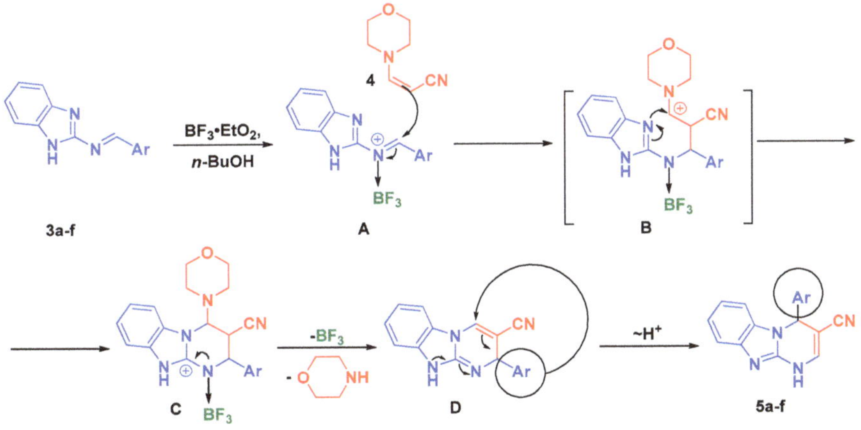

Scheme 6. Plausible reaction mechanisms of dihydropyrimidines **5a–f** formation and [1,3] sigmatropic rearrangement.

As a first stage, the benzimidazole-2-arylimines **3a–f** are activated via the interaction with BF$_3$·Et$_2$O, resulting in the formation of activated complex **A**. At the next stage, there is an asynchronous concerted process interaction of the intermediate **A** with 3-morpholinoacrylonitrile (**4**) through an ephemeral transition state **B** resulting in the formation of a tetrahydropyrimide system **C**. The removal of the morpholine molecule results in system **D**, which undergoes [1,3] sigmatropic rearrangement and yields derivatives **5a–f**.

In addition, we discovered that all of the 4-(aryl)benzo[4,5]imidazo[1,2-a]pyrimidine-3-carbonitriles **6a–f** obtained are fluorescent in solution and solid form. Therefore, photophysical studies of the obtained products **6a–f** were carried out.

2.2. Photophysical Studies

2.2.1. Absorption/Fluorescence Studies in Solution and Solvent Effect

All the obtained fluorophores were soluble in concentrations less than 2×10^{-5} M both in nonpolar (cyclohexane and toluene) and in weakly and strongly apolar aprotonic solvents (THF, acetonitrile, DMSO). Additionally, all the compounds have exhibited an intense fluorescence in solution. Taking into account the subsequent study of the phenomenon of aggregation-induced emission (AIE), THF, which is located at the interface between nonpolar and polar solvents with an average value of orientational polarizability, was chosen as the optimal aprotonic solvent (Δf = 0.21). The results of the photophysical studies are presented below (Table 3).

Table 3. Data of photophysical properties of fluorophores (**6a–f**) (10^{-5} M) in THF solvent.

No.	λ_{abs}^{max}, nm (ε_M, 10^4 M^{-1} cm^{-1}) [1]	λ_{em}^{max}, nm [2]	Stokes Shift, nm/cm^{-1}	τ_{av}, ns [3]	Φ_f, % [4]
6a	268 (4.8) 413 (0.74)	554	141/6162	2.43	7.5
6b	271 (3.53) 312 (0.52) 320 (0.52) 387 (0.11)	540	153/7321	5.12	1.9
6c	254 (14.5) 327 (0.45) 342 (0.45) 371 (0.49) 391 (0.48)	550	159/7394	8.78	<0.1
6d	264 (4.12) 421 (0.76)	567	146/6116	1.59	2.9
6e	266 (5.65) 280 (5.30) 377 (0.14)	520	143/7294	6.24	1.1
6f	255 (13.21) 323 (0.43) 336 (0.47) 372 (0.48) 392 (0.45)	524	132/6426	2.34	4.8

[1] Absorption spectra were measured at r.t. in THF in range from 230 to 500 nm; [2] emission spectra were measured at r.t. in THF; [3] weighted average decay time $\tau_{av} = \Sigma (\tau_i \times \alpha_i)$ in THF (LED 370 nm); and [4] absolute quantum yields were measured using the Integrating Sphere of the Horiba FluoroMax-4 at r.t. in THF.

Emission spectra for all the compounds were measured at low concentrations of 10^{-5} M to avoid any concentration-dependent dimerization and fluorescence quenching. All the graphs were normalized for comparative analysis (Figure 2).

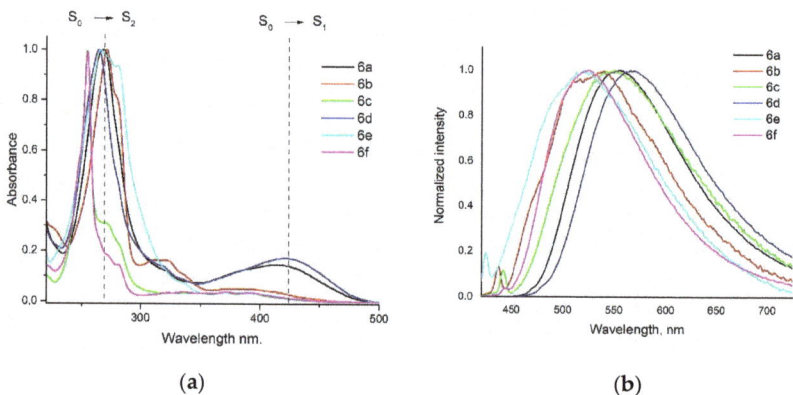

Figure 2. Absorption (**a**) and emission (**b**) spectra of fluorophores **6a–f** in THF (c = 10^{-5} M).

The absorption spectra of the fluorophores **6a–f** are presented by two absorption bands with different intensities at maximum wavelengths in the 220–300 nm and 350–500 nm ranges, which correspond to $S_0 \rightarrow S_2$ and $S_0 \rightarrow S_1$ transitions. In this case, all the compounds show a dominant absorption band due to the transition $S_0 \rightarrow S_2$ with $\varepsilon M < 14.5 \times 10^4$ M^{-1} cm^{-1}.

The emission spectra of the fluorophores **6a–f** are presented by the solid unstructured emission bands with maximums from 520 to 567 nm, referring to the excited ICT-state in a polar aprotic solvent [36]. A significant bathochromic shift was observed for the two 2-dimethylaminophenyl substituted imidazopyrimidine fluorophores **6a,d**, which have some of the most energetically favorable states among the obtained series of fluorophores (3.39 eV for **6a** and 3.24 eV for **6b**) (See Section 2.2.5. Theoretical Calculations). The fluorescence lifetimes of the investigated compounds **6a–f** exhibited a two-exponential decay in THF. The lifetime of the excited state of the fluorophores was measured at r.t. in THF using a nanosecond LED with an excitation wavelength of 370 nm. The average lifetime was calculated using the expression $\tau_{av} = \Sigma\ (\tau_i \times \alpha_i)$ (Table S1). Overall, the average fluorescence lifetime (τ_{av}) ranged from 1.59 ns (lowest for **6d**) to 8.78 ns (highest for **6c**) (Table 3). The compounds were characterized by large Stokes shift values (<140 nm), while the quantum yield values in THF were not higher than 7.5%.

Compounds **6a–f** with variation of electron-donating fragments (4-methoxyphenyl, 4-(dimethylamino)phenyl and anthracen-9-yl) based on the 3-cyanosubstituted benzo [4,5]imidazo [1,2-*a*]pyrimidine, including those substituted with fluorine atoms in positions 7,8 implies that the solvent polarity may influence the electronic state properties of the chromophore (See Section 2.2.5. Theoretical Calculations).

We studied the emission characteristics of **6a–f** compounds in various solvents (Tables S2–S7). Indeed, the effect of the solvent polarity was observed for the chromophores of the entire series with Stokes shifts from 120 to 180 nm. However, only for anthracenyl substituted fluorophores, upon the increasing solvent polarity in a row from nonpolar cyclohexane to the polar DMSO and MeCN, the emission bands of the fluorophores **6c,f** became broad and significantly shifted to the red region, which agrees with the character of strong intramolecular charge transfer (ICT) and is confirmed by the values of theoretically calculated descriptors. Interestingly, in a study of the AIE effect, fluorophore **6f** showed a solvatochromic shift in the THF—water binary system of 10–90% water content in the 520–610 nm wavelength range (Figure S3).

2.2.2. Solid State Fluorescence Studies

The emission spectra of fluorophores **6a–f** in the powder/film as well as the experimental data are presented in Table 4 and Figures 3 and 4. Interestingly, only the dimethoxyphenyl-substituted fluorophores **6a** and **6d** exhibited a redshifted emission in a powder when compared to the spectra in THF solution, implying specific π-π interactions in the solid state.

Table 4. Optical properties of the compounds **6a–f** in the solid state and in PVA film.

No.	In PVA Film		In Powder	
	λ_{em}^{max}, nm	Φ_f, (%) [1]	λ_{em}^{max}, nm	Φ_f, (%) [1]
6a	546	4.8	572	20.5
6b	545	49.6	517	17.8
6c	546	25.6	511	3.9
6d	545	13.9	626	8.3
6e	542	12.0	509	19.3
6f	540	34.5	525	3.4

[1] Absolute quantum yields were measured using the Integrating Sphere of the Horiba FluoroMax-4 at r.t. in film/powder form.

In the manufacture of OLED devices, thin films of compounds are applied in layers; therefore, it is necessary to conduct optical studies with thin films of materials [54]. To examine the emission in the films, thin films of PVA with integrated fluorophores **6** were deposited on quartz plates, and their emission spectra were measured by using the integrating sphere. In all the spectra, the emission maxima were observed at about 545 nm and were quite similar to the ones collected in THF solution. Thus, the absence of an anomalous red shift in the solid emission demonstrates the useful role of the cyano-group in the phenylimidazopyridine chromophore for restraining the formation of heavy J-aggregates in the solid state [55]. In contrast to the emission in powder, the **6b–f** samples in the PVA film showed a significant improvement in fluorescence along with an up to 50% increase in quantum yields, which demonstrates the existence of AIE effects similar to those in solutions.

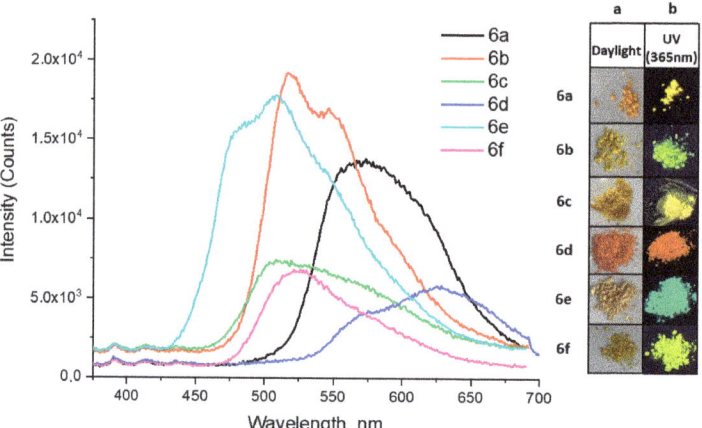

Figure 3. Emission spectra of dyes **6a–f** in powders (**a**) and photographs of the solid samples under daylight and 365 nm UV irradiation (**b**).

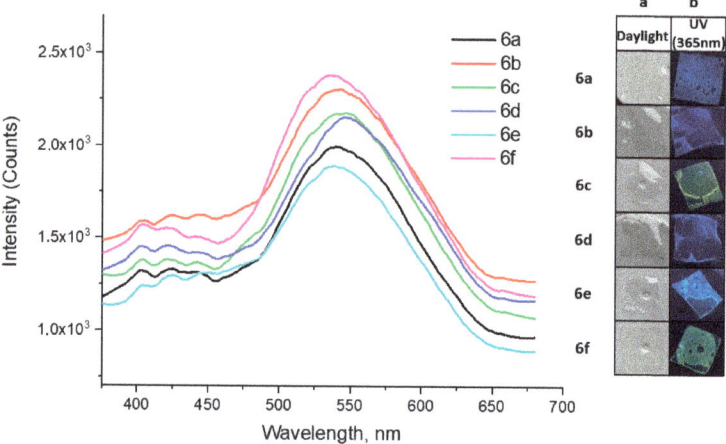

Figure 4. Emission spectra of dyes **6a–f** in PVA films (**a**) and photographs of the samples under daylight and 365 nm UV irradiation (**b**).

2.2.3. Aggregation Studies

The phenomenon of aggregation-induced emission (AIE) is usually associated with the well-known Mie scattering effect and is a signal of nanoaggregate formation [56]. The AIE properties of the **6a–f** dyes were investigated using different water fractions (fw) in THF. As shown in Table 3, anthracenyl substituted fluorophore **6c** almost does not emit in pure THF with a fluorescence quantum yield of less than 0.1%. However, when the water content in the THF solution was increased to 60%, a new green emission band with a maximum at 555 nm was observed for this dye. At the same time, the emission intensity increased approximately two-fold. In addition, the absorption spectra of **6c** with a water fraction of

60% did not coincide with the spectra of pure THF and contained an additional absorption peak in the 425–500 nm range, which may be associated with light scattering due to the formation of nanoaggregates (Figure 5) [57]. In addition, the time-resolved fluorescence curves of **6c** in pure THF and with a water fraction of 60% did not coincide (Table S8, Figure 5). Apparent changes in the mean fluorescence lifetime (τ_{av}) of **6c** from 6.9 ns in THF to 8.8 ns after the addition of water were observed. The experimental results of the effect of the nature of solvents and the values of the theoretically calculated descriptors are consistent with the fluorescence enhancement behavior of **6c** and indicate that the AIE process is accompanied by the formation of molecular aggregates. The optimized **6c** geometries for the ground and excited states in the THF were calculated to interpret the AIE process (See Section 2.2.5. Theoretical Calculations).

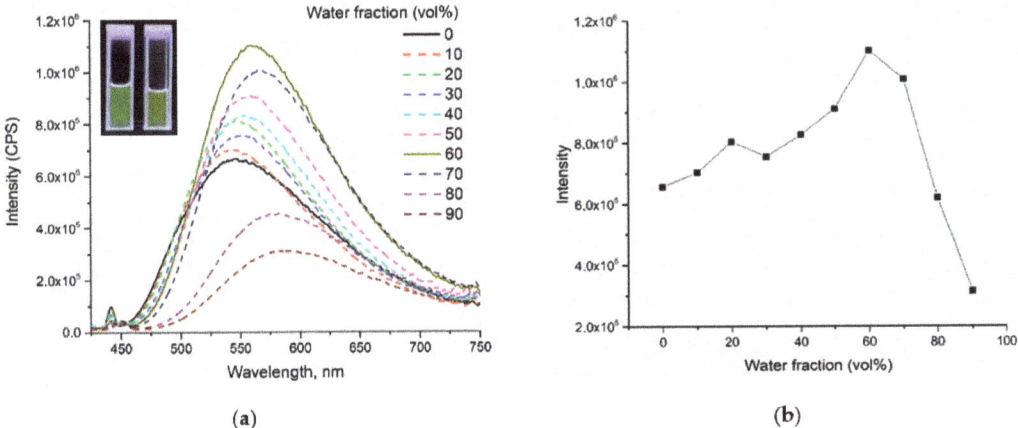

Figure 5. Emission spectra of **6c** in different ratios of THF–water (v/v) mixtures (**a**). Plot of I/I_0 versus water fraction (vol%), where I_0 is the fluorescence intensity in pure THF and emission images of the **6c** in different water fraction mixtures under 365 nm UV illumination (λ_{ex} = 365 nm) with the concentration of 10^{-5} M (**b**).

2.2.4. Mechanochromic Properties

In general, non-planar push-pull luminophores with AIE properties tend to show mechanochromic response [58]. As shown above, fluorophores **6c,f** turned out to be AIE-active; their emission maximums were different in the solid state and in aggregate (Tables 3 and 4); therefore, these two fluorophores were selected as the most suitable candidates for the study of mechanochromic properties. As crystalline samples, anthracenyl substituted fluorophores **6c** and **6f** were obtained with low emission intensities, QYs of 3.9% (**6c**) and 3.4% (**6f**), and emission maxima of 511 and 525 nm, respectively (Table 3).

After grinding with a mortar and pestle, the fluorescence emission of compounds **6c** and **6f** was measured. As it turned out, the compounds demonstrated different responses to mechanical (grinding) stimulation. Thus, the grinding of the yellow powder **6c** led to a red-shift of the fluorescence spectra by 31 nm (the red line) and a decrease in fluorescence intensity (Figures 6a and 7a, Table S9). Additionally, after the resuspension of the sample from CH_2Cl_2, yellow crystals were formed (Figure 7a) and a slight shift of the emission peak to the blue region was recorded.

The **6f** derivative was obtained as yellow crystals with poor emission intensity (Table 3). The grinding of the crystals of 6f resulted in a bright yellow powder (Figure 7b), along with a low red-shift of the fluorescence by 10 nm (the red line) with the same fluorescence intensity. (Figure 6b). Interestingly, after resuspension of the sample in CH_2Cl_2, a mixture of crystals and powder formed, as well as a slightly blue-shifted emission peak that increased with fluorescence intensity (Figures 6b and 7b, Table S9).

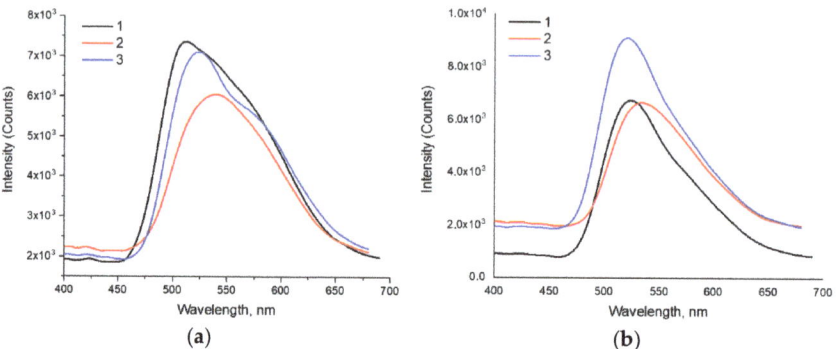

Figure 6. (a) Emission spectra of **6c** in solid states (λ_{ex} = 350 nm): as prepared (1), after grinding (2), and after treatment with CH$_2$Cl$_2$ (3). (b) Emission spectra of **6f** in solid states (λ_{ex} = 350 nm): as prepared (1), after grinding (2), and after treatment with CH$_2$Cl$_2$ (3).

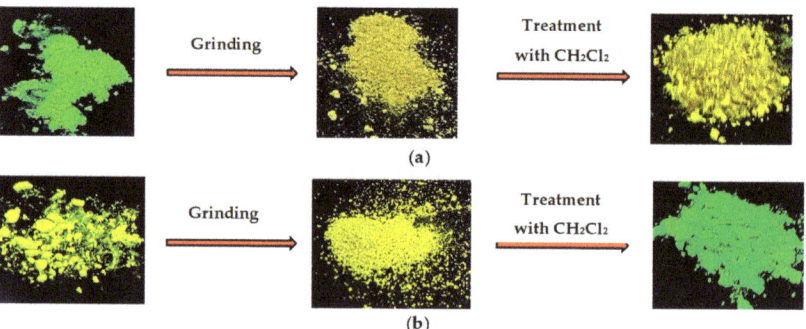

Figure 7. Photographs of **6c** (a) and **6f** (b) taken under 365 nm UV irradiation.

Most probably, the fluorescence response of the samples **6c** and **6f** during grinding depends on the molecular stacking structures in the solid state [59].

2.2.5. Theoretical Calculations

The DFT-calculations were performed in order to evaluate the donor-acceptor properties and the nature of intramolecular charge transfer based on the obtained optimized model structures of fluorophores **6a–f** in the ground and excited states in the solvent phase, energy levels and electron density distribution in frontier molecular orbitals (FMOs), and descriptors—charge-transfer indices (CT-indexes).

The electron density distributions of the boundary molecular orbitals of FMO **6a–f** are shown in Figure 8 and in Table 5.

Table 5. HOMO/LUMO based on the functionality B3LYP/6-311G* in the THF phase.

Compound	HOMO, eV	LUMO, eV	ΔE, eV
6a	−5.43	−2.04	3.39
6b	−5.96	−2.13	3.83
6c	−5.56	−2.16	3.40
6d	−5.38	−2.14	3.24
6e	−6.04	−2.23	3.81
6f	−5.59	−2.21	3.38

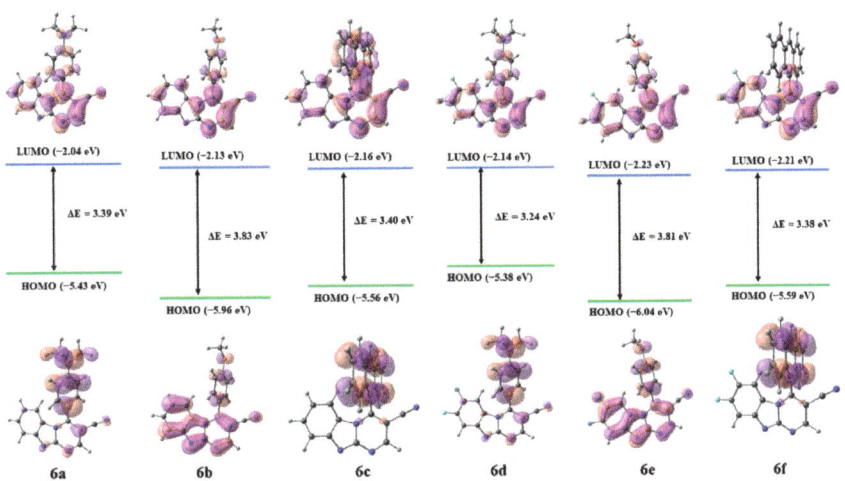

Figure 8. Energy gaps of fluorophores **6a–f** in THF phase.

The highest occupied molecular orbitals (HOMOs) of the anthracenyl substituted dyes **6c,f** delocalized exclusively on the donor group, whereas the acceptor group based on the 3-cyano substituted benzo [4,5]imidazo [1,2-*a*]pyrimidine domain is responsible for the contribution to the lowest unoccupied molecular orbitals (LUMOs). Charge delocalization was less pronounced in the electron density distribution in the FMO for dimethylaminophenyl substituted fluorophores **6a,d**. In fact, there was no delocalization of electron density for methoxyphenyl substituted samples **6b,e**.

Thus, based on theoretical calculations and experimental data, one can present a general model of the studied fluorophores consisting of a donor methoxyphenyl/dimethylaminophenyl/anthracenyl fragment (Ar, blue) and an acceptor 3-cyano-substituted benzo[4,5]imidazo[1,2-*a*]pyrimidine domain (red), including substituted fluorine atoms at positions 7 and 8 (Figure 9).

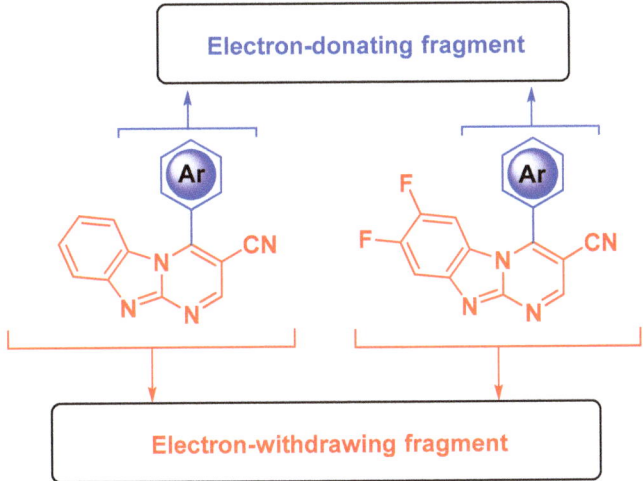

Figure 9. Donor-acceptor structure 4-aryl-substituted benzo[4,5]imidazo[1,2-*a*]pyrimidine chromophors **6a–f**.

To obtain a deeper understanding of the correlation between charge transfer and fluorophore structures, additional calculations of CT-indices were performed [60]. The corresponding indices (D, S_r, and t) presented in Table 6 were calculated for all fluorophores in the Multiwfn program [61].

Table 6. Calculated dipole moments for model structures in ground and excited multiplicity states and estimated indexes related to hole-electron distribution (CT-indexes).

Compound	Dipole Moment in Ground Multiplicity State (Debye)	Dipole Moment in Excited Multiplicity State (Debye)	D(Å)	S_r (a.u.)	t (Å)
6b	3.1676	4.3108	0.978	0.62822	−0.326
6e	4.0313	1.9340	0.927	0.62287	−0.550
6a	3.1463	9.2304	3.722	0.50976	0.617
6d	6.5432	13.7974	3.832	0.50975	0.678
6c	3.8859	1.8867	4.406	0.26547	2.599
6f	1.7563	3.5274	4.523	0.24338	2.707

The highest D index values [60], as the distance between the centers of gravity of the donor and acceptor, were 4.4 and 4.5 Å for anthracenyl substituted **6c** and **6f**, respectively, which result from the highest degree of intramolecular charge transfer. The S_r index introduced by Tozer in 2008 [62] gives a good correlation between the value of the Stokes shift and the CT junction value; that is, the smaller S_r corresponds to the larger Stokes shifts. The lowest values of this index correspond to compounds **6c,f**, as confirmed by studies of the solvatochromic effect with the highest values of the 168–181 nm Stokes shift. The index t > 0 confirms the very fact of charge separation (CD) between the chromophore donor and acceptor due to charge excitation. Thus, the analysis of CT indices confirmed the ICT process for the anthracenyl and dimethylaminophenyl substituted chromophores **6a,d** and **6c,f**, and also made it possible to predict a significant overlap between the centroids of the positive charge of the donor and the negative charge of the acceptor, representing the zones of increase and decrease in electron density upon excitation, based on the calculated values of D at t > 0.

2.3. Crystallography

According to the XRD data, two independent molecules of the compound **6c** crystallize with a molecule of CH_2Cl_2 in the centrosymmetric space group of the triclinic system. In the result, the structurally independent unit $C_{51}H_{30}Cl_2N_8$ (M = 825.73 g/mol) was used for all calculations. The molecule CH_2Cl_2 is disordered and demonstrates the high magnitude of the anisotropic displacement parameters. The geometry of independent heterocyclic molecules differs only slightly, primarily in the dihedral angles between the heterocyclic and anthracene planes. The general geometry of the molecule was shown in Figure 10. The mean bond distances and angles in the molecules are close to expectations. The heterocyclic and anthracene parts of the molecule are non-conjugated due to high dihedral angles between their planes. In the crystal some polar CArH ... NC- contacts are observed with participation of the CN-group, in particular, H(9A) ... N(2) [x − 1, y + 1, z] 2.66 Å (on a scale of 0.09 Å less than the sum of the VdW radii) and N(2A) ... H(19A) [−x, 1 − y, 1 − z] 2.71 Å (on the order of 0.04 Å less than the sum of the VdW radii). The π-π-contacts in the crystal are presented only as shortened π-π-contact between the heterocycle and anthracene moiety C(5A) ... C (17) at a distance of 3.336(4) Å (0.064 Å less than the sum of the VdW radii, Figure 11).

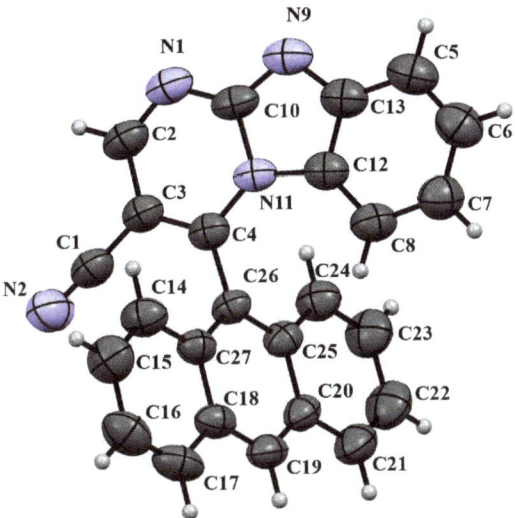

Figure 10. The compound **6c** in the thermal ellipsoid at the 50% probability level.

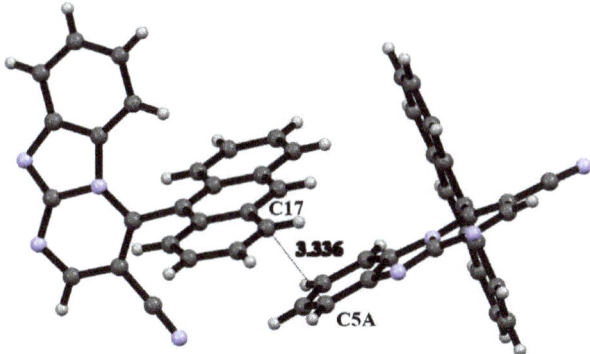

Figure 11. π-π-contacts in the crystal of the compound **6c**.

3. Materials and Methods

3.1. Chemical Experiment

Commercial reagents were obtained from Sigma-Aldrich, Acros Organics, or Alfa Aesar and used without any preprocessing. All workup and purification procedures were carried out using analytical-grade solvents. One-dimensional ^{1}H- and ^{13}C-NMR spectra were acquired on a Bruker DRX-400 instrument (Karlsruhe, Germany) (400 and 101 MHz, respectively), utilizing DMSO-d_6, CDCl$_3$, and CF$_3$COOD as solvents and an external reference, respectively. Chemical shifts are expressed in δ (parts per million, ppm) values, and coupling constants are expressed in hertz (Hz). The following abbreviations are used for the multiplicity of NMR signals: s, singlet; d, doublet; t, triplet; dd, doublet of doublet; m, multiplet; and AN, anthracene. IR spectra were recorded on a Bruker α spectrometer equipped with a ZnSe ATR accessory. Elemental analysis was performed on a PerkinElmer PE 2400 elemental analyzer (Waltham, MA, USA). Melting points were determined on a Stuart SMP3 (Staffordshire, UK) and are uncorrected. The monitoring of the reaction progress was performed using TLC on Sorbfil plates (Imid LTD, Russia, Krasnodar) (the eluent is EtOAc). The spectral characteristics of the compound **3b** correspond to the data [63]. The compound 3-Morpholinoacrylonitrile (**4**) was prepared according to a literature procedure [64].

General procedure for the synthesis of N-(4-arylidene)-1H-benzo[d]imidazol-2-amine (**3a,c** and **3d–f**).

Corresponding 1H-benzo[d]imidazol-2-amine **1a,b** (0.01 mol) was mixed with corresponding aldehydes **2a,c** and **2d–f** (0.0105 mol) and the mixture was heated at 130 °C for 3 h. The reaction mixture was cooled to room temperature and ground up to give the expected pure product.

4-Dimethylaminobenzylidene-1H-benzo[d]imidazol-2-amine (**3a**). Yellow powder (2.37 g, yield 90%), m.p. 245–247 °C. FT-IR (neat) ν_{max} (cm^{-1}): 3051, 1614, 1584, 1443, 1415, 1167. ^{1}H-NMR (400 MHz, DMSO-d_6) δ (ppm) 3.04 (6H, s, -N(CH$_3$)$_2$), 6.82 (2H, d, J = 8.6 Hz, H-2′, H-6′), 7.09–7.15 (2H, m, H-5, H-6), 7.29–7.44 (1H, m, H-4), 7.43–7.57 (1H, m, H-7), 7.86 (2H, d, J = 8.4 Hz, H-3′, H-5′), 9.24 (1H, s, N=CH), 12.36 (1H, s, NH). ^{13}C{^{1}H}-NMR (100 MHz, DMSO-d_6) δ (ppm) 40.1 (2C), 66.8, 111.1, 112.1 (2C), 118.5, 121.8 (2C), 123.0, 132.0, 134.6, 143.1, 153.8, 157.5, 164.9 Calcd for C$_{16}$H$_{16}$N$_4$: C 72.70, H 6.10, N 21.20; found: C 72.63, H 6.15, N 21.22.

N-(Anthracen-9-ylidene)-1H-benzo[d]imidazol-2-amine (**3c**). Orange powder (2.73 g, yield 85%), m.p. 277–279 °C. FT-IR (neat) ν_{max} (cm^{-1}): 3046, 1790, 1620, 1553, 1517, 1338. ^{1}H-NMR (400 MHz, DMSO-d_6) δ (ppm) 7.15–7.22 (2H, m, H-5, H-6), 7.44–7.51 (1H, m, H-4), 7.56–7.65 (3H, m, H-7, 2xH$_{AN}$), 7.67–7.73 (2H, m, 2xH$_{AN}$), 8.17 (2H, d, J = 8.3 Hz, 2xH$_{AN}$), 8.80 (1H, s, H$_{AN}$), 9.13 (2H, d, J = 9.0 Hz, 2xH$_{AN}$), 10.74 (1H, s, N=CH), 12.73 (1H, s, NH). ^{13}C{^{1}H}-NMR (100 MHz, DMSO-d_6) δ (ppm) 111.2, 118.9, 122.1, 124.6, 124.8, 125.8, 128.3, 129.4, 130.9, 131.0, 132.9, 134.4, 142.5, 156.1, 163.9. Calcd for C$_{22}$H$_{15}$N$_3$: C 82.22, H 4.70, N 13.08; found: C 82.25, H 4.66, N 13.03.

5,6-Difluoro-N-(4-dimethylaminobenzylidene)-1H-benzo[d]imidazol-2-amine (**3d**). Yellow powder (2.61 g, yield 87%), m.p. 294–296 °C. FT-IR (neat) ν_{max} (cm^{-1}): 3045, 1636, 1614, 1549, 1353, 1155. ^{1}H-NMR (400 MHz, DMSO-d_6) δ (ppm) 3.07 (6H, s, -N(CH$_3$)$_2$), 6.83 (2H d, J = 8.6 Hz, H-2′, H-6′), 7.31–7.61 (2H, m, H-4, H-7), 7.86 (2H, d, J = 8.5 Hz, H-3′, H-5′), 9.19 (1H, s, N=CH), 12.59 (1H, s, NH). ^{13}C{^{1}H}-NMR (100 MHz, DMSO-d_6) δ (ppm) 31.2 (2C), 99.2, 105.9, 112.1 (2C), 122.7, 129.9 (d, J = 8.9 Hz) 132.2 (2C), 138.7 (d, J = 9.7 Hz), 144.6 (d, J = 249.0 Hz), 145.1 (d, J = 227.2 Hz), 154.0 (2C), 159.2, 165.3. ^{19}F-NMR (376 MHz, DMSO-d_6) δ (ppm) −145.9 (d, J = 22.2 Hz), -145.14 (d, J = 20.8 Hz). Calcd for C$_{16}$H$_{14}$F$_2$N$_4$: C 63.99, H 4.70, N 18.66; found: C 63.81, H 4.73, N 18.53.

5,6-Difluoro-N-(4-methoxybenzylidene)-1H-benzo[d]imidazol-2-amine (**3e**). Yellow powder (2.58 g, yield 90%), m.p. 254–256 °C. FT-IR (neat) ν_{max} (cm^{-1}): 3062, 1593, 1568, 1509, 1453, 1256. ^{1}H-NMR (400 MHz, DMSO-d_6) δ (ppm) 3.86 (3H, s, OCH$_3$), 7.12 (2H d, J = 8.4 Hz, H-2′, H-6′), 7.37–7.62 (2H, m, H-4, H-7), 8.01 (2H, d, J = 8.3 Hz, H-3′, H-5′), 9.32 (1H, s, N=CH), 12.79 (1H, s, NH). ^{13}C{^{1}H}-NMR (100 MHz, DMSO-d_6) δ (ppm) 55.6, 99.0, 105.8, 114.6 (2C), 127.8, 129.7, 137.7 (2C), 138.0, 146.5 (d, J = 237.6 Hz), 146.6 (d, J = 237.2 Hz), 157.8, 163.2, 165.1. ^{19}F-NMR (376 MHz, DMSO-d_6) δ (ppm) -145.3, -144.3. Calcd for C$_{15}$H$_{11}$F$_2$N$_3$O: C 62.72, H 3.86, N 13.23; found: C 62.65, H 3.91, N 13.17.

5,6-Difluoro-N-(Anthracen-9-ylidene)-1H-benzo[d]imidazol-2-amine (**3f**). Orange powder (3.00 g, yield 84%), m.p. 282–284 °C. FT-IR (neat) ν_{max} (cm^{-1}): 3145, 1666, 1553, 1479, 1452, 1199. ^{1}H-NMR (400 MHz, DMSO-d_6) δ (ppm) 7.52–7.76 (6H, m, H-4, 5xH$_{AN}$), 8.20 (2H, d, J = 8.4 Hz, 2xH$_{AN}$), 8.89 (1H, s, H$_{AN}$), 8.96–9.09 (2H, m, H-7, H$_{AN}$), 10.63 (1H, s, N=CH), 13.15 (1H, s, NH). ^{13}C{^{1}H}-NMR (100 MHz, DMSO-d_6) δ (ppm) 99.9 (d, J = 22.5 Hz), 106.7 (d, J = 19.8 Hz), 124.0, 125.0, 126.3, 128.9, 129.76, 129.83, 129.9, 130.4 (d, J = 11.5 Hz), 131.2, 131.4, 131.6, 131.9, 133.7, 135.7, 138.6 (d, J = 10.9 Hz), 147.2 (d, J = 237.7 Hz), 147.4 (d, J = 238.0 Hz), 158.2, 164.9, 194.7. ^{19}F-NMR (376 MHz, DMSO-d_6) δ (ppm) -144.7 (d, J = 21.8 Hz), -143.54 (d, J = 22.1 Hz). Calcd for C$_{22}$H$_{13}$F$_2$N$_3$: C 73.94, H 3.67, N 11.76; found: C 74.03, H 3.53, N 11.46.

General procedure for the synthesis of 4-(aryl)-1,2-dihydrobenzo[4,5]imidazo[1,2-a]pyrimidine-3-carbonitriles (**5a–c**).

To a suspension of the corresponding derivative **3a–c** (0.01 mol, 1 equivalent) in 30 mL of *n*-BuOH, 1.88 mL (0.015 mol., 1.5 equiv.) of BF$_3$·Et$_2$O was added. To the resulting solution, 1.38 g (0.01 mol, 1 equivalent) of 3-morpholinoacrylonitrile (**4**) was added. The reaction mixture was heated in an oil bath at 130 °C for 5 h. The resulting mixture was cooled to room temperature and stirred for 15 min. The obtained precipitate was filtered off and washed with *i*-PrOH, water, and acetone to give the expected pure product.

4-(Dimethylaminophenyl)-1,2-dihydrobenzo[4,5]imidazo[1,2-*a*]pyrimidine-3-carbonitrile (**5a**). White powder (2.33 g, yield 74%), m.p. > 300 °C. FT-IR (neat) ν_{max} (cm^{-1}): 3071, 2805, 2215, 1621, 1578, 1459. ^{1}H-NMR (400 MHz, CDCl$_3$ + 0.1 mL CF$_3$COOD) δ (ppm) 3.36 (6H, s, -N(CH$_3$)$_2$), 6.42 (1H, s, H-4), 6.97 (1H, d, *J* = 8.3 Hz, H-6), 7.35 (1H, t, *J* = 7.9 Hz, H-7), 7.50–7.56 (2H, m, H-8, H-2), 7.64–7.73 (5H, m, H-9, H-2′, H-3′, H-5′ H-6′), 11.30 (1H, s, -NH). ^{13}C{^{1}H}-NMR (100 MHz, CDCl$_3$ + 0.1 mL CF$_3$COOD) δ (ppm) 47.6 (2C), 57.3, 87.6, 111.5, 114.1, 114.4, 122.4 (2C), 126.7, 127.3, 127.8, 128.6, 129.7 (2C), 135.8, 138.4, 141.6, 143.6. Calcd for C$_{19}$H$_{17}$N$_5$: C 72.36, H 5.43, N 22.21; found: C 72.45, H 5.51, N 22.04.

4-(4-Methoxyphenyl)-1,2-dihydrobenzo[4,5]imidazo[1,2-*a*]pyrimidine-3-carbonitrile (**5b**). White powder (1.90 g, yield 63%), m.p. 270–272 °C. FT-IR (neat) ν_{max} (cm^{-1}): 3376, 3109, 2209, 1659, 1624, 1254. ^{1}H-NMR (400 MHz, CDCl$_3$ + 0.1 mL CF$_3$COOD) δ (ppm) 3.82 (3H, s, OCH$_3$), 6.16 (1H, s, H-4), 6.89–7.05 (3H, m, H-6, H-3′, H-5′), 7.25 (1H, t, *J* = 7.8 Hz, H-7), 7.28–7.36 (3H, m, H-2, H-2′, H-6′), 7.40 (1H, t, *J* = 7.8 Hz, H-8), 7.62 (1H, d, *J* = 8.2 Hz, H-9). ^{13}C{^{1}H}-NMR (100 MHz, CDCl$_3$ + 0.1 mL CF$_3$COOD) δ (ppm) 55.6 (2C), 58.1, 89.3, 111.9, 114.0, 115.4 (2C), 125.4, 126.5, 127.5, 128.0, 128.6 (2C), 129.4, 133.8, 142.6, 161.3. Analytical calculated for C$_{18}$H$_{14}$N$_4$O: C 71.51, H 4.67, N 18.53; found: C 71.58, H 4.61, N 18.45.

4-(Anthracen-9-yl)-1,2-dihydrobenzo[4,5]imidazo[1,2-*a*]pyrimidine-3-carbonitrile (**5c**). White powder (2.23 g, yield 60%), m.p. > 300 °C. FT-IR (neat) ν_{max} (cm^{-1}): 3062, 2635, 2202, 1666, 1502, 1447. ^{1}H-NMR (400 MHz, CDCl$_3$ + 0.1 mL CF$_3$COOD) δ (ppm) 6.11 (1H, d, *J* = 8.4 Hz, H$_{AN}$), 6.86 (1H, t, *J* = 8.0 Hz, H-6), 7.21–7.27 (1H, m, H-7), 7.42–7.55 (3H, m, H-2, 2xH$_{AN}$), 7.60 (1H, s, H-4), 7.64–7.70 (1H, m, H-8), 7.76–7.85 (2H, m, 2xH$_{AN}$), 7.97 (1H, s, H$_{AN}$), 8.05–8.12 (1H m, H$_{AN}$), 8.21 (1H, d, *J* = 8.5 Hz, H-9), 8.49 (1H, d, *J* = 9.1 Hz, H$_{AN}$), 8.72 (1H, s, H$_{AN}$). ^{13}C{^{1}H}-NMR (100 MHz, CDCl$_3$ + 0.1 mL CF$_3$COOD) δ (ppm) 53.3, 88.8, 112.0, 113.8, 114.1, 120.3, 120.5, 121.5, 125.9, 126.1, 126.3, 127.0, 128.2, 128.3, 129.1, 129.9, 130.3, 130.9, 131.2, 131.3, 131.5, 132.0, 133.4, 135.5, 141.8. Calcd for C$_{25}$H$_{16}$N$_4$: C 80.63, H 4.33, N 15.04; found: C 80.53, H 4.42, N 15.05.

4-(4-(Dimethylamino)phenyl)-7,8-difluoro-1,2-dihydrobenzo[4,5]imidazo[1,2-*a*]pyrimidine-3-carbonitrile (**5d**). White powder (2.28 g, yield 65%), m.p. > 300 °C. FT-IR (neat) ν_{max} (cm^{-1}): 3106, 2886, 2216, 1658, 1495, 1463. ^{1}H-NMR (400 MHz, DMSO-*d*$_6$) δ (ppm) 2.88 (6H, s, -N(CH$_3$)$_2$), 6.25 (1H, s, H-4), 6.69 (2H, d, *J* = 8.3 Hz, H-3′, H-5′), 6.98 (1H, dd, *J* = 10.5, 7.3 Hz, H-6), 7.20 (2H, d, *J* = 8.3 Hz, H-2′, H-6′), 7.42 (1H, dd, *J* = 11.2, 7.3 Hz, H-9), 7.58 (1H, s, H-2), 11.14 (1H, s, -NH). ^{13}C{^{1}H}-NMR (100 MHz, CDCl$_3$ + 0.1 mL CF$_3$COOD) δ (ppm) 47.4 (2C), 57.5, 87.7, 101.1 (d, *J* = 25.4 Hz), 104.2 (d, *J* = 24.1 Hz), 113.8, 122.7 (2C), 123.1 (d, *J* = 12.7 Hz), 124.6 (d, *J* = 13.1 Hz), 129.8 (2C), 135.6, 137.8, 142.9, 143.9, 149.7 (dd, *J* = 253.1, 15.0 Hz), 150.5 (dd, *J* = 251.6, 15.4 Hz). Calcd for C$_{19}$H$_{15}$F$_2$N$_5$: C 64.95, H 4.30, N 19.93; found: C 64.78, H 4.47, N 19.87.

7,8-Difluoro-4-(4-methoxyphenyl)-1,2-dihydrobenzo[4,5]imidazo[1,2-*a*]pyrimidine-3-carbonitrile (**5e**). White powder (1.99 g, yield 59%), m.p. > 300 °C. FT-IR (neat) ν_{max} (cm^{-1}): 3109, 2213, 1586, 1462, 1374, 1252. ^{1}H-NMR (400 MHz, CDCl$_3$ + 0.1 mL CF$_3$COOD) δ (ppm) 3.84 (3H, s, OCH$_3$), 6.09 (1H, s, H-4), 6.76–6.83 (1H, m, H-6), 6.98 (2H, d, *J* = 8.6 Hz, H-3′, H-5′), 7.28–7.35 (3H, m, H-2, H-2′, H-6′), 7.48–7.55 (1H, m, H-9). ^{13}C{^{1}H}-NMR (100 MHz, CDCl$_3$ + 0.1 mL CF$_3$COOD) δ (ppm) 55.6, 58.4, 89.4, 101.4 (d, *J* = 24.5 Hz), 103.5 (d, *J* = 24.1 Hz), 114.9, 115.7 (2C), 123.7 (d, *J* = 10.2 Hz), 125.5 (d, *J* = 11.8 Hz), 126.7, 128.6 (2C), 133.5, 143.8, 148.7 (dd, *J* = 249.3, 14.4 Hz), 149.6 (dd, *J* = 249.4, 13.7 Hz), 161.6. Calcd for C$_{18}$H$_{12}$F$_2$N$_4$O: C 63.90, H 3.58, N 16.56; found: C 63.83, H 3.61, N 16.38.

4-(Anthracen-9-yl)-7,8-difluoro-1,2-dihydrobenzo[4,5]imidazo[1,2-*a*]pyrimidine-3-carbonitrile (**5f**). White powder (2.57 g, yield 63%), m.p. > 300 °C. FT-IR (neat) ν_{max} (cm^{-1}): 3069,

2204, 1636, 1465, 1384, 1268. ^{1}H-NMR (400 MHz, CDCl$_3$ + 0.1 mL CF$_3$COOD) δ (ppm) 5.80–5.96 (1H, m, H-6), 7.42–7.56 (3H, m, 3x H$_{AN}$), 7.63 (1H, s, H-4), 7.70 (1H, t, J = 7.5 Hz, H$_{AN}$), 7.77–7.87 (2H, m, H-9, H$_{AN}$), 7.95 (1H, s, H$_{AN}$), 8.11–8.18 (1H, m, H$_{AN}$), 8.26 (1H, d, J = 8.5 Hz, H$_{AN}$), 8.46 (1H, d, J = 9.1 Hz, H$_{AN}$), 8.78 (1H, s, H-2). ^{13}C{^{1}H}-NMR (100 MHz, CDCl$_3$ + 0.1 mL CF$_3$COOD) δ (ppm) 53.5, 77.4, 89.0, 101.4 (d, J = 24.8 Hz), 103.5 (d, J = 24.5 Hz), 120.0, 120.2, 120.4, 123.8 (d, J = 10.1 Hz), 124.3 (d, J = 11.1 Hz), 126.0, 126.3, 129.4, 130.2, 130.3, 131.0, 131.3, 131.4, 131.5, 132.0, 133.8, 135.3, 143.0, 149.2 (dd, J = 248.1, 11.8 Hz), 150.0 (dd, J = 256.5, 19.6 Hz). Calcd for C$_{25}$H$_{14}$F$_2$N$_4$: C 73.52, H 3.46, N 13.72; found: C 73.63, H 3.49, N 13.58.

General procedure for the synthesis of 4-(aryl)benzo[4,5]imidazo[1,2-a]pyrimidine-3-carbonitriles (6a–f).

To a stirred solution of the appropriate derivatives 5a–f (0.005 mol, 1 equivalent) in DMF (30 mL), MnO$_2$ (1.74 g, 0.02 mol, 4 equivalent) was added. The resulting mixture was stirred for 2 h at 130 °C (oil bath temperature) in an open air atmosphere until TLC (EtOAc as eluent) indicated total consumption of starting dihydropyrimidines 5a–f. The reaction mixture was filtered through ceolite, the filtrate was poured into 150 mL of water, and the solid product was collected by filtration to give the expected pure product.

4-(4-(Dimethylamino)phenyl)benzo[4,5]imidazo[1,2-a]pyrimidine-3-carbonitrile (6a). Orange powder (1.33 g, yield 85%), m.p. > 300 °C. FT-IR (neat) ν_{max} (cm^{-1}): 2232, 1604, 1538, 1400, 1372, 1189. ^{1}H-NMR (400 MHz, DMSO-d_6) δ (ppm) 3.10 (6H, s, -N(CH$_3$)$_2$), 6.84 (1H, d, J = 8.4 Hz, H-6), 7.00 (2H, d, J = 8.4 Hz, H-3′, H-5′), 7.21 (1H, t, J = 7.9 Hz, H-7), 7.54 (1H, t, J = 7.8 Hz, H-8), 7.60 (2H, d, J = 8.4 Hz, H-2′, H-6′), 7.90 (1H, d, J = 8.2 Hz, H-9), 9.04 (1H, s, H-2). ^{13}C{^{1}H}-NMR (100 MHz, DMSO-d_6) δ (ppm) 39.6 (2C), 94.6, 111.6 (2C), 114.8, 115.1, 116.1, 119.9, 122.2, 126.8, 127.6, 129.6 (2C), 144.6, 150.1, 152.3, 155.3, 156.7. Calcd for C$_{19}$H$_{15}$N$_5$: C 72.83, H 4.82, N 22.35; found: C 72.71, H 5.06, N 22.23.

4-(4-Methoxyphenyl)benzo[4,5]imidazo[1,2-a]pyrimidine-3-carbonitrile (6b). Yellow powder (1.25 g, yield 83%), m.p. 233–235 °C. FT-IR (neat) ν_{max} (cm^{-1}): 2230, 1667, 1473, 1091, 1058, 1020. ^{1}H-NMR (400 MHz, DMSO-d_6) δ (ppm) 3.95 (3H, s, OCH$_3$), 6.54 (1H, d, J = 8.5 Hz, H-6), 7.20 (1H, t, J = 7.8 Hz, H-7), 7.34 (2H, d, J = 8.7 Hz, H-3′, H-5′), 7.55 (1H, t, J = 7.7 Hz, H-8), 7.77 (2H, d, J = 8.5 Hz, H-2′, H-6′), 7.92 (1H, d, J = 8.1 Hz, H-9), 9.11 (1H, s, H-2). ^{13}C{^{1}H}-NMR (100 MHz, DMSO-d_6) δ (ppm) 55.6, 95.0, 114.8, 115.1 (2C), 115.6, 120.0, 121.1, 122.5, 127.0, 127.4, 130.1 (2C), 144.6, 149.8, 155.1, 155.9, 161.9. Calcd for C$_{18}$H$_{12}$N$_4$O: C 71.99, H 4.03, N 18.66; found: C 71.80, H 3.91, N 18.70.

4-(Anthracen-9-yl)benzo[4,5]imidazo[1,2-a]pyrimidine-3-carbonitrile (6c). Yellow powder (1.66 g, yield 90%), m.p. > 300 °C. FT-IR (neat) ν_{max} (cm^{-1}): 3051, 2227, 1621, 1483, 1446, 1350. ^{1}H-NMR (400 MHz, DMSO-d_6) δ (ppm) 5.29 (1H, d, J = 8.4 Hz, H-6), 6.72–6.80 (1H, m, H-7), 7.39 (1H, t, J = 7.3 Hz, H-8), 7.46–7.53 (2H, m, 2xH$_{AN}$), 7.62–7.73 (4H, m, 4xH$_{AN}$), 7.92 (1H, d, J = 8.3 Hz, H-9), 8.38 (2H, d, J = 8.5 Hz, 2xH$_{AN}$), 9.20 (1H, s, H$_{AN}$), 9.37 (1H, s, H-2). ^{13}C{^{1}H}-NMR (100 MHz, DMSO-d_6) δ (ppm) 97.2, 113.5, 115.0, 120.1, 121.0, 123.0, 123.7 (2C), 126.5 (2C), 126.6, 127.0, 128.6 (2C), 128.8 (2C), 129.3 (2C), 130.6 (2C), 132.1, 144.7, 149.9, 153.0, 155.4. Calcd for C$_{25}$H$_{14}$N$_4$: C 81.06, H 3.81, N 15.13; found: C 80.94, H 3.58, N 15.12.

4-(4-(Dimethylamino)phenyl)-7,8-difluorobenzo[4,5]imidazo[1,2-a]pyrimidine-3-carbonitrile (6d). Orange powder (1.41 g, yield 81%), m.p. 284–286 °C. FT-IR (neat) ν_{max} (cm^{-1}): 3082, 2225, 1603, 1438, 1398, 1377. ^{1}H-NMR (400 MHz, DMSO-d_6) δ (ppm) 3.11 (6H, s, -N(CH$_3$)$_2$), 6.63 (1H, t, J = 9.3 Hz, H-6), 7.02 (2H, d, J = 8.4 Hz, H-3′, H-5′), 7.61 (2H, d, J = 8.3 Hz, H-2′, H-6′), 8.02 (1H, t, J = 9.2 Hz, H-9), 9.07 (1H, s, H-2). ^{13}C{^{1}H}-NMR (100 MHz, DMSO-d_6) δ (ppm) 95.0, 103.1 (d, J = 24.4 Hz), 107.1 (d, J = 19.9 Hz), 111.5 (2C), 113.7, 115.4, 122.7 (d, J = 10.7 Hz), 129.5 (2C), 140.8 (d, J = 11.6 Hz), 145.3 (dd, J = 241.7, 15.4 Hz), 149.1 (dd, J = 245.8, 14.8 Hz), 151.2, 152.5, 155.4, 156.1. Calcd for C$_{19}$H$_{13}$F$_2$N$_4$: C 65.32, H 3.75, N 20.05; found: C 65.53, H 3.89, N 19.92.

7,8-Difluoro-4-(4-methoxyphenyl)benzo[4,5]imidazo[1,2-a]pyrimidine-3-carbonitrile (6e). Beige powder (1.34 g, yield 80%), m.p. 246–248 °C. FT-IR (neat) ν_{max} (cm^{-1}): 3046, 2229, 1595, 1530, 1490, 1101. ^{1}H NMR (400 MHz, DMSO-d_6) δ (ppm) 3.96 (3H, s, OCH$_3$), 6.31

(1H, dd, J = 10.8, 7.3 Hz, H-6), 7.37 (2H, d, J = 8.3 Hz, H-3′, H-5′), 7.77 (2H, d, J = 8.3 Hz, H-2′, H-6′), 8.05 (1H, dd, J = 10.8, 7.5 Hz, H-9), 9.15 (1H, s, H-2). ^{13}C{^{1}H}-NMR (100 MHz, DMSO-d_6) δ (ppm) 55.6, 95.6, 103.0 (d, J = 24.4 Hz), 107.5 (d, J = 19.8 Hz), 115.2, 115.3 (2C), 120.3, 122.7 (d, J = 10.9 Hz), 130.2 (2C), 140.9 (d, J = 11.8 Hz), 145.7 (dd, J = 242.2, 15.6 Hz), 149.3 (dd, J = 245.8, 15.0 Hz), 151.0, 155.5, 155.6, 162.2. Calcd for $C_{18}H_{10}F_2N_4O$: C 64.29, H 3.00, N 16.66; found: C 64.35, H 3.19, N 16.52.

4-(Anthracen-9-yl)-7,8-difluorobenzo[4,5]imidazo[1,2-a]pyrimidine-3-carbonitrile (**6f**). Yellow powder (1.71 g, yield 84%), m.p. > 300 °C. FT-IR (neat) ν_{max} (cm^{-1}): 3088, 2230, 1594, 1505, 1465, 1074. ^{1}H NMR (400 MHz, DMSO-d_6) δ (ppm) 5.00 (1H, t, J = 8.9 Hz, H-6), 7.52 (2H, t, J = 7.7 Hz, 2xH$_{AN}$), 7.67 (2H, t, J = 7.5 Hz, 2xH$_{AN}$), 7.75 (2H, d, J = 8.8 Hz, 2xH$_{AN}$), 8.07 (1H, t, J = 9.3 Hz, H-9), 8.38 (2H, d, J = 8.6 Hz, 2xH$_{AN}$), 9.23 (1H, s, H-2), 9.41 (1H, s, H$_{AN}$). ^{13}C{^{1}H}-NMR (100 MHz, DMSO-d_6) δ (ppm) 98.0, 101.5 (d, J = 24.5 Hz), 107.9 (d, J = 19.9 Hz), 114.7, 119.9, 121.9 (d, J = 10.8 Hz), 123.6 (2C), 126.6 (2C), 128.7 (2C), 129.1 (2C), 129.3 (2C), 130.5 (2C), 132.5, 141.1 (d, J = 11.8 Hz), 145.9 (dd, J = 243.2, 15.6 Hz), 149.3 (dd, J = 246.6, 14.9 Hz), 151.3, 152.6, 155.7. Calcd for $C_{25}H_{12}F_2N_4$: C 73.89, H 2.98, N 13.79; found: C 74.05, H 2.85, N 13.62.

3.2. Crystallography Experiment

The XRD analyses were carried out using equipment of the Center for Joint Use "Spectroscopy and Analysis of Organic Compounds" at the Postovsky Institute of Organic Synthesis of the Russian Academy of Sciences (Ural Branch). The experiment was carried out on a standard procedure (MoK$_\alpha$-irradiation, graphite monochromator, ω-scans with 1^0 step at T = 295(2) K) on an automated X-ray diffractometer Xcalibur 3 with a CCD detector. Empirical absorption correction was applied. The solution and refinement of the structures were accomplished using the Olex program package [65]. The structures were solved by the method of the intrinsic phases in the ShelXT program and refined by the ShelXL by full-matrix least-squares method for non-hydrogen atoms [66]. The H atoms were placed in the calculated positions and refined in isotropic approximation.

Crystal Data for $C_{51}H_{30}Cl_2N_8$ (M = 825.73 g/mol): triclinic, space group P-1, a = 8.5135(4) Å, b = 10.4646(5) Å, c = 22.9053(12) Å, α = 88.784(4)°, β = 85.741(4)°, γ = 82.301(4)°, V = 2016.53(17) Å^3, Z = 2, T = 295(2) K, μ(MoK$_\alpha$) = 0.210 mm^{-1}, Dcalc = 1.360 g/cm^3, 20,634 reflections measured (7.384° $\leq$ 2Θ $\leq$ 60.982°), 10,876 unique (R$_{int}$ = 0.0577, R$_{sigma}$ = 0.0845), which were used in all calculations. The final R$_1$ = 0.0767, wR$_2$ = 0.1916 (I > 2σ(I)) and R$_1$ = 0.1463, wR$_2$ = 0.2616 (all data). Largest peak/hole difference is 0.34/−0.35.

The XRD data were deposited in the Cambridge Structural Database with the number CCDC 2215090. This data can be requested free of charge via www.ccdc.cam.ac.uk (accessed on 17 November 2022).

3.3. DFT Calculations

The quantum chemical calculations were performed at the B3LYP/6-31G*//PM6 level of theory using the Gaussian-09 program package (M. J. Frisch, G. W. Trucks, H. B. Schlegel, G. E. Scuseria, M. A. Robb, J. R. Cheeseman, G. Scalmani, V. Barone, B. Mennucci, G. A. Petersson, H. Nakatsuji, M. Caricato, X. Li, H. P. Hratchian, A. F. Izmaylov, J. Bloino, G. Zheng, J. L. Sonnenberg, M. Had DJF. Gaussian 09, Revision C.01. Wallingford, CT 2010). No symmetry restrictions were applied during the geometry optimization procedure. The solvent effects were taken into account using the SMD (solvation model based on density) continuum solvation model suggested by Truhlar et al. [67] for THF. The Hessian matrices were calculated for all optimized model structures to prove the location of correct minima on the potential energy surface (no imaginary frequencies were found in all cases). The Chemcraft program http://www.chemcraftprog.com/ (accessed on 17 November 2022) was used for visualization. The hole-electron analysis was carried out in Multiwfn program (version 3.7) [61]. The Cartesian atomic coordinates for all optimized equilibrium model structures are presented in the attached xyz-files.

4. Conclusions

In summary, we have designed and synthesized a series of novel 4-(aryl)-benzo[4,5]imidazo[1,2-*a*]pyrimidine-3-carbonitriles by successive transformations, including the preparation of benzimidazole-2-arylimines, the Povarov reaction, and the oxidation of dihydrobenzo [4,5]imidazo[1,2-*a*]pyrimidine-3-carbonitriles. Based on the literature data and X-ray diffraction analysis, it was found that during the Povarov reaction, [1,3] sigmatropic rearrangement occurred. The structure of the synthesized compounds is unambiguously confirmed by the set of spectral data. For the derivatives **6a–f**, the ordinary photophysical properties such as absorption, emission, lifetime, and QY in solution, as well as emission and QY in powder, were studied. For the chromophore **6c**, Aggregation-Induced Emission (AIE) has been illustrated using different water fractions (fw) in THF. Finally, the mechanofluorochromic properties of derivatives **6c** and **6f** were investigated, and the response to mechanical stimulation with changing emission maxima or/and intensity was recorded. The significant photophysical properties and availability of 4-(aryl)-benzo[4,5]imidazo[1,2-a]pyrimidine-3-carbonitriles pave the way for future applications in biology, medicine, ecology, and photonics.

Supplementary Materials: The following supporting information can be downloaded at: https://www.mdpi.com/article/10.3390/molecules27228029/s1, Table S1: Fluorescence lifetime of probes **6a–f** (C = 2 × 10^{-6} M) in THF; Table S2: Orientation polarizability for solvents (Δf), absorption and fluorescence emission maxima (λ_{abs}, λ_{em}, nm), and Stokes shift (nm, cm^{-1}) of **6a** in different solvents; Table S3: Orientation polarizability for solvents (Δf), absorption and fluorescence emission maxima (λ_{abs}, λ_{em}, nm), and Stokes shift (nm, cm^{-1}) of **6b** in different solvents; Table S4: Orientation polarizability for solvents (Δf), absorption and fluorescence emission maxima (λ_{abs}, λ_{em}, nm), and Stokes shift (nm, cm^{-1}) of **6c** in different solvents; Table S5: Orientation polarizability for solvents (Δf), absorption and fluorescence emission maxima (λ_{abs}, λ_{em}, nm) and Stokes shift (nm, cm^{-1}) of **6d** in different solvents; Table S6: Orientation polarizability for solvents (Δf), absorption and fluorescence emission maxima (λ_{abs}, λ_{em}, nm) and Stokes shift (nm, cm^{-1}) of **6e** in different solvents; Table S7: Orientation polarizability for solvents (Δf), absorption and fluorescence emission maxima (λ_{abs}, λ_{em}, nm) and Stokes shift (nm, cm^{-1}) of **6f** in different solvents; Table S8: Fluorescence lifetime of probe **6c** (C = 2 × 10^{-6} M) in THF/water mixtures with water fractions 0/60 (vol%); Table S9: Mechanochromic properties of probes **6a–f**; Table S10: Crystal data and structure refinement for **6c**; Table S11: Fractional Atomic Coordinates (×10^4) and Equivalent Isotropic Displacement Parameters ($Å^2 \times 10^3$) for **6c**; Table S12: Anisotropic Displacement Parameters ($Å^2 \times 10^3$) for **6c**; Table S13: Bond Lengths for **6c**; Table S14: Bond Angles for **6c**; Table S15: Torsion Angles for **6c**; Table S16: Hydrogen Atom Coordinates (Å × 10^4) and Isotropic Displacement Parameters ($Å^2 \times 10^3$) Table S17: for **6c**; Atomic Occupancy for 6c; Figure S1: Solvent effect of **6c** and **6f**; Figure S2: UV-Vis absorption spectra of **6c** in THF/water mixtures with water fractions 0/60% (**A**). Time-resolved emission decay curves of **6c** in THF/water mixtures with water fractions 0/60% (**B**); Figure S3: Solvent effect for **6f** in THF/water; Figures S4–S20: ^{1}H- and ^{13}C-NMR spectra of compounds **3a,c**, **3d–f**, **5a–f**, and **6a–f**; Figures S21–S29 IR spectra of compounds **3a,c**, **3d–f**, **5a–f**, and **6a–f**.

Author Contributions: Synthesis, V.V.F., M.A.K. and S.V.A.; methodology, V.V.F., E.N.U. and V.L.R.; writing—original draft preparation, V.V.F., M.I.V. and O.S.T.; writing—review and editing, E.N.U., V.L.R., G.V.Z. and V.N.C.; photophysical studies, M.I.V. and O.S.T.; visualization, A.S.N. and D.S.K.; quantum chemical calculations, A.S.N.; crystallographic investigation, P.A.S.; supervision, V.L.R., G.V.Z. and V.N.C.; project administration, V.L.R. All authors have read and agreed to the published version of the manuscript.

Funding: The research funding from the Ministry of Science and Higher Education of the Russian Federation (Ural Federal University Program of Development within the Priority-2030 Program) is gratefully acknowledged.

Institutional Review Board Statement: Not applicable.

Informed Consent Statement: Not applicable.

Data Availability Statement: Data are contained within the article.

Acknowledgments: The team of authors would like to thank the Laboratory for Comprehensive Research and Expert Evaluation of Organic Materials under the direction of O.S. Eltsov.

Conflicts of Interest: The authors declare no conflict of interest.

Sample Availability: Samples of the compounds **3a–c**, **4**, **5a–f**, and **6a–f** are available from the authors.

References

1. Alqarni, S.; Cooper, L.; Galvan Achi, J.; Bott, R.; Sali, V.K.; Brown, A.; Santarsiero, B.D.; Krunic, A.; Manicassamy, B.; Peet, N.P.; et al. Synthesis, Optimization, and Structure–Activity Relationships of Imidazo[1,2-a]Pyrimidines as Inhibitors of Group 2 Influenza A Viruses. *J. Med. Chem.* **2022**, *65*, 14104–14120. [CrossRef] [PubMed]
2. Massari, S.; Bertagnin, C.; Pismataro, M.C.; Donnadio, A.; Nannetti, G.; Felicetti, T.; Di Bona, S.; Nizi, M.G.; Tensi, L.; Manfroni, G.; et al. Synthesis and Characterization of 1,2,4-Triazolo[1,5-a]Pyrimidine-2-Carboxamide-Based Compounds Targeting the PA-PB1 Interface of Influenza A Virus Polymerase. *Eur. J. Med. Chem.* **2021**, *209*, 112944. [CrossRef] [PubMed]
3. Perlíková, P.; Hocek, M. Pyrrolo[2,3-d]Pyrimidine (7-Deazapurine) as a Privileged Scaffold in Design of Antitumor and Antiviral Nucleosides. *Med. Res. Rev.* **2017**, *37*, 1429–1460. [CrossRef] [PubMed]
4. Jung, E.; Soto-Acosta, R.; Geraghty, R.J.; Chen, L. Zika Virus Inhibitors Based on a 1,3-Disubstituted 1H-Pyrazolo[3,4-d]Pyrimidine-Amine Scaffold. *Molecules* **2022**, *27*, 6109. [CrossRef] [PubMed]
5. Romagnoli, R.; Oliva, P.; Prencipe, F.; Manfredini, S.; Budassi, F.; Brancale, A.; Ferla, S.; Hamel, E.; Corallo, D.; Aveic, S.; et al. Design, Synthesis and Biological Investigation of 2-Anilino Triazolopyrimidines as Tubulin Polymerization Inhibitors with Anticancer Activities. *Pharmaceuticals* **2022**, *15*, 1031. [CrossRef]
6. Yu, G.-X.; Hu, Y.; Zhang, W.-X.; Tian, X.-Y.; Zhang, S.-Y.; Zhang, Y.; Yuan, S.; Song, J. Design, Synthesis and Biological Evaluation of [1,2,4]Triazolo[1,5-a]Pyrimidine Indole Derivatives against Gastric Cancer Cells MGC-803 via the Suppression of ERK Signaling Pathway. *Molecules* **2022**, *27*, 4996. [CrossRef] [PubMed]
7. Shi, X.; Quan, Y.; Wang, Y.; Wang, Y.; Li, Y. Design, Synthesis, and Biological Evaluation of 2,6,7-Substituted Pyrrolo[2,3-d]Pyrimidines as Cyclin Dependent Kinase Inhibitor in Pancreatic Cancer Cells. *Bioorganic Med. Chem. Lett.* **2021**, *33*, 127725. [CrossRef]
8. Ding, R.; Wang, X.; Fu, J.; Chang, Y.; Li, Y.; Liu, Y.; Liu, Y.; Ma, J.; Hu, J. Design, Synthesis and Antibacterial Activity of Novel Pleuromutilin Derivatives with Thieno[2,3-d]Pyrimidine Substitution. *Eur. J. Med. Chem.* **2022**, *237*, 114398. [CrossRef]
9. Sutherland, H.S.; Choi, P.J.; Lu, G.-L.; Giddens, A.C.; Tong, A.S.T.; Franzblau, S.G.; Cooper, C.B.; Palmer, B.D.; Denny, W.A. Synthesis and Structure–Activity Relationships for the Anti-Mycobacterial Activity of 3-Phenyl-N-(Pyridin-2-Ylmethyl)Pyrazolo[1,5-a]Pyrimidin-7-Amines. *Pharmaceuticals* **2022**, *15*, 1125. [CrossRef]
10. Shen, J.; Deng, X.; Sun, R.; Tavallaie, M.S.; Wang, J.; Cai, Q.; Lam, C.; Lei, S.; Fu, L.; Jiang, F. Structural Optimization of Pyrazolo[1,5-a]Pyrimidine Derivatives as Potent and Highly Selective DPP-4 Inhibitors. *Eur. J. Med. Chem.* **2020**, *208*, 112850. [CrossRef]
11. Peytam, F.; Adib, M.; Shourgeshty, R.; Firoozpour, L.; Rahmanian-Jazi, M.; Jahani, M.; Moghimi, S.; Divsalar, K.; Faramarzi, M.A.; Mojtabavi, S.; et al. An Efficient and Targeted Synthetic Approach towards New Highly Substituted 6-Amino-Pyrazolo[1,5-a]Pyrimidines with α-Glucosidase Inhibitory Activity. *Sci. Rep.* **2020**, *10*, 2595. [CrossRef] [PubMed]
12. Tigreros, A.; Macías, M.; Portilla, J. Expeditious Ethanol Quantification Present in Hydrocarbons and Distilled Spirits: Extending Photophysical Usages of the Pyrazolo[1,5-a]Pyrimidines. *Dye. Pigment.* **2022**, *202*, 110299. [CrossRef]
13. Fedotov, V.V.; Ulomsky, E.N.; Belskaya, N.P.; Eltyshev, A.K.; Savateev, K.V.; Voinkov, E.K.; Lyapustin, D.N.; Rusinov, V.L. Benzimidazoazapurines: Design, Synthesis, and Photophysical Study. *J. Org. Chem.* **2021**, *86*, 8319–8332. [CrossRef]
14. Tigreros, A.; Castillo, J.C.; Portilla, J. Cyanide Chemosensors Based on 3-Dicyanovinylpyrazolo[1,5-a]Pyrimidines: Effects of Peripheral 4-Anisyl Group Substitution on the Photophysical Properties. *Talanta* **2020**, *215*, 120905. [CrossRef]
15. Zhang, M.; Cheng, R.; Lan, J.; Zhang, H.; Yan, L.; Pu, X.; Huang, Z.; Wu, D.; You, J. Oxidative C-H/C-H Cross-Coupling of [1,2,4]Triazolo[1,5-a]Pyrimidines with Indoles and Pyrroles: Discovering Excited-State Intramolecular Proton Transfer (ESIPT) Fluorophores. *Org. Lett.* **2019**, *21*, 4058–4062. [CrossRef]
16. Ooyama, Y.; Uenaka, K.; Ohshita, J. Development of a Functionally Separated D–π–A Fluorescent Dye with a Pyrazyl Group as an Electron-Accepting Group for Dye-Sensitized Solar Cells. *Org. Chem. Front.* **2015**, *2*, 552–559. [CrossRef]
17. Dinastiya, E.M.; Verbitskiy, E.V.; Gadirov, R.M.; Samsonova, L.G.; Degtyarenko, K.M.; Grigoryev, D.V.; Kurtcevich, A.E.; Solodova, T.A.; Tel'minov, E.N.; Rusinov, G.L.; et al. Investigation of 4,6-Di(Hetero)Aryl-Substituted Pyrimidines as Emitters for Non-Doped OLED and Laser Dyes. *J. Photochem. Photobiol. A Chem.* **2021**, *408*, 113089. [CrossRef]
18. Fecková, M.; le Poul, P.; Bureš, F.; Robin-le Guen, F.; Achelle, S. Nonlinear Optical Properties of Pyrimidine Chromophores. *Dye. Pigment.* **2020**, *182*, 108659. [CrossRef]
19. Debnath, S.; Parveen, S.; Pradhan, P.; Das, I.; Das, T. Benzo[4,5]Imidazo[1,2-a]Pyridines and Benzo[4,5]Imidazo[1,2-a]Pyrimidines: Recent Advancements in Synthesis of Two Diversely Important Heterocyclic Motifs and Their Derivatives. *New J. Chem.* **2022**, *46*, 10504–10534. [CrossRef]
20. Fedotov, V.V.; Rusinov, V.L.; Ulomsky, E.N.; Mukhin, E.M.; Gorbunov, E.B.; Chupakhin, O.N. Pyrimido[1,2-a]Benzimidazoles: Synthesis and Perspective of Their Pharmacological Use. *Chem. Heterocycl. Comp.* **2021**, *57*, 383–409. [CrossRef]

21. Fedotov, V.V.; Ulomskiy, E.N.; Gorbunov, E.B.; Eltsov, O.S.; Voinkov, E.K.; Savateev, K.V.; Drokin, R.A.; Kotovskaya, S.K.; Rusinov, V.L. 3-Nitropyrimido[1,2-a]Benzimidazol-4-Ones: Synthesis and Study of Alkylation Reaction. *Chem. Heterocycl. Comp.* **2017**, *53*, 582–588. [CrossRef]
22. Manna, S.K.; Das, T.; Samanta, S. Polycyclic Benzimidazole: Synthesis and Photophysical Properties. *ChemistrySelect* **2019**, *4*, 8781–8790. [CrossRef]
23. Vil', V.A.; Grishin, S.S.; Baberkina, E.P.; Alekseenko, A.L.; Glinushkin, A.P.; Kovalenko, A.E.; Terent'ev, A.O. Electrochemical Synthesis of Tetrahydroquinolines from Imines and Cyclic Ethers *via* Oxidation/Aza-Diels-Alder Cycloaddition. *Adv. Synth. Catal.* **2022**, *364*, 1098–1108. [CrossRef]
24. Steinke, T.; Wonner, P.; Gauld, R.M.; Heinrich, S.; Huber, S.M. Catalytic Activation of Imines by Chalcogen Bond Donors in a Povarov [4+2] Cycloaddition Reaction. *Chem. A Eur. J.* **2022**, *28*, e202200917. [CrossRef] [PubMed]
25. Clerigué, J.; Ramos, M.T.; Menéndez, J.C. Mechanochemical Aza-Vinylogous Povarov Reactions for the Synthesis of Highly Functionalized 1,2,3,4-Tetrahydroquinolines and 1,2,3,4-Tetrahydro-1,5-Naphthyridines. *Molecules* **2021**, *26*, 1330. [CrossRef]
26. Cores, Á.; Clerigué, J.; Orocio-Rodríguez, E.; Menéndez, J.C. Multicomponent Reactions for the Synthesis of Active Pharmaceutical Ingredients. *Pharmaceuticals* **2022**, *15*, 1009. [CrossRef]
27. Jiménez-Aberásturi, X.; Palacios, F.; de los Santos, J.M. Sc(OTf)$_3$-Mediated [4 + 2] Annulations of *N*-Carbonyl Aryldiazenes with Cyclopentadiene to Construct Cinnoline Derivatives: Azo-Povarov Reaction. *J. Org. Chem.* **2022**, *87*, 11583–11592. [CrossRef]
28. Ghashghaei, O.; Masdeu, C.; Alonso, C.; Palacios, F.; Lavilla, R. Recent Advances of the Povarov Reaction in Medicinal Chemistry. *Drug Discov. Today Technol.* **2018**, *29*, 71–79. [CrossRef]
29. Sun, Y.; Lei, Z.; Ma, H. Twisted Aggregation-Induced Emission Luminogens (AIEgens) Contribute to Mechanochromism Materials: A Review. *J. Mater. Chem. C* **2022**, *10*, 14834–14867. [CrossRef]
30. Cooper, M.W.; Zhang, X.; Zhang, Y.; Ashokan, A.; Fuentes-Hernandez, C.; Salman, S.; Kippelen, B.; Barlow, S.; Marder, S.R. Delayed Luminescence in 2-Methyl-5-(Penta(9-Carbazolyl)Phenyl)-1,3,4-Oxadiazole Derivatives. *J. Phys. Chem. A* **2022**, *126*, 7480–7490. [CrossRef]
31. Gong, X.; Xiang, Y.; Ning, W.; Zhan, L.; Gong, S.; Xie, G.; Yang, C. A Heterocycle Fusing Strategy for Simple Construction of Efficient Solution-Processable Pure-Red Thermally Activated Delayed Fluorescence Emitters. *J. Mater. Chem. C* **2022**, *10*, 15981–15988. [CrossRef]
32. Wang, X.; Li, Y.; Wu, Y.; Qin, K.; Xu, D.; Wang, D.; Ma, H.; Ning, S.; Wu, Z. A 2-Phenylfuro[2,3-*b*]Quinoxaline-Triphenylamine-Based Emitter: Photophysical Properties and Application in TADF-Sensitized Fluorescence OLEDs. *New J. Chem.* **2022**, *46*, 18854–18864. [CrossRef]
33. Hojo, R.; Mayder, D.M.; Hudson, Z.M. Donor–Acceptor Materials Exhibiting Deep Blue Emission and Thermally Activated Delayed Fluorescence with Tris(Triazolo)Triazine. *J. Mater. Chem. C* **2021**, *9*, 14342–14350. [CrossRef]
34. Rodella, F.; Saxena, R.; Bagnich, S.; Banevičius, D.; Kreiza, G.; Athanasopoulos, S.; Juršėnas, S.; Kazlauskas, K.; Köhler, A.; Strohriegl, P. Low Efficiency Roll-off Blue TADF OLEDs Employing a Novel Acridine–Pyrimidine Based High Triplet Energy Host. *J. Mater. Chem. C* **2021**, *9*, 17471–17482. [CrossRef]
35. Devesing Girase, J.; Rani Nayak, V.; Tagare, J.; Shahnawaz; Ram Nagar, M.; Jou, J.-H.; Vaidyanathan, S. Solution-Processed Deep-Blue (Y~0.06) Fluorophores Based on Triphenylamine-Imidazole (Donor-Acceptor) for OLEDs: Computational and Experimental Exploration. *J. Inf. Disp.* **2022**, *23*, 53–67. [CrossRef]
36. Anupriya; Justin Thomas, K.R; Nagar, M.R.; Jou, J.-H. Effect of Cyano Substituent on the Functional Properties of Blue Emitting Imidazo[1,2-a]Pyridine Derivatives. *Dye. Pigment.* **2022**, *206*, 110658. [CrossRef]
37. Ohsawa, T.; Sasabe, H.; Watanabe, T.; Nakao, K.; Komatsu, R.; Hayashi, Y.; Hayasaka, Y.; Kido, J. A Series of Imidazo[1,2-f]Phenanthridine-Based Sky-Blue TADF Emitters Realizing EQE of over 20%. *Adv. Opt. Mater.* **2019**, *7*, 1801282. [CrossRef]
38. Wang, Z.-Y.; Zhao, J.-W.; Li, P.; Feng, T.; Wang, W.-J.; Tao, S.-L.; Tong, Q.-X. Novel Phenanthroimidazole-Based Blue AIEgens: Reversible Mechanochromism, Bipolar Transporting Properties, and Electroluminescence. *New J. Chem.* **2018**, *42*, 8924–8932. [CrossRef]
39. Godumala, M.; Choi, S.; Cho, M.J.; Choi, D.H. Thermally Activated Delayed Fluorescence Blue Dopants and Hosts: From the Design Strategy to Organic Light-Emitting Diode Applications. *J. Mater. Chem. C* **2016**, *4*, 11355–11381. [CrossRef]
40. Kothavale, S.; Lee, K.H.; Lee, J.Y. Molecular Design Strategy of Thermally Activated Delayed Fluorescent Emitters Using CN-Substituted Imidazopyrazine as a New Electron-Accepting Unit. *Chem. Asian J.* **2020**, *15*, 122–128. [CrossRef]
41. Anupriya; Thomas, K.R.J.; Nagar, M.R.; Shahnawaz; Jou, J.-H. Phenanthroimidazole Substituted Imidazo[1,2-a]Pyridine Derivatives for Deep-Blue Electroluminescence with CIEy ~ 0.08. *J. Photochem. Photobiol. A Chem.* **2022**, *423*, 113600. [CrossRef]
42. Taniya, O.S.; Fedotov, V.V.; Novikov, A.S.; Sadieva, L.K.; Krinochkin, A.P.; Kovalev, I.S.; Kopchuk, D.S.; Zyryanov, G.V.; Liu, Y.; Ulomsky, E.N.; et al. Abnormal Push-Pull Benzo[4,5]Imidazo[1,2-a][1,2,3]Triazolo[4,5-e]Pyrimidine Fluorophores in Planarized Intramolecular Charge Transfer (PLICT) State: Synthesis, Photophysical Studies and Theoretical Calculations. *Dye. Pigment.* **2022**, *204*, 110405. [CrossRef]
43. Barik, S.; Skene, W.G. Turning-on the Quenched Fluorescence of Azomethines through Structural Modifications: Turning-on the Quenched Fluorescence of Azomethines. *Eur. J. Org. Chem.* **2013**, *2013*, 2563–2572. [CrossRef]
44. Barluenga, J.; Aznar, F.; Valdes, C.; Cabal, M.P. Stereoselective Synthesis of 4-Piperidone and 4-Aminotetrahydropyridine Derivatives by the Imino Diels-Alder Reaction of 2-Amino-1,3-Butadienes. *J. Org. Chem.* **1993**, *58*, 3391–3396. [CrossRef]

45. Montalvo-González, R.; Ariza-Castolo, A. Molecular Structure of Di-Aryl-Aldimines by Multinuclear Magnetic Resonance and X-Ray Diffraction. *J. Mol. Struct.* **2003**, *655*, 375–389. [CrossRef]
46. Nowicka, A.; Liszkiewicz, H.; Nawrocka, W.; Wietrzyk, J.; Kempińska, K.; Dryś, A. Synthesis and Antiproliferative Activity in Vitro of New 2-Aminobenzimidazole Derivatives. Reaction of 2-Arylideneaminobenzimidazole with Selected Nitriles Containing Active Methylene Group. *Open Chem.* **2014**, *12*, 1047–1055. [CrossRef]
47. Bogolubsky, A.V.; Moroz, Y.S.; Mykhailiuk, P.K.; Panov, D.M.; Pipko, S.E.; Konovets, A.I.; Tolmachev, A. A One-Pot Parallel Reductive Amination of Aldehydes with Heteroaromatic Amines. *ACS Comb. Sci.* **2014**, *16*, 375–380. [CrossRef]
48. Yu, J.; Hu, P.; Zhou, T.; Xu, Y. Synthesis of Benzimidazole Thiazolinone Derivatives under Microwave Irradiation. *J. Chem. Res.* **2011**, *35*, 672–673. [CrossRef]
49. Chen, C.-H.; Yellol, G.S.; Lin, P.-T.; Sun, C.-M. Base-Catalyzed Povarov Reaction: An Unusual[1,3]Sigmatropic Rearrangement to Dihydropyrimidobenzimidazoles. *Org. Lett.* **2011**, *13*, 5120–5123. [CrossRef]
50. Hsiao, Y.-S.; Narhe, B.D.; Chang, Y.-S.; Sun, C.-M. One-Pot, Two-Step Synthesis of Imidazo[1,2-a]Benzimidazoles via A Multicomponent [4 + 1] Cycloaddition Reaction. *ACS Comb. Sci.* **2013**, *15*, 551–555. [CrossRef]
51. Shah, A.P.; Hura, N.; Kishore Babu, N.; Roy, N.; Krishna Rao, V.; Paul, A.; Kumar Roy, P.; Singh, S.; Guchhait, S.K. A Core-Linker-Polyamine (CLP) Strategy Enables Rapid Discovery of Antileishmanial Aminoalkylquinolinecarboxamides That Target Oxidative Stress Mechanism. *ChemMedChem* **2022**, *17*, e202200109. [CrossRef]
52. Kuznetsova, E.A.; Smolobochkin, A.V.; Rizbayeva, T.S.; Gazizov, A.S.; Voronina, J.K.; Lodochnikova, O.A.; Gerasimova, D.P.; Dobrynin, A.B.; Syakaev, V.V.; Shurpik, D.N.; et al. Diastereoselective Intramolecular Cyclization/Povarov Reaction Cascade for the One-Pot Synthesis of Polycyclic Quinolines. *Org. Biomol. Chem.* **2022**, *20*, 5515–5519. [CrossRef] [PubMed]
53. Vicente-García, E.; Catti, F.; Ramón, R.; Lavilla, R. Unsaturated Lactams: New Inputs for Povarov-Type Multicomponent Reactions. *Org. Lett.* **2010**, *12*, 860–863. [CrossRef]
54. Thakur, D.; Nagar, M.R.; Tomar, A.; Dubey, D.K.; Kumar, S.; Swayamprabha, S.S.; Banik, S.; Jou, J.-H.; Ghosh, S. Through Positional Isomerism: Impact of Molecular Composition on Enhanced Triplet Harvest for Solution-Processed OLED Efficiency Improvement. *ACS Appl. Electron. Mater.* **2021**, *3*, 2317–2332. [CrossRef]
55. Anupriya; Thomas, K.R.J.; Nagar, M.R.; Shahnawaz; Jou, J.-H. Imidazo[1,2-a]Pyridine Based Deep-Blue Emitter: Effect of Donor on the Optoelectronic Properties. *J. Mater. Sci. Mater. Electron.* **2021**, *32*, 26838–26850. [CrossRef]
56. Leung, C.W.T.; Hong, Y.; Chen, S.; Zhao, E.; Lam, J.W.Y.; Tang, B.Z. A Photostable AIE Luminogen for Specific Mitochondrial Imaging and Tracking. *J. Am. Chem. Soc.* **2013**, *135*, 62–65. [CrossRef]
57. Hong, Y.; Lam, J.W.Y.; Tang, B.Z. Aggregation-Induced Emission. *Chem. Soc. Rev.* **2011**, *40*, 5361. [CrossRef]
58. Suman, G.R.; Pandey, M.; Chakravarthy, A.S.J. Review on New Horizons of Aggregation Induced Emission: From Design to Development. *Mater. Chem. Front.* **2021**, *5*, 1541–1584. [CrossRef]
59. Kwon, M.S.; Gierschner, J.; Yoon, S.-J.; Park, S.Y. Unique Piezochromic Fluorescence Behavior of Dicyanodistyrylbenzene Based Donor-Acceptor-Donor Triad: Mechanically Controlled Photo-Induced Electron Transfer (eT) in Molecular Assemblies. *Adv. Mater.* **2012**, *24*, 5487–5492. [CrossRef]
60. Le Bahers, T.; Adamo, C.; Ciofini, I. A Qualitative Index of Spatial Extent in Charge-Transfer Excitations. *J. Chem. Theory Comput.* **2011**, *7*, 2498–2506. [CrossRef]
61. Lu, T.; Chen, F. Multiwfn: A Multifunctional Wavefunction Analyzer. *J. Comput. Chem.* **2012**, *33*, 580–592. [CrossRef] [PubMed]
62. Peach, M.J.G.; Benfield, P.; Helgaker, T.; Tozer, D.J. Excitation Energies in Density Functional Theory: An Evaluation and a Diagnostic Test. *J. Chem. Phys.* **2008**, *128*, 044118. [CrossRef] [PubMed]
63. Nawrocka, W.; Sztuba, B.; Kowalska, M.W.; Liszkiewicz, H.; Wietrzyk, J.; Nasulewicz, A.; Pełczyńska, M.; Opolski, A. Synthesis and Antiproliferative Activity in Vitro of 2-Aminobenzimidazole Derivatives. *Il Farmaco* **2004**, *59*, 83–91. [CrossRef]
64. Rene, L.; Poncet, J.; Auzou, G. A One Pot Synthesis of β-Cyanoenamines. *Synthesis* **1986**, *1986*, 419–420. [CrossRef]
65. Dolomanov, O.V.; Bourhis, L.J.; Gildea, R.J.; Howard, J.A.K.; Puschmann, H. OLEX2: A Complete Structure Solution, Refinement and Analysis Program. *J. Appl. Crystallogr.* **2009**, *42*, 339–341. [CrossRef]
66. Sheldrick, G.M. SHELXT–Integrated Space-Group and Crystal-Structure Determination. *Acta Crystallogr. A Found Adv.* **2015**, *71*, 3–8. [CrossRef] [PubMed]
67. Marenich, A.V.; Cramer, C.J.; Truhlar, D.G. Universal Solvation Model Based on Solute Electron Density and on a Continuum Model of the Solvent Defined by the Bulk Dielectric Constant and Atomic Surface Tensions. *J. Phys. Chem. B* **2009**, *113*, 6378–6396. [CrossRef]

Communication

Synthesis of 2,5-Dialkyl-1,3,4-oxadiazoles Bearing Carboxymethylamino Groups

Marcin Łuczyński, Kornelia Kubiesa and Agnieszka Kudelko *

Department of Chemical Organic Technology and Petrochemistry, The Silesian University of Technology, Krzywoustego 4, PL-44100 Gliwice, Poland
* Correspondence: agnieszka.kudelko@polsl.pl; Tel.: +48-32-237-17-29

Abstract: A series of new symmetrical 2,5-dialkyl-1,3,4-oxadiazoles containing substituted alkyl groups at the terminal positions with substituents, such as bromine, isopropyloxycarbonylmethylamino, and carboxymethylamino, were successfully synthesized. The developed multistep method employed commercially available acid chlorides differing in alkyl chain length and terminal substituent, hydrazine hydrate, and phosphorus oxychloride. The intermediate bromine-containing 2,5-dialkyl-1,3,4-oxadiazoles were easily substituted with diisopropyl iminodiacetate, followed by hydrolysis in aqueous methanol solution giving the corresponding 1,3,4-oxadiazoles bearing carboxymethylamino substituents. The structure of all products was confirmed by conventional spectroscopic methods including ^{1}H NMR, ^{13}C NMR, and HRMS.

Keywords: 1,3,4-oxadiazoles; organic ligands; heterocycles; substitution; diisopropyl iminodiacetate

1. Introduction

Oxadiazoles are five-membered heterocyclic compounds composed of two nitrogen atoms and one oxygen atom. Depending on the heteroatom position, oxadiazoles exist in the form of several isomers [1], including 1,3,4-oxadiazole derivatives that are the most studied due to their high stability and wide range of biological activity [2]. Additionally, they exhibit anti-inflammatory, analgesic [3], antiviral [4], antibacterial [5], antifungal [6], anticancer [7], and blood pressure-lowering effects [8]. Reports have shown that oxadiazoles biological activity characteristics can be employed in agriculture as herbicides, insecticides, and plant protection agents against bacteria, viruses, and fungi [9–12]. Furthermore, oxadiazole compounds possess valuable optical properties. 1,2-Diazole fragment of 1,3,4-oxadiazole derivatives has an electron-accepting effect and contributes to its application in various types of conducting systems, such as organic light-emitting diodes, laser dyes, optical brighteners, and scintillators [13–16]. Certain 2,5-disubstituted 1,3,4-oxadiazole derivatives have high thermal and chemical stability, which is important in materials science, and such compounds are used in the production of heat-resistant polymers, blowing agents, optical brighteners, and anti-corrosion agents [17–20].

One of the most popular methods for 1,3,4-oxadiazole derivatives preparation include the cyclodehydration reaction. The following reagents are commonly employed in the cyclization of N,N'-diacylhydrazines: polyphosphoric acid (PPA) [21], sulfuric acid (H_2SO_4) [22], phosphorus oxychloride ($POCl_3$) [23], thionyl chloride ($SOCl_2$) [24], trifluoromethanesulfonic anhydride (($CF_3SO_2)_2O$) [25], phosphorus pentoxide (P_2O_5) [26], boron trifluoride etherate ($BF_3·OEt_2$) [27], and Burgess reagent [28]. Additionally, it has been shown that the preparation of 1,3,4-oxadiazole derivatives is possible using oxidative cyclization of N-acylhydrazones with oxidizing agents such as cerium ammonium nitrate (CAN) [29], bromine (Br_2) [30], potassium permanganate ($KMnO_4$) [31], lead(IV) oxide (PbO_2) [32], chloramine T [33], 2,3-dichloro-5,6-dicyano-1,4-benzowuinone (DDQ) [34], and hypervalent iodine reagents [35].

1,3,4-Oxadiazoles containing alkyl chains at the 2 and 5 positions and substituted with carboxymethylamino groups are of particular interest. Generally, the presence of aminopolycarboxylic functionalities promote complexing properties in such derivatives, allowing for binding to metal cations. The literature shows a range of organic ligands of this type, some of which have been approved for use in medicine and agriculture. These include ethylenediaminetetraacetic acid (EDTA) [36], N-(hydroxyethyl)ethylenediaminetriacetic acid (HEEDTA) [37], ethylenediamine-N,N'-bis(o-hydroxy-p-methylphenyl)acetic acid (EDDHMA) [38], diethylenetriaminepentaacetic acid (DTPA) [39], nitrilotriacetic acid (NTA) [40], glucoheptanoic acid [41], and citric acid [42,43]. Unfortunately, despite the excellent chelating properties, not all fertilizing chelates show the adequate biodegradation. Some of the most common coordination compounds based on EDTA and DTPA are characterized by very high stability, but they are resistant to biodegradation. Numerous studies revealed that they are present in waters of lakes or ponds, as well as in the soil for a long time, which may result in eutrophication of waters and introduction of metals into the food chain [44]. On the other hand, there has recently been an emphasis on chelating agents based on sugar molecules in order to increase the effectiveness of the micronutrients. There is a possibility that sugar acid derivatives, condensed tannis, and glucohetonates could effectively replace the traditionally used EDTA for the production of chelated micronutrients [43]. Heterocyclic compounds composed of carbon, oxygen, and nitrogen atoms could also constitute another alternative scaffold for the construction of new chelating agents.

According to the above data, we assumed that the combination of the 1,3,4-oxadiazole core containing oxygen and two nitrogen atoms in the ring with aminopolycarboxylic groups allows for the development of a new family of organic ligands with potential applications as complexing agents in medicine and agriculture. Herein, we developed an effective method for the preparation of new symmetrically substituted 1,3,4-oxadiazole derivatives containing carboxymethylaminoalkyl groups at the 2 and 5 positions. Initially, a three-step transformation was conducted using commercially available reagents (Scheme 1). In the first step, the acid chlorides were reacted with hydrazine hydrate to form symmetrical N,N'-diacylhydrazine derivatives with different alkyl chain lengths. In the second step, they were cyclized using POCl$_3$, a known cyclodehydration reagent. Finally, the substitution reaction between bromine-containing 1,3,4-oxadiazole derivatives and iminodiacetic acid was studied.

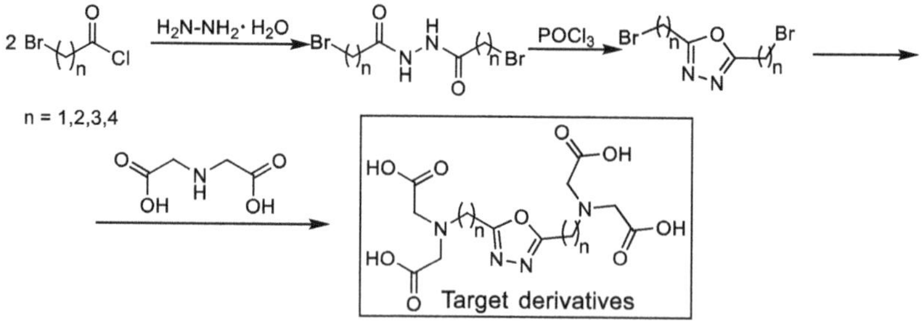

Scheme 1. Initial concept of the synthetic pathway.

2. Results

The target derivatives containing carboxymethylaminoalkyl groups were obtained in a multistep transformation reaction. The first step consisted of the synthesis of symmetrical N,N'-diacylhydrazine derivatives (2a–d) (Scheme 2) using commercially available acid chloride derivatives (1a–d) (Scheme 2) differing in alkyl chain length and bearing a bromine atom at the terminal position. The model reaction employed bromoacetyl chloride (1a) as the starting material. The optimization study consisted of examining the base (triethylamine (TEA) and sodium carbonate), solvent (chloroform, and diethyl ether), and the influence of temperature. The best results were obtained at low temperature (0 °C) in diethyl ether using aqueous sodium carbonate as the base. The products were purified via recrystallization from either methanol or ethanol. The yields of the obtained hydrazine derivatives (**2a–d**) were 65–79% (Table 1, entries 4, 8, 12, and 16).

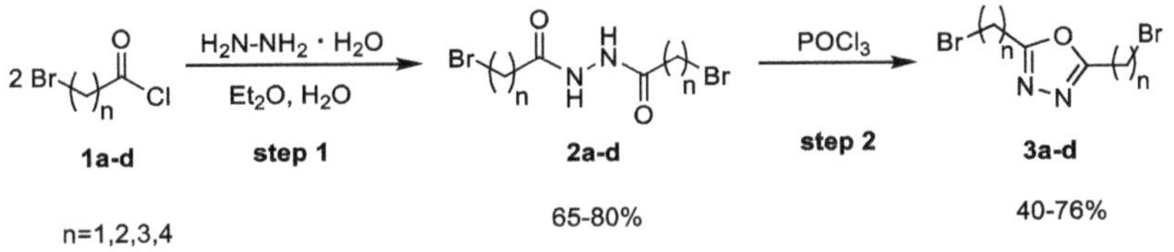

Scheme 2. Synthesis of symmetrical N,N'-diacylhydrazine (**2a–d**) and 1,3,4-oxadiazole derivatives (**3a–d**). Reaction conditions: step 1: acid chloride (**1a–d**, 0.06 mol), hydrazine hydrate (hydrazine 64%, 4.6 mL, 0.06 mol), sodium carbonate (6.36 g, 0.06 mol), diethyl ether (70 mL), water (40 mL), 0 °C, 0.5 h; acid chloride (**1a–d**, 0.06 mol), diethyl ether (10 mL), 25 °C, 2 h; step 2: N,N'-diacylhydrazine (**2a–d**, 0.007 mol), POCl$_3$ (22.4 mL, 0.24 mol), 55 °C, 6–24 h.

Table 1. N,N'-Diacylhydrazines (**2a–d**) derived from acid chlorides.

Entry	Product	n	Base	Solvent	Yield [%]
1	2a	1	TEA	Chloroform	26
2				Diethyl ether	11
3			Na$_2$CO$_3$	Chloroform	47
4				Diethyl ether	65
5	2b	2	TEA	Chloroform	33
6				Diethyl ether	44
7			Na$_2$CO$_3$	Chloroform	59
8				Diethyl ether	73
9	2c	3	TEA	Chloroform	37
10				Diethyl ether	35
11			Na$_2$CO$_3$	Chloroform	66
12				Diethyl ether	76
13	2d	4	TEA	Chloroform	28
14				Diethyl ether	20
15			Na$_2$CO$_3$	Chloroform	74
16				Diethyl ether	79

The synthesis of 1,3,4-oxadiazole derivatives (**3a–d**) (Scheme 2) involved reacting with POCl$_3$, a cyclodehydration reagent. The reaction was conducted in anhydrous toluene or solvent-free conditions and monitored by TLC. In toluene, product was formed in lower yield compared to the solvent-free reaction (Table 2, entries 2, 4, 6, and 8). The final products (**3a–d**) were obtained in 40–76% yield and were used for the subsequent reactions without purification (Table 2).

Table 2. 2,5-Dialkyl-1,3,4-oxadiazole derivatives (**3a–d**) formed by cyclization of N,N'-diacylhydrazine.

Entry	Product	n	Solvent	Yield (%)
1	3a	1	Toluene	36
2			-	51
3	3b	2	Toluene	15
4			-	40
5	3c	3	Toluene	39
6			-	44
7	3d	4	Toluene	59
8			-	76

As part of the synthesis, we planned to use iminodiacetic acid (**4**), which could directly react with the formed 2,5-dialkyl-1,3,4-oxadiazole derivatives (**3a–d**) containing bromine substituents (Scheme 1). After testing a series of bases, including TEA, sodium hydroxide, and sodium carbonate, as well as several organic solvents (chloroform, methanol, acetonitrile) and their mixtures [45], the desired products 7 were not generated. Therefore, the iminodiacetic acid was converted into a more reactive ester, which could be subjected to a substitution reaction with 2,5-bis(bromoalkyl)-1,3,4-oxadiazoles (3a–d), followed by hydrolysis to restore the carboxyl groups. Hence, the esterification reactions were performed using ethanol (a) and isopropanol (b) as substrates, and in the presence of sulfuric acid as the catalyst (Scheme 3). The final esters (5a,b) were obtained in good yields (37–52%).

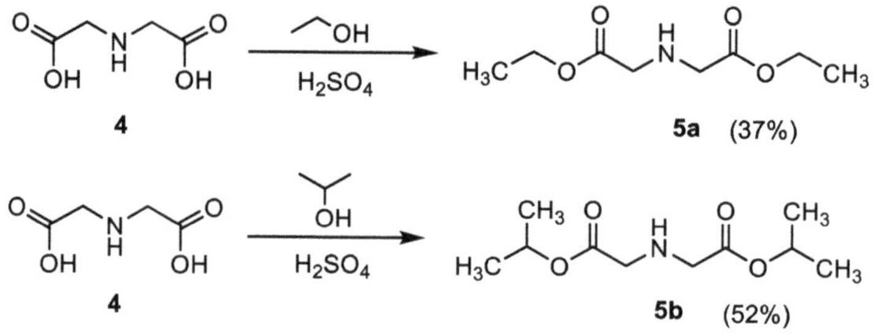

Scheme 3. Synthesis of diethyl iminodiacetate (5a) and diisopropyl iminodiacetate (5b). Reaction conditions: iminodiacetic acid (4, 15.0 g, 0.11 mol), alcohol (120 mL), H_2SO_4 (7.5 mL), reflux, 12 h.

The next step was to perform the substitution reaction involving the appropriate ester (5a,b), oxadiazole derivative (3a–d), and base in an aprotic solvent (Scheme 4). In order to determine the optimal conditions, 2,5-bis(bromomethyl)-1,3,4-oxadiazole (3a) and diethyl iminodiacetate (5a) were first examined. When the reaction was conducted at room temperature, main product 6a was produced but in a low yield, at 23% (Table 3, entry 11). The further study revealed that the optimal temperature was in the range of 50–60 °C, giving the product 6a in a 91% yield (Table 3, entry 12).

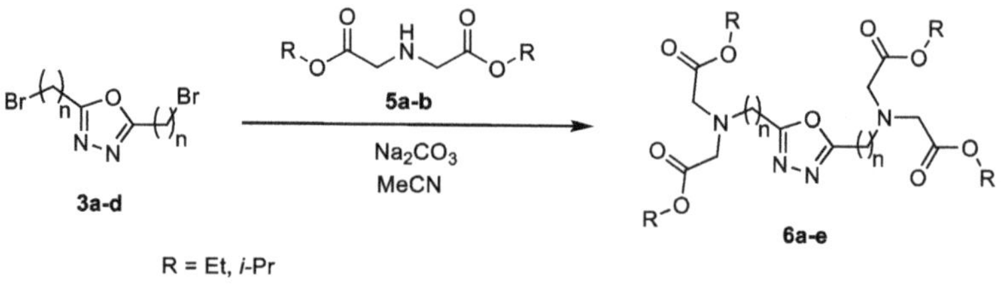

Scheme 4. Synthesis of 2,5-dialkyl-1,3,4-oxadiazole (6a–e) ester derivatives. Reaction conditions: 1,3,4-oxadiazole derivative (3a–d, 0.004 mol), iminodiacetic acid ester (5a,b, 0.01 mol), sodium carbonate (4.24 g, 0.04 mol), acetonitrile (50 mL), 60 °C, 12 h.

Among the solvents tested, acetonitrile gave the best result (Table 3, entry 12) owing to its characteristic aprotic polarity and its relatively low boiling point in relation to dimethylformamide (DMF) or DMSO. The inorganic weak base sodium carbonate enhanced the removal of inorganic compounds during the extraction process. Additionally, the optimal reaction time was 8–12 h, which was determined by TLC. Having the optimized conditions in hand, a series of substitution reactions were conducted using oxadiazoles (3b–d), and producing products (6a–e) in a 71–91% yield (Table 4, entries 1–5). All products (6a–e) were purified by column chromatography on silica gel using ethyl acetate as the eluent.

Table 3. Optimization of the substitution reaction with diethyl iminodiacetate to form derivative **6a**.

Entry	Solvent	Temp. (°C)	Base	Yield (%)
1	DMF	25	TEA	18
2		60		22
3		25	Na_2CO_3	14
4		60		69
5	DMSO	25	TEA	9
6		60		24
7		25	Na_2CO_3	18
8		60		55
9	Acetonitrile	25	TEA	21
10		60		39
11		25	Na_2CO_3	23
12		60		91

Table 4. Obtained yields of 2,5-dialkyl-1,3,4-oxadiazole (**6a–e**) and products of their hydrolysis containing carboxylic groups (**7b,7d**).

Entry	n	R	Product	Yield (%)
1	1	Ethyl	6a	91
2	1	i-Propyl	6b	84
3	2	i-Propyl	6c	71
4	3	i-Propyl	6d	68
5	4	i-Propyl	6e	73
6	1	i-Propyl	7b	54
7	3	i-Propyl	7d	68

The last step in the synthetic pathways was the restoration of the carboxyl groups from esters groups (**6**). Various literature methods were examined, including anhydrous lithium chloride [46] and the classical method of hydrolysis in an acidic and alkaline environment [47]. It was found that the alkaline hydrolysis reaction gave desired 1,3,4-oxadiazole derivatives bearing carboxymethylaminoalkyl groups (**7b, 7d**) at the side alkyl chains (Scheme 5). The lithium chloride method was ineffective, and hydrolysis under acidic conditions gave the desired product, but excessive heating promoted decomposition of the oxadiazole. Optimization of the hydrolysis reaction involved different amounts of NaOH and solvents. The obtained results showed that a significant excess of NaOH and conducting the reaction in a relatively high temperature caused the decomposition. However, we observed the formation of the intended final product (**7b**) when the molar ratio between ester **6** and NaOH was 1:10. Methanol was found to be the best solvent owing to its ability to dissolve the substrate. The final products (**7b, 7d**) were purified by recrystallization from methanol, providing the pure products in a 54–68% yield (Table 4, entries 6 and 7).

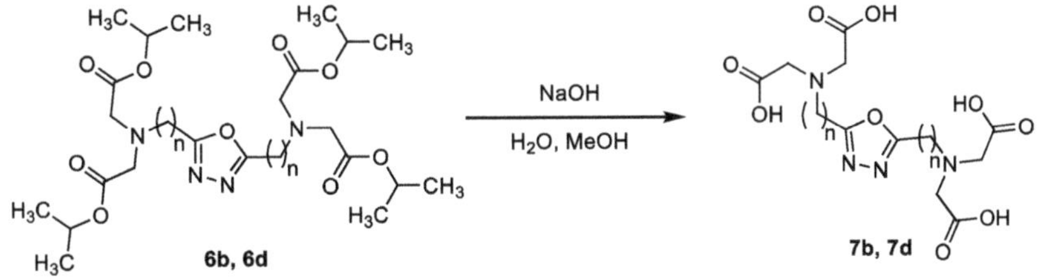

Scheme 5. Hydrolysis reaction of compounds **6b** and **6d** to form 2,5-dialkyl-1,3,4-oxadiazole derivatives containing carboxymethylaminoalkyl moieties (**7b, 7d**). Reaction conditions: ester derivative of 2,5-dialkyl-1,3,4-oxadiazole (**6b, 6d**, 0.18 mmol), NaOH (0.1 g, 1.8 mmol), MeOH (24 mL), H_2O (6 mL), 50 °C, 1 h.

The structure of all obtained intermediates and final products was confirmed by ^{1}H and ^{13}C NMR spectroscopy (see Supplementary Materials). Both ester and acid derivatives were symmetrical molecules; hence, the number of signals was reduced. Among 2,5-dialkyl-1,3,4-oxadiazole derivatives (**6a–e**), and not described so far in the literature, characteristic ^{1}H NMR signals included the doublet at 1.25 ppm and septet at 5.00 ppm corresponding to the isopropyl group. The singlet at 3.50 ppm was related to the iminodiacetate moiety. The remaining signals at 1.00–3.00 ppm corresponded to the alkyl side chain between the oxadiazole and diisopropyl ester. In the case of ^{13}C NMR, the characteristic C2 and C5 signals of the heterocyclic 1,3,4-oxadiazole ring were found at 165.0 ppm. The remaining peaks from carbonyl groups were located at 170.0 ppm, while two signals from isopropyloxy group were found at 22.0 ppm (CH_3) and 68.0 ppm (-CH-), respectively. Signals at 50.0–55.0 ppm corresponded to the iminodiacetate part of the molecule (-N(CH_2)$_2$<). Finally, carbons of the alkyl chain were in the range of 22.0–60.0 ppm. ^{1}H and ^{13}C spectra of the final products, containing carboxyl groups (**7b**, **7d**), showed no visible signals of the ester residue. High-resolution mass spectra further confirmed the structure of the obtained intermediates and final products.

3. Experimental Section

3.1. General Information

All reagents were purchased from commercial sources and used without further purification. Melting points were measured using a Stuart SMP3 melting point apparatus (Staffordshire, UK). NMR spectra were recorded at 25 °C using an Agilent 400-NMR spectrometer (Agilent Technologies, Waldbronn, Germany) at 400 MHz for ^{1}H and 100 MHz for ^{13}C, with CDCl$_3$ or DMSO as solvent, and TMS as the internal standard. High-resolution mass spectra were acquired using a Waters ACQUITY UPLC/Xevo G2QT instrument (Waters Corporation, Milford, MA, USA). Thin-layer chromatography (TLC) was performed using silica gel 60 F254 (Merck, Merck KGaA, Darmstadt, Germany) thin-layer chromatography plates, with ethyl acetate, chloroform/ethyl acetate (5:1 v/v), or methanol/chloroform (4:1 v/v) as the mobile phases.

3.2. Synthesis and Characterization

3.2.1. Synthesis of N,N'-Diacylhydrazine Derivatives (2a–d)

Hydrazine hydrate (4.6 mL, 0.06 mol) was dissolved in diethyl ether (50 mL), and the mixture was cooled to 0 °C. The appropriate amount of acid chloride (**1a–d**, 0.06 mol) was dissolved in diethyl ether (20 mL) and added dropwise to the mixture. The temperature was carefully monitored, keeping it below 35 °C. Then, after 30 min, sodium carbonate (6.36 g, 0.06 mol) dissolved in water (40 mL) was added. After the evolution of carbon dioxide had ceased, acid chloride (**1a–d**, 0.06 mol) dissolved in diethyl ether (10 mL) was added dropwise. The reaction mixture was stirred at room temperature for 2 h. The resulting precipitate was filtered, and dried products was recrystallized from methanol to obtain the desired pure products.

2-Bromo-N'-(2-bromoacetyl)acetohydrazide (**2a**)

The product was obtained as white powder (10.69 g, 65%); m.p. 174–176 °C. ^{1}H-NMR (400 MHz, DMSO): δ 3.92 (s, 4H), 10.58 (s, 2H, NH); ^{13}C-NMR (100 MHz, DMSO): δ 26.9, 164.4.

3-Bromo-N'-(3-bromopropanoyl)propanehydrazide (**2b**)

The product was obtained as white powder (13.23 g, 73%); m.p. 182–183 °C. ^{1}H-NMR (400 MHz, DMSO): δ 2.78 (t, J = 8.0 Hz, 4H), 3.65 (t, J = 8.0 Hz, 4H), 10.08 (s, 2H, NH);); ^{13}C-NMR (100 MHz, DMSO): δ 28.7, 36.3, 167.7.

4-Bromo-N'-(4-bromobutanoyl)butanehydrazide (**2c**)

The product was obtained as white powder (15.05 g, 76%); m.p. 159–160 °C. ^{1}H-NMR (400 MHz, DMSO): δ 2.04 (tt, J = 6,8 Hz, J = 7.2 Hz, 4H), 2.28 (t, J = 7.2 Hz, 4H), 3.55 (t, J = 6.8 Hz, 4H), 9.75 (s, 2H, NH);); ^{13}C-NMR (100 MHz, DMSO): δ 28.3, 31.5, 34.2, 170.1.

5-Bromo-N'-(5-bromopentanoyl)pentanehydrazide (**2d**)

The product was obtained as white powder (16.97 g, 79%); m.p. 149–151 °C. ^{1}H-NMR (400 MHz, DMSO): δ 1.61 (m, 4H), 1.81 (m, 4H), 2.13 (t, J = 8.0 Hz, 4H), 3.52 (t, J = 8.0 Hz, 4H), 9.67 (s, 2H, NH); ^{13}C-NMR (100 MHz, DMSO): δ 23.6, 31.5, 32.1, 34.7, 170.7.

3.2.2. Synthesis of 1,3,4-Oxadiazole Derivatives (**3a–d**)

Phosphorus oxychloride (22.4 mL, 0.24 mol) was added to N,N'-diacylhydrazine (**2a-d**, 0.007 mol). The mixture was heated to reflux for 6–24 h. The progress of the reaction was monitored by TLC using methanol/chloroform (4:1 v/v) as the mobile phase. Excess phosphorus oxychloride was evaporated, and the residue in the flask was dissolved in diethyl ether (40 mL) and poured into water (100 mL). The mixture was neutralized using sodium carbonate, extracted with diethyl ether (40 mL), dried over anhydrous magnesium sulfate, and evaporated to dryness.

2,5-Bis(bromomethyl)-1,3,4-oxadiazole (**3a**)

The product was obtained as yellow oil (0.91 g, 51%). ^{1}H-NMR (400 MHz, CDCl$_3$): δ 4.92 (s, 4H); ^{13}C-NMR (100 MHz, DMSO): δ 17.6, 164.2. HRMS (ESI): m/z calcd for C$_4$H$_4$N$_2$OBr$_2$ + H$^+$: 256.8748; found 256.8756.

2,5-Bis(2-bromoethyl)-1,3,4-oxadiazole (**3b**)

The product was obtained as yellow oil (1.99 g, 40%). ^{1}H-NMR (400 MHz, DMSO): δ 2.96 (t, J = 8.0Hz, 4H), 3.56 (t, J = 8.0 Hz, 4H); ^{13}C-NMR (100 MHz, DMSO): δ 28.7, 36.3, 167.7. HRMS (ESI): m/z calcd for C$_6$H$_8$N$_2$OBr$_2$ + H$^+$: 284.9061; found 284.9064.

2,5-Bis(3-bromopropyl)-1,3,4-oxadiazole (**3c**)

The product was obtained as yellow oil (0.96 g, 44%). ^{1}H-NMR (400 MHz, CDCl$_3$): δ 2.21 (m, 4H), 2.96 (t, J = 8.0 Hz, 4H), 3.63 (t, J = 8.0 Hz, 4H); ^{13}C-NMR (100 MHz, DMSO): δ 28.9, 31.5, 33.5, 165.5. HRMS (ESI): m/z calcd for C$_8$H$_{12}$N$_2$OBr$_2$ + H$^+$: 312.9374; found 312.9386.

2,5-Bis(4-bromobutyl)-1,3,4-oxadiazole (**3d**)

The product was obtained as yellow oil (1.81 g, 76%). ^{1}H-NMR (400 MHz, DMSO): δ 1.77 (m, 4H), 1.83 (m, 4H), 2.81 (t, J = 8.0 Hz, 4H), 3.53 (t, J = 8.0 Hz, 4H); ^{13}C-NMR (100 MHz, DMSO): δ 23.6, 24.4, 31.4, 34.4, 166.1. HRMS (ESI): m/z calcd for C$_{10}$H$_{16}$N$_2$OBr$_2$ + H$^+$: 340.9688; found 340.9692.

3.2.3. Synthesis of Iminodiacetic Acid Ester Derivatives (**5a,b**)

Ethanol (**a**) or isopropanol (**b**) (120 mL) and concentrated H$_2$SO$_4$ (7.5 mL) were added to iminodiacetic acid (**4**) (15.0 g, 0.11 mol). The reaction mixture was heated to reflux for 12 h. Excess alcohol was then evaporated using a rotary evaporator, and the mixture was neutralized with sodium bicarbonate solution. Then, the resulted solution was extracted with ethyl acetate (30 mL), dried over anhydrous magnesium sulfate, and evaporated to dryness.

Diethyl iminodiacetate (**5a**)

The product was obtained as slightly yellow liquid (7.69 g, 37%). ^{1}H-NMR (400 MHz, CDCl$_3$): δ 1.24 (t, J = 8.0 Hz, 6H), 3.42 (s, 4H), 4.15 (m, 4H); ^{13}C-NMR (100 MHz, DMSO): δ 14.1, 50.1, 60.8, 171.6. HRMS (ESI): m/z calcd for C$_8$H$_{15}$NO$_4$ + H$^+$: 190.1079; found 190.1082 [48,49].

Diisopropyl iminodiacetate (**5b**)

The product was obtained as slightly yellow liquid (12.41 g, 52%). ^{1}H-NMR (400 MHz, CDCl$_3$): δ 1.25 (d, J = 6.4 Hz, 12H), 3.42 (s, 4H), 5.07 (m, 2H); ^{13}C-NMR (100 MHz, CDCl$_3$): δ 21.5, 50.3, 68.4, 171.2. HRMS (ESI): m/z calcd for C$_{10}$H$_{19}$NO$_4$ + H$^+$: 218.1392; found 218.1403.

3.2.4. Synthesis of Ester Derivatives of 2,5-Dialkyl-1,3,4-oxadiazole (**6a–e**)

2,5-Bis(bromoalkyl)-1,3,4-oxadiazole (**3a–d**, 0.004 mol), iminodiacetic acid ester (**5a–b**, 0.01 mol), and sodium carbonate (4.24 g, 0.04 mol) were dissolved in acetonitrile (50 mL). The reaction mixture was heated at 60 °C for 12 h. Water (20 mL) was added and extracted using ethyl acetate (50 mL). The organic phase was separated, dried over anhydrous magnesium sulfate, and evaporated to dryness. The crude product was purified using column chromatography with ethyl acetate as the mobile phase.

Tetraethyl 2,2′,2″,2‴-(((1,3,4-oxadiazole-2,5-diyl)bis(methylene))bis(azanetriyl))tetraacetate (**6a**)

The product was obtained as yellow oil (1.72 g, 91%). ^{1}H-NMR (400 MHz, CDCl$_3$): δ 1.26 (t, J = 8.0 Hz, 12H), 3.69 (s, 8H), 4.24 (q, J = 8.0 Hz, 8H); ^{13}C-NMR (100 MHz, CDCl$_3$): δ 14.1, 48.1, 54.5, 60.8, 164.3, 170.5. HRMS (ESI): m/z calcd for C$_{20}$H$_{32}$N$_4$O$_9$ + H$^+$: 473.2248; found 473.2253.

Tetraisopropyl 2,2′,2″,2‴-(((1,3,4-oxadiazole=2,5-diyl)bis(methylene))bis(azanetriyl))tetraacetate (**6b**)

The product was obtained as yellow oil (1.77 g, 84%). ^{1}H-NMR (400 MHz, CDCl$_3$): δ 1.24 (d, J = 6.4 Hz, 24H), 3.65 (s, 8H), 4.24 (s, 4H), 5.02 (sept, J = 6.4 Hz, 4H); ^{13}C-NMR (100 MHz, CDCl$_3$): δ 21.8, 48.1, 54.8, 68.4, 164.4, 170.1. HRMS (ESI): m/z calcd for C$_{24}$H$_{40}$N$_4$O$_9$ + H$^+$: 529.2874; found 529.2870.

Tetraisopropyl 2,2′,2″,2‴-(((1,3,4-oxadiazole-2,5-diyl)bis(ethane-2,1-diyl))bis(azanetriyl))tetraacete (**6c**)

The product was obtained as yellow oil (1.58 g, 71%). ^{1}H-NMR (400 MHz, CDCl$_3$): δ 1.25 (d, J = 4.4 Hz, 24H), 3.34 (t, J = 4.4 Hz, 4H), 3.43 (s, 8H), 3.90 (t, J = 4.4 Hz, 4H), 5.07 (sept, J = 4.4 Hz, 4H); ^{13}C-NMR (100 MHz, CDCl$_3$): δ 21.8, 29.0, 39.4, 50.3, 68.4, 164.1, 171.2. HRMS (ESI): m/z calcd for C$_{26}$H$_{44}$N$_4$O$_9$ + H$^+$: 557.3185; found 557.3185.

Tetraisopropyl 2,2′,2″,2‴-(((1,3,4-oxadiazole-2,5-diyl)bis(propane-3,1-diyl))bis(azanetriyl))-tetraacetate (**6d**)

The product was obtained as yellow oil (1.59 g, 68%). %). ^{1}H-NMR (400 MHz, CDCl$_3$): δ 1.24 (d, J = 4.4 Hz, 24H), 1.94 (m, 4H), 2.84 (t, J = 4.4 Hz, 4H), 2.91 (t, J = 4.8 Hz, 4H), 3.50 (s, 8H), 5.03 (m, 4H); ^{13}C-NMR (100 MHz, CDCl$_3$): δ 21.9, 22.8, 24.7, 53.1, 55.2, 68.0, 166.8, 170.6. HRMS (ESI): m/z calcd for C$_{28}$H$_{48}$N$_4$O$_9$ + H$^+$: 585.3500; found 585.3491.

Tetraisopropyl 2,2′,2″,2‴-(((1,3,4-oxadiazole-2,5-diyl)bis(butane-4,1-diyl))bis(azanetriyl))tetraacetate (**6e**)

The product was obtained as yellow oil (1.79 g, 73%). %). ^{1}H-NMR (400 MHz, CDCl$_3$): δ 1.26 (d, J = 8.0 Hz, 24H), 1.36 (m, 4H), 1.58 (m, 4H), 2.77 (t, J = 8.0 Hz, 4H), 2.84 (t, J = 8.0 Hz, 4H), 3.51 (s, 8H), 5.01 (m, 4H); ^{13}C-NMR (100 MHz, CDCl$_3$): δ 21.7, 23.6, 23.9, 25.1, 49.1, 54.8, 68.4, 164.4, 170.1. HRMS (ESI): m/z calcd for C$_{30}$H$_{52}$N$_4$O$_9$ + H$^+$: 613.3812; found 613.3810.

3.2.5. Synthesis of 2,5-Dialkyl-1,3,4-oxadiazole Derivatives Containing Carboxymethylamino Groups (**7b**, **7d**)

The ester derivatives of 2,5-dialkyl-1,3,4-oxadiazole (**6b**, **6d**) (0.18 mmol) were dissolved in methanol (24 mL), water (6 mL), and NaOH (0.1 g, 1.8 mmol). The reaction mixture was heated at 50 °C for 1 h. The solution was then neutralized with 1 M HCl and evaporated to dryness. The crude product was purified by recrystallization from methanol.

2,2′,2″,2‴-(((1,3,4-oxadiazole-2,5-diyl)bis(methylene))bis(azanetriyl))tetraacetic acid (**7b**)

The product was obtained as white powder (0.04 g, 54%). ^{1}H-NMR (400 MHz, DMSO): δ 3.38 (s, 8H), 4.04 (s, 4H); ^{13}C-NMR (100 MHz, DMSO): δ 47.4, 57.5, 164.5, 173.9. HRMS (ESI): m/z calcd for $C_{12}H_{16}N_4O_9$ + H$^+$ + Na: 384.0893; found 384.0846.

2,2′,2″,2‴-(((1,3,4-oxadiazole-2,5-diyl)bis(propane-3,1-diyl))bis(azanetriyl))tetraacetic acid (**7d**)

The product was obtained as white powder (0.05 g, 68%). ^{1}H-NMR (400 MHz, DMSO): δ 1.79 (m, 4H), 2.71 (t, J = 8.0 Hz, 4H), 2.83 (t, J = 8.0 Hz, 4H), 3.42 (s, 8H); ^{13}C-NMR (100 MHz, DMSO): δ 22.1, 24.1, 52.8, 54.7, 166.3, 172.4. HRMS (ESI): m/z calcd for $C_{16}H_{24}N_4O_9$ + H$^+$: 417.1621; found 417.1638.

4. Conclusions

An interesting methodological process was developed for the synthesis of extended 1,3,4-oxadiazole derivatives containing carboxymethylaminoalkyl groups at positions 2 and 5. The initial synthetic pathway, comprised of nucleophilic substitution of bromine-containing 2,5-dialkyl-1,3,4-oxadiazoles with iminodiacetic acid, was ineffective. However, the replacement of iminodiacetic acid with more reactive diisopropyl iminodiacetate led to the formation of the intermediate esters and final acids in satisfactory yields. The obtained final products constitute a new family of complexing agents with potential applications in various areas, such as agriculture, medicine, or pharmacy.

Supplementary Materials: The following supporting information can be downloaded at: https://www.mdpi.com/article/10.3390/molecules27227687/s1, Figure S1: ^{1}H NMR spectra (400 MHz, dmso) of 2-Bromo-N′-(2-bromoacetyl)acetohydrazide (**2a**); Figure S2: ^{13}C NMR spectra (100 MHz, dmso) of 2-Bromo-N′-(2-bromoacetyl)acetohydrazide (**2a**); Figure S3: ^{1}H NMR spectra (400 MHz, dmso) of 3-Bromo-N′-(3-bromopropanoyl)propanehydrazide (**2b**); Figure S4: ^{13}C NMR spectra (100 MHz, dmso) of 3-Bromo-N′-(3-bromopropanoyl)propanehydrazide (**2b**); Figure S5: ^{1}H NMR spectra (400 MHz, dmso) of 4-Bromo-N′-(4-bromobutanoyl)butanehydrazide (**2c**); Figure S6: ^{13}C NMR spectra (100 MHz, dmso) of 4-Bromo-N′-(4-bromobutanoyl)butanehydrazide (**2c**); Figure S7: ^{1}H NMR spectra (400 MHz, dmso) of 5-Bromo-N′-(5-bromopentanoyl)pentanehydrazide (**2d**); Figure S8: ^{13}C NMR spectra (100 MHz, dmso) of 5-Bromo-N′-(5-bromopentanoyl)pentanehydrazide (**2d**); Figure S9: ^{1}H NMR spectra (400 MHz, CDCl$_3$) of 2,5-Bis(bromomethyl)-1,3,4-oxadiazole (**3a**); Figure S10: ^{13}C NMR spectra (100 MHz, CDCl$_3$) of 2,5-Bis(bromomethyl)-1,3,4-oxadiazole (**3a**); Figure S11: ^{1}H NMR spectra (400 MHz, dmso) of 2,5-Bis(2-bromoethyl)-1,3,4-oxadiazole (**3b**); Figure S12: ^{13}C NMR spectra (100 MHz, dmso) of 2,5-Bis(2-bromoethyl)-1,3,4-oxadiazole (**3b**); Figure S13: ^{1}H NMR spectra (400 MHz, CDCl$_3$) of 2,5-Bis(3-bromopropyl)-1,3,4-oxadiazole (**3c**); Figure S14: ^{13}C NMR spectra (100 MHz, CDCl$_3$) of 2,5-Bis(3-bromopropyl)-1,3,4-oxadiazole (**3c**); Figure S15: ^{1}H NMR spectra (400 MHz, dmso) of 2,5-Bis(4-bromobutyl)-1,3,4-oxadiazole (**3d**); Figure S16: ^{13}C NMR spectra (100 MHz, dmso) of 2,5-Bis(4-bromobutyl)-1,3,4-oxadiazole (**3d**); Figure S17: ^{1}H NMR spectra (400 MHz, CDCl$_3$) of Diethyl iminodiacetate (**5a**); Figure S18: ^{13}C NMR spectra (100 MHz, CDCl$_3$) of Diethyl iminodiacetate (**5a**); Figure S19: ^{1}H NMR spectra (400 MHz, CDCl$_3$) of Diisopropyl iminodiacetate (**5b**); Figure S20: ^{13}C NMR spectra (100 MHz, CDCl$_3$) of Diisopropyl iminodiacetate (**5b**); Figure S21: ^{1}H NMR spectra (400 MHz, CDCl$_3$) of Tetraethyl 2,2′,2″,2‴-(((1,3,4-oxadiazole-2,5-diyl)bis(methylene))bis(azanetriyl))tetraacetate (**6a**); Figure S22: ^{13}C NMR spectra (100 MHz, CDCl$_3$) of Tetraethyl 2,2′,2″,2‴-(((1,3,4-oxadiazole-2,5-diyl)bis(methylene))bis(azanetriyl))tetraacetate (**6a**); Figure S23: ^{1}H NMR spectra (400 MHz, CDCl$_3$) of Tetraisopropyl 2,2′,2″,2‴-(((1,3,4-oxadiazole-2,5-diyl)bis(methylene))bis(azanetriyl))tetraacetate (**6b**); Figure S24: ^{13}C NMR spectra (100 MHz, CDCl$_3$) of Tetraisopropyl 2,2′,2″,2‴-(((1,3,4-oxadiazole=2,5-diyl)bis(methylene))bis(azanetriyl))tetraacetate (**6b**); Figure S25: ^{1}H NMR spectra (400 MHz, CDCl$_3$) of Tetraisopropyl 2,2′,2″,2‴-(((1,3,4-oxadiazole-2,5-diyl)bis(ethane-2,1-diyl))bis(azanetriyl))tetraacete (**6c**); Figure S26: ^{13}C NMR spectra (100 MHz, CDCl$_3$) of Tetraisopropyl 2,2′,2″,2‴-(((1,3,4-oxadiazole-2,5-diyl)bis(ethane-2,1-diyl))bis(azanetriyl))tetraacete (**6c**); Figure S27: ^{1}H NMR spectra (400 MHz, CDCl$_3$) of Tetraisopropyl 2,2′,2″,2‴-(((1,3,4-oxadiazole-2,5-diyl)bis(propane-3,1-diyl))bis(azanetriyl))tetraacetate (**6d**); Figure S28: ^{13}C NMR spectra (100 MHz, CDCl$_3$) of Tetraisopropyl 2,2′,2″,2‴-(((1,3,4-oxadiazole-2,5-diyl)bis(propane-3,1-diyl))bis(azanetriyl))tetraacetate (**6d**); Figure S29: ^{1}H NMR spectra (400 MHz, CDCl$_3$) of Tetraisopropyl 2,2′,2″,2‴-(((1,3,4-oxadiazole-2,5-diyl)bis(butane-4,1-diyl))bis(azanetriyl))tetraacetate (**6e**);

Figure S30: ^{13}C NMR spectra (100 MHz, CDCl$_3$) of Tetraisopropyl 2,2′,2″,2‴-(((1,3,4-oxadiazole-2,5-diyl)bis(butane-4,1-diyl))bis(azanetriyl))tetraacetate (**6e**); Figure S31: ^{1}H NMR spectra (400 MHz, dmso) of 2,2′,2″,2‴-(((1,3,4-oxadiazole-2,5-diyl)bis(methylene))bis(azanetriyl))tetraacetic acid (**7b**); Figure S32: ^{13}C NMR spectra (400 MHz, dmso) of 2,2′,2″,2‴-(((1,3,4-oxadiazole-2,5-diyl)bis(methylene))bis(azanetriyl))tetraacetic acid (**7b**); Figure S33: ^{1}H NMR spectra (100 MHz, dmso) of 2,2′,2″,2‴-(((1,3,4-oxadiazole-2,5-diyl)bis(propane-3,1-diyl))bis(azanetriyl))tetraacetic acid (**7d**); Figure S34: ^{13}C NMR spectra (400 MHz, dmso) of 2,2′,2″,2‴-(((1,3,4-oxadiazole-2,5-diyl)bis(propane-3,1-diyl))bis(azanetriyl))tetraacetic acid (**7d**).

Author Contributions: Conceptualization, methodology, and planning of the experiments, M.Ł. and A.K.; synthesis of chemical compounds, M.Ł. and K.K.; characterization of chemical compounds, M.Ł. and K.K.; analysis and interpretation of the results, M.Ł. and A.K.; writing—original draft, M.Ł.; review and editing of the manuscript, A.K. All authors have read and agreed to the published version of the manuscript.

Funding: The project was financially supported by Silesian University of Technology (Poland) Grant No 04/050/RGP20/0115, and 04/050/BKM22/0152.

Institutional Review Board Statement: Not applicable.

Informed Consent Statement: Not applicable.

Data Availability Statement: Not applicable.

Conflicts of Interest: The authors declare no conflict of interest.

References

1. Shukla, C.; Sanchit, S. Biologically active oxadiazole. *J. Drug Deliv. Ther.* **2015**, *5*, 8–13. [CrossRef]
2. Kumar, K.A.; Jayaroopa, P.; Kumar, V. Comprehensive review on the chemistry of 1,3,4-oxadiazoles and their applications. *Int. J. ChemTech Res.* **2012**, *4*, 1782–1791.
3. Akhter, M.; Husain, A.; Azad, B.; Ajmal, M. Aroylpropionic acid based 2,5-disubstituted-1,3,4-oxadiazoles: Synthesis and their anti-inflammatory and analgesic activities. *Eur. J. Med. Chem.* **2009**, *44*, 2372–2378. [CrossRef] [PubMed]
4. Albratty, M.; El-Sharkawy, K.A.; Alhazmi, H.A. Synthesis and evaluation of some new 1,3,4-oxadiazoles bearing thiophene, thiazole, coumarin, pyridine and pyridazine derivatives as antiviral agents. *Acta Pharm.* **2019**, *69*, 261–276. [CrossRef]
5. Ahsan, M.J.; Samy, J.G.; Khalilullah, H.; Nomani, S.; Saraswat, P.; Gaur, R.; Singh, A. Molecular properties prediction and synthesis of novel 1,3,4-oxadiazole analogues as potent antimicrobial and antitubercular agents. *Bioorganic Med. Chem. Lett.* **2011**, *21*, 7246–7250. [CrossRef] [PubMed]
6. Chen, H.; Li, Z.; Han, Y. Synthesis and fungicidal activity against *Rhizoctonia solani* of 2-alkyl(Alkylthio)-5-pyrazolyl-1,3,4-oxadiazoles (Thiadiazoles). *J. Agric. Food Chem.* **2000**, *48*, 5312–5315. [CrossRef]
7. Ahsan, M.J.; Choupra, A.; Sharma, R.K.; Jadav, S.S.; Padmaja, P.; Hassan, Z.; Al-Tamimi, A.B.S.; Geesi, M.H.; Bakht, M.A. Rationale Design, Synthesis, Cytotoxicity Evaluation, and Molecular Docking Studies of 1,3,4-oxadiazole Analogues. *Anticancer Agents Med. Chem.* **2017**, *18*, 121–138. [CrossRef] [PubMed]
8. Lelyukh, M.; Martynets, M.; Kalytovska, M.; Drapak, I.; Harkov, S.; Chaban, T.; Matiychuk, V. Approaches for synthesis and chemical modification of non-condensed heterocyclic systems based on 1,3,4-oxadiazole ring and their biological activity: A review. *J. Appl. Pharm. Sci.* **2020**, *10*, 151–165. [CrossRef]
9. Zheng, X.; Li, Z.; Wang, Y.; Chen, W.; Huang, Q.; Liu, C.; Song, G. Syntheses and insecticidal activities of novel 2,5-disubstituted 1,3,4-oxadiazoles. *J. Fluor. Chem.* **2003**, *123*, 163–169. [CrossRef]
10. Zou, X.J.; Lai, L.H.; Jin, G.Y.; Zhang, Z.X. Synthesis, Fungicidal Activity, and 3D-QSAR of Pyridazinone-Substituted 1,3,4-Oxadiazoles and 1,3,4-Thiadiazoles. *Agric. Food Chem.* **2002**, *50*, 3757–3760. [CrossRef] [PubMed]
11. Luczynski, M.; Kudelko, A. Synthesis and Biological Activity of 1,3,4-Oxadiazoles Used in Medicine and Agriculture. *Appl. Sci.* **2022**, *12*, 3756. [CrossRef]
12. Tajik, H.; Dadras, A. Synthesis and herbicidal activity of novel 5-chloro-3-fluoro-2-phenoxypyridines with a 1,3,4-oxadiazole ring. *J. Pestic. Sci.* **2011**, *36*, 27–32. [CrossRef]
13. Wróblowska, M.; Kudelko, A.; Kuźnik, N.; Łaba, K.; Łapkowski, M. Synthesis of Extended 1,3,4-Oxadiazole and 1,3,4-Thiadiazole Derivatives in the Suzuki Cross-coupling Reactions. *J. Heterocycl. Chem.* **2017**, *54*, 1550–1557. [CrossRef]
14. Bhujabal, Y.B.; Vadagaonkar, K.S.; Kapdi, A.R. Pd/PTABS: Catalyst for Efficient C-H (Hetero_Arylation of 1,3,4-Oxadiazoles Using Bromo(Hetero)Arenes. *Asian J. Org. Chem.* **2019**, *8*, 289–295. [CrossRef]
15. Zhang, M.; Hu, Z.; He, T. Conducting probe atomic force microscopy investigation of anisotropic charge transport in solution cast PBD single crystals Induced by an external field. *J. Phys. Chem. B* **2004**, *108*, 19198–19204. [CrossRef]
16. Homocianu, M.; Airinei, A. 1,3,4-Oxadiazole Derivatives. Optical Properties in Pure and Mixed Solvents. *J. Fluoresc.* **2016**, *26*, 1617–1635. [CrossRef] [PubMed]

17. Schulz, B.; Orgzall, I.; Freydank, A.; Xu, C. Self-organization of substituted 1,3,4-oxadiazoles in the solid state and at surfaces. *Adv. Colloid Interface Sci.* **2005**, *116*, 143–164. [CrossRef]
18. Tamoto, N.; Adachi, C.; Nagai, K. Electroluminescence of 1,3,4-Oxadiazole and Triphenylamine-Containing Molecules as an Emitter in Organic Multilayer Light Emitting Diodes. *Chem. Mater.* **1997**, *9*, 1077–1085. [CrossRef]
19. Chen, Z.K.; Meng, H.; Lai, Y.H.; Huang, W. Photoluminescent poly(p-phenylenevinylene)s with an aromatic oxadiazole moiety as the side chain: Synthesis, electrochemistry, and spectroscopy study. *Macromolecules* **1999**, *32*, 4351–4358. [CrossRef]
20. Kedzia, A.; Jasiak, K.; Kudelko, A. An Efficient Synthesis of New 2-Aryl-5-phenylazenyl-1,3,4-oxadiazole Derivatives from N,N′-Diarylcarbonohydrazides. *Syn. Lett.* **2018**, *29*, 1745–1748. [CrossRef]
21. Tully, W.R.; Gardner, C.R.; Gillespie, R.J.; Westwood, R. 2-(Oxadiazolyl)- and 2-(Thiazolyl)imidazo[1,2-a]pyrimidines as Agonists and Inverse Agonists at Benzodiazepine Receptors. *J. Med. Chem.* **1991**, *34*, 2060–2067. [CrossRef] [PubMed]
22. Short, F.W.; Long, L.M. Synthesis of 5-aryl-2-oxazolepropionic acids and analogs. Antiinflammatory agents. *J. Heterocycl. Chem.* **1969**, *6*, 707–712. [CrossRef]
23. Theocharis, A.B.; Alexandrou, N.E. Synthesis and spectral data of 4,5-bis[5-aryl-1,3,4-oxadiazol-2-yl]-1-benzyl-1,2,3-triazoles. *J. Heterocycl. Chem.* **1990**, *27*, 1685–1688. [CrossRef]
24. Saeed, A. An expeditious, solvent-free synthesis of some 5-aryl-2-(2-hydroxyphenyl)-1,3,4-oxadiazoles. *Chem. Heterocycl. Compd.* **2007**, *43*, 1072–1075. [CrossRef]
25. Liras, S.; Allen, M.P.; Segelstein, B.E. A mild method for the preparation of 1,3,4-oxadiazoles: Traffic anhydride promoted cyclization of diacylhydrazines. *Synth. Commun.* **2000**, *30*, 437–443. [CrossRef]
26. Carlsen, P.H.J.; Jorgensen, K.B. Synthesis of unsymmetrically substituted 4H-1,2,4-triazoles. *J. Heterocycl. Chem.* **1994**, *31*, 805–807. [CrossRef]
27. Tandon, V.K.; Chhor, R.B. An efficient one pot synthesis of 1,3,4-oxadiazoles. *Synth. Commun.* **2001**, *31*, 1727–1732. [CrossRef]
28. Brain, C.T.; Paul, J.M.; Loong, Y.; Oakley, P.J. Novel procedure for the synthesis of 1,3,4-oxadiazoles from 1,2-diacylhydrazines using polymer-supported Burgess reagent under microwave conditions. *Tetrahedron Lett.* **1999**, *40*, 3275–3278. [CrossRef]
29. Dabiri, M.; Salehi, P.; Baghbanzadeh, M.; Bahramnejad, M. A facile procedure for the one-pot synthesis of unsymmetrical 2,5-disubstituted 1,3,4-oxadiazoles. *Tetrahedron Lett.* **2006**, *47*, 6983–6986. [CrossRef]
30. Majji, G.; Rout, S.K.; Guin, S.; Gogoi, A.; Patel, B.K. Iodine-catalysed oxidative cyclisation of acylhydrazones to 2,5-substituted 1,3,4-oxadiazoles. *RSC Adv.* **2014**, *4*, 5357–5362. [CrossRef]
31. Rostamizadeh, S.; Housaini, A.G. Microwave assisted synthesis of 2,5-disubstituted 1,3,4-oxadiazoles. *Tetrahedron Lett.* **2004**, *45*, 8753–8756. [CrossRef]
32. Milcent, R.; Barbier, G. Oxidation of hydrazones with lead dioxide: New synthesis of 1,3,4-oxadiazoles and 4-amino-1,2,4-triazol-5-one derivatives. *Chem. Informationsd.* **1983**, *14*, 80–81. [CrossRef]
33. Jedlovska, E.; Lesko, J. A simple one-pot procedure for the synthesis of 1,3,4-oxadiazoles. *Synth. Commun.* **1994**, *24*, 1879–1885. [CrossRef]
34. Jasiak, K.; Kudelko, A. Oxidative cyclization of N-aroylhydrazones to 2-(2-arylthenyl)-1,3,4-oxadiazoles using DDQ as an efficient oxidant. *Tetrahedron Lett.* **2015**, *56*, 5878–5881. [CrossRef]
35. Shang, Z.; Reiner, J.; Chang, J.; Zhao, K. Oxidative cyclization of aldazines with bis(trifuloroacetoxy)iodobenzene. *Tetrahedron Lett.* **2005**, *46*, 2701–2704. [CrossRef]
36. Oviedo, C.; Rodriguez, J. EDTA: The chelating agent under environmental scrutiny. *Qumica Nova* **2003**, *26*, 901–905. [CrossRef]
37. Shi, Y.; Campbell, J.A. Study of cyclization of chelating compounds using electrospray ionization mass spectrometry. *J. Radioanal. Nucl. Chem.* **2000**, *245*, 293–300. [CrossRef]
38. Yunta, F.; Garcia-Marco, S.; Lucena, J.J.; Gomez-Gallego, M.; Alcazar, R.; Sierra, M.A. Chelating agents related to ethylenediamine bis(2-hydroxyphenyl)acetic acid (EDDHA): Synthesis, characterization, and equilibrium studies of the free ligands and their Mg^{2+}, Ca^{2+}, Cu^{2+}, Fe^{3+} chelates. *Inorg. Chem.* **2003**, *42*, 5412–5421. [CrossRef]
39. Surgutskaia, N.S.; Di Martino, A.; Zednik, J.; Ozaltin, K.; Lovecka, L.; Domincova-Bergerova, E.; Kimmer, D.; Svoboda, J.; Sedlarik, V. Efficient Cu^{2+}, Pb^{2+} and Ni^{2+} ion removal from wastewater using electrospun DTPA-moified chitosan/polyethylene oxide nanofibers. *Sep. Purif. Technol.* **2020**, *247*, 116914. [CrossRef]
40. Knecht, S.; Ricklin, D.; Eberle, A.N.; Ernst, B. Oligohis-tags: Mechanisms of binding to Ni^{2+}-NTA surfaces. *J. Mol. Recognit.* **2009**, *22*, 270–279. [CrossRef]
41. Shaddox, T.W.; Unruh, J.B.; Kruse, J.K.; Restuccia, N.G. Solubility of Iron, Manganese, and Magnesium Sulfates and Glucoheptonates in Two Alkaline Soils. *Soil Sci. Soc. Am. J.* **2016**, *80*, 765–770. [CrossRef]
42. Pozdnyakov, I.P.; Tyutereva, Y.E.; Mikheilis, A.V.; Grivin, V.P.; Plyusnin, V.F. Primary photoprocesses for Fe(III) complexes with citric and hlycolic acids in aqueous solutions. *J. Photochem. Photobiol. A Chem.* **2022**, *434*, 114274. [CrossRef]
43. Sekhon, B.S. Chelates for Micronutrient Nutrition among Crops. *Resonance* **2003**, *8*, 46–53. [CrossRef]
44. Mrozek-Niecko, A.; Pernak, J. Biodegradacja soli tetrasodowej kwasu N-(1,2-dikarboksyetyleno)-D,L-asparaginowego i jego chelatów. *Przem. Chem.* **2006**, *85*, 635–637.
45. Santos, M.A.; Marques, S.M.; Tuccinardi, T.; Carelli, P.; Panelli, L.; Rossello, A. Design, synthesis and molecular modelling study of iminodiacetyl monohydroxamic acid derivatives as MMP inhibitors. *Bioorganic Med. Chem.* **2006**, *14*, 7539–7550. [CrossRef]
46. Elsinger, F.; Schreiber, J.; Eschenmoser, A. Notiz uber die Selektivitat der Spaltung von Carbonsauremethylestern mit Lithiumjodid. *Chim. Acta* **1960**, *43*, 113–118. [CrossRef]

47. Tanemura, K.; Rohand, T. Activated charcoal as an effective additive for alkaline and acidic hydrolysis of esters in water. *Tetrahedron Lett.* **2020**, *61*, 152467. [CrossRef]
48. Cież, D.; Svetlik, J. A One-Pot Preparation of 5-Oxo 2,4-Disubstituted 2,5-Dihydro-1H-imidazol-2-carboxylates from α-Bromo Esters. *Synlett* **2011**, *3*, 315–318. [CrossRef]
49. Gallagher, J.A.; Levine, L.A.; Williams, M.E. Anion Effects in Cu-Crosslinked Palindromic Artificial Tripeptides with Pendant Bpy Ligands. *Eur. J. Inorg. Chem.* **2011**, *27*, 4168–4174. [CrossRef]

Article

Design, Synthesis and Fungicidal Activity of *N*-(thiophen-2-yl) Nicotinamide Derivatives

Hongfei Wu [1,2], Xingxing Lu [1], Jingbo Xu [2], Xiaoming Zhang [1], Zhinian Li [2], Xinling Yang [1,*] and Yun Ling [1,*]

[1] Innovation Center of Pesticide Research, Department of Applied Chemistry, College of Science, China Agricultural University, Beijing 100193, China
[2] State Key Laboratory of the Discovery and Development of Novel Pesticide, Shenyang Sinochem Agrochemicals R&D Company Ltd., Shenyang 110021, China
* Correspondence: yangxl@cau.edu.cn (X.Y.); lyun@cau.edu.cn (Y.L.); Tel.: +86-10-6273-2223 (X.Y.); +86-010-6273-1446 (Y.L.)

Abstract: Based on the modification of natural products and the active substructure splicing method, a series of new *N*-(thiophen-2-yl) nicotinamide derivatives were designed and synthesized by splicing the nitrogen-containing heterocycle natural molecule nicotinic acid and the sulfur-containing heterocycle thiophene. The structures of the target compounds were identified through ^{1}H NMR, ^{13}C NMR and HRMS spectra. The in vivo bioassay results of all the compounds against cucumber downy mildew (CDM, *Pseudoperonospora cubensis* (Berk.et Curt.) Rostov.) in a greenhouse indicated that compounds **4a** (EC$_{50}$ = 4.69 mg/L) and **4f** (EC$_{50}$ = 1.96 mg/L) exhibited excellent fungicidal activities which were higher than both diflumetorim (EC$_{50}$ = 21.44 mg/L) and flumorph (EC$_{50}$ = 7.55 mg/L). The bioassay results of the field trial against CDM demonstrated that the 10% EC formulation of compound **4f** displayed excellent efficacies (70% and 79% control efficacies, respectively, each at 100 mg/L and 200 mg/L) which were superior to those of the two commercial fungicides flumorph (56% control efficacy at 200 mg/L) and mancozeb (76% control efficacy at 1000 mg/L). *N*-(thiophen-2-yl) nicotinamide derivatives are significant lead compounds that can be used for further structural optimization, and compound **4f** is also a promising fungicide candidate against CDM that can be used for further development.

Keywords: heterocycle; natural product; nicotinic acid; thiophene; fungicidal activity

Citation: Wu, H.; Lu, X.; Xu, J.; Zhang, X.; Li, Z.; Yang, X.; Ling, Y. Design, Synthesis and Fungicidal Activity of *N*-(thiophen-2-yl) Nicotinamide Derivatives. *Molecules* **2022**, *27*, 8700. https://doi.org/10.3390/molecules27248700

Academic Editor: Joseph Sloop

Received: 19 November 2022
Accepted: 6 December 2022
Published: 8 December 2022

Publisher's Note: MDPI stays neutral with regard to jurisdictional claims in published maps and institutional affiliations.

Copyright: © 2022 by the authors. Licensee MDPI, Basel, Switzerland. This article is an open access article distributed under the terms and conditions of the Creative Commons Attribution (CC BY) license (https:// creativecommons.org/licenses/by/ 4.0/).

1. Introduction

The four main classes of fungal phytopathogens, including oomycetes, ascomycetes, basidiomycetes and deuteromycetes, severely threaten human health, food safety, and agriculture [1,2]. Fungicides are the main approaches to control plant diseases and play a critical role in modern agriculture by increasing both crop quality and yield. Nevertheless, with the widespread application of, especially overused, fungicides, the development of resistance is inevitable [3,4]. Therefore, there is an urgent demand to develop efficient, safe, and eco-friendly fungicides with innovative structures.

Natural products bring bioinspiration to laboratories for the discovery of new weeds, plant pathogens and insect pest control agents, and they play an important role in the advancement of crop protection research [5–9]. Heterocycles are present in a large proportion of natural molecules and often contribute significantly to their structural and physical properties as well as to their biological activity [10–13]. Approximately 70% of all the agrochemicals that have been launched within the last 20 years bear at least one heterocyclic ring [14]. The nitrogen-containing heterocycle in the natural molecule nicotinic acid, vitamin B3, is the first lipid-lowering drug used for dyslipidemia treatment, and it has been applied for more than five decades [15]. Nicotinic acid and its derivatives play crucial roles as multifunctional pharmacophores in governing many biological activities related to physiological functions and pharmacological activities [16]. In agriculture, nicotinic acid derivatives display extensive applications as well. Agrochemicals, such as diflufenican as a herbicide, flonicamid as an insecticide and boscalid as a fungicide, have already been widely used for crop protection [17–19]. Currently, a few new nicotinic acid derivatives have been discovered and are under development as agrochemical candidates, for example, aminopyrifen as a fungicide candidate, cyclobutrifluram as a fungicide–nematicide candidate and nicofluprole as an insecticide candidate (Figure 1) [20–22]. Multidisciplinary interest has been focused on the study of nicotinic acid and its derivatives.

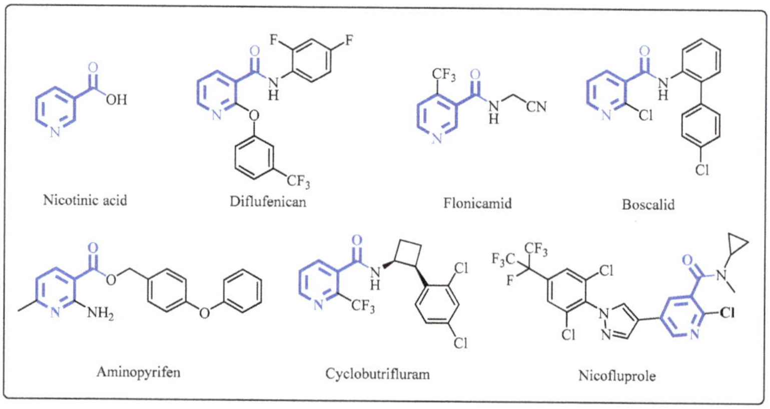

Figure 1. The structures of nicotinic acid and its derivatives.

Meanwhile, thiophene, a widely researched five-membered sulfur heterocycle, exists in commercialized agricultural fungicides, including silthiofam, ethaboxam, penthiopyrad and isofetamid (Figure 2) [23–26]. The extensive literature on thiophenes is indicative of the research on and the commercial interest in the heterocycle. Some hundreds of patents appear each year, many applying thiophene compounds as alternatives to benzenoid products, and, in many cases, the thiophene-containing molecule shows a higher activity than the benzene-containing one [27]. As an attractive small heterocycle molecule, thiophene has been widely studied for the development of novel fungicides because of its wide and satisfactory antifungal activity [28–32].

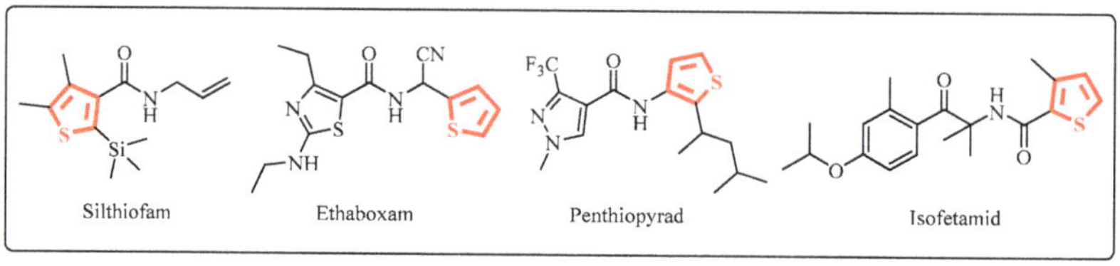

Figure 2. The structures of thiophene-containing agricultural fungicides.

Obviously, based on the modification of natural products and the active substructure splicing method, combinations of the two active substructures of nicotinic acid and thiophene are significant for the discovery of agricultural fungicides with novel molecular structures. A series of N-(thiophen-2-yl) nicotinamide derivatives were designed to generate novel compounds with excellent fungicidal activity (Scheme 1).

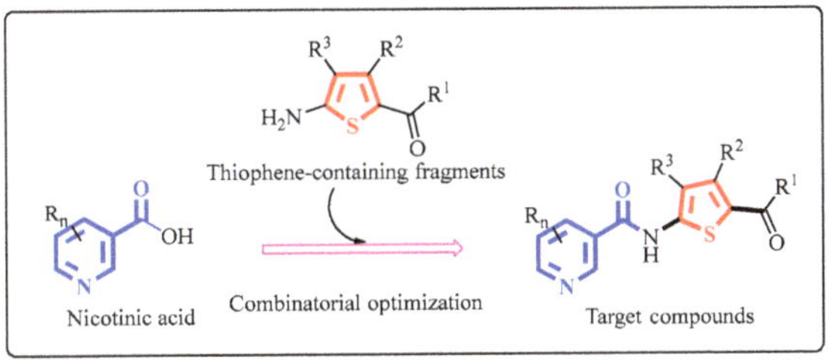

Scheme 1. The design strategy of target compounds.

2. Results and Discussion

2.1. Chemistry

The synthetic pathways used to prepare target compounds **4a~4s** are shown in Scheme 2. The substituted nicotinic acid **1** was acyl chlorinated with oxalyl chloride to obtain acyl chloride **2**. The substituted thiophen-2-amine **3** was converted into the desired *N*-(thiophen-2-yl) nicotinamide derivatives **4a~4s** through acylation with the obtained acyl chloride **2** under basic conditions. Their structures were confirmed with ^{1}H NMR, ^{13}C NMR and HRMS spectra. Meanwhile, compound **4f** was confirmed with IR spectra (supplementary materials).

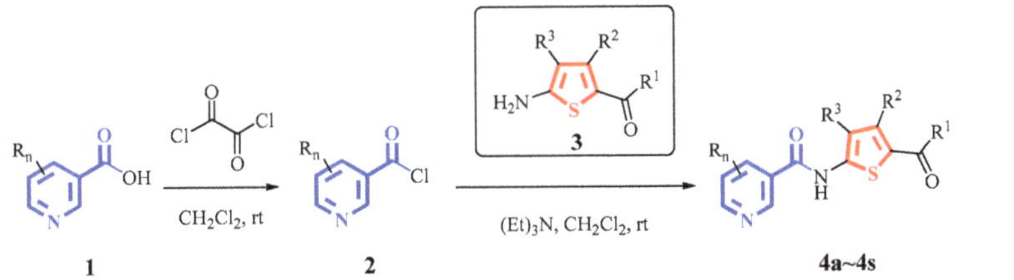

4a: R_n = 2-CH$_3$-5-CN-6-Cl, R^1 = OC$_2$H$_5$, R^2 = CH$_3$, R^3 = CN; **4b**: R_n = 2-CH$_3$-5-Cl-6-Br, R^1 = OC$_2$H$_5$, R^2 = CH$_3$, R^3 = CN;
4c: R_n = 5,6-Br$_2$, R^1 = OC$_2$H$_5$, R^2 = CH$_3$, R^3 = CN; **4d**: R_n = 5-F-6-Br, R^1 = OC$_2$H$_5$, R^2 = CH$_3$, R^3 = CN;
4e: R_n = 5-F-6-Cl, R^1 = OC$_2$H$_5$, R^2 = CH$_3$, R^3 = CN; **4f**: R_n = 5,6-Cl$_2$, R^1 = OC$_2$H$_5$, R^2 = CH$_3$, R^3 = CN;
4g: R_n = 5-Br-6-Cl, R^1 = OCH$_3$, R^2 = CH$_3$, R^3 = CN; **4h**: R_n = 6-Br, R^1 = OCH$_3$, R^2 = CH$_3$, R^3 = CN;
4i: R_n = 5,6-Cl$_2$, R^1 = OCH$_3$, R^2 = CH$_3$, R^3 = CN; **4j**: R_n = 5,6-Cl$_2$, R^1 = OC$_3$H$_7$-i, R^2 = CH$_3$, R^3 = CN;
4k: R_n = 5,6-Cl$_2$, R^1 = OC$_3$H$_7$-n, R^2 = CH$_3$, R^3 = CN; **4l**: R_n = 5,6-Cl$_2$, R^1 = OC$_4$H$_9$-n, R^2 = CH$_3$, R^3 = CN;
4m: R_n = 5,6-Cl$_2$, R^1 = O(CH$_2$)$_2$OCH$_3$, R^2 = CH$_3$, R^3 = CN; **4n**: R_n = 5,6-Cl$_2$, R^1 = OCH$_2$C$_5$H$_6$, R^2 = CH$_3$, R^3 = CN;
4o: R_n = 5,6-Cl$_2$, R^1 = NHCH$_3$, R^2 = CH$_3$, R^3 = CN; **4p**: R_n = 5,6-Cl$_2$, R^1 = cyclopropylamino, R^2 = CH$_3$, R^3 = CN;
4q: R_n = 5,6-Cl$_2$, R^1 = NHPh, R^2 = CH$_3$, R^3 = CN; **4r**: R_n = 5,6-Cl$_2$, R^1 = OC$_2$H$_5$, R^2 = C$_2$H$_5$, R^3 = CN;
4s: R_n = 5,6-Cl$_2$, R^1 = OC$_2$H$_5$, R^2 = CH$_3$, R^3 = H.

Scheme 2. Synthetic routes of target compounds.

Compound **4f** (R_n = 5, 6-Cl$_2$; R^1 = OC$_2$H$_5$; R^2 = CH$_3$; R^3 = CN) exhibited excellent fungicidal activities and was taken as an example to analyze the ^{1}H NMR spectra data (Figure 3). The chemical shift as triplet was observed at δ 1.31 ppm with J = 7.2 Hz due to the protons of the CH$_3$ of OC$_2$H$_5$. A singlet at δ 2.54 ppm was observed due to the protons of the CH$_3$ on the thiophene ring. The chemical shift as quartet was observed at δ 4.29 ppm with J = 7.2 Hz due to the protons of the CH$_2$ of OC$_2$H$_5$. The chemical shifts as doublet were observed at δ 8.64 ppm with J = 1.8 Hz and at δ 8.86 ppm with J = 1.8 Hz due to the protons at the fourth position and second position of pyridine, respectively. The spectrum showed a broad singlet at δ 12.67 ppm due to the proton of the CONH.

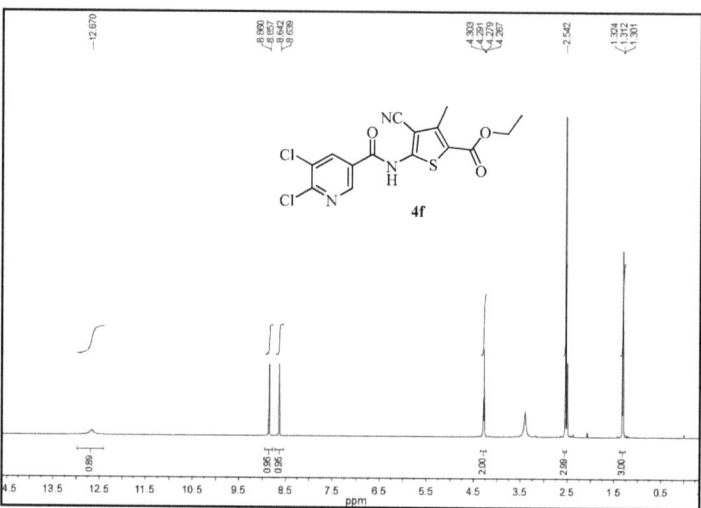

Figure 3. The ^{1}H NMR spectrum of compound **4f**.

2.2. Fungicidal Activities In Vivo in a Greenhouse

The in vivo bioassay results shown in Table 1 indicated that, at 400 mg/L, a few compounds displayed fungicidal activities against wheat powdery mildew (WPM, *Blumeria graminis* (DC.) Speer) and southern corn rust (SCR, *Puccinia sorghi*); a few compounds had good fungicidal activities against cucumber anthracnose (CA, *Colletotrichum orbiculare*); and most compounds exhibited excellent fungicidal activities against cucumber downy mildew (CDM, *Pseudoperonospora cubensis* (Berk.et Curt.) Rostov.), which is one of the most destructive oomycete diseases. Further bioassay results in a greenhouse (Table 2) indicated that more than half of these compounds showed moderate to significant fungicidal activity against CDM, and there were six compounds (**4a**, **4b**, **4c**, **4f**, **4i** and **4r**) that displayed higher activities than diflumetorim (EC_{50} = 21.44 mg/L). Compounds **4a** (EC_{50} = 4.69 mg/L) and **4f** (EC_{50} = 1.96 mg/L) exhibited especially excellent activities, which were higher than flumorph (EC_{50} = 7.55 mg/L). The structure–activity relationships (SAR) were unfolded as follows.

Initially, the structural modification was mostly around the pyridine ring due to the use of the substituted nicotinic acids that were available at hand in compounds **4a~4h**. The testing results shown in Table 2 illustrated that compound **4f**, with a chloro at both the fifth position and sixth position of the pyridine ring, significantly indicated the highest fungicidal activity against CDM with an EC_{50} value of 1.96 mg/L. The other 5,6-dihalosubstitued pyridine compounds, **4c** (EC_{50} = 19.89 mg/L), **4d** (EC_{50} = 32.44 mg/L), **4e** (EC_{50} = 25.61 mg/L) and **4g** (EC_{50} = 34.29 mg/L), also showed good fungicidal activities, but they were much lower than those of **4f**. Additionally, **4a** (R_n = 2-CH_3-5-CN-6-Cl), with a trisubstituted pyridine, had excellent fungicidal activities next to those of **4f**. It was beneficial to the increase of the fungicidal activity to have a chloro at the sixth position because the two best compounds, **4a** and **4f**, had a chloro at the sixth position.

Table 1. The in vivo fungicidal activities of target compounds 4a~4s in green house at 400 mg/L.

Compd.	R_n	R^1	R^2	R^3	WPM	SCR	CA	CDM
4a	2-CH$_3$-5-CN-6-Cl	OC$_2$H$_5$	CH$_3$	CN	0	0	0	100
4b	2-CH$_3$-5-Cl-6-Br	OC$_2$H$_5$	CH$_3$	CN	0	50	0	100
4c	5,6-Br$_2$	OC$_2$H$_5$	CH$_3$	CN	0	0	100	100
4d	5-F-6-Br	OC$_2$H$_5$	CH$_3$	CN	0	0	0	100
4e	5-F-6-Cl	OC$_2$H$_5$	CH$_3$	CN	0	0	0	100
4f	5,6-Cl$_2$	OC$_2$H$_5$	CH$_3$	CN	0	0	0	100
4g	5-Br-6-Cl	OCH$_3$	CH$_3$	CN	0	0	100	100
4h	6-Br	OCH$_3$	CH$_3$	CN	0	0	100	100
4i	5,6-Cl$_2$	OCH$_3$	CH$_3$	CN	0	0	0	98
4j	5,6-Cl$_2$	OC$_3$H$_7$-i	CH$_3$	CN	0	0	40	100
4k	5,6-Cl$_2$	OC$_3$H$_7$-n	CH$_3$	CN	0	0	30	100
4l	5,6-Cl$_2$	OC$_4$H$_9$-n	CH$_3$	CN	0	0	0	100
4m	5,6-Cl$_2$	OC$_2$H$_4$OCH$_3$	CH$_3$	CN	0	0	0	50
4n	5,6-Cl$_2$	OCH$_2$C$_6$H$_5$	CH$_3$	CN	0	0	0	85
4o	5,6-Cl$_2$	NHCH$_3$	CH$_3$	CN	0	0	30	0
4p	5,6-Cl$_2$	Cyclopropylamino	CH$_3$	CN	0	0	0	0
4q	5,6-Cl$_2$	NHPh	CH$_3$	CN	50	100	85	98
4r	5,6-Cl$_2$	OC$_2$H$_5$	C$_2$H$_5$	CN	0	0	0	100
4s	5,6-Cl$_2$	OC$_2$H$_5$	CH$_3$	H	0	0	0	100
Azoxystrobin					100	100	100	/[a]
Diflumetorim					/[a]	/[a]	/[a]	100
Flumorph					/[a]	/[a]	/[a]	100

[a]: Not tested.

Table 2. EC$_{50}$ values of target compounds 4a~4s against cucumber downy mildew in green house.

Compd.	Y = ax + b	EC$_{50}$ (mg/L)	95% CI [a]	r
4a	y = 4.14x + 1.28	4.69	3.27–6.71	0.98
4b	y = 1.71x + 2.92	16.52	13.27–20.58	0.94
4c	y = 1.87x + 2.57	19.89	16.10–24.56	0.95
4d	y = 1.62x + 2.55	32.44	26.12–40.29	0.95
4e	y = 1.58x + 2.78	25.61	20.83–31.48	0.97
4f	y = 1.15x + 4.66	1.96	1.17–3.29	0.95
4g	y = 1.62x + 2.52	34.29	27.46–42.82	0.94
4h	y = 1.62x + 2.72	25.54	20.80–31.35	0.97
4i	y = 1.48x + 3.64	8.31	6.37–10.84	0.97
4j	y = 1.58x + 2.88	21.94	17.58–27.36	0.93
4k	y = 1.58x + 2.66	30.41	24.51–37.73	0.95
4l	/[b]	100–400	/[b]	/[b]
4m	/[b]	>400	/[b]	/[b]
4n	/[b]	100–400	/[b]	/[b]
4o	/[b]	> 400	/[b]	/[b]
4p	/[b]	>400	/[b]	/[b]
4q	/[b]	100–400	/[b]	/[b]
4r	y = 1.48x + 3.70	7.53	5.71–9.94	0.98
4s	y = 1.26x + 2.72	63.81	47.23–86.22	0.92
Diflumetorim	y = 1.43x + 3.10	21.44	17.21–26.70	0.96
Flumorph	y = 1.25x + 3.91	7.55	5.57–10.22	0.98

[a]: Confidence interval. [b]: The value could not be measured accurately.

Next, the 5,6-dichloro on the pyridine ring of compound 4f was maintained, and the substitutions on the thiophene ring were changed to generate compounds 4i~4s. Table 2 also showed that the EC$_{50}$ values of the compounds 4i (R^1 = OCH$_3$), 4f (R^1 = OC$_2$H$_5$), 4j (R^1 = OC$_3$H$_7$-i), 4k (R^1 = OC$_3$H$_7$-n) and 4l (R^1 = OC$_4$H$_9$-n) were 8.31 mg/L, 1.96 mg/L, 21.94 mg/L, 30.41 mg/L and 100–400 mg/L, respectively, which meant that the fungicidal activity first increased and then decreased dramatically with the increase of the carbon chain length of the alkyloxy in the R^1 moiety and that 4f (R^1 = OC$_2$H$_5$) had the highest fungicidal activity. The fungicidal activity got much worse when R^1 had an alkylamine instead of an alkyloxy; for example, the EC_{50} value of compound **4o** (R^1 = NHCH$_3$) was over 400 mg/L, which was much larger than that of compound **4i** (R^1 = OCH$_3$). Furthermore, when the methyl (R^2 = CH$_3$) of compound **4f** was replaced with an ethyl to generate compound **4r** (EC_{50} = 7.53 mg/L), the fungicidal activity reduced. When compound **4f** (R^3 = CN) was decyanated to obtain compound **4s** (R^3 = H), the fungicidal activity decreased significantly. Compound **4f** (R_n = 5,6-Cl$_2$; R^1 = OC$_2$H$_5$; R^2 = CH$_3$; R^3 = CN) was still the most active one in all nineteen compounds.

2.3. Field Trials against CDM

A field trial of compound **4f** was conducted to see its fungicidal activity in the field environment (Table 3), which was a critical assessment indicator for the demonstration of whether the compound was worth further optimization or even commercial development as a fungicide candidate. The results of the field trial in 2021 against CDM demonstrated that compound **4f** displayed better control efficacies than the two commercial fungicides flumorph and mancozeb. Compound **4f** exhibited 70% and 79% control efficacies, respectively, each at a concentration of 100 mg/L and 200 mg/L, whereas flumorph at 200 mg/L and mancozeb at 1000 mg/L showed 56% and 76% control efficacies, respectively. However, the control efficacy of compound **4f** was inferior to that of the commercial fungicide cyazofamid, with a 91% control efficacy at 100 mg/L. Compared with flumorph, compound **4f**, with a much lower EC_{50} value against CDM in the greenhouse tests, was more active against CDM in the field trial.

Table 3. The results of compound **4f** against CDM in field (2021, Shenyang, open field).

Compd.	Concentration (mg/L)	Control Efficacy (%)			
		r1	r2	r3	Mean
10% 4f EC	50	55	53	50	53
	100	74	66	69	70
	200	82	75	79	79
10% Flumorph EC	200	64	52	51	56
80% Mancozeb WP	1000	79	76	72	76
10% Cyazofamid SC	100	91	91	90	91
CK	(disease index)	69	61	60	63

3. Materials and Methods

3.1. Chemicals and Target Compounds

All starting materials and reagents were commercially available and used without further purification except as indicated. (The starting materials **1** and **3** were from Taizhou Jiakang Chemical Co., Ltd., a custom chemicals supplier in Zhejiang in China, and the reagents were from Sinopharm Chemical Reagent Co., Ltd., in Shanghai in China.) ^{1}H nuclear magnetic resonance (NMR) spectra were obtained at 300 MHz using a Varian Mercury 300 spectrometer (Varian, Palo Alto, CA, USA), at 600 MHz using a Varian Unity Plus 600 spectrometer (Varian, Palo Alto, CA, USA) or at 600 MHz using a JEOL JNM-ECZ600R spectrometer (JEOL RESONANCE Inc., Akishima, Tokyo, Japan) with DMSO-d_6 as the solvent and tetramethylsilane (TMS) as the internal standard. ^{13}C nuclear magnetic resonance (NMR) spectra were obtained at 150 MHz using a Varian Unity Plus 600 spectrometer (Varian, Palo Alto, CA, USA) or at 150 MHz using a JEOL JNM-ECZ600R spectrometer (JEOL RESONANCE Inc., Akishima, Tokyo, Japan) with DMSO-d_6 as the solvent and tetramethylsilane (TMS) as the internal standard. Chemical-shift values (δ) were given in parts per million (ppm). Mass spectra were acquired with a Thermo Scientific Q Exactive Focus mass spectrometer system (Thermo Fisher Scientific Inc., Waltham, MA, USA). Melting points were determined on an X-4 precision microscope melting point tester and were uncorrected. Petroleum ether used for column chromatography had a boiling range of 60~90 °C. Chemical names were generated using ChemDraw (Cambridge

Soft, version 15.0). Yields were not optimized. An overview synthesis of N-(thiophen-2-yl)nicotinamide derivatives **4a~4s** is shown in Scheme 2.

3.1.1. General Synthetic Procedures

General procedure for the preparation of compounds **4a~4s**: To a solution of acid **1** (2.3 mmol) in CH_2Cl_2 (20 mL), oxalyl chloride (6.9 mmol) was added dropwise, and then a drop of DMF was added. The mixture was stirred at room temperature for 6 h and concentrated under reduced pressure to obtain acyl chloride **2**. To a mixture of amine **3** (2.0 mmol) and triethylamine (2.4 mmol) in CH_2Cl_2 (20 mL), the solution of acyl chloride **2** in CH_2Cl_2 (10 mL) was added dropwise in an ice-water bath. The resulting mixture was stirred at room temperature until TLC indicated that the reaction was complete. To the mixture, water (10 mL) was added, and then the organic phase was washed with brine, dried over anhydrous $MgSO_4$ and concentrated under reduced pressure to give a crude product. The crude product was purified through chromatography on a column of silica gel with petroleum ether and ethyl acetate to obtain the product amide **4**.

3.1.2. Chemical Property of the Compounds

Ethyl 5-(6-chloro-5-cyano-2-methylnicotinamido)-4-cyano-3-methylthiophene-2-carboxylate (**4a**): White solid (0.58 g, yield 74%). Decomposition temperature > 175 °C. ^{1}H NMR (600 MHz, DMSO-d_6): δ 12.78 (bs, 1H, CONH), 8.77 (s, 1H, pyridine-4-H), 4.29 (q, J = 7.2 Hz, 2H, OCH_2CH$_3$), 2.63 (s, 3H, CH$_3$), 2.56 (s, 3H, CH$_3$) and 1.31 (t, J = 7.2 Hz, 3H, OCH$_2$CH_3). ^{13}C NMR (150 MHz, DMSO-d_6): δ 164.7, 162.3, 161.8, 151.8, 144.2, 144.1, 128.9, 118.1, 118.0, 115.1, 113.8, 106.8, 98.6, 61.4, 23.4, 14.7 and 14.5. HR-MS (m/z): calcd. for $C_{17}H_{12}ClN_4O_3S$ [M-H]$^-$ 387.0324; found, 387.0327.

Ethyl 5-(6-bromo-5-chloro-2-methylnicotinamido)-4-cyano-3-methylthiophene-2-carboxylate (**4b**): White solid (0.66 g, yield 74%). Decomposition temperature > 196 °C. ^{1}H NMR (600 MHz, DMSO-d_6): δ 12.71 (bs, 1H, CONH), 8.44 (s, 1H, pyridine-4-H), 4.24 (q, J = 7.2 Hz, 2H, OCH_2CH$_3$), 2.49 (s, 3H, CH$_3$), 2.45 (s, 3H, CH$_3$) and 1.27 (t, J = 7.2 Hz, 3H, OCH$_2$CH_3). ^{13}C NMR (150 MHz, DMSO-d_6): δ 165.0, 162.0, 156.6, 151.4, 150.3, 144.3, 143.2, 130.0, 118.2, 116.2, 113.9, 98.6, 61.5, 22.4, 14.8 and 14.7. HR-MS (m/z): calcd. for $C_{16}H_{13}BrClN_3NaO_3S$ [M + Na]$^+$ 465.9422; found, 465.9421.

Ethyl 4-cyano-5-(5,6-dibromonicotinamido)-3-methylthiophene-2-carboxylate (**4c**): White solid (0.55g, yield 57%). Decomposition temperature > 231 °C. ^{1}H NMR (300 MHz, DMSO-d_6): δ 12.56 (bs, 1H, CONH), 8.86 (d, J = 2.1 Hz, 1H, pyridine-2-H), 8.69 (d, J = 2.1 Hz, 1H, pyridine-4-H), 4.29 (q, J = 6.9 Hz, 2H, OCH_2CH$_3$), 2.54 (s, 3H, CH$_3$) and 1.34 (t, J = 6.9 Hz, 3H, OCH$_2$CH_3). ^{13}C NMR (150 MHz, DMSO-d_6): δ 163.3, 162.0, 153.2, 151.4, 149.0, 144.3, 143.1, 129.5, 119.1, 118.5, 114.0, 99.0, 61.5, 14.9 and 14.7. HR-MS (m/z): calcd. for $C_{15}H_{10}Br_2N_3O_3S$ [M-H]$^-$ 471.8800; found, 471.8795.

Ethyl 5-(6-bromo-5-fluoronicotinamido)-4-cyano-3-methylthiophene-2-carboxylate (**4d**): White solid (0.53 g, yield 64%). Decomposition temperature > 220 °C. ^{1}H NMR (600 MHz, DMSO-d_6): δ 12.66 (bs, 1H, CONH), 8.75 (s, 1H, pyridine-2-H), 8.42 (d, J = 9.0 Hz, 1H, pyridine-4-H), 4.24 (q, J = 7.2 Hz, 2H, OCH_2CH$_3$), 2.50 (s, 3H, CH$_3$) and 1.27 (t, J = 7.2 Hz, 3H, OCH$_2$CH_3). ^{13}C NMR (150 MHz, DMSO-d_6): δ 163.4, 162.0, 153.9 (d, $^1J_{CF}$ = 258.6 Hz), 151.3, 145.9 (d, $^3J_{CF}$ = 6.3 Hz), 141.4 (d, $^2J_{CF}$ = 19.4 Hz), 144.4, 130.0, 120.2 (d, $^2J_{CF}$ = 20.9 Hz), 118.6, 113.9, 99.1, 61.5, 14.9 and 14.7. HR-MS (m/z): calcd. for $C_{15}H_{10}BrFN_3O_3S$ [M-H]$^-$ 409.9623; found, 409.9616.

Ethyl 5-(6-chloro-5-fluoronicotinamido)-4-cyano-3-methylthiophene-2-carboxylate (**4e**): White solid (0.52 g, yield 70%). Decomposition temperature > 275 °C. ^{1}H NMR (300 MHz, DMSO-d_6): δ 12.59 (bs, 1H, CONH), 8.80 (d, J = 2.1 Hz, 1H, pyridine-2-H), 8.43 (dd, J = 9.0, 2.1 Hz, 1H, pyridine-4-H), 4.30 (q, J = 7.2 Hz, 2H, OCH_2CH$_3$), 2.57 (s, 3H, CH$_3$) and 1.34 (t, J = 7.2 Hz, 3H, OCH$_2$CH_3). ^{13}C NMR (150 MHz, DMSO-d_6): δ 163.5, 162.0, 153.9 (d, $^1J_{CF}$ = 258.6 Hz), 151.5, 145.9 (d, $^3J_{CF}$ = 6.6 Hz), 144.4, 141.3 (d, $^2J_{CF}$ = 19.7 Hz), 130.1, 126.2 (d, $^2J_{CF}$ = 21.6 Hz), 118.5, 114.0, 99.1, 61.5, 14.9 and 14.7. HR-MS (m/z): calcd. for $C_{15}H_{10}ClFN_3O_3S$ [M-H]$^-$ 366.0120; found, 366.0120.

Ethyl 4-cyano-5-(5,6-dichloronicotinamido)-3-methylthiophene-2-carboxylate (**4f**): White solid (0.58 g, yield 75%). Decomposition temperature > 262 °C. ^{1}H NMR (600 MHz, DMSO-d_6): δ 12.67 (bs, 1H, CONH), 8.86 (d, J = 1.8 Hz, 1H, pyridine-2-H), 8.64 (d, J = 1.8 Hz, 1H, pyridine-4-H), 4.29 (q, J = 7.2 Hz, 2H, OCH_2CH$_3$), 2.54 (s, 3H, CH$_3$) and 1.31 (t, J = 7.2 Hz, 3H, OCH$_2$CH_3). ^{13}C NMR (150 MHz, DMSO-d_6): δ 163.2, 161.8, 151.3, 151.2, 148.3, 144.2, 139.7, 129.3, 129.2, 118.3, 113.8, 98.9, 61.3, 14.7 and 14.5. HR-MS (*m/z*): calcd. for $C_{15}H_{11}Cl_2N_3NaO_3S$ [M + Na]$^+$ 405.9791; found, 405.9793.

Methyl 5-(5-bromo-6-chloronicotinamido)-4-cyano-3-methylthiophene-2-carboxylate (**4g**): White solid (0.54 g, yield 64%). Decomposition temperature > 238 °C. ^{1}H NMR (600 MHz, DMSO-d_6): δ 12.62 (bs, 1H, CONH), 8.82 (d, J = 1.8 Hz, 1H, pyridine-2-H), 8.68 (d, J = 1.8 Hz, 1H, pyridine-4-H), 3.78 (s, 3H, OCH$_3$) and 2.49 (s, 3H, CH$_3$). ^{13}C NMR (150 MHz, DMSO-d_6): δ 163.3, 162.4, 153.2, 151.5, 149.0, 144.5, 143.1, 129.1, 119.5, 118.2, 113.9, 99.0, 52.7 and 14.9. HR-MS (*m/z*): calcd. for $C_{14}H_{10}BrClN_3O_3S$ [M + H]$^+$ 415.9284; found, 415.9289.

Methyl 5-(6-bromonicotinamido)-4-cyano-3-methylthiophene-2-carboxylate (**4h**): White solid (0.55 g, yield 72%). Decomposition temperature > 261 °C. ^{1}H NMR (300 MHz, DMSO-d_6): δ 12.60 (bs, 1H, CONH), 8.91 (s, 1H, pyridine-2-H), 8.34 (d, J = 8.1 Hz, 1H, pyridine-4-H), 7.67 (d, J = 8.1 Hz, 1H, pyridine-5-H), 3.83 (s, 3H, OCH$_3$) and 2.56 (s, 3H, CH$_3$). ^{13}C NMR (150 MHz, DMSO-d_6): δ 164.7, 162.5, 154.2, 150.7, 149.8, 144.6, 140.5, 128.0, 124.6, 117.9, 114.0, 99.0, 52.7 and 14.9. HR-MS (*m/z*): calcd. for $C_{14}H_{11}BrN_3O_3S$ [M + H]$^+$ 381.9676; found, 381.9679.

Methyl 4-cyano-5-(5,6-dichloronicotinamido)-3-methylthiophene-2-carboxylate (**4i**): White solid (0.56 g, yield 75%). Decomposition temperature > 263 °C. ^{1}H NMR (300 MHz, DMSO-d_6): δ 12.55 (bs, 1H, CONH), 8.85 (d, J = 2.1 Hz, 1H, pyridine-2-H), 8.59 (d, J = 2.1 Hz, 1H, pyridine-4-H), 3.84 (s, 3H, OCH$_3$) and 2.57 (s, 3H, CH$_3$). ^{13}C NMR (150 MHz, DMSO-d_6): δ 163.4, 162.4, 151.6, 151.4, 148.5, 144.6, 139.9, 129.4, 129.3, 118.1, 114.0, 99.0, 52.7 and 14.9. HR-MS (*m/z*): calcd. for $C_{14}H_8Cl_2N_3O_3S$ [M-H]$^-$ 367.9668; found, 367.9671.

Isopropyl 4-cyano-5-(5,6-dichloronicotinamido)-3-methylthiophene-2-carboxylate (**4j**): White solid (0.57 g, yield 71%). An m.p. of 239~240 °C. ^{1}H NMR (300 MHz, DMSO-d_6): δ 12.58 (bs, 1H, CONH), 8.86 (d, J = 2.1 Hz, 1H, pyridine-2-H), 8.61 (d, J = 2.1 Hz, 1H, pyridine-4-H), 5.11 (hept, J = 6.3 Hz, 1H, OCH(CH$_3$)$_2$), 2.56 (s, 3H, CH$_3$) and 1.34 (d, J = 6.3 Hz, 6H, OCH(CH_3)$_2$). ^{13}C NMR (150 MHz, DMSO-d_6): δ 163.4, 161.6, 151.3 (2 C), 148.5, 144.2, 139.9, 129.4 (2 C), 118.8, 114.1, 99.4, 69.2, 22.2 (2 C) and 14.9. HR-MS (*m/z*): calcd. for $C_{16}H_{12}Cl_2N_3O_3S$ [M-H]$^-$ 395.9981; found, 395.9986.

Propyl 4-cyano-5-(5,6-dichloronicotinamido)-3-methylthiophene-2-carboxylate (**4k**): White solid (0.56 g, yield 70%). An m.p. of 229~230 °C. ^{1}H NMR (300 MHz, DMSO-d_6): δ 12.61 (bs, 1H, CONH), 8.87 (d, J = 2.1 Hz, 1H, pyridine-2-H), 8.62 (d, J = 2.1 Hz, 1H, pyridine-4-H), 4.21 (t, J = 6.6 Hz, 1H, OCH_2CH$_2$CH$_3$), 2.57 (s, 3H, CH$_3$), 1.69–1.81 (m, 2H, OCH$_2$CH_2CH$_3$) and 1.01 (t, J = 6.9 Hz, 3H, OCH$_2$CH$_2$CH_3). ^{13}C NMR (150 MHz, DMSO-d_6): δ 163.5, 162.1, 151.3 (2 C), 148.5, 144.3, 139.9, 129.4 (2 C), 118.4, 114.1, 99.4, 66.8, 22.1, 14.9 and 10.9. HR-MS (*m/z*): calcd. for $C_{16}H_{12}Cl_2N_3O_3S$ [M-H]$^-$ 395.9981; found, 395.9986.

Butyl 4-cyano-5-(5,6-dichloronicotinamido)-3-methylthiophene-2-carboxylate (**4l**): White solid (0.55 g, yield 66%). An m.p. of 215~216 °C. ^{1}H NMR (600 MHz, DMSO-d_6): δ 12.66 (bs, 1H, CONH), 8.83 (d, J = 2.1 Hz, 1H, pyridine-2-H), 8.62 (d, J = 2.1 Hz, 1H, pyridine-4-H), 4.21 (t, J = 6.6 Hz, 1H, OCH_2CH$_2$CH$_2$CH$_3$), 2.51 (s, 3H, CH$_3$), 1.62–1.64 (m, 2H, OCH$_2$CH_2CH$_2$CH$_3$), 1.36–1.39 (m, 2H, OCH$_2$CH$_2$CH_2CH$_3$) and 0.89 (t, J = 7.2 Hz, 3H, OCH$_2$CH$_2$CH$_2$CH_3). ^{13}C NMR (150 MHz, DMSO-d_6): δ 163.5, 162.1, 151.3 (2 C), 148.5, 144.4, 139.9, 129.4 (2 C), 120.6, 114.1, 99.1, 65.1, 30.7, 19.3, 14.9 and 14.1. HR-MS (*m/z*): calcd. for $C_{17}H_{14}Cl_2N_3O_3S$ [M-H]$^-$ 410.0138; found, 410.0145.

2-methoxyethyl 4-cyano-5-(5,6-dichloronicotinamido)-3-methylthiophene-2-carboxylate (**4m**): White solid (0.58 g, yield 69%). An m.p. of 213~214 °C. ^{1}H NMR (300 MHz, DMSO-d_6): δ 12.62 (bs, 1H, CONH), 8.86 (d, J = 2.1 Hz, 1H, pyridine-2-H), 8.62 (d, J = 2.1 Hz, 1H, pyridine-4-H), 4.34–4.38 (m, 2H, OCH_2CH$_2$OCH$_3$), 3.62–3.65 (m, 2H, OCH$_2$CH_2OCH$_3$), 3.33 (s, 3H, OCH$_2$CH$_2$OCH_3) and 2.57 (s, 3H, CH$_3$). ^{13}C NMR (150 MHz,

DMSO-d_6): δ 163.4, 161.9, 151.7, 151.4, 148.5, 144.7, 139.9, 129.5, 129.3, 118.2, 114.0, 99.1, 70.2, 64.4, 58.7 and 14.9. HR-MS (*m/z*): calcd. for $C_{16}H_{12}Cl_2N_3O_4S$ [M-H]$^-$ 411.9931; found, 411.9936.

Benzyl 4-cyano-5-(5,6-dichloronicotinamido)-3-methylthiophene-2-carboxylate (**4n**): White solid (0.57 g, yield 63%). Decomposition temperature > 213 °C. ^{1}H NMR (300 MHz, DMSO-d_6): δ 12.68 (bs, 1H, CONH), 8.85 (d, *J* = 2.1 Hz, 1H, pyridine-2-H), 8.63 (d, *J* = 2.1 Hz, 1H, pyridine-4-H), 7.36–7.45 (m, 5H, Ph-2,3,4,5,6-5H), 5.32 (s, 2H, OCH$_2$) and 2.57 (s, 3H, CH$_3$). ^{13}C NMR (150 MHz, DMSO-d_6): δ 163.5, 161.8, 151.7, 151.4, 148.5, 144.9, 139.9, 136.3, 129.5, 129.3, 129.1 (2 C), 128.8, 128.5 (2 C), 118.0, 113.9, 99.1, 66.8 and 14.9. HR-MS (*m/z*): calcd. for $C_{20}H_{12}Cl_2N_3O_3S$ [M-H]$^-$ 443.9987; found, 443.9981.

5,6-dichloro-*N*-(3-cyano-4-methyl-5-(methylcarbamoyl)thiophen-2-yl)nicotinamide (**4o**): White solid (0.51 g, yield 68%). Decomposition temperature > 265 °C. ^{1}H NMR (300 MHz, DMSO-d_6): δ 12.44 (bs, 1H, CONH), 8.86 (d, *J* = 2.1 Hz, 1H, pyridine-2-H), 8.64 (d, *J* = 2.1 Hz, 1H, pyridine-4-H), 8.05 (q, *J* = 4.5 Hz, 3H, CONH*CH$_3$*), 2.75 (d, *J* = 4.5 Hz, 3H, CON*H*CH$_3$) and 2.47 (s, 3H, CH$_3$). ^{13}C NMR (150 MHz, DMSO-d_6): δ 163.0, 162.4, 151.3, 148.7, 148.4, 139.8, 137.5, 129.5, 129.4, 124.8, 114.3, 98.7, 26.9 and 14.7. HR-MS (*m/z*): calcd. for $C_{14}H_9Cl_2N_4O_2S$ [M-H]$^-$ 366.9828; found, 366.9830.

5,6-dichloro-*N*-(3-cyano-5-(cyclopropylcarbamoyl)-4-methylthiophen-2-yl)nicotinamide (**4p**): White solid (0.55 g, yield 69%). Decomposition temperature > 258 °C; ^{1}H NMR (300 MHz, DMSO-d_6): δ 12.41 (bs, 1H, CONH), 8.86 (d, *J* = 2.1 Hz, 1H, pyridine-2-H), 8.63 (d, *J* = 2.1 Hz, 1H, pyridine-4-H), 8.19 (d, *J* = 6.9 Hz, 1H, CON*H*CH), 2.74–2.83 (m, 1H, cyclopropane-H), 2.44 (s, 3H, CH$_3$) and 0.55–0.72 (m, 4H, cyclopropane-H). ^{13}C NMR (150 MHz, DMSO-d_6): δ 163.3, 163.0, 151.3, 148.6, 148.4, 139.8, 137.7, 129.5 (2 C), 124.6, 114.3, 98.6, 23.6, 14.7 and 6.4. HR-MS (*m/z*): calcd. for $C_{16}H_{13}Cl_2N_4O_2S$ [M + H]$^+$ 395.0131; found, 395.0123.

5,6-dichloro-*N*-(3-cyano-4-methyl-5-(phenylcarbamoyl)thiophen-2-yl)nicotinamide (**4q**): White solid (0.57 g, yield 65%). Decomposition temperature > 243 °C. ^{1}H NMR (600 MHz, DMSO-d_6): δ 12.56 (bs, 1H, CONH), 10.11 (s, 1H, CONH), 8.86 (d, *J* = 2.1 Hz, 1H, pyridine-2-H), 8.64 (d, *J* = 2.1 Hz, 1H, pyridine-4-H), 7.62–7.64 (m, 2H, Ph-H), 7.29–7.32 (m, 2H, Ph-H), 7.06–7.09 (m, 1H, Ph-H) and 2.47 (s, 3H, CH$_3$). ^{13}C NMR (150 MHz, DMSO-d_6): δ 163.1, 160.8, 151.3, 148.5, 139.9, 139.2, 138.7, 129.5, 129.2 (2 C), 129.1, 124.8, 124.5, 121.0 (2 C), 120.8, 114.3, 98.6 and 14.9. HR-MS (*m/z*): calcd. for $C_{19}H_{11}Cl_2N_4O_2S$ [M-H]$^-$ 428.9985; found, 428.9988.

Ethyl 4-cyano-5-(5,6-dichloronicotinamido)-3-ethylthiophene-2-carboxylate (**4r**): White solid (0.55 g, yield 68%). Decomposition temperature > 205 °C. ^{1}H NMR (600 MHz, DMSO-d_6): δ 12.63 (bs, 1H, CONH), 8.80 (d, *J* = 2.1 Hz, 1H, pyridine-2-H), 8.58 (d, *J* = 2.1 Hz, 1H, pyridine-4-H), 4.24 (q, *J* = 7.2 Hz, 2H, OCH$_2$CH$_3$), 2.95 (q, *J* = 7.2 Hz, 2H, *CH$_2$*CH$_3$), 1.26 (t, *J* = 7.2 Hz, 3H, OCH$_2$*CH$_3$*) and 1.12 (t, *J* = 7.2 Hz, 3H, CH$_2$*CH$_3$*). ^{13}C NMR (150 MHz, DMSO-d_6): δ 163.3, 161.6, 151.7, 151.4, 150.5, 148.5, 139.8, 129.5, 129.2, 118.1, 113.8, 98.1, 61.5, 22.0, 14.8 and 14.6. HR-MS (*m/z*): calcd. for $C_{16}H_{14}Cl_2N_3O_3S$ [M + H]$^+$ 398.0128; found, 398.0120.

Ethyl 5-(5,6-dichloronicotinamido)-3-methylthiophene-2-carboxylate (**4s**): White solid (0.53 g, yield 73%). Decomposition temperature > 211 °C. ^{1}H NMR (600 MHz, DMSO-d_6): δ 12.04 (bs, 1H, CONH), 8.85 (d, *J* = 2.4 Hz, 1H, pyridine-2-H), 8.56 (d, *J* = 2.4 Hz, 1H, pyridine-4-H), 6.73 (s, 1H, thiophene-H), 4.18 (q, *J* = 7.2 Hz, 2H, OCH$_2$CH$_3$), 2.40 (s, 3H, CH$_3$) and 1.25 (t, *J* = 7.2 Hz, 3H, OCH$_2$*CH$_3$*). ^{13}C NMR (150 MHz, DMSO-d_6): δ 163.1, 160.9, 151.2, 147.9, 144.5, 143.9, 138.9, 129.9, 129.5, 117.7, 117.5, 60.6, 16.3 and 14.8. HR-MS (*m/z*): calcd. for $C_{14}H_{13}Cl_2N_2O_3S$ [M + H]$^+$ 359.0019; found, 359.0012.

3.2. Fungicidal Activities In Vivo in a Greenhouse

Each of the test compounds (4 mg) were first dissolved in 5 mL of dimethyl sulfoxide, and then 5 mL of water containing 0.1% Tween 80 was added to generate 10 mL stock solutions at a concentration of 400 mg/L. Serial test solutions were prepared by diluting the above solution (testing range of 1.5625~400 mg/L). Evaluations of the in vivo fungicidal activity of the synthesized compounds against *Blumeria graminis* (DC.) Speer, *Puccinia sorghi*, *Colletotrichum orbiculare* and *Pseudoperonospora cubensis* (Berk.et Curt.) Rostov. Were performed as follows: Seeds (wheat: *Triticum aestivum* L., maize: *Zea mays* L., and cucumber: *Cucumis sativus* L.) were grown to the one-leaf, two-leaf, and two- to three-leaf stages, and then the test solutions were sprayed on the host plants with a homemade sprayer. After 24 h, the leaves of the host plants were inoculated with sporangial suspensions of the fungi which were cultured by Shenyang Sinochem Agrochemicals R&D Company Ltd. (Shenyang, China), each at a concentration of 5×10^5 spores/mL, using a PS289 Procon Boy WA double action 0.3 mm airbrush (GSI, Tokyo, Japan), but the fungi *Blumeria graminis* (DC.) Speer was inoculated by shaking off the spores directly onto the wheat leaves. The plants were stored in a humidity chamber (24 ± 1 °C, RH > 90%, dark) and then were transferred to a greenhouse (18~30 °C, RH > 50~60%) 24 h after infection. Three replicates were carried out. The activity of each compound was estimated through visual inspection after 7 d, and the screening results were reported in the range of 0% (no control of the fungus) to 100% (complete control of the fungus). The inhibitory activity (%) was estimated as:

Inhibitory activity (%) = [(viability of the blank control − viability of the treated plant)/viability of the blank control] × 100%

The EC_{50} values were calculated with Duncan's new multiple-range test (DMRT) using DPS version 14.5.

3.3. Field Trials against CDM

Field trials were conducted in the open field owned by Shenyang Sinochem Agrochemicals R&D Company Ltd. in Shenyang, Liaoning province (25 m², at the five-leaf stage). The test solution was sprayed on the host plant (*Cucumis sativus* L.) with a WS-15D knapsack electric sprayer (WishSprayer, Shandong, China). Two spays were carried out with an interval period of 6~7 d. Seven days after the second treatment, the incidence of disease spots in each plot was investigated by randomly selecting 4 samples per plot and 8 plants per sample. The incidence of the whole plant was recorded by counting the number of diseased leaves and by determining the incidence grade [33]. The grade scales were divided into six levels (ratio of leaf-spot area to leaf area): level 0 (no disease), level 1 (<5%), level 3 (6~10%), level 5 (11~25%), level 7 (26~50%), and level 9 (> 51%). The disease index (DI, %) was calculated as

DI (%) = [$\sum$ (number of diseased leaves × relative level)/(total number of investigated leaves × the highest level)] × 100%

The inhibitory activity (%) was calculated as

Inhibitory activity (%) = [(DI of the blank control − DI of the treated plot)/DI of the blank control] × 100%

4. Conclusions

In summary, nineteen *N*-(thiophen-2-yl) nicotinamide derivatives were designed and synthesized. The in vivo bioassay results of all nineteen compounds against WPM, SCR, CA and CDM in a greenhouse indicated that over half of these compounds showed moderate to significant fungicidal activity against CDM, and there were six compounds (**4a**, **4b**, **4c**, **4f**, **4i** and **4r**) that displayed higher activities than diflumetorim (EC_{50} = 21.44 mg/L); compounds **4a** (EC_{50} = 4.69 mg/L) and **4f** (EC_{50} = 1.96 mg/L) exhibited especially excellent activities, which were higher than those of flumorph (EC_{50} = 7.55 mg/L). The bioassay results of the field trial against CDM demonstrated that compound **4f** displayed excellent control efficacies (70% and 79% control efficacies, respectively, each at 100 mg/L and 200 mg/L), which were superior to those of the two commercial fungicides flumorph (56% control efficacy at 200 mg/L) and mancozeb (76% control efficacy at 1000 mg/L). However, the control efficacy of compound **4f** was inferior to that of the commercial fungicide cyazofamid (91% control efficacy at 100 mg/L). This study illustrated that *N*-(thiophen-2-yl) nicotinamide derivatives are significant lead compounds that can be used for the further discovery of new compounds to control the oomycete disease CDM, and it illustrated that compound **4f** is also a promising fungicide candidate against CDM that can be used for further development.

Supplementary Materials: The following supporting information can be downloaded at https://www.mdpi.com/article/10.3390/molecules27248700/s1, ^{1}H NMR, ^{13}C NMR, HRMS and IR of the compound.

Author Contributions: H.W. and Y.L. designed the target compounds and experiments; H.W. and J.X. synthesized all of the compounds; H.W. and Z.L. ran the bioassay evaluation and statistics analysis; H.W. drafted the paper; and X.L., X.Z., X.Y. and Y.L. revised the paper. All authors have read and agreed to the published version of the manuscript.

Funding: This research was funded by the National Natural Science Foundation of China, grant number 22077137.

Institutional Review Board Statement: Not applicable.

Informed Consent Statement: Not applicable.

Data Availability Statement: All data presented in this study are available in the article and in the Supplementary Material.

Conflicts of Interest: The authors declare no conflict of interest.

Sample Availability: Samples of compounds **4a~4s** are available from the authors.

References

1. Rajasekaran, K.; Stromberg, K.D.; Cary, J.W.; Cleveland, T.E. Broad-spectrum antimicrobial activity in vitro of the synthetic peptide D4E1. *J. Agric. Food Chem.* **2001**, *49*, 2799–2803. [CrossRef] [PubMed]
2. Zhang, S.; Liu, S.; Zhang, J.; Reiter, R.J.; Wang, Y.; Qiu, D.; Luo, X.; Khalid, A.R.; Wang, H.; Feng, L.; et al. Synergistic anti-oomycete effect of melatonin with a biofungicide against oomycetic black shank disease. *J. Pineal Res.* **2018**, *65*, e12492. [CrossRef] [PubMed]
3. Gould, F.; Brown, Z.S.; Kuzma, J. Wicked evolution: Can we address the sociobiological dilemma of pesticide resistance? *Science* **2018**, *360*, 728–732. [CrossRef]
4. de Chaves, M.A.; Reginatto, P.; da Costa, B.S.; de Paschoal, R.I.; Teixeira, M.L.; Fuentefria, A.M. Fungicide Resistance in Fusarium Graminearum Species Complex. *Curr. Microbiol.* **2022**, *79*, 1–9. [CrossRef] [PubMed]
5. Wang, T.; Yang, S.; Li, H.; Lu, A.; Wang, Z.; Yao, Y.; Wang, Q. Discovery, structural optimization, and mode of action of essramycin alkaloid and its derivatives as anti-tobacco mosaic virus and anti-phytopathogenic fungus agents. *J. Agric. Food Chem.* **2020**, *68*, 471–484. [CrossRef]
6. Xia, Q.; Dong, J.; Li, L.; Wang, Q.; Liu, Y.; Wang, Q. Discovery of glycosylated genipin derivatives as novel antiviral, insecticidal, and fungicidal agents. *J. Agric. Food Chem.* **2018**, *66*, 1341–1348. [CrossRef]
7. Lei, P.; Xu, Y.; Du, J.; Yang, X.L.; Yuan, H.Z.; Xu, G.F.; Ling, Y. Design, synthesis and fungicidal activity of N-Substituted Benzoyl-1,2,3,4-Tetrahydroquinolyl-1-Carboxamide. *Bioorg. Med. Chem. Lett.* **2016**, *26*, 2544–2546. [CrossRef]
8. Loiseleur, O. Natural products in the discovery of agrochemicals. *Chimia* **2017**, *71*, 810–822. [CrossRef]
9. Sparks, T.C.; Hahn, D.R.; Garizi, N.V. Natural products, their derivatives, mimics and synthetic equivalents: Role in agrochemical discovery. *Pest Manag. Sci.* **2017**, *73*, 700–715. [CrossRef]
10. Davison, E.K.; Sperry, J. Natural products with heteroatom-rich ring systems. *J. Nat. Prod.* **2017**, *80*, 3060–3079. [CrossRef]
11. Taylor, R.D.; MacCoss, M.; Lawson, A.D.G. Rings in drugs. *J. Med. Chem.* **2014**, *57*, 5845–5859. [CrossRef] [PubMed]
12. Pozharskii, A.F.; Soldatenkov, A.T.; Katritzky, A.R. Heterocycles in Agriculture. In *Heterocycles in Life and Society*, 2nd ed.; Wiley-VCH: Weinheim, Germany, 2011; pp. 185–207. [CrossRef]
13. Hemmerling, F.; Hahn, F. Biosynthesis of oxygen and nitrogen-containing heterocycles in polyketides. *Beilstein J. Org. Chem.* **2016**, *12*, 1512–1550. [CrossRef] [PubMed]
14. Lamberth, C. Heterocyclic chemistry in crop protection. *Pest Manag. Sci.* **2013**, *69*, 1106–1114. [CrossRef] [PubMed]
15. Kang, I.; Kim, S.W.; Youn, J.H. Effects of nicotinic acid on gene expression: Potential mechanisms and implications for wanted and unwanted effects of the lipid-lowering drug. *J. Clin. Endocrinol. Metab.* **2011**, *96*, 3048–3055. [CrossRef]
16. Sinthupoom, N.; Prachayasittikul, V.; Prachayasittikul, S.; Ruchirawat, S.; Prachayasittikul, V. Nicotinic acid and derivatives as multifunctional pharmacophores for medical applications. *Eur. Food Res. Technol.* **2015**, *240*, 1–17. [CrossRef]
17. Ashton, I.P.; Abulnaja, K.O.; Pallett, K.E.; Cole, D.J.; Harwood, J.L. Diflufenican, a carotenogenesis inhibitor, also reduces acyl lipid synthesis. *Pestic. Biochem. Physiol.* **1992**, *43*, 14–21. [CrossRef]
18. Morita, M.; Ueda, T.; Yoneda, T.; Koyanagi, T.; Haga, T. Flonicamid, a novel insecticide with a rapid inhibitory effect on aphid feeding. *Pest Manag. Sci.* **2007**, *63*, 969–973. [CrossRef]
19. Ma, S.; Ji, R.; Wang, X.; Yu, C.; Yu, Y.; Yang, X. Fluorescence detection of boscalid pesticide residues in grape juice. *Optik* **2019**, *180*, 236–239. [CrossRef]

20. Hatamoto, M.; Aizawa, R.; Kobayashi, Y.; Fujimura, M. A novel fungicide Aminopyrifen Inhibits GWT-1 protein in Glycosylphosphatidylinositol-Anchor biosynthesis in neurospora crassa. *Pestic. Biochem. Physiol.* **2019**, *156*, 1–8. [CrossRef]
21. O'Sullivan, A.C.; Loiseleur, O.; Staiger, R.; Luksch, T.; Pitterna, T. Preparation of N-Cyclylamides as Nematicides. Patent WO 2,013,143,811, 3 October 2013.
22. Hallenbach, W.; Schwarz, H.G.; Ilg, K.; Goergens, U.; Koebberling, J.; Turberg, A.; Boehnke, N.; Maue, M.; Velten, R.; Harschneck, T.; et al. Preparation of Substituted Benzamides for Treating Arthropodes. Patent WO 2,015,067,646, 14 May 2015.
23. Phillips, G.; Fevig, T.L.; Lau, P.H.; Klemm, G.H.; Mao, M.K.; Ma, C.; Gloeckner, J.A.; Clark, A.S. Process research on the synthesis of silthiofam: A novel fungicide for wheat. *Org. Process. Res. Dev.* **2002**, *6*, 357–366. [CrossRef]
24. Kim, D.S.; Chun, S.J.; Jeon, J.J.; Lee, S.W.; Joe, G.H. Synthesis and fungicidal activity of ethaboxam against oomycetes. *Pest Manag. Sci.* **2004**, *60*, 1007–1012. [CrossRef]
25. Yanase, Y.; Katsuta, H.; Tomiya, K.; Enomoto, M.; Sakamoto, O. Development of a novel fungicide, penthiopyrad. *J. Pestic. Sci.* **2013**, *38*, 167–168. [CrossRef]
26. Zuniga, A.I.; Oliveira, M.S.; Rebello, C.S.; Peres, N.A. Baseline sensitivity of botrytis cinerea isolates from strawberry to isofetamid compared to other SDHIs. *Plant Dis.* **2020**, *104*, 1224–1230. [CrossRef]
27. Swanston, J. Thiophene. In *Ullmann's Encyclopedia of Industrial Chemistry*; Wiley-VCH: Weinheim, Germany, 2006; Volume 36, pp. 657–669. [CrossRef]
28. Wang, B.; Shi, Y.; Zhan, Y.; Zhang, L.; Zhang, Y.; Wang, L.; Zhang, X.; Li, Y.; Li, Z.; Li, B. Synthesis and biological activity of novel furan/thiophene and piperazine-containing (Bis)1,2,4-Triazole mannich bases. *Chin. J. Chem.* **2015**, *33*, 1124–1134. [CrossRef]
29. Zhang, B.; Li, Y.H.; Liu, Y.; Chen, Y.R.; Pan, E.S.; You, W.W.; Zhao, P.L. Design, synthesis and biological evaluation of novel 1,2,4-Triazolo [3,4-b][1,3,4] Thiadiazines bearing furan and Thiophene nucleus. *Eur. J. Med. Chem.* **2015**, *103*, 335–342. [CrossRef]
30. Wang, X.; Ren, Z.; Mei, Y.; Liu, M.; Chen, M.; Si, W.; Yang, C.; Song, Y. Design, synthesis, and antifungal activity of 3-(Thiophen-2-Yl)-1,5-Dihydro-2H-Pyrrol-2-One derivatives bearing a carbonic ester group. *J. Heterocycl. Chem.* **2019**, *56*, 165–171. [CrossRef]
31. Wu, H.B.; Kuang, M.S.; Lan, H.P.; Wen, Y.X.; Liu, T.T. Novel bithiophene dimers from Echinops Latifolius as potential antifungal and nematicidal agents. *J. Agric. Food Chem.* **2020**, *68*, 11939–11945. [CrossRef]
32. Yang, Z.; Sun, Y.; Liu, Q.; Li, A.; Wang, W.; Gu, W. Design, synthesis, and antifungal activity of novel Thiophene/Furan-1,3,4-Oxadiazole carboxamides as potent succinate dehydrogenase inhibitors. *J. Agric. Food Chem.* **2021**, *69*, 13373–13385. [CrossRef]
33. Guan, A.; Wang, M.; Yang, J.; Wang, L.; Xie, Y.; Lan, J.; Liu, C. Discovery of a new fungicide candidate through lead optimization of Pyrimidinamine derivatives and its activity against cucumber downy mildew. *J. Agric. Food Chem.* **2017**, *65*, 10829–10835. [CrossRef]

Article

2-(2-(Dimethylamino)vinyl)-4*H*-pyran-4-ones as Novel and Convenient Building-Blocks for the Synthesis of Conjugated 4-Pyrone Derivatives

Dmitrii L. Obydennov *, Diana I. Nigamatova, Alexander S. Shirinkin, Oleg E. Melnikov, Vladislav V. Fedin, Sergey A. Usachev, Alena E. Simbirtseva, Mikhail Y. Kornev and Vyacheslav Y. Sosnovskikh

Institute of Natural Sciences and Mathematics, Ural Federal University, 620000 Ekaterinburg, Russia
* Correspondence: dobydennov@mail.ru; Tel.: +7-343-3899597

Abstract: A straightforward approach for the construction of the new class of conjugated pyrans based on enamination of 2-methyl-4-pyrones with DMF-DMA was developed. 2-(2-(Dimethylamino)vinyl)-4-pyrones are highly reactive substrates that undergo 1,6-conjugate addition/elimination or 1,3-dipolar cycloaddition/elimination followed by substitution of the dimethylamino group without ring opening. This strategy includes selective transformations leading to conjugated and isoxazolyl-substituted 4-pyrone structures. The photophysical properties of the prepared 4-pyrones were determined in view of further design of novel merocyanine fluorophores. A solvatochromism was found for enamino-substituted 4-pyrones accompanied by a strong increase in fluorescence intensity in alcohols. The prepared conjugated structures demonstrated valuable photophysical properties, such as a large Stokes shift (up to 204 nm) and a good quantum yield (up to 28%).

Keywords: 4-pyrone; DMF-DMA; enamination; cycloaddition; merocyanine; 1,6-conjugate addition; solvatochromism; fluorophore

Citation: Obydennov, D.L.; Nigamatova, D.I.; Shirinkin, A.S.; Melnikov, O.E.; Fedin, V.V.; Usachev, S.A.; Simbirtseva, A.E.; Kornev, M.Y.; Sosnovskikh, V.Y. 2-(2-(Dimethylamino)vinyl)-4*H*-pyran-4-ones as Novel and Convenient Building-Blocks for the Synthesis of Conjugated 4-Pyrone Derivatives. *Molecules* 2022, 27, 8996. https://doi.org/10.3390/molecules27248996

Academic Editor: Joseph Sloop

Received: 20 November 2022
Accepted: 13 December 2022
Published: 16 December 2022

Publisher's Note: MDPI stays neutral with regard to jurisdictional claims in published maps and institutional affiliations.

Copyright: © 2022 by the authors. Licensee MDPI, Basel, Switzerland. This article is an open access article distributed under the terms and conditions of the Creative Commons Attribution (CC BY) license (https://creativecommons.org/licenses/by/4.0/).

1. Introduction

4-Pyrones are an important class of compounds that are widely distributed in nature, exhibiting various beneficial biological activities (e.g., phenoxan (Figure 1), which is active against HIV) [1,2] and are also used as multifunctional building blocks for organic synthesis [3–8]. On the other hand, the presence of conjugated double bonds in the pyrone structure makes it possible to consider these heterocycles in terms of attractive photophysical properties. Hispidin, as an important styryl pyrone, is responsible for bioluminescence in basidiomycete fungi in the result of oxidation [9,10], and Cyercene A is a photoactive protective agent in marine mollusks [11,12]. It has also been shown that styryl-substituted 4-pyrones find applications as fluorophores [13–19], exhibiting mechanochromism and solvatochromism based on the aggregation-induced emission enhancement (AIEE) phenomenon [16–18] and can act as molecular switches [19].

The introduction of an additional electron-withdrawing substituent into the C-4 position of the pyrone ring by means of the Knoevenagel reaction leads to one of the most popular merocyanines, DCM (Figure 1), and other 4-methylene-4*H*-pyrans, which have found wide application due to their important photophysical properties [20–24].

The major methods for the preparation of 2-vinyl-substituted 4-pyrones are based on the functionalization of the active methyl group based on the aldol condensation with aromatic aldehydes (Scheme 1) [13–19]. In addition, approaches are known that include the decarboxylative rearrangement of dehydroacetic acid derivatives [15] or the Horner-Wadsworth–Emmons reaction, which finds particular use in the synthesis of naturally occurring pyrans [1,12]. To the best of our knowledge, the transformation of 2,6-dimethyl-4-pyrone with DMF-DMA resulted in the monoenamination product with poor yield (5%) [24], whereas β-dimethylaminoacrolein aminal acetals led to 6-bis(4-dimethylaminoalka-1,3-dienyl)-4H-pyran-4-ones [13].

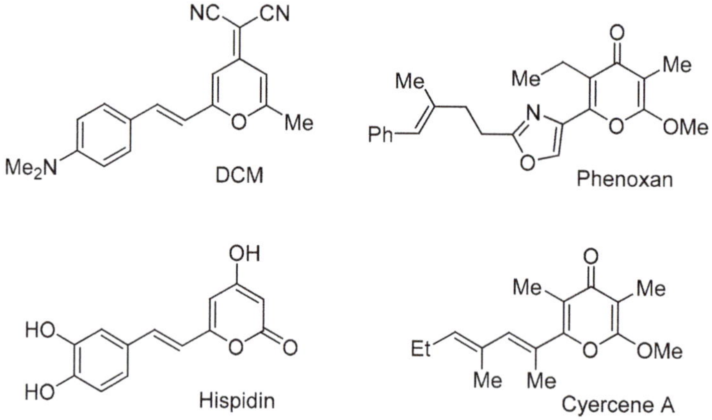

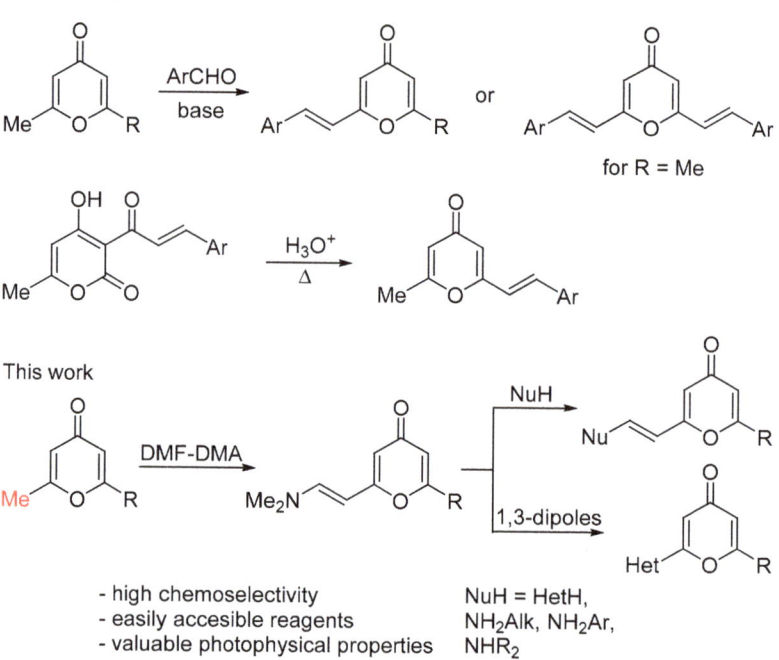

Figure 1. Some important 4-pyrones and their derivatives.

Scheme 1. General strategy for the synthesis of 2-vinyl-4-pyrones.

It is important to note that the methods for functionalizing 4-pyrones using nucleophilic reagents to create strong push-pull systems are scarcely studied. Typically, such reactions proceeded via ring opening transformation to form new cyclic systems [25–27]. The introduction of the enamino moiety [28–34] into 4-pyrone molecules leads to new highly reactive substrates, which can be used for the creation of valuable dyes via modification of 4-pyrone moiety or enamino group. Despite their attractiveness, 4-pyrone-based fluorophores are severely limited and have been described in just a few papers [13–19] due to the modest photophysical properties of the heterocycles and being overshadowed by derived dyes, 4-methylene-4H-pyrans.

In this paper, we describe a general strategy to 4-pyrone-bearing merocyanines based on an enamination with DMF-DMA and subsequent transformation of the dimethylaminovinyl group via a nucleophilic 1,6-addition or cycloaddition reaction. This approach opens straightforward access to a wide range of new promising pyran fluorophores.

2. Results and Discussion
2.1. Synthesis of 2-Enamino-substituted 4-Pyrones and Their Chemical Properties

The functionalization of the pyrone ring was carried out via an enamination reaction at the active methyl group using DMF-DMA as a reagent and a solvent (Scheme 2) [35]. N-Methylimidazole (NMI) was selected as a convenient base for the promotion of the transformation [36]. Enamination of 2-(tert-butyl)-6-methyl-4H-pyran-4-one (**1a**) with DMF-DMA (3 equiv.) and NMI (3 equiv.) at 100 °C in an autoclave afforded enamino-substituted 4-pyrone **2a** in only 15% yield (Table 1, entry 1). We decided to increase the reaction temperature to 120 °C and study the influence of the base amount on both the reaction outcome and time (TLC monitoring). When one equivalent of NMI was used, the reaction was completed in 15 h and the product was prepared in 40% yield (entry 2). We found that a further decrease in the amount of NMI (0.25–0.5 equivalent) allowed the improvement of the enamination reaction outcome until 67–72%, but it required longer heating (20–25 h) (TLC monitoring) (entries 3,4). The best yield (72%) of pyrone **2a** was achieved using 0.25 equivalents of N-methylimidazole though it took 25 h (entry 4). The isolation of the pyrone included simple recrystallization from n-heptane. Interestingly, the reaction also occurred without promotion of the base and gave the product in a lower yield (57%) under heating at 120 °C for 25 h (entry 5). Increasing temperature to 130 °C led to pyrone **2a** in 54% yield (entry 6). The enamination with the use of pyridine as a solvent and DMF-DMA (1.2 equiv.) at 100 °C or 120 °C did not produce the desired product.

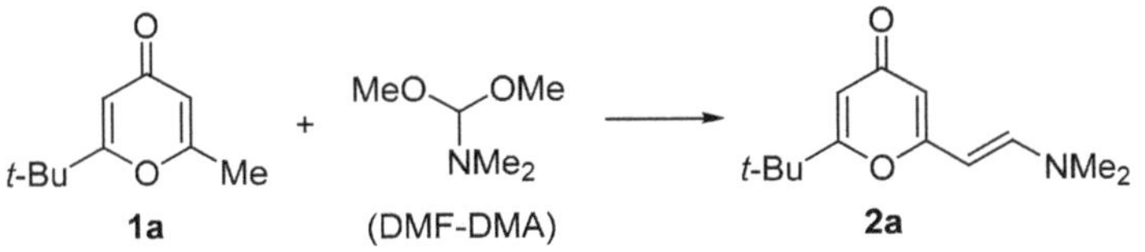

Scheme 2. Reaction of pyrone **1a** with DMF-DMA.

Table 1. Reaction condition optimization for the enamination of **1a**.

Entry	NMI, equiv.	Time, h	Temp., °C	Yield of 2a, %
1	3	8	100	15
2	1	15	120	40
3	0.5	20	120	67
4	0.25	25	120	72
5	–	25	120	57
6	0.25	10	130	54

The enamination reaction conditions were extended for various 2-methyl-4-pyrones (Scheme 3, Table 2), but this transformation turned out to be very sensitive to the nature of substituents at the pyrone ring. The enamination of 2-methyl-6-phenyl-4-pyrone (**1b**) with DMF-DMA proceeded for 12 h under the optimized conditions; as a result, pyrone **2b** was obtained in 53% yield. However, the reaction of 2-methyl-6-trifluoromethyl-4-pyrone (**1c**) was completed in 5 h at 120 °C, leading to the desired product in only low yield (12%). This result can be explained by side processes due to the presence of the trifluoromethyl group and high CH-acidity). Lowering the temperature to 100 °C made it possible to increase the yield up to 43%.

Scheme 3. Enamination of pyrones **1**.

Table 2. The scope of enamino-substituted 4-pyrones **2**.

Entry	R	Product	Temp., °C	Equiv. of NMI	Time, h	Yield, %
1	t-Bu	2a	120	0.25	15	72
2	Ph	2b	120	0.25	12	53
3	CF_3	2c	100	0.25	5	43, 12 [a]
4	CO_2Me	2d	120	4	6	27 [b], 8 [c]
5	PhCH = CH	2e	100	0.25	4	22
6	4-$Me_2NC_6H_4$CH = CH	2f	120	3	3	72
7	4-$MeOC_6H_4$CH = CH	2g	120	3	3	51

[a] The reaction was carried out with 1.2 mmol of **1c** at 120 °C for 5 h in the presence of NMI (24.6 mg, 0.3 mmol) and DMF-DMA (429.0 mg, 3.6 mmol). [b] From ethyl 6-methylcomanate (**1d′**). [c] The reaction was carried out with methyl 6-methylcomanate (**1d**) at 120 °C for 4 h in the presence of NMI (24.6 mg, 0.3 mmol) and DMF-DMA (429.0 mg, 3.6 mmol).

It was found that the enamination of ethyl 6-methylcomanate (**1d′**) with DMF-DMA and an excess of NMI (3 equiv.) at 120 °C was accompanied by the transesterification reaction to produce product **2d** in 27% yield (Scheme 3 Table 2). It is interesting to note that direct enamination of methyl 6-methylcomanate (**1d**) at 100 °C in the presence of NMI (0.25 equiv.) led to the desired product **2d** in only 8% yield.

We tried to extend the enamination on 2-methyl-6-styryl-4-pyrones **1e–g** for the synthesis of unsymmetrical 2,6-divinyl-4-pyrones **2e–g** (Table 2). The reaction of 2-methyl-6-styryl-4-pyrone (**1e**) at 120 °C with a threefold excess of DMF-DMA and different amounts of N-methylimidazole did not lead to the desired product. Lowering the temperature to 100 °C made it possible to obtain product **2e** in 22% yield for 4 h (TLC monitoring). For starting 2-methyl-6-styryl-4-pyrones **1f,g**, the use of the optimized conditions did not give the desired products because of low solubility of the starting materials. The enamination of pyrones **1f,g** with threefold excess of N-methylimidazole and heating at 120 °C for 3 h led to complete conversion of the starting 4-pyrone, and products **2f,g** were isolated by the treatment with diethyl ether in 72 and 51% yields, respectively. Such a difference in the behavior of styrylpyrones may be connected with the presence of electron-donating substituents, which deactivated the double bond of the styryl fragment and reduced the possibility of side reactions.

The reaction of 2,6-dimethyl-4-pyrone **3a** with three equivalent DMF-DMA and 0.5 equiv. of NMI was carried out at 120 °C for 15 h, resulting in a mixture of products of enamination **4a** and **5a** (Scheme 4). Recrystallization from n-heptane easily allowed the separation of pyrones **4a** and **5a** and isolation them in pure form in 17% and 23% yields, respectively. All our attempts to carry out more selective monoenamination by lowering the temperature, variation of reagent amounts, and increasing the reaction time were unsuccessful and accompanied by incomplete conversion and the formation of the bis(enamino) derivative, which did not allow the preparation of product **4a** in pure form directly.

The use of an excess of DMF-DMA (5 equiv.) and heating at 130 °C for 15 h made it possible to increase the yield of bisenamine **5a** to 51%. At the same time, the formation of monoenamino derivative **4a** was not observed. Carrying out the reaction without using NMI or using one equivalent of NMI resulted in product **5a** in 41 and 45% yields, respectively. The transformation of 3-bromo-2,6-dimethyl-4-pyrone (**3b**) led to a double enamination product **5b** in 53% yield. This result can probably be explained by a higher CH-acidity of the methyl groups. We also managed to carry out monoenamination selectively at the methyl group located near the electron-withdrawing bromine atom, and pyrone **4b** was prepared in high yield (80%) (Scheme 4). The structure of the product was assigned on the basis of the chemical shift of the methyl group in comparison with the starting 2,6-dimethyl-3-bromo-4-pyrone (**3b**) and product **4b**.

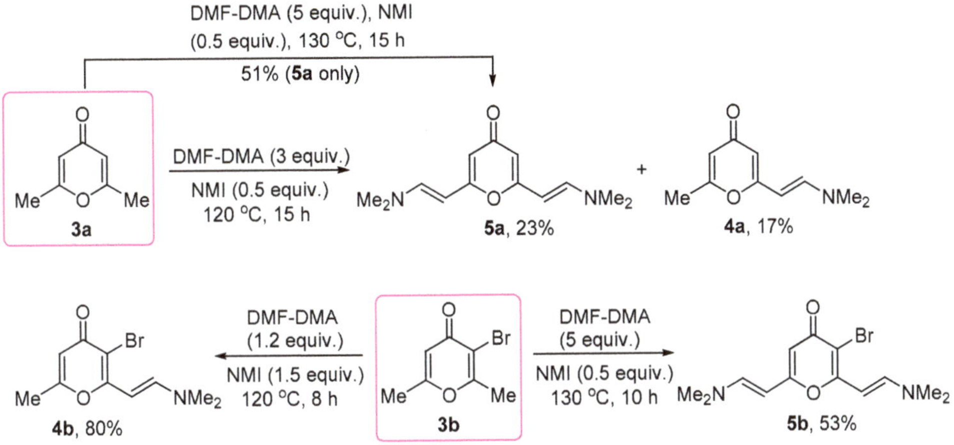

Scheme 4. Enamination of 2,6-dimethyl-4-pyrones **3a,b**.

The ^{1}H NMR spectra of the obtained dimethylamino-substituted pyrones **2,4,5** demonstrate a characteristic set of two doublets of the enamino group with ^3J coupling of 12.6–13.3 Hz, which indicates the E-configuration of the double bond and its partially double order due to the strong push-pull nature [31,36].

To study the chemical properties of monoenamino-substituted 4-pyrones **2** with various nucleophilic reagents, compound **2g** was used as an example to obtain conjugated structures (Scheme 5). We found that heating in AcOH turned out to be convenient conditions for carrying out the reactions. The transformation of pyrone **2g** with aniline or diphenylamine at 90 °C led to the substitution of the dimethylamino group and the formation of products **6a,b** in 82–86% yield. The reaction of substance **2g** with p-phenylenediamine as a binucleophile gave product **6c** as the result of an attack on both amino groups. It was found that pyrone **2g** reacted with benzylamine under reflux in acetonitrile to form product **6d** in 76% yield. The transformation of bis(enamino)pyrone **5a** with aniline was found to proceed at room temperature, leading to product **6e** in 55% yield. Thus, it has been shown that the side chain of γ-pyrone can easily be functionalized with aliphatic and aromatic amines.

Enamino-substituted pyrones **2g** and **5a** were able to react with 2-methylindole as a C-nucleophile to form indolyl-substituted 4-pyrones **7a,b** in 52–62% yields under reflux in AcOH for 7–10 h (Scheme 5). The transformation of bisenamine **5a** with 2-methylindole included the substitution at two enamino fragments and led to bis(indolylvinyl)-4-pyrone **7b** in 52% yield.

Next, we investigated the cycloaddition of enamino-substituted 4-pyrones with 1,3-dipoles (Scheme 6). It was observed that organic azides and diphenylnitrilimine did not give the desired products, which is probably due to the electron-withdrawing properties of the pyrone ring. The reaction with benzonitrile oxide, which was generated in situ from N-hydroxybenzimidoyl chloride [25], in dioxane without the use of a base led to the formation of isoxazolyl-substituted 4-pyrones in 39–80% yields. Although compounds **2f,g** bear two double bonds of different nature, the transformation proceeded chemoselectively at the enamino fragment. In the case of bis(enamino) derivative **5a**, the cycloaddition occurred at both enamino moieties to give 2,6-bis(isoxazolyl)-4-pyrone **9** in 39% yield.

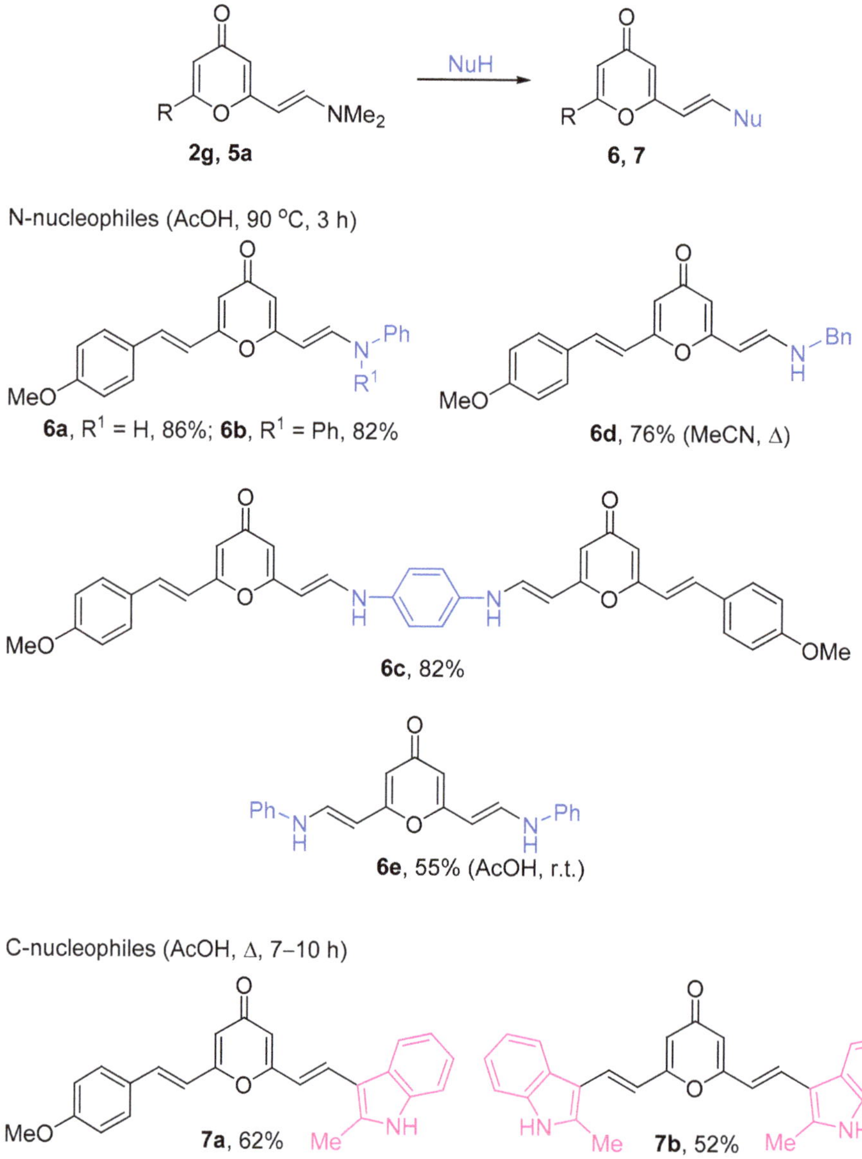

Scheme 5. Reactions of enamino-substituted 4-pyrones **2g** and **5a** with nucleophiles.

2.2. Photophysical Properties of Products

For the series of enamino-substituted 4-pyrones, the photophysical properties were studied to assess the prospects for their use as fluorophores. We started with the study of the influence of the nature of the solvent on the absorption and emission spectra of (E)-2-(2-(dimethylamino)vinyl)-6-methyl-4H-pyran-4-one (**4a**) and 2,6-bis((E)-2-(dimethylamino)vinyl)-4H-pyran-4-one (**5a**).

For the monoenamino-substituted compound **4a**, the absorption spectrum includes one-band at 334–363 nm with an extinction coefficient of 29,200–35,900 $M^{-1}cm^{-1}$ (Figure 2, Table 3). In aprotic solvents, the absorption maximum is observed at 334–350 nm. For alcohol solutions of pyrone **4a**, the absorption maximum shifts slightly to the long-wavelength region and appears at 356 (*i*-PrOH), 361 (EtOH), 363 nm (MeOH) in accordance with the solvent polarity. The emission spectrum demonstrates a single maximum and depends strongly on the nature of the solvent. In alcohols as protic solvents, the fluorescence intensity increases many times over in comparison with aprotic polar solvents, such as DMSO. The highest values of quantum yields are achieved in MeOH (3.6%) and EtOH (1.4%), where the substance exhibits blue fluorescence. The largest Stokes shifts (67–71 nm) are also observed in MeOH and EtOH, while it is equal to 26–59 nm in the other solvents. The peculiarity of fluorescence in protic solvents can be related to the specific solvation of carbonyl oxygen in the excited state due to intramolecular charge transfer (ICT). The solvatochromism of 4-pyrones was previously unknown and distinguishes the studied conjugated 4-pyrones from 4-methylene-4*H*-pyrans, popular merocyanine dyes whose fluorescence is related to the solvent polarity and is most pronounced in DMSO.

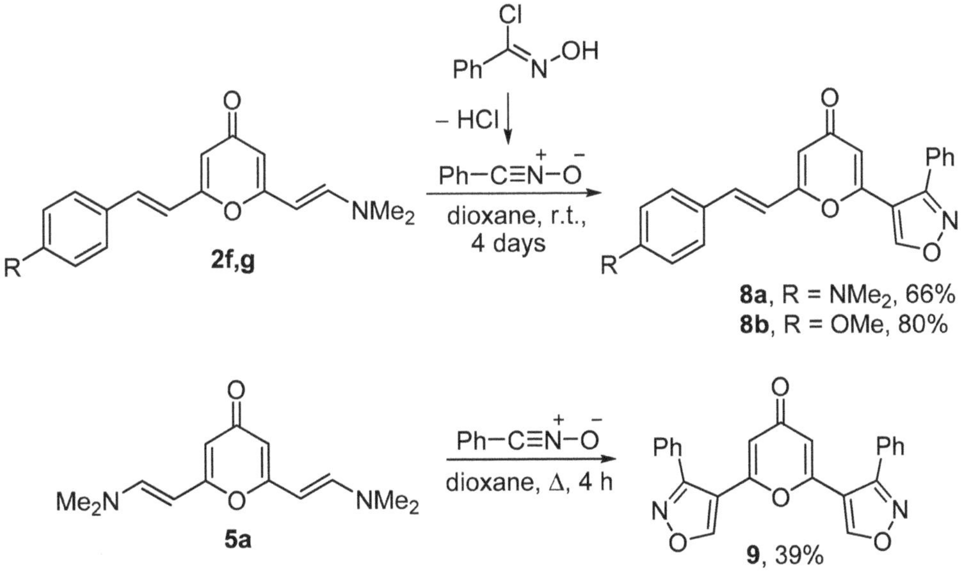

Scheme 6. Cycloaddition of enamino-substituted 4-pyrones **2f,g** and **5a**.

In the case of bis(enamino) derivative **5a**, two maxima are observed in the absorption spectra. The most intense and structured band is in the short wavelength region at 300–304 nm (ε = 41,400–54,400 M^{-1} cm^{-1}), and at 378–408 nm there is a second maximum with an extinction coefficient of 17,400–22,300 M^{-1} cm^{-1} (Figure 3, Table 4). The nature of the solvent most strongly affects the second maximum, which can be associated with intramolecular electron transfer. The strongest redshift of the second band is observed in alcohols (395–408 nm) compared to aprotic solvents (378–381 nm). As in the case of monoenamino derivative **4a**, the fluorescence spectra turned out to be highly sensitive to the nature of the solvent and has one emission maximum located in the range of 455–490 nm. The fluorescence intensity in polar aprotic solvents is observed to be low (QY = 2.3–4.1%). The substance exhibits green fluorescence, and the highest quantum yield is found in methanol (QY = 28%), ethanol (QY = 21%), and isopropanol (QY = 11%). Also, in these solvents, the largest Stokes shifts are observed, which are equal to 83 nm, 82 nm, and 80 nm, respectively.

Figure 2. Absorption (**a**) and emission (**b**) spectra of compound **4a** in various solvents (C = 1.0×10^{-5} M).

Table 3. Absorption and fluorescence spectral data of compound **4a** in various solvents (C = 1.0×10^{-5} M).

Compound	Solvent	λ_{abs}, nm [a]	λ_{em}, nm [b]	ε, M^{-1}cm^{-1}	Stokes Shift, nm	QY, % [c]
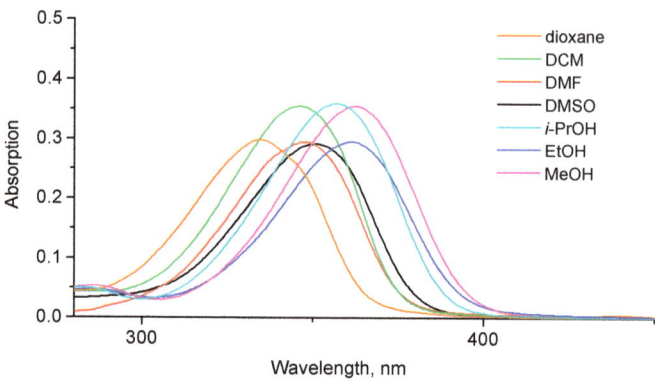 **4a**	dioxane	334	370	29,900	26	<0.1 [d]
	CH$_2$Cl$_2$	345	399	35,500	54	<0.1
	DMF	348	407	29,500	59	<0.1
	DMSO	350	408	29,200	58	0.2
	i-PrOH	356	415	35,900	59	0.3
	EtOH	361	428	29,500	67	1.4
	MeOH	363	434	35,500	71	3.6

[a] Absorption maximum wavelength. [b] Excitation wavelength corresponds to λ_{abs}. [c] The relative fluorescence quantum yield (QY) was estimated using the solution of rhodamine 6G in ethanol as a standard (QY$_{std}$ = 94%, λ_{ex} = 480 nm) according to the described method [35]. [d] The relative fluorescence quantum yield (QY) was estimated using a 0.1 M H$_2$SO$_4$ solution of quinine sulfate solution (QY$_{std}$ = 54%, λ_{ex} = 360 nm) according to the described method [37].

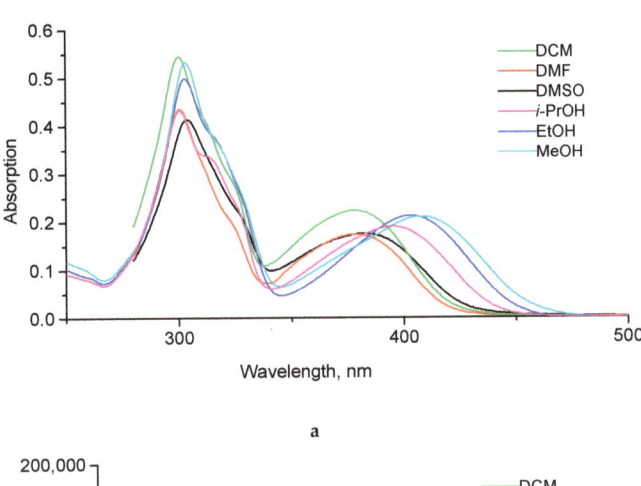

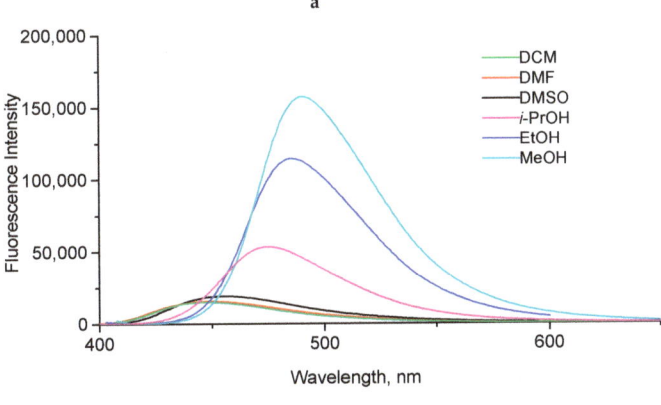

Figure 3. Absorption (**a**) and emission (**b**) spectra of compound **5a** in various solvents (C = 1.0×10^{-5} M).

Table 4. Absorption and fluorescence spectral data of compound **5a** in various solvents (C = 1.0×10^{-5} M).

Compound	Solvent	λ_{abs}, nm [a]	λ_{em}, nm [b]	ε, M^{-1}cm^{-1}	Stokes Shift, nm	QY, % [c]
	CH$_2$Cl$_2$	300 378	447	54,400 22,300	69	3.0
	DMF	300 378	450	43,600 17,400	72	4.1
5a	DMSO	304 381	455	41,400 17,500	74	2.3
	i-PrOH	301 395	475	43,300 19,000	80	11
	EtOH	303 403	485	49,800 21,200	82	21
	MeOH	303 408	490	53,200 21,000	83	28

[a] Absorption maximum wavelengths. [b] Excitation wavelength corresponds to λ_{abs}. [c] The relative fluorescence quantum yield (QY) was estimated using the solution of rhodamine 6G in ethanol as a standard (QY$_{std}$ = 94%, λ_{ex} = 480 nm) according to the described method [37].

For the design of new fluorophores, we studied a number of functionalized enamino-substituted 4-pyrones (Figure 4, Table 5). The introduction of the *tert*-butyl group, compared to the methyl group, has practically no effect on the photophysical properties. The absorption and emission spectra of (*tert*-butyl)-6-(2-(dimethylamino)vinyl)-4*H*-pyran-4-one (**2a**) are very similar for monoenamine derivative **4a**. The introduction of the phenyl group complicates the absorption spectrum, as a result, several maxima are observed. In this case, the major absorption maximum appears at 380 nm with an extinction coefficient of 19,700 $M^{-1}cm^{-1}$. An important feature of fluorescence is the large Stokes shift, which amounts to 147 nm (QY = 3.4%). For dimethylenamino derivative **2f** bearing the *p*-dimethylaminostyryl moiety, the absorption spectra show several maxima with approximately the same value of the extinction coefficient. The fluorescence spectrum exhibits one-band emission at 572 nm (λ_{ex} = 368 nm), which is characterized by higher quantum yield (18%) and a significant Stokes shift (204 nm). Similarly, the *p*-MeO-styryl-substituted compound **2g** has several bands with low intensity in absorption spectra and very weak fluorescence emission.

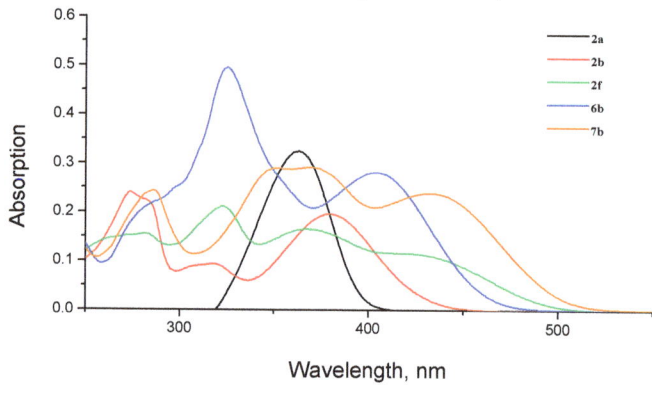

a

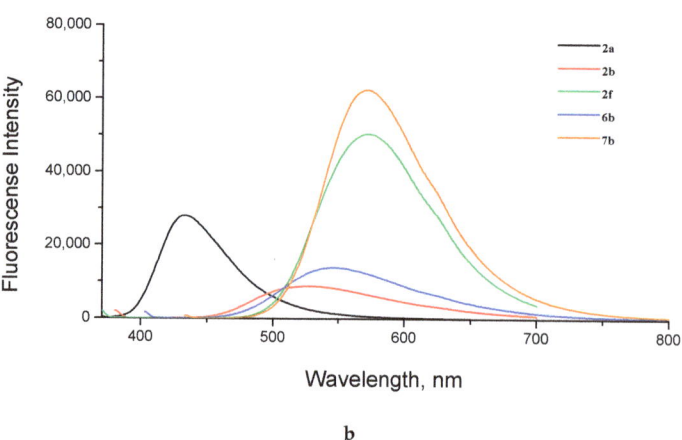

b

Figure 4. Absorption (**a**) and emission (**b**) spectra of compounds **2a,b,f**; **6b**; and **7b** in MeOH (C = 1.0×10^{-5} M).

Table 5. Absorption and fluorescence spectral data of compounds **2a,b,f**; **6b**; and **7b** in MeOH (C = 1.0 × 10^{-5} M).

Compd.	Structure	λ_{abs}, nm [a]	λ_{em}, nm [b]	ε, M^{-1} cm^{-1}	Stokes Shift, nm	QY, % [c]
2a	Me₂N–/=\–O–(C=O)–t-Bu	364	433	32,500	70	3.0
2b	Me₂N–/=\–O–(C=O)–Ph	274 317 380	527	24,000 9400 19,700	147	3.4
2f	Me₂N–/=\–O–(C=O)–/=\–NMe₂	323 368	572	21,200 16,600	204	18
6b	Ph₂N–/=\–O–(C=O)–/=\–OMe	325 404	546	49,700 28,200	142	3.4
7b	(indolyl)–/=\–O–(C=O)–/=\–(indolyl)	286 375–348 (plateau) 433	571	24,400 29,000 23,900	138	15

[a] Absorption maximum wavelengths. [b] Excitation wavelength corresponds to λ_{abs}. [c] The relative fluorescence quantum yield (QY) was estimated using the solution of rhodamine 6G in ethanol as a standard (QY$_{std}$ = 94%, λ_{ex} = 480 nm) according to the described method [37].

Introduction of the diphenylamino substituent allows the improvement of photophysical properties. Thus, for p-MeO-styryl derivative **6b**, two intense maxima are found in the absorption spectrum at 404 nm and 350 nm. This substance shows one maximum in the emission spectrum at 546 nm (λ_{ex} = 368 nm) and a quantum yield of 3.4%. Next, the bis(indolyl) derivative **7b** was investigated as a symmetrical compound with an extended conjugation system. Its absorption spectrum (MeOH) contains a major maximum at 433 nm, a plateau in the range of 375–348 nm, and a minor maximum at 286 nm. The emission spectrum contains one band at 571 nm, while the quantum yield reaches 15%.

2.3. Theoretical Calculations of the Absorption and Emission

We performed a DFT/TD-DFT quantum chemical calculations of absorption/emission maxima for representative compounds **4a** and **5a** in vacuo and in solvated phase (DMSO, EtOH, and MeOH) using the conductor-like polar continuum model (C-PCM). In the series of the ground state (GS), the first singlet excited state (S$_1$) optimizations were made and the energies of the first seven Franck-Condon singlet states were computed. All calculations were carried out at the (TD-)DFT (CAM-)B3LYP/6-31++G** level of theory for the most stable s-trans conformations [13]. Results of this calculations are provided in Table 6; the optimized geometries of GS and S$_1$ are listed in Supplementary Materials.

The optimizations of the ground state (GS) geometry revealed that both molecules have a planar structure of D-π-A conjugation chains. Calculated Stokes shift values in a solvated phase were 27–30 nm for 2-(2-(dimethylamino)vinyl)-6-methyl-4-pyrone (**4a**) and 92–103 nm for 2,6-bis(2-(dimethylamino)vinyl)-4-pyrone (**5a**), which are in agreement with the experimental Stokes shifts. Also, in all solvents the dipole moment values increased under excitation to S$_1$. On the base of Franck-Condon (FC) excitations and S$_1$ optimized geometries energies and their oscillator strengths, we plotted theoretical absorption and emission spectra (See Supplementary Materials Figures S1–S6, absorption/emission maxima are also provided in Table S6). Because the DFT usually overestimates long-wavelength polymethine transitions, all calculated maxima are notably blue-shifted compared to the experimental ones [38]. According to TD-DFT calculations of S_1^{FC}–S_7^{FC} states in all solvents, the most intense absorption corresponds to $\pi \to \pi^*$ transition and $S_0 \to S_1$ from the highest occupied molecular orbital (HOMO) to the lowest unoccupied molecular orbital (LUMO) for compound **4a**, whereas $S_0 \to S_2$ (HOMO-1 $\to$ LUMO) transition prevails for compound **5a** (Table 6). In the case of compound **5a**, the $S_0 \to S_1$ transition is less intense (f_{GS} = 0.527 for MeOH) and redshifted (λ_{abs} = 376 nm for MeOH), which agrees well with the experimental absorption spectrum demonstrating double bands.

Table 6. Calculated absorption and emission properties of **4a** and **5a** in vacuo, DMSO, EtOH, and MeOH [a].

Entry	Compd.	Solvent	λ_{abs}, nm	f_{GS}	λ_{em}, nm	f_{S1}	μ_{GS}, D	μ_{S1}, D	$\mu_{(GS \to S1)}$, D
1		Vacuum	286	0.718	293 [b]	0.744 [b]	8.808	4.246	−4.562
2	**4a**	DMSO	331	1.070	361	1.092	0.097	3.114	3.017
3		EtOH	340	1.072	367	1.082	1.102	4.372	3.270
4		MeOH	342	1.076	369	1.086	1.246	4.407	3.161
5		Vacuum	252	0.631	315	0.363	10.772	13.084	2.312
6	**5a**	DMSO	302	1.891	394	0.507	0.190	2.386	2.196
7		EtOH	307	1.903	408	0.529	1.407	3.882	2.475
8		MeOH	308	1.921	411	0.537	1.556	4.033	2.477

[a] λ_{abs}—absorption maximum, f_{GS}—oscillator strength for the absorption maximum, λ_{em}—emission maximum, f_{S1}—oscillator strength for the emission maximum, μ_{GS}, μ_{S1}—dipole moments of GS and S_1 optimized geometries, $\mu_{(GS \to S1)}$—the difference in dipole moments of GS and S_1. [b] Data provided for S_2, since S_1 (339 nm, 3.662 eV) has oscillator strength very close to zero.

To explain the strong influence of alcohols on fluorescence, the charges on the carbonyl oxygen of pyrones **4a** and **5a** in the ground and excited states were compared. An increase in the electron density on oxygen can cause its specific solvation due to the formation of hydrogen bonds, leading to the improvement of the fluorescence properties. The charge of pyrone **5a** bearing two electron-donating substituents is in all cases higher than that of pyrone **4a** in the corresponding solvent, which can be explained by the stronger push-pull nature of the former. In addition, the maximum negative charge was found in EtOH and MeOH, both in the ground (−0.731 and −0.746 for **4a**; −0.758 and −0.776 for **5a**) and excited (−0.823 and −0.839 for **4a**; −0.905 and −0.925 for **5a**) states (Table 7). For all excited states, an increase in the negative charge on carbonyl oxygen is observed (except for **4a** in vacuum), which is in good agreement with the change in dipole moments. The strongest charge changes during the GS→S_1 transition were found in MeOH (0.093 for **4a**, 0.149 for **5a**). This result is related to the ICT effect in this solvent, which is more pronounced for pyrone **5a** and determines the stabilization of the excited states via the hydrogen bonding interaction between the carbonyl group and alcohol molecules [39]. Besides, the ICT phenomena for the pyrones were confirmed with the use of electron density difference (EDD) maps (See Supplementary Materials Figures S7–S10).

Table 7. Mulliken charges at the carbonyl oxygen atom of pyrones **4a** and **5a**.

Entry	Compd.	Charge	Vacuum	DMSO	EtOH	MeOH
1		GS	−0.537	−0.640	−0.731	−0.746
2	**4a**	S_1	−0.303	−0.711	−0.823	−0.839
3		delta(GS-S_1)	−0.234	0.071	0.092	0.093
4		GS	−0.549	−0.660	−0.758	−0.776
5	**5a**	S_1	−0.634	−0.779	−0.905	−0.925
6		delta(GS-S_1)	0.085	0.119	0.147	0.149

The analysis of the frontier MOs for the ground states in the solvents (DMSO, MeOH, EtOH) and the gas phase showed that both HOMO and LUMO frontier orbitals are localized chiefly on the polymethine chain atoms, which is known to be typical for merocyanine dyes [22] (Table 8, See Supplementary Materials Tables S1–S8). Opposite to mono(enamino) derivative **4a**, bis(enamino) derivative **5a** exhibited a complete absence of the electron density at the carbonyl moiety of the conjugation chain for the HOMO, whereas the LUMO localization involves the whole π-conjugation, indicating the ICT effect under excitation. This feature can be connected with a large Stokes shift of bis(enamino) derivative **5a** and the red-shifted emission compared to **4a**.

Table 8. Ground state frontier orbitals of compounds **4a** and **5a** in MeOH [a].

Entry	Compound	HOMO	LUMO
1	4a	−5.36 eV	−1.36 eV
2	5a	−5.06 eV	−1.41 eV

[a] Isosurfaces correspond to 0.02 psi/0.0004 rho).

Thus, a convenient method of 4-pyrone functionalization is developed via enamination of 2-methyl-4-pyrones with DMF-DMA. Enamino-substituted 4-pyrones were able to react with nucleophiles and 1,3-dipoles with the substitution of the dimethylamino group and the formation of conjugated push-pull and isoxazolyl-substituted 4-pyrones, respectively. The transformations occured with high chemoselectivity without 4-pyrone ring opening and provided a convenient platform for the synthesis and design of 4-pyrone-based fluorophores. For the first time, the solvatochromism of 4-pyrones in protic solvents due to specific solvation was discovered, leading to a strong increase in the fluorescence intensity compared to aprotic solvents. 4-Pyrones bearing two enamino fragments show a higher quantum yield and significant Stokes shift, which can be explained by their stronger push-pull character. The prepared pyrone merocyanines are of interest due to attractive photophysical properties and easy modification of the conjugated chain, which can contribute to the development of the synthesis of new organic fluorophores.

3. Materials and Methods

NMR spectra were recorded on Bruker DRX-400 (Bruker BioSpin GmbH, Ettlingen, Germany, work frequencies: [1]H, 400 MHz; ^{13}C, 101 MHz; ^{19}F, 376 Hz), Bruker Avance-400 (Bruker BioSpin GmbH, Rheinstetten, Germany, work frequencies: [1]H, 400 MHz; ^{13}C, 101 MHz), Bruker Avance III-500 (Bruker BioSpin GmbH, Rheinstetten, Germany, work frequencies: [1]H, 500 MHz; ^{13}C, 126 MHz), and Bruker Avance NEO (Bruker BioSpin GmbH, work frequencies: ^{1}H, 600 MHz; ^{13}C, 151 MHz) spectrometers in DMSO-d_6 or CDCl$_3$. The chemical shifts (δ) are reported in ppm relative to the internal standard TMS (^{1}H NMR), C$_6$F$_6$ (^{19}F NMR), and residual signals of the solvents (^{13}C NMR). IR spectra were recorded on a Shimadzu IRSpirit-T (Shimadzu Corp., Kyoto, Japan) spectrometer using an attenuated total reflectance (ATR) unit (FTIR mode, diamond prism); the absorbance maxima (ν) are reported in cm^{-1}. Electron absorption spectra were obtained with a Shimadzu UV-1900 (Shimadzu Corp.) spectrophotometer; fluorescence spectra were obtained with a Shimadzu RF-6000 (Shimadzu Corp.) fluorescence spectrophotometer.

Mass spectra (ESI-MS) were measured with a Waters Xevo QTof instrument (Waters Corp., Milford, MA, USA). Elemental analyses were performed on an automatic analyzer PerkinElmer PE 2400 (Perkin Elmer Instruments, Waltham, MA, USA). Melting points were determined using a Stuart SMP40 melting point apparatus (Bibby Scientific Ltd., Stone, Staffordshire, UK). Column chromatography was performed on silica gel (Merck 60, 70–230 mesh). All solvents used were dried and distilled by standard procedures. 2-Methyl-6-styryl-4-pyrones [15], esters of 6-methyl-4-pyrone-2-carboxylic acid [40,41], 2-*tert*-butyl-6-methyl-4-pyrone [42], and 2-methyl-6-phenyl-4-pyrone [42] were prepared according to the literature procedure.

3.1. Quantum Mechanical Calculations

The ground state molecular geometry of the compounds under investigation was fully optimized at density functional theory (DFT) level, both in vacuo and in the solvated phase (DMSO, EtOH, MeOH). For all geometry optimizations, the B3LYP hybrid functional [43] coupled with the 6-31G(d,p)++ basis set was chosen. Solvent effects were taken into account via the implicit conductor-like polarizable continuum model (C-PCM). For the evaluation of energetics, Solvation Model Density (SMD) parametrization was employed [44]. The vibrational frequencies and thermochemicals were computed in harmonic approximation at T = 298.15 K and p = 1 atm, and no imaginary frequencies were found.

The UV-vis absorption spectra for the equilibrium geometries were calculated at time-dependent density functional theory (TD-DFT) level, accounting for S$_0 \rightarrow$ S$_n$ (n = 1 to 7). The nature of the vertical excited electronic state was analyzed both in vacuo and in the solvated phase.

The first singlet excited state (S$_1$) geometry was optimized using analytical gradients and the first transitions S$_1 \rightarrow$ S$_0$ of the emission. Properties of the excited states were calculated using the long-range corrected functional CAM-B3LYP [45,46] coupled with the 6-31G(d,p)++ basis set. The non-equilibrium solvation regime was set for vertical excited states calculations in the solvent phase, whereas the equilibrium solvation was used for adiabatic ones. All calculated UV-vis spectra were plotted as Gaussian curves with wavelengths of absorption/emission maxima as an expected value and σ = 0.4 eV.

The integration grid for the calculations was set to 96 radial shells and 302 angular points.

The RMS gradient convergence tolerance was set to 10^{-7} Hartree/Bohr for GS optimizations and to 10^{-5} Hartree/Bohr for S$_1$ optimizations. The density matrix convergence threshold for the self-consistent field was set to 10^{-5} a.u. for all DFT and to 10^{-6} a.u. for all TD-DFT optimizations.

All calculations were performed using the US GAMESS (ver. 30 September 2021, R2 Patch 1) software package for Linux x64 [47]. Frontier MOs were plotted with MacMolPlt software (ver. 7.7) [48]. Electron density difference maps were calculated with the use of Multiwfn v3.8 [49].

3.2. Synthesis of Compounds 2

Corresponding 4-pyrone **1** (1.2 mmol) was heated with DMF-DMA (429.0 mg, 3.6 mmol), *N*-methylimidazole (0.3 mmol for **2a–c,e**; 4.8 mmol for **2d**; 3.6 mmol for **2f,g**) in an autoclave at 100 °C (for **2c,e**) or 120 °C (for **2a,b,d,f,g**) for the needed time. For pyrones **2a–c**, the reaction mixture was treated by boiling *n*-heptane (40 mL). The solvent was decanted and evaporated to 2 mL, and the solid was filtered. For products **2d,f,g**, the reaction mixture was treated with Et$_2$O and the solid that formed was filtered. Pyrone **2e** was isolated by flash-chromatography with the use of CHCl$_3$ as an eluent.

(E)-2-(Tert-butyl)-6-(2-(dimethylamino)vinyl)-4H-pyran-4-one (**2a**). The reaction was carried out for 15 h. Yield 191.2 mg (72%), yellow powder, mp 128–130 °C. IR (ATR) ν 2962, 2907, 2870, 1639, 1564, 1386, 1364, 1104. ^{1}H NMR (500 MHz, DMSO-d_6) δ 1.23 (9H, s, *t*-Bu), 2.90 (6H, br.s, NMe$_2$), 4.82 (1H, d, *J* = 13.2 Hz, =CH(α)), 5.62 (1H, d, *J* = 2.1 Hz, CH), 5.77 (1H, d, *J* = 2.1 Hz, CH), 7.26 (1H, d, *J* = 13.2 Hz, =CH (β)). ^{13}C NMR (126 MHz, CDCl$_3$) δ 28.1, 35.8, 40.8 (br.s. NMe$_2$), 88.0, 104.8, 108.9, 144.8, 165.8, 172.8, 180.5. Anal. Calculated for C$_{13}$H$_{19}$NO$_2$: C 70.56; H 8.65; N 6.33. Found: C 70.58; H 8.76; N 6.33.

(E)-2-(2-(Dimethylamino)vinyl)-6-phenyl-4H-pyran-4-one (**2b**). The reaction was carried out for 12 h. Yield 153.5 mg (53%), yellow powder, mp 200 °C (destr.). IR (ATR) ν 3056, 2910, 1627, 1539, 1381, 1349, 1101. ^{1}H NMR (400 MHz, DMSO-d_6) δ 2.95 (6H, s, NMe$_2$), 4.92 (1H, d, *J* = 13.3, =CH(α)), 5.78 (1H, d, *J* = 2.2 Hz, H-3), 6.61 (1H, d, *J* = 2.2 Hz, H-5), 7.47 (1H, d, *J* = 13.3 Hz, =CH(β)), 7.52 (3H, m, Ph), 7.94 (2H, m, H-2, H-6 Ph). ^{13}C NMR (101 MHz, DMSO-d_6) δ 86.9, 104.3, 110.1, 126.1, 129.4, 131.1, 132.0, 146.7, 160.4, 166.3, 178.3 (NMe$_2$ was not observed). HRMS (ESI) m/z [M + H]$^+$. Calculated for C$_{15}$H$_{16}$NO$_2$: 242.1189. Found: 242.1103.

(E)-2-(2-(Dimethylamino)vinyl)-6-(trifluoromethyl)-4H-pyran-4-one (**2c**). The reaction was carried out for 5 h. Yield 120.3 mg (43%), grey powder, mp 71–72 °C. IR (ATR) ν 3071, 2909, 1669, 1557, 1341, 1267, 1078, 959. ^{1}H NMR (400 MHz, CDCl$_3$) δ 2.99 (6H, s, NMe$_2$), 4.75 (1H, d, *J* = 12.6 Hz, =CH(α)), 5.84 (1H, s), 6.84 (1H, s), 7.18 (1H, d, *J* = 12.6 Hz, =CH(β)). ^{19}F NMR (376 MHz, CDCl$_3$) δ 90.2 (s, CF$_3$). ^{13}C NMR (126 MHz, DMSO-d_6) δ 36.8 (br.s), 44.0 (br.s), 84.9, 103.5, 113.6 (q, *J* = 2.2 Hz, C-3), 118.7 (q, *J* = 273.2 Hz, CF$_3$), 147.6, 148.6 (q, *J* = 38.1 Hz, C-2), 166.9, 175.3. HRMS (ESI) m/z [M + H]$^+$. Calculated for C$_{10}$H$_{11}$F$_3$NO$_2$: 234.0742. Found: 234.0735.

Methyl (E)-6-(2-(dimethylamino)vinyl)-4-oxo-4H-pyran-2-carboxylate (**2d**). The reaction was carried out for 6 h. Yield 72.3 mg (27%), yellow powder, mp 130–132 °C. IR (ATR) ν 3064, 2909, 1745, 1632, 1550, 1351, 1098, 959. ^{1}H NMR (500 MHz, CDCl$_3$) δ 2.97 (6H, s, NMe$_2$), 4.76 (1H, d, *J* = 13.0 Hz, =CH(α)), 5.89 (1H, d, *J* = 2.3 Hz, H-5), 6.88 (1H, d, *J* = 2.3 Hz, H-3), 7.32 (1H, d, *J* = 13.0 Hz, =CH(β)). ^{13}C NMR (126 MHz, CDCl$_3$) δ 40.1 (br.s, NMe$_2$), 53.1, 86.9, 106.9, 118.7, 146.9, 150.0, 161.1, 166.8, 178.7. Anal. Calculated for C$_{11}$H$_{13}$NO$_4$: C 59.41; H 5.72; N 5.91. Found: C 59.19; H 5.87; N 6.27.

2-((E)-2-(Dimethylamino)vinyl)-6-((E)-styryl)-4H-pyran-4-one (**2e**). The reaction was carried out for 4 h. Yield 70.6 mg (22%), yellow powder, mp 75–77 °C. IR (ATR) ν 3055, 1642, 1610, 1524, 1397, 1157, 1103, 749. ^{1}H NMR (500 MHz, CDCl$_3$) δ 2.98 (6H, s, NMe$_2$), 4.79 (1H, d, *J* = 13.1 Hz, -CH=C-N), 5.84 (1H, d, *J* = 1.4 Hz, CH), 6.12 (1H, d, *J* = 1.4 Hz, CH), 6.66 (1H, d, *J* = 16.1 Hz, -CH=C-Ar), 7.24 (1H, d, *J* = 13.1 Hz, =CH-N), 7.27 (1H, d, *J* = 16.1 Hz, =CH-Ar), 7.39 (2H, t, *J* = 7.3 Hz, H-3, H-5 Ph), 7.51 (2H, d, *J* = 7.3 Hz, H-2, H-6 Ph). ^{13}C NMR (126 MHz, CDCl$_3$) δ 40.8, 87.9, 105.6, 113.4, 120.6, 127.3, 128.9, 129.3, 134.1, 135.3, 145.5, 159.5, 165.7, 179.9. HRMS (ESI) m/z [M + H]$^+$. Calculated for C$_{17}$H$_{18}$NO$_2$: 268.1338. Found: 268.1342.

2-((E)-4-(Dimethylamino)styryl)-6-((E)-2-(dimethylamino)vinyl)-4H-pyran-4-one (**2f**). The reaction was carried out for 3 h. Yield 268.2 mg (72%), brown crystals, mp 205–206 °C. IR (ATR) ν 2802, 1596, 1520, 1386, 1357, 1154, 1100. ^{1}H NMR (400 MHz, DMSO-d_6) δ 2.97 (12H, s, 2NMe$_2$), 4.87 (1H, d, *J* = 13.2 Hz, -CH=C-N), 5.65 (1H, d, *J* = 2.1 Hz, CH), 5.93 (1H, d, *J* = 2.1 Hz, CH), 6.69 (1H, d, *J* = 16.2 Hz, -CH=C-Ar), 6.73 (2H, d, *J* = 8.8 Hz, H-3, H-5 Ar), 7.36 (1H, d, *J* = 16.2 Hz, =CH–Ar), 7.48 (1H, d, *J* = 13.2 Hz, =CH–N), 7.53 (2H, d, *J* = 8.8 Hz, H-2, H-6 Ar). ^{13}C NMR (101 MHz, DMSO-d_6) δ 39.7, 86.6, 103.8, 111.1, 111.9, 115.0, 122.9, 128.9, 134.2, 146.1, 150.9, 160.1, 165.2, 177.9 (NMe$_2$ was not observed). Anal. Calculated for C$_{19}$H$_{22}$N$_2$O$_2$: C 73.52; H 7.14; N 9.03. Found: C 73.47; H 7.27; N 9.10.

2-((E)-2-Dimethylamino)vinyl)-6-((E)-4-methoxystyryl)-4H-pyran-4-one (**2g**). The reaction was carried out for 3 h. Yield 185.5 mg (51%), brown crystals, mp 155–157 °C. IR (ATR) ν 3070, 2967, 1637, 1612, 1549, 1383, 1346, 1096, 1020, 937. ^{1}H NMR (400 MHz, DMSO-d_6)

δ 2.96 (6H, s, 2NMe₂), 3.80 (3H, s, OMe), 4.88 (1H, d, *J* = 13.2 Hz, -CH=C-N), 5.67 (1H, d, *J* = 2.1 Hz, CH), 6.00 (1H, d, *J* = 2.1 Hz, CH), 6.85 (1H, d, *J* = 16.2 Hz, -CH=C-Ar), 6.99 (2H, d, *J* = 8.8 Hz, H-3, H-5, Ar), 7.43 (1H, d, *J* = 16.2 Hz, =CH–Ar), 7.51 (1H, d, *J* = 13.2 Hz, =CH–N), 7.65 (2H, d, *J* = 8.8 Hz, H-2, H-6, Ar). ^{13}C NMR (101 MHz, DMSO-d_6) δ 55.2, 86.5, 103.9, 112.2, 114.3, 118.2, 128.1, 129.0, 133.4, 146.3, 159.5, 160.1, 165.4, 177.9 (the NMe₂ group was not observed). HRMS (ESI) m/z [M + H]⁺. Calculated for $C_{18}H_{20}NO_3$: 298.1443. Found: 298.1450.

3.3. Synthesis of Compounds 4

(E)-2-(2-(Dimethylamino)vinyl)-6-methyl-4H-pyran-4-one (**4a**). 2,6-Dimethyl-4-pyrone (**3a**) (0.202 g, 1.63 mmol), DMF-DMA (0.583 g, 4.89 mmol) and *N*-methylimidazole (66.9 mg, 0.815 mmol) were heated at 120 °C in an autoclave for 15 h. The reaction mixture was treated with boiling *n*-heptane (40 mL) to extract product **4a**. The residue was diluted with Et₂O to give 0.0876 g (23%) of brown needles of pyrone **5a** (mp 203–204 °C). The solution of *n*-heptane was evaporated to 2 mL, and the precipitate was filtered. Yield 0.0479 g (17%), yellow powder, mp 109–111 °C. IR (ATR) ν 3058, 2912, 1652. 1557, 1394, 1376, 1098. 913. ^{1}H NMR (400 MHz, CDCl₃) δ 2.21 (3H, s, Me), 2.91 (6H, s, NMe₂), 4.71 (1H, d, *J* = 13.2 Hz, =CH(α)), 5.75 (1H, d, *J* = 2.2 Hz, CH), 5.91 (1H, d, *J* = 2.2 Hz, CH), 7.12 (1H, d, *J* = 13.2 Hz, =CH(β)). ^{13}C NMR (151 MHz, CDCl₃) δ 19.5, 86.8, 103.5, 112.5, 146.3, 163.2, 166.4, 178.5 (NMe₂ was not observed). The NMR spectra are in accordance with the literature data [13].

(E)-3-Bromo-2-(2-(dimethylamino)vinyl)-6-methyl-4H-pyran-4-one (**4b**). 2,6-Dimethyl-3-bromo-4-pyrone (**3b**) (100 mg, 0.493 mmol), DMF-DMA (70.6 mg, 0.592 mmol), and *N*-methylimidazole (60.8 mg, 0.740 mmol) were heated in an autoclave for 8 h at 120 °C. The reaction was monitored by TLC. After completion of the reaction (TLC monitoring), the product was filtered and washed with Et₂O (2 mL). The compound was further purified by column chromatography (CHCl₃:EtOH = 10:0.5). Yield 102 mg (80%), yellow powder, mp 162–163 °C. IR (ATR) ν 3063, 2918, 1661, 1558, 1421, 1270, 1007, 945, 773. ^{1}H NMR (500 MHz, CDCl₃) δ 2.23 (3H, s, Me), 3.01 (6H, s, NMe₂), 5.27 (1H, d, *J* = 13.0 Hz, =CH(α)), 6.00 (1H, s, H-5), 7.24 (1H, d, *J* = 13.0 Hz, =CH(β)). ^{13}C NMR (126 MHz, CDCl₃) δ 19.3, 38.3, 44.0, 87.1, 103.1, 110.7, 147.2, 161.7, 162.8, 173.6. HRMS (ESI) m/z [M + H]⁺. Calculated for $C_{15}H_{16}NO_2$: 258.0085. Found: 258.0130.

3.4. General Method for the Synthesis of Bis(enamino)-substituted 4-Pyrones 5a,b

A mixture of 2,6-dimethyl-4-pyrone **3a** or **3b** (0.806 mmol), DMF-DMA (480 mg, 4.03 mmol), and *N*-methylimidazole (33.0 mg, 0.403 mmol) was heated in an autoclave for 15 h (for **5a**) or 10 h (for **5b**) at 130 °C. Then the reaction mixture was diluted with Et₂O (5 mL) and the product filtered.

2,6-Bis((E)-2-(dimethylamino)vinyl)-4H-pyran-4-one (**5a**). Yield 96.3 mg (51%), brown crystals, mp 203–204 °C. IR (ATR) ν 2990, 2810, 1640, 1615, 1542, 1360, 1335, 1095, 947. ^{1}H NMR (400 MHz, DMSO-d_6) δ 2.89 (12H, s, 2NMe₂), 4.75 (2H, d, *J* = 13.3 Hz, =CH(α)), 5.43 (2H, s, H-3, H-5), 7.29 (2H, d, *J* = 13.3 Hz, =CH(β)). ^{13}C NMR (100 MHz, DMSO-d_6) δ 40.0–41.0 (br.s), 87.2, 103.2, 144,9, 163.5, 177.9. HRMS (ESI) m/z [M + H]⁺. Calculated for $C_{13}H_{19}N_2O_2$: 235.1447. Found: 235.1448.

3-Bromo-2,6-bis((E)-2-(dimethylamino)vinyl)-4H-pyran-4-one (**5b**). The product was additionally purified by column chromatography (CHCl₃:EtOH = 10:0.5). Yield 134 mg (53%), dark orange powder, mp 185–186 °C. IR (ATR) ν 3398, 3020, 2916, 1630, 1487, 1353, 1105, 947, 753. ^{1}H NMR (500 MHz, CDCl₃) δ 2.91 (6H, s, NMe₂), 2.97 (6H, s, NMe₂), 4.79 (1H, d, *J* = 13.2 Hz, CH(α)), 5.24 (1H, d, *J* = 13.0 Hz, =CH(α)), 5.75 (1H, s, H-5), 7.01 (1H, d, *J* = 13.2 Hz, CH(β)), 7.14 (1H, d, *J* = 13.0 Hz, =CH(β)). ^{13}C NMR (126 MHz, CDCl₃) δ 40.7 (br.s, NMe₂), 87.6, 87.9, 102.1, 102.9, 144.7, 146.2, 160.5, 163.0, 173.6. HRMS (ESI) m/z [M + H]⁺. Calculated for $C_{15}H_{16}NO_2$: 313.0507. Found: 313.0550.

3.5. General Method for the Preparation of 2-((E)-4-Methoxystyryl)-6-((E)-2-aminovinyl)-4H-pyran-4-one 6a–c

Enamino-substituted pyrone **2g** (75.7 mg, 0.255 mmol) and N-nucleophile (0.31 mmol) or *p*-phenylenediamine (13.7 mg, 0.127 mmol) were stirred at 90 °C for 3 h in AcOH (1 mL). For **6a,c**, the precipitate was filtered and washed with EtOH. For **6b**, the reaction mixture was diluted with H_2O (5 mL). The solid that formed was recrystallized from EtOH–toluene.

2-((E)-4-Methoxystyryl)-6-((E)-2-(phenylamino)vinyl)-4H-pyran-4-one (**6a**). Yield 75.6 mg (86%), yellow powder, mp 229–231 °C. IR (ATR) ν 3272, 3078, 2958, 1698, 2637, 1596, 1495, 1278, 1269. ^{1}H NMR (400 MHz, DMSO-d_6) δ 3.81 (3H, s, Me), 5.57 (1H, d, *J* = 13.3 Hz, =CH(α)), 5.99 (1H, d, *J* = 1.9 Hz, CH), 6.14 (1H, d, *J* = 1.9 Hz, CH), 6.92 (1H, d, *J* = 16.3 Hz, =CH(α)′), 6.93 (1H, t, *J* = 7.7 Hz, H-4 Ph), 7.01 (2H, d, *J* = 7.9 Hz, H-3, H-5 Ar), 7.16 (2H, d, *J* = 7.9 Hz, H-2, H-6 Ph), 7.31 (2H, t, *J* = 7.8 Hz, H-3, H-5 Ph), 7.50 (1H, d, *J* = 16.3 Hz, =CH(β)′), 7.66 (2H, d, *J* = 8.6 Hz, H-2, H-6 Ar), 7.92 (1H, t, *J* = 12.7 Hz, =CH(β)), 9.58 (1H, d, *J* = 12.2 Hz, NH). ^{13}C NMR (100 MHz, DMSO-d_6) δ 55.7, 94.8, 107.0, 112.6, 114.9, 115.3, 118.5, 121.5, 128.5, 129.6, 129.9, 134.6, 136.3, 142.0, 160.8, 160.7, 164.4, 178.7. HRMS (ESI) m/z [M + H]$^+$. Calculated for $C_{22}H_{20}NO_3$: 346.1456. Found: 346.3980.

2-((E)-2-(Diphenylamino)vinyl)-6-((E)-4-methoxystyryl)-4H-pyran-4-one (**6b**). Yield 88.1 mg (82%), yellow powder, mp 225–226 °C. IR (ATR) ν 3046, 3004, 2836, 1644, 1576, 1489, 1237, 1173. ^{1}H NMR (500 MHz, DMSO-d_6) δ 3.78 (3H, s, Me), 5.07 (1H, d, *J* = 13.5 Hz, =CH(α)), 5.92 (1H, s, CH), 6.14 (1H, s, CH), 6.92 (1H, d, *J* = 16.0 Hz, =CH(α)′), 7.00 (2H, d, *J* = 7.9 Hz, H-3, H-5 Ar), 7.20 (4H, d, *J* = 7.5 Hz, H-2, H-6 Ph), 7.29 (2H, t, *J* = 7.1 Hz, H-4 Ph), 7.43 (1H, d, *J* = 16.0 Hz, =CH(β)′), 7.48 (4H, t, *J* = 7.2 Hz, H-3, H-5 Ph), 7.60 (2H, d, *J* = 8.6, H-2, H-6 Ar), 8.09 (1H, d, *J* = 13.5 Hz, =CH(β)). ^{13}C NMR (125 MHz, DMSO-d_6) δ 55.3, 97.0, 107.7, 112.0, 114.3, 117.9, 123.8, 125.6, 127.9, 129.0, 129.9, 134.3, 139.5, 145.5, 160.2, 160.3, 163.2, 178.2. HRMS (ESI) m/z [M + H]$^+$. Calculated for $C_{28}H_{24}NO_3$: 422.1756. Found: 422.1747.

6,6′-((1E,1′E)-(1,4-Phenylenebis(azanediyl))bis(ethene-2,1-diyl))bis(2-((E)-4-methoxystyryl)-4H-pyran-4-one) (**6c**). Yield 63.5 mg (82%), burgundy powder, mp 204–205 °C. IR (ATR) ν 3040, 2934, 2839, 1637, 1504, 1396, 1253, 1152. ^{1}H NMR (500 MHz, DMSO-d_6) δ 3.78 (6H, s, 2Me), 5.52 (2H, d, *J* = 13.1 Hz, =CH(α)), 5.95 (2H, d, *J* = 2.1 Hz, CH), 6.12 (2H, d, *J* = 2.1 Hz, CH), 6.92 (2H, d, *J* = 16.2 Hz, =CH(α)′), 6.97 (4H, d, *J* = 8.6 Hz, H-3, H-5 Ar), 7.15 (4H, s, Ar), 7.49 (2H, d, *J* = 16.2 Hz, =CH(β)′), 7.66 (4H, d, *J* = 8.6 Hz, H-2, H-6 Ar), 7.87 (2H, t, *J* = 12.0 Hz, =CH(β)), 9.52 (2H, d, *J* = 12.0 Hz, NH). ^{13}C NMR (100 MHz, DMSO-d_6) δ 55.7, 106.4, 112.6, 114.8, 116.9, 118.6, 128.5, 129.6, 136.3, 160.7, 164.7, 172.4, 178.6. HRMS (ESI) m/z [M + H]$^+$. Calculated for $C_{38}H_{33}N_2O_6$: 613.2339. Found: 613.6820.

2-((E)-2-(Benzylamino)vinyl)-6-((E)-4-methoxystyryl)-4H-pyran-4-one (**6d**). Enamino-substituted pyrone **2g** (75.7 mg, 0.255 mmol) and benzylamine (32.8 mg, 0.306 mmol) were refluxed for 7 h in MeCN (1 mL). The reaction mixture was diluted with H_2O (5 mL). The solid that formed was filtered and recrystallized from hexane–toluene. Yield 69.7 mg (76%), yellow powder, mp 144–145 °C. IR (ATR) ν 2900, 1651, 1520, 1394, 1250, 1169, 1027. ^{1}H NMR (400 MHz, DMSO-d_6) δ 3.80 (3H, s, Me), 4.33 (2H, d, *J* = 5.3 Hz, CH_2), 5.04 (1H, d, *J* = 13.4 Hz, =CH(α)), 5.66 (1H, s, *J* = 2.2 Hz, CH), 6.00 (1H, s, *J* = 2.2 Hz, CH), 6.85 (2H, d, *J* = 16.2 Hz, =CH(α)′), 6.99 (2H, d, *J* = 8.7 Hz, H-3, H-5 Ar), 7.25–7.31 (1H, m, Ph), 7.31–7.40 (6H, m), 7.43 (1H, br.s, NH), 7.61 (2H, d, *J* = 8.7 Hz, H-2, H-6 Ar). ^{13}C NMR (101 MHz, DMSO-d_6) δ 31.1, 55.7, 87.5, 104.9, 112.7, 114.8, 118.8, 127.6, 127.8, 128.5, 128.9, 129.5, 133.7, 159.9, 160.6, 165.8, 178.5. HRMS (ESI) m/z [M + H]$^+$. Calculated for $C_{22}H_{20}NO_3$: 346.1456. Found: 346.3980.

2,6-Bis((E)-2-(phenylamino)vinyl)-4H-pyran-4-one (**6e**). Pyrone **5a** (100 mg, 0.427 mmol) and aniline (99.4 g, 1.07 mmol) were stirred at room temperature for 24 h in AcOH (1.5 mL). The precipitate was filtered and washed with EtOH. Yield 77.5 mg (55%), orange powder, mp 217–218 °C. IR (ATR) ν 3221, 3054, 1657, 1645, 1544, 1493, 1278, 1148, 945. ^{1}H NMR (500 MHz, DMSO-d_6) δ 5.52 (2H, d, *J* = 13.3 Hz, =CH(α)), 5.77 (2H, s, CH), 6.92 (2H, t, *J* = 7.3 Hz, H-2, H-6 Ph), 7.13 (4H, d, *J* = 7.9 Hz, H-3, H-5 Ph), 7.87 (2H, t, *J* = 12.8 Hz, =CH(β)), 9.49 (2H, d, *J* = 12.3 Hz, NH). ^{13}C NMR (126 MHz, DMSO-d_6) δ 94.4, 106.0, 114.6,

120.8, 129.4, 135.2, 141.9, 162.5, 178.2. Anal. Calculated for $C_{21}H_{18}N_2O_2$: C 76.34; H 5.49; N 8.48. Found: C 76.42; H 5.56; N 8.43.

3.6. Synthesis of Compounds 7a,b

2-((E)-4-Methoxystyryl)-6-((E)-2-(2-methyl-1H-indol-3-yl)vinyl)-4H-pyran-4-one (**7a**). Styryl-4-pyrone **2g** (105 mg, 0.353 mmol) and 2-methylindole (55.6 mg, 0.424 mmol) were refluxed in AcOH (1 mL) for 7 h. The precipitate formed was filtered and washed with EtOH. Yield 0.0835 g (62%), yellow powder, mp 284–285 °C. IR (ATR) ν 3131, 3035, 2958, 2928, 2834, 1604, 1556, 1394. ^{1}H NMR (400 MHz, DMSO-d_6) δ 2.63 (3H, s, Me), 3.82 (3H, s, OMe), 6.25 (1H, d, *J* = 2.2 Hz, CH), 6.37 (1H, d, *J* = 2.2 Hz, CH), 6.85 (1H, d, *J* = 16.2 Hz, =CH(α)), 7.00 (1H, d, *J* = 16.2 Hz, =CH(α)), 7.03 (2H, d, *J* = 8.9 Hz, H-3, H-5, Ar), 7.12–7.19 (2H, m, H-5, H-6 Ind), 7.35–7.42 (1H, m, H-7 Ind), 7.58 (1H, d, *J* = 16.1 Hz, =CH(β)), 7.68 (2H, d, *J* = 8.7 Hz, H-2, H-6 Ar), 7.79 (1H, d, *J* = 16.2 Hz, =CH(β)), 7.98–8.04 (1H, m, H-4 Ind), 11.65 (1H, s, NH). ^{13}C NMR (151 MHz, DMSO-d_6) δ 12.2, 55.8, 109.4, 111.0, 111.8, 113.1, 113.2, 115.0, 118.5, 120.1, 121.0, 122.2, 126.2, 128.4, 129.7, 135.2, 136.5, 141.4, 161.0, 161.3, 163.3, 179.3. HRMS (ESI) m/z [M + H]$^+$. Calculated for $C_{17}H_{18}NO_2$: 268.1338. Found: 268.1342.

2,6-Bis((E)-2-(2-methyl-1H-indol-3-yl)vinyl)-4H-pyran-4-one (**7b**). Bis(enamino)-substituted 4-pyrone **5a** (100 mg, 0.353 mmol) and 2-methylindole (55.6 mg, 0.424 mmol) was refluxed in AcOH (1 mL) for 10 h. The precipitate formed was filtered and washed with EtOH. Yield 74.6 mg (52%), orange powder, mp >270 °C. IR (ATR) ν 3178, 3056, 1631, 1611, 1539, 1399, 1274, 1153, 936. ^{1}H NMR (500 MHz, DMSO-d_6) δ 2.64 (6H, s, 2Me), 6.28 (2H, s, H-3, H-5), 6.87 (2H, d, *J* = 16.1 Hz, =CH(α)), 7.13–7.19 (4H, m, H-5, H-6 Ind), 7.36–7.41 (2H, m, H-7 Ind), 7.82 (2H, d, *J* = 16.1 Hz, =CH(β)), 7.98–8.04 (2H, m, H-4 Ind), 11.7 (2H, s, NH). ^{13}C NMR (126 MHz, DMSO-d_6) δ 11.6, 108.8, 110.7, 111.3, 112.8, 119.6, 120.5, 121.7, 125.6, 128.4, 136.0, 140.7, 162.1, 179.0. HRMS (ESI) m/z [M + H]$^+$. Calculated for $C_{25}H_{22}NO_3$: 384.1600. Found: 384.1608.

3.7. Synthesis of 6-(3-Phenylisoxazol-4-yl)-4H-pyran-4-ones 8a,b

2-((Dimethylamino)vinyl)-6-styryl-4H-pyran-4-one **2f,g** (0.32 mmol) and *N*-hydroxybenzimidoyl chloride (60.7 mg, 0.390 mmol) were stirred for 4 days in dry 1,4-dioxane at room temperature. The precipitate that formed was filtered and washed with toluene.

(E)-2-(4-(Dimethylamino)styryl)-6-(3-phenylisoxazol-4-yl)-4H-pyran-4-one (**8a**). Yield 81.2 mg (66%), brown crystals, mp 199–200 °C. IR (ATR) ν 3078, 2909, 1587, 1523, 1350, 1127, 940. ^{1}H NMR (400 MHz, DMSO-d_6) δ 2.97 (6H, s, NMe$_2$), 6.20 (1H, s, CH), 6.33–6.48 (2H, m), 6.65 (1H, d, *J* = 16.5 Hz, =CH(β)), 6.70 (2H, d, *J* = 7.6 Hz, Ar), 7.24 (2H, d, *J* = 7.6 Hz, Ar), 7.47–7.71 (5H, m, Ph), 9.82 (1H, s, H-5 Isox). ^{13}C NMR (100 MHz, DMSO-d_6) δ 112.3, 113.4, 113.7, 114.0, 122.7, 128.2, 129.2, 129.4, 129.5, 130.8, 136.3, 151.7, 154.7, 159.9, 162.2, 162.9, 178.5 (NMe$_2$ + 1C were not observed). HRMS (ESI) m/z [M + H]$^+$. Calculated for $C_{24}H_{21}N_2O_3$: 388.1522. Found: 388.1524.

((E)-2-(4-Methoxystyryl)-6-(3-phenylisoxazol-4-yl)-4H-pyran-4-one (**8b**). Yield 95.1 mg (80%), light yellow crystals, mp 221–222 °C. IR (ATR) ν 3048, 2837, 1657, 1626, 1512, 1379, 823. ^{1}H NMR (400 MHz, DMSO-d_6) δ 3.80 (3H, s, OMe), 6.27 (1H, s, CH), 6.42 (1H, s, CH), 6.46 (1H, d, *J* = 15.3 Hz, =CH(α)), 6.83 (1H, d, *J* = 15.3 Hz, =CH(β)), 6.97 (2H, br.s, Ar), 7.37 (2H, br.s, Ar), 7.45–7.78 (5H, m, Ph), 9.82 (1H, s, H-5 Isox). ^{13}C NMR (126 MHz, DMSO-d_6) δ 54.9, 112.5, 113.0, 113.3, 114.5, 127.3, 128.2, 128.5, 128.9, 129.7, 130.3, 154.8, 157.9, 169.1, 161.1, 166.4, 177.4 (2C were not observed). Anal. Calculated for $C_{27}H_{17}NO_4$: C 74.38; H 4.61; N 3.77. Found: C 74.15; H 4.56; N 3.69.

3.8. Synthesis of compound 9

2,6-Bis(3-phenylisoxazol-4-yl)-4H-pyran-4-one (**9**). Bis(enamino)-substituted pyrone **5a** (100 mg, 0.427 mmol) and *N*-hydroxybenzimidoyl chloride (146 mg, 0.938 mmol) were refluxed in dry dioxane (2 mL) for 4 h. The precipitate formed was filtered and washed with EtOH. Yield 64.0 mg (39%), beige powder, mp 238–239 °C. IR (ATR) ν 3066, 2890, 1657, 1612, 1549, 1445, 1404, 1143, 913. ^{1}H NMR (500 MHz, DMSO-d_6) δ 6.19 (2H, s, H-3, H-5), 7.51–7.55

(m, 8H, Ph); 7.55–7.60 (m, 2H, Ph), 9.13 (s, 2H, Isox). ^{13}C NMR (126 MHz, DMSO-d_6) δ 112.5, 113.1, 127.1, 128.6, 128.9, 130.3, 155.1, 159.1, 161.3, 177.0. Anal. Calculated for $C_{23}H_{14}N_2O_4$: C 72.25; H 3.69; N 7.33. Found: C 72.13; H 3.84; N 7.45.

Supplementary Materials: The following supporting information can be downloaded at: https://www.mdpi.com/xxx/s1; Table S1: ground state frontier orbitals for compounds **4a** and **5a** in vacuo; Table S2: Ground state frontier orbitals for compounds **4a** and **5a** in DMSO; Table S3: ground state frontier orbitals for compounds **4a** and **5a** in ethanol; Table S4: ground state frontier orbitals for compounds **4a** and **5a** in methanol; Table S5: frontier orbitals of the first singlet excited state for the relaxed geometry of compounds **4a** and **5a** in vacuo; Table S6: frontier orbitals of the first singlet excited state for the relaxed geometry of compounds **4a** and **5a** in DMSO; Table S7: frontier orbitals of the first singlet-excited state for the relaxed geometry of compounds **4a** and **5a** in ethanol; Table S8: frontier orbitals of the first singlet-excited state for the relaxed geometry of compounds **4a** and **5a** in methanol; calculated normalized UV-vis spectra for compounds **4a** and **5a** in DMSO, methanol, and ethanol; electron density difference maps for compounds **4a** and **5a**; the method of preparation of 2-methyl-6-trifluoromethyl-4-pyrone; full ^{1}H, ^{19}F, and ^{13}C NMR spectra of all synthesized compounds; Figure S1: Normalized absorption and emission spectra of **4a** in DMSO at a CAM-B3LYP level; Figure S2: Normalized absorption and emission spectra of **4a** in ethanol at a CAM-B3LYP level; Figure S3: Normalized absorption and emission spectra of **4a** in methanol at a CAM-B3LYP level; Figure S4: Normalized absorption and emission spectra of **5a** in DMSO at a CAM-B3LYP level; Figure S5: Normalized absorption and emission spectra of **5a** in ethanol at a CAM-B3LYP level; Figure S6: Normalized absorption and emission spectra of **5a** in methanol at a CAM-B3LYP level; Figure S7: An electron density difference map for the S0→S1 transition of compound **4a** in vacuo; Figure S8: An electron density difference map for the S0→S1 transition of compound **4a** in methanol; Figure S9: An electron density difference map for the S0→S1 transition of compound **5a** in vacuo; Figure S10: An electron density difference map for the S0→S1 transition of compound 5a in methanol.

Author Contributions: Conceptualization and methodology were provided by V.Y.S. and D.L.O.; D.L.O. conceived and designed the experiments. The experimental work was conducted by D.I.N., O.E.M., A.S.S., V.V.F., S.A.U. and A.E.S.; A.E.S. carried out photophysical experiments. D.L.O. and A.E.S. analyzed the results. D.L.O. studied and systemized the spectral data. Theoretical calculation was carried out by M.Y.K. D.L.O., M.Y.K. and V.Y.S. wrote the paper. Project administration and funding acquisition were carried out by V.Y.S. and D.L.O. All authors have read and agreed to the published version of the manuscript.

Funding: This research was funded by the Russian Science Foundation, grant number 18-13-00186 (https://rscf.ru/project/18-13-00186/ (accessed on 12 December 2022)).

Institutional Review Board Statement: Not applicable.

Informed Consent Statement: Not applicable.

Data Availability Statement: Data is contained within the article and Supplementary Materials.

Acknowledgments: Analytical studies were carried out using equipment at the Center for Joint Use 'Spectroscopy and Analysis of Organic Compounds' at the Postovsky Institute of Organic Synthesis of the Russian Academy of Sciences (Ural Branch) and the Laboratory of Complex Investigations and Expert Evaluation of Organic Materials of the Center for Joint Use at the Ural Federal University.

Conflicts of Interest: The authors declare no conflict of interest.

Sample Availability: Samples of the compounds are not available from the authors.

References

1. Singh, K.S. Pyrone-derived marine natural products: A review on isolation, bio-activities and synthesis. *Curr. Org. Chem.* **2020**, *24*, 354–401. [CrossRef]
2. Ishibashi, Y.; Ohba, S.; Nishiyama, S.; Yamamura, S. Total synthesis of phenoxan and a related pyrone derivative. *Tetrahedron Lett.* **1996**, *37*, 2997–3000. [CrossRef]

3. Schiavone, D.V.; Kapkayeva, D.M.; Murelli, R.P. Investigations into a stoichiometrically equivalent intermolecular oxidopyrylium [5 + 2] cycloaddition reaction leveraging 3-hydroxy-4-pyrone-based oxidopyrylium dimers. *J. Org. Chem.* **2021**, *86*, 3826–3835. [CrossRef]
4. Usachev, S.A.; Nigamatova, D.I.; Mysik, D.K.; Naumov, N.A.; Obydennov, D.L.; Sosnovskikh, V.Y. 2-Aryl-6-polyfluoroalkyl-4-pyrones as promising R^F-building-blocks: Synthesis and application for construction of fluorinated azaheterocycles. *Molecules* **2021**, *26*, 4415. [CrossRef] [PubMed]
5. Milyutin, C.V.; Komogortsev, A.N.; Lichitsky, B.V.; Melekhina, V.G.; Minyaev, M.E. Construction of spiro-γ-butyrolactone core via cascade photochemical reaction of 3-hydroxypyran-4-one derivatives. *Org. Lett.* **2021**, *23*, 5266–5270. [CrossRef]
6. Chen, J.; Wu, L.; Wu, J. Kojic acid and maltol: The "transformers" in organic synthesis. *Chin. Chem. Lett.* **2020**, *31*, 2993–2995. [CrossRef]
7. Yasukata, T.; Masui, M.; Ikarashi, F.; Okamoto, K.; Kurita, T.; Nagai, M.; Sugata, Y.; Miyake, N.; Hara, S.; Adachi, Y.; et al. Practical synthetic method for the preparation of pyrone diesters: An efficient synthetic route for the synthesis of dolutegravir sodium. *Org. Process Res. Dev.* **2019**, *23*, 565–570. [CrossRef]
8. Obydennov, D.L.; Khammatova, L.R.; Eltsov, O.S.; Sosnovskikh, V.Y. A chemo- and regiocontrolled approach to bipyrazoles and pyridones via the reaction of ethyl 5-acyl-4-pyrone-2-carboxylates with hydrazines. *Org. Biomol. Chem.* **2018**, *16*, 1692–1707. [CrossRef]
9. Palkina, K.A.; Ipatova, D.A.; Shakhova, E.S.; Balakireva, A.V.; Markina, N.M. Therapeutic potential of hispidin—Fungal and plant polyketide. *J. Fungi* **2021**, *7*, 323. [CrossRef]
10. Purtov, K.V.; Petushkov, V.N.; Baranov, M.S.; Mineev, K.S.; Rodionova, N.S.; Kaskova, Z.M.; Tsarkova, A.S.; Petunin, A.I.; Bondar, V.S.; Rodicheva, E.K.; et al. The chemical basis of fungal bioluminescence. *Angew. Chem. Int. Ed.* **2015**, *54*, 8124–8128. [CrossRef]
11. Zuidema, D.R.; Jones, P.B. Triplet photosensitization in cyercene A and related pyrones. *J. Photochem. Photobiol. B Biol.* **2006**, *83*, 137–145. [CrossRef] [PubMed]
12. Moses, J.E.; Baldwin, J.E.; Adlington, R.M. An efficient synthesis of cyercene A. *Tetrahedron Lett.* **2004**, *45*, 6447–6448. [CrossRef]
13. Krasnaya, Z.A.; Smirnova, Y.V.; Tatikolov, A.S.; Kuz'min, V.A. Synthesis and photonics of ketocyanine dyes, 2,6-bis(4-dimethylaminoalka-1,3-dienyl)-4H-pyran-4-ones and ethoxytridecamethine salts based on them. *Russ. Chem. Bull.* **1999**, *48*, 1329–1334. [CrossRef]
14. Obydennov, D.L.; Simbirtseva, A.E.; Sosnovskikh, V.Y. Synthesis of 4-oxo-6-styryl-4H-pyran-2-carbonitriles and their application for the construction of new 4-pyrone derivatives. *Res. Chem. Intermed.* **2022**, *48*, 2155–2179. [CrossRef]
15. Rahimpour, K.; Zarenezhad, H.; Teimuri-Mofrad, R. Transition metal-catalyzed C–N cross-coupling reaction of bromine-substituted pyranylidene derivatives: Synthesis, characterization, and optical properties study of pyran-based chromophores. *J. Iran. Chem. Soc.* **2020**, *17*, 2627–2636. [CrossRef]
16. Cao, Y.; Xi, Y.; Teng, X.; Li, Y.; Yan, X.; Chen, L. Alkoxy substituted D-p-A dimethyl-4-pyrone derivatives: Aggregation induced emission enhancement, mechanochromic and solvatochromic properties. *Dyes Pigm.* **2017**, *137*, 75–83. [CrossRef]
17. Cao, Y.; Chen, L.; Xi, Y.; Li, Y.; Yan, X. Stimuli-responsive 2,6-diarylethene-4H-pyran-4-one derivatives: Aggregation induced emission enhancement, mechanochromism and solvatochromism. *Mater. Lett.* **2018**, *212*, 225–230. [CrossRef]
18. Liu, D.; Cao, Y.; Yan, X.; Wang, B. Two stimulus responsive carbazole substituted D–π–A pyrone compounds exhibiting mechanochromism and solvatochromism. *Res. Chem. Intermed.* **2019**, *45*, 2429–2439. [CrossRef]
19. Pecourneau, J.; Losantos, R.; Monari, A.; Parant, S.; Pasc, A.; Mourer, M. Synthesis and photoswitching properties of bioinspired dissymmetric γ-pyrone, an analogue of cyclocurcumin. *J. Org. Chem.* **2021**, *86*, 8112–8126. [CrossRef]
20. Guo, Z.; Zhu, W.; Tian, H. Dicyanomethylene-4H-pyranchromophores for OLED emitters, logic gates and optical chemosensors. *Chem. Commun.* **2012**, *48*, 6073–6084. [CrossRef]
21. Casimiro, L.; Maisonneuve, S.; Retailleau, P.; Silvi, S.; Xie, J.; Métivier, R. Photophysical properties of 4-dicyanomethylene-2-methyl-6-(p-dimethylamino-styryl)-4H-pyran revisited: Fluorescence versus photoisomerization. *Chem. Eur. J.* **2020**, *26*, 14341–14350. [CrossRef] [PubMed]
22. Hoche, J.; Schulz, A.; Dietrich, L.M.; Humeniuk, A.; Stolte, M.; Schmidt, D.; Brixner, T.; Würthner, F.; Mitric, R. The origin of the solvent dependence of fluorescence quantum yields in dipolar merocyanine dyes. *Chem. Sci.* **2019**, *10*, 11013–11022. [CrossRef] [PubMed]
23. Gao, M.-J.; Hua, Y.; Xu, J.-Q.; Zhang, L.-X.; Wang, S.; Kang, Y.-F. Near-infrared fluorescence probe with a large Stokes shift for selectively imaging of hydrogen peroxide in living cells and in vivo. *Dyes Pigm.* **2022**, *197*, 109930. [CrossRef]
24. Krasnaya, Z.A.; Smirnova, Y.V.; Shvedova, L.A.; Tatikolov, A.S.; Kuz'min, V.A. Synthesis and spectroscopic properties of cross-conjugated ketones and meso-substituted tridecamethine salts containing the pyran or pyridone fragment. *Russ. Chem. Bull.* **2003**, *52*, 2029–2037. [CrossRef]
25. Obydennov, D.L.; Simbirtseva, A.E.; Piksin, S.E.; Sosnovskikh, V.Y. 2,6-Dicyano-4-pyrone as a novel and multifarious building block for the synthesis of 2,6-bis(hetaryl)-4-pyrones and 2,6-bis(hetaryl)-4-pyridinols. *ACS Omega* **2020**, *5*, 33406–33420. [CrossRef]
26. Obydennov, D.L.; Usachev, B.I.; Sosnovskikh, V.Y. Reactions of 2-mono- and 2,6-disubstituted 4-pyrones with phenylhydrazine as general method for the synthesis of 3-(N-phenylpyrazolyl)indoles. *Chem. Heterocycl. Compd.* **2015**, *50*, 1388–1403. [CrossRef]
27. Usachev, B. 2-(Trifluoromethyl)-4H-pyran-4-ones: Convenient, available and versatile building-blocks for regioselective syntheses of trifluoromethylated organic compounds. *J. Fluor. Chem.* **2015**, *172*, 80–91. [CrossRef]

28. Zatsikha, Y.V.; Yakubovskyi, V.P.; Shandura, M.P.; Kovtun, Y.P. Functionalized bispyridoneannelated BODIPY—Bright long-wavelength fluorophores. *Dyes Pigm.* **2015**, *114*, 215–221. [CrossRef]
29. Bosson, J.; Labrador, G.M.; Besnard, C.; Jacquemin, D.; Lacour, J. Chiral near-infrared fluorophores by self-promoted oxidative coupling of cationic helicenes with amines/enamines. *Angew. Chem.* **2021**, *133*, 8815–8820. [CrossRef]
30. Wang, D.; Guo, X.; Wu, H.; Wu, Q.; Wang, H.; Zhang, X.; Hao, E.; Jiao, L. Visible light excitation of BODIPYs enables dehydrogenative enamination at their α-positions with aliphatic amines. *J. Org. Chem.* **2020**, *85*, 8360–8370. [CrossRef]
31. Baleeva, N.S.; Zaitseva, S.O.; Mineev, K.S.; Khavroshechkina, A.V.; Zagudaylova, M.B.; Baranov, M.S. Enamine–azide [2+3]-cycloaddition as a method to introduce functional groups into fluorescent dyes. *Tetrahedron Lett.* **2019**, *60*, 456–459. [CrossRef]
32. Lugovik, K.I.; Popova, A.V.; Eltyshev, A.K.; Benassi, E.; Belskaya, N.P. Synthesis of thiazoles bearing aryl enamine/aza-enamine side chains: Effect of the π-conjugated spacer structure and hydrogen bonding on photophysical properties. *Eur. J. Org. Chem.* **2017**, *2017*, 4175–4187. [CrossRef]
33. Stanovnik, B.; Svete, J. Synthesis of heterocycles from alkyl 3-(dimethylamino)propenoates and related enaminones. *Chem. Rev.* **2004**, *104*, 2433–2480. [CrossRef]
34. Ghosh, C.K.; Bhattacharyya, S.; Ghosh, C.; Patra, A. Benzopyrans. Part 41. Reactions of 2-(2-dimethylaminovinyl)-1-benzopyran-4-ones with various dienophiles. *J. Chem. Soc. Perkin Trans.* **1999**, *1*, 3005–3013. [CrossRef]
35. Gümüş, M.; Koca, İ. Enamines and dimethylamino imines as building blocks in heterocyclic synthesis: Reactions of DMF-DMA reagent with different functional groups. *Chemistryselect* **2020**, *5*, 12377–12397. [CrossRef]
36. Beliaev, N.A.; Shafikov, M.Z.; Efimov, I.V.; Beryozkina, T.V.; Lubec, G.; Dehaen, W.; Bakulev, V.A. Design and synthesis of imidazoles linearly connected to carbocyclic and heterocyclic rings via a 1,2,3-triazole linker. Reactivity of β-azolyl enamines towards heteroaromatic azides. *New J. Chem.* **2018**, *42*, 7049–7059. [CrossRef]
37. Brouwer, A.M. Standards for photoluminescence quantum yield measurements in solution (IUPAC Technical Report). *Pure Appl. Chem.* **2011**, *83*, 2213–2228. [CrossRef]
38. Fabian, J. TDDFT-calculations of Vis/NIR absorbing compounds. *Dyes Pigm.* **2010**, *84*, 36–53. [CrossRef]
39. Chipem, F.A.S.; Mishra, A.; Krishnamoorthy, G. The role of hydrogen bonding in excited state intramolecular charge transfer. *Phys. Chem. Chem. Phys.* **2012**, *14*, 8775–8790. [CrossRef]
40. Clark, B.P.; Ross, W.J.; Todd, A. 6-Substituted Pyranone Compounds and Their Use as Pharmaceuticals. U.S. Patent 4471129, 30 July 1984.
41. Coffin, A.; Ready, J.M. Selective synthesis of (+)-dysoline. *Org. Lett.* **2019**, *21*, 648–651. [CrossRef]
42. Chen, C.H.; Klubek, K.P.; Shi, J. Red Organic Electroluminescent Materials. U.S. Patent 5908581, 1 June 1999.
43. Becke, A.D. Density-functional thermochemistry. III. The role of exact exchange. *J. Chem. Phys.* **1993**, *98*, 5648–5652. [CrossRef]
44. Marenich, A.V.; Cramer, C.J.; Truhlar, D.G. Universal solvation model based on solute electron density and on a continuum model of the solvent defined by the bulk dielectric constant and atomic surface tensions. *J. Phys. Chem. B* **2009**, *113*, 6378–6396. [CrossRef] [PubMed]
45. Yanai, T.; Tew, D.P.; Handy, N.C. A new hybrid exchange–correlation functional using the Coulomb-attenuating method (CAM-B3LYP). *Chem. Phys. Lett.* **2004**, *393*, 51–57. [CrossRef]
46. Parthasarathy, V.; Castet, F.; Pandey, R.; Mongin, O.; Das, P.K.; Blanchard-Desce, M. Unprecedented intramolecular cyclization in strongly dipolar extended merocyanine dyes: A route to novel dyes with improved transparency, nonlinear optical properties and thermal stability. *Dyes Pigm.* **2016**, *130*, 70–78. [CrossRef]
47. Barca, G.M.J.; Bertoni, C.; Carrington, L.; Datta, D.; De Silva, N.; Deustua, J.E.; Fedorov, D.G.; Gour, J.R.; Gunina, A.O.; Guidez, E.; et al. Recent developments in the general atomic and molecular electronic structure system. *J. Chem. Phys.* **2020**, *152*, 154102. [CrossRef]
48. Bode, B.M.; Gordon, M.S. Macmolplt: A graphical user interface for GAMESS. *J. Mol. Graph. Model.* **1998**, *16*, 133–138. [CrossRef] [PubMed]
49. Lu, T.; Chen, F. Multiwfn: A multifunctional wavefunction analyzer. *J. Comput. Chem.* **2012**, *33*, 580–592. [CrossRef] [PubMed]

Article

Synthesis and Molecular Docking of Some Novel 3-Thiazolyl-Coumarins as Inhibitors of VEGFR-2 Kinase

Tariq Z. Abolibda [1], Maher Fathalla [1,2], Basant Farag [2], Magdi E. A. Zaki [3] and Sobhi M. Gomha [1,4,*]

[1] Department of Chemistry, Faculty of Science, Islamic University of Madinah, Madinah 42351, Saudi Arabia
[2] Department of Chemistry, Faculty of Science, Zagazig University, Zagazig 44519, Egypt
[3] Department of Chemistry, Faculty of Science, Imam Mohammad Ibn Saud Islamic University (IMSIU), Riyadh 11623, Saudi Arabia
[4] Department of Chemistry, Faculty of Science, Cairo University, Cairo 12613, Egypt
* Correspondence: smgomha@iu.edu.sa or s.m.gomha@cu.edu.eg or s.m.gomha@gmail.com

Abstract: One crucial strategy for the treatment of breast cancer involves focusing on the Vascular Endothelial Growth Factor Receptor (VEGFR-2) signaling system. Consequently, the development of new (VEGFR-2) inhibitors is of the utmost importance. In this study, novel 3-thiazolhydrazinylcoumarins were designed and synthesized via the reaction of phenylazoacetylcoumarin with various hydrazonoyl halides and α-bromoketones. By using elemental and spectral analysis data (IR, ^{1}H-NMR, ^{13}C-NMR, and Mass), the ascribed structures for all newly synthesized compounds were clarified, and the mechanisms underlying their formation were delineated. The molecular docking studies of the resulting 6-(phenyldiazenyl)-2H-chromen-2-one (**3**, **6a–e**, **10a–c** and **12a–c**) derivatives were assessed against VEGFR-2 and demonstrated comparable activities to that of Sorafenib (approved medicine) with compounds **6d** and **6b** showing the highest binding scores (−9.900 and −9.819 kcal/mol, respectively). The cytotoxicity of the most active thiazole derivatives **6d**, **6b**, **6c**, **10c** and **10a** were investigated for their human breast cancer (MCF-7) cell line and normal cell line LLC-Mk2 using MTT assay and Sorafenib as the reference drug. The results revealed that compounds **6d** and **6b** exhibited greater anticancer activities (IC$_{50}$ = 10.5 ± 0.71 and 11.2 ± 0.80 μM, respectively) than the Sorafenib reference drug (IC$_{50}$ = 5.10 ± 0.49 μM). Therefore, the present study demonstrated that thiazolyl coumarins are potential (VEGFR-2) inhibitors and pave the way for the synthesis of additional libraries based on the reported scaffold, which could eventually lead to the development of efficient treatment for breast cancer.

Keywords: acetylcoumarin; hydrazonoyl halides; thiazoles; molecular docking; VEGFR-2

Citation: Abolibda, T.Z.; Fathalla, M.; Farag, B.; Zaki, M.E.A.; Gomha, S.M. Synthesis and Molecular Docking of Some Novel 3-Thiazolyl-Coumarins as Inhibitors of VEGFR-2 Kinase. *Molecules* **2023**, *28*, 689. https://doi.org/10.3390/molecules28020689

Academic Editor: Joseph Sloop

Received: 10 December 2022
Revised: 4 January 2023
Accepted: 6 January 2023
Published: 10 January 2023

Copyright: © 2023 by the authors. Licensee MDPI, Basel, Switzerland. This article is an open access article distributed under the terms and conditions of the Creative Commons Attribution (CC BY) license (https:// creativecommons.org/licenses/by/ 4.0/).

1. Introduction

According to statistics, among all cancers affecting women, breast cancer accounts for 33.1 percent, making it the most prevalent disease of either gender [1]. Breast cancer early detection programs were a valuable resource for the tens of thousands of women who were diagnosed with cervical cancer or malignant or premalignant breast cancer [2]. The percentages of incidence and fatality, however, remain at historically high levels [3,4]. Invasive breast cancer has been used to express a number of angiogenic factors at all tumor stages [5]. Additionally, vascular endothelial growth factor receptor-2 (VEGFR-2) has been discovered to be significantly expressed in both primary and metastatic invasive breast carcinomas, suggesting a role for the VEGF signaling pathway in the regulation of breast tumor angiogenesis [6]. The C-terminal and N-terminal lobes each contributed residues to the region of the VEGFR-2 of protein kinases, which actively bind adenosine triphosphate (ATP), which is placed in the gap between the two lobes [7]. At the lobe, the C-terminal is an activation loop that has a conserved aspartate-phenylalanine-glycine (DFG) motif at the start of it [8]. Type I through III inhibitors are the three classes of VEGFR-2 inhibitors. Type II inhibitors maintain the DFG motif-containing DFG-out conformation of inactive

VEGFR2 kinase; Type II inhibitors create a hydrophobic allosteric pocket next to the ATP-binding site. Improved kinase selectivity and high cellular potency are just two benefits of type II inhibitors [9]. Through the suppression of the Ras/MAPK pathway, VEGFR-2 inhibitors also delayed the development of selective estrogen receptor modulator (SERM) resistance in breast cancer [10]. An important field of study in the fight against cancer is the discovery and creation of novel anticancer drugs with high efficacy and low toxicity. According to reports in the literature, compounds comprising coumarins, thiazoles, or thiazolylcoumarins have drawn a lot of attention from drug research due to their potential anticancer action with good IC_{50} [11,12] (Figure 1).

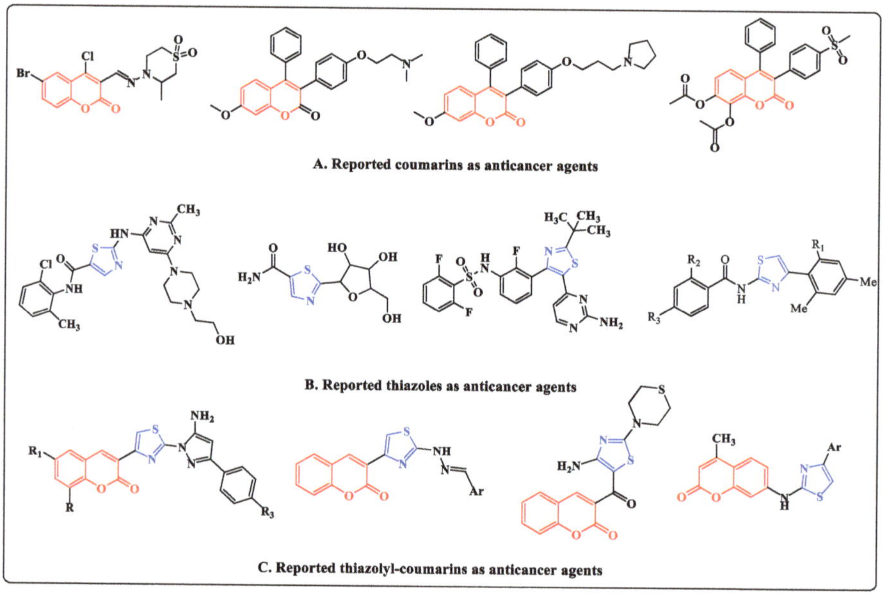

Figure 1. Examples of some reported coumarins, thiazoles and thiazolylcoumarins as anticancer agents.

Coumarin is a versatile molecule that serves as the pharmacological and biological building block for a wide range of naturally occurring chemicals [13–15]. It has been regarded as an intriguing framework for the development of anticancer drugs [16–19]. Furthermore, recent research revealed that a variety of coumarin compounds, both natural and synthetic, have antiproliferative properties via inhibiting VEGFR-2-mediated signaling pathways (Figure 2) [20–23]. On the other hand, thiazoles are considered to be important chemical synthons found in a variety of pharmacologically active compounds [24]. They possess a wide range of biological activities as anticancer, antimicrobial and anti-inflammatory agents [25–27]. Some thiazole derivatives were reported as type II VEGFR-2 inhibitors with similar activity compared with the Sorafenib reference drug (Figure 2) [28–38].

In light of our previous work on the synthesis of novel antitumor heterocycles [39–46] and with consideration of the aforementioned results, a new sort of VEGFR-2 inhibitors has been developed as prospective anti-breast cancer agents by hybridizing the coumarin and 1,3-thiazole moieties, which have been found to inhibit kinases and have antiproliferative properties. In this study, we developed and synthesized new 3-thiazolhydrazinylcoumarins in an effort to enhance the target compounds' synergistic pharmacological significance and assess their anti-breast cancer activity targeting VEGFR-2. Finally, in order to occupy the hydrophobic back pocket of VEGFR-2, a side phenyl ring was maintained, either mono or di, substituted with a wide variety consisting of hydrophobic groups, such as chloro, methyl, and phenyl. The molecular docking studies of these compounds were performed to confirm their ability to satisfy the pharmacophoric features. Moreover, it also determines the binding mode interaction that occurred with the desired VEGFR-2 inhibition.

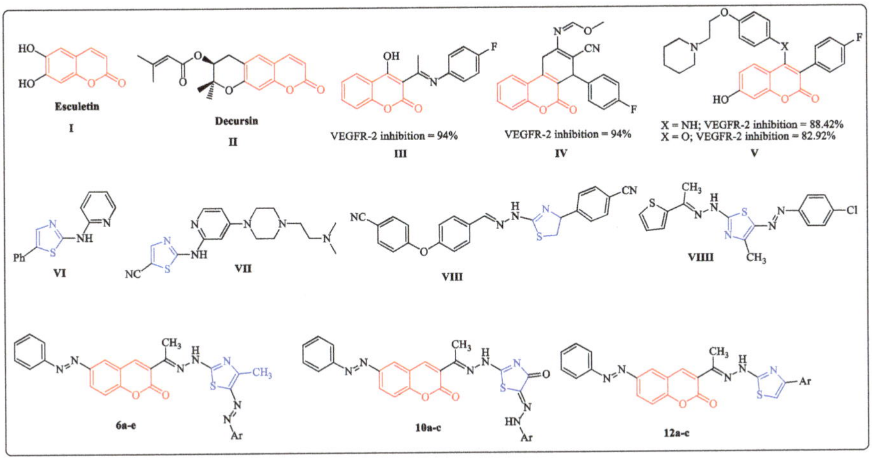

Figure 2. Examples of some reported coumarins (I–V) and thiazoles (VI–VIIII) based VEGFR-2 inhibitors and the target compounds **6a–e**, **10a–c** and **12a–c**.

2. Results and Discussion

2.1. Chemistry

Our research aims to synthesize a new series of bioactive thiazole derivatives, and this may be performed by synthesizing the starting derivative 2-(1-(2-oxo-6-(phenyldiazenyl)-2H-chromen-3-yl)ethylidene)hydrazine-1-carbothioamide (**3**) via the reaction of 3-acetyl-6-(phenyldiazenyl)-2H-chromen-2-one (**1**) [47] and hydrazinecarbothioamide **2** in EtOH in the presence of a catalytic amount of HCl under reflux as depicted in Scheme 1. Element and spectral data techniques (IR, ^{1}H-NMR, mass) were used to determine structure **3** (see Section 3 experimental section).

The reaction of compound **3** with hydrazonoyl chlorides **4a–e** [48] in EtOH containing Et$_3$N yielded the thiazole derivatives **6a–e** via cyclization with the removal of the H$_2$O molecule from intermediate **5** (Scheme 1). The structure of product **6** was proved by spectral (IR, mass, ^{1}H-NMR, ^{13}C-NMR) and elemental data. The ^{1}H-NMR spectrum of product **6a** showed a singlet signal at δ = 10.36 ppm assigned to the -NH proton, in addition to the usual signals of the fourteen aromatic protons and the two CH$_3$ group protons. Furthermore, the IR spectrum showed two stretching bands at υ = 1725 and 3347 cm^{-1} due to the C=O and the NH groups. Its ^{13}C-NMR spectrum showed δ = 8.4, 14.7 (2CH$_3$), 116.3–162.1 (20Ar-C and C=N) and 163.5 (C=O) ppm. Moreover, the mass spectra of all derivatives of compound **6** showed molecular ion peaks at the right molecular weight for the corresponding molecules (see the Section 3 experimental section, supporting information).

It is envisioned that the nucleophilic attack of the thiol group of compound **3** at the electron-deficient carbon of the hydrazone group of compound **4** creates an intermediate **5**, which would then undergo a dehydrative cyclization to produce the final products of compound **6**.

Alternative synthetic techniques might be used to create authentic samples of **6a**. Product **6a** was obtained as a result of the reaction of phenyl diazonium chloride and 3-(1-(2-(4-methyl-5-(phenyldiazenyl)thiazol-2-yl)hydrazineylidene)ethyl)-2H-chromen-2-one (**7**) [49] in pyridine (Scheme 1).

On the other hand, the reaction of compound **3** with ethyl 2-chloro-2-(2-arylhydrazineylidene)acetate **8** [50] in refluxing EtOH containing TEA as a basic catalyst afforded products **10** (Scheme 1). The structures of product **10** were confirmed based on spectral (IR, mass, ^{1}H-NMR, ^{13}C-NMR) and elemental data. The ^{1}H-NMR spectrum of **10a** revealed the expected signals at δ = 2.26, 2.27 (2s, 2CH$_3$), 7.02–7.92 (m, 13 Ar-H), 10.54, 10.79 and (2 br s, 2 NH) ppm. Also, the ^{13}C-NMR spectrum showed the expected signals at δ = 8.9, 14.2 (2CH$_3$), 117.0–156.7 (19 Ar-C and C=N), and 163.0, 171.3 (2C=O) ppm. In addition, its IR spectrum showed the expected characteristic stretching bands at υ = 3413, 3289 (2NH) and 1724, 1692 (2C=O) cm^{-1} (see Section 3 experimental section, supporting information).

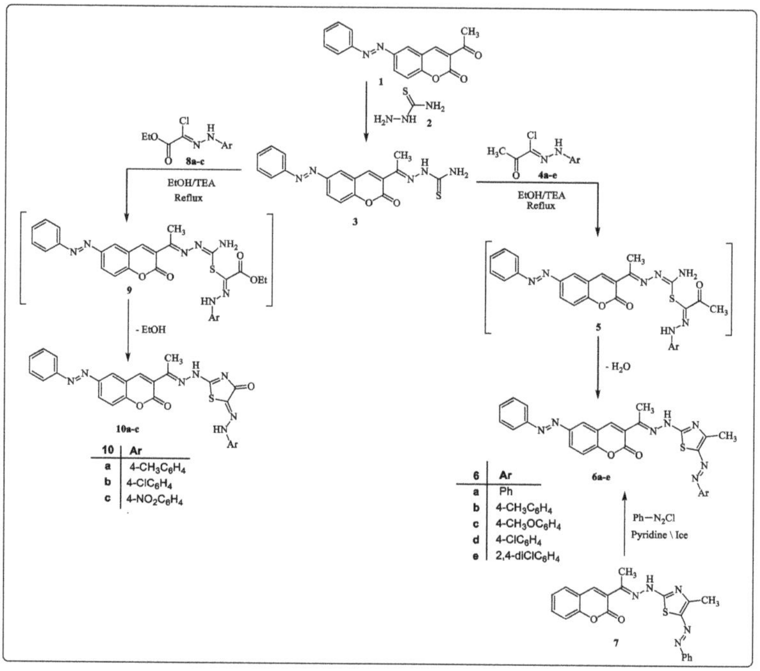

Scheme 1. Synthesis of arylazothiazole derivatives **6a–e** and arylhydrazothiazolone derivatives **10a–c**.

Furthermore, the reaction of compound **3** with α-bromoketones was used to investigate the potential of compound **3** as a building block for the production of another series of predicted physiologically active thiazoles. Thus, the reaction of compound **3** with substituted phenacyl bromides **11a–c** in EtOH afforded thiazoles **12a–c** (Scheme 2). The ^{1}H-NMR spectrum of product **12a**, considered an example of product **12**, revealed the predicted signals assigned for the CH$_3$, fourteen aromatic protons, and NH at δ = 2.31, 6.94–8.42, and 11.32 ppm, respectively. In addition, its ^{13}C-NMR spectrum showed δ 10.1(CH$_3$), 106.1–159.6 (20Ar-C and C=N) and 164.5 (C=O) ppm. The mass spectra of compounds **12a–c** exhibited a peak that matched their molecular ions in each case. Infrared spectra showed four bands, each corresponding to the carbonyl and NH groups, at υ = 1722 and 3327 cm^{-1}.

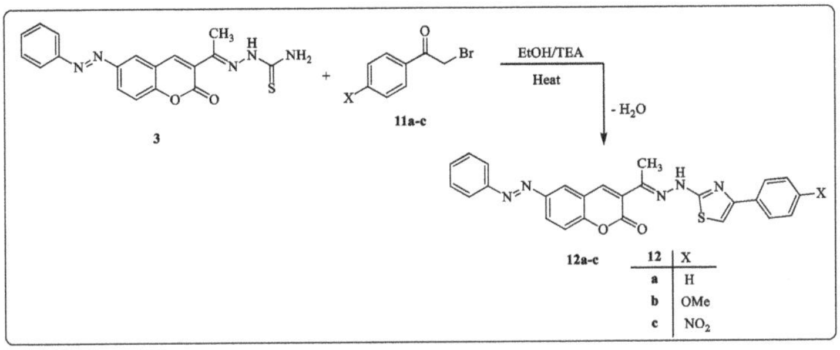

Scheme 2. Synthesis of thiazole derivatives **12a–c**.

2.2. Molecular Docking

The molecular docking was performed using MOE (Molecular Operating Environment) software version 2015.10 and was confirmed by re-docking the native ligand Sorafenib in the VEGFR-2 active regions giving a docking score of −9.284 kcal/mol. The modes of Sorafenib's interactions with VEGFR-2's active site residues are demonstrated in Figure 3. It is evident that the urea linker of Sorafenib plays a significant role in its binding ability for the enzyme VEGFR2 as it fits into the enzyme's allosteric site to form four significant hydrogen bonds with four essential residues (one *H*-bond acceptor between chloro atom with Ile1025, one *H*-bond acceptor between the oxygen of urea moiety and Asp1046, and finally two *H*-bonds donors between two NH of urea side chain with Glu885). The hydrophobic 4-chloro-3-trifluoromethylphenyl is directed toward the hydrophobic back pocket by the urea linker's binding mechanism. In addition, one *H*-bond acceptor between *N* atom of pyridine with essential amino acid Cys919 (inserted in Figure 3 as a 2D view).

Molecular docking studies demonstrated that the synthesized coumarin-based derivatives (**3**, **6a–e**, **10a–c** and **12a–c**) interact with the VEGFR-2 enzyme active site in similar ways to those of Sorafenib with binding energy values between −7.723 to −9.900 kcal/mol. The energy scores and receptor interactions of type II VEGFR-2 inhibitor for the synthesized compounds compared to the native ligand (Sorafenib) are summarized in Tables 1 and 2.

For example, compound **6a** in type II VEGFR-2 inhibitor showed two *H*-bond; one *H*-bond donor between the carbon phenyl of 2*H*-chromen-2-one skeleton with Asp814, and another *H*-bond acceptor between the nitrogen of thiazole ring with Asp1046, respectively, but compounds **6a, b** in type I VEGFR-2 inhibitor showed *H*-bond acceptor between the nitrogen atom with Lys868 (inserted in Figures 4 and 5 as a 2D view).

Table 1. Energy scores and receptor interactions of type II VEGFR-2 inhibitors for the synthesized compounds (**3**, **6a–e**, **10a–c** and **12a–c**), compared to Sorafenib.

Compound	Energy Score (S) (kcal/mol)	Interacting Residues
3	−7.72	Asp814, Asp1046, Lys868, and Leu889
6a	−9.15	Leu1035 and Lys868
6b	−9.81	Asp1046, Arg1027 and Lys868
6c	−9.58	Lys868
6d	−9.90	Asp1046 and Lys868
6e	−9.37	Asp1046 and Lys868
10a	−9.67	Asp1046
10b	−9.42	Asp1046 and Phe1047
10c	−9.69	Cys919, Gly922 and Lys868
12a	−8.52	Asp814 and Arg1027
12b	−8.90	Asp814, Asp1046, Lys868 and Leu889
12c	−8.95	Asp814
Sorafenib	−8.65	Cys1045, Asp1046, Glu885 and Cys919

Table 2. Energy scores and receptor interactions of active VEGFR-2 with DFG-in motif for the synthesized compounds (**3**, **6a–e**, **10a–c** and **12a–c**), compared to Sorafenib.

Compound	Energy Score (S) (kcal/mol)	Receptor Interactions
3	−7.319	Leu840/π-H
6a	−9.155	Lys868/π-H Lys868/H-bond acceptor
6b	−9.089	Lys868/π-H Leu1035/π-H Lys868/H-bond acceptor Leu1035/π-H
6c	−9.580	-
6d	−9.857	-
6e	−9.254	Lys868/π-H
10a	−9.205	Lys868/π-H
10b	−8.728	Asp814/H-bond donor
10c	−9.572	Cyss919/H-bond acceptor Cly922/H-bond acceptor Lys868/π-H
12a	−8.135	-
12b	−8.363	Lys868/π-H
12c	−8.735	Asp814/H-bond donor
Sorafenib	−8.214	Asp814

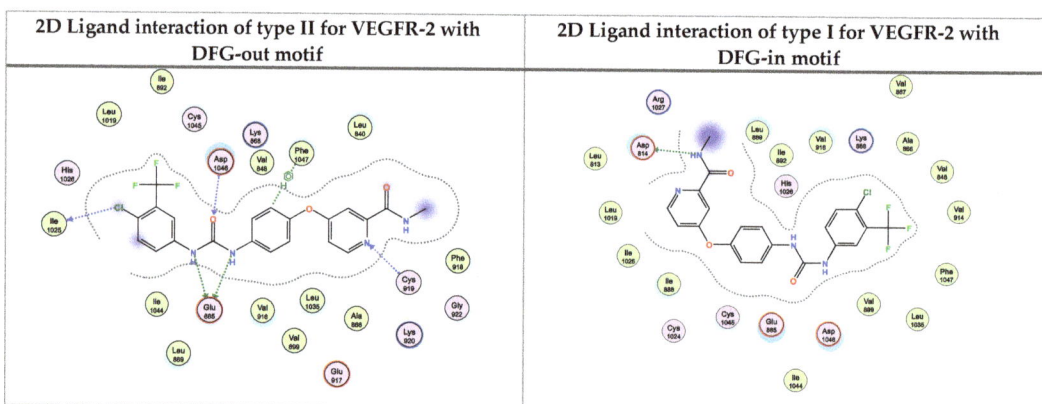

Figure 3. 2D Binding of sorafenib (a reference ligand) with DFG-out and DFG-in motif of type II and I for VEGFR-2, respectively.

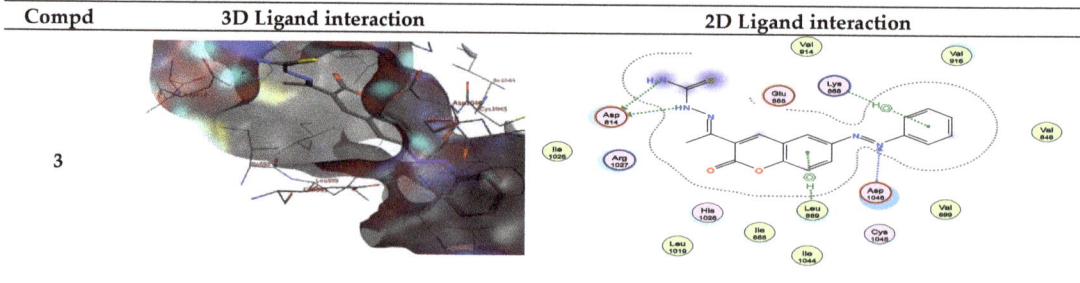

Figure 4. *Cont.*

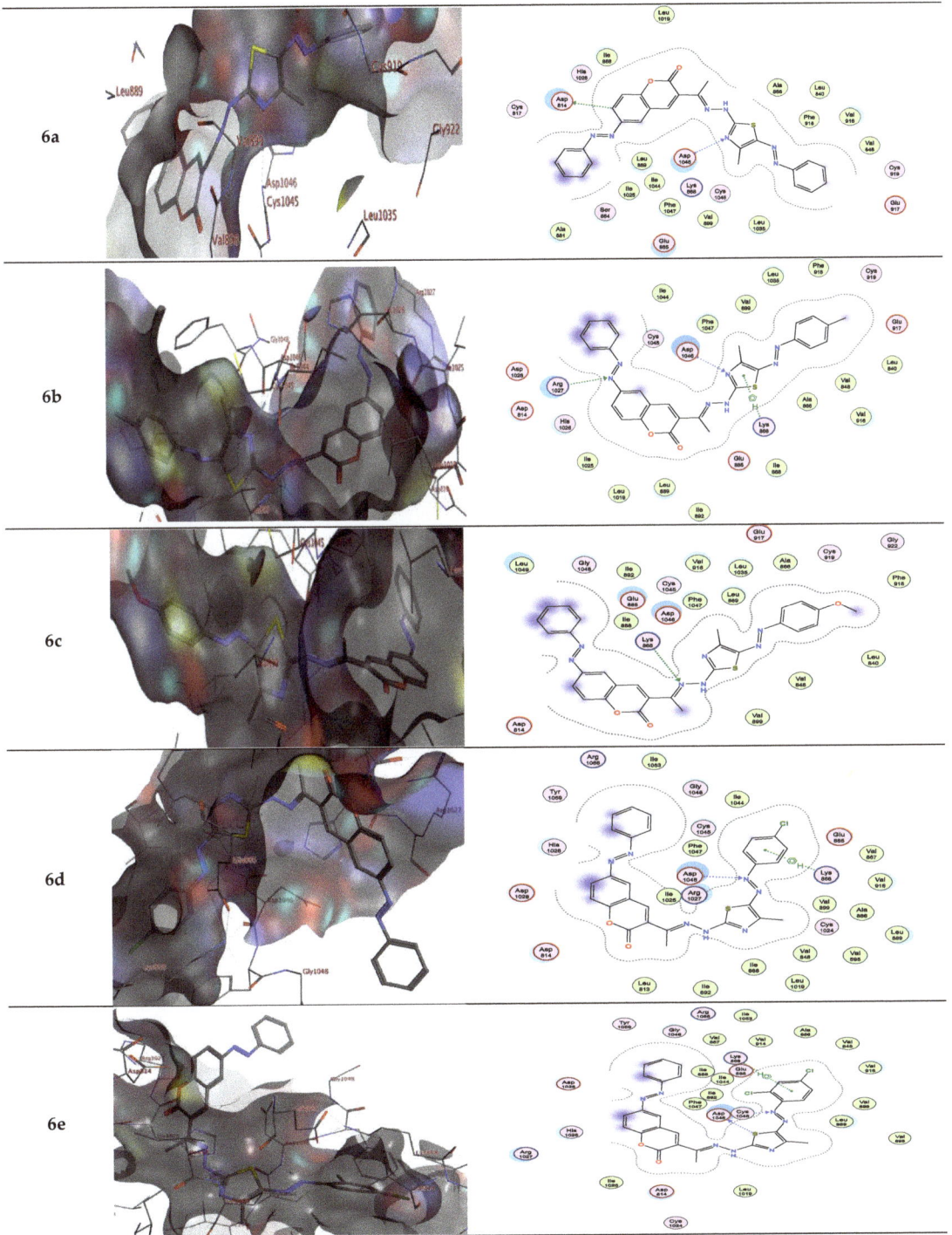

Figure 4. *Cont.*

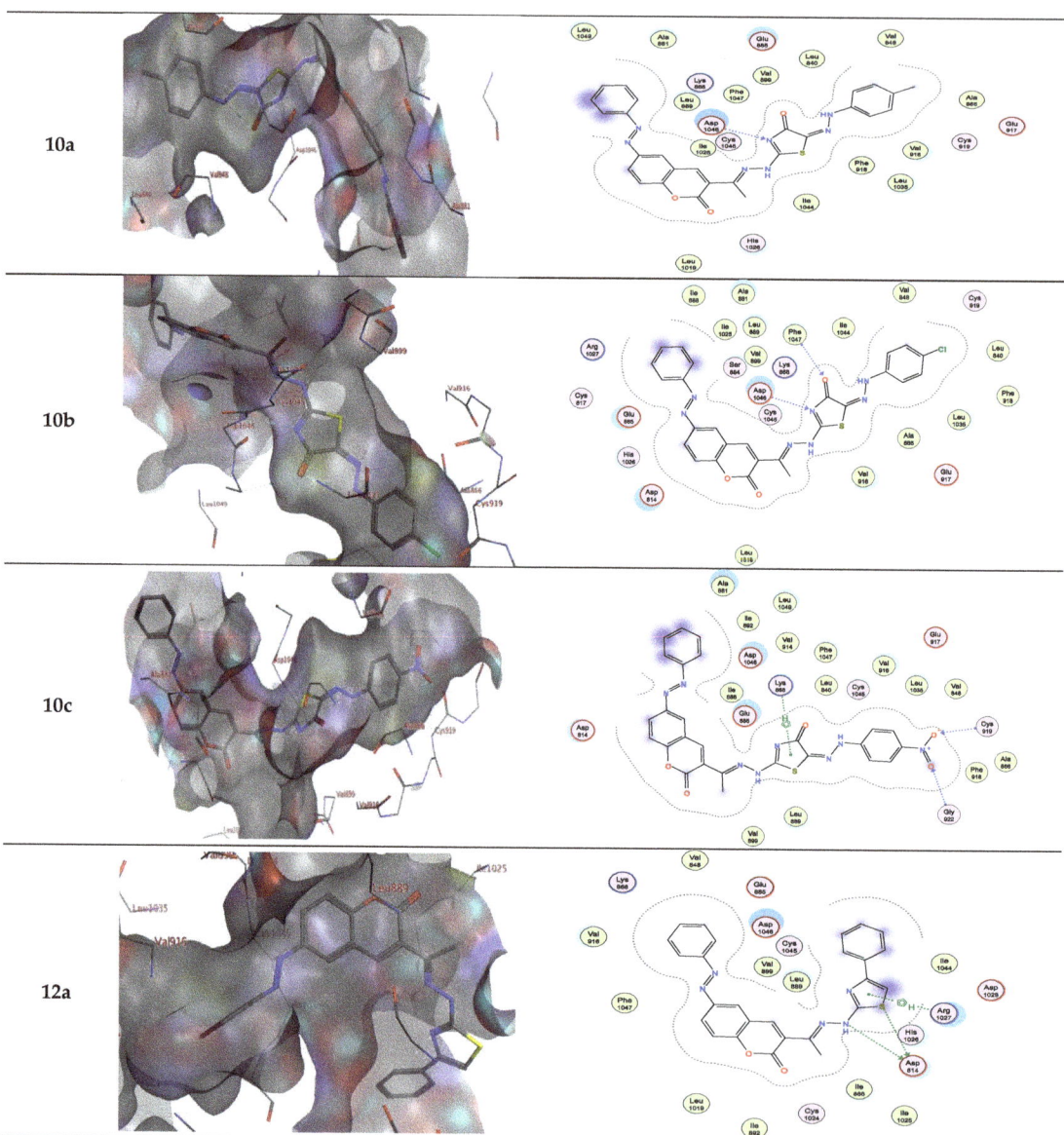

Figure 4. *Cont.*

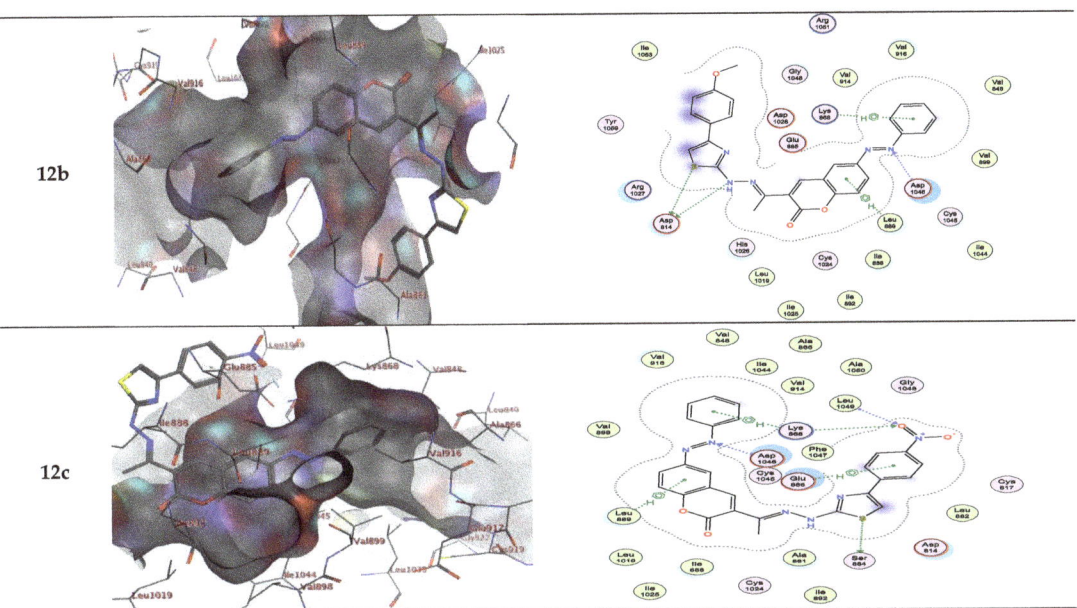

Figure 4. Interactions between ligands in three dimensions (left) and two dimensions (right) within a compound's binding pocket of 4ASD (type II VEGFR-2 inhibitor) for compounds 3, 6a–e, 10a–c and 12a–c.

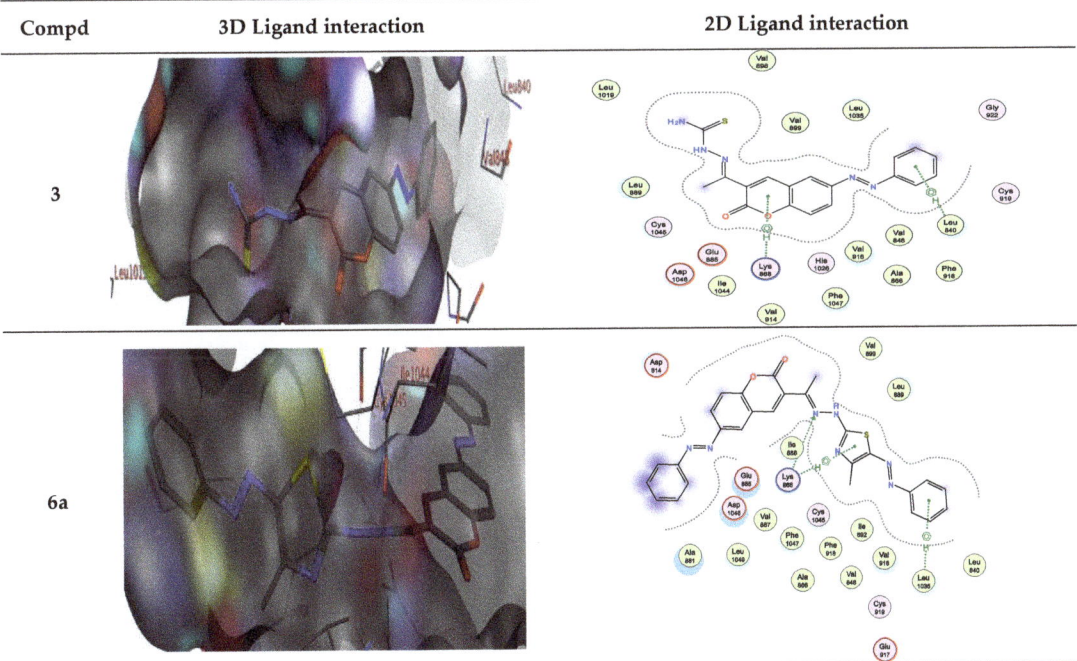

Figure 5. *Cont.*

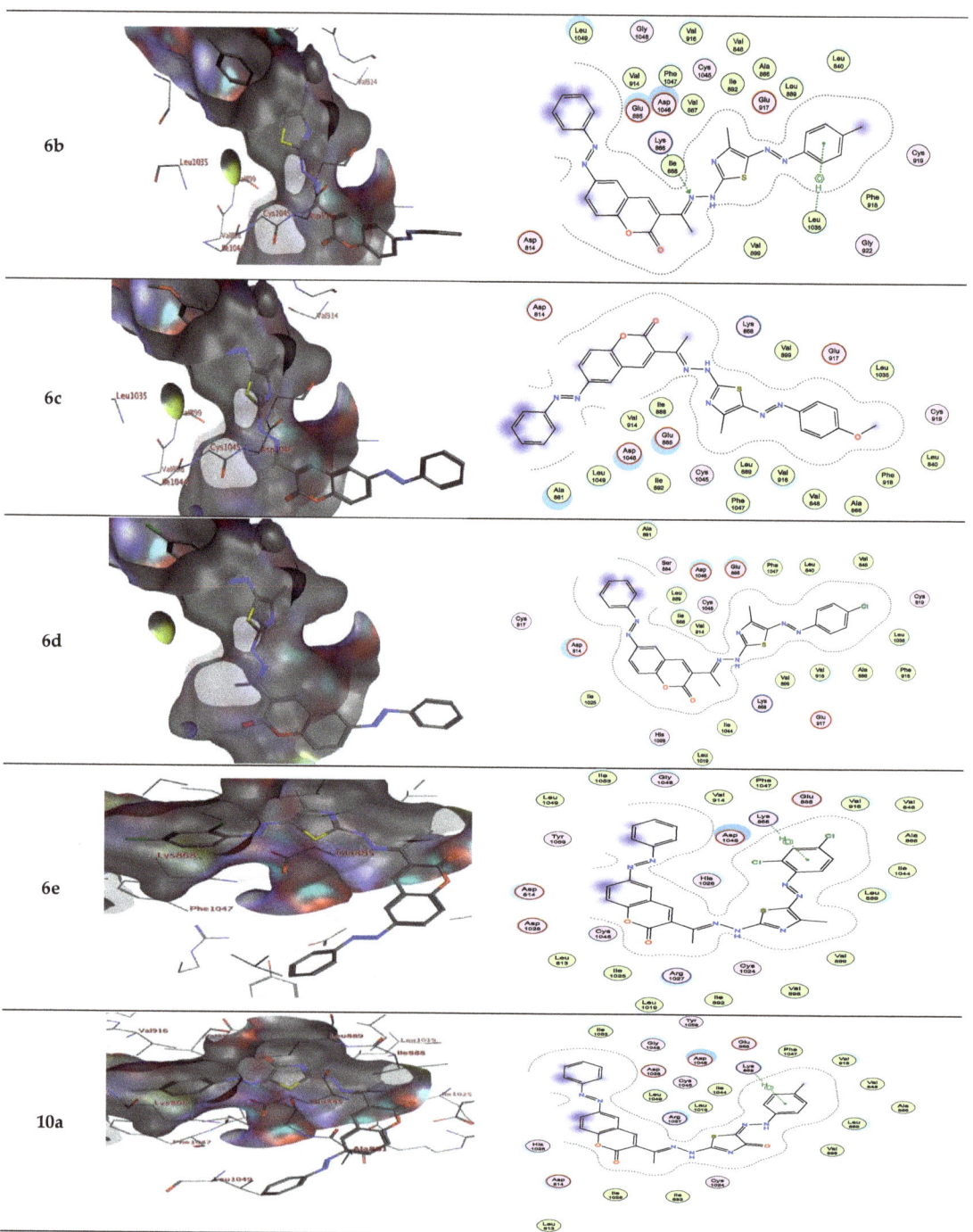

Figure 5. *Cont.*

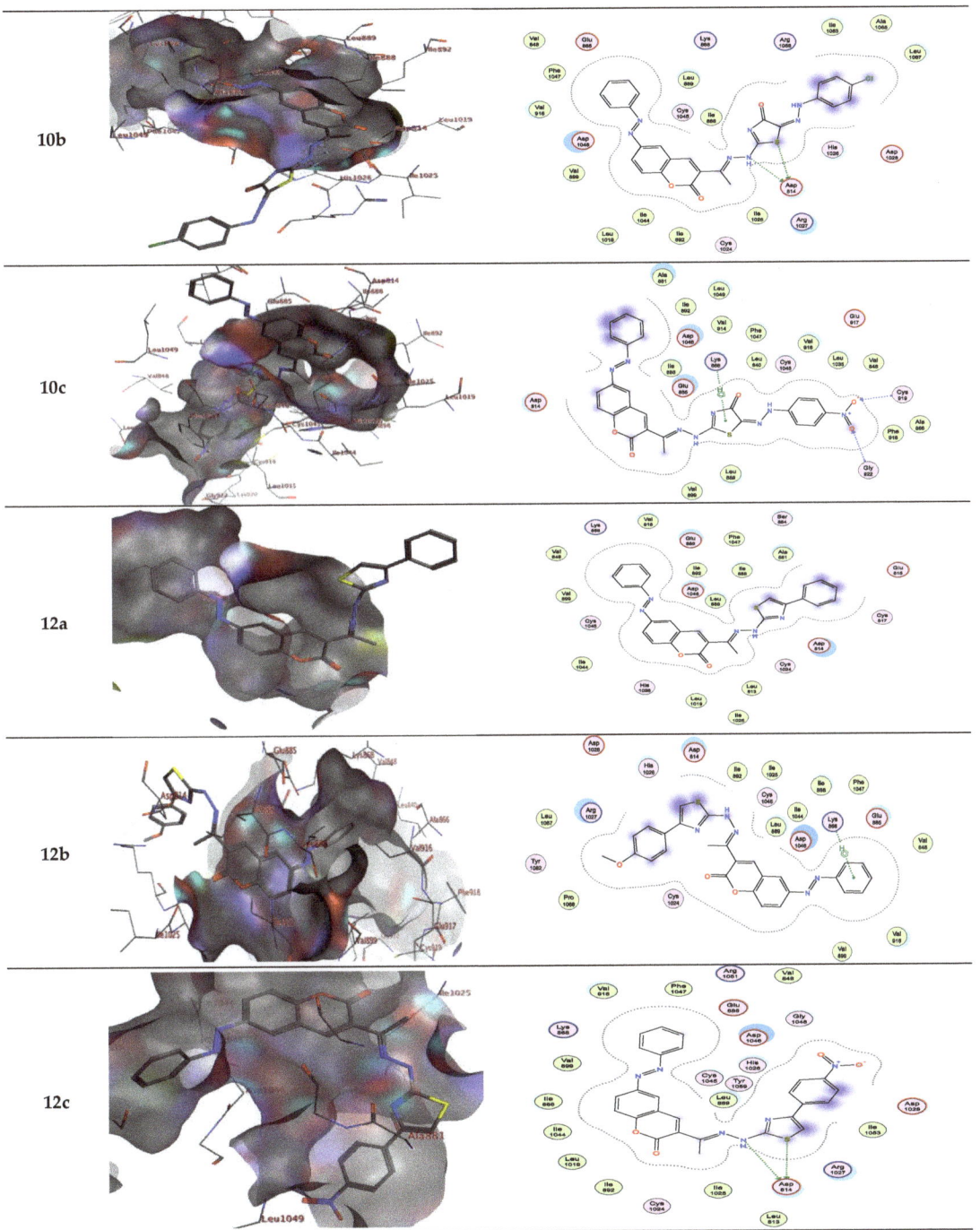

Figure 5. Interactions between ligands in three dimensions (left) and two dimensions (right) within a compound's binding pocket of 4ASD (type I VEGFR-2 inhibitor) for compounds **3**, **6a–e**, **10a–c** and **12a–c**.

Compound **6b** in type II VEGFR-2 inhibitor form two *H*-bond acceptors; one between the nitrogen atom of thiazole moiety and Asp1046 and another with Arg1027. In addition, π-*H* bond interaction with Lys868. Compound **6c** forms an *H*-bond acceptor with the Lys868. As depicted in Figure 4, compounds **6d** and **6e** form one *H*-bond acceptor between the nitrogen atom of the azo moiety and the side chain of Asp1046. Moreover, there was π-*H* interaction between the phenyl scaffold and Lys868. Moreover, compound **6e** showed an additional *H*- bond donor between the sulfur atom of the thiazole ring and Asp1046.

On the other hand, compounds **10a** and **b** form an *H*-bond acceptor with the Asp1046 via the nitrogen atom of the thiazole ring. Moreover, compound **10b** in type II VEGFR-2 inhibitor showed an additional *H*-bond acceptor between the carbonyl group of the thiazole ring and Phe1047. However, this compound in type I VEGFR-2 inhibitor showed two *H*-bond donors between the nitrogen atom and the sulfur atom of the thiazole ring with Asp814 (inserted in Figure 5 as a 2D view). Compound **10c** exhibits two *H*-bond acceptors between the two oxygen atoms of the nitro group with the side chain of Cys919 and Gly922, respectively. Moreover, there was one π-*H* interaction between the thiazole ring and Lys868 (inserted in Figures 4 and 5 as a 3D view).

Compounds **12a** and **b** form two *H*-bond donors with Asp814 via NH and the sulfur atom of thiazole moiety. Moreover, compound **12a** exhibits an additional π-*H* interaction between the thiazole ring and Arg1027. Compound **12b** shows an additional *H*-bond acceptor between the nitrogen atom of the azo group and Asp1046. In addition, two π-*H* interactions, one between the phenyl ring and Lys868, and another π-*H* interaction between the phenyl of 2*H*-chromen-2-one skeleton and Leu889. These latter π-*H* interactions of **12b** are present in compound **3** in addition to two *H*-bond donors with Asp814 and one *H*-bond acceptor with Asp1046. Compound **12c** in type II VEGFR-2 inhibitor showed three *H*-bonds; one *H*-bond donor between the sulfur atom of the thiazole ring and Ser884, and two *H*-bond acceptors between the oxygen atom of the nitro group between Lys868 with Leu1049, respectively. In addition, three π-*H* interactions; one between the phenyl ring with Lys868, another π-*H* interaction between the phenyl of 2*H*-chromen-2-one skeleton with Leu889, and finally, π-*H* interaction between the phenyl of the 4-nitrophenyl skeleton with Glu885. However, this compound in type I VEGFR-2 inhibitor exhibits two *H*-bond donors between the nitrogen atom and the sulfur atom of the thiazole ring with the side chain of Asp814 (inserted in Figure 5 as a 3D view). The results implied that the ligands under investigation occupy similar positions and orientations within Sorafenib's hypothesized binding sites.

2.3. Cytotoxic Potential

Using the MTT test and Sorafenib as a reference medication, the cytotoxicity of the most active synthesized thiazole derivatives **6b**, **6c**, **6d**, **10a**, and **10c** for their human breast cancer (MCF-7) cell line and normal cell line LLC-Mk2 was examined. Afterward, the determination of the tested sample concentrations that were sufficient to kill 50 percent of the cell population (IC_{50}) was done by using the cytotoxicity results in plotting a dose-response curve.

In addition, cytotoxic activities were reported as the average IC_{50} from three separate tests. Table 3 and Figure 6 demonstrate that most of the evaluated compounds had very varied activity when compared to the reference drug.

Examination of the SAR leads to the following conclusions:

The 1,3-thiazoles **6d** and **6b** (IC_{50} = 10.5 ± 0.71 and 11.2 ± 0.80 µM, respectively) demonstrated promising anticancer activity against MCF-7 and outperformed the reference drug (IC_{50} = 5.10 ± 0.69 µM).

For 1,3-thiazoles **6**: Substitution of the phenyl group at position 5 in the 1,3-thiazole ring with Cl atom (electron-withdrawing atom) enhances the anticancer activity (**6d** > **6b**, **6c**).

Table 3. In vitro cytotoxic activity of thiazoles **6b**, **6c**, **6d**, **10a** and **10c** against MCF-7 and LLC-MK2.

Tested Compounds	MCF-7 IC$_{50}$ (µM) *	LLC-MK2 * CC$_{50}$ (µM)
6b	11.2 ± 0.80	152.3 ± 9.30
6c	14.1 ± 0.63	141.0 ± 5.15
6d	10.5 ± 0.71	132.2 ± 8.26
10a	28.6 ± 0.64	140.3 ± 10.29
10c	15.7 ± 0.67	163.1 ± 9.12
Sorafenib	5.10 ± 0.49	135.3 ± 4.08

* The data are presented as mean, standard error (SE).

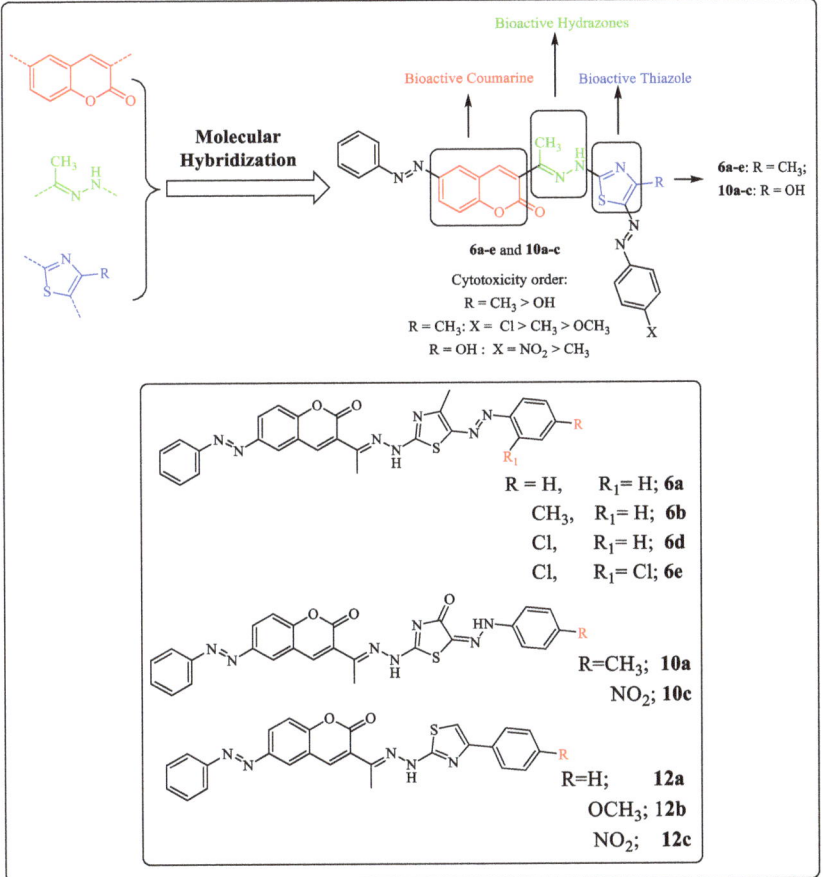

Figure 6. SAR of the synthesized thiazoles.

It was observed that the chloro **6d** derivative is more active than its methyl **6b** counterpart, which may be owing to the influence of substituent lipophilicity and the atomic size; hence, **6d** is the most potent derivative. Thus, the orientation of compound **6d** in the pocket allowed effective hydrophobic interactions as compared to compound **6b** (see the 3D models of both compounds, Figure 3). Additionally, the electron-withdrawing effect of the chloro substituent of **6d** has a greater stimulatory effect on activity than the electron-donating methyl substituent of **6b**. On the other hand, 2,4-dichloro derivative **6e** was more active than the unsubstituted one **6a**.

For 1,3-thiazolones **10**: The introduction of an electron-withdrawing group (e.g., NO_2) at the para-position of the phenyl group at position 5 in the 1,3-thiazole ring enhances the antitumor activity (**10c** > **10a**).

The introduction of the p-substitution of the electron-withdrawing group at derivative **12** has a similar activity enhancement effect. Thus, compound **12c** was the most reactive in this series, and the unsubstituted derivative **12a** exhibited the lowest activity. In addition, derivatives with thiazole moiety exhibited higher activity as compared to compound **3**, thus highlighting the positive effect of the thiazole ring on the activity of the reported compounds.

The 1,3-thiazole derivatives **6** have higher anticancer activity towards MCF-7 cell lines as compared to thiazolone derivatives **10**.

The cytotoxic activity of the Sorafenib standard drug and most active compounds **6b**, **6c**, **6d**, **10a** and **10c** were also estimated on LLC-Mk2 (rhesus monkey kidney epithelial normal cells). The outcomes of these measurements demonstrated the non-toxic effect of the tested derivatives because their CC_{50} toward normal cell lines is higher than 100 µM, as shown in Table 3.

3. Experimental Section

See the supporting information file S1.

Synthesis of *2-(1-(2-oxo-6-(phenyldiazenyl)-2H-chromen-3-yl)ethylidene) hydrazine-1-carbothioamide* (**3**).

A mixture of 3-acetyl-6-(phenyldiazenyl)-2H-chromen-2-one (**1**) (10 mmol, 2.92 g) and hydrazinecarbothioamide (**2**) (10 mmol, 0.91 g) in EtOH (40 mL) containing a few drops of concentrated HCl was heated under reflux for 4 h. The formed precipitate was recrystallized from EtOH to give product **3** as a yellowish-white solid in 73% yield; m.p. 177–179 °C; ^{1}H-NMR (DMSO-d_6) δ: 2.41 (s, 3H, CH_3), 7.15–8.32 (m, 11H, Ar-H and NH_2), 10.39 (s, br, 1H, NH) ppm; IR (KBr) ν cm^{-1}: 3419, 3372, 3284 (NH_2 and NH), 1725 (C=O), 1606 (C=N); MS m/z (%): 365 (M+, 37). Anal. Calcd: for $C_{18}H_{15}N_5O_2S$ (365.41): C, 59.17; H, 4.14; N, 19.17. Found: C, 59.04; H, 4.05; N, 19.00%.

General procedure for the synthesis of 1,3-thiazole derivatives **6a–e**, **10a–c** and **12a–c**.

Catalytic amounts of TEA were added into a solution of compound **3** (1 mmol, 0.365 g) and the appropriate hydrazonoyl chlorides **4a–e** or **8a–c** or α-bromoketones **11a–c** (1 mmol for each) in EtOH (20 mL), and the reaction mixture was refluxed for 4–6 h (monitored by TLC). Finally, the formed precipitate was recrystallized to give thiazoles **6a–e** or **10a–c**, or **12a–c**, respectively.

Alternate synthesis of **6a**.

A cold aqueous solution of benzenediazonium salt was added portion wise to a cold solution of compound **7** (1 mmol, 0.403 g) in pyridine (15 mL) under stirring then the precipitated product was recrystallized from DMF to give compound **6a** in 72% yield.

The physical properties and spectral data of the isolated products are listed in the supporting information file S1.

3.1. Molecular Docking

The most active compounds were sketched using Chemdraw 12.0, and their molecular modeling was performed using molecular operating environment software. The results were refined using the London DG force and force field energy. All minimizations were performed until a root mean square deviation (RMSD) gradient 0.1 kcal·mol^{-1}Å^{-1} using MMFF 94× (Merck molecular force field 94×), and the partial charges were determined automatically. The binding affinity of the ligand was evaluated using the scoring function and dock function (S, Kcal/mol) created by the MOE software. The enzyme's X-ray crystal structure (PDB ID: 4ASD, resolution: 2.03 Å) was downloaded in PDB format according to the protein data bank [51,52]. The enzyme was ready for studies using docking: (i) the water was eliminated from the protein; (ii) hydrogen atoms were added to the structure in their characteristic geometries, then reconnected the bonds broken and fixing the potential; (iii) as the large site, dummy atoms were used to execute a site search using MOE Alpha Site Finder upon this enzyme structure [53]; (iv) analyzing the ligand's interaction with the active site's amino acids. The best docking score is obtained as the most negative value for the active ligands. All docking procedures and scoring were recorded according to established protocols [54–56]. Triangle Matcher placement method and London dG score tool were used for docking.

3.2. Cytotoxic Assay

The cytotoxicity of the investigated substances was assessed and determined by MTT assay, and the detailed cytotoxicity assay is included in the Supplementary File S1 [57,58].

4. Conclusions

This study disclosed the design and synthesis of novel 3-thiazolhydrazinyl coumarins utilizing 3-acetyl-6-methyl-2H-chromen-2-one. Spectroscopy and elemental analyses were utilized to confirm the hypothesized product's structures. Moreover, a molecular docking study of synthesized 2H-chromen-2-one derivatives was performed to investigate their interactions with VEGFR-2's active region. In addition, the most active derivatives of the designed compounds were tested in vitro against the MCF-7 and LLC-Mk2 cell lines using MTT assay and Sorafenib as a reference drug. The results demonstrated the potential antitumor activities of compounds **6d** and **6b** (IC_{50} = 10.5 ± 0.71 and 11.2 ± 0.80 µM, respectively). Therefore, the present study demonstrated that the reported thiazolyl coumarins are potential (VEGFR-2) inhibitors and pave the way for the synthesis of additional libraries based on the reported scaffold, which could eventually lead to the development of efficient treatment for breast cancer

Supplementary Materials: The following supporting information can be downloaded at: https://www.mdpi.com/article/10.3390/molecules28020689/s1. Supplementary File S1: Physical and spectral data of the synthesized compounds and their ^{1}H- and ^{13}C-NMR spectra.

Author Contributions: T.Z.A., M.F., B.F., M.E.A.Z. and S.M.G.: Supervision, Investigation, Methodology, Resources, Formal analysis, Data curation, Funding acquisition, Writing-original draft, Writing-review and editing. All authors have read and agreed to the published version of the manuscript.

Funding: This research received no external funding.

Institutional Review Board Statement: Not applicable.

Informed Consent Statement: Not applicable.

Data Availability Statement: The data presented in this study are available on request.

Acknowledgments: The authors express their recognition to Islamic University of Madinah, Saudi Arabia, for supporting and providing Labs and chemicals for the research.

Conflicts of Interest: The authors declare no conflict of interest.

Sample Availability: Samples of the compounds **3**, **6a–e**, **10a–c** and **12a–c** are available from the authors.

References

1. Stewart, B.W.; Wild, C.P. *World Cancer Report 2014*; World Health Organization, International Agency for Research on Cancer [IARC]: Lyon, France, 2014.
2. Lee, N.C.; Wong, F.L.; Jamison, P.M.; Jones, S.F.; Galaska, L.; Brady, K.T.; Wethers, B.; Stokes-Townsend, G.A. Implementation of the National Breast and Cervical Cancer Early Detection Program. *Cancer* **2014**, *120*, 2540–2548. [CrossRef] [PubMed]
3. National Cancer Institute. SEER Stat Fact Sheet: Breastcancer. 2014. Available online: http://seer.cancer.gov/statfacts/html/breast.html (accessed on 22 August 2014).
4. National Cancer Institute. SEER Stat Fact Sheet: Cervix Uteri Cancer. 2014. Available online: http://seer.cancer.gov/statfacts/html/cervix.html (accessed on 22 August 2014).

5. Fox, S.B.; Generali, D.G.; Harris, A.L. Breast tumour angiogenesis. *Breast Cancer Res.* **2007**, *9*, 216. [CrossRef] [PubMed]
6. Ryden, L.; Stendahl, M.; Jonsson, H.; Emdin, S.; Bengtsson, N.O.; Landberg, G. Tumor-specific VEGF-A and VEGFR2 in postmenopausal breast cancer patients with long-term follow-up. Implication of a link between VEGF pathway and tamoxifen response. *Breast Cancer Res. Treat.* **2005**, *89*, 135–143. [CrossRef]
7. Roskoski, R.J. VEGF receptor protein-tyrosine kinases: Structure and regulation. *Biochem. Biophys. Res. Commun.* **2008**, *37*, 287–291. [CrossRef]
8. Zuccotto, F.; Ardini, E.; Casale, E.; Angiolini, M. Through the "gatekeeper door": Exploiting the active kinase conformation. *J. Med. Chem.* **2010**, *53*, 2681–2694. [CrossRef]
9. Regan, J.; Pargellis, C.A.; Cirillo, P.F.; Gilmore, T.; Hickey, E.R.; Peet, G.W.; Proto, A.; Swinamer, A.; Moss, N. The kinetics of binding to p38MAP kinase by analogues of BIRB 796. *Bioorg. Med. Chem. Lett.* **2003**, *13*, 3101. [CrossRef]
10. Patel, R.R.; Sengupta, S.; Kim, H.R.; Klein-Szanto, A.J.; Pyle, J.R.; Zhu, F.; Li, T.; Ross, E.A.; Oseni, S.; Fargnoli, J.; et al. Experimental treatment of oestrogen receptor (ER) positive breast cancer with tamoxifen and brivanib alaninate, a VEGFR-2/FGFR-1 kinase inhibitor: A potential clinical application of angiogenesis inhibitors. *Eur. J. Cancer* **2010**, *46*, 1537–1553. [CrossRef]
11. Rawat, A.; Reddy, A.V.B. Recent advances on anticancer activity of coumarin derivative. *Eur. J. Med. Chem. Rep.* **2022**, *5*, 100038. [CrossRef]
12. Morigi, R.; Locatelli, A.; Leoni, A.; Rambaldi, M. Recent patents on thiazole derivatives endowed with antitumor activity. *Recent Pat. Anticancer. Drug Discov.* **2015**, *10*, 280–297. [CrossRef]
13. Batran, R.Z.; Dawood, D.H.; El-Seginy, S.A.; Maher, T.J.; Gugnani, K.S.; Rondon-Ortiz, A.N. Coumarinyl pyranopyrimidines as new neuropeptide S receptor antagonists; design, synthesis, homology and molecular docking. *Bioorg. Chem.* **2017**, *75*, 274–290. [CrossRef]
14. Abdelhafez, O.M.; Amin, K.M.; Ali, H.I.; Abdalla, M.M.; Batran, R.Z. Synthesis of new 7-oxycoumarin derivatives as potent and selective monoamine oxidase A inhibitors. *J. Med. Chem.* **2012**, *55*, 10424–10436. [CrossRef]
15. Batran, R.Z.; Kassem, A.F.; Abbas, E.M.H.; Elseginy, S.A.; Mounier, M.M. Design, synthesis and molecular modeling of new 4-phenylcoumarin derivatives as tubulin polymerization inhibitors targeting MCF-7 breast cancer cells. *Bioorg. Med. Chem.* **2018**, *26*, 3474–3490. [CrossRef] [PubMed]
16. Alshabanah, L.A.; Al-Mutabagani, L.A.; Gomha, S.M.; Ahmed, H.A. Three-component synthesis of some new coumarin derivatives as anti-cancer agents. *Front. Chem.* **2022**, *9*, 762248. [CrossRef]
17. Zhao, P.; Chen, L.; Li, L.; Wei, Z.; Tong, B.; Jia, Y.; Kong, L.; Xia, Y.; Dai, Y. SC-III3, a novel scopoletin derivative, induces cytotoxicity in hepatocellular cancer cells through oxidative DNA damage and ataxia telangiectasia-mutated nuclear protein kinase activa-tion. *BMC Cancer* **2014**, *14*, 987. [CrossRef] [PubMed]
18. Batran, R.Z.; Dawood, D.H.; El-Seginy, S.A.; Ali, M.M.; Maher, T.J.; Gugnani, K.S.; Rondon-Ortiz, A.N. New Coumarin Derivatives as Anti-Breast and Anti-Cervical Cancer Agents Targeting VEGFR-2 and p38α MAPK. *Arch. Pharm. (Weinheim)* **2017**, *350*, e1700064. [CrossRef]
19. Gomha, S.M.; Abdel-aziz, H.M. Synthesis and antitumor activity of 1,3,4-thiadiazole derivatives bearing coumarine ring. *Heterocycles* **2015**, *91*, 583–592. [CrossRef]
20. Luo, G.; Li, X.; Zhang, G.; Wu, C.; Tang, Z.; Liu, L.; You, Q.; Xiang, H. Novel SERMs based on 3-aryl-4-aryloxy-2H-chromen-2-one skeleton—A possible way to dual ERα/VEGFR-2 ligands for treatment of breast cancer. *Eur. J. Med. Chem.* **2017**, *140*, 252–273. [CrossRef]
21. Pan, R.; Dai, Y.; Gao, X.H.; Lu, D.; Xia, Y.F. Inhibition of vascular endothelial growth factorinduced angiogenesis by scopoletin through interrupting the autophosphorylation of VEGF receptor 2 and its downstream signaling pathways. *Vascul. Pharmacol.* **2011**, *54*, 18–28. [CrossRef] [PubMed]
22. Park, S.L.; Won, S.Y.; Song, J.H.; Lee, S.Y.; Kim, W.J.; Moon, S.K. Esculetin inhibits VEGF-Induced angiogenesis both in vitro and in vivo. *Am. J. Chin. Med.* **2016**, *44*, 61–76. [CrossRef]
23. Franchin, M.; Rosalen, P.L.; da Cunha, M.G.; Silva, R.L.; Colon, D.F.; Bassi, G.S.; de Alenca, S.M.; Ikegaki, M.; Alves-Filho, J.C.; Cunha, F.Q.; et al. Cinnamoyloxy-mammeisin Isolated from Geopropolis Attenuates Inflammatory Process by Inhibiting Cytokine Production: Involvement of MAPK, AP-1, and NF-κB. *J. Nat. Prod.* **2016**, *79*, 1828–1833. [CrossRef]
24. Arshad, M.F.; Alam, A.; Alshammari, A.A.; Alhazza, M.B.; Alzimam, I.M.; Alam, M.A.; Mustafa, G.; Ansari, M.S.; Alotaibi, A.M.; Alotaibi, A.A.; et al. Thiazole: A Versatile Standalone Moiety Contributing to the Development of Various Drugs and Biologically Active Agents. *Molecules* **2022**, *27*, 3994. [CrossRef] [PubMed]
25. Raveesha, R.; Anusuya, A.M.; Raghu, A.V.; Kumar, K.Y.; Kumar, M.G.D.; Prasad, S.B.B.; Prashanth, M.K. Synthesis and characterization of novel thiazole derivatives as potential anticancer agents: Molecular docking and DFT studies. *Comput. Toxicol.* **2022**, *21*, 100202–100219. [CrossRef]
26. Gomha, S.M.; Riyadh, S.M.; Huwaimel, B.; Zayed, M.E.M.; Abdellattif, M.H. Synthesis, molecular docking study and cytotoxic activity on MCF cells of some new thiazole clubbed thiophene scaffolds. *Molecules* **2022**, *27*, 4639. [CrossRef]
27. Abdel-Aziz, S.A.; Taher, E.S.; Lan, P.; El-Koussi, N.A.; Salem, O.I.A.; Gomaa, H.A.M.; Youssif, B.G.M. New pyrimidine/thiazole hybrids endowed with analgesic, anti-inflammatory, and lower cardiotoxic activities: Design, synthesis, and COX-2/sEH dual inhibition. *Arch. Pharm.* **2022**, *355*, e2200024. [CrossRef]
28. Altıntop, M.D.; Sever, B.; Çiftçi, G.A.; Özdemir, A. Design, Synthesis, and Evaluation of a New Series of Thiazole-Based Anticancer Agents as Potent Akt Inhibitors. *Molecules* **2018**, *23*, 1318. [CrossRef]

29. Hassan, A.; Badr, M.; Hassan, H.A.; Abdelhamid, D.; Abuo-Rahma, G.E.D.A. Novel 4-(piperazin-1-yl)quinolin-2(1H)-one bearing thiazoles with antiproliferative activity through VEGFR-2-TK inhibition. *Bioorg. Med. Chem.* **2021**, *40*, 116168–116181. [CrossRef]
30. Perrot-Applanat, M.; Di Benedetto, M. Autocrine functions of VEGF in breast tumor cells: Adhesion, survival, migration and invasion. *Cell Adh. Migr.* **2012**, *6*, 547–553. [CrossRef]
31. Song, G.; Li, Y.; Jiang, G. Role of VEGF/VEGFR in the pathogenesis of leukemias and as treatment targets. *Oncol. Rep.* **2012**, *28*, 1935–1944. [CrossRef]
32. Bilodeau, M.T.; Rodman, L.D.; McGaughey, G.B.; Coll, K.E.; Koester, T.J.; Hoffman, W.F.; Thomas, K.A. The discovery of N-(1,3-thiazol-2-yl) pyridin-2-amines as potent inhibitors of KDR kinase. *Bioorg. Med. Chem. Lett.* **2004**, *14*, 2941–2945. [CrossRef]
33. Bilodeau, M.T.; Balitza, A.E.; Koester, T.J.; Manley, P.J.; Rodman, L.D.; Buser-Doepner, C.; Hartman, G.D. Potent N-(1,3-thiazol-2-yl)pyridin-2-amine vascular endothelial growth factor receptor tyrosine kinase inhibitors with excellent pharmacokinetics and low affinity for the hERG ion channel. *J. Med. Chem.* **2004**, *47*, 6363–6372. [CrossRef]
34. Sisko, J.T.; Tucker, T.J.; Bilodeau, M.T.; Buser, C.A.; Ciecko, P.A.; Coll, K.E.; Hartman, G.D. Potent 2-[(pyrimidin-4-yl)amine]-1,3-thiazole-5-carbonitrile-based inhibitors of VEGFR-2 (KDR) kinase. *Bioorg. Med. Chem. Lett.* **2006**, *16*, 1146–1150. [CrossRef]
35. Kiselyov, A.S.; Piatnitski, E.; Semenov, V.V. N-(aryl)-4-(azolylethyl)thiazole-5-carboxamides: Novel potent inhibitors of VEGF receptors I and II. *Bioorg. Med. Chem. Lett.* **2006**, *16*, 602–606. [CrossRef]
36. Wickens, P.; Kluender, H.; Dixon, J.; Brennan, C.; Achebe, F.; Bacchiocchi, A.; Levy, J. SAR of a novel "anthranilamide like" series of VEGFR-2, multi protein kinase inhibitors for the treatment of cancer. *Bioorg. Med. Chem. Lett.* **2007**, *17*, 4378–4381. [CrossRef]
37. Abou-Seri, S.M.; Eldehna, W.M.; Ali, M.M.; Abou El Ella, D.A. 1-piperazinylphthalazines as potential VEGFR-2 inhibitors and anticancer agents: Synthesis and in vitro biological evaluation. *Eur. J. Med. Chem.* **2016**, *107*, 165–179. [CrossRef]
38. El-Miligy, M.M.; Abd El Razik, H.A.; Abu-Serie, M.M. Synthesis of piperazine-based thiazolidinones as VEGFR2 tyrosine kinase inhibitors inducing apoptosis. *Fut. Med. Chem.* **2017**, *9*, 1709–1729. [CrossRef] [PubMed]
39. Gomha, S.M.; Abdelhady, H.A.; Hassain, D.Z.H.; Abdelmonsef, A.H.; El-Naggar, M.; Elaasser, M.M.; Mahmoud, H.K. Thiazole based thiosemicarbazones: Synthesis, cytotoxicity evaluation and molecular docking study. *Drug Des. Dev. Ther.* **2021**, *15*, 659–677. [CrossRef] [PubMed]
40. Aljohani, G.F.; Abolibda, T.Z.; Alhilal, M.; Al-Humaidi, J.Y.; Alhilal, S.; Ahmed, H.A.; Gomha, S.M. Novel thiadiazole-thiazole hybrids: Synthesis, molecular docking, and cytotoxicity evaluation against liver cancer cell lines. *J. Taibah Uni. Sci.* **2022**, *16*, 1005–1015. [CrossRef]
41. Abouzied, A.S.; Al-Humaidi, J.Y.; Bazaid, A.S.; Qanash, H.; Binsaleh, N.K.; Alamri, A.; Ibrahim, S.M.; Gomha, S.M. Synthesis, molecular docking study, and cytotoxicity evaluation of some novel 1,3,4-thiadiazole as well as 1,3-thiazole derivatives bearing a pyridine moiety. *Molecules* **2022**, *27*, 6368. [CrossRef]
42. Nayl, A.A.; Arafa, W.A.A.; Ahmed, M.; Abd-Elhamid, A.I.; El-Fakharany, E.M.; Abdelgawad, M.A.; Gomha, S.M.; Ibrahim, H.M.; Aly, A.A.; Bräse, S.; et al. Novel pyridinium based ionic liquid promoter for aqueous knoevenagel condensation: Green and efficient synthesis of new derivatives with their anticancer evaluation. *Molecules* **2022**, *27*, 2940. [CrossRef] [PubMed]
43. Gomha, S.M.; Edrees, M.M.; Muhammad, Z.A.; El-Reedy, A.A. 5-(Thiophen-2-yl)-1,3,4-thiadiazole derivatives: Synthesis, molecular docking and in-vitro cytotoxicity evaluation as potential anti-cancer agents. *Drug Des. Devel. Ther.* **2018**, *12*, 1511–1523. [CrossRef]
44. Gomha, S.M.; Edrees, M.M.; Faty, R.A.M.; Muhammad, Z.A.; Mabkhot, Y.N. Microwave-assisted one pot three-component synthesis of some novel pyrazole scaffolds as potent anticancer agents. *Chem. Central J.* **2017**, *11*, 37. [CrossRef] [PubMed]
45. Gomha, S.M.; Abdelaziz, M.R.; Kheder, N.A.; Abdel-aziz, H.M.; Alteraiy, S.; Mabkhot, Y.N. A Facile access and evaluation of some novel thiazole and 1,3,4-thiadiazole derivatives incorporating thiazole moiety as potent anticancer agents. *Chem. Central J.* **2017**, *11*, 105. [CrossRef]
46. Edrees, M.M.; Abu-Melha, S.; Saad, A.M.; Kheder, N.A.; Gomha, S.M.; Muhammad, Z.A. Eco-friendly synthesis, characterization and biological evaluation of some new pyrazolines containing thiazole moiety as potential anticancer and antimicrobial agents. *Molecules* **2018**, *23*, 2970. [CrossRef]
47. Sivaguru, P.; Sandhiya, R.; Adhiyaman, M.; Lalitha, A. Synthesis and antioxidant properties of novel 2H-chromene-3-carboxylate and 3-acetyl-2H-chromene derivatives. *Tetrahedron Lett.* **2016**, *57*, 2496–2501. [CrossRef]
48. Badrey, M.G.; Gomha, S.M. 3-Amino-8-hydroxy-4-imino-6-methyl-5-phenyl-4,5-dihydro-3H-chromeno[2,3-d]pyrimidine: An efficient key precursor for novel synthesis of some interesting triazines and triazepines as potential anti-tumor agents. *Molecules* **2012**, *17*, 11538–11553. [CrossRef] [PubMed]
49. Gomha, S.M.; Khalil, K.D. A Convenient ultrasound-promoted synthesis and cytotoxic activity of some new thiazole derivatives bearing a coumarin nucleus. *Molecules* **2012**, *17*, 9335–9347. [CrossRef]
50. Gomha, S.M. A facile *one-pot* synthesis of 6,7,8,9-tetrahydrobenzo[4,5]thieno[2,3-d]-1,2,4-triazolo[4,5-a]pyrimidin-5-ones. *Monatsh. Chem.* **2009**, *140*, 213–220. [CrossRef]
51. McTigue, M.; Murray, B.W.; Chen, J.H.; Deng, Y.; Solowiej, J.; Kania, R.S. Molecular Conformations, Interactions, and Properties Associated with Drug Efficiency and Clinical Performance Among Vegfr Tk Inhibitors molecular conformations, interactions, and properties associated with drug efficiency and clinical performance among Vegfr Tk Inhibitors. *Proc. Natl. Acad. Sci. USA* **2012**, *109*, 18281–18289.

52. Othman, I.M.M.; Alamshany, Z.M.; Tashkandi, N.Y.; Gad-Elkareem, M.A.; Anwar, M.M.; Nossier, E.S. New pyrimidine and pyrazole-based compounds as potential EGFR inhibitors: Synthesis, anticancer, antimicrobial evaluation and computational studies. *Bioorg. Chem.* **2021**, *114*, 105078. [CrossRef]
53. Labute, P. Protonate3D: Assignment of Ionization States and Hydrogen Coordinates to Macromolecular Structures. *Proteins* **2008**, *75*, 187–205. [CrossRef]
54. Kattan, S.W.; Nafie, M.S.; Elmgeed, G.A.; Alelwani, W.; Badar, M.; Tantawy, M.A. Molecular docking, anti-proliferative activity and induction of apoptosis in human liver cancer cells treated with androstane derivatives: Implication of PI3K/AKT/mTOR pathway. *J. Steroid Biochem. Mol. Biol.* **2020**, *198*, 105604. [CrossRef]
55. Tantawy, M.A.; Sroor, F.M.; Mohamed, M.F.; El-Naggar, M.E.; Saleh, F.M.; Hassaneen, H.M.; Abdelhamid, I.A. Molecular Docking Study, Cytotoxicity, Cell Cycle Arrest and Apoptotic Induction of Novel Chalcones Incorporating Thiadiazolyl Isoquinoline in Cervical Cancer. *Anti-Cancer Agents Med. Chem.* **2020**, *20*, 70–83. [CrossRef] [PubMed]
56. Nafie, M.S.; Tantawy, M.A.; Elmgeed, G.A. Screening of different drug design tools to predict the mode of action of steroidal derivatives as anti-cancer agents. *Steroids* **2019**, *152*, 108485. [CrossRef] [PubMed]
57. Gomha, S.M.; Riyadh, S.M.; Mahmmoud, E.A.; Elaasser, M.M. Synthesis and anticancer activities of thiazoles, 1,3-thiazines, and thiazolidine using chitosan-grafted-poly(vinylpyridine) as basic catalyst. *Heterocycles* **2015**, *91*, 1227–1243.
58. Mosmann, T. Rapid colorimetric assay for cellular growth and survival: Application to proliferation and cytotoxicity assays. *J. Immunol. Methods* **1983**, *65*, 55–63. [CrossRef] [PubMed]

Disclaimer/Publisher's Note: The statements, opinions and data contained in all publications are solely those of the individual author(s) and contributor(s) and not of MDPI and/or the editor(s). MDPI and/or the editor(s) disclaim responsibility for any injury to people or property resulting from any ideas, methods, instructions or products referred to in the content.

Article

Synthesis of Mono- and Polyazole Hybrids Based on Polyfluoroflavones

Mariya A. Panova, Konstantin V. Shcherbakov, Ekaterina F. Zhilina, Yanina V. Burgart and Victor I. Saloutin *

Postovsky Institute of Organic Synthesis, Ural Branch of the Russian Academy of Sciences, 22/20 Kovalevskoy St, Yekaterinburg 620108, Russia
* Correspondence: victor.saloutin@yandex.ru or saloutin@ios.uran.ru

Abstract: The possibility of functionalization of 2-(polyfluorophenyl)-4H-chromen-4-ones, with them having different numbers of fluorine atoms, with 1,2,4-triazole or imidazole under conditions of base-promoted nucleophilic aromatic substitution has been shown. A high selectivity of monosubstitution was found with the use of an azole (1.5 equiv.)/NaOBu^t (1.5 equiv.)/MeCN system. The structural features of fluorinated mono(azolyl)-substituted flavones in crystals were established using XRD analysis. The ability of penta- and tetrafluoroflavones to form persubstituted products with triazole under azole (6 equiv.)/NaOBu^t (6 equiv.)/DMF conditions was found in contrast to similar transformations with imidazole. On the basis of mono(azolyl)-containing polyfluoroflavones in reactions with triazole and pyrazole, polynuclear hybrid compounds containing various azole fragments were obtained. For poly(pyrazolyl)-substituted flavones, green emission in the solid state under UV-irradiation was found, and for some derivatives, weak fungistatic activity was found.

Keywords: polyfluoroflavones; 1H-1,2,4-triazole; imidazole; nucleophilic aromatic substitution; regioselectivity; azolyl-substituted flavones; photoluminescence

1. Introduction

Flavones based on a 2-phenylchromen-4-one backbone are important heteroaromatic scaffolds in organic chemistry due to their availability and significant synthetic and biological potential [1–5]. The uniqueness of this heterocyclic backbone is also due to the fact that its derivatives are widely represented in the plant world, which often determines their diverse biological action [6–11]. Isolation, identification, and chemical modification of flavones of plant origin is one of the rapidly developing areas in drug design [12,13]. Another equally important area of progress in the chemistry of flavones is the development of synthetic strategies for their modification. In addition, here, certain successes have recently been achieved, for example, its modification has been proposed by the Buchwald–Hartwig reaction [14], metal-catalyzed cascade rearrangements [15,16], electrochemical dimerization [17], CH-functionalization [18], etc. However, all of these transformations most often require expensive catalysts or complex installations.

In turn, fluorinated flavones offer extra possibilities for their functionalization in reactions with nucleophilic reagents. Fluoroaromatic compounds are well known to be perspective frameworks for their modification by different methods for formation of new C–C and C–heteroatom bonds [19–21], including C–N bond formation in reactions with azole-type heterocycles [22–24]. There are known approaches to the synthesis of chromone–azole dyads [25]; however, data on direct functionalization of fluorine-containing flavones with azoles under S$_N$Ar reaction conditions have only been only in publications from our research team [26,27], although the reaction of nucleophilic aromatic substitution of fluorine atoms is a quite simple, economically and environmentally friendly process, which offers the possibility of substitution for fluorinated substrates with advantages over reactions that use expensive catalysts [28]. To date, we have considerable practice in the synthesis and modification of polyfluoroflavones [26,27,29–32]. Previously, we proposed a convenient and efficient method for the synthesis of B-ring polyfluorinated flavones (2-(polyfluorophenyl)chromen-4-ones) [29], which can be involved in S_NAr reactions to obtain polynuclear heterocyclic compounds based on flavones and azoles. We have shown the possibility of controlling the fluorine atoms substitution of 2-(polyfluorophenyl)chromen-4-ones by pyrazole and assumed the mechanism for sequential fluorine substitution on an example of 2-pentafluorophenyl-4H-chromen-4-one [26].

Within this work in continuation of our research in this area, the features of the functionalization of 2-(polyfluoroaryl)-4H-chromen-4-ones **1–3**, with them having different numbers of fluorine atoms in the aryl substituent (Figure 1), by 1H-1,2,4-triazole and imidazole under conditions of base-promoted nucleophilic aromatic substitution were studied.

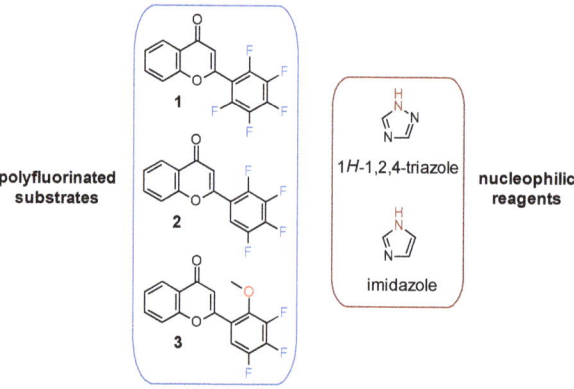

Figure 1. Polyfluorinated flavones in the S_NAr reaction with the 1H-1,2,4-triazole and imidazole under study in this work.

The introduction of azole fragments has a wide perspective due to their ability to form various noncovalent interactions with different therapeutic targets, which is valuable for drug design. Different azole derivatives have significant potential for medicinal chemistry [33–40]. Of particular interest are 1H-1,2,4-triazole and imidazole derivatives, which are known to possess therapeutic effect against drug-resistant pathogens [40–42]. The way to more effective medicines is through the synthesis of azole hybrids with other pharmacophores [27,43–45], which might be flavones, with them having a pyran framework that determines their great potential as antiviral antibacterial agents. Polyazole hybrid derivatives are also given special attention due to their potential applications as electron-transporting materials, emitters, and host materials in OLEDs, the most attractive products of organic electronics [46–48]. In this regard, the synthesis of hybrid compounds based on the flavone and azole cycles seems to be a prominent problem, which can be solved by the S_NAr reaction of polyfluorinated flavones with azoles.

2. Results

The study of the reaction of polyfluoroflavones **1–3** with 1H-1,2,4-triazole and imidazole under base-promoted nucleophilic aromatic substitution was carried out according to three synthetic protocols developed during the research of transformations with pyrazole [26]. The application of the Cs_2CO_3-promoted conditions will allow for observation of the spectrum of possible substituted products, while the application of the NaOBu^t-promoted conditions should facilitate for selective mono- and persubstitution of fluorine, which depends on variation of the nucleophile and base loading. The most convenient method for identifying fluorine-containing compounds in their mixtures is ^{19}F NMR spectroscopy data.

The reaction of flavone **1** with 3 equiv. of triazole and Cs$_2$CO$_3$ in MeCN led to a mono-, tri-, tetra-, and penta(1H-1,2,4-triazol-1-yl)-substituted products mixture, from which only mono- and penta-substituted flavones **4** and **7** were isolated by column chromatography (Scheme 1). The ^{19}F NMR data of fluorine-containing products **4–7** and their ratio in the reaction mixture are shown in Table 1. Under optimized conditions, selective synthesis of mono- and penta(1H-1,2,4-triazol-1-yl)-substituted products **4** and **7** with a good preparative yield was achieved (Scheme 1).

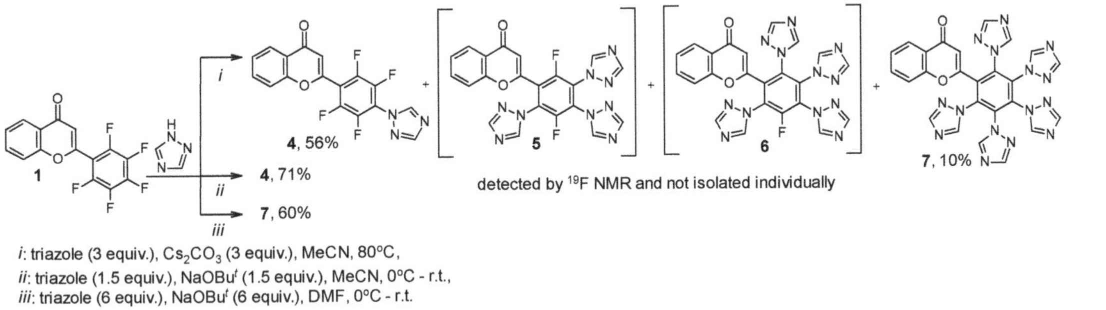

i: triazole (3 equiv.), Cs$_2$CO$_3$ (3 equiv.), MeCN, 80°C,
ii: triazole (1.5 equiv.), NaOBu^t (1.5 equiv.), MeCN, 0°C - r.t.,
iii: triazole (6 equiv.), NaOBu^t (6 equiv.), DMF, 0°C - r.t.

Scheme 1. Reaction of 2-(pentafluorophenyl)-4H-chromen-4-one **1** with 1H-1,2,4-triazole.

Table 1. ^{19}F NMR data of mixture of Cs$_2$CO$_3$-promoted reaction of flavone **1** with 1H-1,2,4-triazole.

Compound	^{19}F NMR Data	
	δ, ppm	Products Ratio, %
4	16.66, m; 24.82, m	97
5	15.95, m; 26.03, m; 26.86, m	1
6	31.53, m	2

In contrast to the transformations with triazole, the reaction of flavone **1** with 3 equiv. of imidazole under the same conditions resulted in the formation of mono-, di-, and tri(1H-imidazol-1-yl)-substituted products **8–10**, which were isolated with a poor preparative yield. The ^{19}F NMR data of fluorine-containing products **8–10** and their ratio in the mixture are shown in Table 2. Under conditions preferable to monosubstitution, 2-[2,3,5,6-tetrafluoro-4-(1H-imidazol-1-yl)phenyl]-4H-chromen-4-one **8** was synthesized with a good yield (Scheme 2). In addition, under conditions conducive to the formation of a persubstituted product, the reaction of these reagents is extremely nonselective.

Table 2. ^{19}F NMR data of a mixture of Cs$_2$CO$_3$-promoted reaction of flavone **1** with imidazole.

Compound	^{19}F NMR Data	
	δ, ppm, J, Hz	Products Ratio, %
8	14.88, m; 24.52, m	63
9	23.57, m; 26.31, m; 30.56, m	6
10	31.46, dm, J = 14.2 Hz; 41.17, dd, J = 14.9, 1.1 Hz	31

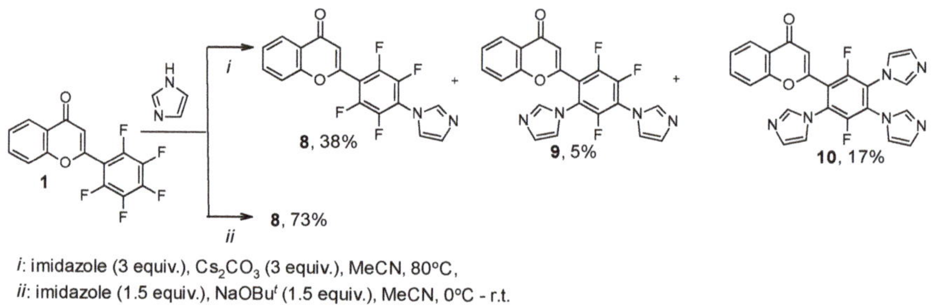

Scheme 2. Reaction of 2-(pentafluorophenyl)-4H-chromen-4-one **1** with imidazole.

It should be noted that for 2-[2,5-difluoro-3,4,6-tri(1H-imidazol-1-yl)phenyl]-4H-chromen-4-one **10**, an isomeric structure—2-[3,5-difluoro-2,4,6-tri(1H-imidazol-1-yl)phenyl]-substituted product can be proposed. The structures of **9** and **10** were solved on the basis of a comparative analysis of ^{1}H, ^{19}F NMR data of these compounds **9** and **10**, and NMR data of formerly synthesized [26] tri(1H-pyrazol-1-yl)-substituted analogue **A** (Appendix A, Table A1).

The crystal structures of 2-[2,3,5,6-tetrafluoro-4-(1H-azol-1-yl)phenyl]-4H-chromen-4-ones flavones **4** and **8** were confirmed by XRD analysis (Figures 2 and 3). Both homologues **4** and **8** have similar structural characteristics, in contrast to the pyrazolyl-substituted analogue [26], which does not have intramolecular interactions coordinating both pyrone and aryl moieties, which are therefore aplanar. In addition, the introduction of triazole and imidazole fragments has a critical effect on the structure of their unit cell in the crystal. Thus, a cell of compound **4** has a rhombic syngony and compound **8** has a monoclinic syngony. It is important to note that for two systems of atoms C16C11C2C3 and C17N1C14C15 of product **4** and similar systems C3C2C11C12 and C13C14N1C17 of analogue **8**, the torsion angles have values close in absolute value, but pairwise opposite in sign, equal to 42.70(0.76), −47.50(0.85), and −58.89(0.31), 56.50(0.34) degrees for **4** and **8**, respectively. The unit cell of the crystal **4** consists of four molecules due to the formation of O⋯H, F⋯F, and C⋯C sp2 short intermolecular contacts (O3⋯H17 2.373, O3⋯H18 2.634, C5⋯C14 3.331(8), C9⋯C12 3.303(7), F2⋯F4 2.889(5) Å) (Figure 2). The unit cell of crystal **8** also consists of four molecules stabilized by C⋯H, C⋯F, and C⋯C sp2 short intermolecular contacts (C18⋯H5 2.80(2), C7⋯F3 3.117(3), C9⋯C16 3.396(3) Å) (Figure 3).

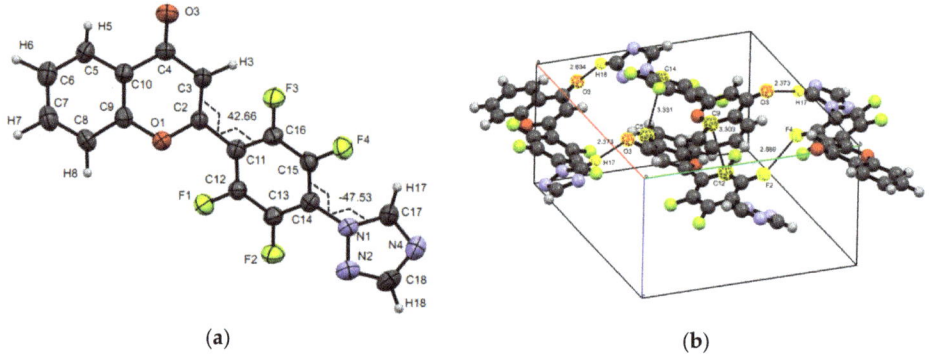

(a) (b)

Figure 2. Molecular structure and selected torsions (**a**), and unit cell (**b**) of compound **4** with atoms represented as thermal ellipsoids of thermal vibrations with a 50% probability.

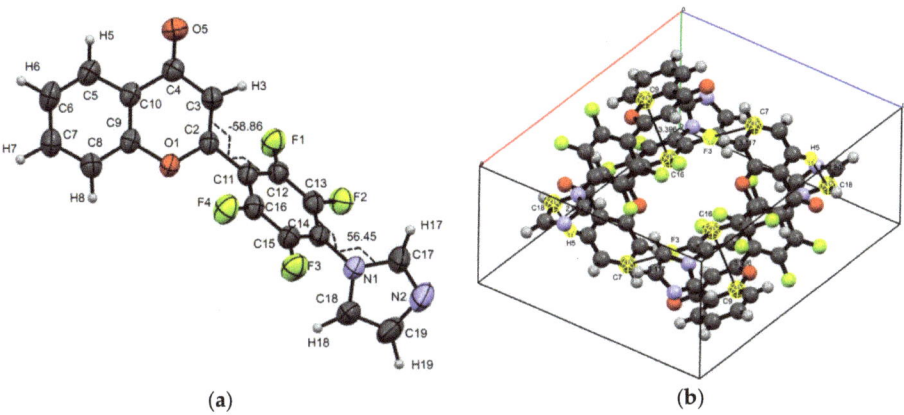

Figure 3. Molecular structure and selected torsions (**a**), and unit cell (**b**) of compound **8** with atoms represented as thermal ellipsoids of thermal vibrations with a 50% probability.

Furthermore, we introduced flavone **2**, with it having 2,3,4,5-tetrafluorophenyl substituent, in the S_NAr reaction with triazole and imidazole. The reaction of compound **2** with 3 equiv. of triazole (Scheme 3) and imidazole (Scheme 4) in the presence of Cs_2CO_3 in MeCN led to mono-, di-, and tri(azol-1-yl)-substituted products mixture **11–13** and **15–17**, respectively. The ^{19}F NMR data of fluorine-containing flavones **11–13** and **15–17** and their ratio in the mixture are shown in Table 3. Under optimized conditions, selective synthesis of mono- **11** and tetra(triazol-1-yl)-substituted products **14** was conducted with a good preparative yield (Scheme 3). However, under similar conditions, only mono(imidazolyl)-substituted flavone **15** was obtained in a good yield from the reaction with imidazole, and it was not possible to isolate the persubstituted product (Scheme 4) as in the reaction of flavone **1** (Scheme 2). This may indicate a lower reactivity of imidazole compared to triazole and the previously studied pyrazole under the conditions used [26].

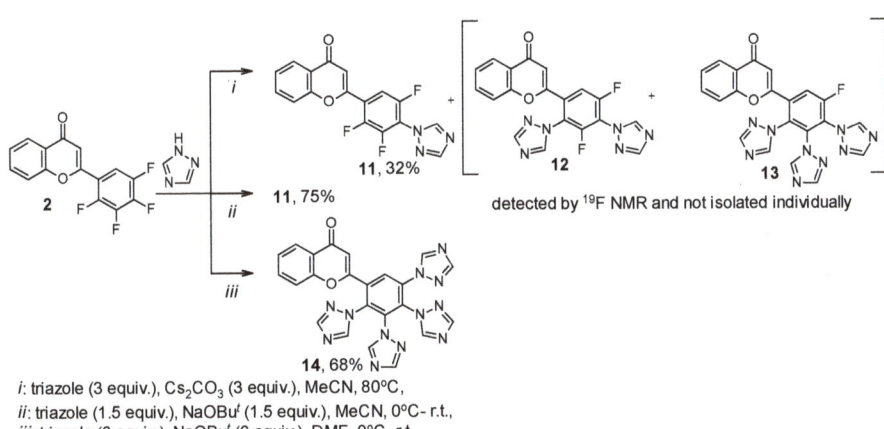

i: triazole (3 equiv.), Cs_2CO_3 (3 equiv.), MeCN, 80 °C,
ii: triazole (1.5 equiv.), NaOBu^t (1.5 equiv.), MeCN, 0 °C- r.t.,
iii: triazole (6 equiv.), NaOBu^t (6 equiv.), DMF, 0 °C- r.t.

Scheme 3. Reaction of 2-(2,3,4,5-tetrafluorophenyl)-4*H*-chromen-4-one **2** with 1*H*-1,2,4-triazole.

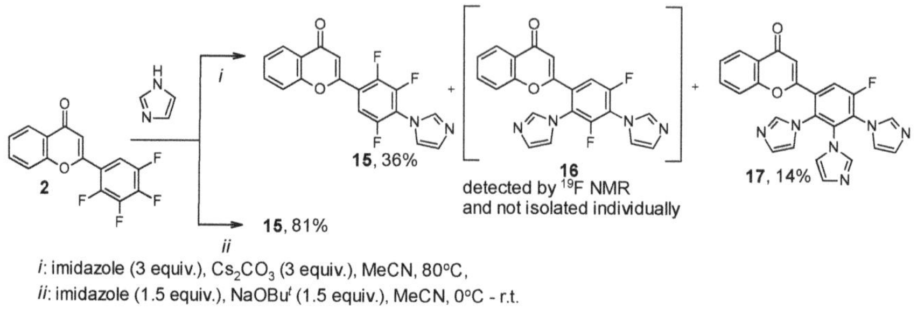

Scheme 4. Reaction of 2-(2,3,4,5-tetrafluorophenyl)-4H-chromen-4-one **2** with imidazole.

i: imidazole (3 equiv.), Cs$_2$CO$_3$ (3 equiv.), MeCN, 80°C,
ii: imidazole (1.5 equiv.), NaOBu^t (1.5 equiv.), MeCN, 0°C - r.t.

Table 3. ^{19}F NMR data of mixtures of Cs$_2$CO$_3$-promoted reactions of flavone **2** with 1H-1,2,4-triazole and imidazole.

Compound	^{19}F NMR Data	
	δ, ppm	Products Ratio, %
11	23.72, d, J = 19.9 Hz; 24.92, ddd, J = 20.1, 14.7, 5.6 Hz; 39.24, dd, J = 14.1, 10.7 Hz	56
12	39.54, dd, J = 15.2, 9.8 Hz; 39.86, dd, J = 16.0, 5.7 Hz	34
13	49.03, d, J = 9.2 Hz	10
15	21.76, dd, J = 20.0, 1.9 Hz; 24.80, ddd, J = 20.1, 14.4, 6.1 Hz, 38.16, m	35
16	39.41, m; 39.71, m	20
17	47.34, d, J = 9.1 Hz	45

The structure of 2-[2,3,5-trifluoro-4-(1H-imidazol-1-yl)phenyl]-4H-chromen-4-one **15** was confirmed by XRD analysis (Figure 4). As well as products **4** and **8**, compound **15** has no intramolecular interactions, which coordinated both pyrone and aryl moieties. However, the replacement of fluorine at the C6′ site by hydrogen allows the mutual position of these two moieties of flavone **15** in crystal close to coplanar. Torsions C3C2C11C12 and C1N1C14C13 are −7.10(0.30) and −48.14(0.27) degrees. The unit cell of the crystal **15** is monoclinic, consists of four molecules, stabilized by pairs of O ⋯ H, C ⋯ O short intermolecular contacts (O4 ⋯ H17 2.33(3), C17 ⋯ O4 3.148(3), and C2 ⋯ O4 3.127(2) Å).

The reaction of 2-(3,4,5-trifluoro-2-methoxy-phenyl)-4H-chromen-4-one **3** with 3 equiv. of triazole in the presence of Cs$_2$CO$_3$ in MeCN led to the formation of mono- and di(1H-1,2,4-triazol-1-yl)-substituted products **18**, **19**, which were isolated from the mixture (Scheme 5). In the ^{19}F NMR spectrum of the mixture, in addition to the signals corresponding to products **18** and **19**, two pairs of signals were recorded, presumably assigned by us to compounds **20** and **21** (Table 4). Due to the low intensity and insufficient resolution of these signals, the possibility of their detection is difficult. We believe that the formation of product **20** is possible due to the demethylation of flavone **18** under the reaction conditions used. The formed phenolic group in compound **20** can be in equilibrium between keto and enol forms, and keto form may undergo a nucleophilic addition reaction with triazole followed by aromatization to provide flavone **21**.

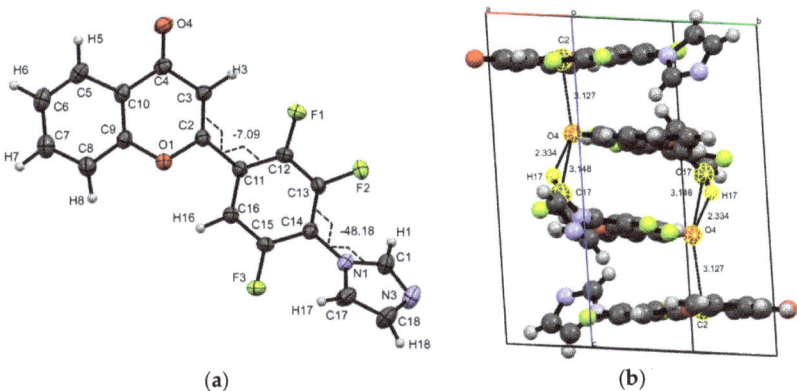

Figure 4. Molecular structure and selected torsions (**a**), and unit cell (**b**) of compound **15** with atoms represented as thermal ellipsoids of thermal vibrations with a 50% probability.

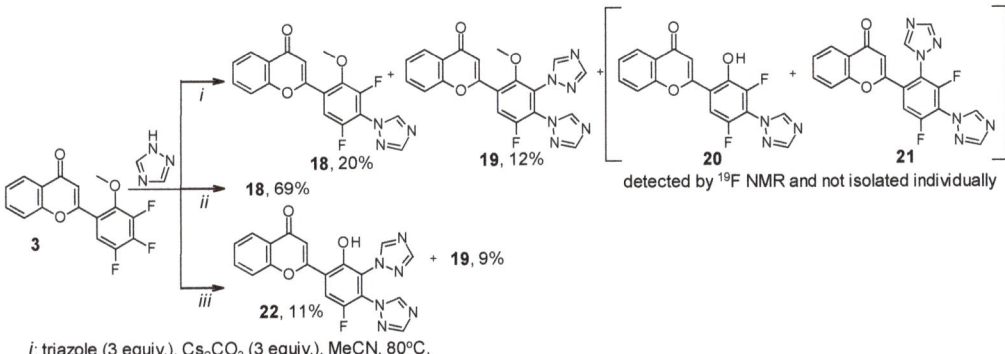

i: triazole (3 equiv.), Cs_2CO_3 (3 equiv.), MeCN, 80°C,
ii: triazole (1.5 equiv.), NaOBu^t (1.5 equiv.), MeCN, 0°C - r.t.,
iii: triazole (4 equiv.), NaOBu^t (4 equiv.), DMF, 0°C - r.t.

Scheme 5. Reaction of 2-(3,4,5-trifluoro-2-methoxyphenyl)-4*H*-chromen-4-one **3** with 1*H*-1,2,4-triazole.

Table 4. ^{19}F NMR data of mixture of Cs_2CO_3-promoted reaction of flavone **3** and 1*H*-1,2,4-triazole.

Compound	^{19}F NMR Data	
	δ, ppm	Products Ratio, %
18	29.29, m; 37.18, m	31
19	38.83, dd, *J* = 9.9, 1.9 Hz	64
20	38.77, m; 39.82, m	2
21	37.13, m; 39.63, m	3

Under NaOBu^t-promoted conditions for selective monosubstitution corresponding 2-[2,3,5,6-tetrafluoro-4-(1*H*-1,2,4-triazol-1-yl)phenyl]-4*H*-chromen-4-one **18** was synthesized with a good preparative yield. The reaction with four-fold excess of triazole and NaOBu^t led to flavone **19** and 2-[2-hydroxy-3,4-di(1*H*-1,2,4-triazol-1-yl)phenyl]-4*H*-chromen-4-one **22**, which was obtained by demethylation of 2-methoxy-3,4-di(1*H*-1,2,4-triazol-1-yl)-substituted precursor **19**, with poor yields, similar to pyrazolyl-substituted analogues [26]. The use of six-fold excess of triazole and NaOBu^t gave the same result as above with four-fold excess.

From the reaction of flavone **3** with 3 equiv. of imidazole in the presence of Cs_2CO_3 in MeCN, only mono-substituted product **23** was isolated individually. The ^{19}F NMR spectrum of the mixture (Table 5) also contained signals assigned to flavones **24** and **25** based on the conclusions given above for similar transformations with triazole. Under optimized conditions, 2-[3,5-difluoro-2-methoxy-4-(1*H*-imidazol-1-yl)phenyl]-4*H*-chromen-4-one **23** was synthesized (Scheme 6). Using an excess of this azole has not been successful in isolating any individual products.

Table 5. ^{19}F NMR data of mixture of Cs_2CO_3-promoted reaction of flavone **3** and imidazole.

Compound	^{19}F NMR Data	
	δ, ppm	Products Ratio, %
23	27.60, m; 36.46, m	31
24	38.70, m	67
25	28.22, m; 38.63, m	2

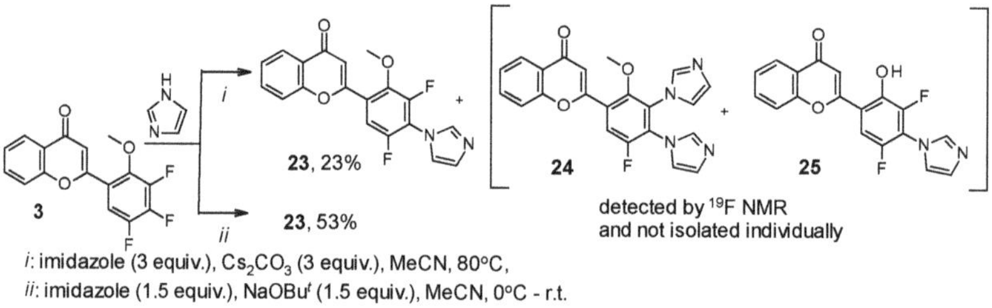

i: imidazole (3 equiv.), Cs_2CO_3 (3 equiv.), MeCN, 80 °C,
ii: imidazole (1.5 equiv.), NaOBu^t (1.5 equiv.), MeCN, 0 °C - r.t.

Scheme 6. Reaction of 2-(3,4,5-trifluoro-2-methoxyphenyl)-4*H*-chromen-4-one **3** with imidazole.

The structures of 2-[5-fluoro-2-methoxy-3,4-di(1*H*-1,2,4-triazol-1-yl)phenyl]-4*H*-chromen-4-one **19** and 2-[4,5-difluoro-2-methoxy-4-(1*H*-1,2,4-imidazol-1-yl)phenyl]-4*H*-chromen-4-one **23** were confirmed by XRD analysis (Figures 5 and 6). Compounds **19** and **23** also do not have any intramolecular interactions. It was found that the introduction of one or two azole fragments affects the geometric parameters of crystals **19** and **23**. The unit cell of compound **19** has a triclinic syngony and consists of two molecules stabilized by a pair of short intramolecular contacts C ··· F (C10 ··· F1 3.159(6) Å). The cell of compound **23** of the monoclinic system consists of four molecules forming short intermolecular contacts O ··· H, C ··· F (O3 ··· H8 2.568, O3 ··· H16 2.32(3), C3 ··· F001 3.148(3) Å). It should be noted that the torsion angles of two similar systems of atoms C16C11C2C3 and C3C2C11C12 of di(triazolyl)- and mono(imidazolyl)-substituted flavones **19** and **23** have a significant difference both in absolute value and in sign and are equal to −22.81(0.72) and 8.64(0.33) degrees, respectively. The geometry of the azole fragments of molecules **19** and **23** does not reveal fundamental differences. The torsion angles of the atomic systems C18N1C15C16, C20N4C14C15, and C17N1C14C13 are 56.12(0.58), 56.96(3.50), and 49.82(0.37) degrees, respectively.

It should be noted that a common characteristic of the crystals of both mono(1*H*-imidazol-1-yl)-substituted products **15** and **23** is the formation of a unit cell of the monoclinic system. However, the torsion angles of the systems of atoms C3C2C11C12, C1N1C14C13 of trifluoroflavone **15** and C3C2C11C12 and C17N1C14C13 of difluoroflavone **23** have practically equal absolute values and opposite values (Figures 4 and 6).

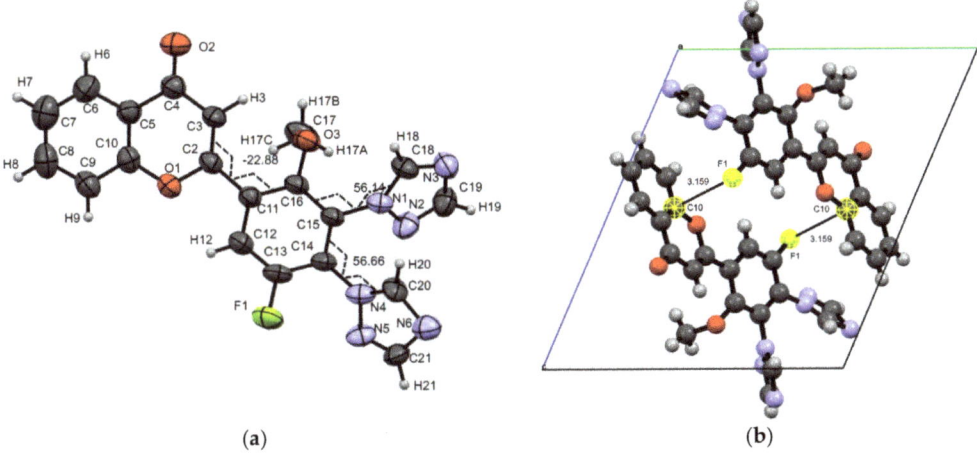

Figure 5. Molecular structure and selected torsions (**a**) and unit cell (**b**) of compound **19** with atoms represented as thermal ellipsoids of thermal vibrations with a 50% probability.

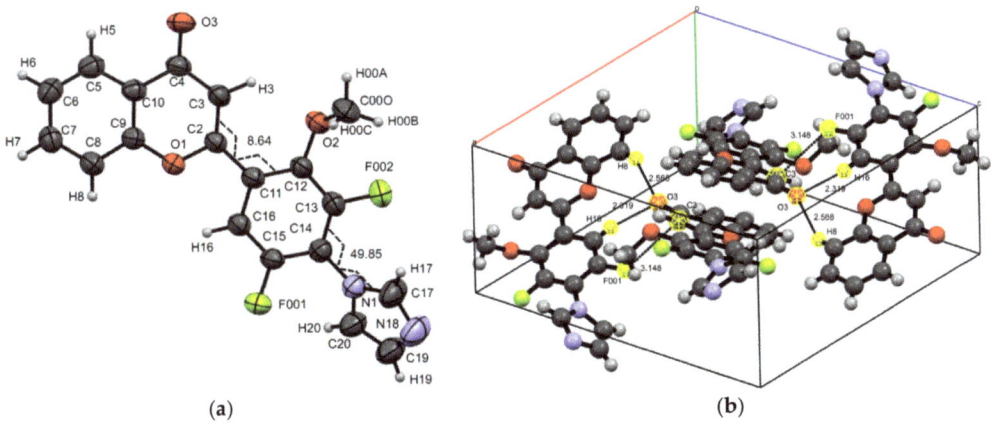

Figure 6. Molecular structure and selected torsions (**a**) and unit cell (**b**) of compound **23** with atoms represented as thermal ellipsoids of thermal vibrations with a 50% probability.

In this work, we also studied the possibility of the synthesis of 2-[poly(1*H*-azol-1-yl)phenyl]-4*H*-chromen-4-ones, involving two different azole-type heterocycles in their structure by the NaOBu^t-promoted S$_N$Ar reaction of mono(1*H*-azol-1-yl)-substituted flavones with azoles.

Thus, the reaction of mono(triazolyl)-substituted flavone **4** with four-fold excess of pyrazole and NaOBu^t led to 2-[2,5-difluoro-3,6-di(1*H*-pyrazol-1-yl)-4-(1*H*-1,2,4-triazol-1-yl)phenyl]- and 2-[2,3,5,6-tetra(1*H*-pyrazol-1-yl)-4-(1*H*-1,2,4-triazol-1-yl)phenyl]-4*H*-chromen-4-ones **26** and **27**. Flavone **26** was shown to react with two-fold excess of pyrazole and NaOBu^t to form product of complete substitution **27** (Scheme 7). Under conditions preferable for persubstitution, flavone **8** reacts with pyrazole and triazole, leading to the formation of compounds **28** and **29**. Similarly, 2-(2,3,5,6-tetrafluoro-4-(1*H*-pyrazol-1-yl)phenyl)-4*H*-chromen-4-one **30** obtained earlier [26] forms product **31** with triazole (Scheme 8). Unfortunately, we have not yet succeeded in growing a suitable single crystal for XRD due to the limited solubility of polynuclear products **27–29** and **31** in organic solvents.

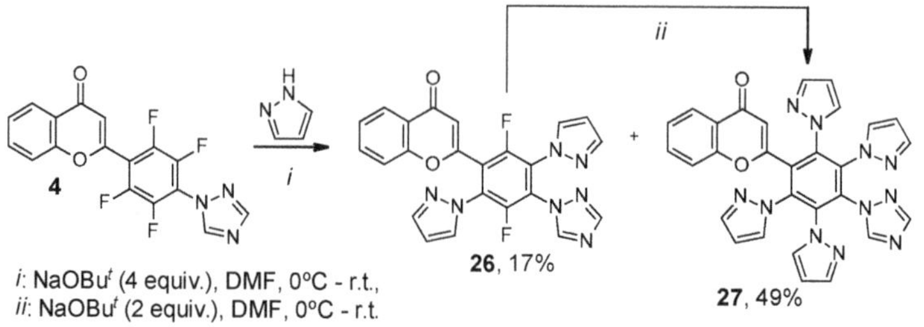
ii
4
i
26, 17%
27, 49%
i: NaOBu^t (4 equiv.), DMF, 0°C - r.t.,
ii: NaOBu^t (2 equiv.), DMF, 0°C - r.t.

Scheme 7. Reaction of flavone **4** with pyrazole.

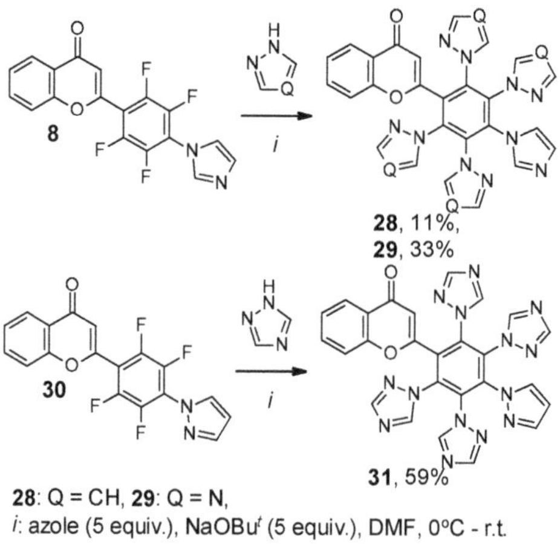
8
i
28, 11%,
29, 33%
30
i
31, 59%
28: Q = CH, 29: Q = N,
i: azole (5 equiv.), NaOBu^t (5 equiv.), DMF, 0°C - r.t.

Scheme 8. Reaction of monoazolyl-substituted flavone **4, 30** with azoles.

As is known, aryl-azole scaffolds are considered as promising materials for OLEDs [46–49]; therefore, we recorded the photoluminescence spectra for poly(azolyl)-substituted flavones **28–31** and formerly synthesized penta(pyrazolyl)-substituted analogue **32** [26]. It was found that 2-[4-(imidazolyl)-2,3,5,6-tetra(pyrazolyl)- and penta(pyrazolyl)-substituted flavones **28** and **32** exhibit emission in the solid state under UV-irradiation, in contrast to poly(triazolyl)-containing derivatives **29** and **31**. The emission spectra of these compounds were recorded, and the data are presented in Table 6 and Figure 7. Flavones **28** and **32** possess green emission with maxima 504 nm. The commission international de L'Eclairage (CIE) coordinates were (0.255; 0.528) and (0.255; 0.526) for **28** and **32**, respectively. Substitution of pyrazole fragment at C4′ by imidazole results in a 1.6-fold increase in quantum yield (0.18 and 0.29 for **28** and **32**, correspondingly). Detailed data of the fluorescence lifetime measurements of flavones **28** and **32** are given in Appendix B.

Table 6. Photoluminescent data for compounds **28** and **32** in powder at r.t.

Compound	Emission, λ_{em}, nm	τ_{avg}, [ns]/χ^2	Φ_F
28	504	7.72/ 1.109	0.29
32	504	6.81/ 1.206	0.18

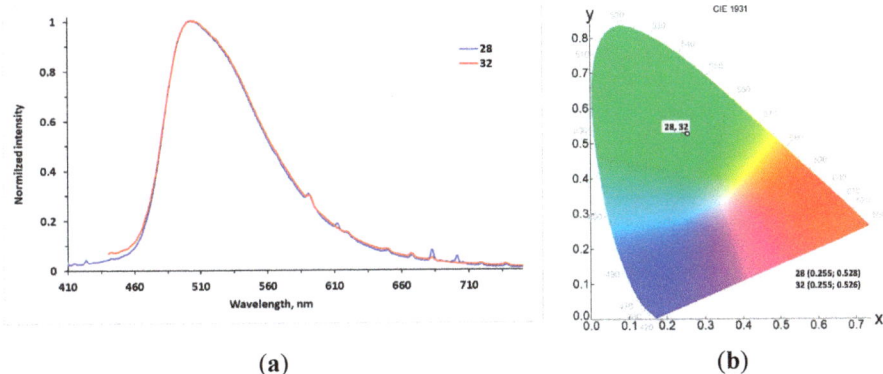

(a) (b)

Figure 7. Emission (–) spectra (**a**) and CIE 1931 chromaticity diagram (**b**) for chromophores **28** and **32** in solid state.

Azolyl-substituted flavones are of great interest in the search for the bioactive compounds among them, and, in particular, for antimycotic agents [50]; therefore, we screened the fungistatic activity of a number of compounds (Appendix C) obtained in this work and earlier [29] in relation to four control strains of clinically significant species of pathogenic fungi *Trichophyton rubrum, Epidermophyton floccosum, Microsporum canis,* and *Candida parapsilosis*. It was found that flavones **8** and **14** have a weak inhibitory effect against *T. rubrum* and *M. canis* (MIC 100 mg/mL), and derivatives **23** and **33**, combined methoxy and azole substituents, in the absence of fungistatic activity (MIC > 100 μg/mL) showed high and moderate activity in inhibiting the growth of 50% fungal culture (MIC_{50} 1.56–12.5 μg/mL).

3. Materials and Methods

3.1. Chemistry: General Information and Synthetic Techniques

Solvents and reagents except fluorine-containing flavones are commercially available and were used without purification. The NMR spectra of the synthesized compounds (see Supplementary Materials) were recorded on Bruker DRX-400 and Bruker AVANCE III 500 spectrometers (^{1}H, 400.13 (DRX400) and 500.13 (AV500) MHz, ^{13}C, 125.76 MHz, Me_4Si as an internal standard, ^{19}F, 376.44 (DRX400) and 470.52 (AV500) MHz, C_6F_6 as an internal standard, chemical shifts were not converted to CCl_3F)). IR spectra were recorded on a Perkin Elmer Spectrum Two FT-IR spectrometer (UATR) in the range of 4000–400 cm^{-1}. Elemental (C, H, N) analysis was performed on a Perkin Elmer PE 2400 Series II CHNS-O EA 1108 elemental analyzer. The melting points were measured on a Stuart SMP3 in open capillaries. The reaction progress was monitored by TLC on ALUGRAM Xtra SIL G/UV$_{254}$ sheets. The starting flavones (**1–3**) were synthesized by a procedure [29].

Synthetic technique A for the synthesis of azolyl-substituted flavones using Cs_2CO_3. flavone (0.5 or 1 mmol), azole (3 equiv.), and Cs_2CO_3 (3 equiv.) were suspended in 10 mL of MeCN. The reaction mixture was heated to 80 °C. The reaction progress was monitored by TLC. At the end of the reaction, the mixture was diluted with water (10 mL) and extracted with $CHCl_3$ or DCM (2 × 10 mL). Organic layers were combined, and the solvent was removed. The residue was immobilized on silica gel and purified by column chromatography using an appropriate eluent mixture (2:1 *v/v*).

Synthetic technique B for the synthesis of azolyl-substituted flavones using NaOBu^t. flavone (0.5 or 1 mmol) was dissolved in dry DMF (5 mL), placed in a sealed vial and cooled to 0 °C. Azole (from 1.5 to 6 equiv.) and NaOBu^t (from 1.5 to 6 equiv.) were suspended in dry DMF (5 mL) and stirred at room temperature for 5 min. The mixture was cooled to 0 °C and was added to a flavone solution in DMF while stirring. After 10 min. in a cooling bath (0 °C), the vial was removed and the reaction continued at room temperature. The reaction progress was monitored by TLC. At the end, the reaction mixture was diluted with water (10 mL), stirred, and cooled. The formed precipitate was filtered off and washed with water. The water solution was neutralized with 0.1M HCl and extracted with CHCl$_3$ or DCM. The organic layer was separated, and the solvent was removed. Organic residues were combined, immobilized on silica gel and purified by column chromatography using an appropriate eluent mixture (2:1 v/v).

3.2. Spectral and Elemental Analysis Data of Synthesized Compounds

2-[2,3,5,6-Tetrafluoro-4-(1*H*-1,2,4-triazol-1-yl)phenyl]-4*H*-chromen-4-one (**4**). Yield 202 mg (56% according to technique A), 256 mg (71% according to technique B); white powder; mp 192–194 °C; IR ν 3108, 3043 (C–H^{Ar}), 1639 (C=O), 1528, 1485, 1381 (C=C^{Ar}, C–H^{Ar}, C–N), 1139, 1110 (C–F) cm^{-1}; ^{1}H NMR (500.13 MHz, CDCl$_3$) δ 6.68 (s, 1H, CH^{Pyranone}), 7.50 (m, 1H, CH^{Ar}), 7.53 (d, *J* = 8.6 Hz, 1H, CH^{Ar}), 7.76 (ddd, *J* = 8.6, 7.1, 1.7 Hz, 1H, CH^{Ar}), 8.26 (dd, *J* = 8.0, 1.5 Hz, 1H, CH^{Ar}), 8.29 (s, 1H, CH^{Triazole}), 8.52 (s, 1H, CH^{Triazole}) ppm; ^{13}C NMR (125.76 MHz, CDCl$_3$) δ 113.7 (t, *J* = 15 Hz, C^{Ar}), 116.1 (t, *J* = 3 Hz, C^{Ar}), 118.2 (s, C^{Ar}), 118.8 (s, C^{Ar}), 123.9 (s, C^{Pyranone}), 125.9 (s, C^{Ar}), 126.0 (s, C^{Ar}), 134.5 (s, C^{Ar}), 141.7 (m, 2C^{ArF}), 144.7 (ddt, *J* = 257, 14, 5 Hz, 2C^{ArF}), 145.3 (t, *J* = 3 Hz, C^{Triazole}), 151.7 (s, C^{Ar}), 153.5 (s, C^{Triazole}), 156.6 (s, C^{Pyranone}), 177.0 (s, C^{Pyranone}) ppm; ^{19}F NMR (470.52 MHz, CDCl$_3$) δ 16.66 (m, 2F), 24.83 (m, 2F) ppm; Anal. calcd. for C$_{17}$H$_7$F$_4$N$_3$O$_2$: C 56.52, H 1.95, N 11.63, found: C 56.26, H 1.84, N 11.38.

2-[2,3,4,5,6-Penta(1*H*-1,2,4-triazol-1-yl)phenyl]-4*H*-chromen-4-one (**7**). Yield 39 mg (7% according to technique A), 335 mg (60% according to technique B); light-yellow powder; mp 322–323 °C; IR ν 3118, 3086 (C–H^{Ar}), 1651 (C=O), 1514, 1459, 1379, 1272 (C=C^{Ar}, C–H^{Ar}, C–N) cm^{-1}; ^{1}H NMR (500.13 MHz, CDCl$_3$) δ 6.20 (s, 1H, CH^{Pyranone}), 7.16 (d, *J* = 8.4 Hz, 1H, CH^{Ar}), 7.42 (m, 1H, CH^{Ar}), 7.65 (ddd, *J* = 8.5, 7.2, 1.7 Hz, 1H, CH^{Ar}), 7.87 (s, 1H, CH^{Triazole}), 7.88 (s, 2H, CH^{Triazole}), 7.90 (s, 2H, CH^{Triazole}), 8.08 (dd, *J* = 8, 1.6 Hz, 1H, CH^{Ar}), 8.10 (s, 1H, CH^{Triazole}), 8.15 (s, 2H, CH^{Triazole}), 8.22 (s, 2H, CH^{Triazole}) ppm; ^{13}C NMR (125.76 MHz, CDCl$_3$) δ 114.9, 117.4 (s, 2C^{Ar}), 123.2, 126.1, 126.5, 133.1, 134.3 (s, 2C^{Ar}), 134.9, 135.3, 136.1, 145.3 (s, 2C^{Triazole}), 145.7, 146.0 (s, 2C^{Triazole}), 153.4, 153.7 (s, 2C^{Triazole}), 153.7 (s, 2C^{Triazole}), 155.7, 175.8 ppm; Anal. Calcd. For C$_{25}$H$_{15}$N$_{15}$O$_2$: C 53.86, 2.71, N 37.69, found: C 53.60, H 2.68, N 37.69.

2-[2,3,5,6-Tetrafluoro-4-(1*H*-imidazol-1-yl)phenyl]-4*H*-chromen-4-one (**8**). Yield 137 mg (38% according to technique A), 263 mg (73% according to technique B); white powder, mp 172–174 °C; IR ν 3110 (C–H^{Ar}), 1653 (C=O), 1489, 1459, 1376 (C=C^{Ar}, C–H^{Ar}, C–N), 1224 (C–F) cm^{-1}; ^{1}H NMR (400.13 MHz, CDCl$_3$) δ 6.67 (s, 1H, CH^{Pyranone}), 7.31 (br.s., 1H, CH^{Imidazole}), 7.33 (br.s., 1H, CH^{Imidazole}), 7.48–7.53 (m, 2H, 2CH^{Ar}), 7.76 (ddd, *J* = 8.6, 7.3, 1.5 Hz, 1H, CH^{Ar}), 7.88 (br.s., 1H, CH^{Imidazole}), 8.27 (dd, *J* = 7.9, 1.5 Hz, 1H, CH^{Ar}) ppm; ^{13}C NMR (125.76 MHz, CDCl$_3$) δ 112.0 (t, *J* = 15 Hz, C^{Ar}), 115.9 (t, *J* = 3 Hz, C^{Ar}), 118.2, 119.3 (t, *J* = 13 Hz, C^{Ar}), 119.7 (t, *J* = 2 Hz, C^{Imidazole}), 123.9, 125.9, 126.0, 130.6, 134.5, 137.5 (t, *J* = 4 Hz, C^{Imidazole}), 140.1–142.3 (m, 2C^{ArF}), 144.9 (ddt, *J* = 257, 14, 5 Hz, 2C^{ArF}), 152.0, 156.6, 177.1 ppm; ^{19}F NMR (376.44 MHz, CDCl$_3$) δ 14.88 (dqd, *J* = 6.0, 4.4, 2.4 Hz, 2F), 24.51 (dq, *J* = 7.3, 4.2 Hz, 2F) ppm; Anal. calcd. for C$_{18}$H$_8$F$_4$N$_2$O$_2$: C 60.01, H 2.24, N 7.78, found: C 59.89, H 2.30, N 7.96.

2-[2,3,5-Trifluoro-4,6-di(1*H*-imidazol-1-yl)phenyl]-4*H*-chromen-4-one (**9**). Yield 20 mg (5%); light-yellow powder, mp 200–203 °C; ^{1}H NMR (400.13 MHz, CDCl$_3$) δ 6.52 (d, *J* = 1.5 Hz, 1H, CH^{Pyranone}), 7.05 (br.s., 1H, CH^{Imidazole}), 7.13 (br.s., 1H, CH^{Imidazole}), 7.18 (d, *J* = 8.4 Hz, 1H, CH^{Ar}), 7.33 (br.s., 1H, CH^{Imidazole}), 7.35 (br.s., 1H, CH^{Imidazole}), 7.41–7.45 (m, 1H, CH^{Ar}), 7.65 (br.s., 1H, CH^{Imidazole}), 7.65 (dd, *J* = 15.7, 1.7 Hz, 1H, CH^{Ar}), 7.90 (br.s., 1H, CH^{Imidazole}), 8.17 (dd, *J* = 7.9, 1.5 Hz, 1H, CH^{Ar}) ppm; ^{19}F NMR (376.44 MHz, CDCl$_3$) δ 23.60 (d, *J* = 22.2 Hz, 1F), 26.33 (ddd, *J* = 22.6, 13.0, 1.3 Hz, 1F), 30.57 (d, *J* = 13.0 Hz, 1F) ppm. Anal. calcd. for C$_{21}$H$_{11}$F$_3$N$_4$O$_2$: C 61.77, H 2.72, N 13.72, found: C 61.89, H 2.79, N 13.66.

2-[2,5-Difluoro-3,4,6-tri(1*H*-imidazol-1-yl)phenyl]-4*H*-chromen-4-one (**10**). Yield 95 mg (17%); light-yellow powder; mp 227–228 °C; ^{1}H NMR (400.13 MHz, CDCl$_3$) δ 6.51 (d, *J* = 1.3 Hz, 1H, CH^{Pyranone}), 6.81 (d, *J* = 1.3 Hz, 1H, CH^{Imidazole}), 6.83 (d, *J* = 1.2 Hz, 1H,

CH^{Imidazole}), 7.10 (d, J = 1.2 Hz, 1H, CH^{Imidazole}), 7.15–7.20 (m, 1H, CH^{Ar}), 7.21–7.23 (m, 3H, 3CH^{Imidazole}), 7.43–7.47 (m, 1H, CH^{Ar}), 7.50 (br.s, 2H, 2CH^{Imidazole}), 7.65–7.69 (m, 1H, CH^{Ar}), 7.70 (br.s, 1H, 1CH^{Imidazole}), 8.18 (dd, J = 8.0, 1.6 Hz, 1H, CH^{Ar}) ppm; ^{13}C NMR (125.76 MHz, CDCl$_3$) δ 115.8 (d, J = 3 Hz, CH^{Pyranone}), 117.8, 119.0, 119.2, 120.0, 120.5 (d, J = 17 Hz, C^{ArF}), 123.3 (d, J = 16 Hz, C^{ArF}), 123.6, 125.6 (dd, J = 15, 2 Hz, C^{ArF}), 125.8–126.0 (m, C^{ArF}), 125.9, 126.2, 131.0, 131.46, 131.47, 134.7, 136.9 (d, J = 3 Hz, C^{ArF}), 137.0 (d, J = 2 Hz, C^{Imidazole}), 137.5 (d, J = 2 Hz, C^{Imidazole}), 148.3 (dd, J = 256, 4 Hz, C^{ArF}), 150.9 (dd, J = 257, 4 Hz, C^{ArF}), 152.7 (d, J = 2 Hz, C^{Imidazole}), 156.2, 176.6 ppm; ^{19}F NMR (376.44 MHz, CDCl$_3$) δ 31.47 (dm, J = 14.4 Hz, 1F), 41.18 (dd, J = 14.7, 1.2 Hz, 1F) ppm. Anal. calcd. for C$_{24}$H$_{14}$F$_2$N$_6$O$_2$: C 63.16, H 3.09, N 18.41, found: C 62.90, H 2.82, N 18.24.

2-[2,3,5-Trifluoro-4-(1H-1,2,4-triazol-1-yl)phenyl]-4H-chromen-4-one (**11**). Yield 257 mg (75%); white powder; mp 210–211 °C; IR ν 3138, 3118, 3070 (C–H^{Ar}), 1634 (C=O), 1531, 1462, 1369 (C=C^{Ar}, C–H^{Ar}, C–N), 1040 (C–F) cm^{-1}; ^{1}H NMR (500.13 MHz, CDCl$_3$) δ 7.02 (s, 1H, CH^{Pyranone}), 7.49 (t, J = 7.5 Hz, 1H, CH^{Ar}), 7.58 (d, J = 8.4 Hz, 1H, CH^{Ar}), 7.74–7.79 (m, 2H, 2CH^{Ar}), 8.25 (dd, J = 8.0, 1.6 Hz, 1H, CH^{Ar}), 8.27 (s, 1H, CH^{Triazole}), 8.49 (s, 1H, CH^{Triazole}) ppm; ^{13}C NMR (125.76 MHz, CDCl$_3$) δ 110.6 (dd, J = 24, 4 Hz, C^{Ar}), 113.9, 114.0, 118.0 (m, C^{Ar}, C^{Triazole}), 122.7 (t, J = 9 Hz, C^{Ar}), 123.8, 125.9, 134.5, 144.8–145.1 (m, C^{ArF}), 145.3 (t, J = 3 Hz, C^{Triazole}), 146.9–147.2 (m, C^{ArF}), 151.7 (ddd, J = 254, 3, 2 Hz, C^{ArF}), 153.3, 154.9 (td, J = 4, 2 Hz, C^{Pyranone}), 156.1, 177.6 ppm; ^{19}F NMR (470.52 MHz, CDCl$_3$) δ 23.72 (d, J = 20.1 Hz, 1F), 24.92 (ddd, J = 20.3, 14.5, 5.9 Hz, 1F), 39.24 (dd, J = 14.6, 10.8 Hz, 1F) ppm; Anal. calcd. for C$_{17}$H$_8$F$_3$N$_3$O$_2$: C 59.48, H 2.35, N 12.24, found: C 54.40, H 2.32, N 12.29.

2-[2,3,4,5-Tetra(1H-1,2,4-triazol-1-yl)phenyl]-4H-chromen-4-one (**14**). Yield 333 mg (68%); light-yellow powder; mp 295–296 °C; ; IR ν 3134, 3108 (C–H^{Ar}), 1644 (C=O), 1508, 1467 (C=C^{Ar}, C–H^{Ar}, C–N) cm^{-1}; ^{1}H NMR (500.13 MHz, (CD$_3$)$_2$SO) δ 6.87 (s, 1H, CH^{Pyranone}), 7.22 (d, J = 8.3 Hz, 1H, CH^{Ar}), 7.49–7.53 (m, 1H, CH^{Ar}), 7.80 (ddd, J = 8.7, 7.4, 1.6 Hz, 1H, CH^{Ar}), 8.00 (s, 1H, CH^{Triazole}), 8.03 (dd, J = 7.9, 1.5 Hz, 1H, CH^{Ar}), 8.05 (s, 1H, CH^{Triazole}), 8.10 (s, 1H, CH^{Triazole}), 8.17 (s, 1H, CH^{Triazole}), 8.61 (s, 1H, CH^{Triazole}), 8.68 (s, 1H, CH^{Triazole}), 8.86 (s, 1H, CH^{Triazole}), 8.87 (s, 1H, CH^{Triazole}), 9.04 (s, 1H, CH^{Triazole}) ppm; ^{13}C NMR (125.76 MHz, CDCl$_3$) δ 112.8, 118.1 (s, 2C^{Ar}), 122.9, 124.9, 126.1, 129.3, 131.1, 132.9 (d, J = 2 Hz, C^{Ar}), 133.6, 134.9, 135.2, 146.0, 146.8, 146.9, 147.2, 152.6, 152.7, 152.8, 153.1, 155.4, 158.6, 176.5 ppm; Anal. calcd. for C$_{23}$H$_{14}$N$_{12}$O$_2$: C 56.33, H 2.88, N 34.27, found: C 56.11, H 2.90, N 34.22.

2-[2,3,5-Trifluoro-4-(1H-imidazol-1-yl)phenyl]-4H-chromen-4-one (**15**). Yield 277 mg (81%); white powder; mp 204–206 °C; IR ν 3160, 3133, 3077 (C–H^{Ar}), 1625, 1606 (C=O), 1515, 1465, 1351 (C=C^{Ar}, C–H^{Ar}, C–N), 1007 (C–F) cm^{-1}; ^{1}H NMR (500.13 MHz, CDCl$_3$) δ 7.00 (s, 1H, CH^{Pyranone}), 7.31 (br.s, 2H, 2CH^{Imidazole}), 7.47–7.50 (m, 1H, 1CH^{Ar}), 7.58 (d, J = 8.2 Hz, 1H, CH^{Ar}), 7.71–7.74 (m, 1H, 1CH^{Ar}), 7.75–7.78 (m, 1H, 1CH^{Ar}), 7.88 (br.s, 1H, CH^{Imidazole}), 8.25 (dd, J = 8.0, 1.6 Hz, 1H, CH^{Ar}) ppm; ^{13}C NMR (125.76 MHz, CDCl$_3$) δ 110.6 (dd, J = 25, 4 Hz, C^{Ar}), 113.5, 113.6, 118.0, 118.5 (dd, J = 17, 12 Hz, C^{ArF}), 119.8 (t, J = 2 Hz, C^{Imidazole}), 120.7 (t, J = 9 Hz, C^{ArF}), 123.7, 125.9, 130.2, 134.4, 137.6 (t, J = 4 Hz, C^{Imidazole}), 145.3 (ddd, J = 256, 17, 4 Hz, C^{ArF}), 146.2 (ddd, J = 258, 14, 4 Hz, C^{ArF}), 151.2 (dt, J = 251, 3 Hz, C^{ArF}), 155.1 (m, C^{ArF}), 156.1, 177.7 ppm; ^{19}F NMR (470.52 MHz, CDCl$_3$) δ 21.78 (dd, J = 19.8, 1.2 Hz, 1F), 24.83 (ddd, J = 20.1, 14.3, 1.2 Hz, 1F), 38.13–38.18 (m, 1F) ppm; Anal. calcd. for C$_{18}$H$_9$F$_3$N$_2$O$_2$: C 63.16, H 2.65, N 8.18, found: C 63.16, H 2.81, N 8.29.

2-[5-Fluoro-2,3,4-tri(1H-imidazol-1-yl)phenyl]-4H-chromen-4-one (**17**). Yield 31 mg (14%); yellow powder; mp 168–170 °C; ^{1}H NMR (400.13 MHz, CDCl$_3$) δ 6.40 (s, 1H, CH^{Pyranone}), 6.57 (t, J = 1.3 Hz, 1H, CH^{Imidazole}), 6.73 (d, J = 1.1 Hz, 1H, CH^{Ar}), 6.78 (t, J = 1.3 Hz, 1H, CH^{Imidazole}), 7.03–7.04 (m, 2H, CH^{Imidazole}), 7.13 (br.s, 1H, CH^{Imidazole}), 7.15–7.17 (m, 2H, CH^{Imidazole}), 7.39 (br.s, 1H, CH^{Imidazole}), 7.40–7.44 (m, 1H, CH^{Ar}), 7.47 (br.s, 1H, CH^{Imidazole}), 7.65 (ddd, J = 8.7, 7.3, 1.6 Hz, 1H, CH^{Ar}), 7.85 (d, J = 9.1 Hz, 1H, CH^{Ar}), 8.15 (dd, J = 8.0, 1.5 Hz, 1H, CH^{Ar}) ppm; ^{13}C NMR (125.76 MHz, CDCl$_3$) δ 112.8, 117.9, 118.5 (d, J = 23 Hz, C^{ArF}), 119.0, 119.1, 120.0, 123.4, 125.7–125.8 (m, C^{ArF}, C^{Ar}), 126.1, 129.6 (d, J = 4 Hz, C^{ArF}), 131.0, 131.1, 131.5, 132.8 (d, J = 2 Hz, C^{ArF}), 132.9 (d, J = 9 Hz, C^{ArF}), 134.7, 136.5, 137.0 (d, J = 2 Hz, C^{Imidazole}), 137.4, 155.9, 156.2 (d, J = 259 Hz, C^{ArF}), 158.6

(d, J = 2Hz, $C^{Pyranone}$), 177.1 ppm; ^{19}F NMR (376.44 MHz, CDCl$_3$) δ 47.37 (d, J = 9.2 Hz, 1F) ppm; Anal. calcd. for C$_{24}$H$_{15}$FN$_6$O$_2$: C 65.75, H 3.45, N 19.17, found: C 65.95, H 3.32, N 18.84.

2-[3,5-Difluoro-4-(1H-1,2,4-triazol-1-yl)-2-methoxyphenyl]-4H-chromen-4-one (**18**). Yield 35 mg (20% according to technique A), 123 mg (69% according to technique B); white powder; mp 172–174 °C; IR ν 3143, 3108, 3063 (C–H^{Ar}, C–H^{Alk}), 1665 (C=O), 1523, 1464, 1377 (C=C^{Ar}, C–H^{Ar}, C–N), 1034 (C–F) cm^{-1}; ^{1}H NMR (400.13 MHz, CDCl$_3$) δ 4.06 (d, J = 1.9 Hz, 3H, OCH$_3$), 7.15 (s, 1H, CH^{Pyranone}), 7.45–7.49 (m, 1H, CH^{Ar}), 7.56 (dd, J = 8.4, 0.5 Hz, CH^{Ar}), 7.66 (dd, J = 10.4, 2.3 Hz, CH^{Ar}), 7.75 (ddd, J = 8.7, 7.2, 1.7 Hz, CH^{Ar}), 8.25 (s, 1H, CH^{Triazole}), 8.26 (dd, J = 7.9, 1.6 Hz, CH^{Ar}), 8.45 (br.s, 1H, CH^{Triazole}) ppm; ^{13}C NMR (125.76 MHz, CDCl$_3$) δ 62.1 (d, J = 6 Hz, C^{OMe}), 111.2 (dd, J = 23, 4 Hz, C^{ArF}), 113.6, 117.6 (dd, J = 16, 14 Hz, C^{ArF}), 118.0, 123.8, 125.6, 125.8, 127.3 (dd, J = 9, 4 Hz, C^{ArF}), 134.2, 144.0 (dd, J = 12, 4 Hz, C^{ArF}), 145.4 (br.s, C^{Triazole}), 150.5 (dd, J = 257, 4 Hz, C^{ArF}), 151.4 (dd, J = 252, 3 Hz, C^{ArF}), 153.1, 156.2, 157.2 (dd, J = 4, 2 Hz, C^{Pyranone}), 178.2 ppm; ^{19}F NMR (376.44 MHz, CDCl$_3$) δ 29.29–29.30 (m, 1F), 37.19 (dd, J = 10.5, 1.3 Hz, 1F) ppm; Anal. calcd. for C$_{18}$H$_{11}$F$_2$N$_3$O$_3$: C 60.85, H 3.12, N 11.83, found: C 60.92, H 3.26, N 11.72.

2-[5-Fluoro-2-methoxy-3,4-di(1H-1,2,4-triazol-1-yl)phenyl]-4H-chromen-4-one (**19**). Yield 24 mg (12%); white powder; mp 203–204 °C; IR ν 3113, 3100, 2953, 2924 (C–H^{Ar}, C–H^{Alk}), 1640 (C=O), 1510, 1467, 1374 (C=C^{Ar}, C–H^{Ar}, C–N), 1008 (C–F) cm^{-1}; ^{1}H NMR (400.13 MHz, CDCl$_3$) δ 3.49 (s, 3H, OCH$_3$), 7.16 (s, 1H, CH^{Pyranone}), 7.48–7.52 (m, 1H, CH^{Ar}), 7.59 (d, J = 8.4 Hz, 1H, CH^{Ar}), 7.76–7.80 (m, 1H, CH^{Ar}), 7.96–8.01 (m, 3H, CH^{Triazole}), 8.27 (dd, J = 8.0, 1.4 Hz, 1H, CH^{Ar}), 8.37 (d, J = 1.5 Hz, 1H, CH^{Ar}), 8.42 (s, 1H, CH^{Triazole}) ppm; ^{13}C NMR (125.76 MHz, CDCl$_3$) δ 62.4 (s, C^{OMe}), 113.4, 118.0 (d, J = 24 Hz, C^{ArF}), 118.0, 123.8, 124.9 (d, J = 15 Hz, C^{ArF}), 125.9, 126.0, 129.0 (d, J = 8 Hz, C^{ArF}), 129.6 (d, J = 2 Hz, C^{Triazole}), 134.5, 145.7 (d, J = 2 Hz, C^{Triazole}), 146.3, 151.0 (d, J = 4 Hz, C^{ArF}), 152.6 (d, J = 253 Hz, C^{ArF}), 152.8, 152.9, 156.3, 157.1 (d, J = 2 Hz, C^{Pyranone}), 177.9 ppm; ^{19}F NMR (376.44 MHz, CDCl$_3$) δ 38.81 (dd, J = 9.9, 1.6 Hz, 1F) ppm; Anal. calcd. for C$_{20}$H$_{13}$FN$_6$O$_3$: C 59.41, H 3.24, N 20.78, found: C 59.51, H 3.23, N 20.79.

2-[5-Fluoro-2-hydroxy-3,4-di(1H-1,2,4-triazol-1-yl)phenyl]-4H-chromen-4-one (**22**). Yield 22 mg (11%); yellow powder; mp 313–316 °C; ^{1}H NMR (400.13 MHz, (CD$_3$)$_2$SO) δ 7.04 (s, 1H, CH^{Pyranone}), 7.52–7.56 (m, 1H, CH^{Ar}), 7.78–7.80 (m, 1H, CH^{Ar}), 7.88 (ddd, J = 8.6, 7.2, 1.6 Hz, 1H, CH^{Ar}), 8.07 (s, 1H, CH^{Triazole}), 8.09 (dd, J = 8.1, 1.5 Hz, 1H, CH^{Ar}), 8.11 (s, 1H, CH^{Triazole}), 8.26 (d, J = 10.4 Hz, 1H, CH^{Ar}), 8.76 (s, 1H, CH^{Triazole}), 8.79 (s, 1H, CH^{Triazole}), 10.98 (s, 1H, OH) ppm; ^{13}C NMR (125.76 MHz, (CD$_3$)$_2$SO) δ 112.6, 118.3 (d, J = 23 Hz, C^{ArF}), 118.7, 123.3 (d, J = 9 Hz, C^{ArF}), 124.0, 124.7, 124.8 (d, J = 15 Hz, C^{ArF}), 125.7, 134.5, 146.6, 147.3, 148.7 (d, J = 2 Hz, C^{Triazole}), 149.1 (d, J = 244 Hz, C^{ArF}), 152.3, 152.5, 156.0, 159.0, 177.1 ppm; ^{19}F NMR (376.44 MHz, (CD$_3$)SO) δ 30.74 (d, J = 9.6 Hz, 1F) ppm.

2-[3,5-Difluoro-4-(1H-imidazol-1-yl)-2-methoxyphenyl]-4H-chromen-4-one (**23**). Yield 39 mg (23% according to technique A), 94 mg (53% according to technique B); white powder; mp 176–177 °C; IR ν 3108, 3043, 2999, 2950 (OMe, C–H^{Ar}), 1630 (C=O), 1524, 1480, 1370 (C=C^{Ar}, C–H^{Ar}, C–O, C–N), 1112 (C–F) cm^{-1}; °C; ^{1}H NMR (400.13 MHz, CDCl$_3$) δ 4.04 (d, J = 1.7 Hz, 3H, OMe), 7.14 (s, 1H, CH^{Pyranone}), 7.27–7.29 (m, 2H, 2CH^{Imidazole}), 7.45–7.49 (m, 1H, 1CH^{Ar}), 7.55–7.57 (m, 1H, 1CH^{Ar}), 7.64 (dd, J = 10.8, 2.2 Hz, 1H, CH^{Ar}), 7.75 (ddd, J = 8.6, 7.2, 1.7 Hz, 1H, CH^{Ar}), 7.83 (br.s, 1H, CH^{Imidazole}), 8.25 (dd, J = 8.0, 1.5 Hz, 1H, CH^{Ar}) ppm; ^{13}C NMR (125.76 MHz, CDCl$_3$) δ 62.1 (d, J = 6 Hz, C^{OMe}), 111.2 (dd, J = 24, 4 Hz, C^{Ar}), 113.3, 118.0, 118.2 (dd, J = 14, 2 Hz, C^{Ar}), 120.0 (t, J = 2 Hz, C^{Imidazole}), 123.8, 125.5–125.6 (m, C^{Ar}, C^{Imidazole}), 125.8, 129.9, 134.2, 137.7 (t, J = 3 Hz, C^{Imidazole}), 144.2 (dd, J = 12, 4 Hz, C^{Ar}), 150.1 (dd, J = 254, 4 Hz, C^{ArF}), 151.1 (dd, J = 249, 4 Hz, C^{ArF}), 156.2, 157.4 (dd, J = 4, 2 Hz, C^{Pyranone}), 178.2 ppm; ^{19}F NMR (376.44 MHz, CDCl$_3$) δ 27.59–27.60 (m, 1F), 36.45–36.48 (m, 1F)ppm; Anal. calcd. for C$_{19}$H$_{12}$F$_2$N$_2$O$_3$: C 64.41, H 3.41, N 7.91, found: C 64.37, H 3.40, N 8.02.

2-[2,5-Difluoro-3,6-di(1H-pyrazol-1-yl)-4-(1H-1,2,4-triazol-1-yl)phenyl]-4H-chromen-4-one (**26**). Yield 39 mg (17%); white powder; mp 256–258 °C; IR ν 3130, 3106, 3071 (C–H^{Ar}), 1642 (C=O), 1529, 1483, 1390 (C=C^{Ar}, C–H^{Ar}, C–N), 1139, 1127 (C–F) cm^{-1}; ^{1}H NMR

(400.13 MHz, CDCl$_3$) δ 6.44 (d, J = 1.2 Hz, 1H, C^{Pyranone}), 6.45–6.46 (m, 1H, CH^{Imidazole}), 6.48–6.49 (m, 1H, CH^{Imidazole}), 7.19–7.21 (m, 1H, CH^{Ar}), 7.40–7.44 (m, 1H, CH^{Ar}), 7.58 (d, J = 1.6 Hz, 1H, C^{Ar}), 7.63–7.66 (m, 3H, CH^{Imidazole}), 7.84 (t, J = 2.6 Hz, 1H, C^{Imidazole}), 8.05 (s, 1H, CH^{Triazole}), 8.19 (dd, J = 8.0, 1.5 Hz, 1H, C^{Ar}), 8.24 (s, 1H, CH^{Triazole}) ppm; ^{13}C NMR (125.76 MHz, CDCl$_3$) δ 108.4, 108.6, 115.0 (d, J = 3 Hz, C^{Pyrazole}), 118.0, 121.4 (d, J = 17 Hz, C^{ArF}), 123.7, 125.5 (dd, J = 14, 3 Hz, C^{ArF}), 125.6, 125.8, 126.3 (d, J = 17 Hz, C^{ArF}), 128.7 (dd, J = 14, 3 Hz, C^{ArF}), 132.0 (d, J = 4 Hz, C^{Pyrazole}), 132.3 (d, J = 2 Hz, C^{Pyrazole}), 134.1, 142.7, 142.9, 145.7 (d, J = 1 Hz, C^{Pyranone}), 148.0 (dd, J = 258, 4 Hz, C^{ArF}), 151.1 (dd, J = 257, 4 Hz, C^{ArF}), 153.1, 154.3 (d, J = 2 Hz, C^{Pyranone}), 177.2 ppm; ^{19}F NMR (376.44 MHz, CDCl$_3$) δ 30.62 (dd, J = 14.4, 1.8 Hz, 1F), 30.75–39.80 (m, 1F) ppm; Anal. calcd. for C$_{23}$H$_{13}$F$_2$N$_7$O$_2$: C 60.40, 2.86, N 21.44, found: C 60.15, H 2.68, N 21.54.

2-[2,3,5,6-Tetra(1H-pyrazol-1-yl)-4-(1H-1,2,4-triazol-1-yl)phenyl]-4H-chromen-4-one (**27**). Yield 136 mg (49%); yellow powder; mp 299–301 °C; IR ν 3126, 3092 (C–H^{Ar}), 1646 (C=O), 1525, 1468, 1389 (C=C^{Ar}, C–H^{Ar}, C–N) cm^{-1}; ^{1}H NMR (400.13 MHz, CDCl$_3$) δ 5.30 (s, 1H, CH^{Pyrazole}), 6.08 (s, 1H, CH^{Pyrazole}), 6.16–6.18 (m, 5H, CH^{Pyranone}, CH^{Pyrazole}), 7.11 (d, J = 8.4 Hz, 1H, CH^{Ar}), 7.24 (d, J = 2.5 Hz, 1H, CH^{Pyrazole}), 7.29 (d, J = 2.5 Hz, 1H, CH^{Pyrazole}), 7.32–7.36 (m, 1H, CH^{Ar}), 7.43 (d, J = 1.6 Hz, 2H, CH^{Pyrazole}), 7.49 (d, J = 1.6 Hz, 2H, CH^{Pyrazole}), 7.54–7.59 (m, 1H, CH^{Ar}), 7.74 (s, 1H, CH^{Triazole}), 8.05 (s, 1H, CH^{Triazole}), 8.06–8.08 (m, 1H, CH^{Ar}) ppm; ^{13}C NMR (125.76 MHz, CDCl$_3$) δ 108.0 (s, 2C^{Pyrazole}), 108.2 (s, 2C^{Pyrazole}), 113.3, 117.8, 123.3, 125.4, 125.7 (s, 2C^{Ar}), 131.8 (s, 2C^{Pyrazole}), 132.1 (s, 2C^{Pyrazole}), 132.7, 133.8 (s, 2C^{Ar}), 135.2, 135.9, 138.2, 142.2 (s, 2C^{Pyrazole}), 142.3 (s, 2C^{Pyrazole}), 145.9, 152.3, 156.0, 156.6, 177.0 ppm; Anal. calcd. for C$_{29}$H$_{19}$N$_{11}$O$_2$: C 62.92, 3.46, N 27.83, found: C 63.15, H 3.68, N 27.69.

2-[4-(1H-Imidazol-1-yl)-2,3,5,6-tetra(1H-pyrazol-1-yl)phenyl]-4H-chromen-4-one (**28**). Yield 31 mg (11%); yellow powder; mp 320–321 °C; IR ν 3127 (C–H^{Ar}), 1649 (C=O), 1525, 1467, 1388 (C=C^{Ar}, C–H^{Ar}, C–N) cm^{-1}; ^{1}H NMR (400.13 MHz, CDCl$_3$) δ 6.03 (s, 1H, CH^{Pyranone}), 6.12–6.13 (m, 1H, CH^{Pyrazole}), 6.15 (dd, J = 4.3, 2.2 Hz, 4H, CH^{Pyrazole}), 7.11 (d, J = 8.3 Hz, 1H, CH^{Ar}), 7.20–7.23 (m, 3H, CH^{Pyrazole}, CH^{Imidazole}), 7.31–7.34 (m, 3H, CH^{Ar}, CH^{Pyrazole}, CH^{Imidazole}), 7.39 (d, J = 1.6 Hz, 2H, CH^{Pyrazole}), 7.41 (d, J = 1.6 Hz, 1H, CH^{Imidazole}), 7.47 (d, J = 1.5 Hz, 2H, CH^{Pyrazole}), 7.55 (ddd, J = 8.6, 7.3, 1.6 Hz, 1H, CH^{Ar}), 8.06 (dd, J = 7.9, 1.5 Hz, 1H, CH^{Ar}) ppm; ^{13}C NMR (125.76 MHz, CDCl$_3$) δ 107.3, 107.5 (s, 2C^{Pyrazole}), 107.8 (s, 2C^{Pyrazole}), 113.2, 117.9, 123.3, 125.3, 125.6, 131.5, 131.9 (s, 2C^{Ar}), 132.0 (s, 2C^{Pyrazole}), 132.2 (s, 2C^{Pyrazole}), 133.7, 135.8, 138.3, 138.4, 141.7 (s, 2C^{Ar}), 141.7 (s, 2C^{Pyrazole}), 141.9 (s, 2C^{Pyrazole}), 156.0, 157.1, 177.1 ppm; Anal. calcd. for C$_{30}$H$_{20}$N$_{10}$O$_2$: C 65.21, 3.65, N 25.35, found: C 65.15, H 3.57, N 25.55.

2-[4-(1H-Imidazol-1-yl)-2,3,5,6-tetra(1H-1,2,4-triazol-1-yl)phenyl]-4H-chromen-4-one (**29**). Yield 92 mg (33%); pale pink powder; mp 299–301 °C; IR ν 3123, 3105 (C–H^{Ar}), 1647 (C=O), 1510, 1464, 1378 (C=C^{Ar}, C–H^{Ar}, C–N) cm^{-1}; ^{1}H NMR (500.13 MHz, (CD$_3$)$_2$SO) δ 6.30 (s, 1H, CH^{Pyranone}), 6.84 (s, 1H, CH^{Imidazole}), 7.01 (t, J = 1.3 Hz, 1H, CH^{Imidazole}), 7.34 (d, J = 8.2 Hz, 1H, CH^{Ar}), 7.47–7.49 (m, 1H, CH^{Ar}), 7.52 (s, 1H, CH^{Imidazole}), 7.80 (ddd, J = 8.7, 7.2, 1.7 Hz, 1H, CH^{Ar}), 7.92 (dd, J = 8.0, 1.6 Hz, 1H, CH^{Ar}), 8.08 (s, 2H, CH^{Triazole}), 8.11 (s, 2H, CH^{Triazole}), 8.63 (s, 2H, CH^{Triazole}), 8.82 (s, 2H, CH^{Triazole}) ppm; ^{13}C NMR (125.76 MHz, CDCl$_3$) δ 113.7, 117.9, 121.0, 122.4, 124.9, 126.3, 129.3, 130.6, 134.0 (s, 2C^{Ar}), 135.0, 135.5 (s, 2C^{Ar}), 136.0, 137.9, 146.7 (s, 2C^{Triazole}), 146.8 (s, 2C^{Triazole}), 152.9 (s, 2C^{Triazole}), 152.9 (s, 2C^{Triazole}), 154.8, 155.4, 175.4 ppm; Anal. calcd. for C$_{26}$H$_{16}$N$_{14}$O$_2$: C 56.11, 2.90, N 35.24, found: C 56.30, H 2.98, N 35.04.

2-[4-(1H-Pyrazol-1-yl)-2,3,5,6-tetra(1H-1,2,4-triazol-1-yl)phenyl]-4H-chromen-4-one (**31**). Yield 159 mg (59%); pale pink powder; mp 309–310 °C; IR ν 3113 (C–H^{Ar}), 1651 (C=O), 1511, 1466, 1386 (C=C^{Ar}, C–H^{Ar}, C–N) cm^{-1}; ^{1}H NMR (500.13 MHz, (CD$_3$)$_2$SO) δ 6.31–6.32 (m, 1H, CH^{Pyrazole}), 6.32 (s, 1H, CH^{Pyranone}), 7.34 (d, J = 8.2 Hz, 1H, CH^{Ar}), 7.46–7.49 (m, 1H, CH^{Ar}), 7.56 (d, J = 1.7 Hz, 1H, CH^{Pyrazole}), 7.67 (d, J = 2.5 Hz, 1H, CH^{Pyrazole}), 7.79 (ddd, J = 8.7, 7.2, 1.7 Hz, 1H, CH^{Ar}), 7.92 (dd, J = 8.0, 1.6 Hz, 1H, CH^{Ar}), 8.04 (s, 2H, CH^{Triazole}), 8.05 (s, 2H, CH^{Triazole}), 8.53 (s, 2H, CH^{Triazole}), 8.82 (s, 2H, CH^{Triazole}) ppm; ^{13}C NMR (125.76 MHz, (CD$_3$)$_2$SO) δ 107.9, 113.7, 118.0, 122.4, 124.9, 126.3, 130.8, 133.1, 133.9 (s, 2C^{Ar}), 135.1,

135.4 (s, 2C^{Ar}), 137.6, 142.5, 146.8 (s, 2C^{Triazole}), 146.9 (s, 2C^{Triazole}), 152.7 (s, 2C^{Triazole}), 152.9 (s, 2C^{Triazole}), 155.0, 155.4, 175.5 ppm; Anal. calcd. for $C_{26}H_{16}N_{14}O_2$: C 56.11, 2.90, N 35.24, found: C 56.26, H 2.99, N 35.11.

3.3. XRD Experiments

The X-ray studies were performed on an Xcalibur 3 CCD (Oxford Diffraction Ltd., Abingdon, UK) diffractometer with a graphite monochromator, λ(MoKα) 0.71073 Å radiation and T 295(2) K. An empirical absorption correction was applied. Using Olex2 [51], the structure was solved with the Superflip [52] structure solution program using charge flipping and refined with the ShelXL [53] refinement package using Least Squares minimization. All non-hydrogen atoms were refined in the anisotropic approximation; H-atoms at the C–H bonds were refined in the "rider" model with dependent displacement parameters. An empirical absorption correction was carried out through spherical harmonics, implemented in the SCALE3 ABSPACK scaling algorithm by the program "CrysAlisPro" (Rigaku Oxford Diffraction).

The main crystallographic data for **4**: $C_{17}H_7F_4N_3O_2$, M 361.26, orthorhombic, *a* 15.8944(12), *b* 12.7694(11), *c* 7.3245(6) Å, V 1486.6(2) Å^3, space group Pna2$_1$, Z 4, μ(Mo Kα) 0.125 mm^{-1}, 256 refinement parameters, 3609 reflections measured, and 2493 unique (R_{int} = 0.0617), which were used in all calculations. CCDC 2225826 contains the supplementary crystallographic data for this compound.

The main crystallographic data for **8**: $C_{18}H_8F_4N_2O_2$, M 360.26, monoclinic, *a* 15.0164(11), *b* 7.9494(7), *c* 12.8592(10) Å, β 99.256(7)°, V 1515.0(2) Å^3, space group P2$_1$/c, Z 4, μ(Mo Kα) 0.125 mm^{-1}, 268 refinement parameters, 4162 reflections measured, and 2246 unique (R_{int} = 0.0633), which were used in all calculations. CCDC 2,225,827 contains the supplementary crystallographic data for this compound.

The main crystallographic data for **15**: $C_{18}H_9F_3N_2O_2$, M 342.27, monoclinic, *a* 13.3565(10), *b* 7.8477(5), *c* 14.7075(12) Å, β 113.251(9)°, V 1416.4(2) Å^3, space group P2$_1$/n, Z 4, μ(Mo Kα) 0.125 mm^{-1}, 262 refinement parameters, 3865 reflections measured, and 2692 unique (R_{int} = 0.0597), which were used in all calculations. CCDC 2,225,828 contains the supplementary crystallographic data for this compound.

The main crystallographic data for **19**: $C_{20}H_{13}FN_6O_3$, M 404.36, triclinic, *a* 13.3565(10), *b* 7.8477(5), *c* 14.7075(12) Å, α 111.535(14), β 94.311(13), γ 101.586(12)°, V 1416.4(2) Å^3, space group P$\bar{1}$, Z 2, μ(Mo Kα) 0.125 mm^{-1}, 288 refinement parameters, 3662 reflections measured, 1504 unique (R_{int} = 0.0711) which were used in all calculations. CCDC 2,225,829 contains the supplementary crystallographic data for this compound.

The main crystallographic data for **23**: $C_{19}H_{12}F_2N_2O_3$, M 354,31, monoclinic, *a* 14.0716(10), *b* 7.7151(5), *c* 14.5964(12) Å, β 100.453(7)°, V 1553.0(2) Å^3, space group P2$_1$/c, Z 4, μ(Mo Kα) 0.125 mm^{-1}, 257 refinement parameters, 4236 reflections measured, and 2349 unique (R_{int} = 0.0568) which were used in all calculations. CCDC 2,225,830 contains the supplementary crystallographic data for this compound.

3.4. Fungistatic Activity Evaluation

The following dermatophyte fungal strains were used: *Trichophyton rubrum* (RCPF F 1408), *Epidermophyton floccosum* (RCPF F 1659/17), and *Microsporum canis* (RCPF F 1643/1585), as well as yeast-like fungus *Candida parapsilosis* (RCPF 1245/ ATCC 22019). The fungi cultures were obtained from the Russian Collection of Pathogenic Fungi (Kashkin Research Institute of Medical Mycology; Mechnikov Northwest State Medical University, St.-Petersburg). Saburo agar and Saburo broth were used for the fungi. The microorganisms were identified as matrix-extracted bacterial proteins with an accuracy of 99.9% using a BioMerieux VITEK MS MALDI-TOF analyzer. The test cultures were prepared to an optical density of 0.5 according to McFarland (1.5 × 10^8 CFU/ mL) using a BioMerieux DensiCHEK densimeter. The suspensions of *C. parapsilosis* were prepared from 24 h cultures, and dermatophyte inocula were prepared after incubation for 2 weeks and preliminary homogenization in sterile saline. The fungi were inoculated at a concentration of 10^5 CFU/mL.

The antimycotic activity was evaluated by a micro method [54]. The agar nutrient medium was maintained in liquid by heating to 52 °C. The chemical compounds to be tested were dissolved in DMSO to a concentration of 1000 µg/ mL, and the stock solutions were diluted with distilled sterilized water; serial dilutions (from 250–200 µg/ mL) were made using nutrient media. Dermatophytes were incubated at 27 °C for up to 7–10 days and *C. parapsilosis* for 24 h in a moist 5.0% CO_2 chamber. In each case, positive and negative controls were used. The minimum inhibitory concentration was determined visually as the lowest concentration at which a test culture no longer grows. Chemically pure fluconazole was used as a reference drug.

4. Conclusions

The data obtained in this work and earlier in the study of transformations with pyrazole [29] thus indicate that base-promoted reactions of nucleophilic aromatic substitution are a convenient method for the functionalization of polyfluoroflavones **1–3** with azoles with different numbers of nitrogen atoms. At the same time, it was found that monosubstitution of the para-fluorine atom successfully and selectively occurs while using the system (azole (1.5 equiv.)/NaOBu^t (1.5 equiv.)/MeCN) regardless of the structure and properties of the used polyfluorinated substrates and nucleophilic reagents, since in all cases, mono(azolyl)-substituted flavones were obtained in good yields. Under the conditions (azole (6 equiv.)/NaOBu^t (6 equiv.)/DMF), which promote the formation of persubstituted products, the interactions of polyfluoroflavones **1–3** with pyrazole are distinguished by high selectivity [26], while similar reactions with triazole produced productively only for penta- and tetrafluoroflavones **1** and **2**, and the same transformations with imidazole in general are extremely non-selective.

Comparing the conversion of polyfluoroflavones **1–3** in reactions with azoles under conditions that do not provide selective substitution (azole (3 equiv.)/Cs_2CO_3 (3 equiv.)/MeCN)), it can be noted that transformations with pyrazole [26] are characterized by easier formation of polysubstituted products, and under conditions conducive to persubstitution, it is possible to build the following series of azoles according to reactivity: pyrazole $\geq$ triazole > imidazole. Obviously, in both cases, the reactivity of azoles does not correspond to their basicity, and therefore to some extent, nucleophilicity, since imidazole is known to be the strongest base among them [55]. According to the literature data [56,57], polyfluoroaromatic compounds generally react with nucleophiles via Meisenheimer complexes. For flavones **1–3**, we assume an analogous mechanism [26]; first nucleophile attack occurs on the activated C4' site of flavones **1–3** with the generation of an intermediate of a stable quinoid structure, followed by formation of a mono(azolyl)-substituted product. Sequential substitution is coordinated by the joint activating effect of the substituents, and the resulting intermediate complexes are stabilized both by O,N-bidentate coordination between the azole and the pyrone fragment of the molecule with Na$^+$ and by the coordination of neighboring azole moieties with Na$^+$. It is likely that the higher reactivity of pyrazole, triazole, and their intermediates compared to imidazole and its derivatives in S_NAr poly- and per-substitution reactions is due to the possibility of participation of their imine nitrogen atoms in the N–N=C function in coordination during the formation of transitional complexes.

In addition, it was shown that the per-substitution conditions can be successfully used for the synthesis of polynuclear hybrid compounds containing two different azole fragments by the reaction of mono(azolyl)-substituted flavones with pyrazole and triazole.

Using XRD data, the structural features of triazolyl- and imidazolyl-substituted flavones in crystal form were established. For example, in contrast to previously synthesized pyrazole analogues [26], new azole derivatives do not contain an intramolecular H-bond.

In terms of possible practical applications, it has been established that the resulting poly(pyrazolyl)-substituted flavones have luminescent properties, which makes further development of research in this area promising. In addition, weak antimycotic activity was found for some azolyl-containing flavones.

Supplementary Materials: NMR data of the synthesized compounds can be downloaded at: https://www.mdpi.com/article/10.3390/molecules28020869/s1. Figure S1. ^{1}H NMR spectrum of compound **4**; Figure S2. ^{13}C NMR spectrum of compound **4**; Figure S3. ^{19}F NMR spectrum of compound **4**; Figure S4. ^{1}H NMR spectrum of compound **7**; Figure S5. ^{13}C NMR spectrum of compound **7**; Figure S6. ^{1}H NMR spectrum of compound **8**; Figure S7. ^{13}C NMR spectrum of compound **8**; Figure S8. ^{19}F NMR spectrum of compound **8**; Figure S9. ^{1}H NMR spectrum of compound **9**; Figure S10. ^{19}F NMR spectrum of compound **9**; Figure S11. ^{1}H NMR spectrum of compound **10**; Figure S12. ^{13}C NMR spectrum of compound **10**; Figure S13. ^{19}F NMR spectrum of compound **10**; Figure S14. ^{1}H NMR spectrum of compound **11**; Figure S15. ^{13}C NMR spectrum of compound **11**; Figure S16. ^{19}F NMR spectrum of compound **11**; Figure S17. ^{1}H NMR spectrum of compound **14**; Figure S18. ^{13}C NMR spectrum of compound **14**; Figure S19. ^{1}H NMR spectrum of compound **15**; Figure S20. ^{13}C NMR spectrum of compound **15**; Figure S21. ^{19}F NMR spectrum of compound **15**; Figure S22. ^{1}H NMR spectrum of compound **17**; Figure S23. ^{13}C NMR spectrum of compound **17**; Figure S24. ^{19}F NMR spectrum of compound **17**; Figure S25. ^{1}H NMR spectrum of compound **18**; Figure S26. ^{13}C NMR spectrum of compound **18**; Figure S27. ^{19}F NMR spectrum of compound **18**; Figure S28. ^{1}H NMR spectrum of compound **19**; Figure S29. ^{13}C NMR spectrum of compound **19**; Figure S30. ^{19}F NMR spectrum of compound **19**; Figure S31. ^{1}H NMR spectrum of compound **22**; Figure S32. ^{13}C NMR spectrum of compound **22**; Figure S33. ^{19}F NMR spectrum of compound **22**; Figure S34. ^{1}H NMR spectrum of compound **23**; Figure S35. ^{13}C NMR spectrum of compound **23**; Figure S36. ^{19}F NMR spectrum of compound **23**; Figure S37. ^{1}H NMR spectrum of compound **26**; Figure S38. ^{13}C NMR spectrum of compound **26**; Figure S39. ^{19}F NMR spectrum of compound **26**; Figure S40. ^{1}H NMR spectrum of compound **27**; Figure S41. ^{13}C NMR spectrum of compound **27**; Figure S42. ^{1}H NMR spectrum of compound **28**; Figure S43. ^{13}C NMR spectrum of compound **28**; Figure S44. ^{1}H NMR spectrum of compound **29**; Figure S45. ^{13}C NMR spectrum of compound **29**; Figure S46. ^{1}H NMR spectrum of compound **31**; Figure S47. ^{13}C NMR spectrum of compound **31**.

Author Contributions: Conceptualization Y.V.B.; methodology and synthesis, validation, and interpretation of analysis data, K.V.S. and M.A.P.; analytical experiments E.F.Z.; writing—original draft preparation, K.V.S. and M.A.P.; writing—review and editing, Y.V.B.; supervision, V.I.S. All authors have read and agreed to the published version of the manuscript.

Funding: This work was financially supported by the Ministry of Science and Higher Education of the Russian Federation (State task AAAA-A19-119012290115-2).

Institutional Review Board Statement: Not applicable.

Informed Consent Statement: Not applicable.

Data Availability Statement: Not applicable.

Acknowledgments: The analytical data analysis (IR and NMR spectroscopy, elemental analysis, XRD analysis, and photoluminescent data) was performed on equipment belonging to the Center for Joint Use of Scientific Equipment "Spectroscopy and Analysis of Organic Compounds" IOS UB RAS. Fungistatic activity evaluation was performed on equipment belonging to the Ural Research Institute of Dermatovenereology and Immunopathology, Yekaterinburg, Russian Federation. The authors acknowledge Natalia A. Gerasimova for her contribution.

Conflicts of Interest: The authors declare no conflict of interest.

Sample Availability: Samples of the compounds are available from the authors.

Appendix A

Table A1. NMR data of characteristic F and H nuclei of flavones **9**, **10**, and **A**.

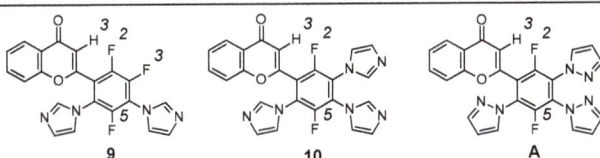

Compound	H3		F2		F3		F5	
	δ	J	δ	J	δ	J	δ	J
9	6.52, d	$^5J_{HF}$ 1.4	26.33 ddd	$^3J_{FF}$ 22.6 $^5J_{FF}$ 13.0 $^5J_{FH}$ 1.4	23.60 d	$^3J_{FF}$ 22.4	30.57 d	$^5J_{FF}$ 13.1
10	6.51, d	$^5J_{HF}$ 1.3	41.18 dd	$^5J_{FF}$ 14.7 $^5J_{FH}$ 1.3	–	–	31.47 d	$^5J_{FF}$ 14.7
A	6.44, d	$^5J_{HF}$ 1.1	29.51 dd	$^5J_{FF}$ 14.4 $^5J_{FH}$ 1.1	–	–	39.83 d	$^5J_{FF}$ 14.3

Appendix B

Table A2. Detailed data of the fluorescence lifetime measurements of **28** and **32**: τ—lifetime, f—fractional contribution, τ_{avg}—average lifetime, and χ^2—chi-squared distribution.

Compound	Solid					
	τ_1, ns	f_1, %	τ_2, ns	f_2, %	τ_{avg}, ns	χ^2
28	3.57	16.5	8.53	83.5	7.72	1.109
32	2.84	21.3	7.88	78.7	6.81	1.206

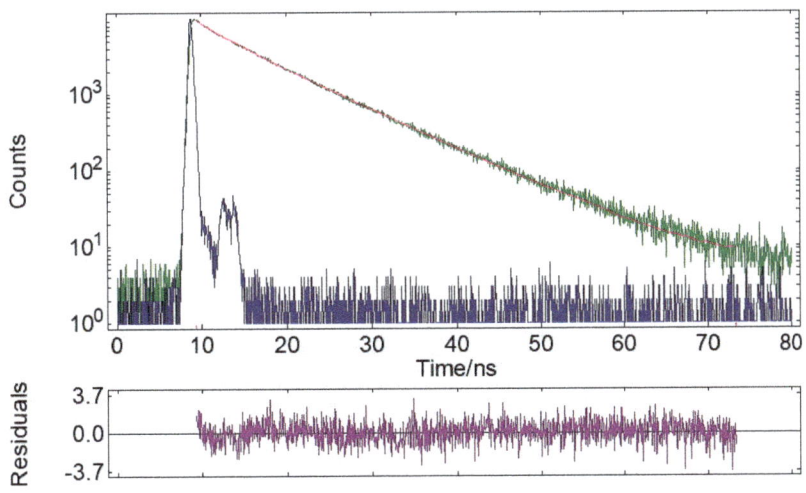

Figure A1. Time−resolved fluorescence lifetime decay profile of solid powder **28** (green) and instrumental response function (IRF, blue). λ_{ex} = 375 nm and λ_{em} = 504 nm.

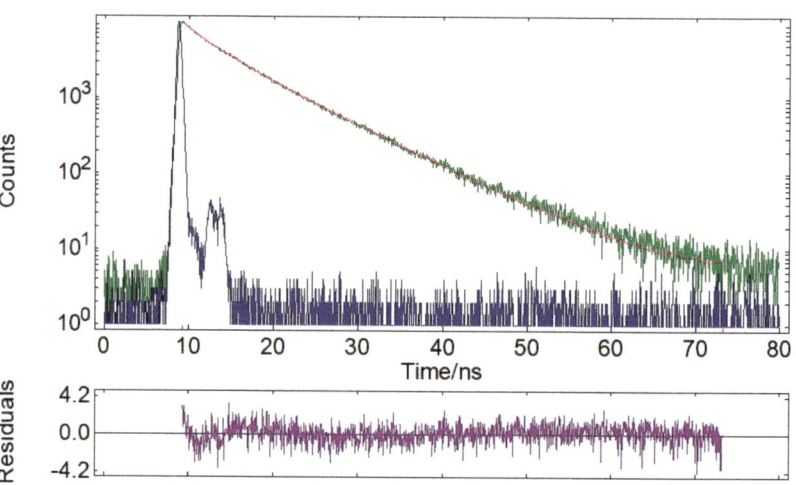

Figure A2. Time−resolved fluorescence lifetime decay profile of solid powder **32** (green), instrumental response function (IRF, blue). λ_{ex} = 375 nm and λ_{em} = 504 nm.

Appendix C

Table A3. Fungistatic activity of fluorinated flavones and their azolyl-substituted derivatives.

No	R^1	R^2	R^3	R^4	R^5	T. rubrum	E. floccosum	M. canis	C. parapsilosis
4	F	F	trz	F	F	>200	>200	>200	>200
8	F	F	imz	F	F	100/**25**	200/100	100/**50**	>200
11	H	F	trz	F	F	>200	>200	>200	>200
14	H	trz	trz	trz	trz	100/**25**	200/100	100/**50**	>200
18	OMe	F	trz	F	H	>200	>200	>200	>200
19	OMe	trz	trz	F	H	-	>200/50	>200/200	>200/200
23	OMe	F	imz	F	H	>200/**12.5**	>200/**12.5**	>200/**6.25**	>200
26	F	pz	trz	F	pz	>200	>200	>200	>200
33	OMe	pz	pz	F	H	>100/**1.56**	>100/**1.56**	>100	>100
34	F	pz	pz	F	pz	>100	>100	>100	>100
35	pz	pz	pz	pz	pz	>100	>100	>100	>100
Fluconazole						3.12	1.56	3.12	0.5–2

References

1. Tian, S.; Luo, T.; Zhu, Y.; Wan, J.-P. Recent advances in diversification of chromones and flavones by direct C–H bond activation or functionalization. *Chin. Chem. Lett.* **2020**, *31*, 3073–3082. [CrossRef]

2. Kshatriya, R.; Jejurkar, V.P.; Saha, S. Recent advances in the synthetic methodologies of flavones. *Tetrahedron* **2018**, *74*, 811–833. [CrossRef]
3. Spagnuolo, C.; Moccia, S.; Russo, G.L. Anti-inflammatory effects of flavonoids in neurodegenerative disorders. *Eur. J. Med. Chem.* **2018**, *153*, 105–115. [CrossRef]
4. Singh, M.; Kaur, M.; Silakari, O. Flavones: An important scaffold for medicinal chemistry. *Eur. J. Med. Chem.* **2014**, *84*, 206–239. [CrossRef] [PubMed]
5. Verma, A.K.; Singh, H.; Satyanarayana, M.; Srivastava, S.P.; Tiwari, P.; Singh, A.B.; Dwivedi, A.K.; Singh, S.K.; Srivastava, M.; Nath, C.; et al. Flavone-based novel antidiabetic and antidyslipidemic agents. *J. Med. Chem.* **2012**, *55*, 4551–4567. [CrossRef] [PubMed]
6. Lo, M.-M.; Benfodda, Z.; Dunyach-Rémy, C.; Bénimélis, D.; Roulard, R.; Fontaine, J.-X.; Mathiron, D.; Quéro, A.; Molinié, R.; Meffre, P. Isolation and identification of flavones responsible for the antibacterial activities of *Tillandsia bergeri* extracts. *ACS Omega* **2022**, *7*, 35851–35862. [CrossRef] [PubMed]
7. Dong, H.; Wu, M.; Xiang, S.; Song, T.; Li, Y.; Long, B.; Feng, C.; Shi, Z. Total syntheses and antibacterial evaluations of neocyclomorusin and related flavones. *J. Nat. Prod.* **2022**, *85*, 2217–2225. [CrossRef] [PubMed]
8. Rubin, D.; Sansom, C.E.; Lucas, N.T.; McAdam, J.C.; Simpson, J.; Lord, J.M.; Perry, N.B. O-Acylated flavones in the alpine daisy *Celmisia viscosa*: Intraspecific variation. *J. Nat. Prod.* **2022**, *85*, 1904–1911. [CrossRef] [PubMed]
9. Wang, X.; Cao, Y.; Chen, S.; Lin, J.; Yang, X.; Huang, D. Structure–activity relationship (SAR) of flavones on their anti-inflammatory activity in murine macrophages in culture through the NF-κB pathway and c-Src kinase receptor. *J. Agric. Food Chem.* **2022**, *70*, 8788–8798. [CrossRef] [PubMed]
10. Tsai, H.-Y.; Chen, M.-Y.; Hsu, C.; Kuan, K.-Y.; Chang, C.-F.; Wang, C.-W.; Hsu, C.-P.; Su, N.-W. Luteolin phosphate derivatives generated by cultivating *Bacillus subtilis* var. Natto BCRC 80517 with luteolin. *J. Agric. Food Chem.* **2022**, *70*, 8738–8745. [CrossRef] [PubMed]
11. Dong, H.; Wu, M.; Li, Y.; Lu, L.; Qin, J.; He, Y.; Shi, Z. Total syntheses and anti-inflammatory evaluations of pongamosides A–C, natural furanoflavonoid glucosides from fruit of *Pongamia pinnata* (L.) Pierre. *J. Nat. Prod.* **2022**, *85*, 1118–1127. [CrossRef]
12. Li, J.; Tan, L.-H.; Zou, H.; Zou, Z.-X.; Long, H.-P.; Wang, W.-X.; Xu, P.-S.; Liu, L.-F.; Xu, K.-P.; Tan, G.-S. Palhinosides A–H: Flavone glucosidic truxinate esters with neuroprotective activities from *Palhinhaea cernua*. *J. Nat. Prod.* **2020**, *83*, 216–222. [CrossRef] [PubMed]
13. Lin, S.; Koh, J.-J.; Aung, T.T.; Sin, W.L.W.; Lim, F.; Wang, L.; Lakshminarayanan, R.; Zhou, L.; Tan, D.T.H.; Cao, D.; et al. Semisynthetic flavone-derived antimicrobials with therapeutic potential against methicillin-resistant *Staphylococcus aureus* (MRSA). *J. Med. Chem.* **2017**, *60*, 6152–6165. [CrossRef]
14. Pajtás, D.; Kónya, K.; Kiss-Szikszai, A.; Džubák, P.; Pethő, Z.; Varga, Z.; Panyi, G.; Patonay, T. Optimization of the synthesis of flavone–amino acid and flavone–dipeptide hybrids via Buchwald–Hartwig reaction. *J. Org. Chem.* **2017**, *82*, 4578–4587. [CrossRef] [PubMed]
15. Byun, Y.; Moon, K.; Park, J.; Ghosh, P.; Mishra, N.K.; Kim, I.S. Methylene thiazolidinediones as alkylation reagents in catalytic C–H functionalization: Rapid access to glitazones. *Org. Lett.* **2022**, *24*, 8578–8583. [CrossRef] [PubMed]
16. Xiong, Y.; Schaus, S.E.; Porco, J.A., Jr. Metal-catalyzed cascade rearrangements of 3-alkynyl flavone ethers. *Org. Lett.* **2013**, *15*, 1962–1965. [CrossRef]
17. Hosseini, S.; Thapa, B.; Medeiros, M.J.; Pasciak, E.M.; Pence, M.A.; Twum, E.B.; Karty, J.A.; Gao, X.; Raghavachari, K.; Peters, D.G.; et al. Electrosynthesis of a baurone by controlled dimerization of flavone: Mechanistic insight and large-scale application. *J. Org. Chem.* **2020**, *85*, 10658–10669. [CrossRef]
18. Tang, Q.; Bian, Z.; Wu, W.; Wang, J.; Xie, P.; Pittman, C.U., Jr.; Zhou, A. Making flavone thi-oethers using halides and powdered sulfur or $Na_2S_2O_3$. *J. Org. Chem.* **2017**, *82*, 10617–10622. [CrossRef]
19. Tokárová, Z.; Balogh, R.; Tisovský, P.; Hrnčariková, K.; Végh, D. Direct nucleophilic substitution of polyfluorobenzenes with pyrrole and 2,5-dimethylpyrrole. *J. Fluorine Chem.* **2017**, *204*, 59–64. [CrossRef]
20. Gerencsér, J.; Balázs, A.; Dormán, G. Synthesis and modification of heterocycles by metal-catalyzed cross-coupling reactions. In *Topics in Heterocyclic Chemistry*; Patonay, T., Kónya, K., Eds.; Springer International Publishing AG: Cham, Switzerland, 2016; Volume 45.
21. Kong, X.; Zhang, H.; Cao, C.; Shi, Y.; Pang, G. Effective transition metal free and selective C–F activation under mild conditions. *RSC Adv.* **2015**, *5*, 7035–7048. [CrossRef]
22. Boelke, A.; Sadat, S.; Lork, E.; Nachtsheim, B.J. Pseudocyclic bis-N-heterocycle-stabilized iodanes—Synthesis, characterization and applications. *Chem. Commun.* **2021**, *57*, 7434–7437. [CrossRef] [PubMed]
23. Zhong, W.; Wang, L.; Qin, D.; Zhou, J.; Duan, H. Two novel fluorescent probes as systematic sensors for multiple metal ions: Focus on detection of Hg^{2+}. *ACS Omega* **2020**, *5*, 24285–24295. [CrossRef] [PubMed]
24. Zhou, Q.; Hong, X.; Cui, H.-Z.; Huang, S.; Yi, Y.; Hou, X.-F. The construction of C–N, C–O, and $C(sp^2)$–$C(sp^3)$ bonds from fluorine-substituted 2-aryl benzazoles for direct synthesis of N-, O-, C-functionalized 2-aryl benzazole derivatives. *J. Org. Chem.* **2018**, *83*, 6363–6372. [CrossRef] [PubMed]
25. Santos, C.M.M.; Silva, V.L.M.; Silva, A.M.S. Synthesis of chromone-related pyrazole compounds. *Molecules* **2017**, *22*, 1665–1713. [CrossRef] [PubMed]

26. Shcherbakov, K.V.; Panova, M.A.; Burgart, Y.V.; Saloutin, V.I. Selective nucleophilic aromatic substitution of 2-(polyfluorophenyl)-4*H*-chromen-4-ones with pyrazole. *J. Fluorine Chem.* **2022**, *263*, 110034. [CrossRef]
27. Shcherbakov, K.V.; Artemyeva, M.A.; Burgart, Y.V.; Saloutin, V.I.; Volobueva, A.S.; Misiurina, M.A.; Esaulkova, Y.L.; Sinegubova, E.O.; Zarubaev, V.V. 7-Imidazolylsubstituted 4′-methoxy and 3′,4′-dimethoxy-containing polyfluoroflavones as promising antiviral agents. *J. Fluorine Chem.* **2020**, *240*, 109657. [CrossRef]
28. Podlech, J. Elimination of fluorine to form C–N bonds. In *Organo-Fluorine Compounds*, 4th ed.; Baasner, B., Hagemann, H., Tatlow, J.C., Eds.; Georg Thieme Verlag: Stuttgart, Germany, 1999; pp. 449–464.
29. Shcherbakov, K.V.; Panova, M.A.; Burgart, Y.V.; Zarubaev, V.V.; Gerasimova, N.A.; Evstigneeva, N.P.; Saloutin, V.I. The synthesis and biological evaluation of A- and B-ring fluorinated flavones and their key intermediates. *J. Fluorine Chem.* **2021**, *249*, 109857. [CrossRef]
30. Shcherbakov, K.V.; Artemyeva, M.A.; Burgart, Y.V.; Evstigneeva, N.P.; Gerasimova, N.A.; Zilberberg, N.V.; Kungurov, N.V.; Saloutin, V.I.; Chupakhin, O.N. Transformations of 3-acyl-4*H*-polyfluorochromen-4-ones under the action of amino acids and biogenic amines. *J. Fluorine Chem.* **2019**, *226*, 109354. [CrossRef]
31. Shcherbakov, K.V.; Burgart, Y.V.; Saloutin, V.I.; Chupakhin, O.N. Modification of polyfluoro-containing 3-(ethoxycarbonyl)flavones by biogenic amines and amino acids. *Curr. Org. Synth.* **2018**, *15*, 707–714. [CrossRef]
32. Shcherbakov, K.V.; Burgart, Y.V.; Saloutin, V.I.; Chupakhin, O.N. Polyfluorinecontaining chromen-4-ones: Synthesis and transformations. *Russ. Chem. Bull.* **2016**, *65*, 2151–2162. [CrossRef]
33. Ahmadi, A.; Mohammadnejadi, E.; Karami, P.; Razzaghi-Asi, N. Current status and structure activity relationship of privileged azoles as antifungal agents. *Int. J. Antimicrob. Agents* **2022**, *59*, 106518. [CrossRef]
34. Seck, I.; Nguemo, F. Triazole, imidazole, and thiazole-based compounds as potential agents against coronavirus. *Results Chem.* **2021**, *3*, 100132. [CrossRef]
35. Kerru, N.; Gummidi, L.; Maddila, S.; Gangu, K.K.; Jonnalagadda, S.B. A review on recent advances in nitrogen-containing molecules and their biological applications. *Molecules* **2020**, *25*, 1909–1951. [CrossRef] [PubMed]
36. Aggarwal, R.; Sumran, G. An insight on medicinal attributes of 1,2,4-triazoles. *Eur. J. Med. Chem.* **2020**, *205*, 112652. [CrossRef] [PubMed]
37. Prasher, P.; Sharma, M. "Azole" as privileged heterocycle for targeting the inducible cyclooxygenase enzyme. *Drug Dev. Res.* **2020**, *82*, 167–197. [CrossRef]
38. Hou, Y.; Shang, C.; Wang, H.; Yun, J. Isatin-azole hybrids and their anticancer activities. *Arch. Pharm. Chem. Life Sci.* **2019**, *353*, e1900272. [CrossRef] [PubMed]
39. Xu, M.; Peng, Y.; Wang, S.; Ji, J.; Rakesh, K.P. Triazole derivatives as inhibitors of Alzheimer's disease: Current developments and structure-activity relationships. *Eur. J. Med. Chem.* **2019**, *180*, 656–672. [CrossRef] [PubMed]
40. Fan, Y.-L.; Jin, X.-H.; Huang, Z.-P.; Yu, H.-F.; Zeng, Z.-G.; Gao, T.; Feng, L.-S. Resent advances of imidazole-containing derivatives as anti-tubercular agents. *Eur. J. Med. Chem.* **2018**, *150*, 347–365. [CrossRef]
41. Yan, M.; Xu, L.; Wang, Y.; Wan, J.; Liu, T.; Liu, W.; Wan, Y.; Zhang, B.; Wang, R. Opportunities and challenges of using five-membered ring compounds as promising antitubercular agents. *Drug Dev. Res.* **2020**, *81*, 402–418. [CrossRef]
42. Gao, F.; Wang, T.; Xiao, J.; Huang, G. Antibacterial activity study of 1,2,4-triazole derivatives. *Eur. J. Med. Chem.* **2019**, *173*, 274–281. [CrossRef]
43. Teli, G.; Chawla, P.A. Hybridization of imidazole with various heterocycles in targeting cancer. *ChemistrySelect* **2021**, *6*, 4803–4836. [CrossRef]
44. Zhang, J.; Wang, S.; Ba, Y.; Xu, Z. 1,2,4-Triazole-quinoline/ quinolone hybrids as potential anti-bacterial agents. *Eur. J. Med. Chem.* **2019**, *174*, 1–8. [CrossRef]
45. Fan, Y.-L.; Liu. M. Coumarin-triazole hybrids and their biological activities. *J. Heterocycl. Chem.* **2018**, *55*, 791–802. [CrossRef]
46. Zhang, T.; Zhu, M.; Li, J.; Zhang, Y.; Wang, X. Bipolar host materials comprising carbazole, pyridine and triazole moieties for efficient and stable phosphorescent OLEDs. *Dyes Pigm.* **2021**, *192*, 109426. [CrossRef]
47. Ye, S.; Zhuang, S.; Pan, B.; Guo, R.; Wang, L. Imidazole derivatives for efficient organic light-emitting diodes. *J. Inf. Disp.* **2020**, *21*, 173–196. [CrossRef]
48. Xu, H.; Zhao, Y.; Zhang, J.; Zhang, D.; Miao, Y.; Shinar, J.; Shinar, R.; Wang, H.; Xu, B. Low efficiency rol-off phosphorescent organic light-emitting devices using thermally activated delayed fluorescence hosts materials based 1,2,4-triazole acceptor. *Org. Electron.* **2019**, *74*, 13–22. [CrossRef]
49. Tao, Y.; Yang, C.; Qin, J. Organic host materials for phosphorescent organic light-emitting diodes. *Chem. Soc. Rev.* **2011**, *40*, 2943–2970. [CrossRef]
50. Emami, L.; Faghih, Z.; Ataollahi, E. Azole derivatives: Recent advances as potent antibacterial and antifungal agents. *Curr. Med. Chem.* **2022**, *129*, 220–249. [CrossRef]
51. Dolomanov, O.V.; Bourhis, L.J.; Gildea, R.J.; Howard, J.A.K.; Puschmann, H. OLEX2: A complete structure solution, refinement and analysis program. *J. Appl. Crystallogr.* **2009**, *42*, 339–341. [CrossRef]
52. Palatinus, L.; Chapuis, G. SUPERFLIP–A computer program for the solution of crystal structures by charge flipping in arbitrary dimensions. *J. Appl. Crystallogr.* **2007**, *40*, 786–790. [CrossRef]
53. Sheldrick, G.M. A short history of SHELX. *Acta Crystallogr. Sect. A Found. Crystallogr.* **2007**, *64*, 112–122. [CrossRef] [PubMed]

54. Kubanova, A.A.; Stepanova, Z.V.; Gus'kova, T.A.; Pushkina, T.V.; Krylova, L.Y.; Shilova, I.B.; Trenin, A.S. Metodicheskie rekomendacii po izucheniyu protivogribkovoi aktivnosti lekarstvennykh sredstv (Guidelines for the study of the antifungal activity of medicines). In *Rukovodstvo po Provedeniyu Doklinicheskikh Issledovanii Lekarstvennykh Sredstv (A Guide to Preclinical Trials of Medicines)*; Mironov, A.N., Ed.; Grif i K: Moscow, Russia, 2012; pp. 578–586.
55. Koldobskii, G.I.; Ostrovskii, V.A. Acid-base properties of five-membered nitrogen-containing heterocycles. *Chem. Heterocycl. Compds.* **1988**, *24*, 469–480. [CrossRef]
56. Krishnan, R.; Parthiban, A. Regioselective preparation of functional aryl ethers and esters by stepwise nucleophilic aromatic substitution reaction. *J. Fluorine Chem.* **2014**, *162*, 17–25. [CrossRef]
57. Chambers, R.D.; Martin, P.A.; Sandford, G.; Williams, L.H. Mechanisms of reactions of halogenated compounds: Part 7. Effects of fluorine and other groups as substituents on nucleophilic aromatic substitution. *J. Fluorine Chem.* **2008**, *129*, 998–1002. [CrossRef]

Disclaimer/Publisher's Note: The statements, opinions and data contained in all publications are solely those of the individual author(s) and contributor(s) and not of MDPI and/or the editor(s). MDPI and/or the editor(s) disclaim responsibility for any injury to people or property resulting from any ideas, methods, instructions or products referred to in the content.

Article

The Construction of Polycyclic Pyridones via Ring-Opening Transformations of 3-hydroxy-3,4-dihydropyrido[2,1-*c*][1,4]oxazine-1,8-diones

Viktoria V. Viktorova, Elena V. Steparuk, Dmitrii L. Obydennov *, and Vyacheslav Y. Sosnovskikh *

Institute of Natural Sciences and Mathematics, Ural Federal University, 51 Lenina Ave., 620000 Ekaterinburg, Russia
* Correspondence: dobydennov@mail.ru (D.L.O.); vy.sosnovskikh@urfu.ru (V.Y.S.)

Abstract: This work describes the synthesis of 3-hydroxy-3,4-dihydropyrido[2,1-*c*][1,4]oxazine-1,8-diones, their tautomerism, and reactivity towards binucleophiles. These molecules are novel and convenient building-blocks for the direct construction of biologically important polycyclic pyridones via an oxazinone ring-opening transformation promoted with ammonium acetate or acetic acid. In the case of *o*-phenylenediamine, partial aromatization of the obtained heterocycles proceeded to form polycyclic benzimidazole-fused pyridones (33–91%).

Keywords: 4-pyridone; oxazinone; ring-opening; benzimidazole; aldehyde-lactol tautomerism; ammonium acetate

1. Introduction

4-Pyridones are important nitrogen-containing heterocycles, which have recently attracted much attention as biologically active [1–4] and natural compounds [5,6]. Polycyclic structures, such as dolutegravir, bictegravir, and cabotegravir, are used as modern inhibitors of HIV integrase for antiretroviral therapy [1,3] (Figure 1). Baloxavir marboxil also belongs to this class of these compounds and is applied as the first cap-dependent endonuclease inhibitor for the treatment of influenza [2].

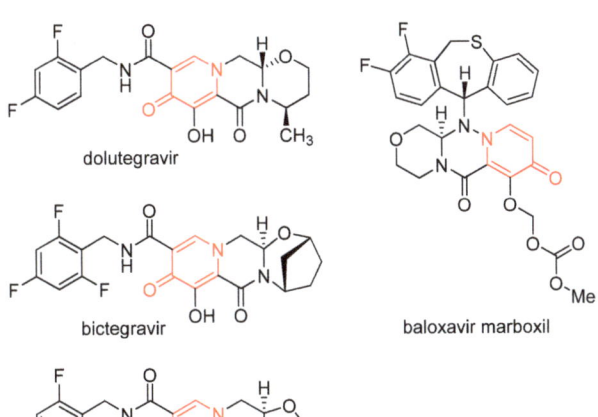

Figure 1. Some important polycyclic 4-pyridones.

The chemistry of these heterocycles is actively developed not only for the design of biologically important compounds [7–9], but also for effective preparation in the industry [1–3,10–15]. At the same time, there is a need to search for new multifarious pyridone building blocks [16–25] and convenient synthetic tools for the construction of polycyclic pyridones [1–3,15], including CH functionalization [26].

The general approach is well known in the literature based on the reaction of alkyl 1-(2,2-dimethoxyethyl)-4-oxo-1,4-dihydropyridine-2-carboxylates with binucleophiles to obtain the fused heterocycles [1,3,10–14] (Scheme 1). In this case, the acid-catalyzed deprotection of the dimethyl acetal group led to 2-(4-oxopyridin-1(4H)-yl)acetaldehydes, which are usually considered as intermediates of polycyclic pyridone formation. Substituted 3-hydroxy-3,4-dihydropyrido[2,1-c][1,4]oxazine-1,8-dione was also detected as the result of hydrolysis of the carboethoxy and dimethylacetal groups as a by-product in the synthesis of dolutegravir [10]. Such structure also can be suggested as a possible intermediate for this heterocyclization [1,10]. To the best of our knowledge, only unsubstituted 3-hydroxy-3,4-dihydropyrido[2,1-c][1,4]oxazine-1,8-dione was prepared from comanic acid in pure form and used for the transformation with (R)-3-aminobutan-1-ol [3,11]. Moreover, there are data on the Ugi reaction of a pyridone-bearing aldoacid with isonitriles for the synthesis of various piperazinone-fused pyridones [27].

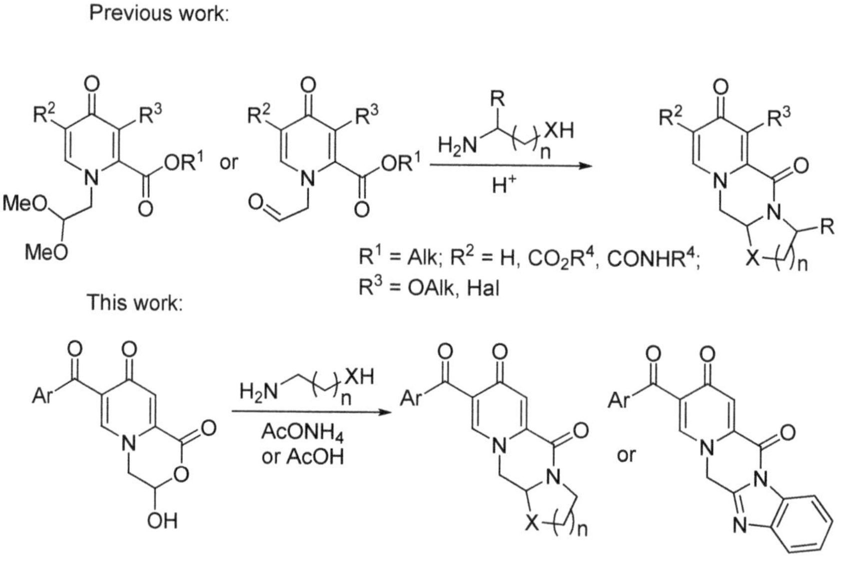

Scheme 1. General strategy for the synthesis of polycyclic pyridones.

We decided to study 3-hydroxy-3,4-dihydropyrido[2,1-c][1,4]oxazine-1,8-diones in more detail in order to find new directions for the construction of polycyclic pyridones through morpholinone ring-opening reactions. These molecules bear the hidden aldehyde moiety, which can determine their high reactivity towards nucleophiles via the tautomeric equilibrium. This strategy based on the transformation with diamines can open access to new cyclic fused pyridones, which are of interest for the further design of biologically active compounds.

2. Results and Discussion

Synthesis of 3-hydroxy-3,4-dihydropyrido[2,1-c][1,4]oxazine-1,8-diones 3 and Their Chemical Properties

We started with the ANRORC reaction of 5-acyl-4-pyrone-2-carboxylate **1** with 2,2-dimethoxyethylamine as the effective method for the preparation of 4-pyridones [3,17,28,29]

(Scheme 2, Table 1). The ring-opening transformations proceeded under reflux in toluene for 4 h to produce pyridones **2a–d,f** in 30–90% yields. Pivaloyl-substituted pyrone **1e** did not provide the desired product, and the reaction was carried in more polar MeCN, leading to pyridone **2e** in 42% yield. Compounds **2a–e** underwent the deprotection of the dimethyl acetal moiety in aqueous HCl to form 3-hydroxy-3,4-dihydropyrido[2,1-c][1,4]oxazine-1,8-diones **3a–e**. For 2,5-dicarbethoxy-4-pyridone **2f**, the heating in formic acid was used for the selective hydrolysis of the COOEt group at the C-2 position as a result of the promotion by the presence of the adjacent aldehyde fragment.

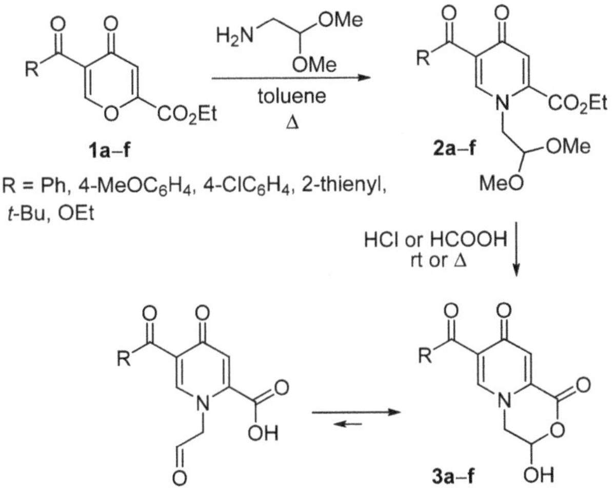

Scheme 2. Synthesis of 3-hydroxy-3,4-dihydropyrido[2,1-c][1,4]oxazine-1,8-diones **3**.

Table 1. The scope of products **2** and **3**.

Entry	Compound 2,3	R	Yield of 2, %	Yield of 3, %
1	a	Ph	90	53
2	b	4-MeOC$_6$H$_4$	63	85
3	c	4-ClC$_6$H$_4$	73	59
4	d	2-Th	72	84
5	e	t-Bu	42	41
6	f	OEt	30	74

Pyridones **3** can undergo aldehyde-lactol tautomerism [30] and exist as acyclic aldoacid or a cyclic lactol form (3-hydroxy-3,4-dihydropyrido[2,1-c][1,4]oxazine-1,8-diones) (Scheme 2). It is interesting to note that this type of the ring-chain tautomerism for morpholinones has not been studied before.

The ^{1}H NMR spectra of products **3** in DMSO-d_6 demonstrates the existence of only the lactol form. The spectral feature of the tautomer is the presence of a downfield signal of the OH group at δ 8.13–8.72 ppm and an ABX system of the morpholinone moiety. For the pivaloyl-substituted compound **3e**, a singlet of the methylene group and a strongly broadened singlet of the CH proton were observed probably due to the rapid interconversion between different forms.

Pyridones **3** bear the carbonyl group at the C-5 position, which can be used for further modifications of the heterocyclic fragment. Therefore, the important task included the search for selective transformations on the morpholinone fragment. We have studied the detailed influence of conditions on the ring-opening reaction of 3-hydroxy-3,4-dihydropyrido[2,1-c][1,4]oxazine-1,8-diones (**3b**) with 3-aminopropan-1-ol (Table 2). A mixture of methanol–toluene was used as a solvent to increase the solubility of pyridone **3b**. The reaction did not proceed without the use of catalysts even under the prolonged reflux. It was found that the transformation in the presence of acetic acid as an additive led to product **4a** in 48% yield. We suggested that the formation of 3-hydroxypropylammonium acetate occurred, which acts as a nucleophile and activator of the morpholinone moiety. However, this transformation did not proceed at room temperature, as well as with the use of the 0.2 equiv. of acetic acid.

Table 2. The optimization of reactions conditions for the synthesis of **4a** from **3b** and 3-aminopropan-1-ol [a].

Entry	Catalyst, Equiv.	Time, h	Temp., °C	Yield of 4a, %
1	–	12	reflux	–
2	AcOH, 1.2	12	reflux	48
3	AcOH, 0.2	12	reflux	–
4	AcOH, 1.2	12	room temperature	–
5	**AcONH₄, 1.0**	**12**	**reflux**	**69**
6	AcONH₄, 1.0	12	room temperature	–
7	NH₄Cl, 1.0	12	reflux	–
8	NBu₄Br, 1.0	12	reflux	–
9	NH₂Et₂OAc, 1.0	12	reflux	–

[a] Pyridone **3b** (99.9 mg, 0.317 mmol) and 3-aminopropan-1-ol (28.6 mg, 0.381 mmol) were stirred in a mixture of toluene and methanol (1:1, 2 mL).

Taking into account the effect of AcOH, we tried to use AcONH$_4$ as a bifunctional catalyst [31,32] for this process. To our delight, the product was obtained in a good yield (69%) under reflux for 12 h (TLC monitoring) (Table 2). The variation of the nature of the ammonium salt or temperature did not allow for the improvement of the reaction yield.

We tried to extend the optimized conditions with the use of ammonium acetate (Method A) for other 5-acylpyridones **3** and binucleophiles for the synthesis of polycyclic 4-pyridones (Table 3). In most cases, the acyl fragment strongly influenced the reaction selectivity, and an alternative method included the use of acetic acid (Method B). Pyridones **3b,c** bearing para-substituted benzoyl fragments underwent the transformation in the presence of ammonium acetate and led to the formation of products **4a,b** in 69–75% yields. Benzoyl- and thienoyl-substituted compounds **3a,d** reacted more effectively in the conditions of method B and provided products **4c,d** in 31–52% yields. When propane-1,3-diamine was used as a binucleophile, we were not able to isolate the desired polycyclic products in a pure form directly. The precipitates that formed always contained the starting diamine in significant amounts. Next, binucleophiles bearing two carbon atoms in the linker was used for the heterocyclization. The reaction with ethylenediamine proceeded in good yields and led to the formation of imidazo[1,2-a]pyrido[1,2-d]pyrazine-5,7-diones **5a,b** in 78–84% yields (Method A). Thienoyl-substituted pyridone **3d** underwent the ring-opening process in the presence of acetic acid (Method B) to produce compound **5c** in 48% yield. At the same time, we failed to isolate any products in the pure form in the reaction with ethanolamine.

Table 3. Reactions of compounds **3** with binucleophiles [a].

4a: Ar = 4-MeOC$_6$H$_4$ (69%, Method A)
b: Ar = 4-ClC$_6$H$_4$ (75%, Method A)[b]
c: Ar = 2-Th (52%, Method B)
d: Ar = Ph (31%, Method B)

5a: Ar = 4-MeOC$_6$H$_4$ (84%, Method A)
b: Ar = 4-ClC$_6$H$_4$ (78%, Method A)
c: Ar = 2-Th (48%, Method B)

[a] *Method A.* Compound **3** (0.317 mmol) was stirred with amine (0.381 mmol) and ammonium acetate (0.0245 g, 0.317 mmol) in a mixture of toluene (1 mL) and methanol (1 mL) at 90 °C for 8–12 h. *Method B.* Compound **3** (0.317 mmol) was stirred with amine (0.381 mmol) and acetic acid (22.8 mg, 0.380 mmol) in a mixture of toluene (1 mL) and methanol (1 mL) at 90 °C for 7–12 h. [b] 2 equiv. of the amine was used.

The peculiarity of ammonium acetate is probably associated with its solubility in a methanol–toluene mixture and the ability to promote the ring-opening process of the morpholinone ring, which leads to the formation of the aldoacid (Scheme 3). Subsequent stages, including intermolecular attack of a binucleophile and intramolecular cyclization, can be catalyzed by both the ammonium cation and acetic acid. An experiment was carried out to study the reaction of ammonium acetate with pyridone **3b** under reflux. According to the ^{1}H NMR spectrum of the obtained precipitate, it was found that the formation of an open-chain structure occurred (see Supplementary Materials). Although we did not detect the aldehyde group, a singlet of the methylene group and absence of the 3-CH proton of the lactol form were observed in the ^{1}H NMR spectrum.

The reaction of pyridone **3** with *o*-phenylenediamine proceeded in the presence of ammonium acetate at room temperature or under reflux and was accompanied by aromatization under the action of atmospheric oxygen (Scheme 4, Table 4). The intermediate **C** was not isolated in pure form, but was detected as by-products in all cases. Carrying out the reaction under argon did not allow the selective formation of compound **C**. This result can indicate that the oxidation additionally promotes the reaction leading to the most stable product **6**.

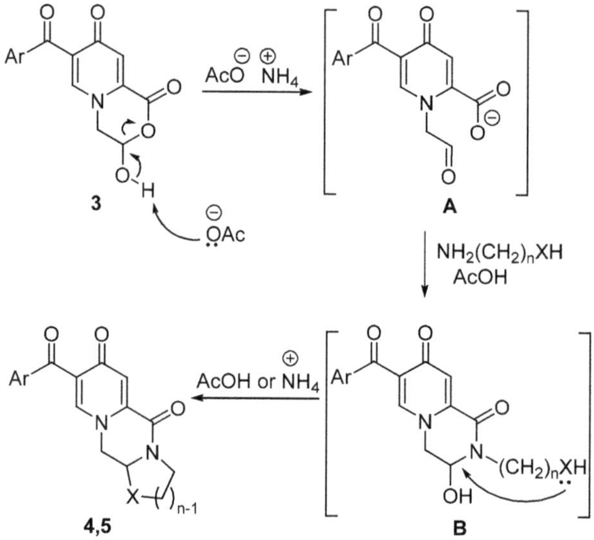

Scheme 3. The proposed mechanism of the ring-opening transformation.

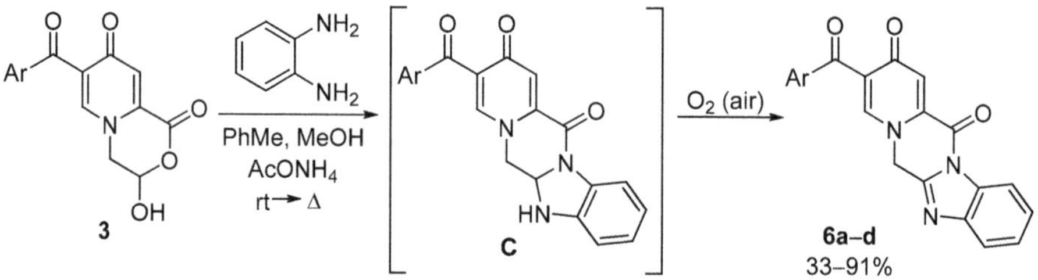

Scheme 4. Reactions of compounds 3 with o-phenylenediamine.

Table 4. The scope of products 6.

Entry	Compound 6	Ar	Yield of 6, %
1	a	Ph	71
2	b	4-MeOC$_6$H$_4$	91
3	c	4-ClC$_6$H$_4$	33
4	d	2-Th	42

To obtain compounds **6** in a pure form directly, the reaction was carried out at room temperature for 12 h and subsequent reflux for 2 h. In these conditions, the aromatization proceeded completely and pyridones **6** bearing the benzimidazole fragment were isolated in 33–91% yields. The reaction turned out to be sensitive to the nature of the acyl moiety, which probably determined the occurrence of side reactions. Pivaloyl pyridone **3e** led to degradation products, which did not bear the *t*-Bu group. In the ^{1}H NMR spectra of compounds **6**, the downfield singlet of methylene group was observed at δ 5.81–5.84 ppm due to the presence of two adjacent aromatic systems.

Thus, 3-hydroxy-3,4-dihydropyrido[2,1-*c*][1,4]oxazine-1,8-diones have been synthesized and demonstrated to exist predominantly in the lactol tautomeric form. The new and convenient approach has been developed for the preparation of polycyclic pyridones based on the pyridomorpholinones via ring-opening reactions. The binucleophile linker and the nature of nucleophile centers strongly influence the reaction outcome. The most active binucleophiles in this heterocyclization process are ethylenediamine and 3-aminopropan-1-ol. It has been demonstrated that the reaction with *o*-phenylenediamine is followed by oxidation and the formation of benzimidazole-fused 4-pyridones.

3. Materials and Methods

NMR spectra were recorded on Bruker DRX-400 (Bruker BioSpin GmbH, Ettlingen, Germany, work frequencies: ^{1}H, 400 MHz; ^{13}C, 101 MHz), Bruker Avance-400 (Bruker BioSpin GmbH, Rheinstetten, Germany, work frequencies: ^{1}H, 400 MHz; ^{13}C, 101 MHz), Bruker Avance III-500 (Bruker BioSpin GmbH, Rheinstetten, Germany, work frequencies: ^{1}H, 500 MHz; ^{13}C, 126 MHz), and Bruker Avance NEO (Bruker BioSpin GmbH, Rheinstetten, Germany, work frequencies: ^{1}H, 600 MHz; ^{13}C, 151 MHz) spectrometers in DMSO-d_6 or CDCl$_3$. The chemical shifts (δ) are reported in ppm relative to the internal standard TMS (^{1}H NMR) and residual signals of the solvents (^{13}C NMR). IR spectra were recorded on a Shimadzu IRSpirit-T (Shimadzu Corp., Kyoto, Japan) spectrometer using an attenuated total reflectance (ATR) unit (FTIR mode, diamond prism); the absorbance maxima (ν) are reported in cm^{-1}. Mass spectra (ESI-MS) were measured with a Waters Xevo QTof instrument (Waters Corp., Milford, MA, USA). Elemental analyses were performed on an automatic analyzer PerkinElmer PE 2400 (Perkin Elmer Instruments, Waltham, MA, USA). Melting points were determined using a Stuart SMP40 melting point apparatus (Bibby Scientific Ltd., Stone, Staffordshire, UK). Column chromatography was performed on silica gel (Merck 60, 70–230 mesh). All solvents that were used were dried and distilled by standard procedures. 4-Pyrones **1** were prepared according to the literature methods [28,33,34].

*3.1. General Procedure for the Preparation of 1-(2,2-dimethoxyethyl)-4-pyridones **2***

Ethyl 5-acyl-4-oxo-4*H*-pyran-2-carboxylate **1** (0.330 mmol) was added to a cooled solution of 2,2-dimethoxyethanamine (0.0380 g, 0.361 mmol) in toluene (1 mL). The resulting mixture was stirred for 20 min at room temperature (the precipitation was observed) and heated under reflux for 4 h. The solvent was evaporated under reduced pressure, and the product was isolated by flash chromatography using ethyl acetate as an eluent. For pyridone **2e**, acetonitrile was used instead of toluene (reflux for 3 h).

Ethyl 5-benzoyl-1-(2,2-dimethoxyethyl)-4-oxo-1,4-dihydropyridine-2-carboxylate (**2a**). Yield 0.1067 g (90%), yellow crystals. IR (ATR) ν 3053, 2932, 2836, 1720, 1652, 1630, 1475, 1294, 829, 701. ^{1}H NMR (400 MHz, CDCl$_3$) δ 1.41 (t, *J* = 7.1 Hz, 3H, Me), 3.42 (s, 6H, 2MeO), 4.34 (d, *J* = 4.5 Hz, 2H, CH$_2$N), 4.40 (q, *J* = 7.1 Hz, 2H, CH$_2$), 4.53 (t, *J* = 4.5 Hz, 1H, CH), 7.07 (s, 1H, H-3 Py), 7.43 (t, *J* = 7.6 Hz, 2H, H-3, H-5 Ph), 7.55 (t, *J* = 8.0 Hz, 2H, H-4 Ph), 7.79 (s, 1H, H-6 Py), 7.86 (dd, *J* = 8.0 Hz, *J* = 1.4 Hz, 2H, H-2, H-6 Ph). ^{13}C NMR (126 MHz, DMSO-d_6) δ 13.7, 54.2, 55.4, 62.5, 102.7, 122.3, 127.9, 128.4 (2C), 129.1 (2C), 133.2, 136.9, 141.6, 147.2, 162.0, 174.4, 193.2. HRMS (ESI) m/z [M + H]$^+$. Calculated for C$_{19}$H$_{22}$NO$_6$: 360.1453. Found: 360.1447.

Ethyl 1-(2,2-dimethoxyethyl)-5-(4-methoxybenzoyl)-4-oxo-1,4-dihydropyridine-2-carboxylate (**2b**). Yield 0.0828 g (63%), light yellow oil. ^{1}H NMR (500 MHz, CDCl$_3$) δ 1.41 (t, *J* = 7.1 Hz, 3H, Me), 3.42 (s, 6H, 2MeO), 4.36 (d, *J* = 4.5 Hz, 2H, CH$_2$N), 4.40 (q, *J* = 7.1 Hz, 2H, CH$_2$), 4.53 (t, *J* = 4.5 Hz, 1H, CH), 7.06 (s, 1H, H-3 Py), 7.40 (d, *J* = 8.7 Hz, 2H, H-3, H-5 Ar), 7.79 (d, *J* = 8.7 Hz, 2H, H-2, H-6 Ar), 7.82 (s, 1H, H-6 Py). ^{13}C NMR (151 MHz, CDCl$_3$) δ 14.0, 55.5, 55.7, 55.9, 62.8, 103.3, 113.6, 124.8, 129.4, 129.8, 132.3, 140.2, 147.3, 162.4, 163.8, 175.7, 191.6. Anal. Calculated for C$_{20}$H$_{23}$NO$_7$·0.5H$_2$O: C 60.29; H 6.07; N 3.52. Found: C 59.98; H 6.20; N 3.47.

Ethyl 5-(4-chlorobenzoyl)-1-(2,2-dimethoxyethyl)-4-oxo-1,4-dihydropyridine-2-carboxylate (**2c**). Yield 0.0949 g (73%), brown viscous liquid. IR (ATR) ν 3033, 2978, 2862, 1729, 1660, 1630, 1481, 1246, 851, 767. ^{1}H NMR (400 MHz, CDCl$_3$) δ 1.41 (t, *J* = 7.1 Hz, 3H, Me), 3.42 (s, 6H, 2MeO), 4.36 (d, *J* = 4.5 Hz, 2H, CH$_2$N), 4.40 (q, *J* = 7.1 Hz, 2H, CH$_2$), 4.53 (t, *J* = 4.5 Hz, 1H, CH), 7.06 (s, 1H, H-3 Py), 7.40 (d, *J* = 8.7 Hz, 2H, H-3, H-5 Ar), 7.79 (d, *J* = 8.7 Hz,

2H, H-2, H-6 Ar), 7.82 (s, 1H, H-6 Py). ^{13}C NMR (126 MHz, CDCl$_3$) δ 14.0, 55.7, 55.8, 62.8, 103.2, 125.3, 128.2, 128.5 (2C), 131.0 (2C), 135.4, 139.4, 140.3, 148.1, 162.2, 175.6, 192.1. Anal. Calculated for C$_{19}$H$_{20}$ClNO$_6$: C 57.95; H 5.12; N 3.56. Found: C 58.35; H 5.42; N 3.48.

Ethyl 1-(2,2-dimethoxyethyl)-4-oxo-5-(thiophene-2-carbonyl)-1,4-dihydropyridine-2-carboxylate (**2d**). Yield 0.0868 g (72%), dark yellow viscous liquid. IR (ATR) ν 3066, 2985, 2933, 2841, 1637, 1620, 1478, 1250, 849, 719. ^{1}H NMR (400 MHz, CDCl$_3$) δ 1.41 (t, *J* = 7.1 Hz, 3H, Me), 3.42 (s, 6H, 2MeO), 4.34 (d, *J* = 4.5 Hz, 1H, CH$_2$N), 4.40 (q, *J* = 7.1 Hz, 2H, CH$_2$), 4.53 (t, *J* = 4.5 Hz, 1H, CH), 7.09 (s, 1H, H-3 Py), 7.12 (dd, *J* = 4.9 Hz, *J* = 3.9 Hz, 1H, H-4 Th), 7.68 (dd, *J* = 4.9, *J* = 1.1 Hz, 1H, H-5 Th), 7.81 (s, 1H, H-6 Py), 7.85 (dd, *J* = 3.9, *J* = 1.1 Hz, 1H, H-3 Th). ^{13}C NMR (126 MHz, CDCl$_3$) δ 13.9, 55.7, 55.8, 62.8, 103.2, 125.1, 128.1, 128.8, 134.7, 135.1, 140.0, 143.7, 147.4, 162.2, 175.3, 184.2. Anal. Calculated for C$_{17}$H$_{19}$NO$_5$S: C 55.88; H 5.24; N 3.83. Found: C 56.09; H 5.15; N 3.90.

Ethyl 1-(2,2-dimethoxyethyl)-4-oxo-5-pivaloyl-1,4-dihydropyridine-2-carboxylate (**2e**). Yield 0.0470 g (42%), brown viscous liquid. IR (ATR) ν 2960, 2837, 1731, 1628, 1480, 1247, 952, 805. ^{1}H NMR (400 MHz, CDCl$_3$) δ 1.28 (s, 9H, *t*-Bu), 1.40 (t, *J* = 7.1 Hz, 3H, Me), 3.41 (s, 6H, 2MeO), 4.26 (d, *J* = 4.7 Hz, 2H, CH$_2$N), 4.38 (q, *J* = 7.1 Hz, 2H, CH$_2$), 4.50 (t, *J* = 4.7 Hz, 1H, CH), 6.97 (s, 1H, H-3 Py), 7.41 (s, 1H, H-6). ^{13}C NMR (126 MHz, CDCl$_3$) δ 13.9, 26.2, 44.7, 55.6, 55.8, 62.7, 103.4, 123.9, 132.1, 139.7, 144.1, 162.4, 175.4, 209.8. HRMS (ESI) m/z [M + H]$^+$. Calculated for C$_{17}$H$_{26}$NO$_6$: 340.1766. Found: 340.1760.

Diethyl 1-(2,2-dimethoxyethyl)-4-oxo-1,4-dihydropyridine-2,5-dicarboxylate (**2f**). Yield 0.0324 g (30%), brown viscous liquid. IR (ATR) ν 2978, 2839, 1725, 1664, 1631, 1474, 1299, 867, 776. ^{1}H NMR (400 MHz, CDCl$_3$) δ 1.38 (t, *J* = 7.1 Hz, 3H, Me), 1.39 (t, *J* = 7.1 Hz, 3H, Me), 3.41 (s, 6H, 2MeO), 4.33 (d, *J* = 4.7 Hz, 2H, CH$_2$N), 4.37 (q, *J* = 7.1 Hz, 2H, CH$_2$), 4.38 (q, *J* = 7.1 Hz, 2H, CH$_2$), 4.49 (t, *J* = 4.7 Hz, 1H, CH), 7.07 (s, 1H, H-3 Py), 8.18 (s, 1H, H-6 Py). HRMS (ESI) m/z [M + H]$^+$. Calculated for C$_{15}$H$_{22}$NO$_7$: 328.1406. Found: 328.1396. The spectral data are in accordance with the patent literature [11].

3.2. General Method for the Preparation of dihydropyrido[2,1-c][1,4]oxazine-1,8-diones 3

1-(2,2-Dimethoxyethyl)-4-pyridone **2** (0.332 mmol) was stirred in hydrochloric acid (1: 1, 2 mL) for 24 h or, for **2b**, in concentrated hydrochloric acid (2 mL) for 3 h at room temperature and for 3 h under reflux. The precipitate formed was filtered. For compound **2f**, formic acid (85%, 2 mL) was used. The reaction mixture was stirred at room temperature for 3 h and heated at 85 °C for 4 h. After evaporation of the solvent, the product was isolated by flash chromatography using ethyl acetate as an eluent.

7-Benzoyl-3-hydroxy-3,4-dihydropyrido[2,1-c][1,4]oxazine-1,8-dione (**3a**). Yield 0.0506 g (53%), white powder, mp 187–188 °C. IR (ATR) ν 3228, 3078, 2489, 1665, 1637, 1470, 1213, 854, 748. ^{1}H NMR (400 MHz, DMSO-*d*$_6$) δ 4.29 (dd, *J* = 13.5 Hz, *J* = 3.3 Hz, 1H, C*H*H), 4.47 (dd, *J* = 13.5 Hz, *J* = 1.6 Hz, 1H, CH*H*), 6.08 (unresolved m, 1H, H-3), 6.99 (s, 1H, H-9), 7.50 (d, *J* = 7.7 Hz, 2H, H-3, H-5 Ph), 7.64 (t, *J* = 7.4 Hz, *J* = 1.0 Hz, 1H, H-4 Ph), 7.76 (dd, *J* = 8.4 Hz, *J* = 1.3 Hz, 2H, H-2, H-6 Ph), 8.28 (br.s, 1H, H-6), 8.72 (br.s, 1H, OH). ^{13}C NMR (126 MHz, DMSO-*d*$_6$) δ 52.4, 93.7, 121.1, 128.6 (2C), 129.3 (2C), 133.6, 136.5, 144.4, 158.1, 174.0, 192.9. HRMS (ESI) m/z [M + H]$^+$. Calculated for C$_{15}$H$_{12}$NO$_5$: 286.0714. Found: 286.0715.

3-Hydroxy-7-(4-methoxybenzoyl)-3,4-dihydropyrido[2,1-c][1,4]oxazine-1,8-dione (**3b**). Yield 0.0902 g (85%), beige powder, mp 205–206 °C. IR (ATR) ν 3060, 2934, 1684, 1599, 1268, 1153, 1021, 912, 843. ^{1}H NMR (500 MHz, DMSO-*d*$_6$) δ 3.84 (s, 3H, OMe), 4.25 (dd, *J* = 13.5 Hz, *J* = 3.6 Hz, 1H, C*H*H), 4.44 (dd, *J* = 13.5 Hz, *J* = 1.5 Hz, 1H, CH*H*), 6.07 (unresolved m, 1H, H-3), 6.94 (s, 1H, H-9), 7.03 (d, *J* = 8.9 Hz, 2H, H-3, H-5 Ar), 7.75 (d, *J* = 8.8 Hz, 2H, H-2, H-6 Ar), 8.18 (s, 1H, H-6), 8.66 (br.s, 1H, OH). ^{13}C NMR (126 MHz, DMSO-*d*$_6$) δ 52.1, 55.6, 93.6, 113.8 (2C), 121.2, 129.3, 129.9, 131.8 (2C), 135.9, 143.5, 158.2, 163.5, 174.2, 191.3. Anal. Calculated for C$_{16}$H$_{13}$NO$_6$·0.25H$_2$O: C 60.10; H 4.26; N 4.38. Found: C 60.17; H 3.98; 4.42.

7-(4-Chlorobenzoyl)-3-hydroxy-3,4-dihydropyrido[2,1-c][1,4]oxazine-1,8-dione (**3c**). Yield 0.0626 g (59%), white powder, mp 209–210 °C. IR (ATR) ν 3077, 1728, 1645, 1635, 1557, 1533, 1404, 1216, 1047, 732. ^{1}H NMR (500 MHz, DMSO-*d*$_6$) δ 4.32 (br.s, 1H, C*H*H), 4.45 (br.s, 1H, CH*H*), 6.10 (unresolved m, 1H, H-3), 6.90 (br.s, 1H, H-9), 7.57 (d, *J* = 8.9 Hz, 2H, H-3, H-5

Ar), 7.75 (d, *J* = 8.9 Hz, 2H, H-2, H-6 Ar), 8.28 (s, 1H, H-6), 8.67 (s, 1H, OH). ^{13}C NMR (126 MHz, DMSO-d_6) δ 52.0, 122.5, 128.6 (2C),128.7, 130.9 (2C), 135.5, 138.0, 144.6, 163.0, 174.8, 192.5 (2C were not observed). HRMS (ESI) m/z [M + H]$^+$. Calculated for $C_{15}H_{11}ClNO_5$: 320.0330. Found: 320.0326.

3-Hydroxy-7-(thiophene-2-carbonyl)-3,4-dihydropyrido[2,1-c][1,4]oxazine-1,8-dione (**3d**). Yield 0.0812 g (84%), beige powder, mp 206–207 °C. IR (ATR) ν 3071, 2904, 1730, 1638, 1407, 1209, 852, 749. ^{1}H NMR (500 MHz, DMSO-d_6) δ 4.26 (dd, *J* = 13.3 Hz, *J* =2.7 Hz, 1H, C*HH*), 4.46 (d, *J* = 13.3 Hz, 1H, CH*H*), 6.06 (s, 1H, H-3), 6.95 (s, 1H, H-9), 7.23 (dd, *J* = 4.9 Hz, *J* = 3.9 Hz, 1H, H-4 Th), 7.72 (dd, *J* = 3.9 Hz, *J* = 0.9 Hz, 1H, H-3 Th), 8.08 (dd, *J* = 4.9 Hz, *J* = 0.9 Hz, 1H, H-5 Th), 8.25 (s, 1H, H-6), 8.65 (d, *J* = 4.5 Hz, 1H, OH). ^{13}C NMR (126 MHz, DMSO-d_6) δ 51.8, 93.5, 122.1, 128.7, 129.3, 135.5, 135.7, 143.2, 143.3, 158.4, 174.4, 184.7 (1C was not observed). Anal. Calculated for $C_{13}H_9NO_5S$: C 53.61; H 3.11; N 4.81. Found: C 53.64; H 3.39; N 4.90.

3-Hydroxy-7-pivaloyl-3,4-dihydropyrido[2,1-c][1,4]oxazine-1,8-dione (**3e**). Yield 0.0361 g (41%), brown liquid product. IR (ATR) ν 2972, 2935, 2909, 1697, 1479, 1364, 1216, 1060. ^{1}H NMR (500 MHz, DMSO-d_6) δ 1.18 (s, 9H, *t*-Bu), 4.33 (s, 2H, 4-CH$_2$), 6.13 (br.s, 1H, H-3), 6.83 (s, 1H, H-9), 7.90 (s, 1H, H-6), 8.13 (s, 1H, OH). ^{13}C NMR (126 MHz, DMSO-d_6) δ 26.0, 44.0, 54.6, 88.4, 120.1, 131.9, 140.9, 159.7, 163.1, 174.9, 210.1. HRMS (ESI) m/z [M + H]$^+$. Calculated for $C_{13}H_{16}NO_5$: 266.1033. Found: 266.1028.

Ethyl 3-hydroxy-1,8-dioxo-1,3,4,8-tetrahydropyrido[2,1-c][1,4]oxazine-7-carboxylate (**3f**). Yield 0.0622 g (74%), brown powder, mp 193–194 °C. IR (ATR) ν 2985, 1734, 1695, 1575, 1450, 1306, 1200, 1112, 1012, 876, 797. ^{1}H NMR (500 MHz, DMSO-d_6) δ 1.26 (t, *J* = 7.1 Hz, 2H, Me), 4.21 (q, *J* = 7.1 Hz, 2H, CH$_2$), 4.31 (br.s, 1H, CH), 4.41 (s, 1H, CH), 6.03 (s, 1H, H-3), 6.87 (s, 1H, H-9), 8.47 (s, 1H, H-6), 8.61 (s, 1H, OH). ^{13}C NMR (126 MHz, DMSO-d_6) δ 14.2, 52.0, 60.2, 93.6, 119.8, 123.1, 135.4, 146.5, 158.6, 163.8, 173.7. HRMS (ESI) m/z [M + H]$^+$. Calculated for $C_{11}H_{12}NO_6$: 254.0669. Found: 254.0665.

3.3. General Method for the Preparation of Compounds **4** *and* **5**

Method A. 3-Hydroxy-7-acyl-3,4-dihydropyrido[2,1-c][1,4]oxazine-1,8-dione **3** (0.317 mmol) was stirred with amine (0.381 mmol) and ammonium acetate (0.0245 g, 0.317 mmol) in a mixture of toluene (1 mL) and methanol (1 mL) at 90 °C of an oil bath for 7–12 h. The precipitate obtained was filtered and recrystallized in a mixture of toluene and ethanol.

Method B. 3-Hydroxy-7-acyl-3,4-dihydropyrido[2,1-c][1,4]oxazine-1,8-dione **3** (0.317 mmol) was stirred with amine (0.381 mmol) and acetic acid (22.8 mg, 0.380 mmol) in a mixture of toluene (1 mL) and methanol (1 mL) at 90 °C of an oil bath for 7–12 h. The precipitate obtained was filtered and recrystallized in a mixture of toluene and ethanol.

9-(4-Methoxybenzoyl)-3,4,12,12a-tetrahydro-2H-pyrido[1′,2′:4,5]pyrazino[2,1-b][1,3]oxazine-6,8-dion (**4a**). The reaction was carried out for 12 h. Method A. Yield 0.0775 g (69%), yellow powder, mp 222–223 °C. IR (ATR) ν 2956, 2875, 1673, 1640, 1573, 1467, 1384, 783. ^{1}H NMR (400 MHz, DMSO-d_6) δ 1.61 (dm, *J* = 13.2 Hz, 1H, H-3′), 1.64–1.84 (m, 1H, H-3), 3.22 (td, *J* = 12.9 Hz, *J* = 3.2 Hz, 1H, CH), 3.84 (s, 3H, OMe), 3.90 (td, *J* = 11.8 Hz, *J* = 2.7 Hz, 1H, CH), 4.07 (dd, *J* = 11.3 Hz, *J* = 4.7 Hz, 1H, CH), 4.28 (dd, *J* = 14.1 Hz, *J* = 3.5 Hz, 1H, CH), 4.46 (dd, *J* = 14.0 Hz, *J* = 4.2 Hz, 2H, CH), 5.28 (t, *J* = 3.6 Hz, 1H, H-12a), 6.94 (s, 1H, H-7), 7.02 (d, *J* = 8.9 Hz, 2H, H-3, H-5 Ar), 7.75 (d, *J* = 8.9 Hz, 2H, H-2, H-6 Ar), 8.11 (s, 1H, H-10). ^{13}C NMR (126 MHz, DMSO-d_6) δ 25.0, 43.0, 50.2, 55.5, 67.1, 81.5, 113.7 (2C), 120.1, 129.6, 129.7, 131.7, 137.9 (2C), 142.9, 157.0, 163.3, 175.0, 192.0. HRMS (ESI) m/z [M + H]$^+$. Calculated for $C_{19}H_{19}N_2O_5$: 355.1288. Found: 355.1294.

9-(4-Chlorobenzoyl)-3,4,12,12a-tetrahydro-2H-pyrido[1′,2′:4,5]pyrazino[2,1-b][1,3]oxazine-6,8-dione (**4b**). Method A. The reaction was carried out for 12 h. 3-Aminopropan-1-ol (47.6 mg, 0.634 mmol) was used. Yield 0.0853 g (75%), yellow powder, mp 259–261 °C. IR (ATR) ν 3020, 2950, 1674, 1627, 1572, 1452, 1357, 789. ^{1}H NMR (500 MHz, DMSO-d_6) δ 1.61 (dm, *J* = 12.5 Hz, 1H, H-3′), 1.70–1.82 (m, 1H, H-3), 3.22 (td, *J* = 13.0 Hz, *J* = 3.2 Hz, 1H, CH), 3.90 (td, *J* = 12.0 Hz, *J* = 2.4 Hz, 1H, CH), 4.07 (dd, *J* = 11.4 Hz, *J* = 4.8 Hz, 1H, CH), 4.32 (dd,

J = 14.1 Hz, J = 3.5 Hz, 1H, CH), 4.43–4.50 (m, 2H, CH), 5.29 (t, J = 3.6 Hz, 1H, H-12a), 6.96 (s, 1H, H-7), 7.55 (d, J = 8.5 Hz, 2H, H-3, H-5 Ar), 7.75 (d, J = 8.5 Hz, 2H, H-2, H-6 Ar), 8.22 (s, 1H, H-10). ^{13}C NMR (126 MHz, DMSO-d_6) δ 25.0, 43.0, 50.3, 67.1, 81.4, 120.7, 128.4, 128.5 (2C), 131.0 (2C), 135.6, 137.9, 138.1, 144.2, 156.9, 175.0, 192.7. HRMS (ESI) m/z [M + H]$^+$. Calculated for $C_{18}H_{15}ClN_2O_4$: 359.0789. Found: 359.0799.

9-(Thiophene-2-carbonyl)-3,4,12,12a-tetrahydro-2H-pyrido[1',2':4,5]pyrazino[2,1-b][1,3]oxazine-6,8-dione (**4c**). Method B. The reaction was carried out for 8h. Yield 0.0545 g (52%), yellow powder, mp 210–211 °C. IR (ATR) ν 2953, 1634, 1581, 1468, 1373, 1047, 714. ^{1}H NMR (500 MHz, DMSO-d_6) δ 1.61 (dm, J = 13.5 Hz, 1H, H-3'), 1.70–1.82 (m, 1H, H-3), 3.22 (td, J = 12.8 Hz, J = 3.2 Hz, 1H, CH), 3.89 (td, J = 11.9 Hz, J = 2.3 Hz, 1H, CH), 4.05 (dd, J = 11.3 Hz, J = 4.7 Hz, 1H, CH), 4.28 (dd, J = 14.2 Hz, J = 3.2 Hz, 1H, CH), 4.43–4.50 (m, 2H, CH), 5.27 (t, J = 3.6 Hz, 1H, H-12a), 6.97 (s, 1H, H-7), 7.22 (dd, J = 4.9 Hz, J = 4.0 Hz, 1H, H-4 Th), 7.74 (dd, J = 4.0 Hz, J = 0.7 Hz, 1H, H-3 Th), 8.05 (dd, J = 4.9 Hz, J = 0.7 Hz, 1H, H-5 Th), 8.19 (s, 1H, H-10). ^{13}C NMR (126 MHz, DMSO-d_6) δ 25.0, 43.0, 52.0, 67.1, 81.5, 120.4, 128.6, 128.9, 135.6, 137.9, 143.1, 143.4, 156.9, 174.6, 184.9 (1C was not observed). HRMS (ESI) m/z [M + H]$^+$. Calculated for $C_{16}H_{15}N_2O_4S$: 331.0745. Found: 331.0753.

9-Benzoyl-3,4,12,12a-tetrahydro-2H-pyrido[1',2':4,5]pyrazino[2,1-b][1,3]oxazine-6,8-dione (**4d**). Method B. The reaction was carried out for 10 h. Yield 0.0319 g (31%), yellow powder, mp 245–246 °C. ^{1}H NMR (400 MHz, DMSO-d_6) δ 1.61 (dm, J = 14.6 Hz, 1H, H-3'), 1.67–1.84 (m, 1H, H-3), 3.22 (td, J = 13.0 Hz, J = 3.7 Hz, 1H, CH), 3.90 (td, J = 11.3 Hz, J = 2.5 Hz, 1H, CH), 4.07 (dd, J = 11.3 Hz, J = 4.5 Hz, 1H, CH), 4.31 (dd, J = 14.5 Hz, J = 3.5 Hz, 1H, CH), 4.42–4.51 (m, 2H, CH), 5.29 (t, J = 3.5 Hz, 1H, H-12a), 6.95 (s, 1H, H-7), 7.49 (t, J = 7.8 Hz, 2H, H-3, H-5 Ph), 7.62 (t, J = 7.8 Hz, 1H, H-4 Ph), 7.75 (d, J = 8.0 Hz, 2H, H-2, H-6 Ph), 8.18 (s, 1H, H-10). ^{13}C NMR (126 MHz, DMSO-d_6) δ 25.0, 50.2, 67.1, 81.5, 120.4, 128.4 (2C), 129.0, 129.1 (2C), 133.1, 136.8, 138.1, 143.6, 156.9, 175.0, 193.8. HRMS (ESI) m/z [M + H]$^+$. Calculated for $C_{18}H_{17}N_2O_4$: 325.1188. Found: 325.1177.

8-(4-Methoxybenzoyl)-2,3,11,11a-tetrahydro-1H-imidazo[1,2-a]pyrido[1,2-d]pyrazine-5,7-dione (**5a**). Method A. The reaction was carried out for 12 h. Yield 0.0904 g (84%), brown powder, mp 249–250 °C;. IR (ATR) ν 2996, 2837, 1650, 1629, 1577, 1454, 1250, 786. ^{1}H NMR (500 MHz, DMSO-d_6) δ 2.99 (dt, J = 11.5 Hz, J = 7.4 Hz, 1H, CH), 3.20 (ddd, J = 11.2 Hz, J = 7.0 Hz, J = 3.9 Hz, 1H, CH), 3.39 (ddd, J = 11.1 Hz, J = 7.2 Hz, J = 3.8 Hz, 1H, CH), 3.54 (dt, J = 11.2 Hz, J = 7.3 Hz, 1H, CH), 3.84 (s, 3H, OMe), 3.92 (t, J = 11.8 Hz, 1H, CH), 4.52 (dd, J = 12.4 Hz, J = 3.7 Hz, 1H, CH), 4.75 (dd, J = 11.1 Hz, J = 3.6 Hz, 1H, CH), 6.79 (s, 1H, H-6), 7.01 (d, J = 8.8 Hz, 2H, H-3, H-5 Ar), 7.75 (d, J = 8.8 Hz, 2H, H-2, H-6 Ar), 8.09 (s, 1H, H-9) (NH was not observed). ^{13}C NMR (126 MHz, DMSO-d_6) δ 44.3, 44.8, 53.0, 55.5, 69.2, 113.6 (2C), 119.3, 129.0, 129.7, 131.7 (2C), 139.3, 143.3, 154.9, 163.2, 175.2, 192.2. HRMS (ESI) m/z [M + H]$^+$. Calculated for $C_{18}H_{18}N_3O_4$: 340.1285. Found: 340.1297.

8-(4-Chlorobenzoyl)-2,3,11,11a-tetrahydro-1H-imidazo[1,2-a]pyrido[1,2-d]pyrazine-5,7-dione (**5b**). The reaction was carried out for 12 h. Method A. Yield 0.0850 g (78%), beige powder, mp 264–265 °C. IR (ATR) ν 2901, 1657, 1630, 1562, 1456, 1251, 769. ^{1}H NMR (500 MHz, DMSO-d_6) δ 2.95–3.03 (m, 1H, CH), 3.21 (ddd, J = 11.5 Hz, J = 7.1 Hz, J = 3.8 Hz, 1H, CH), 3.40 (ddd, J = 11.2 Hz, J = 7.1 Hz, J = 3.9 Hz, 1H, CH), 3.54 (dt, J = 11.4 Hz, J = 7.3 Hz, 1H, CH), 3.94 (t, J = 11.8 Hz, 1H, CH), 4.57 (dd, J = 12.4 Hz, J = 3.7 Hz, 1H, CH), 4.76 (dd, J = 11.1 Hz, J = 3.3 Hz, 1H, CH), 6.80 (s, 1H, H-6), 7.56 (d, J = 8.5 Hz, 2H, H-2, H-6 Ar), 7.75 (d, J = 8.5 Hz, 2H, H-2, H-6 Ar), 8.22 (s, 1H, H-9) (NH was not observed). ^{13}C NMR (126 MHz, DMSO-d_6) δ 44.3, 44.8, 53.1, 69.2, 119.9, 127.6, 128.5 (2C), 131.0 (2C), 135.8, 137.8, 139.5, 144.6, 154.7, 175.2, 192.8. HRMS (ESI) m/z [M + H]$^+$. Calculated for $C_{17}H_{15}ClN_3O_3$: 344.0813. Found: 344.0802.

8-(Thiophene-2-carbonyl)-2,3,11,11a-tetrahydro-1H-imidazo[1,2-a]pyrido[1,2-d]pyrazine-5,7-dione (**5c**). Method B. The reaction was carried out for 7h. Yield 0.0480 g (48%), beige powder, mp 267–268 °C. ^{1}H NMR (500 MHz, DMSO-d_6) δ 2.99 (dt, J = 11.5 Hz, J = 7.4 Hz, 1H, CH), 3.20 (ddd, J = 11.1 Hz, J = 6.9 Hz, J = 3.9 Hz, 1H, CH), 3.39 (ddd, J = 11.2 Hz, J = 7.2 Hz, J = 3.8 Hz, 1H, CH), 3.54 (dt, J = 11.2 Hz, J = 7.3 Hz, 1H, CH), 3.92 (t, J = 11.8 Hz, 1H, CH), 4.51 (dd, J = 12.4 Hz, J = 3.7 Hz, 1H, CH), 4.75 (dd, J = 11.0 Hz, J = 3.6 Hz, 1H, CH), 6.81 (s,

1H, H-6), 7.22 (dd, J = 5.1 Hz, J = 3.9 Hz, 1H, H-4 Th), 7.76 (dd, J = 3.9 Hz, J = 0.8 Hz, 1H, H-3 Th), 7.76 (dd, J = 5.1 Hz, J = 0.8 Hz, 1H, H-5 Th), 8.18 (s, 1H, H-9). ^{13}C NMR (126 MHz, DMSO-d_6) δ 44.3, 44.8, 53.1, 69.2, 119.5, 128.2, 128.6, 135.5, 135.6, 139.3, 143.5, 154.8, 174.8, 185.0 (1C was not observed). HRMS (ESI) m/z [M + H]$^+$. Calculated for $C_{15}H_{14}N_3O_3S$: 316.0756. Found: 316.0761.

3.4. General Method for the Preparation of Compound 6

3-Hydroxy-7-acyl-3,4-dihydropyrido[2,1-c][1,4]oxazine-1,8-dione 3 (0.317 mmol) was stirred with benzene-1,2-diamine (41.2 mg, 0.381 mmol) and ammonium acetate (0.0245 g, 0.317 mmol) in a mixture of toluene (1 mL) and methanol (1 mL) at room temperature for 12 h. Then, the reaction mixture was refluxed for 2 h. The precipitate obtained was filtered and washed with H$_2$O and EtOH. The product was recrystallized in a mixture of toluene and ethanol.

3-Benzoyl-6H-benzo[4,5]imidazo[1,2-a]pyrido[1,2-d]pyrazine-2,13-dione (**6a**). Yield 0.0800 g (71%), beige powder, mp 252–253 °C. IR (ATR) ν 3037, 2808, 1685, 1443, 1211, 1114, 1027, 807. ^{1}H NMR (500 MHz, DMSO-d_6) δ 5.84 (s, 2H, CH$_2$), 6.80 (s, 1H, H-1), 7.18 (dd, J = 6.0 Hz, J = 3.2 Hz, 2H, Ar), 7.50 (t, J = 7.7 Hz, 2H, H-3 H-5 Ph), 7.55 (dd, J = 6.0 Hz, J = 3.2 Hz, 2H, Ar), 7.63 (t, J = 7.4 Hz, 1H, H-4 Ph), 7.81 (d, J = 7.4 Hz, 2H, H-2, H-6 Ph), 8.33 (s, 1H, H-4). ^{13}C NMR (126 MHz, DMSO-d_6) δ 51.7, 114.9, 122.0 (2C), 122.6, 128.2, 128.3 (2C), 129.2 (2C), 133.0, 137.0, 138.2, 142.4, 147.0, 150.2, 168.1, 175.1, 193.5 (2C were not observed). HRMS (ESI) m/z [M + H]$^+$. Calculated for $C_{21}H_{14}N_3O_3$: 356.1035. Found: 356.1048.

3-(4-Methoxybenzoyl)-6a,7-dihydro-6H-benzo[4,5]imidazo[1,2-a]pyrido[1,2-d]pyrazine-2,13-dione (**6b**). Yield 0.1112 g (91%), yellow powder, mp 234–235 °C. IR (ATR) ν 2970, 2751, 1632, 1556, 1466, 1329, 1214, 1028, 840. ^{1}H NMR (500 MHz; DMSO-d_6) δ 3.84 (s, 3H, OMe), 5.82 (s, 2H, CH$_2$), 6.82 (s, 1H, H-1), 7.02 (d, J = 8.8 Hz, 2H, H-3, H-5 Ar), 7.18 (dd, J = 6.1 Hz, J = 3.1 Hz, 2H, Ar'), 7.54 (dd, J = 6.1 Hz, J = 3.1 Hz, Ar'), 7.80 (d, J = 8.8 Hz, 2H, H-2, H-6 Ar), 8.26 (s, 1H, H-4). ^{13}C NMR (126 MHz, DMSO-d_6) δ 51.5, 55.5, 113.6 (2C), 115.0, 121.9 (2C), 125.2, 128.1, 128.7, 128.8, 129.7, 131.7 (2C), 137.3, 138.3, 146.1, 150.2, 163.1, 163.2, 175.1, 191.8. HRMS (ESI) m/z [M + H]$^+$. Calculated for $C_{22}H_{16}N_3O_4$: 386.1141. Found: 386.1141.

3-(4-Chlorobenzoyl)-6H-benzo[4,5]imidazo[1,2-a]pyrido[1,2-d]pyrazine-2,13-dione (**6c**). Yield 0.0408 g (33%), beige powder, mp 251–252 °C. IR (ATR) ν 2971, 2756, 1633, 1597, 1466, 1256, 779. ^{1}H NMR (500 MHz, DMSO-d_6) δ 5.84 (s, 2H, CH$_2$), 6.82 (s, 1H, H-1), 7.18 (dd, J = 6.0 Hz, J = 3.2 Hz, 2H, Ar'), 7.54 (dd, J = 6.0 Hz, J = 3.2 Hz, 2H, Ar'), 7.57 (d, J = 8.5 Hz, 2H, H-3, H-5 Ar), 7.80 (d, J = 8.5 Hz, 2H, H-2, H-6 Ar), 8.37 (s, 1H, H-4). ^{13}C NMR (126 MHz, DMSO-d_6) δ 55.5, 113.5, 115.0, 122.1 (2C), 123.1, 125.2, 127.7, 128.5 (2C), 131.1 (2C), 135.8, 137.9, 138.0, 142.0, 147.9, 150.2, 163.0, 175.1, 192.5. Anal. Calculated for $C_{21}H_{12}ClN_3O_3$: C 64.71; H 3.10; N 10.78. Found: C 64.64; H 3.73; N 10.60.

3-(Thiophene-2-carbonyl)-6H-benzo[4,5]imidazo[1,2-a]pyrido[1,2-d]pyrazine-2,13-dione (**6d**). Yield 0.0481 g (42%), orange powder, mp 243–244 °C. IR (ATR) ν 2875, 2760, 1635, 1557, 1464, 1326, 1211, 1061, 878. ^{1}H NMR (500 MHz, DMSO-d_6) δ 5.81 (s, 2H, CH$_2$), 6.81 (s, 1H, H-1), 7.18 (dd, J = 6.0 Hz, J = 3.2 Hz, 2H, Ar), 7.23 (dd, J = 4.9 Hz, J = 3.9 Hz, 1H, H-4 Th), 7.54 (dd, J = 6.0 Hz, J = 3.2 Hz, 2H, Ar), 7.84 (dd, J = 3.9 Hz, J = 1.1 Hz, 1H, H-3 Th), 8.05 (dd, J = 4.9 Hz, J = 1.1 Hz, 1H, H-5 Th), 8.35 (s, 1H, H-4). ^{13}C NMR (126 MHz, DMSO-d_6) δ 51.7, 115.0, 122.1 (2C), 122.8, 128.1, 128.6 (2C), 135.7 (2C), 138.0, 141.9, 143.6, 146.8, 150.3, 163.1, 174.7, 184.6 (1C was not observed). Anal. Calculated for $C_{19}H_{11}N_3O_3S$: C 63.15; H 3.07; N 11.63. Found: C 62.83; H 3.33; N 11.25.

Supplementary Materials: The following supporting information can be downloaded at https://www.mdpi.com/article/10.3390/molecules28031285/s1, Full ^{1}H, and ^{13}C NMR spectra of all synthesized compounds; HRMS spectra of compounds **2a,e**, **3a,c,e,f**, **4a–d**, **5a–c**, **6a,b**.

Author Contributions: Conceptualization and methodology were provided by V.Y.S. and D.L.O.; D.L.O. conceived and designed the experiments. The experimental work was conducted by V.V.V. and E.V.S.; D.L.O. and V.V.V. analyzed the results. D.L.O. studied and systemized the spectral data. Project administration and funding acquisition were carried out by V.Y.S. and D.L.O. All authors have read and agreed to the published version of the manuscript.

Funding: The research funding from the Ministry of Science and Higher Education of the Russian Federation (Ural Federal University Program of Development within the Priority-2030 Program) is gratefully acknowledged.

Institutional Review Board Statement: Not applicable.

Informed Consent Statement: Not applicable.

Data Availability Statement: Data are contained within the article and Supplementary Materials.

Acknowledgments: Analytical studies were carried out using equipment at the Center for Joint Use 'Spectroscopy and Analysis of Organic Compounds' at the Postovsky Institute of Organic Synthesis of the Russian Academy of Sciences (Ural Branch) and the Laboratory of Complex Investigations and Expert Evaluation of Organic Materials of the Center for Joint Use at the Ural Federal University.

Conflicts of Interest: The authors declare no conflict of interest.

Sample Availability: The samples of the compounds are not available from the authors.

References

1. Hughes, D.L. Review of synthetic routes and final forms of integrase inhibitors dolutegravir, cabotegravir, and bictegravir. *Org. Process Res. Dev.* **2019**, *23*, 716–729. [CrossRef]
2. Hughes, D.L. Review of the patent literature: Synthesis and final forms of antiviral drugs tecovirimat and baloxavir marboxil. *Org. Process Res. Dev.* **2019**, *23*, 1298–1307. [CrossRef]
3. Schreiner, E.; Richter, F.; Nerdinger, S. Development of synthetic routes to dolutegravir. *Top. Heterocycl. Chem.* **2016**, *44*, 187–208. [CrossRef]
4. He, M.; Fan, M.; Peng, Z.; Wang, G. An overview of hydroxypyranone and hydroxypyridinone as privileged scaffolds for novel drug discovery. *Eur. J. Med. Chem.* **2021**, *221*, 113546. [CrossRef]
5. Hayat, F.; Sonavane, M.; Makarov, M.V.; Trammell, S.A.J.; McPherson, P.; Gassman, N.R.; Migaud, M.E. The biochemical pathways of nicotinamide-derived pyridones. *Int. J. Mol. Sci.* **2021**, *22*, 1145. [CrossRef] [PubMed]
6. Yang, X.-L.; Zhang, J.-Z.; Luo, D.-Q. The taxonomy, biology and chemistry of the fungal Pestalotiopsis genus. *Nat. Prod. Rep.* **2012**, *29*, 622–641. [CrossRef] [PubMed]
7. Wang, H.; Kowalski, M.D.; Lakdawala, A.S.; Vogt, F.G.; Wu, L. An efficient and highly diastereoselective synthesis of GSK1265744, a potent HIV integrase inhibitor. *Org. Lett.* **2015**, *17*, 564–567. [CrossRef]
8. Johns, B.A.; Kawasuji, T.; Weatherhead, J.G.; Taishi, T.; Temelkoff, D.P.; Yoshida, H.; Akiyama, T.; Taoda, Y.; Murai, H.; Kiyama, R.; et al. Carbamoyl pyridone HIV-1 integrase inhibitors 3. A diastereomeric approach to chiral nonracemic tricyclic ring systems and the discovery of dolutegravir (S/GSK1349572) and (S/GSK1265744). *J. Med. Chem.* **2013**, *56*, 5901–5916. [CrossRef]
9. Kawasuji, T.; Johns, B.A.; Yoshida, H.; Weatherhead, J.G.; Akiyama, T.; Taishi, T.; Taoda, Y.; Mikamiyama-Iwata, M.; Murai, H.; Kiyama, R.; et al. Carbamoyl pyridone HIV-1 integrase inhibitors. 2. Bi- and tricyclic derivatives result in superior antiviral and pharmacokinetic profiles. *J. Med. Chem.* **2013**, *56*, 1124–1135. [CrossRef] [PubMed]
10. Sankareswaran, S.; Mannam, M.; Chakka, V.; Mandapati, S.R.; Kumar, P. Identification and control of critical process impurities: An improved process for the preparation of dolutegravir sodium. *Org. Process Res. Dev.* **2016**, *20*, 1461–1468. [CrossRef]
11. Sumino, Y.; Okamoto, K.; Masui, M.; Yamada, D.; Ikarashi, F. Process for Preparing Compound Having HIV Integrase Inhibitory Activity. WO Patent 018065, 9 February 2012.
12. Maras, N.; Selic, L.; Cusak, A. Processes for Preparing Dolutegravir and Cabotegravir and Analogues Thereof. U.S. Patent 0368040, 11 July 2017.
13. Srinivasachary, K.; Subbareddy, D.; Ramadas, C.; Balaji, S.K.K.; Somannavar, Y.S.; Ramadevi, B. Practical and efficient route to dolutegravir sodium via one-pot synthesis of key intermediate with controlled formation of impurities. *Russ. J. Org. Chem.* **2022**, *58*, 526–535. [CrossRef]
14. Ziegler, R.E.; Desai, B.K.; Jee, J.-A.; Gupton, B.F.; Roper, T.D.; Jamison, T.F. 7-Step flow synthesis of the HIV integrase inhibitor dolutegravir. *Angew. Chem. Int. Ed.* **2018**, *57*, 7181–7185. [CrossRef] [PubMed]
15. Dietz, J.-P.; Lucas, T.; Groß, J.; Seitel, S.; Brauer, J.; Ferenc, D.; Gupton, B.F.; Opatz, T. Six-step gram-scale synthesis of the human immunodeficiency virus integrase inhibitor dolutegravir sodium. *Org. Process Res. Dev.* **2021**, *25*, 1898–1910. [CrossRef]
16. Kong, J.; Xia, H.; He, R.; Chen, H.; Yu, Y. Preparation of the key dolutegravir intermediate via $MgBr_2$-promoted cyclization. *Molecules* **2021**, *26*, 2850. [CrossRef] [PubMed]

17. Yasukata, T.; Masui, M.; Ikarashi, F.; Okamoto, K.; Kurita, T.; Nagai, M.; Sugata, Y.; Miyake, N.; Hara, S.; Adachi, Y.; et al. Practical synthetic method for the preparation of pyrone diesters: An efficient synthetic route for the synthesis of dolutegravir sodium. *Org. Process Res. Dev.* **2019**, *23*, 565–570. [CrossRef]
18. Stojanović, M.; Bugarski, S.; Baranac-Stojanović, M. Synthesis of 2,3-dihydro-4-pyridones and 4-pyridones by the cyclization reaction of ester-tethered enaminones. *J. Org. Chem.* **2020**, *85*, 13495–13507. [CrossRef] [PubMed]
19. Yu, Y.; Zhang, Y.; Xiao, L.-Y.; Peng, Q.-Q.; Zhao, Y.-L. Thermally induced formal [4+2] cycloaddition of 3-aminocyclobutenones with electron-deficient alkynes: Facile and efficient synthesis of 4-pyridones. *Chem. Commun.* **2018**, *54*, 8229–8232. [CrossRef] [PubMed]
20. Liu, S.; Li, J.; Lin, J.; Liu, F.; Liu, T.; Huang, C. Substituent-controlled chemoselective synthesis of multi-substituted pyridones *via* a one-pot three-component cascade reaction. *Org. Biomol. Chem.* **2020**, *18*, 1130–1134. [CrossRef]
21. Fedin, V.V.; Usachev, S.A.; Obydennov, D.L.; Sosnovskikh, V.Y. Reactions of trifluorotriacetic acid lactone and hexafluorodehydroacetic acid with amines: Synthesis of trifluoromethylated 4-pyridones and aminoenones. *Molecules* **2022**, *27*, 7098. [CrossRef]
22. Zantioti-Chatzouda, E.-M.; Kotzabasaki, V.; Stratakis, M. Synthesis of γ-pyrones and *N*-methyl-4-pyridones via the Au nanoparticle-catalyzed cyclization of skipped diynones in the presence of water or aqueous methylamine. *J. Org. Chem.* **2022**, *87*, 8525–8533. [CrossRef]
23. Ropero, B.P.F.D.; Elsegood, M.R.J.; Fairley, G.; Pritchard, G.J.; Weaver, G.W. Pyridone functionalization: Regioselective deprotonation of 6-methylpyridin-2(1*H*)- and -4(1*H*)-one derivatives. *Eur. J. Org. Chem.* **2016**, *2016*, 5238–5242. [CrossRef]
24. Obydennov, D.L.; El-Tantawy, A.I.; Sosnovskikh, V.Y. Synthesis of multifunctionalized 2,3-dihydro-4-pyridones and 4-pyridones *via* the reaction of carbamoylated enaminones with aldehydes. *J. Org. Chem.* **2018**, *83*, 13776–13786. [CrossRef] [PubMed]
25. Obydennov, D.L.; Chernyshova, E.V.; Sosnovskikh, V.Y. Acyclic enaminodiones in the synthesis of heterocyclic compounds. *Chem. Heterocycl. Comp.* **2020**, *56*, 1241–1253. [CrossRef]
26. Diesel, J.; Finogenova, A.M.; Cramer, N. Nickel-catalyzed enantioselective pyridone C–H functionalizations enabled by a bulky N-heterocyclic carbene ligand. *J. Am. Chem. Soc.* **2018**, *140*, 4489–4493. [CrossRef] [PubMed]
27. Kea, D.; Wu, Y.; Zhang, L.; Shao, J.; Yu, Y.; Chen, W. Group-assisted-purification chemistry strategy for the efficient assembly of cyclic fused pyridinones. *Synthesis* **2022**, *54*, 1765–1774. [CrossRef]
28. Obydennov, D.L.; Roeschenthaler, G.-V.; Sosnovskikh, V.Y. An improved synthesis and some reactions of diethyl 4-oxo-4*H*-pyran-2,5-dicarboxylate. *Tetrahedron Lett.* **2013**, *54*, 6545–6548. [CrossRef]
29. Obydennov, D.L.; Khammatova, L.R.; Steben'kov, V.D.; Sosnovskikh, V.Y. Synthesis of novel polycarbonyl Schiff bases by ring-opening reaction of ethyl 5-acyl-4-pyrone-2-carboxylates with primary mono- and diamines. *RSC Adv.* **2019**, *9*, 40072–40083. [CrossRef] [PubMed]
30. Li, Z.; Zhang, L.; Zhou, Y.; Zha, D.; Hai, Y.; You, L. Dynamic covalent reactions controlled by ring-chain tautomerism of 2-formylbenzoic acid. *Eur. J. Org. Chem.* **2022**, *2022*, e202101461. [CrossRef]
31. Osyanin, V.A.; Osipov, D.V.; Semenova, I.A.; Korzhenko, K.S.; Lukashenko, A.V.; Demidov, O.P.; Klimochkin, Y.N. Eco-friendly synthesis of fused pyrano[2,3-*b*]pyrans via ammonium acetate-mediated formal oxa-[3 + 3] cycloaddition of 4*H*-chromene-3-carbaldehydes and cyclic 1,3-dicarbonyl compounds. *RSC Adv.* **2020**, *10*, 34344–34354. [CrossRef]
32. Zhang, Z.; Gao, X.; Wan, Y.; Huang, Y.; Huang, G.; Zhang, G. Ammonium acetate-promoted one-pot tandem aldol condensation/aza-addition reactions: Synthesis of 2,3,6,7-tetrahydro-1*H*-pyrrolo[3,2-*c*]pyridin-4(5*H*)-ones. *ACS Omega* **2017**, *2*, 6844–6851. [CrossRef]
33. Obydennov, D.L.; Roeschenthaler, G.-V.; Sosnovskikh, V.Y. Synthesis of 6-aryl- and 5-aroylcomanic acids from 5-aroyl-2-carbethoxy-4-pyrones *via* a deformylative rearrangement and ring-opening/ring-closure sequence. *Tetrahedron Lett.* **2014**, *55*, 472–474. [CrossRef]
34. Obydennov, D.L.; Goncharov, A.O.; Sosnovskikh, V.Y. Preparative synthesis of ethyl 5-acyl-pyrone-2-carboxylates and 6-aryl-, 6-alkyl-, and 5-acylcomanic acids on their basis. *Russ. Chem. Bull.* **2016**, *65*, 2233–2242. [CrossRef]

Disclaimer/Publisher's Note: The statements, opinions and data contained in all publications are solely those of the individual author(s) and contributor(s) and not of MDPI and/or the editor(s). MDPI and/or the editor(s) disclaim responsibility for any injury to people or property resulting from any ideas, methods, instructions or products referred to in the content.

Review

Heteroaromatic Diazirines Are Essential Building Blocks for Material and Medicinal Chemistry

Yuta Murai [1,2,*] and Makoto Hashimoto [3,*]

[1] Graduate School of Life Science, Hokkaido University, Kita 21, Nishi 11, Kita-ku, Sapporo 001-0021, Japan
[2] Faculty of Advanced Life Science, Hokkaido University, Kita 21, Nishi 11, Kita-ku, Sapporo 001-0021, Japan
[3] Division of Applied Bioscience, Graduate School of Agriculture, Hokkaido University, Kita 9, Nishi 9, Kita-ku, Sapporo 060-8589, Japan
* Correspondence: ymurai@sci.hokudai.ac.jp (Y.M.); hasimoto@abs.agr.hokudai.ac.jp (M.H.); Tel.: +81-11-706-9030 (Y.M.); +81-11-706-3849 (M.H.)

Abstract: In materials (polymer) science and medicinal chemistry, heteroaromatic derivatives play the role of the central skeleton in development of novel devices and discovery of new drugs. On the other hand, (3-trifluoromethyl)phenyldiazirine (TPD) is a crucial chemical method for understanding biological processes such as ligand–receptor, nucleic acid–protein, lipid–protein, and protein–protein interactions. In particular, use of TPD has increased in recent materials science to create novel electric and polymer devices with comparative ease and reduced costs. Therefore, a combination of heteroaromatics and (3-trifluoromethyl)diazirine is a promising option for creating better materials and elucidating the unknown mechanisms of action of bioactive heteroaromatic compounds. In this review, a comprehensive synthesis of (3-trifluoromethyl)diazirine-substituted heteroaromatics is described.

Keywords: diazirine; heteroaromatics; medicinal chemistry; photoaffinity labeling

1. Introduction

Heteroaromatic compounds are essential building blocks in a wide variety of functional molecules, including organic materials [1], electric devices [2], pharmaceuticals [3], and natural products [4]. For example, installation of crosslinks in (heteroaromatic) polymers can improve mechanical strength and corrosion resistance. However, it is difficult to install crosslinks into ingredients of a functional polymer in a well-controlled manner due to the lack of functional groups required for coupling. In recent materials science, photocrosslinking has been one of the essential chemical techniques used to make polymer networks into functional materials in order to increase thermal stability, shock resistance, corrosion resistance, and tensile strength without the particular functional groups required for coupling [5,6]. On the other hand, photoaffinity labeling (PAL) has also been demonstrated to identify ligand-binding biomolecules in a variety of drug-discovery fields. Its site mapping and protein–protein interaction are visualized through live-cell imaging of photoinduced crosslinkage [7–10]. Especially in PAL experiments, selection of photophores is an important factor for the performance of an efficient photolabeling reaction. Three common photophores, arylazide [11,12], benzophenone [13], and trifluoromethyldiazirine [14], have been used and contributed to PAL. To the best of our knowledge, trifluoromethyldiazirine is a useful photophore because it has advantages such as its relatively small size, stability, and low rate of rearrangement and generates a highly reactive carbene with a longer wavelength (≈365 nm), which reduces damage to polymer materials and biomolecules. Therefore, a combination of heteroaromatics and trifluoromethyldiazirine is a rational strategy for preparation of functional polymer materials and for revealing the unexplained mechanisms and target biomolecules of bioactive heteroaromatic compounds.

In this review, a comprehensive synthesis of (3-trifluoromethyl)diazirine-substituted heteroaromatics and their applications are described.

2. Diazirinyl-Substituted Pyridines and Pyrimidines

Recently, use of (3-trifluoromethyl)phenyldiazirine (TPD) has been widely increasing in materials science for creation of crosslinking of component parts of organic light-emitting diodes [15,16], for primers for fiber-reinforced polymer composites [17], and for patterning of wearable elastic circuits [18]. Despite the success of TPD as a photocross-coupling agent, in order to establish functional materials more efficiently and easily with versatile photocrosslinking, there is a need to improve the photolabeling performance of TPD. This includes factors such as stability against ambient light and water solubility to enable reactions in aqueous solutions. Kumar et al. designed and synthesized novel 3-pyridyl and 3-pyrimidyl-substituted 3-trifluoromethyl-diazirines 11 and 12, which possess stability under ambient light and have aqueous solubility [19]. As shown in Scheme 1, 5-bromo pyridyl compound 1 and pyrimidyl compound 2 were subjected to the protection of primary alcohol with silyl groups (*tert*-butyldimethylsilyl, TBS; *tert*-butyldiphenylsilyl, TBDPS); subsequently, trifluoroacetylation of compounds 3 and 4 was conducted with *n*-BuLi and methyl trifluoroacetate to obtain 5 and 6 with a moderate yield. Next, the trifluoroacetyl groups were converted to oximes, followed by tosylation in an appropriate manner for each compound. Tosyl oximes 7 and 8 underwent the addition of ammonia (liquid) to give diaziridines 9 and 10 with a good yield. Finally, the desired products, diazirines 11 and 12, were obtained through oxidation of the diaziridinyl moiety with silver oxide and deprotection of the silyl group with tetrabutylammonium fluoride (TBAF).

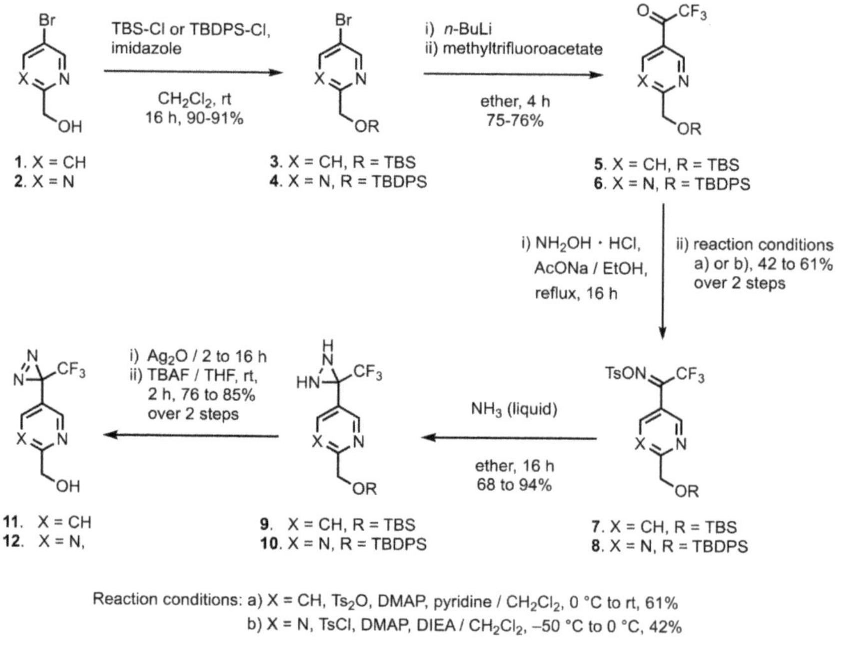

Reaction conditions: a) X = CH, Ts$_2$O, DMAP, pyridine / CH$_2$Cl$_2$, 0 °C to rt, 61%
b) X = N, TsCl, DMAP, DIEA / CH$_2$Cl$_2$, −50 °C to 0 °C, 42%

Scheme 1. The synthesis of 3-trifluoromethyl-diazirinyl pyridine 11 and pyrimidine 12.

The photoactivation and the ambient light stability of synthesized diazirinyl compounds 11 and 12 were tested and compared to those of (4-(3-(trifluoromethyl)-3*H*-diazirin-3-yl)phenyl)methanol 13. The photoreactive kinetics of these three compounds were calculated through photoactivation of a solution of the photolabel in methanol-d_4 with UV light and measurement of the compound change using ^{19}F NMR over time. Neither electron-withdrawing pyridine nor pyrimidine rings affected the ratio of generation of carbene from a diazirine, and they indicated the same photoreactive efficiency as in compound 13. Moreover, the ambient light stability of 11 and 12 was investigated. Compound 13 had already demonstrated significant photodecomposition after seven days with ambient light exposure. In contrast, compounds 11 and 12 were negligibly photodecomposed. After exposure of the probes to ambient light for a period of one month, only 27% of compound 13 remained, whereas 79% of compound 11 and 90% of compound 12 remained (Table 1).

Table 1. Comparison of the ambient light stability of 11 and 12 versus that of (4-(3-(trifluoromethyl)-3H-diazirin-3-yl)phenyl)methanol 13.

11. X = CH, Y = N
12. X = N, Y = N
13. X = CH, Y = CH

Duration of Ambient Light Exposure (Days)	Amounts of Unreacted Compound Yield (%)		
	11	12	13
0	100	100	100
4	97.1	99	87.7
7	95.2	98	78.1
14	90.1	95.2	58.1
18	87.7	94.3	49
26	82.6	92.6	35
31	79.4	90.1	26.8

Green chemistry in materials science is one of the most vital challenges: for example, replacement of organic solvents, success of organic reactions in aqueous solutions, etc. The aqueous-solubility enhancements of compounds 11 and 12 were demonstrated, and it was confirmed, using a HPLC-based assay, that the 11 and 12 derivatives were 100–7500 times more soluble than the 13 derivatives at pH = 7.4 and 5.0 [20] (Table 2).

Table 2. Comparison of aqueous solubility of photoaffinity probes derivatized with modified compounds 11 and 12 with that of compound 13.

Compound	Aqueous Solubility (mM)	
	pH 7.4	pH 5.0
(CF$_3$-diazirinyl phenyl derivative with AcHN, indole)	<0.02	<0.02
(CF$_3$-diazirinyl pyridyl derivative with AcHN, indole)	4.09 ± 0.19	4.29 ± 0.14

Table 2. Cont.

Compound	Aqueous Solubility (mM)	
	pH 7.4	pH 5.0
[structure: CF3-pyrimidyl-diazirine, CH2-O-C(=O)-CH(NHAc)-CH2-indole]	133 ± 0.5	131 ± 1.7
[structure: CF3-diazirine-phenyl-CH2-O-quinoline]	11.3 ± 0.55	14.1 ± 0.43
[structure: CF3-diazirine-pyridyl-CH2-O-quinoline]	374 ± 2.7	422 ± 4.9
[structure: CF3-diazirine-pyrimidyl-CH2-O-quinoline]	≥1000	≥1000

Therefore, 3-pyridyl and 3-pyrimidyl-substituted 3-trifluoromethyl-diazirines 11 and 12 demonstrated significant ambient-light-stability and aqueous-solubility improvements over the conventional 3-trifluoromethyl-3-aryldiazirines. These physicochemical properties could contribute huge advantages, and not only for photolabeling experiments in materials science.

3. Diazirinyl-Substituted Benzimidazoles

Benzimidazole and imidazole derivatives are medicinally used as anticancer [21–25], antioxidant [26,27], anti-inflammatory [28,29], anticoagulant [30,31], anthelmintic, or opioid products, and for analgesic activity [32]. Therefore, in order to discover better medicinal reagents or drugs, it is important to identify the target biomolecules of benzimidazole and imidazole derivatives and understand the mechanism of action of their pharmacological effects. Diazirinyl-substituted benzimidazole was reported by Raimer et al., and its chemical properties, thermal stability, and photoreactivity have been demonstrated [33]. The starting material, 14, was prepared following a previous report [34], then converted to oxime, tosyloxime, diaziridine 15, and diazirine 16 in four steps according to standard protocol (Scheme 2). Moreover, the synthesis of diazirinyl-substituted imidazoles was also attempted, but only diaziridines were able to be reached due to the violent decomposition of the diazirinyl imidazoles with self-ignition.

Next, the physical properties of diazirinyl-substituted benzimidazole 16 were investigated. The photoreaction of 16 in EtOH was slow, and ethoxy adduct 17 was formed, with a moderate yield (46%). Interestingly, in the case of decreasing the ethanol ratio to one equivalent of 16 in the presence of CH_2Cl_2, the reaction proceeded more quickly and gave a higher yield (60%; Scheme 3). This observation was already reported, identifying which solvent effect on the carbene reaction explained its different reactivity [35–38]. Furthermore, compound 16 was proved to have thermal stability at up to 88 °C in differential-scanning calorimetry measurement. Therefore, diazirinyl-substituted benzimidazole 16 could be the backbone of a reliable PAL probe and contribute to new drug discovery through identification of target molecules of benzimidazole-containing bioactive substances.

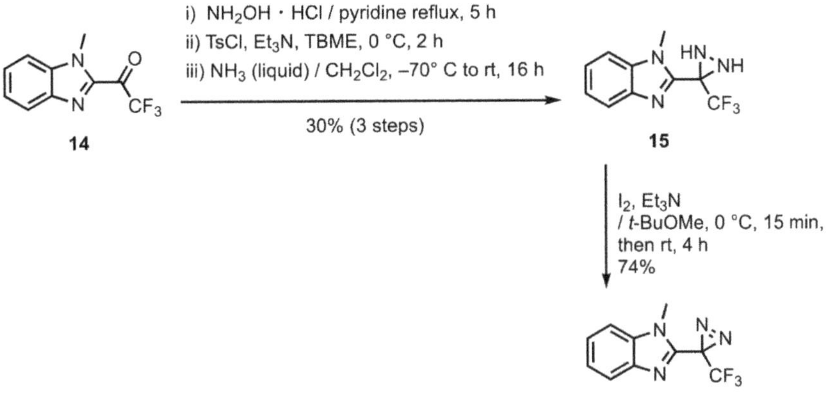

Scheme 2. The synthesis of diazirinyl benzimidazole 16.

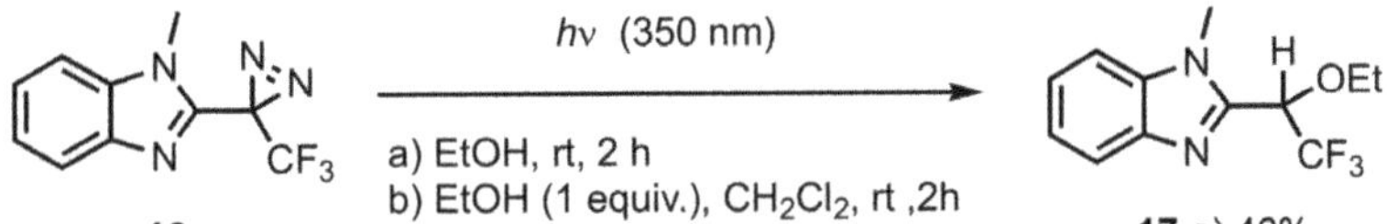

Scheme 3. Photoreaction of diazirinyl-substituted benzimidazole 16 (10 mM) in the presence of EtOH and EtOH/CH_2Cl_2.

4. Diazirinyl-Substituted Pyrazoles

Pyrazoles, five-membered heterocyclic compounds including two nitrogen atoms, are an important class of compounds for drug development, with a wide application in medicinal chemistry [39–41]. Pyrazole derivatives exhibit diverse targets and resistance: for example, to cancer [42,43], acquired immunodeficiency syndrome [44,45], tubercular [46], insects [47], etc. Furthermore, 5-Amino-1-[2,6-dichloro-4-(trifluoromethyl)phenyl]-3-cyano-4-(trifluoromethyl)sulfinylpyrazole (fipronil, Scheme 4), a chemical pesticide, is a widely used broad-spectrum insecticide. Fipronil, as a noncompetitive blocker, inhibits the γ-aminobutyric acid (GABA)-gated chloride-ion channel receptor. Furthermore, within the GABA receptor family, dysfunction of the Type A GABA receptor leads to neurological disorders and mental illnesses in humans; therefore, the GABA receptor is a drug-discovery target for these disorders [48–50]. Moreover, understanding fipronil's noncompetitive binding site in the GABA-gated chloride channel might be a fascinating matter for medicinal chemistry.

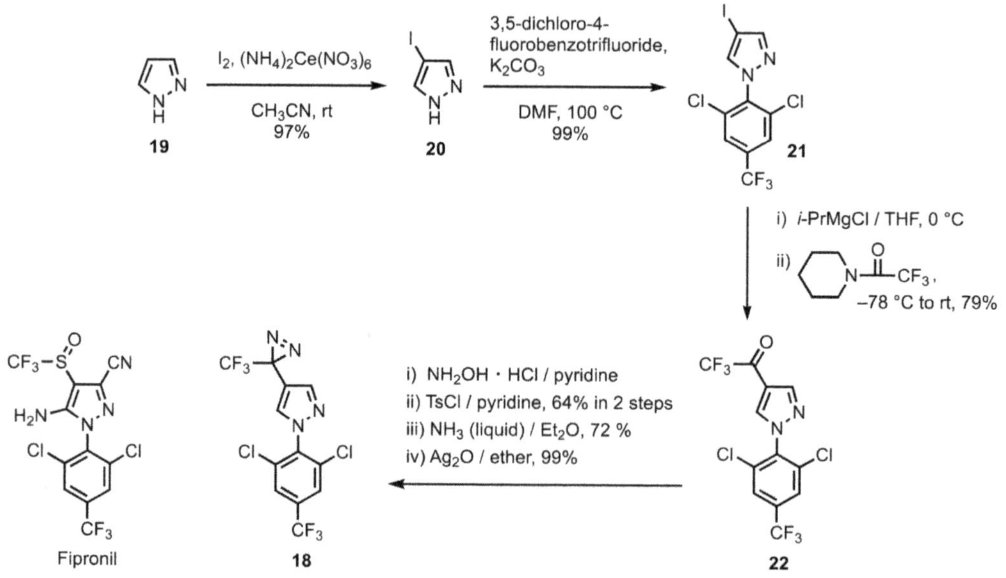

Scheme 4. The synthesis of diazirinyl fipronil based on benzimidazole skeleton 18.

For this purpose, fipronil-based photoaffinity probe 18 (Scheme 4) was prepared by the Casida J.E. group [51,52]. Its synthesis was started from commercially available pyrazole 19, reacted with iodine and ceric ammonium nitrate to produce 4-iodopyrazole 20. Compound 21 was obtained via the nucleophilic aromatic substitution of 20 with potassium carbonate in DMF at 100 °C with a quantitative yield. Subsequently, an iodine–magnesium exchange of 21 was performed with *i*-propylmagnesium chloride, followed by nucleophilic substitution with *N*-(trifluoroacetyl)piperidine to obtain 4-trifluoroacetylpyrazole 22. Finally, compound 18 was prepared according to a previous synthetic method [53–55] (Scheme 4).

The use of radioisotopes incorporated in a PAL probe is an efficient method for detection of labeled molecules because of the radioisotopes high sensitivity. Thus, the incorporation of tritium into fipronil-based photoaffinity probe 18 was also attempted. Initially, iodine incorporation into compound 18 or 22 was attempted several times via ortho lithiation; however, the desired products could not be obtained. Conversely, compound 25, which was prepared through the reduction of 22 with NaBH$_4$, was smoothly subjected to iodination under standard lithiation conditions [56]. Compound 24 was reoxidized back to trifluoroacetyl 25, with the Dess–Martin reagent [57,58], which was followed by conversion to diazirine 26. Next, reduction of diiodoarene 26 with tritium gas, 10% Pd/C, and triethylamine in ethyl acetate was conducted to obtain tritium-labeled, fipronil-based photoaffinity probe 27 (Scheme 5). The binding potency of 18 for the GABA receptor was also evaluated through competitive inhibition with 4′-ethynyl-4-[2,3-^{3}H$_2$]propylbicycloorthobenzoate, and the molar concentrations for 50% inhibition via 18 indicated approximately 2 nM. Therefore, compounds 18 and 27 could be suitable mimics of fipronil for use in photoaffinity labeling.

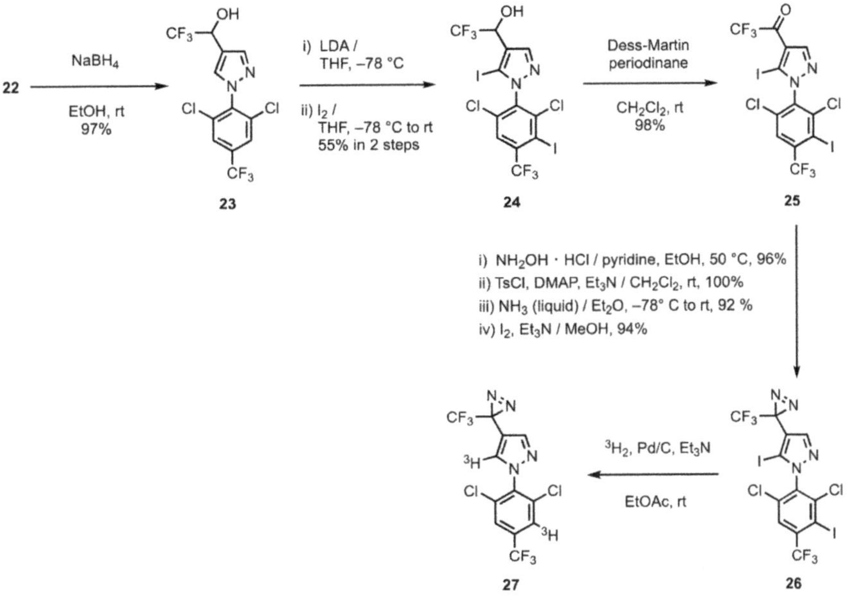

Scheme 5. The synthesis of tritium-labeled, fipronil-based photoaffinity probe 27.

5. Diazirinyl-Substituted Benzoxazolinone

Benzoxazolinones 28, 29, and 30 (Table 3), isolated from light-grown maize (*Zea mays* L.) shoots, have the potential to interfere with auxin behavior or inhibit the auxin receptor [59]. Kosemura, S. et al. reported that the structure–activity relationships of benzoxazolinones 31 and 32 (Table 3) were related to auxin-induced growth and auxin-binding protein [60]. The precise mechanism through which this bioactivity is triggered via benzoxazolinones has remained unknown. To address this subject, Kosemura, S. et al. tried synthesizing two photolabile benzoxazolinone analogues, 33 and 34, and evaluated their photoreactivities [61].

Table 3. Structure–activity relationship study of benzoxazolinones regarding auxin-induced growth and membrane-bound auxin-binding proteins.

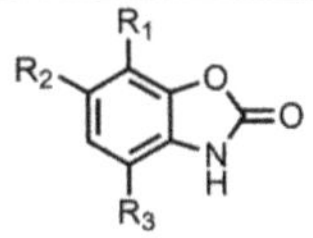

Compound	R_1	R_2	R_3
28	H	OMe	H
29	OMe	OMe	H
30	OMe	OMe	Cl
31	H	OCH$_2$CHMe$_2$	H
32	H	Ac	H

Photoreactive compound 42 was prepared from starting material 3-bromoanisole 35 according to a previous method [62,63]. Briefly, compound 35 was converted to the Grignard reagent treated with magnesium, followed by nucleophilic substitution with *N*-(trifluoroacetyl)piperidine to obtain 3-trifluoroacetylanisole 36. Compound 40 was prepared from 36, following the general diazirine synthetic method. Subsequently, nitration of 40 with fuming nitric acid and demethylation of 41 with BBr$_3$ were conducted to obtain compound 42 with a moderate yield, 42%, in two steps. The reduction of the nitro group in 42 was carried out with Na$_2$S$_2$O$_4$ while diazirine reduction was avoided as much as possible. Finally, Compound 43 was subjected to phenyl carbamation 44, followed by intramolecular cyclization to obtain the desired product, 33, with 14% in three steps (Scheme 6).

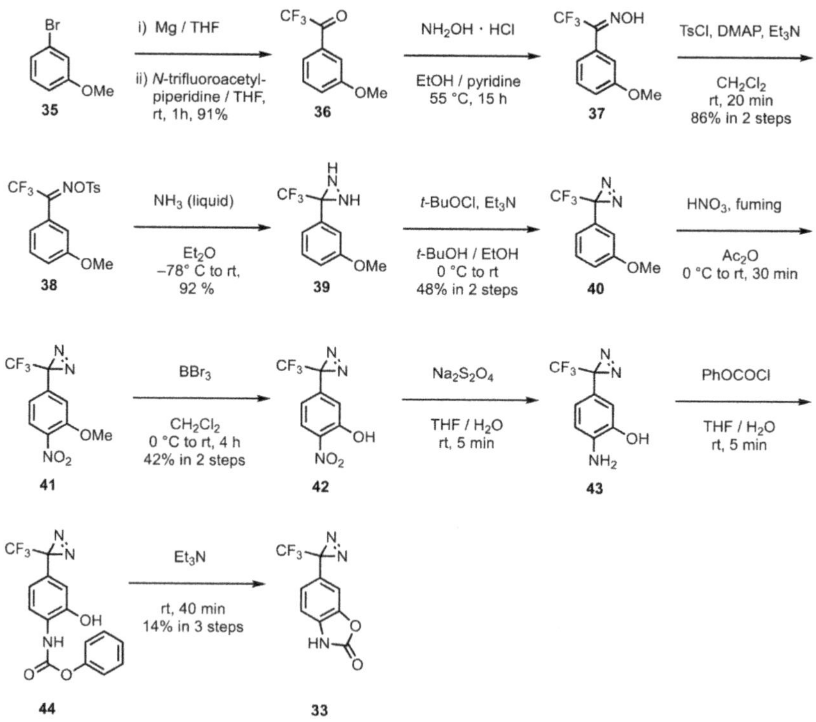

Scheme 6. The synthesis of diazirinyl benzoxazolinone analogue 33.

Moreover, photoreactive compound 34 was synthesized with the following procedure: Phenolic alcohol 45 was protected with chloromethyl methyl ether (MOMCl), followed by bromination with *N*-bromosuccinimide (NBS), to obtain compound 46. Trifluoroacetylation of 46 was carried out through a bromine–lithium exchange with *n*-BuLi, followed by nucleophilic substitution with ethyl trifluoroacetate, to obtain compound 47. Diazirine 48 was prepared according to the same procedure to construct 40 from 36. Nitration of 48 with fuming nitric acid was conducted at −72 °C to avoid the decomposition of the diazirine moiety. Compound 49 was deprotected with acidic hydrolysis 50 and then protected again with acetic anhydride to obtain compound 51. The nitro group of 51 was subjected to reduction with Na$_2$S$_2$O$_4$ within 5 min, followed by carbamation with phenyl chloroformate, then treated with sodium hydroxide to obtain the desired diazirine, 34 (32% in three steps) (Scheme 7).

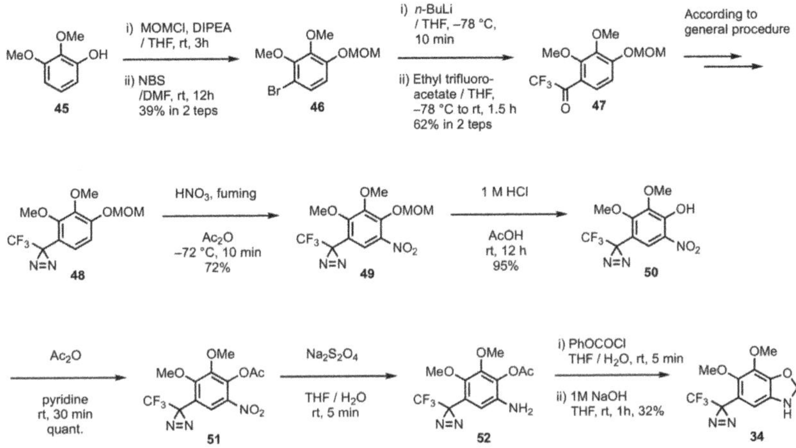

Scheme 7. The synthesis of diazirinyl-substituted benzoxazolinone analogue 34.

Next, diazirinyl benzoxazolinones 33 and 34 were evaluated as to whether they had suitable characteristics for photoaffinity-labeling reagents. The photoirradiation of compound 33 (1 mM in methanol) was smoothly carried out with a black light (12 W) to produce methanol adduct 53 with a moderate yield. The half-life ($t_{1/2}$) of 33 was calculated to be approximately 16 min, whereas the photoreactive kinetics of 34 with the black light (12 W) were much slower than those of 33. Using a 500 W high-pressure mercury lamp for photolysis of 34, compound 54 was produced, and the half-life ($t_{1/2}$) of 34 was 6 min (Scheme 8). Therefore, both of the compounds have suitable characteristics for photoaffinity-labeling reagents.

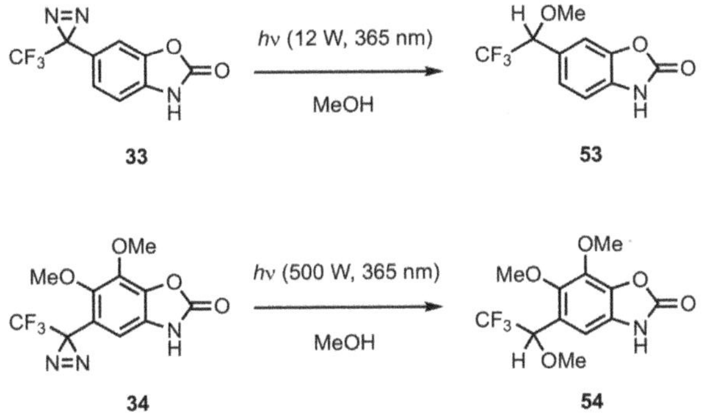

Scheme 8. Photoreaction of diazirinyl-substituted benzoxazolinones 33 and 34 (1 mM) in the presence of MeOH.

6. Diazirinyl-Substituted Benzoxazole

Duchenne muscular dystrophy (DMD), caused by loss-of-function mutations in the dystrophin gene, leads to progressive muscle degeneration and results in heart and respiratory failure [64,65]. Although quality of life and longevity have been improved with developments in the clinical standard of care [66–68], there is no complete treatment available for DMD. Recently, ezutromid (55) has been developed as a utrophin modulator and demonstrated reduced muscle-fiber damage and increased levels of utrophin after 24 trial weeks [69]. However, due to an administration effect, ezutromid could not retain this effect for a long period, and the mechanism of action of ezutromid is unknown. The development of ezutromid was discontinued. Therefore, to elucidate the mechanism of action of ezutromid for these effects and discover new drugs for DMD, Wilkinson, I.V.L. et al. synthesized photoreactive ezutromid derivatives 56 and 57 [70] for use in affinity-based protein profiling (ABPP) [71–73]. As a photophore of 56 and 57, 3-trifluoromethyldiazirine was chosen to replace the ethylsulfonyl group of ezutromid because of the similarity in the electronics of the diazirine and sulfonyl groups. The naphthyl moiety was replaced with alkynyl-phenyl substituents for installation of detection tags with click chemistry. The synthetic scheme of photoreactive compounds 56 and 57 is shown in Scheme 9, beginning with microwave-assisted benzoxazole cyclization 60 from 3-amino-4-hydroxybenzoic acid 58, and the corresponding acid chloride, 59, without purification [74]. The carboxylic acid was subjected to conversion of Weinreb amides 61 and 62, followed by trifluoroacetylation with the Ruppert–Prakash reagent to obtain 63 [75]. Diazirine derivatives 64 and 65 were introduced according to the general method [76]. Finally, Sonogashira couplings with TMS-acetylene were followed by TMS deprotection to obtain the desired products, 56 and 57, with a moderate yield (50–70%).

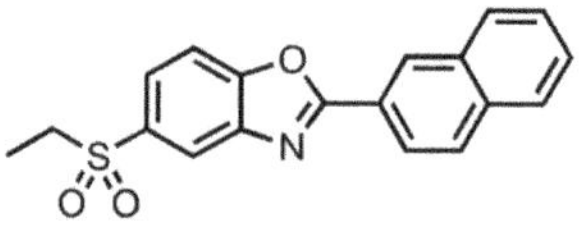

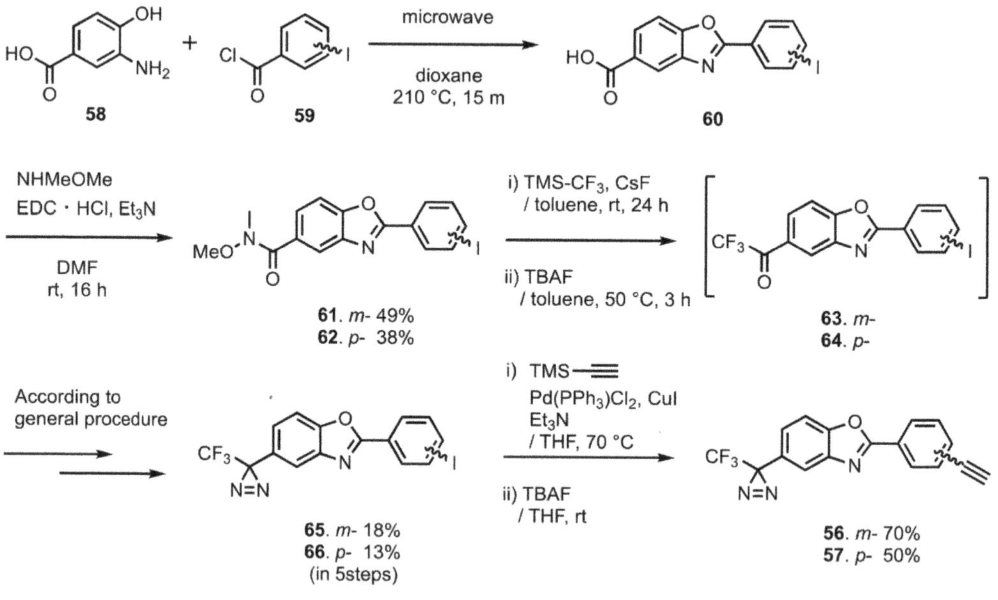

Scheme 9. The synthesis of diazirinyl-substituted benzoxazole analogues 56 and 57.

Furthermore, compounds 56 and 57 were applied to ABPP and found to bind the aryl hydrocarbon receptor (AhR). As a result, ezutromid was revealed as a novel AhR antagonist, inhibiting nuclear translocation and downregulating AhR-responsive genes such as AhRR and Cyp1b1. Therefore, this study could pave the way for the first target-based drug discovery in DMD treatment as well as provide a biomarker for future clinical trials.

7. Diazirinyl-Substituted (Benzo)thiophene

Thiophene is a five-membered, sulfur-containing heteroaromatic ring commonly used as a building block in the field of medicinal chemistry. In particular, thiophene derivatives are expected to possess a wide range of therapeutic properties, such as antitumor [77], anti-inflammatory [78], antiarrhythmic, antianxiety [79], and antifungal [80] effects and kinase inhibition [81,82]. Thus, evaluation and mechanism elucidation of novel thiophene moieties with wider therapeutic activity are a topic of interest for the medicinal chemist to synthesize and investigate new structural prototypes with more effective pharmacological activity. Oncodazole (methyl [5-(2-thienylcarbonyl)-1H-benzimidazol-2-yl]carbamate) 66, composed of a thiophen skeleton, is a tubulin-binding agent that has potent anthelmintic [83] and antifungal [84] activities. In particular, tubulin binding via drugs that cause disruption or hyperstabilization of the mitotic apparatus is presently an area of great interest. Thus, in order to characterize the interaction of 66 with tubulin, Ladd, D.L. et al. evaluated diazirinyl oncodazole (methyl [5-(2-thienylcarbonyl)-1H-benzimidazol-2-yl]carbamate) 67 [85]. They placed the diazirine function in the 4′-position on the basis of a systematic study of oncodazole derivatives because the position could be substituted without loss of biological activity.

Firstly, compound 68 was converted to acid chloride 69 with thionyl chloride, followed by a Friedel-Crafts reaction with anisole and without purification to obtain compound 70 with a quantitative yield. After protection of the carbonyl group of 70 with ethylene glycol, nucleophilic substitution was carried out with N-(trifluoroacetyl)piperidine through a lithium–bromo exchange on 71 to obtain trifluoroacetyl derivative 72. Subsequently, compound 73 was produced via acid hydrolysis of ethylene ketal; then, a reaction of 73 with nitric and sulfuric acids produced the appropriately substituted nitro derivative 74.

Next, compound 74 was converted into diazirinyl derivative 75 through the general TPD synthetic procedure. The methoxy group of 75 was substituted to an amine group with ammonia (liquid) with a good yield; then, the nitro group of 76 was reduced with sodium hydrosulfite, with diazirine moiety degradation avoided as much as possible. Finally, both diamine groups were subjected to condensation with bis(methoxycarbonyl)-S-methylisothiourea catalyzed with p-toluenesulfonic acid to obtain the desired compound, 67 (Scheme 10). Furthermore, the photoreactivity and affinity of diazirinyl derivative 67 to tubulin were evaluated. The photolysis of compound 67 (0.5 mM in methanol) was carried out with a black light under 45 °C for 20 min. The half-life ($t_{1/2}$) of 67 was calculated to be approximately 21 s. In addition, the relative affinity of 67 to tubulin was determined through competitive-equilibrium binding-assay displacement of ^{3}H-labeled oncodazole. In the result, IC$_{50}$ values of 5.7 ± 1.9 µM for 66 and 8.3 ± 3.0 µM for 67 were obtained; therefore, compound 67 has potential as a photoaffinity-labeling reagent.

Furthermore, the benzothiophene scaffold is a capable moiety in drug discovery because it exhibits various biological activities such as antimicrobial [86], anticancer [87,88], and anti-inflammatory [89] effects and many more. Therefore, diazirinyl benzothiophene (81) was synthesized, according to a general TPD synthetic method, by Wang, J. and Sheridan, R.S. [90]. Briefly, benzothiophene-2-trifluoromethyl ketone 77 was oximated with hydroxylamine hydrochloride in pyridine 78, followed by tosylation with tosyl chloride in CH$_2$Cl$_2$ with triethylamine 79. Subsequently, the tosyl oxime moiety was subjected to diaziridinylation with liquid ammonia under −65 °C. Finally, diaziridine 80 was oxidized to the desired diazirine, 81, with iodine, which resulted in a total yield of 26% for proceeding (Scheme 11).

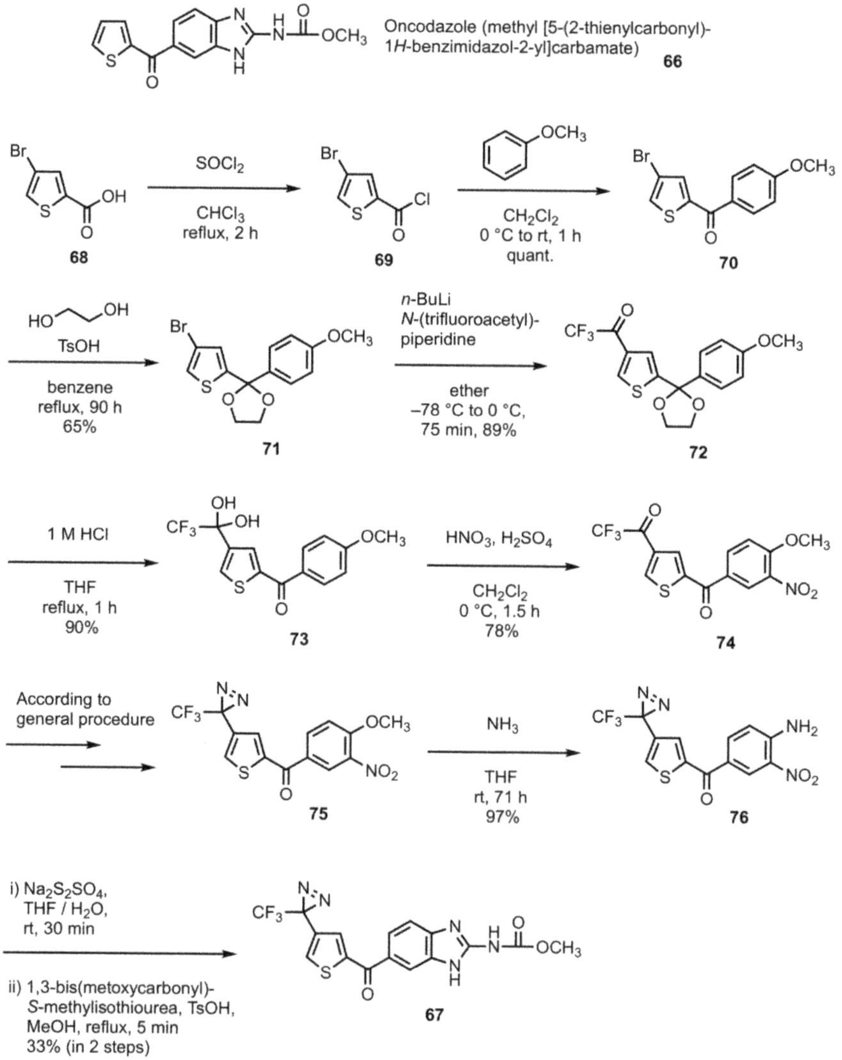

Scheme 10. The synthesis of diazirinyl-substituted thiophene analogue 67.

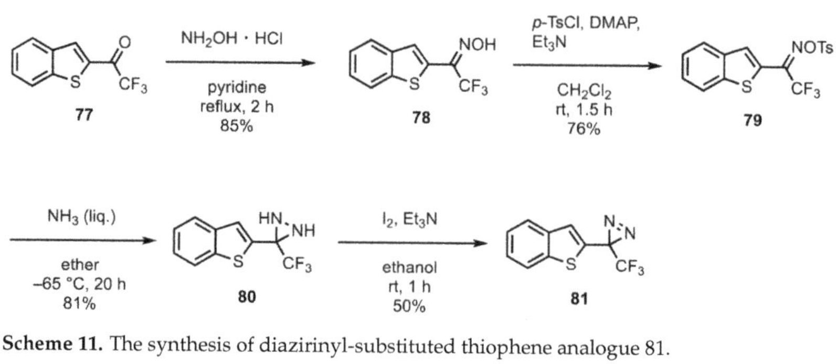

Scheme 11. The synthesis of diazirinyl-substituted thiophene analogue 81.

8. Diazirinyl-Substituted Coumarin

Coumarin (1,2-benzopyrone) and its derivatives are widely available in nature and exhibit various biological activities such as antitumor [91], anti-HIV [92–95], and anti-inflammatory effects [96,97] as well as resistance to triglyceride accumulation [98] and central-nervous-system stimulant effects [99]. In addition to this, coumarin–metal complexes have also attracted attention due to their interesting fluorescent properties as sensors or probes [100,101]. Therefore, there are broad possible uses of coumarin derivatives in materials and medicinal-chemistry scientific research.

In PAL, generally, there are some problems while identification of the cross-linked target molecules due to involves complexity of reaction mixtures and resulting in small quantity of the photo-labelled molecules. Therefore, photolabeled molecules require enrichment with detection tags such as the biotin–avidin system, fluorous tagging [102,103], or clickable groups [104]. However, these PAL probes are each composed of a photophore, a ligand, and a detection tag via a branching structure, and the affinity of the PAL probe towards target molecules tends to reduce because of bulky and unstable fluorophores. Recently, to overcome these problems, Tomohiro, T. et al. have developed a diazirinyl coumarin as a photocrosslinker with a masked fluorogenic beacon [105]. In this section, synthesis of diazirinyl coumarin derivatives and applications for PAL are presented. The synthetic schemes of diazirinyl coumarin derivatives 90a and 90b are described in Scheme 12. Initially, 1-chloro-3,5-dimethoxyphenylbenzene was converted to the Grignard reagent through treatment with magnesium and 1,2-dibromoethane, followed by nucleophilic substitution with 2,2,2-trifluoro-1-(piperidin-1-yl)ethanone to obtain compound 83 with a good yield. Diazirinyl compound 87 was prepared according to the general diazirine synthetic method. Subsequently, compound 87 was subjected to formylation with TiCl$_4$ and dichloromethyl methyl ether to obtain two regioisomers, 88a and 88b, which were able to be separated with silica column chromatography. The methoxy group of each compound 88 was removed with BBr$_3$; then, compounds 89a and 89b were treated with Meldrum's acid (2,2-dimethyl-1,3-dioxane-4,6-dione) to obtain the desired diazirinyl coumarin derivatives, 90a and 90b (Scheme 12).

Next, derivatives of both compounds 90a and 90b (20 mM in CD$_3$OD) were subjected to photoreaction with a 250 W black-light lamp at room temperature for 1 h, monitored, with ^{1}H and ^{19}F-NMR. The compound-90a derivative was smoothly photolyzed to generate a CD$_3$OD adduct without production of a diazo compound, whereas the compound-90b derivative exhibited different phenomena. The rate constant of photolysis of the 90b derivative was calculated as less than one twentieth of the value of the compound-90a derivative. In addition to this, the coumarin fluorescence of the photolyzed 90b derivative was weaker than that of the 90a derivative. On the basis of these data, compound 90a was determined to be suitable as a crosslinker unit for further fluorogenic labeling of proteins.

Furthermore, the versatility of compound 90a as a photoaffinity probe with geldanamycin (GA), which is a potent inhibitor of heat-shock protein 90 (Hsp90), was demonstrated. A photoactivatable GA probe 91 (Figure 1) was prepared according to the synthetic method in a previous report [106] and demonstrated an inhibitory effect on ATPase activity through competition for ATP binding (dissociation constant (K_d) = 1.2 µM) [107]. As a result of Hsp90 PAL with probe 91, photolabeled Hsp90 was detected through measurement of coumarin fluorescence, which increased in an irradiation-time- or probe-concentration-dependent manner. Therefore, compound 90a can be expected to be a an essential photophore for photolabeling and fluorescent imaging of target molecules without the necessity of preinstalling a large, unstable fluorophore in the probe.

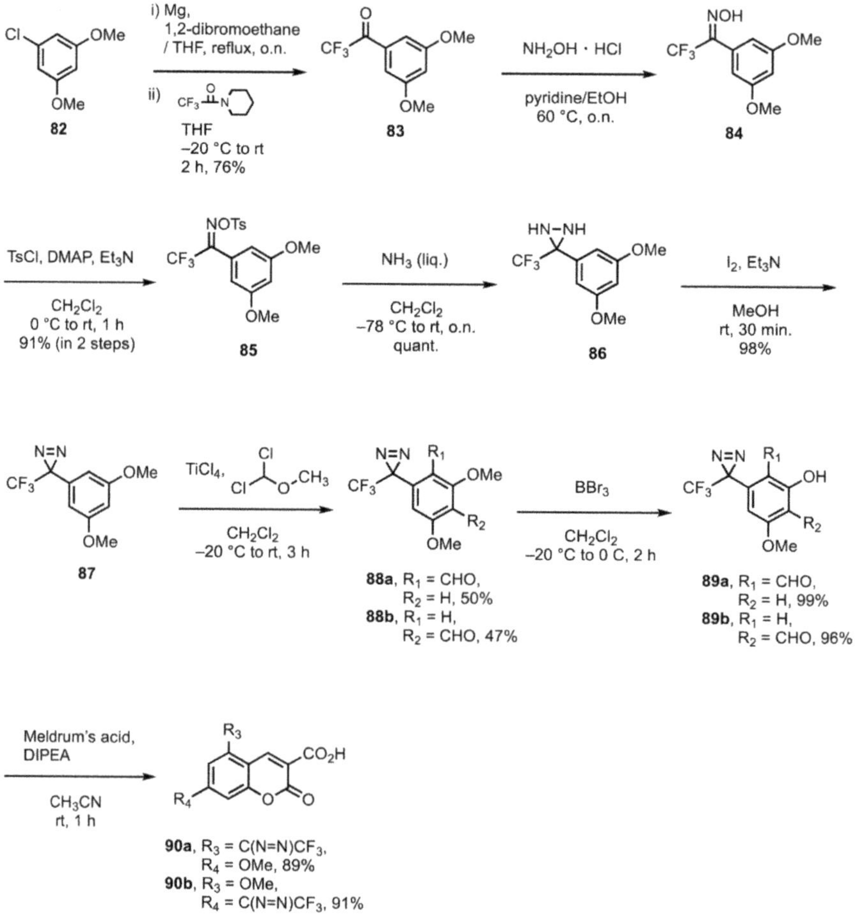

Scheme 12. The synthesis of diazirinyl-substituted diazirinyl coumarin derivatives 90a and 90b.

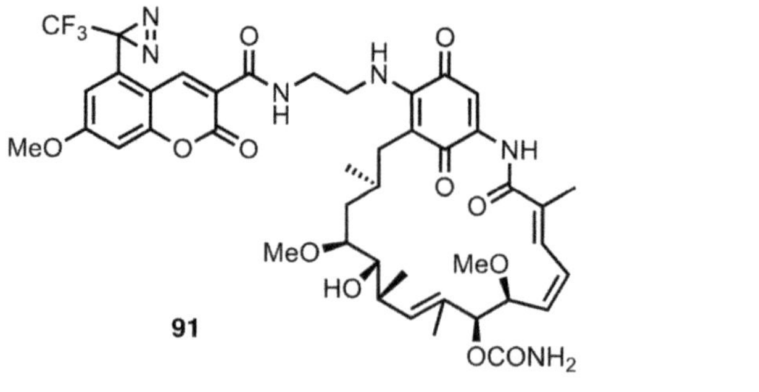

Figure 1. Diazirinyl coumarin-substituted GA probe 91.

Hotta, Y. et al. developed a novel diazirinyl-(coumarin-4-yl)methyl ester that possesses multiple photochemical properties in a single molecule: crosslinking, fluorogenicity, and cleavage functions regulated through photoinduced electron transfer (PeT) [108]. The cleavage system especially has a large advantage for higher-resolution mass spectroscopic analysis of labeled peptides because a ligand and a purification tag in the PAL probe often complicate analysis (Figure 2).

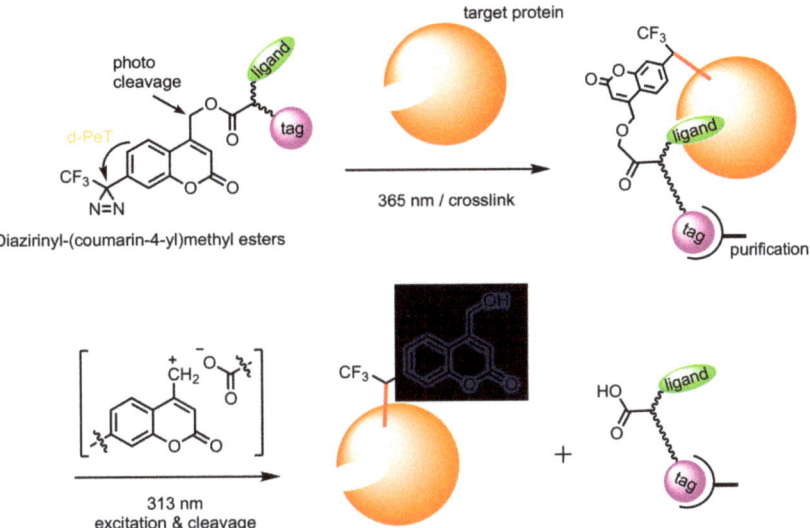

Figure 2. Concept of PAL with novel diazirinyl-coumarin based on (coumarin-4-yl)methyl esters.

Therefore, to evaluate the potential of diazirinyl-(coumarin-4-yl)methyl esters for PAL using carbonic anhydrase II (CA-II), PAL probe 97 was synthesized, and benzenesulfonamide was incorporated as a potent inhibitor of CA-II enzyme activity [109] and biotin. Firstly, compound 92 was subjected to cyclization with ethyl-4-chloroacetoacetate to form coumarin 93, followed by installation of Fmoc-Lys(Boc)-OH with ester linkage to give 94. Next, the Fmoc group of 94 was removed using acidic conditions; then, the amino group of 95 was coupled with NHS-biotin to generate 96 with a moderate yield. Finally, the Boc group in the side chain of 96 was deprotected with TFA/CHCl$_3$, followed by the incorporation of 4-sulfamoylbenzoic acid with 1-ethyl-3-(3-dimethylaminopropyl)carbodiimide hydrochloride (EDC-HCl) and 1-hydroxybenzotriazole (HOBt) to yield the desired 97 (Scheme 13). The inhibitory activity of 97 towards CA-II was calculated with a p-nitrophenyl acetate (4-NPA) esterase assay [110], and the IC$_{50}$ value was 3.0 µM. The PAL probe 97 was furthermore applied for PAL experiments with CA-II. The photocrosslinked CA-II with a 365 nm wavelength was subjected to purification with the avidin–biotin system, followed by photocleavage with a 313 nm wavelength, and the captured protein was observed at approximately 29 kDa, as was detected with coumarin fluorescence. Therefore, diazirinyl-(coumarin-4-yl)methyl esters succeeded in protein identification with controlled photocrosslinking and photocleavage using different photoirradiation wavelengths.

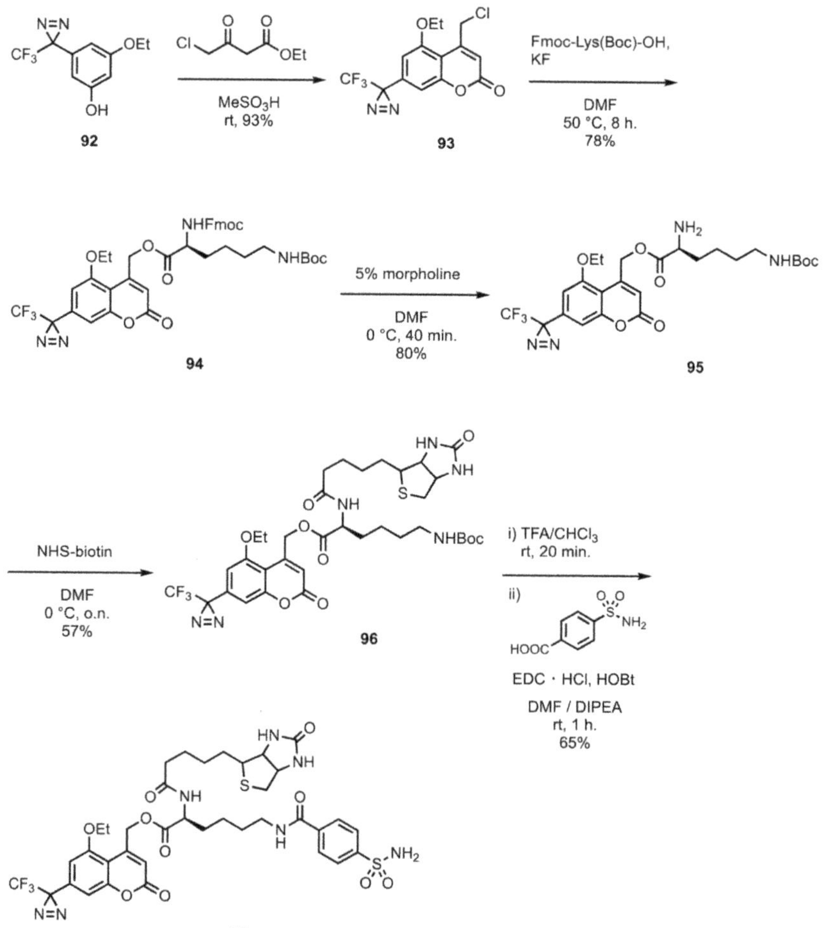

Scheme 13. The synthesis of PAL probe 97, including diazirinyl-(coumarin-4-yl)methyl esters (photophore), 4-sulfamoylbenzoic acid (ligand) and biotin (detection tag).

9. Diazirinyl-Substituted Indoles

Indole, an electron-rich heteroaromatic compound, exhibits a wide range of pharmacological activities and is a versatile pharmacophore. In particular, indole derivatives occur broadly in nature; for example, indole-3-acetic acid (auxin) is a cell-growth hormone essential for cellular division and expansion in plants [111], and 2-amino-3-(1H-indol-3-yl)propanoic acid (tryptophan) is an essential amino acid used as a building block in biomolecules. Furthermore, natural indole alkaloids are crucial molecules for drug discovery, and their actions include muscle-contraction, migraine-relief [112], cytotoxic, antibacterial, antimicrobial and antineoplastic activity [113]. Moreover, synthetic indole alkaloids exhibit anti-HIV [114], anti-Alzheimer's-disease [115], anti-inflammatory, and analgesic effects. Indole and its derivatives are essential; therefore, using them with PAL will contribute more to medicinal chemistry and drug discovery. We have succeeded for the first time in synthesis and postfunctional synthesis of diazirine indoles, especially for diazirinyl indole-3-acetic acid, and we have also evaluated their bioactivity experiments [116]. Subsequently, 5- and 6- bromoindoles 98a and 98b were deprotonated with KH, followed by a lithium–bromide exchange with t-BuLi, and nucleophilic substitution with trifluoroacetyl donors afforded trifluoroacetyl indole derivatives 99a and 99b with an approximately 80% yield. The trifluoroacetyl group was converted to oximes 100a and 100b with hydroxylamine hydrochloride in pyridine, followed by tosylation with tosyl chloride in triethylamine and acetone at 0 °C. However, tosyl oximes 101a and 101b were too unstable under silica-gel conditions to be hydrolyzed to the ketone moiety. In order to avoid a decrease in yield, these tosyl oximes were directly converted to diaziridines 102a and 102b without purification. The isolated yield for 102a and 102b was maintained at over 85%. Finally, oxidation of the diaziridine moiety with iodine in triethylamine occurred simultaneously with iodination at the 3-position of the indole skeleton (103a and 103b). Moreover, oxidation with activated MnO_2 was able to proceed without any side reactions to obtain 104a and 104b with a good yield (80–92%) (Scheme 14). Soon after, 6-diazirinyl indole was reported, with a similar synthetic scheme, by Wartmann, T. and Lindel, T. [117].

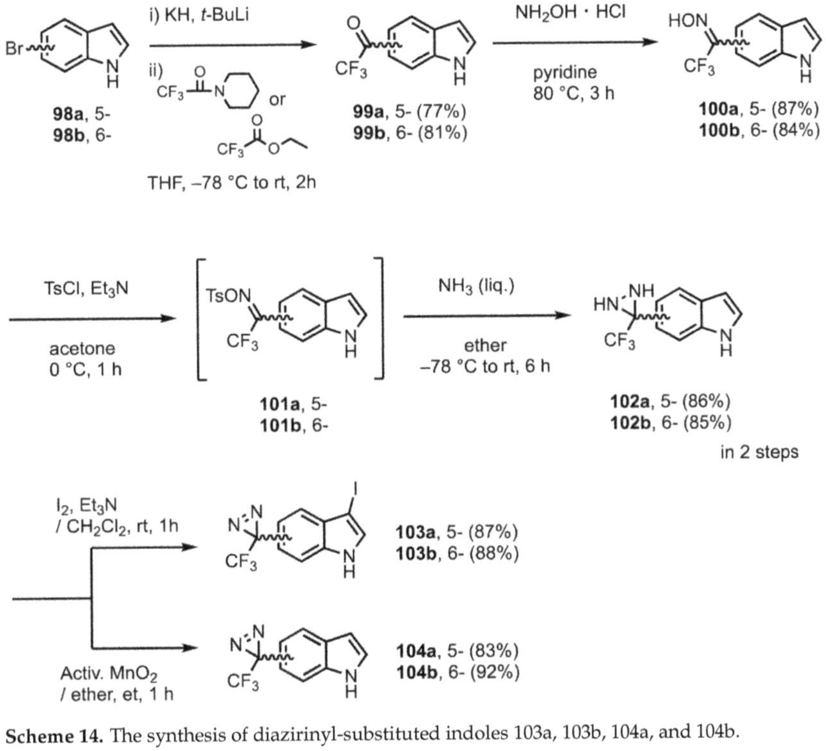

Scheme 14. The synthesis of diazirinyl-substituted indoles 103a, 103b, 104a, and 104b.

Furthermore, we developed the postfunctional derivatization of diazirinyl indole for expansion of the use of its derivatives in PAL. Tryptophan is one of the most biologically significant metabolites synthesized from indole. Once, we tried to synthesize diazirinyl tryptophan from TPD according to reported information [118–120]; however, the diazirinyl tryptophan could not be obtained due to decomposition of the diazirinyl moiety under the harsh conditions required for its construction. Moreover, tryptophan has been prepared through condensation of indole and serine in acetic acid and acetic anhydride under reflux conditions [121]. However, this reaction was conducted under reflux conditions in the presence of all reagents, which the diazirinyl moiety cannot tolerate. Thus, we modified the reaction conditions; initially, active varieties were generated from serine, acetic anhydride, and acetic acid in reflux conditions without indole, followed by the addition of diazirinyl indoles 104a and 104b to the mixture at a low temperature to yield N-acetyl tryptophan derivatives 105a and 105b without decomposition of the diazirinyl moiety. Subsequently, racemates of diazirinyl *N*-acetyltryptophans 105a and 105b were subjected to enzymatic resolution with L-acylase to afford optically pure diazirinyl L-tryptophans 106a and 106b without decomposition of the diazirinyl moiety (Scheme 15A). In addition to tryptophan, diazirinyl carbinols 108a and 108b have shown anticarcinogenic, antioxidant, and antiatherogenic effects, and diazirinyl gramines 109a and 109b, which play a defensive role in plants, were able to be synthesized (Scheme 15B,C).

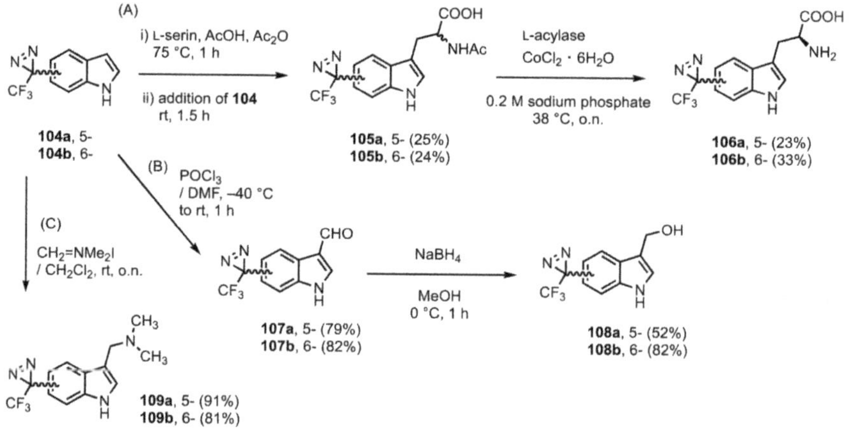

Scheme 15. Postfunctional synthesis of diazirinyl indole derivatives (**A**) tryptophan, (**B**) carbinol and (**C**) gramine from diazirinyl indoles.

Indole-3-acetic acid (IAA), well-known as the plant hormone auxin, is essential throughout cellular division and expansion in plants. However, deeper understanding of the biological mechanism of IAA is required. Thus, diazirinyl IAA is a promising chemical tool for understanding those veiled mechanisms in PAL. Diazirinyl indoles 104a and 104b were reacted with oxalyl chloride, followed by methanolysis, to obtain methyl indole 3-oxoacetates 110a and 110b. Reduction with triethylsilane in trifluoroacetic acid [122] occurred not only on the aromatic α-keto moiety of 110 but between the 2- and 3-positions of the indole skeleton, producing 111a and 111b. Next, dehydrogenation at the 2,3-positions of 111a and 111b with MnO_2 produced only 113b because 111a was unstable under these conditions. Therefore, to construct 5-diazirinyl IAA methyl ester 113a, compound 110a was subjected to reduction with sodium borohydride to obtain α-hydroxy ester 112a, followed by dehydration with P_2I_4 [123] to obtain 5-diazirinyl IAA methyl ester 113a with a moderate yield. This method was also able to be applied to the 6-diazirinyl indoles (from 112b to 113b). Finally, hydrolysis of the methyl ester under alkaline conditions produced the desired 5- and 6-diazirinyl IAA derivatives, 114a and 114b (Scheme 16). We also performed oat coleoptile segment growth bioassays using 114a and 114b and compared them to IAA [124]. Typical auxin responses were observed in the presence of 114a and of 114b. Especially, photoreactive IAA-substituted diazirin at the 5-position (114a) exhibited a higher effect than did IAA. Therefore, both photoreactive IAA 114a and 114b will contribute to future elucidation of the role of IAA and its target proteins via PAL.

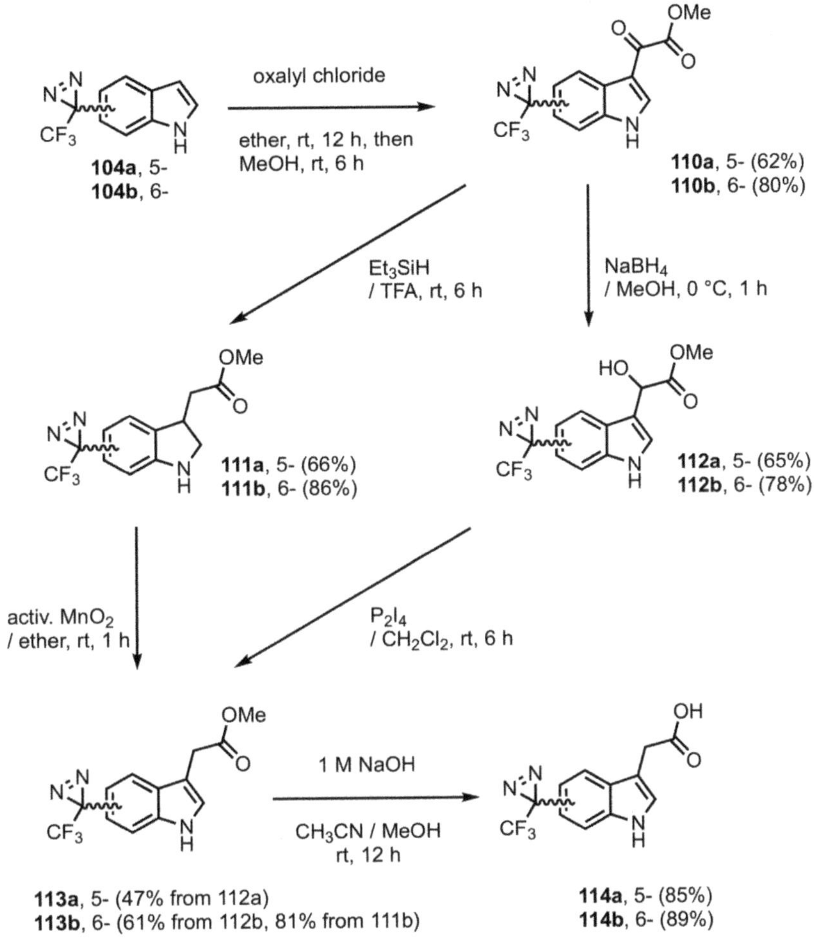

Scheme 16. The synthesis of diazirinyl indole-3-acetic acid derivatives from diazirinyl indole.

10. Conclusions

Heterocyclic aromatics are very important mother skeletons in the fields of materials science and medicinal chemistry. In synthetic chemistry, for heteroaromatic polymers, diazirines have the potential to overcome previous challenges. In particular, application of heteroaromatic diazirines to materials science requires a thorough understanding of their physical properties and characteristics, such as thermal properties and photostability. Although the application of TPD to materials science is known to some extent, there are few reports on heteroaromatic diazirines, and a review regarding the total synthesis and applications of heteroaromatic diazirines has not been published thus far. As shown in this review, heteroaromatic diazirines exhibit different physical properties to those of TPD, which may lead to breakthroughs in novel-polymer-material creation and other challenges. Therefore, it is suggested that the heteroaromatic diazirines presented here could also contribute to development of sustainable new materials. Furthermore, over the past decades, advances in heteroaromatic medicinal chemistry have stimulated progress in the fields of chemical biology and led to significant improvement in quality of life and an increase in the length of human life. However, there are still many unmet medical needs that cannot be addressed with existing therapeutics. In particular, current drug discovery is a search for hit compounds from huge compound libraries and repositioning of existing drugs.

Heteroaromatic compounds are often included in these hit compounds, and understanding their mechanisms of action is a crucial issue in drug discovery. Therefore, heteroaromatic diazirines can make a significant contribution to elucidation of such functional mechanisms in medicinal chemistry. This review provides a comprehensive overview of them. Diazirine-substituted heterocyclic aromatics are expected to be one of the most powerful tools to solve these issues, and various studies that will investigate them are expected and will expand greatly in the future. Therefore, this review will greatly contribute to academia as well as the industry.

Author Contributions: Y.M. and M.H. proposed this review article concept and wrote this manuscript. All authors have read and agreed to the published version of the manuscript.

Funding: This research received no external funding.

Institutional Review Board Statement: Not applicable.

Informed Consent Statement: Not applicable.

Data Availability Statement: Not applicable.

Conflicts of Interest: The authors declare no conflict of interest.

References

1. Borissov, A.; Maurya, Y.K.; Moshniaha, L.; Wong, W.S.; Żyła-Karwowska, M.; Stępień, M. Recent Advances in Heterocyclic Nanographenes and Other Polycyclic Heteroaromatic Compounds. *Chem. Rev.* **2022**, *122*, 565–788. [CrossRef] [PubMed]
2. Lv, M.; Zhang, F.; Wu, Y.; Chen, M.; Yao, C.; Nan, J.; Shu, D.; Zeng, R.; Zeng, H.; Chou, S.L. Heteroaromatic Organic Compound with Conjugated Multi-Carbonyl as Cathode Material for Rechargeable Lithium Batteries. *Sci. Rep.* **2016**, *11*, 23515. [CrossRef] [PubMed]
3. Kabir, E.; Uzzaman, M. A Review on Biological and Medicinal Impact of Heterocyclic Compounds. *Results Chem.* **2022**, *4*, 100606. [CrossRef]
4. Davison, E.K.; Sperry, J. Natural Products with Heteroatom-Rich Ring Systems. *J. Nat. Prod.* **2017**, *80*, 3060–3079. [CrossRef]
5. Lepage, M.L.; Simhadri, C.; Liu, C.; Takaffoli, M.; Bi, L.; Crawford, B.; Milani, A.S.; Wulff, J.E. A Broadly Applicable Cross-Linker for Aliphatic Polymers Containing C-H Bonds. *Science* **2019**, *366*, 875–878. [CrossRef]
6. Musolino, S.F.; Mahbod, M.; Nazir, R.; Bi, L.; Graham, H.A.; Milani, A.S.; Wulff, J.E. Electronically Optimized Diazirine-based Polymer Crosslinkers. *Polym. Chem.* **2022**, *13*, 3833–3839. [CrossRef]
7. Brunner, J. New hotolabeling and crosslinking methods. *Annu. Rev. Biochem.* **1993**, *62*, 483–514. [CrossRef]
8. Tomohiro, T.; Hashimoto, M.; Hatanaka, Y. Cross-Linking Chemistry and Biology: Development of Multifunctional Photoaffinity Probes. *Chem. Rec.* **2005**, *5*, 385–395. [CrossRef]
9. Hashimoto, M.; Hatanaka, Y. Recent Progress in Diazirine-Based Photoaffinity Labeling. *Eur. J. Org. Chem.* **2008**, *2008*, 2513–2523. [CrossRef]
10. Hatanaka, Y. Development and Leading-Edge Application of Innovative Photoaffinity Labeling. *Chem. Pharm. Bull.* **2015**, *63*, 1–12. [CrossRef]
11. Platz, M.S. Comparison of Phenylcarbene and Phenylnitrene. *Acc. Chem. Res.* **1995**, *28*, 487–492. [CrossRef]
12. Karney, W.L.; Borden, W.T. Why Does o-Fluorine Substitution Raise the Barrier to Ring Expansion of Phenylnitrene? *J. Am. Chem. Soc.* **1997**, *119*, 3347–3350. [CrossRef]
13. Galardy, R.E.; Craig, L.C.; Printz, M.P. Benzophenone Triplet: A New Photochemical Probe of Biological Ligand-Receptor Interactions. *Nat. New Biol.* **1973**, *242*, 127–128. [CrossRef]
14. Brunner, J.; Senn, H.; Richards, F.M. 3-Trifluoromethyl-3-Phenyldiazirine. A New Carbene Generating Group for Photolabeling Reagents. *J. Biol. Chem.* **1980**, *255*, 3313–3318. [CrossRef]
15. Dey, K.; Roy Chowdhury, S.; Dykstra, E.; Koronatov, A.; Lu, H.P.; Shinar, R.; Shinar, J.; Anzenbacher, P., Jr. Diazirine-Based Photo-Crosslinkers for Defect Free Fabrication of Solution Processed Organic Light-Emitting Diodes. *J. Mater. Chem. C* **2020**, *8*, 11988–11996. [CrossRef]
16. Chowdhury, S.; Jana, A.; Mandal, A.K.; Choudhary, R.J.; Phase, D.M. Time Evolution of the Structural, Electronic, and Magnetic Phases in Relaxed $SrCoO_3$ Thin Films. *ACS Appl. Electron. Mater.* **2021**, *3*, 3060–3071. [CrossRef]
17. Nazir, R.; Bi, L.; Musolino, S.F.; Margoto, O.H.; Çelebi, K.; Mobuchon, C.; Takaffoli, M.; Milani, A.S.; Falck, G.; Wulf, J.E. Polyamine–Diazirine Conjugates for Use as Primers in UHMWPE–Epoxy Composite Materials. *ACS Appl. Polym. Mater.* **2022**, *4*, 1728–1742. [CrossRef]
18. Yan, Z.; Xu, D.; Lin, Z.; Wang, P.; Cao, B.; Ren, H.; Song, F.; Wan, C.; Wang, L.; Zhou, J.; et al. Highly Stretchable van der Waals Thin Films for Adaptable and Breathable Electronic Membranes. *Science* **2022**, *375*, 852–859. [CrossRef]
19. Kumar, A.B.; Tipton, J.D.; Manetsch, R. 3-Trifluoromethyl-3-Aryldiazirine Photolabels with Enhanced Ambient Light Stability. *Chem. Commun.* **2016**, *52*, 2729–2732. [CrossRef]

20. Cross, R.M.; Monastyrskyi, A.; Mutka, T.S.; Burrows, J.N.; Kyle, D.E.; Manetsch, R. Endochin Optimization: Structure-Activity and Structure-Property Relationship Studies of 3-Substituted 2-Methyl-4(1H)-quinolones with Antimalarial Activity. *J. Med. Chem.* **2010**, *53*, 7076–7094. [CrossRef]
21. Wang, Z.; Deng, X.; Xiong, S.; Xiong, R.; Liu, J.; Zou, L.; Lei, X.; Cao, X.; Xie, Z.; Chen, Y.; et al. Design, Synthesis and Biological Evaluation of Chrysin Benzimidazole Derivatives as Potential Anticancer Agents. *Nat. Prod. Res.* **2018**, *32*, 2900–2909. [CrossRef] [PubMed]
22. Morais, G.R.; Palma, E.; Marques, F.; Gano, L.; Oliveira, M.C.; Abrunhosa, A.; Miranda, H.V.; Outeiro, T.F.; Santos, I.; Paulo, A. Synthesis and Biological Evaluation of Novel 2-Aryl Benzimidazoles as Chemotherapeutic Agents. *J. Heterocycl. Chem.* **2017**, *54*, 255–267. [CrossRef]
23. Shaker, Y.M.; Omar, M.A.; Mahmoud, K.; Elhallouty, S.M.; El-Senousy, W.M.; Ali, M.M.; Mahmoud, A.E.; Abdel-Halim, A.H.; Soliman, S.M.; El Diwani, H.I. Synthesis, in Vitro and in Vivo Aantitumor and Antiviral Activity of Novel 1-Substituted Benzimidazole Derivatives. *J. Enzyme Inhib. Med. Chem.* **2015**, *30*, 826–845. [CrossRef] [PubMed]
24. Onnis, V.; Demurtas, M.; Deplano, A.; Balboni, G.; Baldisserotto, A.; Manfredini, S.; Pacifico, S.; Liekens, S.; Balzarini, J. Design, Synthesis and Evaluation of Antiproliferative Activity of New Benzimidazolehydrazones. *Molecules* **2016**, *21*, 579. [CrossRef] [PubMed]
25. Çevik, U.A.; Sağlık, B.N.; Korkut, B.; Özkay, Y.; Ilgın, S. Antiproliferative, Cytotoxic, and Apoptotic Effects of New Benzimidazole Derivatives Bearing Hydrazone Moiety. *J. Heterocycl. Chem.* **2018**, *55*, 138–148. [CrossRef]
26. Abd el Al, S.N.; Soliman, F.M.A. Synthesis, Some Reactions, Cytotoxic Evaluation and Antioxidant Study of Novel Benzimidazole Derivatives. *Der Pharma Chem.* **2015**, *7*, 71–84.
27. Bellam, M.; Gundluru, M.; Sarva, S.; Chadive, S.; Netala, V.R.; Tartte, V.; Cirandur, S.R. Synthesis and Antioxidant Activity of Some New N-Alkylated Pyrazole Containing Benzimidazoles. *Chem. Heterocycl. Compd.* **2017**, *53*, 173–178. [CrossRef]
28. Sharma, R.; Bali, A.; Chaudhari, B.B. Synthesis of Methanesulphonamido-Benzimidazole Derivatives as Gastro-Sparing Antiinflammatory Agents with Antioxidant Effect. *Bioorg. Med. Chem. Lett.* **2017**, *27*, 3007–3013. [CrossRef]
29. Sethi, P.; Bansal, Y.; Bansal, G. Synthesis and PASS-Assisted Evaluation of Coumarin–Benzimidazole Derivatives as Potential Anti-Inflammatory and Anthelmintic Agents. *Med. Chem. Res.* **2018**, *27*, 61–71. [CrossRef]
30. Yang, H.; Ren, Y.; Gao, X.; Gao, Y. Synthesis and Anticoagulant Bioactivity Evaluation of 1,2,5-Trisubstituted Benzimidazole Fluorinated Derivatives. *Chem. Res. Chin. Univ.* **2016**, *32*, 973–978. [CrossRef]
31. Wang, F.; Ren, Y.-J. Design, Synthesis, Biological Evaluation and Molecular Docking of Novel Substituted 1-Ethyl-1H-Benzimidazole Fluorinated Derivatives as Thrombin Inhibitors. *J. Iran. Chem. Soc.* **2016**, *13*, 1155–1166. [CrossRef]
32. Vandeputte, M.M.; Van Uytfanghe, K.; Layle, N.K.; St Germaine, D.M.; Iula, D.M.; Stove, C.P. Synthesis, Chemical Characterization, and μ-Opioid Receptor Activity Assessment of the Emerging Group of "Nitazene" 2-Benzylbenzimidazole Synthetic Opioids. *ACS Chem. Neurosci.* **2021**, *12*, 1241–1251. [CrossRef]
33. Raimer, B.; Wartmann, T.; Jones, P.G.; Lindel, T. Synthesis, Stability, and Photoreactivity of Diazirinyl-Substituted N-Heterocycles Based on Indole, Benzimidazole, and Imidazole. *Eur. J. Org. Chem.* **2014**, *2014*, 5509–5520. [CrossRef]
34. Cheng, H.; Pei, Y.; Leng, F.; Li, J.; Liang, A.; Zou, D.; Wu, Y.; Wu, Y. Highly Efficient Synthesis of Aryl and Heteroaryl Trifluoromethyl Ketones via o-iodobenzoic acid (IBX). *Tetrahedron Lett.* **2013**, *54*, 4483–4486. [CrossRef]
35. Moss, R.A.; Perez, L.A.; Turro, N.J.; Gould, I.R.; Hacker, N.P. Hammett Analysis of Absolute Carbene Addition Rate Constants. *Tetrahedron Lett.* **1983**, *24*, 685–688. [CrossRef]
36. Mueller, P.H.; Rondan, N.G.; Houk, K.N.; Harrison, J.F.; Hooper, D.; Willen, B.H.; Liebman, J.F. Carbene Singlet-Triplet Gaps. Linear Correlations with Substituent Donation. *J. Am. Chem. Soc.* **1981**, *103*, 5049–5052. [CrossRef]
37. Creary, X. Regioselectivity in the Addition of Singlet and Triplet Carbenes to 1,1-Dimethylallene. A Probe for Carbene Multiplicity. *J. Am. Chem. Soc.* **1980**, *102*, 1611–1618. [CrossRef]
38. Wang, J.; Kubicki, J.; Peng, H.; Platz, M.S. Influence of Solvent on Carbene Intersystem Crossing Rates. *J. Am. Chem. Soc.* **2008**, *130*, 6604–6609. [CrossRef]
39. Faisal, M.; Saeed, A.; Hussain, S.; Dar, P.; Larik, F.A. Recent Developments in Synthetic Chemistry and Biological Activities of Pyrazole Derivatives. *J. Chem. Sci.* **2019**, *131*, 70. [CrossRef]
40. Costa, R.F.; Turones, L.C.; Cavalcante, K.V.N.; Rosa Júnior, I.A.; Xavier, C.H.; Rosseto, L.P.; Napolitano, H.B.; Castro, P.F.D.S.; Neto, M.L.F.; Galvão, G.M.; et al. Heterocyclic Compounds: Pharmacology of Pyrazole Analogs from Rational Structural Considerations. *Front. Pharmacol.* **2021**, *12*, 666725. [CrossRef]
41. Ebenezer, O.; Shapi, M.; Tuszynski, J.A. A Review of the Recent Development in the Synthesis and Biological Evaluations of Pyrazole Derivatives. *Biomedicines* **2022**, *10*, 1124. [CrossRef] [PubMed]
42. Faisal, M.; Hussain, S.; Haider, A.; Saeed, A.; Laril, F.A. Assessing the Effectiveness of Oxidative Approaches for the Synthesis of Aldehydes and Ketones from Oxidation of Iodomethyl Group. *Chem. Pap.* **2019**, *73*, 1053–1067. [CrossRef]
43. Ran, F.; Liu, Y.; Zhang, D.; Liu, M.; Zhao, G. Discovery of Novel Pyrazole Derivatives as Potential Anticancer Agents in MCL. *Bioorg. Med. Chem. Lett.* **2019**, *29*, 1060–1064. [CrossRef] [PubMed]
44. Sony Jacob, K.; Swastika, G. A Battle Against Aids: New Pyrazole Key to an Older Lock-reverse Transcriptase. *Int. J. Pharm. Pharm. Sci.* **2016**, *8*, 75–79.
45. Kumar, S.; Gupta, S.; Rani, V.; Sharma, P. Pyrazole Containing Anti-HIV Agents: An Update. *Med. Chem.* **2022**, *18*, 831–846. [CrossRef]

46. Bekhit, A.A.; Hassan, A.M.; Abd El Razik, H.A.; El-Miligy, M.M.; El-Agroudy, E.J.; Bekhit, A.-D. New Heterocyclic Hybrids of Pyrazole and Its Bioisosteres: Design, Synthesis and Biological Evaluation as Dual Acting Antimalarial-Antileishmanial Agents. *Eur. J. Med. Chem.* **2015**, *94*, 30–44. [CrossRef]
47. Heller, S.T.; Natarajan, S.R. 1,3-Diketones from Acid Chlorides and Ketones: A Rapid and General One-Pot Synthesis of Pyrazoles. *Org. Lett.* **2006**, *8*, 2675–2678. [CrossRef]
48. Chebib, M.; Johnston, G.A. GABA-Activated Ligand Gated Ion Channels: Medicinal Chemistry and Molecular Biology. *J. Med. Chem.* **2000**, *43*, 1427–1447. [CrossRef]
49. Zhu, S.; Noviello, C.M.; Teng, J.; Walsh, R.M., Jr.; Kim, J.J.; Hibbs, R.E. Structure of a Human Synaptic GABAA Receptor. *Nature* **2018**, *559*, 67–72. [CrossRef]
50. Singh, N.S.; Sharma, R.; Singh, S.K.; Singh, D.K. A Comprehensive Review of Environmental Fate and Degradation of Fipronil and Its Toxic Metabolites. *Environ. Res.* **2021**, *199*, 111316. [CrossRef]
51. Sirisoma, N.S.; Ratra, G.S.; Tomizawa, M.; Casida, J.E. Fipronil-Based Photoaffinity Probe for Drosophila and Human Beta 3 GABA Receptors. *Bioorg. Med. Chem. Lett.* **2001**, *11*, 2979–2981. [CrossRef]
52. Sammelson, R.E.; Casida, J.E. Synthesis of a Tritium-Labeled, Fipronil-Based, Highly Potent, Photoaffinity Probe for the GABA Receptor. *J. Org. Chem.* **2003**, *68*, 8075–8079. [CrossRef]
53. Delfino, J.M.; Schreiber, S.L.; Richards, F.M. Design, Synthesis, and Properties of a Photoactivatable Membrane-Spanning Phospholipidic Probe. *J. Am. Chem. Soc.* **1993**, *115*, 3458–3474. [CrossRef]
54. Weber, T.; Brunner, J. 2-(Tributylstannyl)-4-[3-(trifluoromethyl)-3H-diazirin-3-yl]benzyl Alcohol: A Building Block for Photolabeling and Cross-Linking Reagents of Very High Specific Radioactivity. *J. Am. Chem. Soc.* **1995**, *117*, 3084–3095. [CrossRef]
55. Li, G.; Samadder, P.; Arthur, G.; Bittmana, R. Synthesis and Antiproliferative Properties of a Photoactivatable Analogue of ET-18-OCH$_3$. *Tetrahedron* **2001**, *57*, 8925–8932. [CrossRef]
56. Leroux, F.; Schlosser, M. The "Aryne" Route to Biaryls Featuring Uncommon Substituent Patterns. *Angew. Chem. Int. Ed.* **2002**, *41*, 4272–4274. [CrossRef]
57. Dess, D.B.; Martin, J.C. Readily Accessible 12-I-5 Oxidant for the Conversion of Primary and Secondary Alcohols to Aldehydes and Ketones. *J. Org. Chem.* **1983**, *48*, 4155–4156. [CrossRef]
58. Linderman, R.J.; Graves, D.M. Oxidation of Fluoroalkyl-Substituted Carbinols by the Dess-Martin. *J. Org. Chem.* **1989**, *54*, 661–668. [CrossRef]
59. Kosemura, S.; Emori, H.; Yamamura, S.; Anai, T.; Aizawa, H.; Ohtake, N.; Hasegawa, K. Isolation and Characterization of 4-Chloro-6,7-dimethoxybenzoxazolin-2-one, A New Auxin-inhibiting Benzoxazolinone from *Zea mays*. *Chem. Lett.* **1995**, *24*, 1053–1054. [CrossRef]
60. Hoshi-Sakoda, M.; Usui, K.; Ishizuka, K.; Kosemura, S.; Yamamura, S.; Hasegawa, K. Structure-Activity Relationships of Benzoxazolinones with Respect to Auxin-Induced Growth and Auxin-Binding Protein. *Phytochemistry* **1994**, *37*, 297–300. [CrossRef]
61. Kosemura, S.; Emori, H.; Yamamura, S.; Anai, T.; Tomita, K.; Hasegawa, K. Design of Photoaflinity Reagents for Labeling the Auxin Receptor in Maize. *Tetrahedron Lett.* **1997**, *38*, 2125–2128. [CrossRef]
62. Hatanaka, Y.; Hashimoto, M.; Nakayama, H.; Kanaoka, Y. Synthesis of Nitro-Substituted Aryl Diazirines. An Entry to Chromogenic Carbene Precursors for Photoaffinity Labeling. *Chem. Pharm. Bull.* **1994**, *42*, 826–831. [CrossRef]
63. Hatanaka, Y.; Hashimoto, M.; Kurihara, H.; Nakayama, H.; Kanaoka, Y. A Novel Family of Aromatic Diazirines for Photoaffinity Labeling. *J. Org. Chem.* **1994**, *59*, 383–387. [CrossRef]
64. Kohler, M.; Clarenbach, C.F.; Bahler, C.; Brack, T.; Russi, E.W.; Bloch, K.E. Disability and Survival in Duchenne Muscular Dystrophy. *J. Neurol. Neurosurg. Psychiatry* **2009**, *80*, 320–325. [CrossRef] [PubMed]
65. McDonald, C.M.; Henricson, E.K.; Abresch, R.T.; Han, J.J.; Escolar, D.M.; Florence, J.M.; Duong, T.; Arrieta, A.; Clemens, P.R.; Hoffman, E.P.; et al. Cinrg Investigators. The Cooperative International Neuromuscular Research Group Duchenne Natural History Study—A Longitudinal Investigation in the Era of Glucocorticoid Therapy: Design of Protocol and the Methods Used. *Muscle Nerve* **2013**, *48*, 32–54. [CrossRef]
66. Bushby, K.; Finkel, R.; Birnkrant, D.J.; Case, L.E.; Clemens, P.R.; Cripe, L.; Kaul, A.; Kinnett, K.; McDonald, C.; Pandya, S.; et al. Diagnosis and Management of Duchenne Muscular Dystrophy, Part 1: Diagnosis, and Pharmacological and Psychosocial Management. *Lancet Neurol.* **2010**, *9*, 77–93. [CrossRef]
67. Bushby, K.; Finkel, R.; Birnkrant, D.J.; Case, L.E.; Clemens, P.R.; Cripe, L.; Kaul, A.; Kinnett, K.; McDonald, C.; Pandya, S.; et al. Diagnosis and Management of Duchenne Muscular Dystrophy, Part 2: Implementation of Multidisciplinary Care. *Lancet Neurol.* **2010**, *9*, 177–189. [CrossRef]
68. Moxley, R.T., III; Pandya, S.; Ciafaloni, E.; Fox, D.J.; Campbell, K. Change in Natural History of Duchenne Muscular Dystrophy with Long-Term Corticosteroid Treatment: Implications for Management. *J. Child Neurol.* **2010**, *25*, 1116–1129. [CrossRef]
69. Guiraud, S.; Roblin, D.; Kay, D.E. The Potential of Utrophin Modulators for the Treatment of Duchenne Muscular Dystrophy. *Expert Opin. Orphan. Drugs* **2018**, *6*, 179–192. [CrossRef]
70. Wilkinson, I.V.L.; Perkins, K.J.; Dugdale, H.; Moir, L.; Vuorinen, A.; Chatzopoulou, M.; Squire, S.E.; Monecke, S.; Lomow, A.; Geese, M.; et al. Chemical Proteomics and Phenotypic Profiling Identifies the Aryl Hydrocarbon Receptor as a Molecular Target of the Utrophin Modulator Ezutromid. *Angew. Chem. Int. Ed.* **2020**, *59*, 2420–2428. [CrossRef]
71. Medvedev, A.; Kopylov, A.; Buneeva, O.; Zgoda, V.; Archakov, A. Affinity-Based Proteomic Profiling: Problems and Achievements. *Proteomics* **2012**, *12*, 621–637. [CrossRef]

72. Shi, H.; Zhang, C.J.; Chen, G.Y.; Yao, S.Q. Cell-Based Proteome Profiling of Potential Dasatinib Targets by Use of Affinity-Based Probes. *J. Am. Chem. Soc.* **2012**, *134*, 3001–3014. [CrossRef]
73. Won, S.J.; Eschweiler, J.D.; Majmudar, J.D.; Chong, F.S.; Hwang, S.Y.; Ruotolo, B.T.; Martin, B.R. Affinity-Based Selectivity Profiling of an In-Class Selective Competitive Inhibitor of Acyl Protein Thioesterase 2. *ACS Med. Chem. Lett.* **2016**, *8*, 215–220. [CrossRef]
74. Chancellor, D.R.; Davies, K.E.; De Moor, O.; Dorgan, C.R.; Johnson, P.D.; Lambert, A.G.; Lawrence, D.; Lecci, C.; Maillol, C.; Middleton, P.J.; et al. Discovery of 2-Arylbenzoxazoles as Upregulators of Utrophin Production for the Treatment of Duchenne Muscular Dystrophy. *J. Med. Chem.* **2011**, *54*, 3241–3250. [CrossRef]
75. Rudzinski, D.M.; Kelly, C.B.; Leadbeater, N.E. A Weinreb Amide Approach to the Synthesis of Trifluoromethylketones. *Chem. Commun.* **2012**, *48*, 9610–9612. [CrossRef]
76. Hill, J.R.; Robertson, A.A.B. Fishing for Drug Targets: A Focus on Diazirine Photoaffinity Probe Synthesis. *J. Med. Chem.* **2018**, *61*, 6945–6963. [CrossRef]
77. Chen, Z.; Ku, T.C.; Seley-Radtke, K.L. Thiophene-Expanded Guanosine Analogues of Gemcitabine. *Bioorg. Med. Chem. Lett.* **2015**, *25*, 4274–4276. [CrossRef]
78. Pillai, A.D.; Rathod, P.D.; Xavier, F.P.; Padh, H.; Sudarsanam, V.; Vasu, K.K. Tetra Substituted Thiophenes as Anti-Inflammatory Agents: Exploitation of Analogue-Based Drug Design. *Bioorg. Med. Chem.* **2005**, *13*, 6685–6692. [CrossRef]
79. Amr, A.-G.; Sherif, M.H.; Assy, M.G.; Al-Omar, M.A.; Ragab, I. Antiarrhythmic, Serotonin Antagonist and Antianxiety Activities of Novel Substituted Thiophene Derivatives Synthesized from 2-Amino-4,5,6,7-Tetrahydro-N-Phenylbenzo[b]thiophene-3-Carboxamide. *Eur. J. Med. Chem.* **2010**, *45*, 5935–5942. [CrossRef]
80. Alomar, K.; Landreau, A.; Allain, M.; Bouet, G.; Larcher, G. Synthesis, Structure and Antifungal Activity of Thiophene-2,3-Dicarboxaldehyde Bis(thiosemicarbazone) and Nickel(II), Copper(II) and Cadmium(II) Complexes: Unsymmetrical Coordination Mode of Nickel Complex. *J. Inorg. Biochem.* **2013**, *126*, 76–83. [CrossRef]
81. Emmitte, K.A.; Andrews, C.W.; Badiang, J.G.; Davis-Ward, R.G.; Dickson, H.D.; Drewry, D.H.; Emerson, H.K.; Epperly, A.H.; Hassler, D.F.; Knick, V.B.; et al. Discovery of Thiophene Inhibitors of Polo-Like Kinase. *Bioorg. Med. Chem. Lett.* **2009**, *19*, 1018–1021. [CrossRef] [PubMed]
82. Feng, Y.; Park, H.; Bauer, M.; Ryu, J.C.; Yoon, S.O. Thiophene-Pyrazolourea Derivatives as Potent, Orally Bioavailable, and Isoform-Selective JNK3 Inhibitors. *ACS Med. Chem. Lett.* **2020**, *12*, 24–29. [CrossRef] [PubMed]
83. Vanparijs, O.; Hermans, L.; Thienpont, D. Anthelmintic Activity of Flubendazole against Trichinella Spiralis in Rats. *Vet. Parasitol.* **1979**, *5*, 237–242. [CrossRef]
84. Davidse, L.C.; Flach, W. Interaction of Thiabendazole with Fungal Tubulin. *Biochim. Biophys. Acta* **1978**, *543*, 82–90. [CrossRef] [PubMed]
85. Ladd, D.L.; Harrsch, P.B.; Kruse, L.I. Synthesis and Tubulin Binding of 4′-(l-Azi-2,2,2-trifluoroethyl)oncodazole, a Photolabile Analogue of Oncodazole. *J. Org. Chem.* **1988**, *53*, 417–420. [CrossRef]
86. Liao, X.; Liu, L.; Tan, Y.; Jiang, G.; Fang, H.; Xiong, Y.; Duan, X.; Jiang, G.; Wang, J. Synthesis of Ruthenium Complexes Functionalized with Benzothiophene and Their Antibacterial Activity against *Staphylococcus aureus*. *Dalton Trans.* **2021**, *50*, 5607–5616. [CrossRef]
87. Sweidan, K.; Engelmann, J.; Rayyan, W.A.; Sabbah, D.; Zarga, M.A.; Sabbah, D.; Al-Qirim, T.; Al-Hiari, Y.; Sheikha, G.A.; Shattat, G. Synthesis and Preliminary Biological Evaluation of New Heterocyclic Carboxamide Models. *Lett. Drug Des. Discov.* **2015**, *12*, 417–429. [CrossRef]
88. Martorana, A.; Gentile, C.; Perricone, U.; Piccionello, A.P.; Bartolotta, R.; Terenzi, A.; Pace, A.; Mingoia, F.; Almerico, A.M.; Lauria, A. Synthesis, Antiproliferative Activity, and in Silico Insights of New 3-Benzoylamino-Benzo[b]thiophene Derivatives. *Eur. J. Med. Chem.* **2015**, *90*, 537–546. [CrossRef]
89. Fakhr, I.M.; Radwan, M.A.; el-Batran, S.; Abd el-Salam, O.M.; el-Shenawy, S.M. Synthesis and Pharmacological Evaluation of 2-Substituted Benzo[b]thiophenes as Anti-inflammatory and Analgesic Agents. *Eur. J. Med. Chem.* **2009**, *44*, 1718–1725. [CrossRef]
90. Wang, J.; Sheridan, R.S. A Singlet Aryl-CF$_3$ Carbene: 2-Benzothienyl(trifluoromethyl)carbene and Interconversion with a Strained Cyclic Allene. *Org. Lett.* **2007**, *9*, 3177–3180. [CrossRef]
91. Liu, H.; Wang, Y.; Sharma, A.; Mao, R.; Jiang, N.; Dun, B.; She, J.X. Derivatives Containing Both Coumarin and Benzimidazole Potently Induce Caspase-Dependent Apoptosis of Cancer Cells through Inhibition of PI3K-AKT-mTOR Signaling. *Anticancer Drugs* **2015**, *26*, 667–677. [CrossRef]
92. Yeung, K.S.; Meanwell, N.A.; Qiu, Z.; Hernandez, D.; Zhang, S.; McPhee, F.; Weinheimer, S.; Clark, J.M.; Janc, J.W. Structure-Activity Relationship Studies of a Bisbenzimidazole-Based, Zn^{2+}-Dependent Inhibitor of HCV NS3 Serine Protease. *Bioorg. Med. Chem. Lett.* **2001**, *11*, 2355–2359. [CrossRef]
93. Beaulieu, P.L.; Bousquet, Y.; Gauthier, J.; Gillard, J.; Marquis, M.; McKercher, G.; Pellerin, C.; Valois, S.; Kukolj, G. Non-Nucleoside Benzimidazole-Based Allosteric Inhibitors of the Hepatitis C Virus NS5B Polymerase: Inhibition of Subgenomic Hepatitis C Virus RNA Replicons in Huh-7 Cells. *J. Med. Chem.* **2004**, *47*, 6884–6892. [CrossRef]
94. Hirashima, S.; Suzuki, T.; Ishida, T.; Noji, S.; Yata, S.; Ando, I.; Komatsu, M.; Ikeda, S.; Hashimoto, H. Benzimidazole Derivatives Bearing Substituted Biphenyls as Hepatitis C Virus NS5B RNA-Dependent RNA Polymerase Inhibitors: Structure-Activity Relationship Studies and Identification of a Potent and Highly Selective Inhibitor JTK-109. *J. Med. Chem.* **2006**, *49*, 4721–4736. [CrossRef]

95. Ishida, T.; Suzuki, T.; Hirashima, S.; Mizutani, K.; Yoshida, A.; Ando, I.; Ikeda, S.; Adachi, T.; Hashimoto, H. Benzimidazole Inhibitors of Hepatitis C Virus NS5B Polymerase: Identification of 2-[(4-Diarylmethoxy)phenyl]-Benzimidazole. *Bioorg. Med. Chem. Lett.* **2006**, *16*, 1859–1863. [CrossRef]
96. Hunter, P. The Inflammation Theory of Disease. *EMBO Rep.* **2012**, *13*, 968–970. [CrossRef]
97. Liu, C.H.; Abrams, N.D.; Carrick, D.M.; Chander, P.; Dwyer, J.; Hamlet, M.R.J.; Macchiarini, F.; PrabhuDas, M.; Shen, G.L.; Tandon, P.; et al. Biomarkers of Chronic Inflammation in Disease Development and Prevention: Challenges and Opportunities. *Nat. Immunol.* **2017**, *18*, 1175–1180. [CrossRef]
98. Taşdemir, E.; Atmaca, M.; Yıldırım, Y.; Bilgin, H.M.; Demirtaş, B.; Obay, B.D.; Kelle, M.; Oflazoğlu, H.D. Influence of Coumarin and Some coumarin Derivatives on Serum Lipid Profiles in Carbontetrachloride-Exposed Rats. *Hum. Exp. Toxicol.* **2017**, *36*, 295–301. [CrossRef]
99. Skalicka-Woźniak, K.; Orhan, I.E.; Cordell, G.A.; Nabavi, S.M.; Budzyńska, B. Implication of Coumarins towards Central Nervous System Disorders. *Pharmacol. Res.* **2016**, *103*, 188–203. [CrossRef]
100. Sun, X.-Y.; Liu, T.; Sun, J.; Wang, X.-J. Synthesis and Application of Coumarin Fluorescence Probes. *RSC Adv.* **2020**, *10*, 10826–10847. [CrossRef]
101. Balewski, Ł.; Szulta, S.; Jalińska, A.; Kornicka, A. A Mini-Review: Recent Advances in Coumarin-Metal Complexes with Biological Properties. *Front. Chem.* **2021**, *9*, 781779. [CrossRef] [PubMed]
102. Song, Z.; Zhang, Q. Fluorous Aryldiazirine Photoaffinity Labeling Reagents. *Org. Lett.* **2009**, *11*, 4882–4885. [CrossRef] [PubMed]
103. Song, Z.; Huang, W.; Zhang, Q. Isotope-Coded, Fluorous Photoaffinity Labeling Reagents. *Chem. Commun.* **2012**, *48*, 3339–3341. [CrossRef] [PubMed]
104. Mackinnon, A.L.; Taunton, J. Target Identification by Diazirine Photo-Cross-linking and Click Chemistry. *Curr. Protoc. Chem. Biol.* **2009**, *1*, 55–73. [CrossRef] [PubMed]
105. Tomohiro, T.; Yamamoto, A.; Tatsumi, Y.; Hatanaka, Y. [3-(Trifluoromethyl)-3H-Diazirin-3-yl]coumarin as a Carbene-Generating Photocross-Linker with Masked Fluorogenic Beacon. *Chem. Commun.* **2013**, *49*, 11551–11553. [CrossRef]
106. Shen, Y.; Xie, Q.; Norberg, M.; Sausville, E.; Vande Woude, G.; Wenkert, D. Geldanamycin Derivative Inhibition of HGF/SF-Mediated Met Tyrosine Kinase Receptor-Dependent Urokinase-Plasminogen Activation. *Bioorg. Med. Chem.* **2005**, *13*, 4960–4971. [CrossRef]
107. Roe, S.M.; Prodromou, C.; O'Brien, R.; Ladbury, J.E.; Piper, P.W.; Pearl, L.H. Structural Basis for Inhibition of the Hsp90 Molecular Chaperone by the Antitumor Antibiotics Radicicol and Geldanamycin. *J. Med. Chem.* **1999**, *42*, 260–266. [CrossRef]
108. Hotta, Y.; Kaneko, T.; Hayashi, R.; Yamamoto, A.; Morimoto, S.; Chiba, J.; Tomohiro, T. Photoinduced Electron Transfer-Regulated Protein Labeling with a Coumarin-Based Multifunctional Photocrosslinker. *Chem. Asian. J.* **2019**, *14*, 398–402. [CrossRef]
109. Alterio, V.; Di Fiore, A.; D'Ambrosio, K.; Supuran, C.T.; De Simone, G. Multiple Binding Modes of Iinhibitors to Carbonic Anhydrases: How to Design Specific Drugs Targeting 15 Different Isoforms? *Chem. Rev.* **2012**, *112*, 4421–4468. [CrossRef]
110. Bergmann, F.; Rimon, S.; Segal, R. Effect of pH on the activity of eel esterase towards different substrates. *Biochem. J.* **1958**, *68*, 493–499. [CrossRef]
111. Zhao, Y. Auxin Biosynthesis and Its Role in Plant Development. *Annu. Rev. Plant Biol.* **2010**, *61*, 49–64. [CrossRef]
112. Negård, M.; Uhlig, S.; Kauserud, H.; Andersen, T.; Høiland, K.; Vrålstad, T. Links between Genetic Groups, Indole Alkaloid Profiles and Ecology within the Grass-Parasitic Claviceps purpurea Species Complex. *Toxins* **2015**, *7*, 1431–1456. [CrossRef]
113. Netz, N.; Opatz, T. Marine Indole Alkaloids. *Mar. Drugs* **2015**, *13*, 4814–4914. [CrossRef]
114. Chen, X.; Zhan, P.; Li, D.; De Clercq, E.; Liu, X. Recent Advances in DAPYs and Related Analogues as HIV-1 NNRTIs. *Curr. Med. Chem.* **2011**, *18*, 359–376. [CrossRef]
115. Goyal, D.; Kaur, A.; Goyal, B. Benzofuran and Indole: Promising Scaffolds for Drug Development in Alzheimer's Disease. *ChemMedChem* **2018**, *13*, 1275–1299. [CrossRef]
116. Murai, Y.; Masuda, K.; Sakihama, Y.; Hashidoko, Y.; Hatanaka, Y.; Hashimoto, M. Comprehensive Synthesis of Photoreactive (3-Trifluoromethyl)diazirinyl Indole Derivatives from 5- and 6- Trifluoroacetylindoles for Photoaffinity Labeling. *J. Org. Chem.* **2012**, *77*, 8581–8587. [CrossRef]
117. Wartmann, T.; Lindel, T. L-Phototryptophan. *Eur. J. Org. Chem.* **2013**, *2013*, 1649–1652. [CrossRef]
118. Eto, H.; Eguchi, C.; Kagawa, T. Production of L-Tryptophan from γ-Glutamaldehydic Acid. *Bull. Chem. Soc. Jpn.* **1989**, *62*, 961–963. [CrossRef]
119. Baran, P.S.; Hafensteiner, B.D.; Ambhaikar, N.B.; Guerrero, C.A.; Gallagher, J.D. Enantioselective Total Synthesis of Avrainvillamide and the Stephacidins. *J. Am. Chem. Soc.* **2006**, *128*, 8678–8693. [CrossRef]
120. Ma, J.; Yin, W.; Zhou, H.; Liao, X.; Cook, J.M. General Approach to the Total Synthesis of 9-Methoxy-Substituted Indole Alkaloids: Synthesis of Mitragynine, as well as 9-Methoxygeissoschizol and 9-Methoxy-N(b)-Methylgeissoschizol. *J. Org. Chem.* **2009**, *74*, 264–273. [CrossRef]
121. Blaser, G.; Sanderson, J.M.; Batsanov, A.S.; Howard, J.A.K. The Facile Synthesis of a Series of Tryptophan Derivatives. *Tetrahedron Lett.* **2008**, *49*, 2795–2798. [CrossRef]
122. Hashimoto, M.; Hatanaka, Y.; Nabeta, K. Effective Synthesis of a Carbon-linked Diazirinyl Fatty Acid Derivative via Reduction of the Carbonyl Group to Methylene with Triethylsilane and Trifluoroacetic Acid. *Heterocycles* **2003**, *59*, 395–398. [CrossRef]

123. Collot, V.; Schmitt, M.; Marwah, P.; Bourguignon, J.-J. Regiospecific Functionalization of Indole-2-carboxylates and Diastereoselective Preparation of the Corresponding Indolines. *Heterocycles* **1999**, *51*, 2823–2847. [CrossRef]
124. Leonard, N.J.; Greenfield, J.C.; Schmitz, R.Y.; Skoog, F. Photoaffinity-labeled Auxins: Synthesis and Biological Activity. *Plant Physiol.* **1975**, *55*, 1057–1061. [CrossRef] [PubMed]

Disclaimer/Publisher's Note: The statements, opinions and data contained in all publications are solely those of the individual author(s) and contributor(s) and not of MDPI and/or the editor(s). MDPI and/or the editor(s) disclaim responsibility for any injury to people or property resulting from any ideas, methods, instructions or products referred to in the content.

Review

Application of Olefin Metathesis in the Synthesis of Carbo- and Heteroaromatic Compounds—Recent Advances

Szymon Rogalski and Cezary Pietraszuk *

Faculty of Chemistry, Adam Mickiewicz University, Poznań, Uniwersytetu Poznańskiego 8, 61-614 Poznań, Poland
* Correspondence: cezary.pietraszuk@amu.edu.pl

Abstract: The olefin metathesis reaction has found numerous applications in organic synthesis. This is due to a number of advantages, such as the tolerance of most functional groups and sterically demanding olefins. This article reviews recent advances in the application of the metathesis reaction, particularly the metathetic cyclization of dienes and enynes, in synthesis protocols leading to (hetero)aromatic compounds.

Keywords: metathesis; heteroaromatics; carboaromatics

1. Introduction

Olefin metathesis is currently an invaluable tool for advanced organic and polymer synthesis [1–3]. The reaction is frequently used as a step in a sequence of processes leading to complex multifunctional molecules, such as natural products, and biologically active compounds [4]. The great importance of (hetero)arenes in nature, scientific research and the chemical industry does not require justification. Aromatic compounds are essential for the industry and vital to biochemistry [5,6]. More recent applications of aromatics include the synthesis of functional materials. The greatest interest among research groups in using various types of metathetic transformations of olefins, dienes, and enynes in organic synthesis, including the synthesis of aromatic compounds, followed by an explosion of development in the design and synthesis of well-defined alkylidene complexes of molybdenum and ruthenium catalysts of olefin metathesis. The application of olefin metathesis in the synthesis of carbo- and heteroaromatic compounds has been described in numerous reviews. The most comprehensive overview of the literature up to 2009 was provided by Otterlo and Koning [7]. The state-of-the-art of synthetic routes to carbocyclic aromatic compounds via ring-closing alkene and ene metathesis up to 2014 was reviewed by de Koning and van Otterlo [8]. In 2013, a monograph *Transition-Metal-Mediated Aromatic Ring Construction* was published with a chapter by Yoshida on olefin metathesis [9]. The work of Donohoe et al. [10] is an earlier review in this area. Race and Bower [11] described the state-of-the-art in the synthesis of heteroaromatic compounds by alkene and enyne metathesis up to and including 2014. An earlier review describing the possibilities offered by RCM in the synthesis of aromatic heterocycles was written by Donohoe, Fishlock, and Procopiou [12]. Moreover, in 2016, Potukuchi, Colomer, and Donohoe reviewed the synthesis of heteroaromatic compounds, with a focus on the achievements of their own group [13]. Reviews describing the synthesis of specific groups of compounds using metathetic transformations are indicated in the relevant sections of this review. The aim of this review is to highlight new developments in the use of olefin metathesis in the synthesis of aromatic compounds that were reported between the years 2015 and 2022. Procedures involving the de novo formation of aromatic rings, rather than the modification of aromatic starting compounds, are described.

Citation: Rogalski, S.; Pietraszuk, C. Application of Olefin Metathesis in the Synthesis of Carbo- and Heteroaromatic Compounds—Recent Advances. *Molecules* **2023**, *28*, 1680. https://doi.org/10.3390/molecules28041680

Academic Editor: Joseph Sloop

Received: 29 December 2022
Revised: 4 February 2023
Accepted: 8 February 2023
Published: 9 February 2023

Copyright: © 2023 by the authors. Licensee MDPI, Basel, Switzerland. This article is an open access article distributed under the terms and conditions of the Creative Commons Attribution (CC BY) license (https://creativecommons.org/licenses/by/4.0/).

2. Olefin Metathesis

Olefin metathesis is a catalytic transformation of olefins which involves the exchange of double bonds between carbon atoms. The reaction is catalyzed by alkylidene complexes, and its mechanism consists of a sequence of cycloaddition and productive cycloreversion steps proceeding via metallacyclobutane intermediates (Scheme 1) [14].

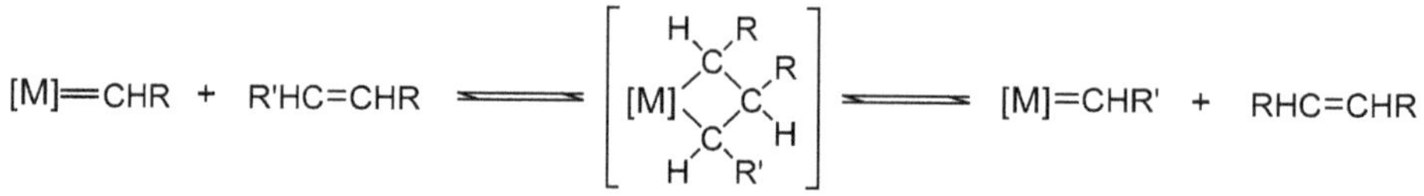

$$[M]=CHR + R'HC=CHR \rightleftharpoons \left[[M]\underset{H\ R'}{\overset{H\ R}{\underset{|}{\overset{|}{C}}\underset{C}{\overset{C}{\underset{|}{\overset{|}{C}}}}}} \right] \rightleftharpoons [M]=CHR' + RHC=CHR$$

Scheme 1. Metallacarbene mechanism of olefin metathesis.

Of the many types of olefin metathesis, the most useful in organic synthesis are the cross-metathesis of olefins (CM) and the metathetic cyclization of dienes (RCM) (Schemes 2 and 3).

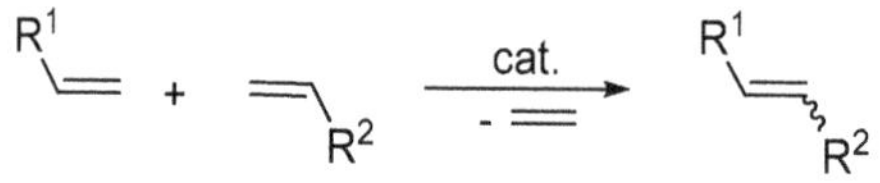

Scheme 2. Olefin cross-metathesis (CM).

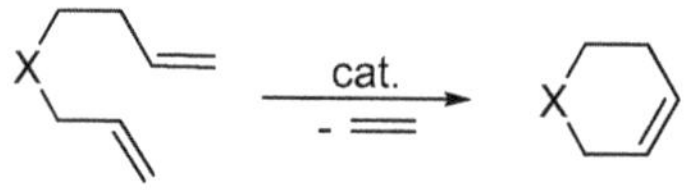

Scheme 3. Ring-closing olefin metathesis (RCM).

The mechanism of olefin metathesis also allows for transformations involving triple bonds, i.e., ene-yne cross-metathesis (EYCM) and enyne ring-closing metathesis (RCEM) (Schemes 4 and 5, respectively).

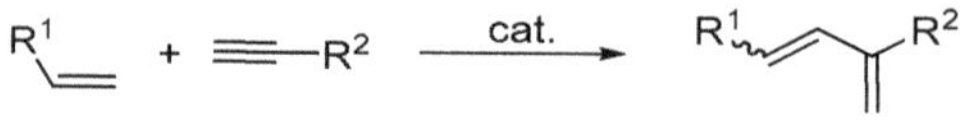

Scheme 4. Ene-yne cross-metathesis (EYCM).

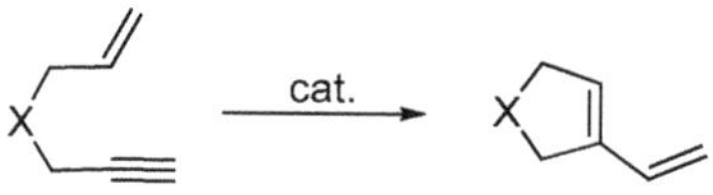

Scheme 5. Ring-closing enyne metathesis (RCEM).

Metathetic transformations of the triple bond can lead to a mixture of regio- and stereoisomers [15]. Fortunately, in the case of RCEM, the synthesis of small and medium rings in the presence of ruthenium catalysts selectively leads to the formation of an exocyclic product, as shown in Scheme 5 [16]. RCEM should not be confused with skeletal reorganization, which is catalyzed by late-transition metals, such as Pd, Pt, Ru, Au, Ir, and Rh, and can lead to similar products but does not follow the metallacarbene mechanism [17,18]. EYCM may also be used for the synthesis of 1,3-dienes. Although the reaction can lead to the formation of stereo- and regioisomers, a number of examples of the selective course of the reaction have been described [19–21]. Cross-metathesis between a 1-alkene and a 1-alkyne leads to 1,3-substituted 1,3-dienes, as shown in Scheme 4. Moreover, the CM of internal alkynes with ethene selectively leads to the formation of 2,3-substituted 1,3-butadienes. Metathetic transformations involving triple bonds are characterized by total atom efficiency.

The olefin metathesis has become an important tool in the synthesis of organic compounds. This is a consequence of the dynamic progress in the synthesis and design of well-defined catalysts such as the alkylidene complexes of tungsten, molybdenum and ruthenium. Especially the family of ruthenium-based catalysts tolerant of normal organic and polymer processing conditions and preserving their catalytic properties in water and in the presence of the majority of functional groups has enabled a great number of applications in organic (and polymer) synthesis (see [22], for example). Tungsten and molybdenum complexes, including the most commonly used, commercially available, highly active molybdenum complex, [(NAr){OC(CH$_3$)(CF$_3$)$_2$}$_2$M=CH(2,6-i-Pr$_2$C$_6$H$_3$)] (Mo-1) [23], suffer from the high oxophilicity of the metal centers, making them extremely sensitive to oxygen, moisture, and numerous functional groups [22]. Although many well-defined ruthenium olefin metathesis catalysts have been described, most of the literature procedures use only a few commercially available complexes (Ru-1–Ru-3, Figure 1).

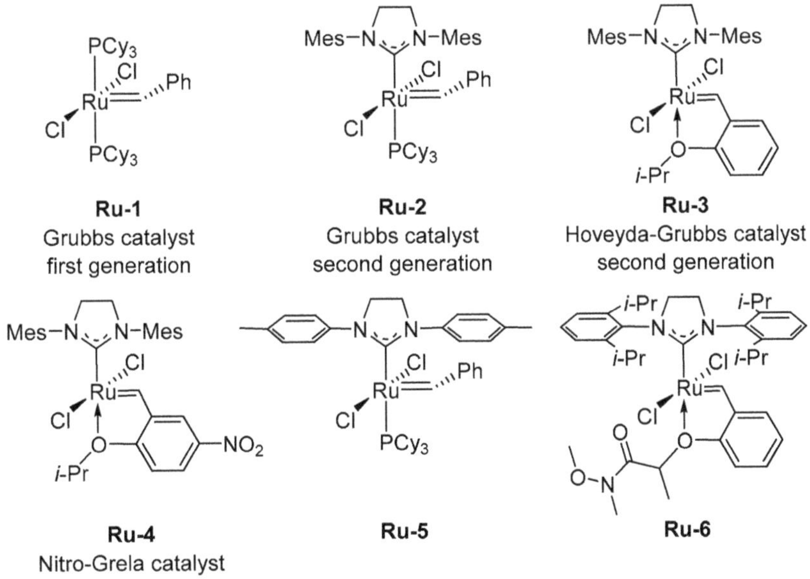

Ru-1
Grubbs catalyst
first generation

Ru-2
Grubbs catalyst
second generation

Ru-3
Hoveyda-Grubbs catalyst
second generation

Ru-4
Nitro-Grela catalyst

Ru-5

Ru-6

Figure 1. Ruthenium-based olefin metathesis catalysts referred to in this review.

Olefin metathesis has gained a number of applications in advanced organic synthesis, chemical biology, medicinal chemistry, the synthesis of simple molecules, and the total synthesis of natural products [1–4].

3. Synthesis of Aromatic Heterocycles

Olefin metathesis, especially ring-closing metathesis (RCM), has been extensively used in the synthesis of heteroaromatic compounds of divergent types. It has been exploited efficiently as a key step for the preparation of the pyrroles, furans, pyridines, imidazoles, and related systems. All of these compounds are largely important as structural units in natural products and are used as synthons in further transformations. In this section, different approaches to the synthesis of heteroaromatic compounds using olefin metathesis as a key step are reviewed. The syntheses of heteroaromatics involving olefin metathesis as one of the key steps has been summarized in several reviews [24–27].

This section presents various strategies for the synthesis of heteroaromatic compounds by using olefin metathesis as a key step. Knowledge on the synthesis of heteroaromatic compounds involving olefin metathesis as one of the key steps has been summarized in several review papers [24–27]. In the synthesis of (hetero)aromatic compounds, the metathesis step is used in a sequence of reactions involving the formation of a heterocyclic ring by RCM of a diene or enyne and its subsequent aromatization.

3.1. Synthesis of Five-Membered Rings

Since pyrrole and furan synthons are commonly present in many pharmaceuticals and biologically active molecules, the efficient synthesis of these compounds continues to be of interest to organic chemists. In 2015, Castagnolo and co-workers described an elegant methodology that represents the first example of one-pot synthesis of substituted pyrroles via enyne cross-metathesis (CM) cyclization reaction, using propargylamines and ethyl vinyl ether as reagents. The described CM/cyclization protocol is performed in the presence of Ru-2, is microwave-assisted, and offers a convenient approach to the synthetically challenging 1,2,3-substituted pyrroles (Scheme 6) [28].

Scheme 6. Synthesis of 1,2,3-substituted pyrroles.

In 2015, Spring reported the development of a new strategy for the synthesis of biologically interesting indolizin-5(3H)-ones [29]. Performed experiments showed that indolizin-5(3H)-ones are a relatively unstable class of compounds and have a tendency to tautomerize to indolizin-5-ols. This observation was further exploited to prepare other useful compounds based on 6,5-azabicyclic scaffolds, which are difficult to obtain with the use of typical methods. The authors developed a procedure based on the ring-closing metathesis reaction effectively applied to construct the azabicyclic heteroaromatic ring system present in indolizin-5-ols (Scheme 7).

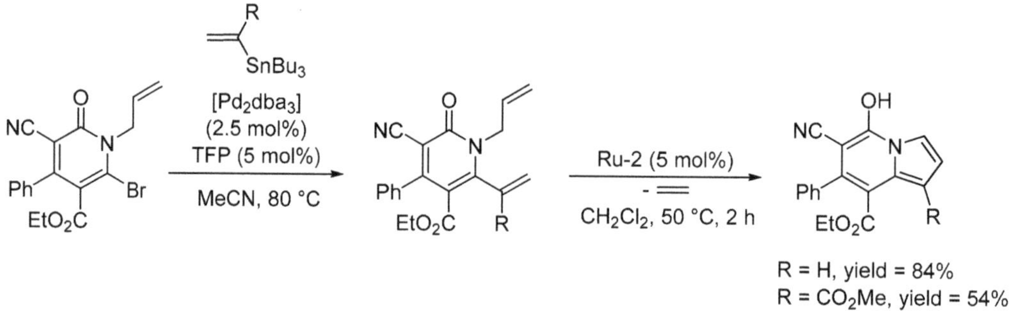

Scheme 7. Synthesis of indolizin-5-ol derivatives.

The proposed strategy represents an unprecedented approach toward this type of scaffold. Several heteroaromatic derivatives were obtained with good-to-high yields (50–84%). The procedure is step-efficient, and a number of novel analogues could be readily accessed through the adaptation of the substituents around the heterocyclic core in the starting pyridine molecule.

Obtained indolizin-5(3H)-ones and indolizin-5-ols are relatively rare; therefore, they can potentially be of interest as new synthons in drug and agrochemical synthesis.

Another route to pyrroles substituted in the β-position was developed by Samec and co-workers, using four high-yielding catalytic steps in the presence of Pd, Ru, and Fe catalysts [30]. The authors described a Pd-catalyzed effective method for the synthesis of unsymmetrical diallylated aromatic amines. In the final key step of the procedure, Ru-Catalyzed RCM, followed by aromatization in the presence of Fe complexes, was employed (Scheme 8).

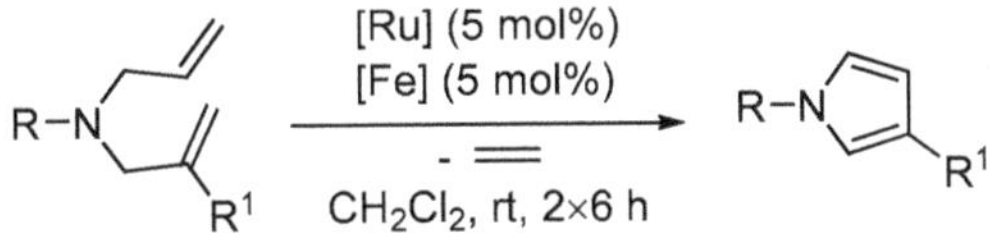

R = Ph, Bn, Cy, C_6H_4-p-F, C_6H_4-p-Cl, C_6H_4-p-Br, C_6H_4-p-OMe
R^1 = n-Pen, i-Pr, Bn, Ph, Np, C_6H_4-p-F, C_6H_4-p-Cl, C_6H_4-p-OMe, C_6H_4-p-Me

Scheme 8. Synthesis of substituted pyrroles via Ru-catalyzed RCM and Fe-catalyzed aromatization.

The method is atom-efficient, with the formation of water and ethene as side-products. The procedure is general and gives pyrroles substituted in the β-position with alkyl, benzyl, or aryl groups with good overall yields.

An efficient and selective approach for the synthesis of polyfunctionalized 3-fluoropyrroles starting from commercial aldehydes was described by Marquez and co-workers [31]. The key step of the described methodology is RCM of diene in the presence of Ru-2, followed by an aromatization process based on the alkylation of the obtained fluorolactams with methyllithium. The procedure proceeded smoothly to generate the desired pyrrole units in excellent yield (73–92%). The authors proposed that the aromatization process in the presence of MeLi proceeded through the formation of a hemiaminal, followed by the elimination of water and the isomerization of double bonds (Scheme 9).

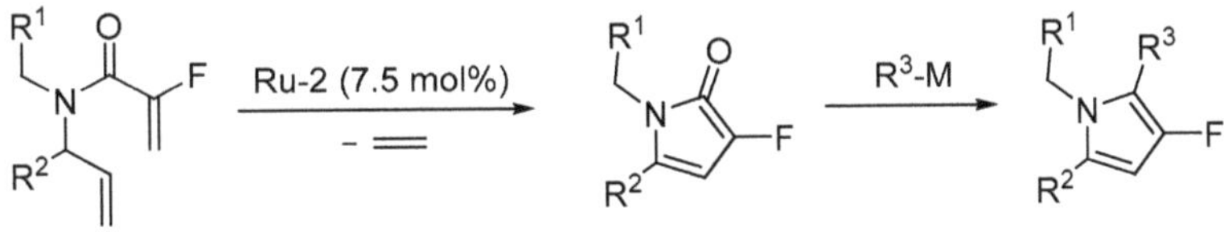

R^1 = Ph, Cy, C_6H_4-p-Br, pyrrole
R^2 = Cy, Bz, C_6H_4-p-Br, C_6H_4-p-OMe, C_6H_4-p-CF_3
R^3 = Me, Ph, n-Bu, allyl

Scheme 9. Synthesis of fluorinated polysubstituted pyrroles.

N-sulfonyl pyrroles were synthesized via the combination of ring-closing, or enyne metathesis, with oxidation [32]. A range of different N-substituted diallyl amines were subjected to RCM with Grubbs catalyst. After the first metathetical step, heterogeneous MnO_2 was used as an effective oxidant (Scheme 10). Reasonable-to-good yields were obtained for a variety of substituted amines even when the process was performed in one-pot procedure.

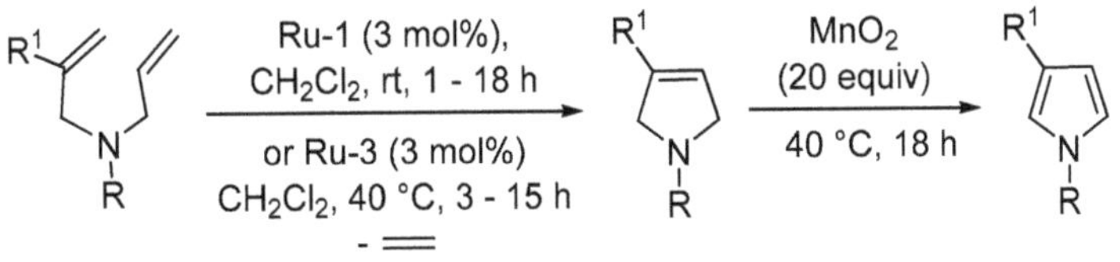

Scheme 10. One-pot RCM-oxidation for the synthesis of substituted pyrroles.

Lamaty and co-workers described an easily scalable pyrrole synthesis strategy involving RCM in the first step and subsequent aromatization by base-induced nitrogen deprotection (Scheme 11) [33].

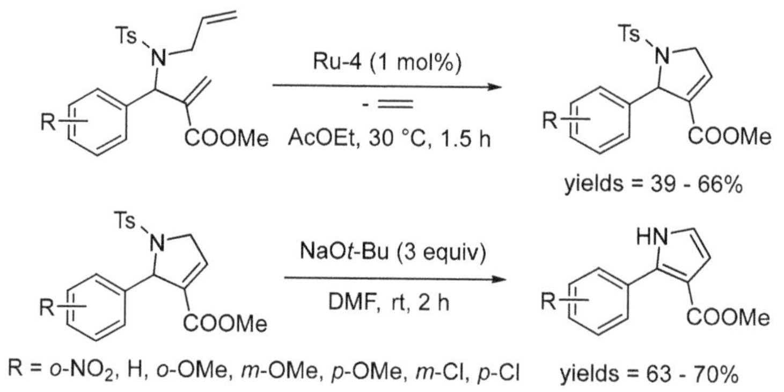

Scheme 11. RCM/deprotection–aromatization sequence for the preparation of pyrroles.

The authors developed the procedure of synthesis of 2-aryl-1H-pyrrole-3-carboxylates, using ring-closing metathesis of the corresponding β-amino esters as a key step. In a two-step procedure, the described methodology allowed for an efficient formation of 2-aryl substituted pyrroles, especially unprecedented 2-aryl-1H-pyrrole-3-carboxylates. The products were obtained in good yields, ranging from 63 to 70%, in the presence of the relatively low loading (1 mol%) of Ru-4 (Figure 1) in green solvent, such as ethyl acetate, except for DCM.

In 2017, the Castagnolo group reported the first chemoenzymatic cascade process for the sustainable preparation of substituted pyrroles in which the ring-closing metathesis reaction was applied for the preparation of 3-pyrrolines [34]. The protocol used, for the first time, the unexplored aromatizing activity of monoamine oxidase enzymes (MAO-N and 6-HDNO), which are able to convert a wide range of N-aryl- and N-alkyl-3-pyrrolines into pyrroles under mild conditions and in high yields (Scheme 12). The authors showed that MAO-N can work in combination with Ru-2, leading to the formation of a series of substituted pyrroles derived from diallylamines, as well as anilines, in a one-pot metathesis−aromatization sequence.

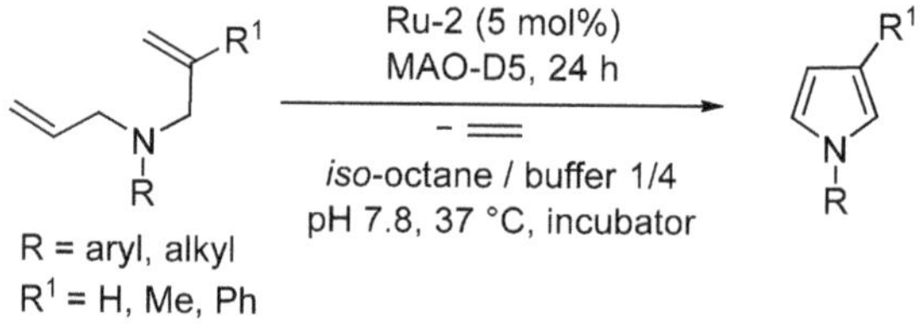

Scheme 12. Chemoenzymatic synthesis of pyrroles.

Kotha and co-workers demonstrated three divergent synthetic strategies to construct pyrrole-based C_3-symmetric molecule [35]. One of the synthetic protocols involved ring-closing metathesis as a key step in the presence of Ru-1. The star-shaped pyrrole derivative with C_3-symmetry was prepared in good yield (84%), without the involvement of additional reagents, in one-pot RCM and a subsequent aromatization protocol, starting from hexa-allyl derivative. The hexa-allyl substituted compound could be synthesized from the corresponding di-allyl ketone by trimerization, starting from the readily available 4-aminoacetophenone (Scheme 13).

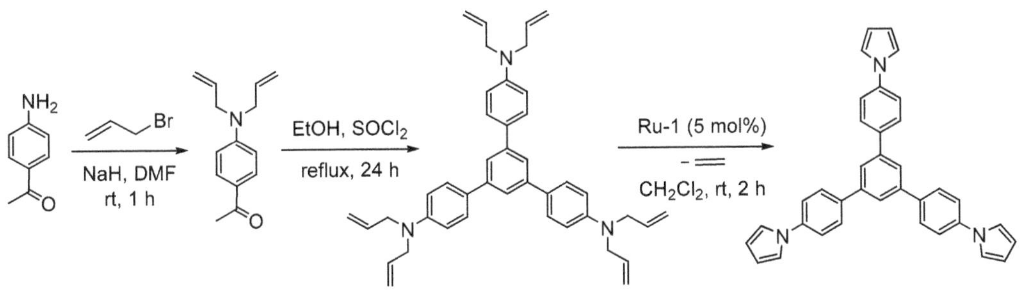

Scheme 13. Synthesis of star-shaped pyrrole derivative via RCM key step.

The described facile RCM/dehydrogenation (aromatization) sequence was preceded by the non-metathetic behavior of the Grubbs catalyst previously reported by the authors. An earlier approach to pyrroles involving RCM required a separate step for aromatization. Thus, the approach reported by Kotha, which does not require additional aromatization step, is more sustainable.

Wu and Li proposed another convenient method for preparation of N-sulfonyl- and N-acylpyrroles via the olefin ring-closing metathesis of diallylamines in the presence of Ru-2 combined with in situ oxidative aromatization with atmospheric O_2 and suitable copper catalyst such as $Cu(OTf)_2$ or $CuBr_2$ (Scheme 14) [36]. The reaction was performed via a one-pot tandem RCM/dehydrogenation procedure and afforded N-sulfonyl- and N-acylpyrroles with moderate-to-good yields (45–93%).

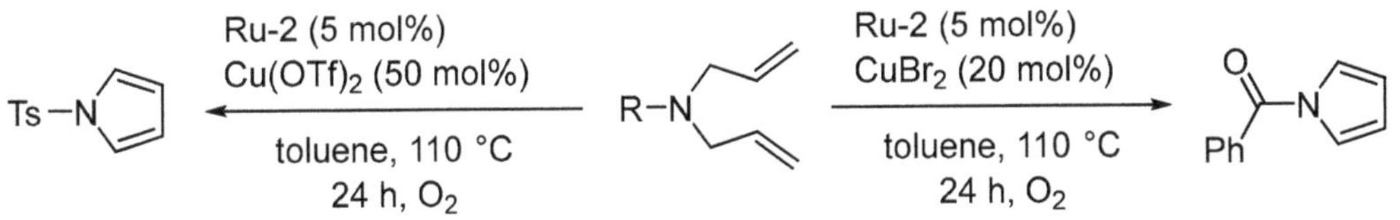

Scheme 14. RCM/dehydrogenation sequence of sulfonyl- and acyldiallylamines.

Arisawa and co-workers synthesized 5-methylisoindolo[2,1-a]quinoline derivatives as novel near-infrared absorption dyes [37]. The authors applied a previously developed one-pot ring-closing metathesis (RCM)/oxidation/1,3-dipolar cycloaddition cascade protocol to prepare various isoindolo[2,1-a]quinoline derivatives from substituted N-allyl-N-benzylanilines (Scheme 15 and Figure 2). They proposed that the key intermediate in this reaction is the azomethine ylide derived from 1,2-dihydroquinoline.

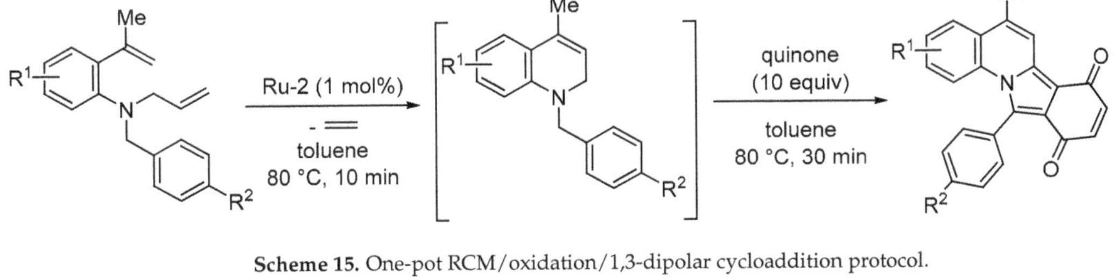

Scheme 15. One-pot RCM/oxidation/1,3-dipolar cycloaddition protocol.

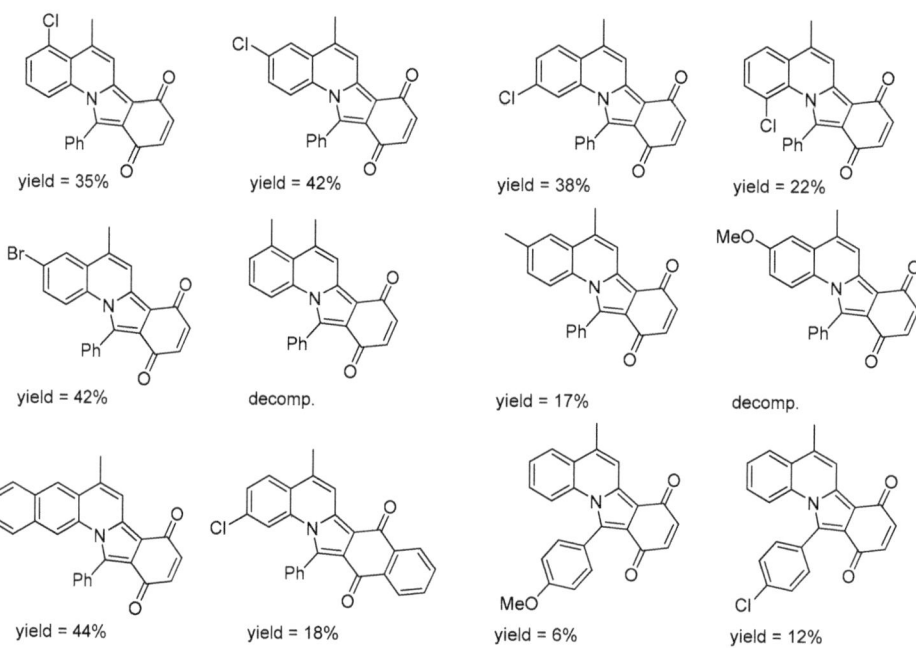

Figure 2. Scope of products obtained by one-pot RCM/oxidation/1,3-dipolar cycloaddition protocol.

The first RCM step was performed effectively in the presence of Ru-2 during 10 min, with relatively low catalyst loading (1 mol%) giving the intermediates nearly quantitative yields. Unfortunately, the electronic effects of substituents on the aromatic ring were responsible for the low substrate reactivity during the 1,3-dipolar cycloaddition step. It was observed that substrates bearing an electron-withdrawing group in the aromatic ring afforded higher yields in the 1,3-dipolar cycloaddition, thus giving the target molecules overall yields ranging from 22 to 55%.

In order to understand a structure and absorption−wavelength relationships, the authors performed calculations by using a time-dependent density functional theory (TD-DFT).

Whiting and Carboni showed a different approach to the synthesis of pyrroles [38]. In the first step, they performed ring-closing enyne metathesis (RCEM), followed by cross-metathesis (CM) with vinylboronic esters to form a large spectrum of cyclic dienyl boronic esters (Scheme 16).

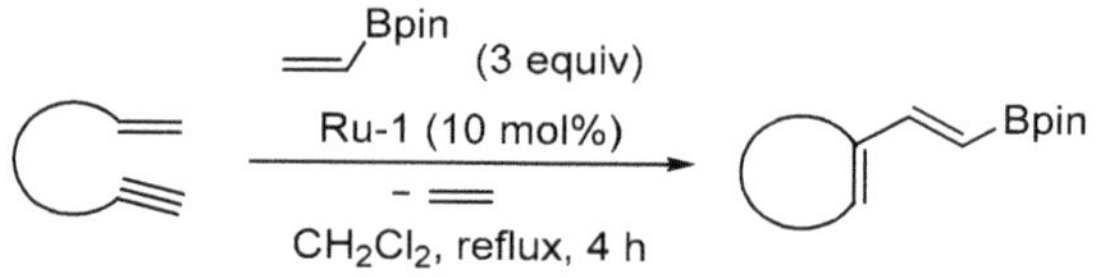

Scheme 16. Synthesis of cyclic dienyl boronic esters via an RCEM/CM sequence.

Second step involved a nitroso-Diels–Alder reaction between a series of previously obtained dienyl boronic esters and nitrosoarenes with the formation of a heteroaromatic ring (Scheme 17).

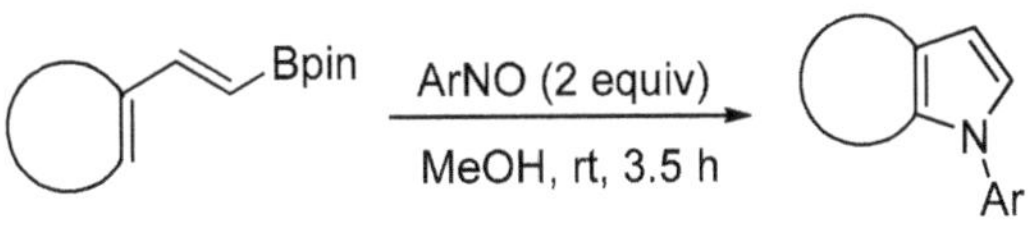

Scheme 17. Synthesis of fused pyrroles from cyclic dienyl boronic esters.

The authors developed a new and efficient route to a series of cyclic 1,3-dienyl boronic esters via diene or enyne metathesis that were further transformed with good yields to fused pyrroles, using a one-pot nitroso-Diels−Alder/ring contraction sequence.

The unprecedented cascade procedure based on the Grubbs-catalyzed ring-closing metathesis of diallyl ethers, followed by laccase/TEMPO-catalyzed aromatization, was successfully developed by Castagnolo (Scheme 18) [39]. The catalytic activity of the Trametes versicolor laccase/TEMPO system was demonstrated for the first time in the aromatization of 2,5-dihydrofurans to the corresponding furans. The synthesis of furans from diallylethers was performed in moderate-to-high yields (31–76%), using a two-step RCM process in the presence of Ru-2 and subsequent laccase/TEMPO-catalyzed chemo-enzymatic aromatization.

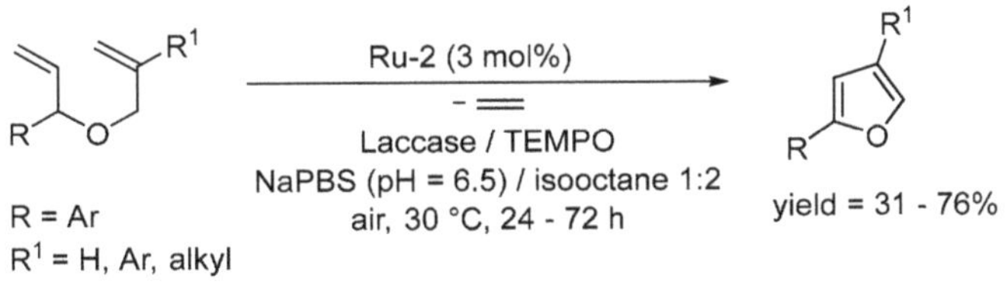

Scheme 18. One-pot chemoenzymatic cascade for the synthesis of furans from diallyl ethers.

Kotha reported a new synthetic strategy leading to C_3-symmetric star-shaped phenyl and triazine central cores bearing oxepine and benzofuran ring systems [40]. The authors exploited a three-step procedure to construct C_3-symmetric molecules, starting with the commercially available *p*-hydroxyacetophenone and *p*-hydroxybenzonitrile, through cyclotrimerization, double-bond isomerization, and ring-closing metathesis sequence as key steps. A series of star-shaped allyl-ether derivatives were obtained through the sequence: cyclotrimerization, Claisen rearrangement and allylation. Then allyl substituted ethers were subjected to Ru-hydride-catalyzed double-bond isomerization performed in toluene at 80 °C. The reaction resulted in the formation of a double-bond isomerized product with 91% yield as a mixture of diastereomers. The obtained compounds were subjected to metathesis with a second-generation Grubbs catalyst in methylene chloride at 40 °C to give C_3-symmetric star-shaped phenyl and triazine core derivatives bearing benzofuran moieties with yields of the metathesis step reaching 80% (Scheme 19).

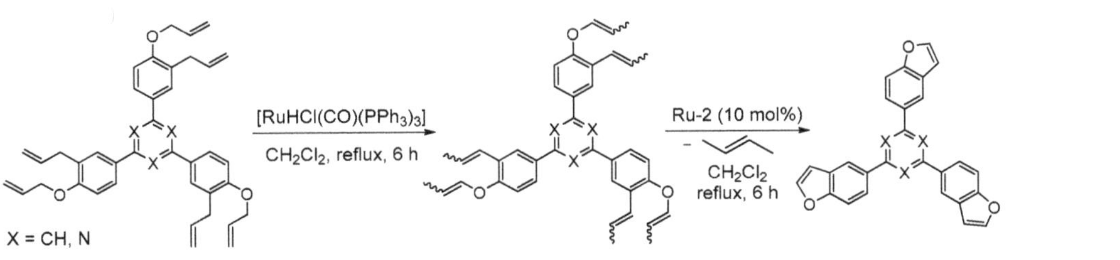

Scheme 19. Synthesis of star-shaped phenyl or triazine central core benzofuran derivatives via isomerization and RCM sequence.

All the compounds obtained therein showed moderate-to-strong fluorescence, with a strong emission band around 325–350 nm. The described methodology is the first successful attempt to introduce oxepine ring systems into C_3-symmetric star-shaped derivatives of a phenyl or triazine core by means of an olefin metathesis reaction.

A multistep strategy involving the synthesis of benzofurans, 2H-chromenes, and benzoxepines from various phenols was recently proposed by Kotha (Scheme 20) [41]. The described procedure employed Claisen rearrangement and followed by RCM as key steps. At first, a number of substituted allyl aryl ethers were synthesized by alkylation of corresponding phenols with allyl bromide in the presence of potassium carbonate. The resulting allyl aryl ethers were then subjected to a Claisen rearrangement in the absence of a solvent to produce allylphenols, which were then converted to diallyl compounds by O-allylation in 97–98% yield.

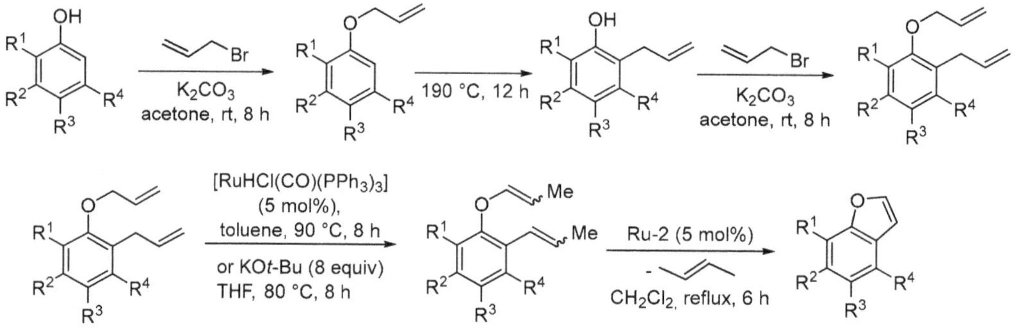

Scheme 20. Synthesis of benzofurans from variously substituted phenols.

The authors then carried out a standard isomerization reaction catalyzed by [RuHCl(CO)(PPh$_3$)$_3$] and investigated an alternative method of double-bond migration, using potassium tert-butoxide (KOt-Bu) in THF. Isomerization products were obtained as a mixture of isomers in 89–94% yields. The final step in the construction of the heteroaromatic ring involved an RCM step, providing the corresponding benzofurans, with a yield of 84–85%. Using the proposed procedure, a wide range of products differing in the substituents on the 6-membered ring was obtained. In addition, naturally occurring benzofurans, such as 7-methoxywutaifuranate, were also synthesized with good yields (80%), using the described protocol (Scheme 21).

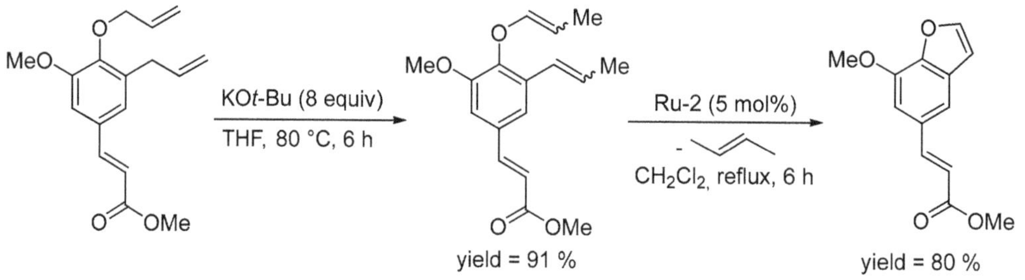

Scheme 21. Synthesis of 7-methoxywutaifuranate.

Another interesting example of a multistep synthesis of 7-methoxywutaifural using olefin metathesis was recently reported by Schmidt and co-workers [42]. The authors showed that isoeugenol can undergo efficient cross-metathesis with acrolein and crotonaldehyde in the presence of Hoveyda–Grubbs catalyst (Ru-3). This observation was used in the multistep synthesis of the natural product-7-methoxywutaifuranal (Scheme 22).

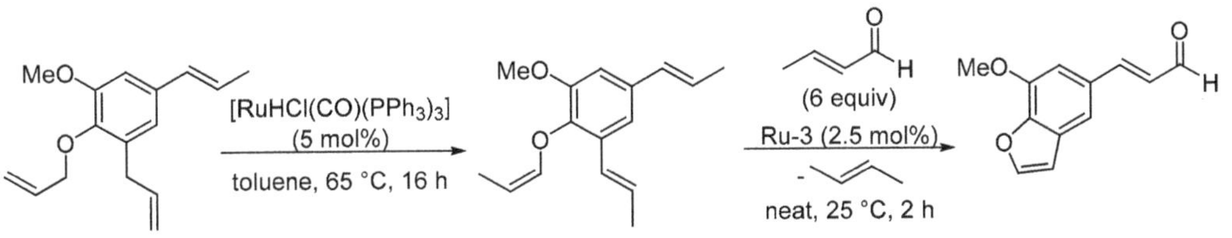

Scheme 22. Key isomerization and metathesis steps in the synthesis of 7-methoxywutaifuranal.

3.2. Synthesis of Six-Membered Rings

Six-membered ring heteroaromatics, such as pyridines and pyridazines, are useful scaffolds that are present in various drugs and natural products. This section reviews strategies for the synthesis of six-membered heteroaromatics that use RCM as a key step. Progress in the synthesis of azonic aromatic heterocycles was summarized by Vaquero [43].

In 2015, Cuadro and Vaquero reported new approach to synthesize the 1-azaquinolizinium (pyrido[1,2-*a*]pyrimidin-5-ium) and its derivatives based on a ring-closing metathesis reaction of pyridinium azadienes in the presence of the second-generation Grubbs catalyst (Ru-2) or Hoveyda–Grubbs catalyst (Ru-3) (Scheme 23) [44]. The described strategy was also successfully applied to the unprecedented synthesis of benzo-1-azaquinolizinium (pyrimido[2,1-*a*]isoquinolinium) cation by RCM (Scheme 24). The proposed synthetic route involved the formation of the new C3–C4 bond to give the pyrimidinium ring. The authors prepared protected pyridinium azadienes derivative with the use of protection/allylation/dehydrohalogenation cascade, starting from commercially available 2-aminopyridine. Then obtained diene was subjected to RCM, affording the expected product. The subsequent deprotection of the RCM product gave the corresponding 1,2-dihydro-1-azaquinolizinium triflate, which was further dehydrogenated by heating at 200 °C in the presence of Pd/C (10%) without solvent to give the 1-azaquinolizinium triflate.

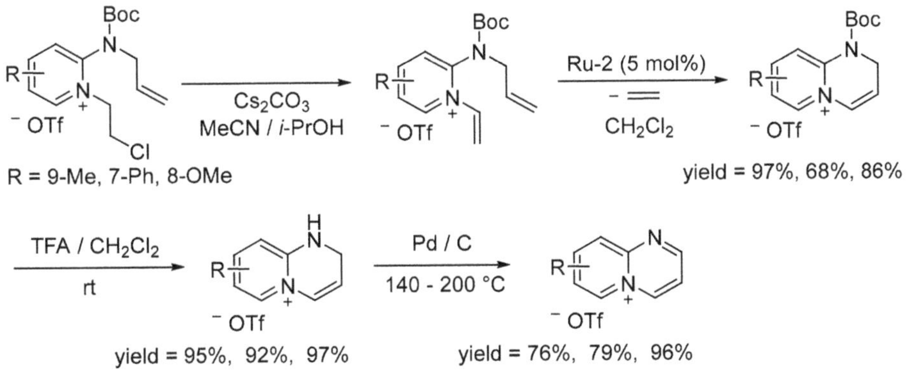

Scheme 23. Application of RCM for the construction of the pyrimidinium ring.

A simple three-step synthesis of substituted pyridines based on an alkylation/ring-closing metathesis/aromatization sequence was described by Opatz and colleagues [45]. The synthetic strategy involved the preparation of diallylic α-aminonitrile in a two-step approach. At first Strecker reaction starting from benzaldehyde, allylamine, and potassium cyanide was performed to afford 2-(allylamino)-2-phenylacetonitrile. Then deprotonation with the use of KHMDS, followed by alkylation with allyl bromide, was applied to give corresponding diallyl-substituted α-aminonitrile. Finally, the RCM of unprotected allylic amine and subsequent spontaneous dehydrocyanation and oxidation were employed as the key steps of the developed sequence (Scheme 25). The desired substituted pyridines were obtained with moderate yields ranging from 24 to 48%.

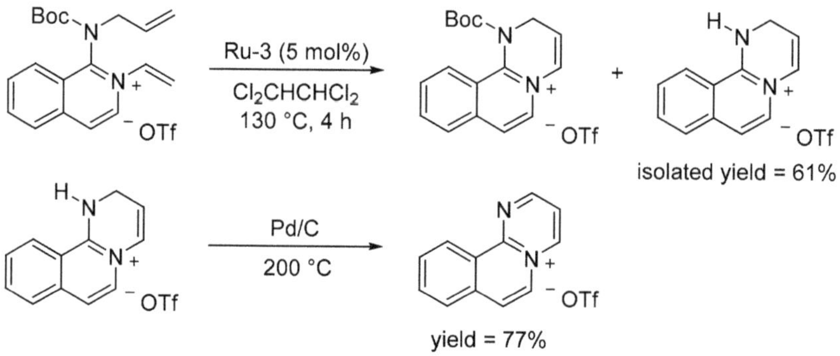

Scheme 24. Synthesis of pyrimido[2,1-a]isoquinolinium, using RCM/aromatization sequence.

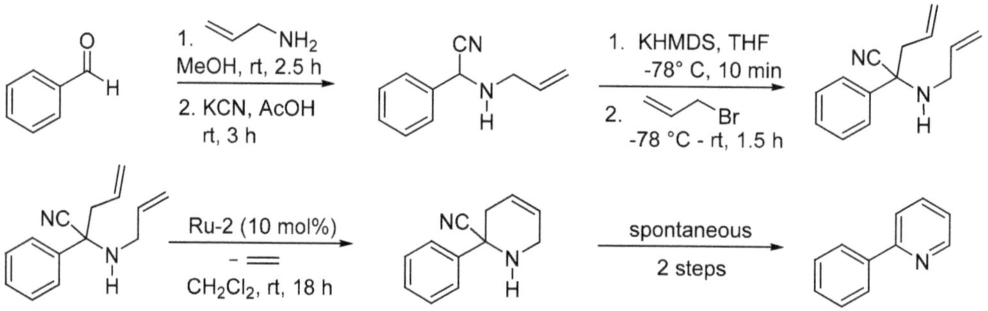

Scheme 25. Synthesis of phenylpyridines, using RCM as a key step.

Suresh studied [2+2+2] cyclotrimerization of diynes and nitriles leading to efficient synthesis of pyridine and its derivatives by DFT calculations. It was shown that two mechanisms of the reaction are plausible—metallacarbene mechanism of metathesis and non-metallacarbene route. The strategy can be potentially used in the total synthesis of natural products [46].

Dash and co-workers reported the addition of Grignard reagent to oxindoles. This previously unexplored strategy represented a regiospecific approach to 2- and 2,3-disubstituted indole derivatives obtained via a one-pot aromatization driven dehydration pathway [47]. The described method allowed for the convenient preparation of diallylindoles, which can be used as ring-closing-metathesis (RCM) precursors for the orthogonal synthesis of pyridyl[1,2-a]indoles (Scheme 26).

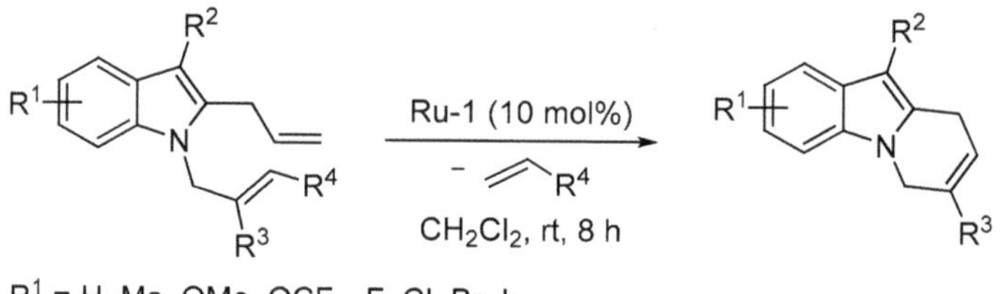

R^1 = H, Me, OMe, OCF$_3$, F, Cl, Br, I;
R^2 = Me, Et, Bn, Ph, p-OMeC$_6$H$_4$, indolyl
R^3 = Me, Ph; R^4 = H, Me, Ph

Scheme 26. RCM of 1,2-diallylindoles.

The authors synthesized a variety of N-allyl 2-oxindoles and effectively transformed them to N-substituted 2-allyl indole derivatives by treatment with allylmagnesium bromide in yields up to 86%. Then the diallylindoles were used for the orthogonal synthesis of conjugated nitrogen-containing heterocycles, using RCM. In this approach, a series of 1,2-diallyl indolyl derivatives were subjected to metathesis cyclization and yielded the desired substituted dihydropyridylindoles in good yields (72–86%). Finally, dihydropyridoindoles were effectively transformed into aromatic derivatives via a DDQ-mediated oxidative aromatization protocol (Scheme 27 and Figure 3).

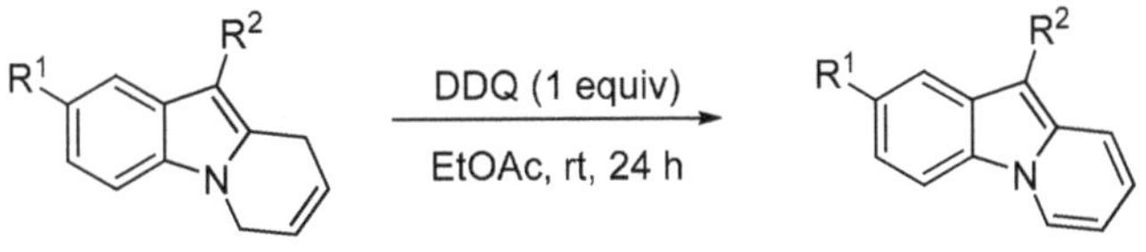

Scheme 27. DDQ-mediated aromatization of dihydropyridoindoles.

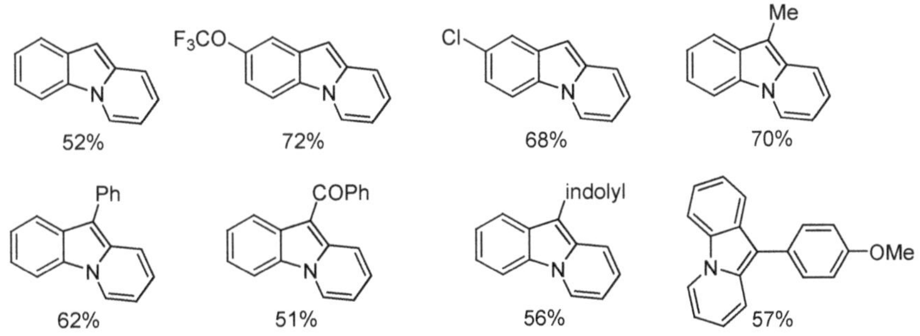

Figure 3. Scope of pyridoindoles synthesized via RCM/aromatization sequence.

The optimized reaction sequence was employed in the synthesis of dimeric pyridoindoles (Scheme 28)

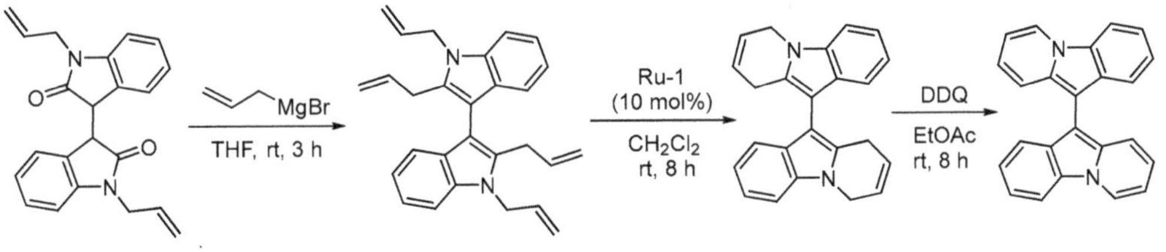

Scheme 28. Synthesis of dimeric pyridoindoles.

4. Synthesis of Aromatic Carbocycles

Dienes or enynes RCM can lead directly to the formation of new aromatic ring systems. Such a strategy has been used in the synthesis of phenanthrenes (Scheme 29) [48,49].

Metathesis cyclization, however, does not usually lead to the formation of an aromatic ring. Thus, the synthesis of (hetero)aromatic compounds uses a metathesis step in a sequence of reactions involving the formation of a carbo- or heterocyclic ring and its subsequent aromatization. The sequence using the diene RCM step is shown in Scheme 30.

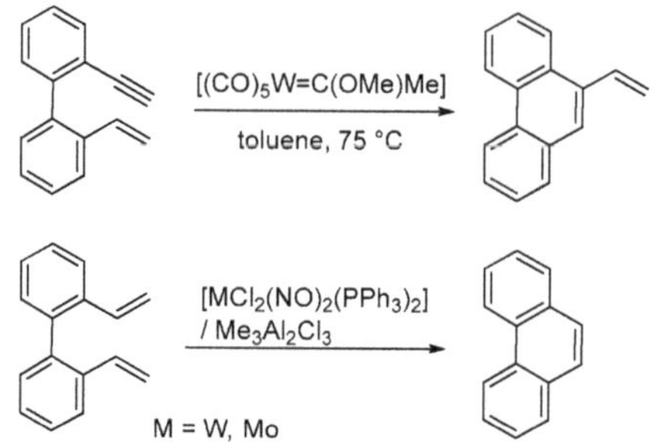

Scheme 29. Direct formation of aromatic compounds via RCEM or RCM.

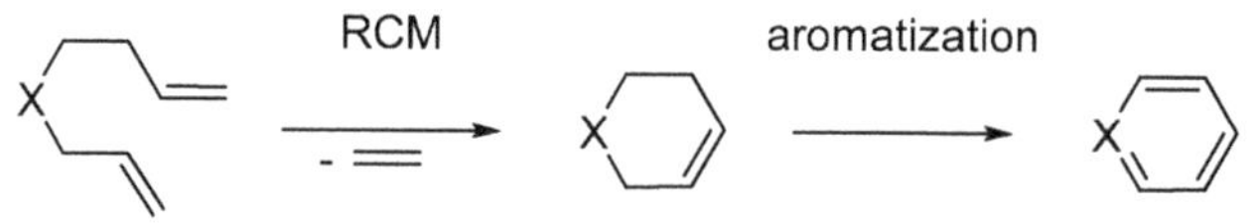

Scheme 30. A strategy for the synthesis of aromatic rings, RCM/aromatization sequence.

RCM enynes (RCEM) and ene-yne cross-metathesis (Scheme 31) are used in a similar manner.

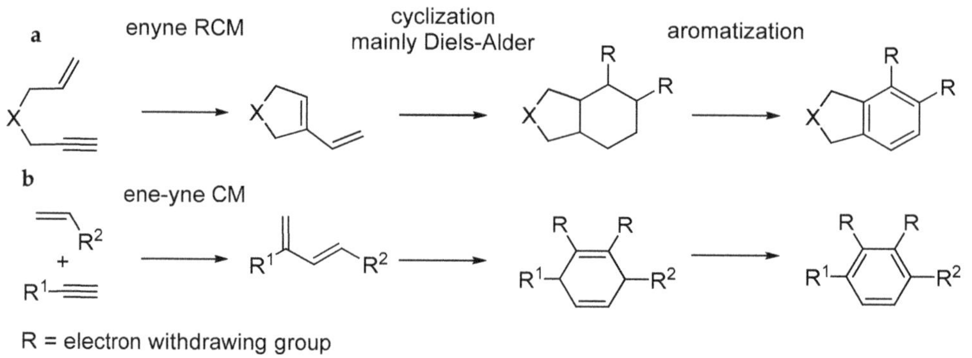

Scheme 31. A strategy for the synthesis of aromatic compounds. Reaction sequence, RCEM (**a**), or EYCM; (**b**) Diels–Alder cycloaddition–aromatization.

4.1. Synthesis of Aromatic Carbocycles Fused to Heterocyclic Rings

Recently, Dash overviewed applications of RCM for the synthesis of carbazoles and indole-fused heterocycles [50].

A new method for the synthesis of carbazoles via ring-closing metathesis of 2,2-diallyl-3-oxindoles, followed by ring-rearrangement aromatization of the resulting spirocyclic indoles, was proposed by Dash and co-workers (Scheme 32) [51]. The method is based on the novel ring-rearrangement aromatization that was observed when spirocyclic indoles were treated with Bronsted acids (TsOH, TfOH, and TFA) in hot toluene (80 °C).

Key substrates were prepared by addition of an excess of allylmagnesium bromide to N-substituted isatins (Schemes 33 and 34).

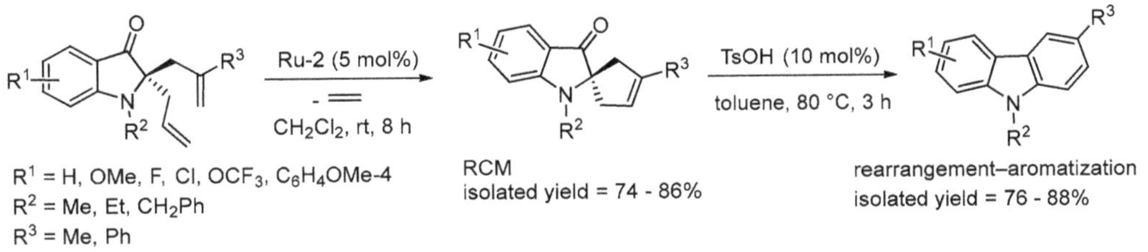

Scheme 32. A strategy for the synthesis of carbazoles from isatins based on RCM and ring-rearrangement aromatization.

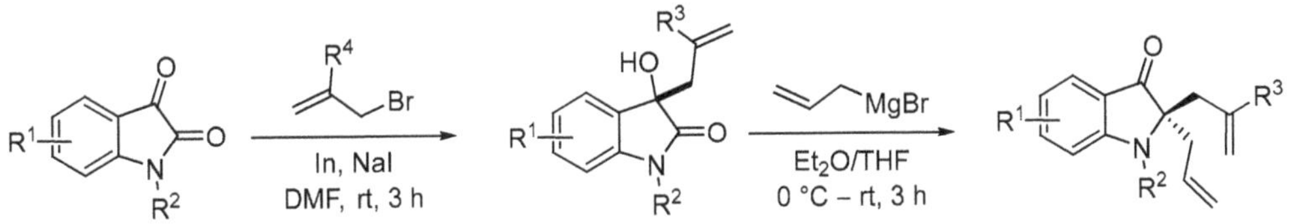

Scheme 33. Introduction of two different allyl groups into the isatin molecule.

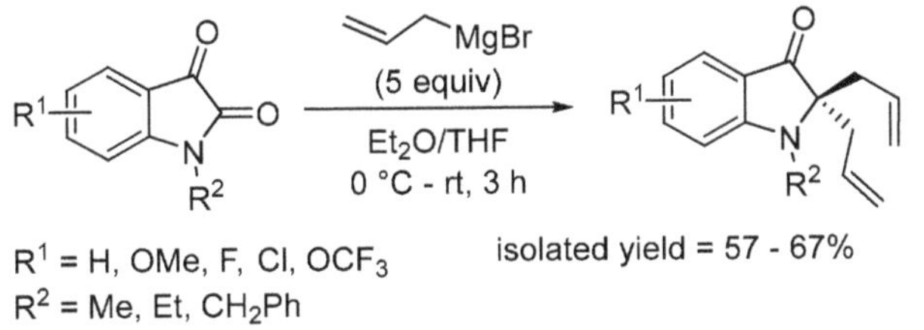

Scheme 34. Addition of allylmagnesium bromide to N-substituted isatins.

The procedure enables the convenient synthesis of carbazole from isatin derivatives and can be successfully used for the synthesis of carbazole natural products. Continued research on the addition of the allylic Grignard reagent to oxindoles resulted in the development of a procedure to obtain 2-substituted and 2,3-disubstituted indoles via a one-pot dehydrative aromatization protocol [47]. The synthesized 2,3-diallyl indoles were used to synthesize carbazole derivatives via RCM and aromatization sequence (Scheme 35).

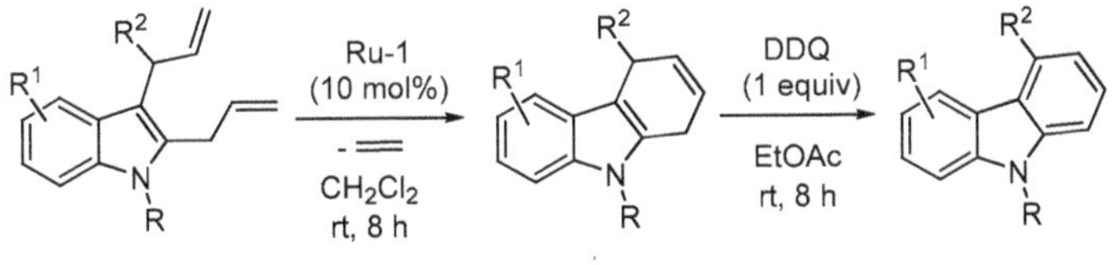

Scheme 35. Synthesis of substituted carbazoles.

The addition of Grignard reagents to substituted 2-oxoindoles was proven to be a tool for generating nitrogen-containing heterocycles by using simple organic transformations. This strategy was used in the synthesis of various substituted carbazoles, as well as several naturally occurring carbazole alkaloids.

The ring-closing metathesis reaction and aromatization were proposed as key steps in the formation of the tricyclic backbone of cyclopenta[b]benzofuran, a key intermediate of Beraprost (a prostacyclin analog) (Figure 4a), starting with (-)-Corey lactone diol (Figure 4b) [52].

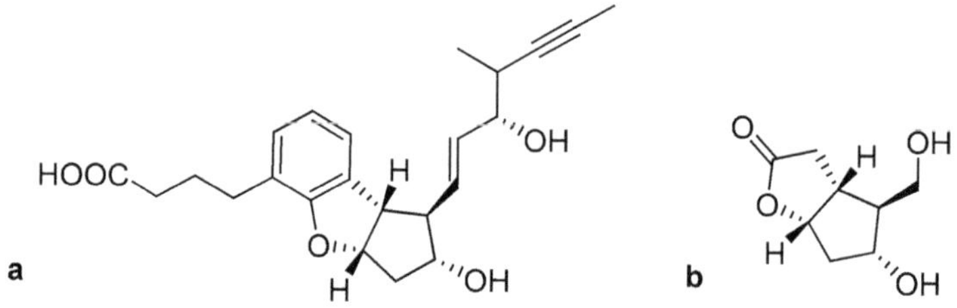

Figure 4. Structures of Beraprost (**a**) and (-)-Corey lactone diol (**b**).

RCM proceeded efficiently in the presence of a second-generation Grubbs catalyst under mild conditions and was followed by dehydration step performed in the presence of pyridine hydrochloride (Scheme 36).

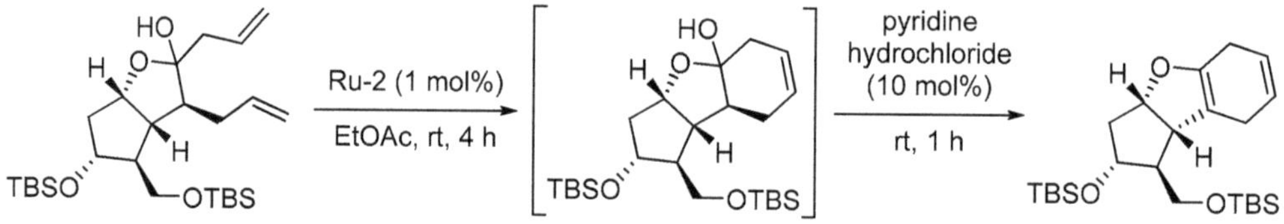

Scheme 36. RCM of diene lactol, followed by spontaneous formation of cyclic diene.

The benzofuran core was obtained by dehydrogenative aromatization of diene ring, using 2,3-dicyano-5,6-dichlorobenzoquinone (DDQ) (Scheme 37).

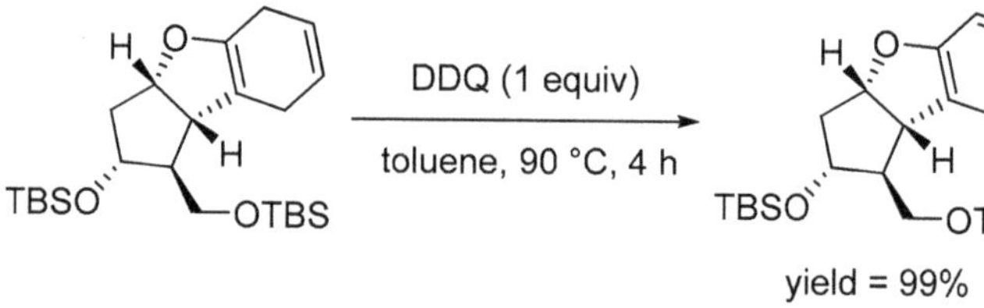

Scheme 37. Synthesis of benzofuran skeleton via oxidative aromatization.

Tokuyama and co-workers reported the catalytic activity of ruthenium alkylidene complexes in the aerobic catalytic dehydrogenation of nitrogen-containing heterocycles with dioxygen as an oxidant [53]. The reaction can be used for the dehydrogenative aromatization of various nitrogen-containing heterocycles. Moreover, using the catalytic activity of alkylidene ruthenium complexes in the RCM of dienes, the authors developed the novel assisted tandem catalysis, using the ring-closing metathesis and dehydrogenative aromatization sequence (Scheme 38). Molecular oxygen was used as a trigger and an oxidation agent. Conditions to enable the highly efficient course of the process were proposed. The procedure involves the use of a ruthenium alkylidene complex Ru-5 (Figure 1), which is a modified version of the Ru-2 that bears less steric substituents at the nitrogen atoms of the imidazol-2-ylidene ligand. This catalyst was found to be optimal for the RCM/aerobic oxidation sequence.

The usefulness of the Tokuyama procedure was demonstrated by its application to the synthesis of the natural antibiotic pyocyanin.

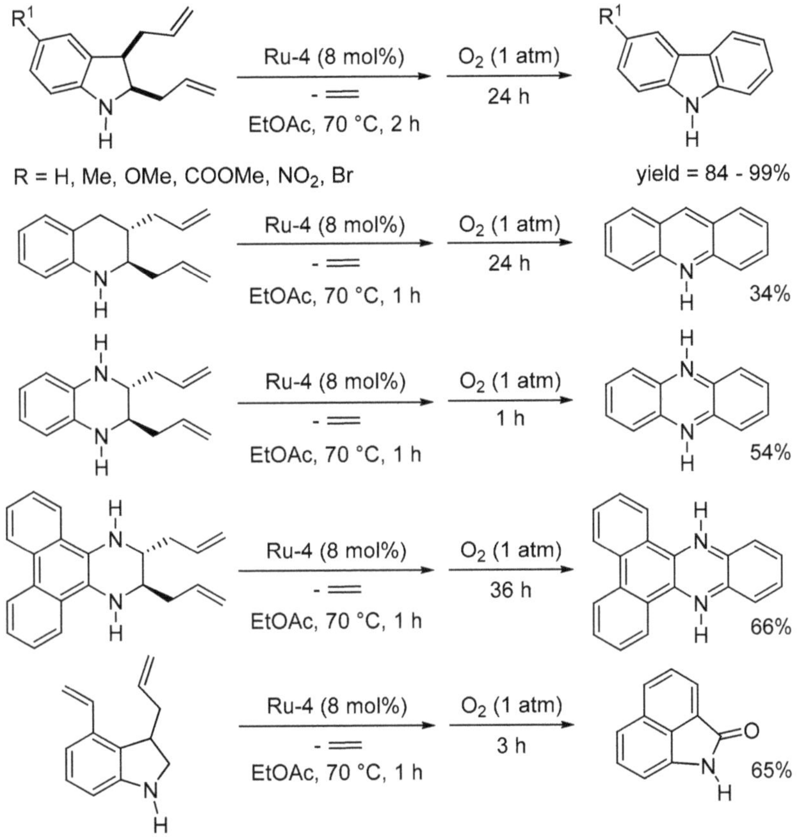

Scheme 38. Scope of the assisted-tandem RCM/aerobic dehydrogenative aromatization. The optimal reaction conditions are indicated.

4.2. Synthesis of Benzene Derivatives

Most of synthetic effort in the synthesis of simple hetero- and carboarenes via metathesis was performed during the huge development of olefin metathesis applications in organic synthesis and is described in previous reviews [7,8,10]. Currently, the research intensity, as measured by the number of publications, is much lower, and research groups are focusing on using metathesis to obtain more complex systems.

Lobo and co-workers reported a procedure for obtaining terephthalates, which includes the use of the CM, Diels–Alder cycloaddition and dehydrogenative aromatization sequence (Scheme 39) [54]. The proposed procedure is eco-friendly in nature due to the use of fully biodegradable reagents (sorbate and acrylic esters).

The highest activity of the cross-metathesis step was achieved in the presence of modified Hoveyda–Grubbs catalyst Ru-6 (Figure 1).

Ward described a procedure of the generation of monosubstituted aromatic rings that uses a ring-closing metathesis and spontaneous 1,2-elimination sequence (Scheme 40) [55]. The method enables the synthesis of benzene derivatives with moderate-to-high yields. The advantage of the procedure is that there is no need to use strong oxidants in the aromatization step.

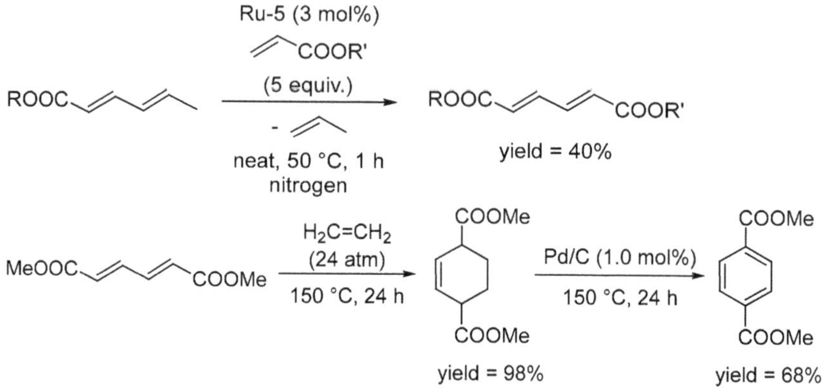

Scheme 39. A one-pot, two-step process for the synthesis of dimethyl terephthalate based on the Diels–Alder reaction of dimethyl muconate with ethylene and subsequent aromatization in the presence of a Pd/C catalyst.

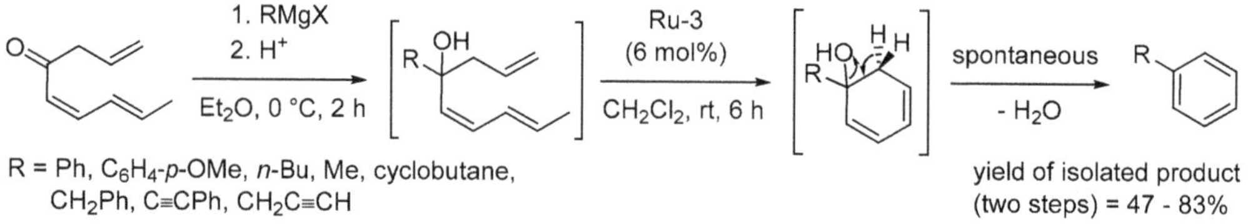

Scheme 40. Synthesis of monosubstituted aromatic ring via ring-closing metathesis, followed by spontaneous 1,2-elimination.

The sequence ene-yne cross-metathesis, Diels−Alder cycloaddition, and aromatization were applied to synthesize a series of biaryls, starting from alkynes [56]. In the ene-yne cross-metathesis step, monosubstituted phenylacetylenes were treated with 1,5-hexadiene (Scheme 41) or ethylene (Scheme 42) in the presence of a Grubbs first- or a second-generation catalyst. Cycloaddition and subsequent aromatization afforded the desired biphenyls in good yields. Two exemplary ethynylated amino acid derivatives were also shown to undergo the proposed sequence.

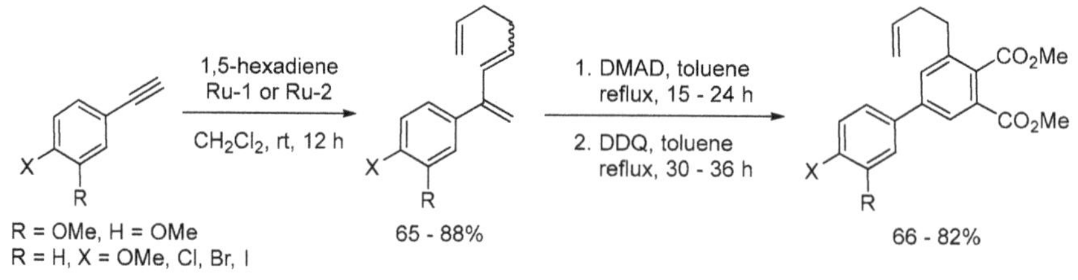

Scheme 41. Strategy for the synthesis of biphenyls from arylalkynes.

Dash and co-workers demonstrated that o-terphenyls can be synthesized from easily accessible substituted benzils. The proposed synthetic pathway involves RCM for the synthesis of a tricyclic skeleton and subsequent aromatization step. It enables the synthesis of symmetrical and unsymmetrical o-terphenyls containing both electron-rich and electron-deficient groups (Scheme 43) [57].

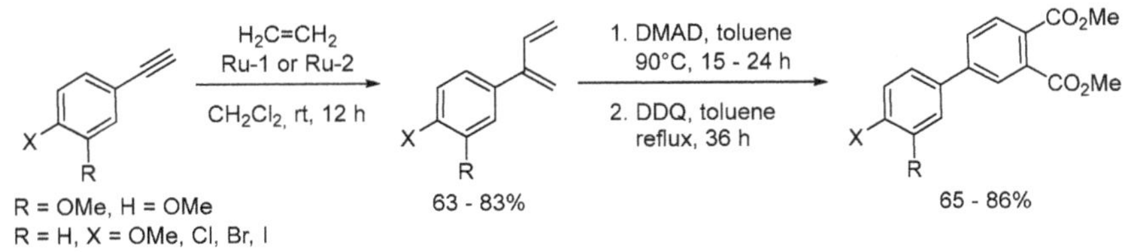

Scheme 42. Strategy for the synthesis of biphenyls from arylalkynes. Ethene was used as the olefinic reaction partner in ene-yne CM.

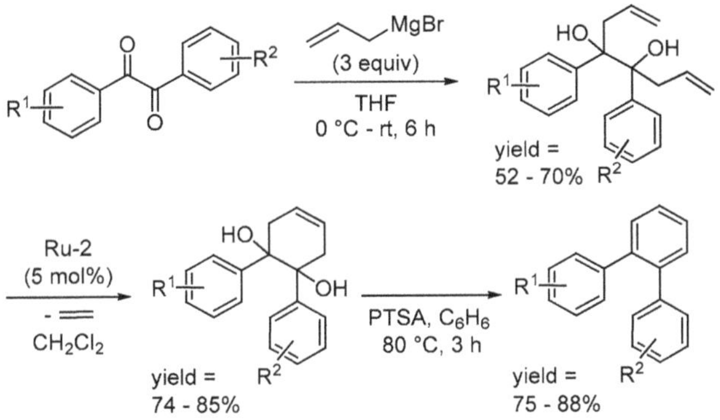

Scheme 43. RCM approach for the synthesis of o-terphenyls.

Electron-rich o-terphenyls were successfully transformed to triphenylenes via the Sholl reaction (Scheme 44).

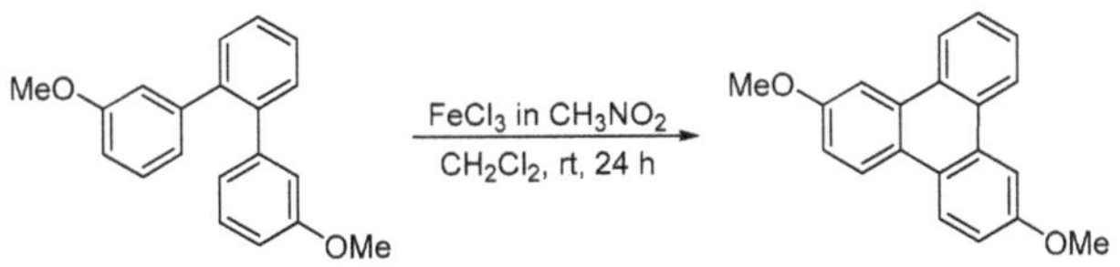

Scheme 44. Synthesis of triphenylene derivatives from o-terphenyls.

4.3. Synthesis of Aromatic Ring in Complex Molecules

Kaliappan reported the synthesis of a new class of sugar–oxasteroid–quinone hybrid molecules via a sequential enyne-metathesis/Diels–Alder cycloaddition/aromatization (Scheme 45) [58].

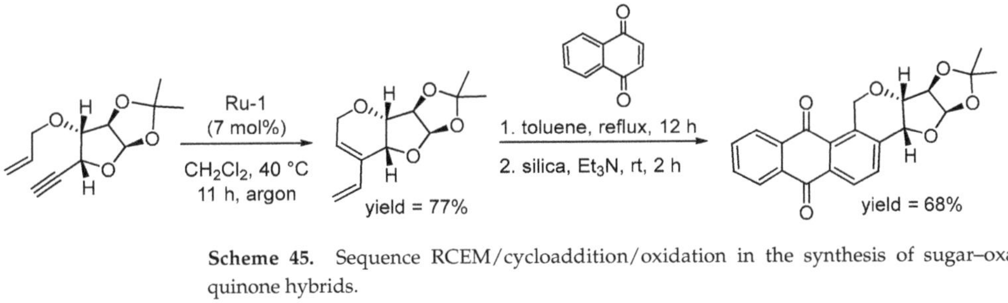

Scheme 45. Sequence RCEM/cycloaddition/oxidation in the synthesis of sugar–oxasteroid–quinone hybrids.

The obtained Diels–Alder cycloadducts were unstable, and their tendency to undergo aromatization (aerobic oxidation on silica gel) was observed. For maximum aromatization efficiency, the crude cycloaddition product was immediately treated with trimethylamine and silica gel. The method yielded a library of sugar–oxasteroid–quinone hybrids.

De Koning and co-workers proposed novel methods for the assembly of benz[a]anthracenes and related compounds [59]. Their strategy involves the Suzuki–Miyaura cross-coupling, isomerization, and ring-closing metathesis. The RCM step in this case results in the formation of an aromatic ring, which can be easily oxidized to the corresponding quinone (Scheme 46).

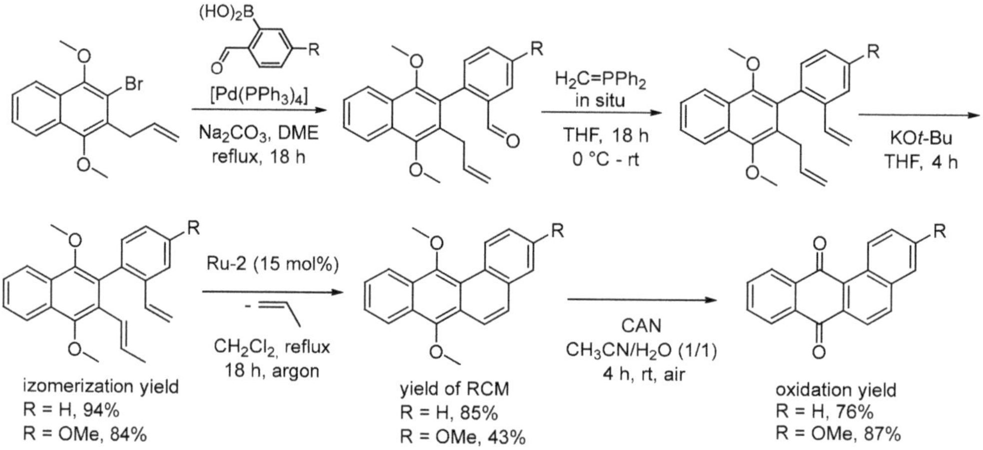

Scheme 46. Key steps in the synthesis of the angucycline core.

Examples of structures of antibiotics from the angucycline group are shown in Figure 5.

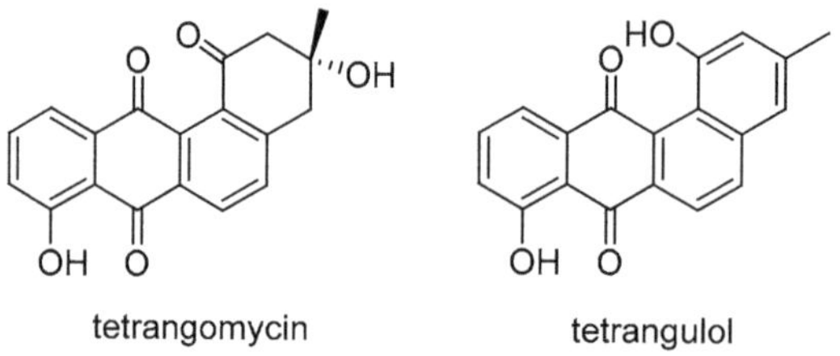

Figure 5. Structures of exemplary angucyclines.

The solid-phase synthesis of numerous polycyclic tetrahydroisoquinolines and tetrahydrobenzo[d]azepines starting from Wang resin-immobilized allylglycine was reported by Soural [60]. Subsequent Fmoc cleavage, sulfonylation with nitrobenzenesulfonyl chlorides, and Mitsunobu alkylation with phenylalkynols afforded corresponding enynes. Further ring-closing metathesis, cycloaddition with functionalized 1,4-naphthoquinones, and finally spontaneous aromatization enables the formation of a variety of tetrahydroisoquinolines and tetrahydrobenzo[d]azepines (Scheme 47). The proposed further functionalization of the obtained derivatives by heterocyclization enables the rapid synthesis of relevant bioactive scaffolds.

Recently, Sparr and co-workers reported on the great potential of metathesis in the atroposelective formation of arenes [61]. The authors demonstrated that, in the presence of enantiopure molybdenum metathesis catalysts, stereodynamic trienes were enantioselectively converted into the corresponding binaphthalene atropisomers (Scheme 48).

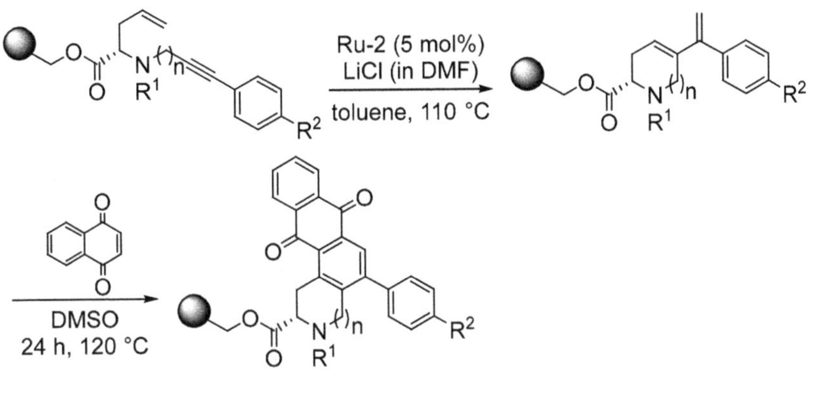

Scheme 47. Key steps in solid-phase synthesis of tetrahydroisoquinolines.

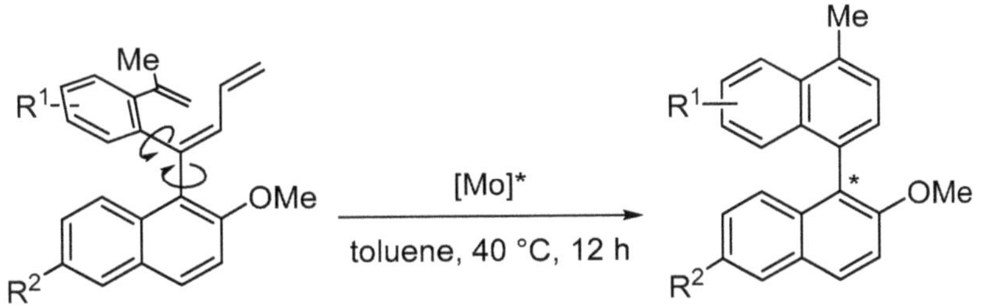

R^1 = 2-nitrobenzenesulfonyl, 4-nitrobenzenesulfonyl;
R^2 = H, CF$_3$; n = 1, 2

Scheme 48. Asymmetric synthesis of binaphthalene atropisomers from stereodynamic trienes.

In the presence of commercially available molybdenum catalysts (Figure 6), enantioselectivity, e.r. = 83:17, was observed for the model triene.

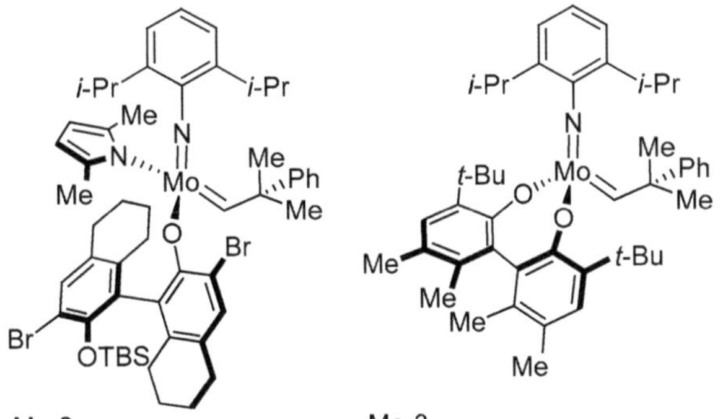

Figure 6. Commercially available chiral molybdenum alkylidene complexes tested in atroposelective synthesis of binaphthalenes.

In order to optimize the yield and enantioselectivity, tests were carried out in the presence of a series of catalysts obtained in situ by reacting a dipyrrolyl molybdenum precursor with binaphthol ligands differing in stereoelectronic properties (Scheme 49).

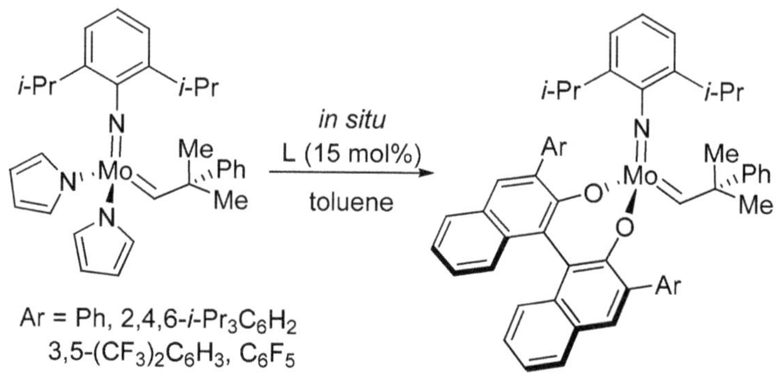
i-Pr
i-Pr
N
Mo
Me
Ph
Me
N
N
in situ
L (15 mol%)
toluene
i-Pr
i-Pr
Ar
N
Mo
O
O
Me
Ph
Me
Ar
Ar = Ph, 2,4,6-i-Pr3C6H2
3,5-(CF3)2C6H3, C6F5

Scheme 49. In situ generation of chiral, non-racemic molybdenum binaphthol complexes.

It was shown that the most enantioselective catalyst was the binaphthyl complex containing pentafluorophenyl substituents. Under optimized conditions, a variety of conformationally dynamic triene substrates were efficiently and enantioselectively converted into binaphthyl atropisomers (Figure 7).

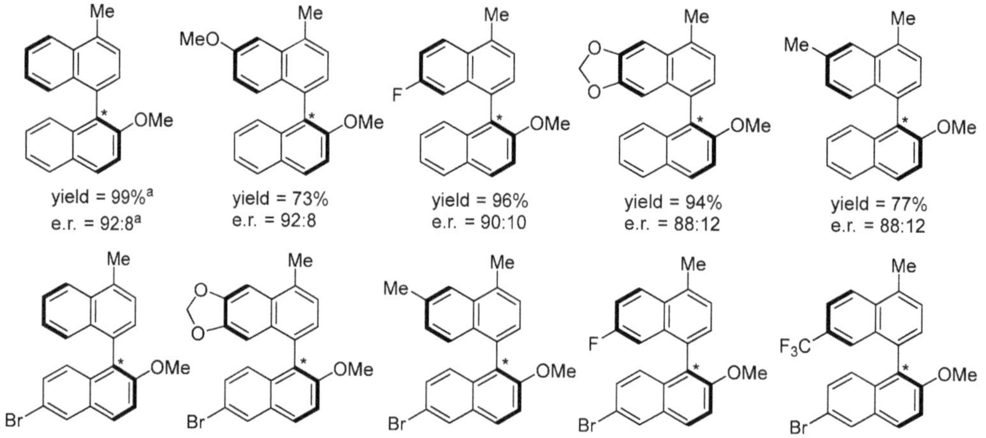
yield = 99%a
e.r. = 92:8a
yield = 73%
e.r. = 92:8
yield = 96%
e.r. = 90:10
yield = 94%
e.r. = 88:12
yield = 77%
e.r. = 88:12
yield = 83%
e.r. = 91:9
yield = 78%
e.r. = 98:2
yield = 71%
e.r. = 89:11
yield = 89%
e.r. = 93:7
yield = 99%
e.r. = 91:9

Figure 7. The scope of synthesized binaphthyl atropisomers.

Finally, Tanaka and co-workers discussed the possibility of using RCM for the in vivo synthesis of cytostatic agents [62]. Thus, cytostatic agents, naphthyl analogues of combretastatin, can be synthesized by the sequential ring-closing metathesis and aromatization of relevant prodrugs (Scheme 50). It was shown that, in the presence of glycosylated Hoveyda–Grubbs-type artificial metalloenzyme Alb-Ru (Figure 8) and in biologically relevant concentrations of prodrugs (dienes), the proposed reaction sequence proceeds readily for all ester-based prodrugs, with pivalate being the most active. Therapy against cancer-cell growth was studied in cellulo and in vivo in mice.

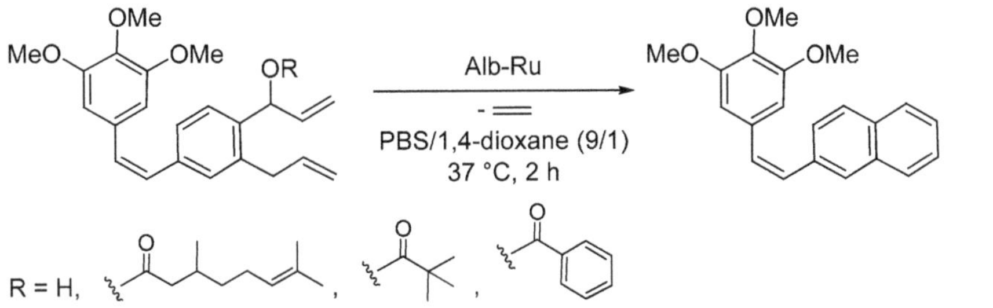

Scheme 50. Naphthyl analogues of combretastatin synthesized by sequential ring-closing metathesis and aromatization.

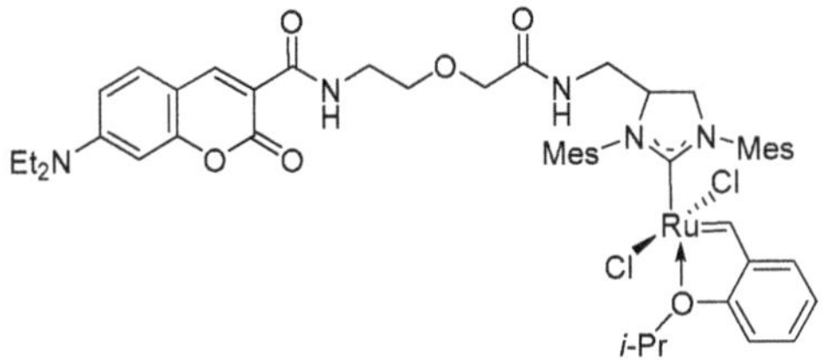

Figure 8. Glycosylated Hoveyda–Grubbs-type artificial metalloenzyme (Alb-Ru).

The promising results show new possibilities for the use of artificial metalloenzymes and metathesis cyclization in pro-drug therapy.

4.4. Synthesis of Polyarenes

Cycloparaphenylenes have gained the interest of many research groups in the hope that these systems can be used for the bottom-up chemical synthesis of carbon nanotubes [63,64]. A series of papers on the synthetic strategy that employs 1,4-diketo-bridged macrocycle as a precursor to a strained 1,4-arene-bridged (bent *para*-phenylene) macrocycle were reported by Merner and co-workers [65–68]. The described strategy involves a metathesis cyclization step, followed by aromatization (Scheme 51). Efficient metathesis cyclization was observed only for the *syn* diastereomer.

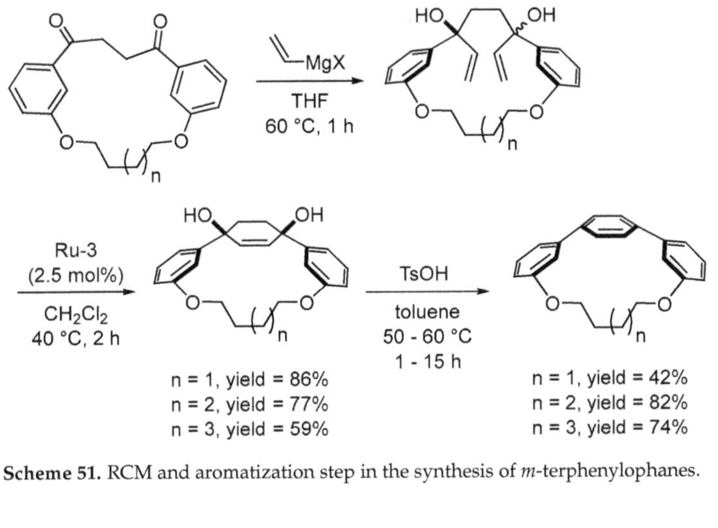

n = 1, yield = 86%
n = 2, yield = 77%
n = 3, yield = 59%

n = 1, yield = 42%
n = 2, yield = 82%
n = 3, yield = 74%

Scheme 51. RCM and aromatization step in the synthesis of *m*-terphenylophanes.

The aromatization reaction may be accompanied by a rearrangement of the desired *p*-terphenylophane to *m*-terphenylophane, which was observed for the arene-bridged macrocycle (Scheme 52) [66,67]. Such a strain-relief-driven rearrangement is not observed for n = 2 (Scheme 51). The selectivity of the aromatization process strongly depends on the reaction conditions and the reagents used. It was found that, in the presence of the Burgess reagent, the reaction selectively leads to the expected product.

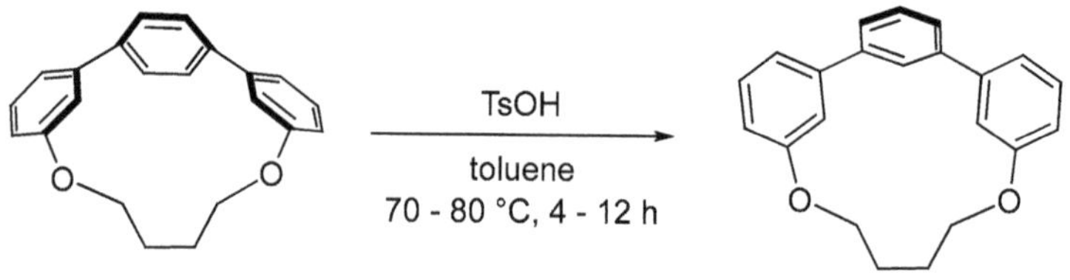

Scheme 52. Strain-relief-driven rearrangement of *p*-terphenylophane to *m*-terphenylophane.

Merner's strategy enabled the synthesis of a *p*-terphenyl-based macrocycle containing a *p*-phenylene unit with 42.6 kcal/mol of strain energy [64]. The sequence uses metathesis cyclization (Scheme 53), but aromatization is crucial to obtain a stretched ring. It involves the use of iterative elimination reactions of the two hydroxyl groups present in the macrocyclic precursor [66,67].

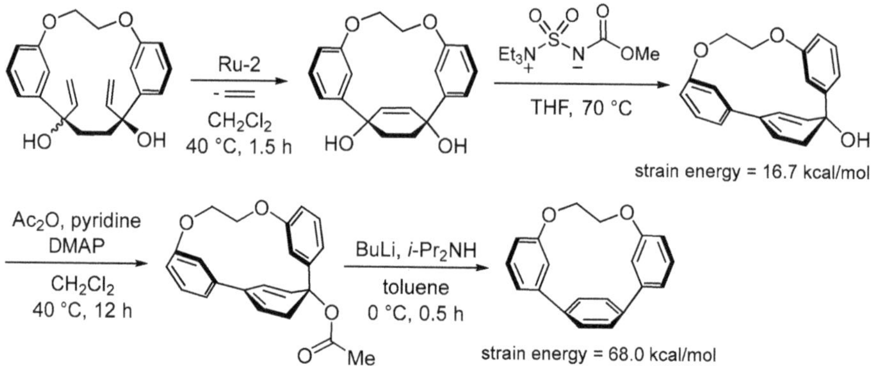

Scheme 53. Merner's strategy to obtain a strained aromatic ring.

The synthesis of double-stranded aromatic macrocycles remains a challenge. Jasti and co-workers developed a reductive aromatization/ring-closing metathesis sequence for the synthesis of aromatic belt fragments (Scheme 54) [69].

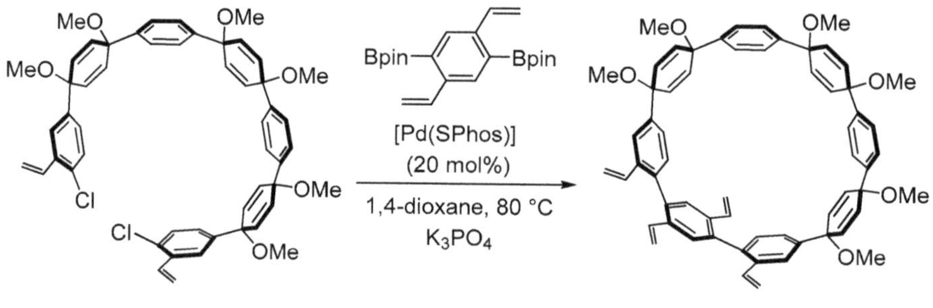

Scheme 54. Synthesis of macrocycles via Suzuki cross-coupling.

For the macrocycle presented in Scheme 55, the sequence starting from RCM was used in order to avoid potential problems of interaction of strong reductant (sodium naphthalenide) with styrene functionality. RCM proceeded readily in the presence of Ru-2 in dichloromethane at 40 °C. No side-products were observed. As shown in Scheme 55, the RCM strategy used leads to the formation of aromatic rings. A final aromatization step was made by subjecting macrocycle to sodium naphthalenide at −78 °C.

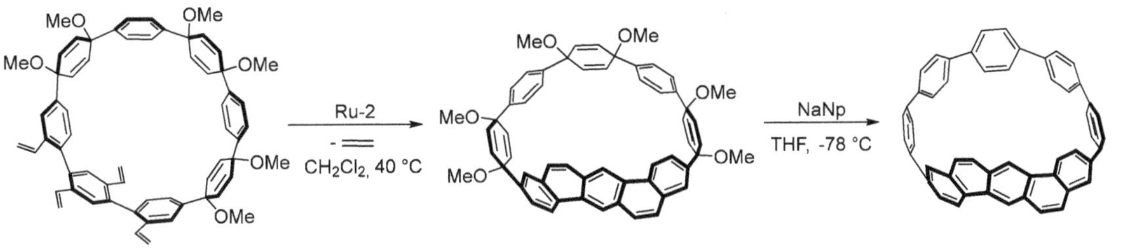

Scheme 55. RCM/reductive aromatization sequence.

The proposed strategy, which is a combination of reductive aromatization and ring-closing metathesis, was tested for the synthesis of smaller, more strained belts. In this case, the order of steps was changed, and the reductive aromatization was carried out prior to the RCM (Scheme 56). Deprotection of hydroxy groups, followed by reductive aromatization under mild conditions, yields strained ring. In the last step, the RCM enabled the formation of a strongly strained fragment of aromatic belt. Jasti demonstrated that the RCM reaction leading to the formation of aromatic rings can be carried out in highly strained systems, without undesirable side-processes [69].

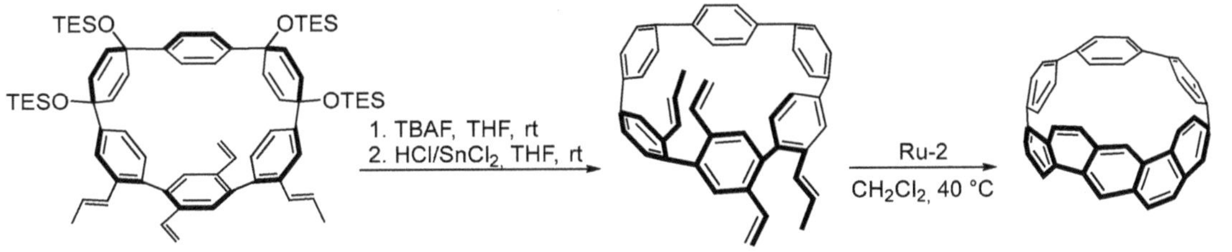

Scheme 56. Reductive aromatization/RCM sequence in the synthesis of highly strained aromatic belt fragment.

Fang and co-workers proposed efficient synthesis of donor–acceptor ladder polymer by the sequence of Suzuki polycondensation and ring-closing olefin metathesis (Scheme 57) [70]. The authors assumed that the use of RCM as a ladderization step, which, in this case, is also an aromatization step, will afford the desired ladder polymer in good yield and without structural defects.

Preliminary tests of polymer properties were performed. As expected, the rigid coplanar nature of the ladder polymer was found to exhibit improved physicochemical properties that were not observed with non-ladder type analogs.

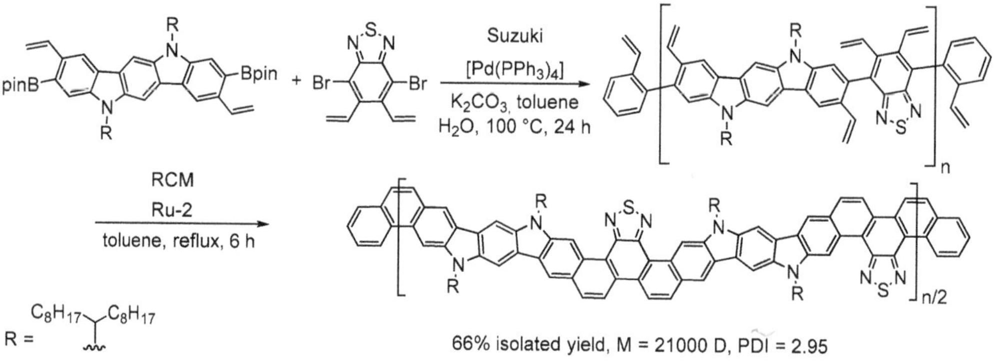

Scheme 57. Synthesis of donor–acceptor ladder polymer by the sequence of Suzuki polycondensation and ring-closing olefin metathesis.

5. Summary and Outlook

The synthesis of aromatic compounds using olefins and enyne metathesis employs several previously described strategies. The synthesis of fused aromatic compounds can be accomplished directly through diene or enyne RCM. Other strategies include a cyclization step via RCM, followed by aromatization, which can be implemented by oxidation or elimination of suitable leaving groups. Finally, enyne RCM or ene-yne CM allows the formation of conjugated dienes that can undergo successive cyclization, mainly through Diels–Alder cycloaddition, and aromatization. Based on the results described and previous reports, it seems that further progress is possible, particularly in the aromatization step. The use of (spontaneous) aromatization through aerobic oxidation eliminates the need for strong oxidants, which can expand the reaction scope and/or improve the yields achieved. A fascinating idea is to use RCM for the synthesis of atropisomers, using metathetic cyclization of conformationally dynamic trienes. The development of the method may open up a new area of application of metathetic cyclization in the asymmetric synthesis of aromatic compounds. Exciting reports have been made on the syntheses of macrocyclic aromatic strips with high ring strain. The efficient synthesis of linear ladder polymers with a stiff coplanar structure has been achieved using the strong thermodynamic driving force of aromatization. Due to the stiffened structure, such polymers exhibit improved physicochemical properties compared to non-ladder analogues. Finally, the possibility of using metathesis/aromatization sequences in the synthesis of prodrugs is described. This opens up a new and poorly explored area of research.

The survey clearly shows that the field of research will continue to develop in the coming years. In particular, material and medicinal chemistry, in the broadest sense, remain an important and underexplored area of research.

Author Contributions: Conceptualization, C.P.; writing—original draft preparation, C.P. and S.R.; writing—review and editing, C.P. and S.R.; visualization, C.P.; supervision, C.P.; funding acquisition, C.P. and S.R. All authors have read and agreed to the published version of the manuscript.

Funding: Authors acknowledge the financial support by the Faculty of Chemistry, Adam Mickiewicz University, Poznan, from the funds awarded by the Ministry of Education and Science in the form of a subsidy for the maintenance and development of research potential in 2022.

Institutional Review Board Statement: Not applicable.

Informed Consent Statement: Not applicable.

Data Availability Statement: Not applicable.

Conflicts of Interest: The authors declare no conflict of interest.

References

1. *Handbook of Metathesis*; Grubbs, R.H. (Ed.) Wiley-VCH: Weinheim, Germany, 2003.
2. *Handbook of Metathesis*; Grubbs, R.H.; Wenzel, A.G.; O'Leary, D.J.; Khosravi, E. (Eds.) Wiley-VCH: Weinheim, Germany, 2015.
3. *Olefin Metathesis. Theory and Practice*; Grela, K., Ed.; Wiley: Hoboken, NJ, USA, 2014.
4. *Metathesis in Natural Product Synthesis*; Cossy, J.; Arseniyadis, S.; Meyer, C. (Eds.) Wiley-VCH: Weinheim, Germany, 2010.
5. Franc, H.-G.; Stadelhofer, J.W. *Industrial Aromatic Chemistry*; Springer: Berlin, Germany, 2011.
6. Quin, L.D.; Tyrell, J.A. *Fundamentals of Heterocyclic Chemistry. Importance in Nature and in the Synthesis of Pharmaceuticals*; Wiley: Hoboken, NJ, USA, 2010.
7. Van Otterlo, W.A.L.; de Koning, C.B. Metathesis in the synthesis of aromatic compounds. *Chem. Rev.* **2009**, *109*, 3743–3782. [CrossRef] [PubMed]
8. De Koning, C.B.; van Otterlo, W.A.L. Ring-closing metathesis: Synthetic routes to carbocyclic aromatic compounds using ring-closing alkene and enyne metathesis. In *Arene Chemistry. Reaction Mechanisms and Methods for Aromatic Compounds*; Mortier, J., Ed.; Wiley: Hoboken, NJ, USA, 2016; pp. 485–510.
9. Yoshida, K. Metathesis reactions. In *Transition-Metal-Mediated Aromatic Ring Construction*; Tanaka, K., Ed.; Wiley: Hoboken, NJ, USA, 2013; pp. 721–742.
10. Donohoe, T.J.; Orr, A.J.; Bingham, M. Ring-Closing Metathesis as a basis for the construction of aromatic compounds. *Angew. Chem. Int. Ed.* **2006**, *45*, 2664–2670. [CrossRef] [PubMed]
11. Race, N.J.; Bower, J.F. Synthesis of heteroaromatic compounds by alkene and enyne metathesis. *Top. Heterocycl. Chem.* **2017**, *47*, 1–32.
12. Donohoe, T.J.; Fishlock, L.P.; Procopiou, P.A. Ring-Closing Metathesis: Novel routes to aromatic heterocycles. *Chem. Eur. J.* **2008**, *14*, 5716–5726. [CrossRef]
13. Potukuchi, H.K.; Colomer, I.; Donohoe, T.J. Synthesis of aromatic heterocycles using ring-closing metathesis. *Adv. Heterocycl. Chem.* **2016**, *120*, 43–65.
14. Herisson, J.-L.; Chauvin, Y. Catalyse de transformation des olefines par les complexes du tungsten. II. Telomerisation des olefines cycliques en presence d'olefines acycliques. *Macromol. Chem.* **1971**, *141*, 161–176. [CrossRef]
15. Hansen, E.C.; Lee, D. Search for solutions to the reactivity and selectivity problems in enyne metathesis. *Acc. Chem. Res.* **2006**, *39*, 509–519. [CrossRef]
16. Hansen, E.C.; Lee, D. Ring-closing enyne metathesis: Control over mode selectivity and stereoselectivity. *J. Am. Chem. Soc.* **2004**, *126*, 15074–15080. [CrossRef]
17. Trost, B.M.; Krische, M.J. Transition metal catalyzed cycloisomerizations. *Synlett* **1998**, *1998*, 1–16. [CrossRef]
18. Zhang, Z.; Zhu, G.; Tong, X.; Wang, F.; Xie, X.; Wang, J.; Jiang, L. Transition metal-catalyzed intramolecular enyne cyclization reaction. *Curr. Org. Chem.* **2006**, *10*, 1457–1478. [CrossRef]
19. Diver, S.T.; Giessert, A.J. Enyne metathesis (enyne bond reorganization). *Chem. Rev.* **2004**, *104*, 1317–1382. [CrossRef] [PubMed]
20. Villar, H.; Fringsa, M.; Bolm, C. Ring-closing enyne metathesis: A powerful tool for the synthesis of heterocycles. *Chem. Soc. Rev.* **2007**, *36*, 55–66. [CrossRef]
21. Li, J.; Lee, D. Enyne metathesis. In *Handbook of Metathesis*; Grubbs, R.H., O'Leary, D.J., Eds.; Wiley-VCH: Weinheim, Germany, 2015; Volume 2, pp. 381–444.
22. Trnka, T.N.; Grubbs, R.H. The development of L_2X_2Ru=CHR olefin metathesis catalysts: An organometallic success story. *Acc. Chem. Res.* **2001**, *34*, 18–29. [CrossRef] [PubMed]
23. Schrock, R.R. Olefin metathesis by molybdenum imido alkylidene catalysts. *Tetrahedron* **1999**, *55*, 8141–8153. [CrossRef]
24. Zielinski, G.K.; Grela, K. Tandem catalysis utilizing olefin metathesis reactions. *Chem. Eur. J.* **2016**, *22*, 9440–9454. [CrossRef] [PubMed]
25. Raviola, C.; Protti, S.; Ravelli, D.; Fagnoni, M. (Hetero)aromatics from dienynes, enediynes and enyne–allenes. *Chem. Soc. Rev.* **2016**, *45*, 4364–4390. [CrossRef] [PubMed]
26. Donohoe, T.J.; Bower, J.F.; Chan, L.K.M. Olefin cross-metathesis for the synthesis of heteroaromatic compounds. *Org. Biomol. Chem.* **2012**, *10*, 1322–1328. [CrossRef]
27. Georgiades, S.N.; Nicolaou, P.G. Recent advances in carbazole syntheses. *Adv. Heterocycl. Chem.* **2019**, *129*, 1–88.
28. Chachignon, H.; Scalacci, N.; Petricci, E.; Castagnolo, D. Synthesis of 1,2,3-substituted pyrroles from propargylamines via a one-pot tandem enyne cross metathesis–cyclization reaction. *J. Org. Chem.* **2015**, *80*, 5287–5295. [CrossRef]
29. Frei, M.S.; Bilyard, M.K.; Alanine, T.A.; Galloway, W.R.J.D.; Stokes, J.E.; Spring, D.R. Studies towards the synthesis of indolizin-5(3H)-one derivatives and related 6,5-azabicyclic scaffolds by ring-closing metathesis. *Bioorg. Med. Chem.* **2015**, *23*, 2666–2679. [CrossRef]
30. Bunrit, A.; Sawadjoon, S.; Tsupova, S.; Sjöberg, P.J.R.; Samec, J.S.M. A general route to β-substituted pyrroles by transition-metal catalysis. *J. Org. Chem.* **2016**, *81*, 1450–1460. [CrossRef] [PubMed]
31. Cogswell, T.J.; Donald, C.S.; Marquez, R. Flexible synthesis of polyfunctionalised 3-fluoropyrroles. *Org. Biomol. Chem.* **2016**, *14*, 183–190. [CrossRef] [PubMed]
32. Keeley, A.; McCauley, S.; Evans, P. A ring-closing metathesis-manganese dioxide oxidation sequence for the synthesis of substituted pyrroles. *Tetrahedron* **2016**, *72*, 2552–2559. [CrossRef]

33. Grychowska, K.; Kubica, B.; Drop, M.; Colacino, E.; Bantreil, X.; Pawlowski, M.; Martinez, J.; Subra, G.; Zajdel, P.; Lamaty, F. Application of the ring-closing metathesis to the formation of 2-aryl-1H-pyrrole-3-carboxylates as building blocks for biologically active compounds. *Tetrahedron* 2016, 72, 7462–7469. [CrossRef]
34. Scalacci, N.; Black, G.W.; Mattedi, G.; Brown, N.L.; Turner, N.J.; Castagnolo, D. Unveiling the biocatalytic aromatizing activity of monoamine oxidases MAO-N and 6-HDNO: Development of chemoenzymatic cascades for the synthesis of pyrroles. *ACS Catal.* 2017, 7, 1295–1300. [CrossRef]
35. Kotha, S.; Todeti, S.; Das, T.; Datta, A. Synthesis of star-shaped pyrrole-based C_3-symmetric molecules via ring-closing metathesis, Buchwald–Hartwig cross-coupling and Clauson–Kaas pyrrole synthesis as key steps. *Tetrahedron Lett.* 2018, 59, 1023–1027. [CrossRef]
36. Chena, W.; Zhanga, Y.-L.; Li, H.-J.; Nana, X.; Liua, Y.; Wu, Y.-C. Synthesis of N-sulfonyl- and N-acylpyrroles via a ring-closing metathesis/dehydrogenation tandem reaction. *Synthesis* 2019, 51, 3651–3666. [CrossRef]
37. Fujii, Y.; Suwa, Y.; Wada, Y.; Takehara, T.; Suzuki, T.; Kawashima, Y.; Kawashita, N.; Takagi, T.; Fujioka, H.; Arisawa, M. Metal-free nitrogen-containing polyheterocyclic near-infrared (NIR) absorption dyes: Synthesis, absorption properties, and theoretical calculation of substituted 5-methylisoindolo[2,1-a]quinolones. *ACS Omega* 2019, 4, 5064–5075. [CrossRef]
38. Francois, B.; Eberlin, L.; Berrée, F.; Whiting, A.; Carboni, B. Access to fused pyrroles from cyclic 1,3-dienyl boronic esters and arylnitroso compounds. *J. Org. Chem.* 2020, 85, 5173–5182. [CrossRef]
39. Risi, C.; Zhao, F.; Castagnolo, D. Chemo-enzymatic metathesis/aromatization cascades for the synthesis of furans: Disclosing the aromatizing activity of Laccase/TEMPO in oxygen-containing heterocycles. *ACS Catal.* 2019, 9, 7264–7269. [CrossRef]
40. Kotha, S.; Solanke, B.U.; Gupta, N.K. Design and synthesis of C_3-symmetric molecules containing oxepine and benzofuran moieties via metathesis. *J. Mol. Struct.* 2021, 1244, 130907. [CrossRef]
41. Kotha, S.; Solanke, B.U. Modular approach to benzofurans, 2H-chromenes and benzoxepines via Claisen rearrangement and ring-closing metathesis: Access to phenylpropanoids. *Chem. Asian J.* 2022, 17, e202200084. [CrossRef]
42. Lood, K.; Tikk, T.; Krüger, M.; Schmidt, B. Methylene capping facilitates cross-metathesis reactions of enals: A short synthesis of 7-methoxywutaifuranal from the xylochemical isoeugenol. *J. Org. Chem.* 2022, 87, 3079–3088. [CrossRef]
43. Sucunza, D.; Cuadro, A.M.; Alvarez-Builla, J.; Vaquero, J.J. Recent advances in the synthesis of azonia aromatic heterocycles. *J. Org. Chem.* 2016, 81, 10126–10135. [CrossRef]
44. Abengózar, A.; Abarca, B.; Cuadro, A.M.; Sucunza, D.; Álvarez-Builla, J.; Vaquero, J.J. Azonia aromatic cations by ring-closing metathesis: Synthesis of azaquinolizinium cations. *Eur. J. Org. Chem.* 2015, 2015, 4214–4223. [CrossRef]
45. Weber, C.; Nebe, M.M.; Kaluza, L.P.V.; Opatz, T. Short synthesis of pyridines from deprotonated α-aminonitriles by an alkylation/RCM sequence. *Z. Naturforsch.* 2016, 71b, 633–641. [CrossRef]
46. Remya, P.R.; Suresh, C.H. Grubbs and Hoveyda-Grubbs catalysts for pyridine derivative synthesis: Probing the mechanistic pathways using DFT. *Mol. Catal.* 2018, 450, 29–38. [CrossRef]
47. Mandal, T.; Chakraborti, G.; Karmakar, S.; Dash, J. Divergent and orthogonal approach to carbazoles and pyridoindoles from oxindoles via indole intermediates. *Org. Lett.* 2018, 20, 4759–4763. [CrossRef]
48. Katz, T.J.; Sivavec, T.M. Metal-catalyzed rearrangement of alkene-alkynes and the stereochemistry of metallacyclobutene ring opening. *J. Am. Chem. Soc.* 1985, 107, 737–738. [CrossRef]
49. Katz, T.J.; Rothchild, R. Mechanism of the olefin metathesis of 2,2'-divinylbiphenyl. *J. Am. Chem. Soc.* 1976, 98, 2519–2526. [CrossRef]
50. Mandal, T.; Dash, J. Ring-closing metathesis for the construction of carbazole and indole-fused natural products. *Org. Biomol. Chem.* 2021, 19, 9797–9808. [CrossRef] [PubMed]
51. Dhara, K.; Mandal, T.; Das, J.; Dash, J. Synthesis of carbazole alkaloids by ring-closing metathesis and ring rearrangement–aromatization. *Angew. Chem. Int. Ed.* 2015, 54, 15831–15835. [CrossRef] [PubMed]
52. Liu, X.; Tian, C.; Jiao, X.; Li, X.; Yang, H.; Yaoa, Y.; Xie, P. Practical synthesis of chiral tricyclic cyclopenta[b]benzofuran, a key intermediate of Beraprost. *Org. Biomol. Chem.* 2016, 14, 7715–7721. [CrossRef] [PubMed]
53. Kawauchi, D.; Noda, K.; Komatsu, Y.; Yoshida, K.; Ueda, H.; Tokuyama, H. Aerobic dehydrogenation of N-heterocycles with Grubbs catalyst: Its application to assisted-tandem catalysis to construct N-containing fused heteroarenes. *Chem. Eur. J.* 2020, 26, 15793–15798. [CrossRef] [PubMed]
54. Saraci, E.; Wang, L.; Theopold, K.H.; Lobo, R.F. Bioderived muconates by cross-metathesis and their conversion into terephthalates. *ChemSusChem* 2018, 11, 773–780. [CrossRef]
55. Lozhkin, B.; Ward, T.R. A Close-to-aromatize approach for the late-stage functionalization through ring-closing metathesis. *Helv. Chim. Acta* 2021, 104, e2100024. [CrossRef]
56. Kotha, S.; Seema, V.; Banerjee, S.; Dipak, M.K. Diversity oriented approach to polycyclics via cross-enyne metathesis and Diels–Alder reaction as key steps. *J. Chem. Sci. (India)* 2015, 127, 155–162. [CrossRef]
57. Karmakar, S.; Mandal, T.; Dash, J. Ring-closing metathesis approach for the synthesis of o-terphenyl derivatives. *Eur. J. Org. Chem.* 2019, 2019, 5916–5924. [CrossRef]
58. Sayyad, A.; Kaliappan, K.P. Sequential enyne-metathesis/Diels–Alder strategy: Rapid access to sugar–oxasteroid–quinone hybrids. *Eur. J. Org. Chem.* 2017, 2017, 5055–5065. [CrossRef]

59. Johnson, M.M.; Ngwira, K.J.; Rousseau, A.L.; Lemmerer, A.; de Koning, C.B. Novel methodology for the synthesis of the benz[a]anthracene skeleton of the angucyclines using a Suzuki-Miyaura/isomerization/ring-closing metathesis strategy. *Tetrahedron* **2018**, *74*, 12–18. [CrossRef]
60. Králová, P.; Soural, M. Synthesis of polycyclic tetrahydroisoquinolines and tetrahydrobenzo[d]azepines from polymer-supported allylglycine. *J. Org. Chem.* **2022**, *87*, 5242–5256. [CrossRef]
61. Jončev, Z.; Sparr, C. Atroposelective arene-forming alkene metathesis. *Angew. Chem. Int. Ed.* **2022**, *61*, e202211168. [CrossRef]
62. Nasibullin, I.; Smirnov, I.; Ahmadi, P.; Vong, K.; Kurbangalieva, A.; Tanaka, K. Synthetic prodrug design enables biocatalytic activation in mice to elicit tumor growth suppression. *Nat. Commun.* **2022**, *13*, 39. [CrossRef]
63. Wu, D.; Cheng, W.; Ban, X.; Xia, J. Cycloparaphenylenes (CPPs): An overview of synthesis, properties, and potential applications. *Asian J. Org. Chem.* **2018**, *7*, 2161–2181. [CrossRef]
64. Lewis, S.E. Cycloparaphenylenes and related nanohoops. *Chem. Soc. Rev.* **2015**, *44*, 2221–2304. [CrossRef]
65. Mitra, N.K.; Meudom, R.; Gorden, J.D.; Merner, B.L. A Non-Cross-Coupling approach to arene-bridged macrocycles: Synthesis, structure, and direct, regioselective functionalization of a cycloparaphenylene fragment. *Org. Lett.* **2015**, *17*, 2700–2703. [CrossRef] [PubMed]
66. Mitra, N.K.; Corzo, H.H.; Merner, B.L. A macrocyclic 1,4-diketone enables the synthesis of a p-phenylene ring that is more strained than a monomer unit of [4]cycloparaphenylene. *Org. Lett.* **2016**, *18*, 3278–3281. [CrossRef] [PubMed]
67. Mitra, N.K.; Meudom, R.; Corzo, H.H.; Gorden, J.D.; Merner, B.L. Overcoming strain-induced rearrangement reactions: A mild dehydrative aromatization protocol for synthesis of highly distorted p-phenylenes. *J. Am. Chem. Soc.* **2016**, *138*, 3235–3240. [CrossRef] [PubMed]
68. Mitra, N.K.; Merryman, C.P.; Merner, L.B. Highly Strained para-phenylene-bridged macrocycles from unstrained 1,4-diketo macrocycles. *Synlett* **2017**, *28*, 2205–2211.
69. Golder, M.R.; Colwell, C.E.; Wong, B.M.; Zakharov, L.N.; Zhen, J.; Jasti, R. Iterative reductive aromatization/ring-closing metathesis. Strategy toward the synthesis of strained aromatic belts. *J. Am. Chem. Soc.* **2016**, *138*, 6577–6582. [CrossRef]
70. Lee, J.; Kalin, A.J.; Wang, C.; Early, J.T.; Al-Hashimi, M.; Fang, L. Donor–acceptor conjugated ladder polymer via aromatization-driven thermodynamic annulation. *Polym. Chem.* **2018**, *9*, 1603–1609. [CrossRef]

Disclaimer/Publisher's Note: The statements, opinions and data contained in all publications are solely those of the individual author(s) and contributor(s) and not of MDPI and/or the editor(s). MDPI and/or the editor(s) disclaim responsibility for any injury to people or property resulting from any ideas, methods, instructions or products referred to in the content.

Article

4,5-Dihydro-5-Oxo-Pyrazolo[1,5-a]Thieno[2,3-c]Pyrimidine: A Novel Scaffold Containing Thiophene Ring. Chemical Reactivity and In Silico Studies to Predict the Profile to GABA$_A$ Receptor Subtype

Letizia Crocetti, Gabriella Guerrini *, Fabrizio Melani, Claudia Vergelli and Maria Paola Giovannoni

Neurofarba, Pharmaceutical and Nutraceutical Section, University of Florence, 50019 Sesto Fiorentino, Italy; letizia.crocetti@unifi.it (L.C.); fabrizio.melani@unifi.it (F.M.); claudia.vergelli@unifi.it (C.V.); mariapaola.giovannoni@unifi.it (M.P.G.)
* Correspondence: gabriella.guerrini@unifi.it; Tel.: +39-055-457-3766

Abstract: The isosteric replacement of the benzene with thiophene ring is a chemical modification widely applied in medicinal chemistry. Several drugs containing the thiophene ring are marketed for treating various pathologies (osteoporosis, peripheral artery disorder, psychosis, anxiety and convulsion). Taking into account this evidence and as a continuation of our study in the GABA$_A$ receptor modulators field, we designed and synthesized new compounds containing the thiophene ring with 4,5-dihydro-5-oxo-pyrazolo[1,5-a]thieno[2,3-c]pyrimidine and pyrazolo[1,5-a]thieno[2,3-c] pyrimidine scaffold. Moreover, these cores, never reported in the literature, are isosteres of pyrazolo[1,5-a]quinazolines (PQ), previously published by us as GABA$_A$R subtype ligands. We introduced in the new scaffold those functions and groups (esters, ketones, alpha/beta-thiophene) that in our PQ derivatives were responsible for the activity, and at the same time, we have extensively investigated the reactivity of the new nucleus regarding the alkylation, reduction, halogenation and hydrolyses. On the six final designed compounds (**12c–f**, **22a,b**) molecular docking and dynamic simulation studies have been performed. The analysis of dynamic simulation, applying our reported model 'Proximity Frequencies', collocates with high probability **12c**, **22b**, in the agonist class towards α1β2γ2-GABA$_A$R.

Citation: Crocetti, L.; Guerrini, G.; Melani, F.; Vergelli, C.; Giovannoni, M.P. 4,5-Dihydro-5-Oxo-Pyrazolo[1,5-a]Thieno[2,3-c]Pyrimidine: A Novel Scaffold Containing Thiophene Ring. Chemical Reactivity and In Silico Studies to Predict the Profile to GABA$_A$ Receptor Subtype. *Molecules* **2023**, *28*, 3054. https://doi.org/10.3390/molecules28073054

Academic Editor: Joseph Sloop

Received: 9 March 2023
Revised: 27 March 2023
Accepted: 28 March 2023
Published: 29 March 2023

Copyright: © 2023 by the authors. Licensee MDPI, Basel, Switzerland. This article is an open access article distributed under the terms and conditions of the Creative Commons Attribution (CC BY) license (https:// creativecommons.org/licenses/by/ 4.0/).

Keywords: 4,5-dihydro-5-oxo-pyrazolo[1,5-a]thieno[2,3-c]pyrimidines; novel scaffold; thiophene; molecular dynamic studies; proximity frequencies (PF); GABA$_A$ receptor subtype; GABA$_A$ receptor ligands

1. Introduction

Thiophene is known to be an isostere of the benzene ring, thus representing a good tool in the field of medicinal chemistry to develop new drugs. The concept of isosterism applied to the thiophene ring was established in 1932 by Erlenmeyer, which proposed the equivalence between -CH = CH- and -S- (in benzene and thiophene, respectively) in terms of size, mass and capacity to provide an aromatic lone pair. Over time, compounds containing thiophene have been extensively investigated in medicinal chemistry, since they show a variety of pharmacological activities, such as anti-inflammatory, antioxidant, antimicrobial, antitumor, and antidepressant action [1–4]. Moreover, several thiophene-heterocycles-fused compounds have been approved by FDA as therapeutic agents for the treatment of osteoporosis (Raloxifene), peripheral artery disorder (Ticlopidine), psychosis (Olanzapine), anxiety and convulsion (Etizolam); see Figure 1.

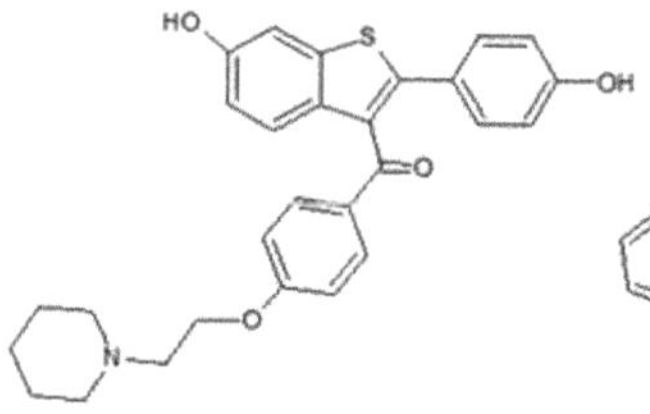

Figure 1. Thiophene-heterocycles-fused compounds known in literature.

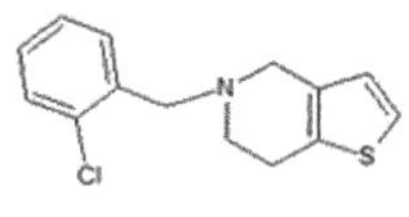

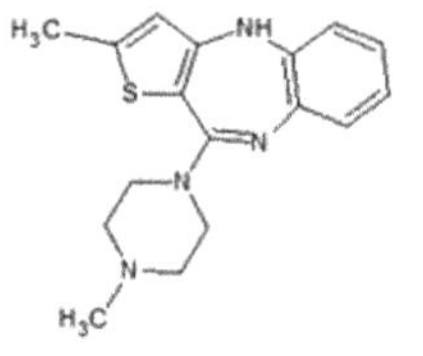

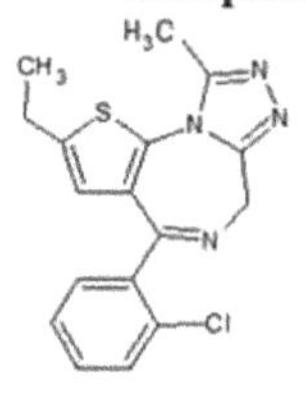

Starting from these evidences, we addressed our research towards the synthesis of heterocyclic compounds containing the thiophene ring. We applied this strategy to the synthesis of potential GABA$_A$R subtypes ligands, which represent a field of research extensively investigated by us for many years, by obtaining some interesting derivatives with pyrazolobenzotriazine, pyrazolopyrimidine and pirazoloquinazoline scaffold [5–8]. This choice is also supported by the fact that in the literature are present some GABA$_A$R ligands (Ro 19-4603, TB21007, Comp. 4, Comp. 16), in which the thiophene ring is fused or bonded to other cycles (Figure 2) [9–11]. In particular, we report here the synthesis of new compounds with 4,5-dihydro-5-oxo-pyrazolo[1,5-a]thieno[2,3-c]pyrimidine and pyrazolo[1,5-a]thieno[2,3-c]pyrimidine scaffold as result of the isosteric replacement of the benzene with thiophene ring in our pyrazolo[1,5-a]quinazolines previously published as GABA$_A$R subtype ligands [8,12] (Figure 3). Moreover, these new compounds can be considered as analogues of the abovementioned GABA$_A$R ligand Ro 19-4603, in which we formally operated a contraction of the central diazepine ring. As a first approach, in this new scaffold, we tried to introduce at position 3 those functions and groups (esters, ketons, alpha/beta-thiophene) responsible for activity in our pyrazolo[1,5-a]quinazolines and in compounds of the literature (i.e., Ro 19-4603 and Comp 16) (Figure 2).

The pyrazolo[1,5-a]thieno[2,3-c]pyrimidine core is not found in the literature (SciFinder, Reaxys), with the exception of the ethyl 4,5-dihydro-5-oxo-pyrazolo [1,5-a]thieno[2,3-c]pyrimidine-3-carboxylate (RN 942034-93-7), of which neither the synthesis nor the characterization is reported, however. In addition, this single compound is not mentioned in any published work. Thus, it was very intriguing to investigate the feasibility and the reactivity of this nucleus toward the most common reactions, such as alkylation, reduction, halogenation, and hydrolysis. Finally, molecular docking studies and evaluation of the 'Proximity Frequencies' (exploiting our reported model) [8,13] were performed on all the final compounds to predict their profile on the α1β2γ2-GABAAR subtype.

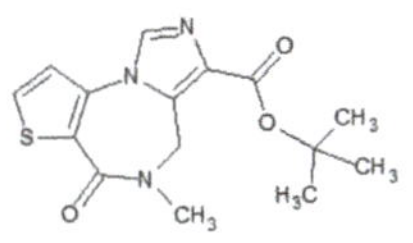

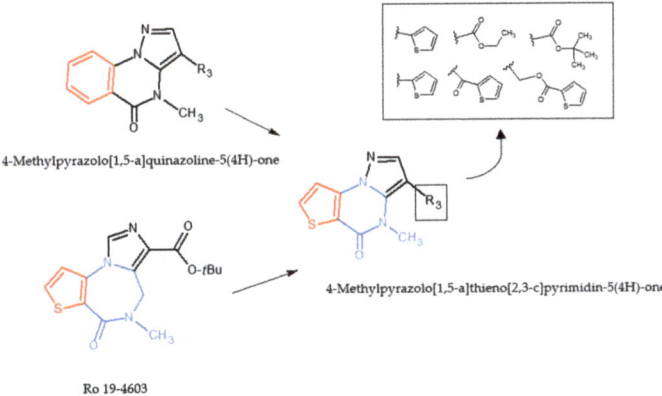

Figure 2. GABA$_A$R ligands with a thiophene ring reported in the literature (numbering is that of the original manuscript).

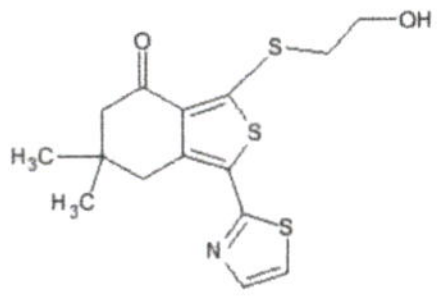

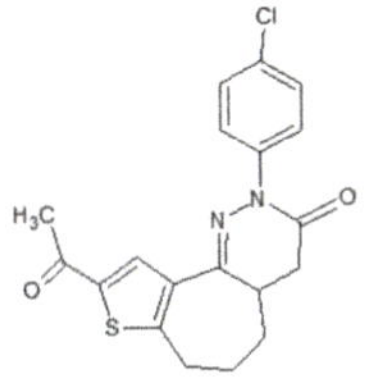

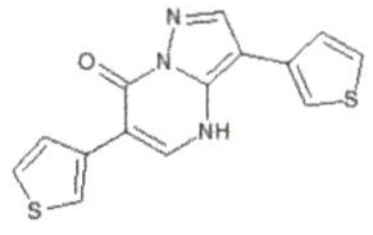

Figure 3. Aim of the work.

2. Results

2.1. Chemistry

The synthetic pathways for obtaining derivatives with pyrazolo[1,5-a]thieno[2,3-e]pyrimidine scaffold are depicted in Schemes 1–6. In this synthetic section, we report not only the procedures for obtaining the final designed products mentioned in Figure 3, but also some reactivity studies on this new scaffold which, furthermore, have produced interesting results. NMR spectra, elemental analysis and other structural information are reported in Supplementary Materials, Table S1.

The first step in building the pyrazolo[1,5-a]thieno[2,3-c]pyrimidine core is a diazotization reaction in the usual manner followed by reduction with SnCl$_2$, on the commercial material methyl 3-aminothiophene-2-carboxylate **1**, obtaining the corresponding 3-hydrazino hydrochloride derivative **2** [14]. This latter was then reacted with ethoxymethylenmalononitrile and ethyl 2-cyano-3-ethoxyacrylate, affording the pyrazolo[1,5-a]thieno[2,3-e]pyrimidin-5(4H)-ones 3-carbonitrile and 3-ethoxycarbonyl derivative **3** and **4**, respectively; further hydrolysis of **3** with H$_2$SO$_4$ conc. gave the 3-carboxamide derivative **5**. Compounds **6–8** were instead obtained by treatment of the same 3-hydrazino hydrochloride derivative **2** with 3-(dimethylamino)-2- (thien-2-carbonyl)acrylonitrile (compound **6**), 3-oxo-2-

(thien-3-yl)propionitrile (compound **7**), and 3-oxo-2-(thien-2-yl)propionitrile (compound **8**), respectively, following the procedure reported in our references [15,16] (Scheme 1).

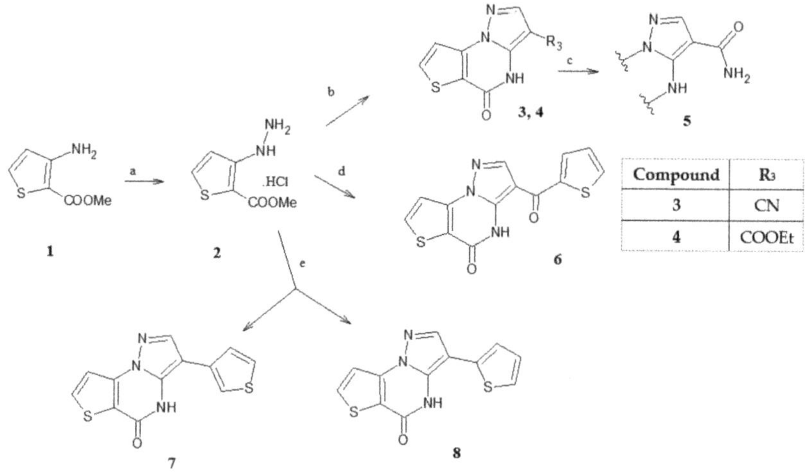

Scheme 1. Reagent and conditions: (a) NaNO$_2$/H$_2$O, HCl, SnCl$_2$, 0 °C; (b) ethoxymethylenmalononitrile for **3** and ethyl 2-cyanoethoxyacrylate for **5**, DMF/AcONa, refluxing temperature; (c) H$_2$SO$_4$, 80 °C; (d) 3-(dimethylamino)-2-(thien-2-ylcarbonyl)acrylonitrile, DMF/AcONa, refluxing temperature; (e) 3-oxo-2-(thien-3-yl)propionitrile for **7** and 3-oxo-2-(thien-2-yl)propionitrile for **8**, AcOH, refluxing temperature.

All the **3–8** pyrazolo[1,5-a]thieno[2,3-e]pyrimidin-5(4H)-ones intermediates synthesized in Scheme 1 were then subjected to halogenation, reduction and alkylation reactions (Scheme 2) and the results differed depending on the function/group bound at position 3 (**3**, R$_3$ = CN; **4**, R$_3$ = COOEt; **5**, R$_3$ = CONH$_2$; **6**, R$_3$ = thien-2-yl carbonyl; **7**, R$_3$ = 3-thienyl; **8**, R$_3$ = 2-thienyl). Starting from the halogenation reaction, the treatment with POCl$_3$/PCl$_5$ only for compounds **5** and **6** gave the corresponding 5-chloro derivatives **9c** and **9d**, which were effortlessly isolated and purified. From compound **7**, the intermediate 5-chloro derivative results as one spot in TLC but was not isolated (**9e**) and used as such for the subsequent reduction reaction. Differently, for compounds bearing a cyano group (**3**) or a thienyl ring at the 3-position (**7** and **8**), the 5-chloro derivatives were not obtained or were not easily isolable. In particular, from compounds **3** and **8** dark tares resulted in TLC many spots of complex purification. For the next step, involving the C-Cl bond cleavage and the resulting reduction to the 4,5-dihydro derivative **10c,d**, we choose NaBH$_4$ in the EtOH/CH$_2$Cl$_2$ mixture. The ethyl 5-chloropyrazolo[1,5-a]thieno [2,3-e]pyrimidine-3-carboxylate **9c** and the 3-(thien-2-ylcarbonyl)-5-chloropyrazolo [1,5-a]thieno[2,3-e]pyrimidine **9d** were rapidly transformed into the final compounds **10c,d**, which were easily recovered and purified; the (hetero) aromatization was then realized by treating **10c** and **10d** in toluene with Pd/C at refluxing temperature, obtaining the final products **11c** and **11d**.

A particular reactivity was evidenced for not isolated intermediate **9e**, the 3-(thien-3-yl)-5-chloropyrazolo[1,5-a]thieno[2,3-e]pyrimidine. In fact, when the reduction is performed with NaBH$_4$, starting and final products mixture was anyway recovered, also largely changing the reaction conditions. On the other hand, performing a catalytic transfer hydrogenation (CTH) with HCOONH$_4$ and Pd/C in EtOH [17], it was possible to highlight in TLC a spot that could be the 4,5-dihydro derivate, which rapidly and spontaneously converts into the 3-(thien-3-yl)pyrazolo[1,5-a]thieno[2,3-e]pyrimidine **11e**. Moving to the alkylation reactions, which afforded the final desired compounds **12c–f**, the lactam derivatives (**3–8**) were all treated following the classical method (DMF/K$_2$CO$_3$/CH$_3$I), but depending on the substituent at the 3-position, different reactivity was evidenced, specifically:

- Compounds **3**, **4** and **6** gave in good yield the 4-methyl derivatives **12a,c,d** (4-methyl-5-oxo-4,5-dihydropyrazolo[1,5-a]thieno[2,3-e]pyrimidine-3-carbonitrile **12a**, 3-ethyl carboxylate **12c** and 3-(thien-2-ylcarbonyl) **12d**).
- The 3-carboxamide derivative **5** gave different reaction products depending on the reaction conditions, in particular on the temperature. In fact, at 80 °C (standard condition), only the O-methyl derivative **13b** (5-methoxypyrazolo[1,5-a]thieno[2,3-e]pyrimidine-3-carboxamide) was obtained, while when maintaining the reaction at reflux temperature, the N-methyl isomer **12b** was obtained as a single product. The 3-carboxyamide derivative **12b** was also obtained by treatment of 4-methyl-5-oxo-4,5-dihydropyrazolo[1,5-a]thieno[2,3-e]pyrimidine-3-carbonitrile **12a** with conc. H_2SO_4, and thus, confirming the N-alkylation.
- Finally, the alkylation of **7** and **8** in the usual manner afforded a mixture of the N-methyl and O-methyl derivatives (**12e** and **13e**; **12f** and **13f**), which were separated by flash chromatography. The assignment of the exact structure was carried out with the use of ^{1}H-NMR technique, whose results are in agreement with our previous data [8]: when methyl is bound to N-4, the peak falls between 3.33 and 4.00 ppm, depending on the substituent at 3-position, while the O-methyl is unchanged for all compounds at 4.20 ppm.

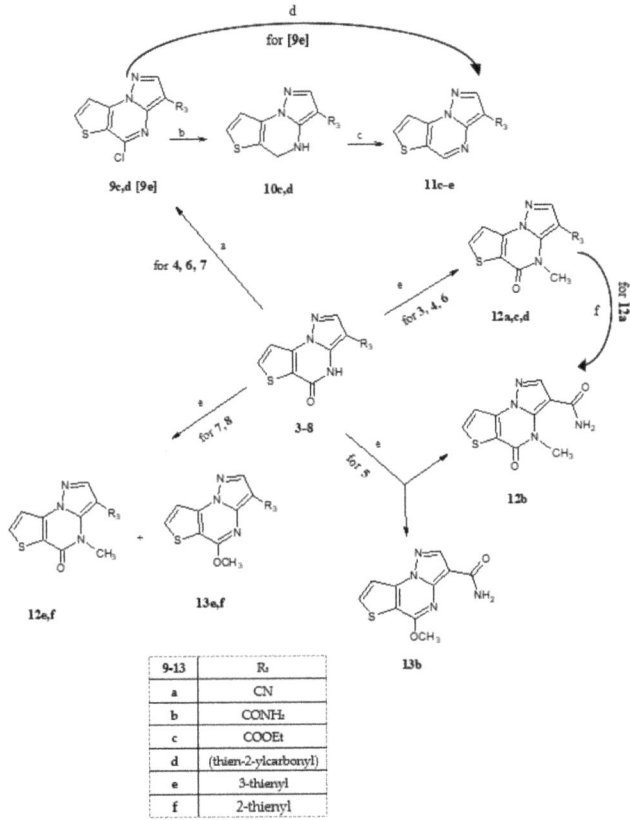

Scheme 2. Reagent and conditions: (a) $POCl_3/PCl_5$ refluxing temperature; (b) $NaBH_4$/EtOH/CH_2Cl_2, RT°, 40 min; (c) Toluene, Pd/C, refluxing temperature for **10c,d**; (d) $HCOONH_4$/EtOH, Pd/C, refluxing temperature for **9e**; (e) DMF/K_2CO_3/CH_3I, 80 °C for **12a,c,d–f** and **13b,e,f**; at reflux temperature for **12b**; (f) H_2SO_4 conc.

The regioselective O-alkylation observed for compound **5** could be due to a prevalence of the tautomeric form −N = C-OH with respect −NH-C = O; the predominance of the first one could indeed be associated with the amide function at position 3, since the $CONH_2$ group could form H-bonds with N-4, thus favoring the tricyclic heteroaromatic structure (see Figure 4). On the other hand, the rising temperature (reflux) can promote a free rotation of the C3-CO bond of the amide group, no longer involved in an H-bond, and thus, allowing alkylation at N-4 (compound **12b**).

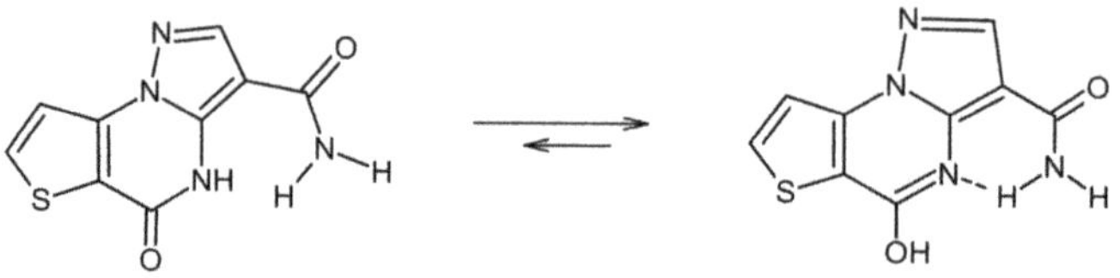

Figure 4. A possible explanation for O-alkylation.

Scheme 3 describes the different reactivity of the 3-ethyl carboxylate derivatives **4** and **12c** towards the alkaline or acid hydrolysis. The starting ethyl 5-oxo-4,5-dihydropyrazolo [1,5-a]thieno[2,3-e]pyrimidine-3-carboxylate **4** behaves in the usual manner to alkaline or acid hydrolysis, giving the corresponding 3-carboxylic acid **14**, which in turn undergoes decarboxylation in HCl 12M, at reflux temperature, yielding compound **15**. Instead, starting from the ester **12c**, also using different reaction conditions (NaOH 10% or LiOH in THF/water or AcOH/HCl or conc. HCl), it has never been possible to obtain the 3-carboxylic acid, but only the 3-decarboxylate derivative **16** is recovered.

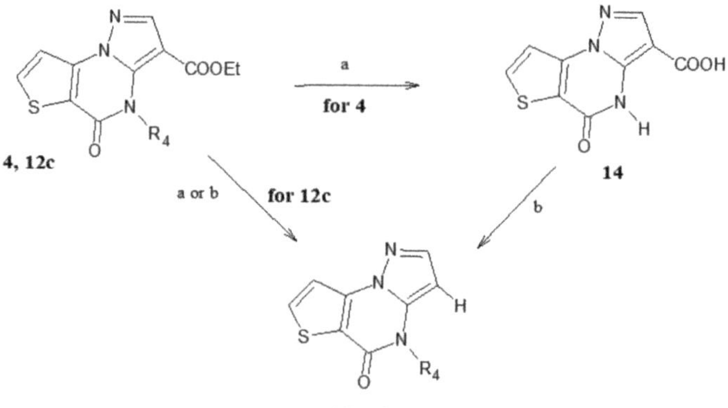

Compounds	R₄
4, 15	H
12c, 16	CH₃

Scheme 3. Reagent and conditions: (a) for **4** NaOH 10% solution or HCl conc.; for **12c** NaOH 10% solution or LiOH or HCl conc.; (b) HCl conc. reflux temperature.

Thus, to get the 4-methyl-5-oxo-4,5-dihydropyrazolo[1,5-a]thieno[2,3-e] pyrimidine-3-carboxylic acid, as a key intermediate for obtaining the final designed esters, we followed a reported procedure [18], which involves diazotization of the 3-carboxamide **12b** (Scheme 4); the reaction did not afford the desired 3-carboxylic acid, but a mixture of two compounds, one identified as the 3-decarboxylate **16** and a possible mechanism of decarboxylation process is reported. The second compound, colored green, was assumed to be a 3-nitroso derivative **17**; this hypothesis could be supported by the fact that, after decarboxylation, a high concentration of nitrosonium ion (NO^+) in the chemical environment could be able to make an electrophilic attach at position 3 of the pyrazolothienoquinazoline scaffold. The ^{1}H-NMR spectrum of the supposed compound **17**, in addition to the absence of the proton at position 3, shows a shift of the methyl bound to N-4 (4.23 ppm) that is not consistent with the chemical shift values of the 4-methyl-5-oxo-4,5-dihydropyrazolo[1,5-a]thieno[2,3-e]pyrimidine derivatives, whose N-methyl group constantly falls in the 3.3–3.7 ppm range. Moreover, the chemical shift of the proton in position 2 is lower than in other products with different substituents (CN, COOEt, $CONH_2$, COOH), suggesting a different electronic/steric environment resulting from the substituent in position 3. The mass analysis confirmed the structure of compound **17**. The same reaction performed on the 5-methoxypyrazolo[1,5-a]thieno[2,3-e]pyrimidine-3-carboxyamide **13b** also gave a mixture of two products, but in this case, the green 3-nitroso derivative **19** was obtained together with the desired 5-methoxypyrazolo[1,5-a]thieno[2,3-e]pyrimidine-3-carboxylic acid (**18**). The mass analysis again confirmed the structure. A possible mechanism of decarboxylation is reported in Figure 5.

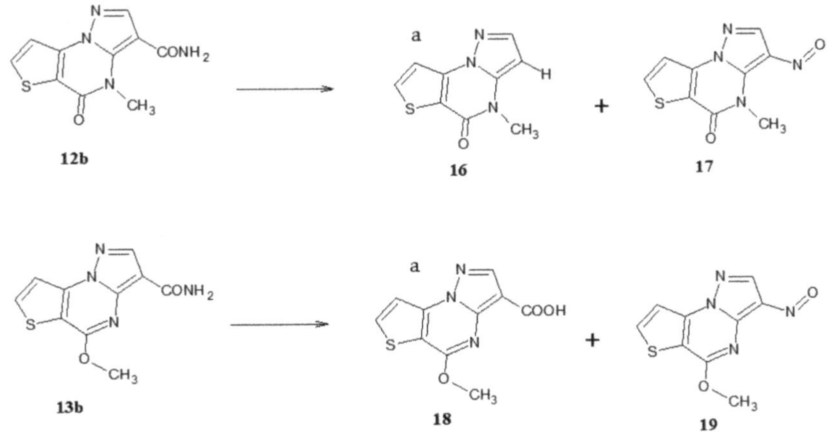

Scheme 4. Reagent and conditions: (a) H_2SO_4 conc., $NaNO_2/H_2O$, RT for 12 h.

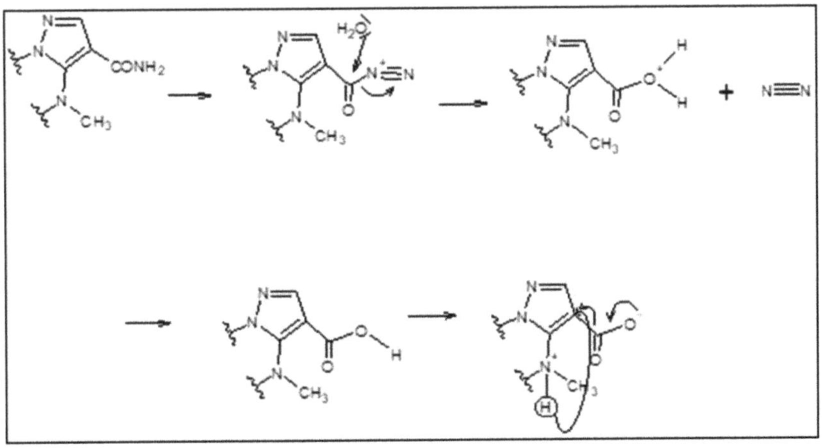

Figure 5. A possible mechanism of decarboxylation.

In order to obtain the desired 4-methyl-5-oxo-4,5-dihydropyrazolo[1,5-a]thieno[2,3-e]pyrimidine-3-carboxylic acid **21**, we explored a further synthetic strategy reported in Scheme 5. The 3-unsubstituted compound **16** was treated with HMTA obtaining the 4-methyl-5-oxo-4,5-dihydropyrazolo[1,5-a]thieno[2,3-e]pyrimidin- 3-carboxaldehyde derivative **20** and its further oxidation (KMnO$_4$/water/acetone/sodium hydroxide) finally gave the desired 3-carboxylic acid **21**. From acid **21**, the final desired esters **22a,b** were obtained by treatment with thionyl chloride and further addition of the suitable alcohol (t-BuOH and 2-thiophenemethanol respectively) in CH$_2$Cl$_2$.

The hydrolysis of the ester function to carboxylic acid also created problems on the ethyl pyrazolo[1,5-a]thieno[2,3-e]pyrimidine-3-carboxylate **11c** (Scheme 6). In fact, the desired product **23** was recovered in a meagre yield together with a big amount of the 3-decarboxylate **24** only if the hydrolysis of **11c** was performed in an alkaline medium; the reaction then evolved spontaneously towards the total formation of compound **24**. On the contrary, in acid medium, compound **11c** underwent a decarboxylation, directly giving the 3-unsubstituted derivative **24**.

Therefore, we explored a different synthetic way, starting from the 5-oxo-4,5-dihydropyrazolo[1,5-a]thieno[2,3-e]pyrimidin-3-carboxylic acid **14**, by using LiAlH$_4$ as reducing agent for the lactam function. Thus, after quenching the reaction and performing the standard workup, the residue was refluxed in toluene and Pd/C. The presence in ^{1}H-NMR spectrum of H5 at 9.06 ppm confirmed that the pyrazolothienopyrimidine core was indeed dehydrogenated but, at the same time, the carboxylic function was transformed in a methyl group, as evidenced by the peak at 2.33 ppm, compound **25**, again preventing the desired **23**.

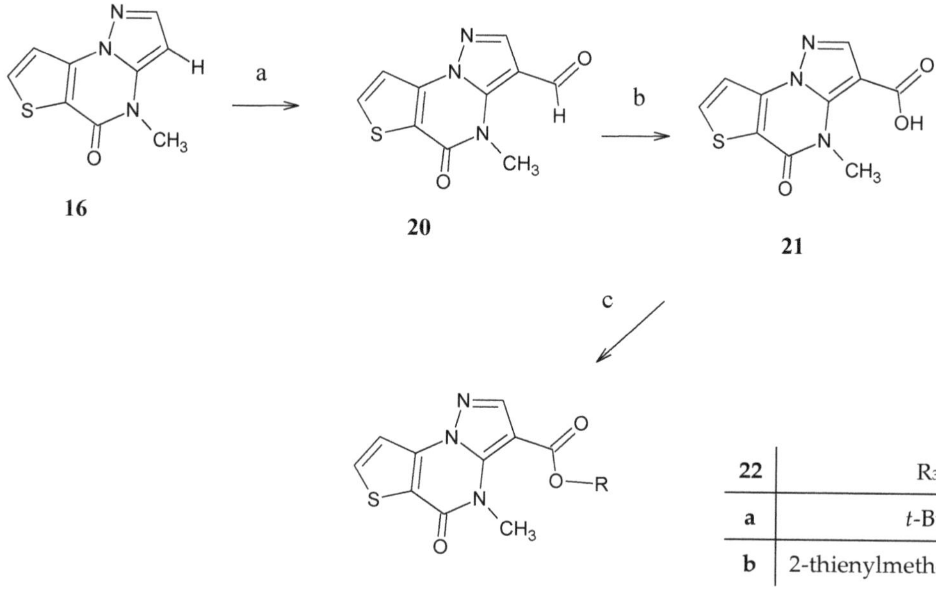

22	R$_3$
a	t-Bu
b	2-thienylmethoxycarbonyl

Scheme 5. Reagent and conditions: (a) AcOH/HMTA, 80 °C; (b) aceton/water, KMnO$_4$, NaOH 10%, 80 °C; (c) SOCl$_2$, CH$_2$Cl$_2$, t-BuOH for **22a** and 2-thiophenemethanol for **22b**, 50 °C.

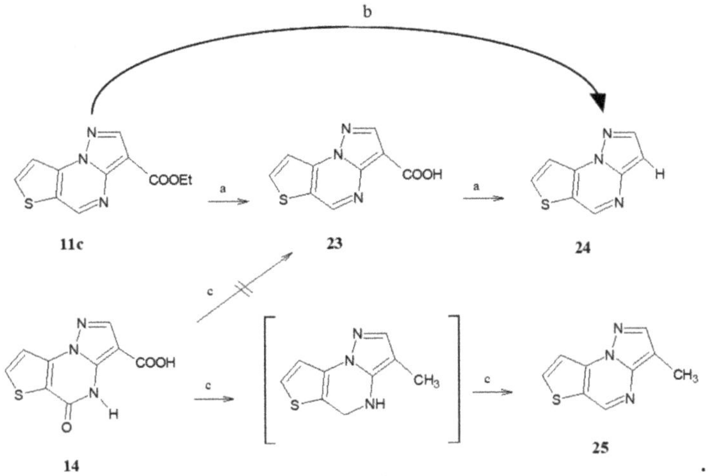

Scheme 6. Reagent and conditions: (a) NaOH 10% solution, or LiOH; (b) HCl or AcOH; (c) LiAlH4, THF abs., then: toluene, Pd/C, refluxing temperature.

2.2. Molecular Dynamic Studies

On the six final designed compounds (**12c–f** and **22a,b**), a molecular docking study and an evaluation of the 'Proximity Frequencies' [8,13] were performed to predict their profile on the $\alpha 1\beta 2\gamma 2$-GABA$_A$R subtype.

The value of Proximity Frequencies (PFs), used in a linear discriminant function (LDA), was able to correctly collocate 70.6% of agonists and 72.7% of antagonists by combining a double PF (αVal203-γThr142) with a triple PF (αHis102-αTyr160-γTyr58). The predictive capacity was evaluated on an appropriate training set of molecules with a cross-validation 'leave one out' (LOO) procedure. During a molecular dynamic simulation (60 ns), the agonist compounds were simultaneously close to the αVal203 and γThr142 amino acids, with a frequency of 37% compared to the frequency of 16% found by the antagonist compounds, while the antagonist compounds were simultaneously close to the αHis102, αTyr160, and γTyr58 amino acids, with a frequency of 35% against a frequency of 13% for agonist compounds. All the 3D structures of the molecules, as a training set and new final compounds, were designed [19] (DS ViewerPro 6.0 Accelrys Software Inc., San Diego, CA, USA) and placed in the binding site of the BDZs with the AUTODOCK 4.2 [20] docking program. The structure of the BDZ binding site was obtained from the recently solved GABA$_A$R structure (PDB ID 6D6T) [21].

The docking program performed on the selected compounds (**12c–f** and **22a,b**) gave a number of clusters of conformation(s) for each compound (rmsd 2.0). The evaluation of trajectories in the dynamic simulation was performed on the conformations that covered at least 90% of poses; the dynamic simulations were performed on an isolated portion of the protein between the α and γ chains comprising all amino acids within a radius of 2 nm from the center of the benzodiazepine binding site. Applying the PF model to the new selected compounds, it emerges that all six ones are collocated in the agonist class, **12d–f** with low probability, while **12c** and **22b**, the 3-ethyl and the 3-(2-thienylmethyl)carboxylate, respectively, with a percentage of prediction of 74% and 78%, are more probable. The 3-t-buthylcarboxylate **22a** shows a percentage of prediction slightly lower (69%), but always in the agonist class; see Table 1.

Table 1. Profile prediction obtained by PF model (LDA results).

N°	Prediction	% Ago	% Anta
12d	ago	66	34
12e	ago	64	36
12f	ago	64	36
12c	AGO	78	22
22a	AGO	69	31
22b	AGO	74	26

Compounds **12c** and **22b** have in the position 3 an ester group which is able to engage a strong hydrogen bond interaction with γThr142 through the carbonyl moiety. Additionally, **22a** has an ester group in the 3 position but the steric hindrance of the *t*-butyloxycarbonyl fragment makes less probable the hydrogen bond interaction of the carbonyl group with the γThr142 residue; see Figures 6–8.

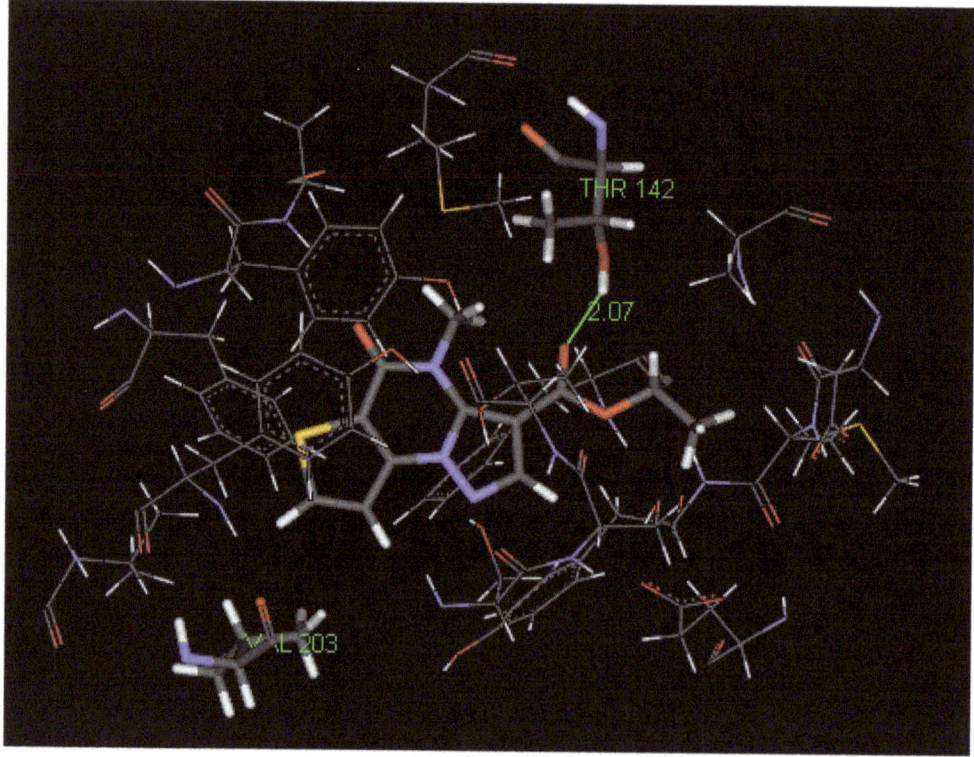

Figure 6. Hydrogen bond interaction of compound **12c** in the last frame of dynamic simulation.

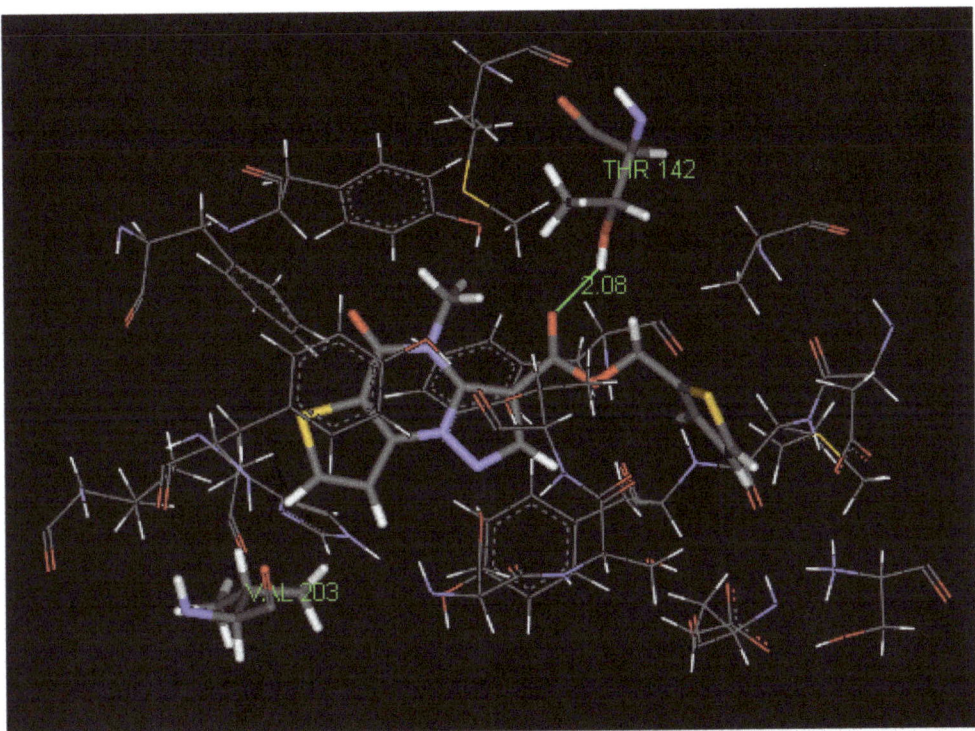

Figure 7. Hydrogen bond interaction of compound 22b in the last frame of dynamic simulation.

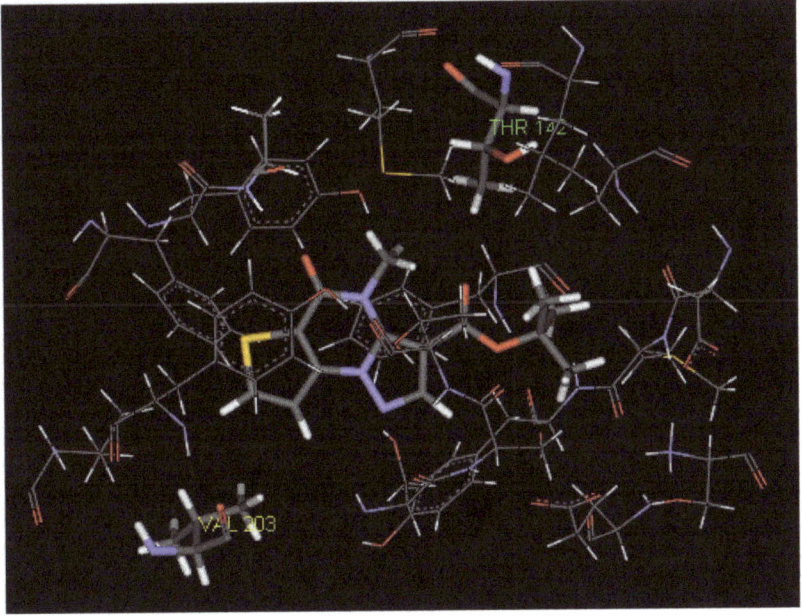

Figure 8. Hydrogen bond interaction of compound **22a** in the last frame of dynamic simulation.

These results are in accordance with our previously reported data [6,12], which evidenced the importance of the carbonyl group of the ester moiety to engage a strong hydrogen bond interaction with receptor protein. Compounds missing the ester group (**12d–f**) show a weak interaction with γThr142 in agreement with the low prediction percentage.

3. Experimental Section

3.1. Chemistry

General procedure for the synthesis of compounds 3,4,6-8. To a solution of **2** (1 mmol, 0.170 g) in DMF abs. and sodium acetate (1.3 mmol) was added 2-ethoxymethylenmalononitrile or ethyl-2-cyano-3-ethoxyacrylate, 3-(dimethylamino)- 2-(thien-2-carbonyl)acrylonitrile [15] (1.12 mmol) to obtain **3**, **4** and **6**. The solvent was AcOH when 3-oxo-2-(thien-3-yl)propionitrile and 3-oxo-2-(thien-2-yl)propionitrile [16] were used to obtain **7** and **8**. The reaction was refluxed for three hours, and after cooling, the addition of water and ice gave a precipitate that was filtered and purified by a suitable solvent

5-Oxo-4,5-dihydropyrazolo[1,5-a]thieno[2,3-e]pyrimidine-3-carbonitrile (3). From **2** and 2-ethoxymethylenmalononitrile. Recrystallized by *i*-propanol, yield 60%, cream crystals, mp > 300 °C. TLC: toluene/ethyl acetate/methanol 8:2:1.5 $v/v/v$ (Rf: 0.4); ^{1}H-NMR (400 MHz, DMSO-d_6) δ 13.47 (bs, 1H, NH, exch.); 8.28 (m, 2H, H-2 and H-7); 7.67 (d, 1H, H-8, J = 5.2 Hz). ^{13}C-NMR (100 MHz, DMSO-d_6) δ 163.70, 156.56, 148.45, 144.35, 138.87, 135.86, 129.09, 117.24, 115.45. Anal. $C_9H_4N_4OS$ (C, H, N).

Ethyl 5-oxo-4,5-dihydropyrazolo[1,5-a]thieno[2,3-e]pyrimidine-3-carboxylate (4). From **2** and ethyl 2-cyano-3-ethoxyacrylate. Recrystallized by ethanol, yield 58%, cream crystals, mp 215–216 °C. TLC: toluene/ethyl acetate/methanol 8:2:1.5 $v/v/v$ (Rf: 0.5); ^{1}H-NMR (400 MHz, DMSO-d_6) δ 11.74 (bs, 1H, NH, exch.); 8.35 (d, 1H, H-7, J = 5.2 Hz); 8.17 (s, 1H, H-2); 7.68 (d, 1H, H-8, J = 5.2 Hz); 4.29 (q, 2H, CH_2, J = 7.2 Hz); 1.29 (t, 3H, CH_3, J = 7.2 Hz). ^{1}H-NMR (400 MHz, CDCl$_3$) δ 9.63 (bs, 1H, NH, exch.); 8.08 (s, 1H, H-2); 7.93 (d, 1H, H-7, J = 5.2 Hz); 7.67 (d, 1H, H-8, J = 5.2 Hz); 4.39 (q, 2H, CH_2, J = 6.8 Hz); 1.41 (t, 3H, CH_3, J = 6.8 Hz). ^{13}C-NMR (100 MHz, DMSO-d_6) δ 165.29, 161.50, 156.56, 144.35, 143.95, 135.87, 133.26, 129.05, 117.25, 60.94, 14.15. Anal. $C_{11}H_9N_3O_3S$ (C, H, N).

3-(Thiophene-2-carbonyl)pyrazolo[1,5-a]thieno[2,3-e]pyrimidin-5(4H)-one (6). From **2** and 3-(dimethylamino)-2-(thien-2-carbonyl)acrylonitrile [15]. Recrystallized by ethanol, yield 95%, cream crystals, mp > 300 °C. TLC: toluene/ethyl acetate/methanol 8:2:1.5 $v/v/v$ (Rf: 0.7); ^{1}H-NMR (400 MHz, DMSO-d_6) δ 11.59 (bs, 1H, NH, exch.); 8.63 (s, 1H, H-2); 8.37 (d, 1H, H-7, J = 5.2 Hz); 8.17 (s, 1H, H-5′); 8.05 (d, 1H, H-3′, J = 4.4 Hz); 7.74 (d, 1H, H-8, J = 5.2 Hz); 7.29 (m, 1H, H-4′). ^{13}C-NMR (100 MHz, DMSO-d_6) δ 157.90, 142.76, 142.25, 138.37, 134.86, 133.09, 129.39, 117.34. Anal. $C_{13}H_7N_3O_2S_2$ (C, H, N).

3-(Thiophene-3-yl)pyrazolo[1,5-a]thieno[2,3-e]pyrimidin-5(4H)-one (7). From **2** and 3-oxo-2-(thien-3-yl)propionitrile [16]. Recrystallized by ethanol, yield 94%, cream crystals, mp 274–276 °C. TLC: toluene/ethyl acetate/methanol 8:2:1.5 $v/v/v$ (Rf: 0.7); ^{1}H-NMR (400 MHz, DMSO-d_6) δ 12.06 (bs, 1H, NH, exch.); 8.29 (d, 1H, H-7, J = 5.2 Hz); 8.18 (s, 1H, H-2); 7.86 (s, 1H, H-2′); 7.67 (d, 1H, H-8, J = 5.2 Hz); 7.06 (m, 1H, H-4′); 7.52 (m, 1H, H-5′). ^{13}C-NMR (100 MHz, DMSO-d_6) δ 165.10, 144.35, 141.71, 139.60, 129.95, 128.30, 127.90, 124.75, 117.20. Anal. $C_{12}H_7N_3OS_2$ (C, H, N).

3-(Thiophene-2-yl)pyrazolo[1,5-a]thieno[2,3-e]pyrimidin-5(4H)-one (8). From **2** and 3-oxo-2-(thien-2-yl)propionitrile [16]. Recrystallized by ethanol, yield 89%, green light crystals, mp 248–250 °C. TLC: toluene/ethyl acetate/methanol 8:2:1.5 $v/v/v$ (Rf: 0.7); ^{1}H-NMR (400 MHz, DMSO-d_6) δ 12.18 (bs, 1H, NH, exch.); 8.30 (d, 1H, H-7, J = 5.2 Hz); 8.00 (s, 1H, H-2); 7.67 (d, 1H, H-8, J = 5.2 Hz); 7.46 (d, 1H, H-5′, J = 4.8 Hz); 7.40 (d, 1H, H-3′, J = 2.8 Hz); 7.11 (dd, 1H, H-4′, J_1 = 4.8 Hz, J_2 = 4.0 Hz). ^{13}C-NMR (100 MHz, DMSO-d_6) δ 157.90, 141.71, 137.66, 128.40, 124.61, 117.19Anal. $C_{12}H_7N_3OS_2$ (C, H, N).

5-Oxo-4,5-dihydropyrazolo[1,5-a]thieno[2,3-e]pyrimidine-3-carboxamide (5). Compound **3** (0.23 mmol, 0.05 g) was suspended in H_2SO_4 conc., 1 mL and heated at 80 °C under stirring. After the starting material disappeared in TLC (toluene/ethyl acetate/methanol 8:2:1.5 $v/v/v$, as eluent, Rf: 0.2), the reaction was stopped; the addition of ice/water gave a precipitate which was filtered and purified by recrystallization with ethanol. Yield 92%, white crystals, mp > 300 °C. TLC: ^{1}H-NMR (400 MHz, DMSO-d_6) δ 10.75 (bs, 1H, NH, exch.); 8.33 (d, 1H, H-7, J = 4.8 Hz); 8.28 (s, 1H, H-2); 7.83 (bs, 1H, CONH, exch.); 7.68 (d, 1H, H-8, J = 4.8 Hz); 7.32 (bs, 1H, CONH, exch.).^{13}C-NMR (100 MHz, DMSO-d_6) δ 165.22, 163.70, 156.56, 150.55, 144.35, 143.91, 135.85, 132.06, 129.00. Anal. $C_9H_6N_4O_2S$ (C, H, N).

General procedure for the synthesis of compounds 9c,d. Compounds **4**, **6** and **7** (0.6 mmol) were suspended in a mixture of POCl$_3$ (5.5 mL) and PCl$_5$ (0.91 mmol, 0.190 g) and refluxed for three hours. The evaporation to dryness to eliminate the excess of POCl$_3$ gave a residue recuperated with ice/water filtered and purified with a suitable solvent, obtaining **9c,d**, starting from **4** and **6**, respectively. From compound **7**, the 5-chloro intermediate **9e** was not isolated but used as such, see below.

Ethyl 5-chloropyrazolo[1,5-a]thieno[2,3-e]pyrimidine-3-carboxylate (9c). From **5**. Recrystallized by ethanol, yield 90%, cream crystals, mp > 300 °C. TLC: toluene/ethyl acetate/methanol 8:2:1.5 $v/v/v$ (Rf: 0.7); ^{1}H-NMR (400 MHz, DMSO-d_6) δ 8.61 (m, 2H, H-2 and H-7); 8.02 (d, 1H, H-8, J = 5.2 Hz); 4.29 (q, 2H, CH$_2$, J = 7.2 Hz); 1.30 (t, 3H, CH$_3$, J = 7.2 Hz). ^{1}H-NMR (400 MHz, CDCl$_3$) δ 8.55 (s, 1H, H-2); 8.08 (d, 1H, H-7, J = 5.6 Hz); 7.95 (d, 1H, H-8, J = 5.6 Hz); 4.45 (q, 2H, CH$_2$, J = 7.2 Hz); 1.44 (t, 3H, CH$_3$, J = 7.2 Hz). ^{13}C-NMR (100 MHz, DMSO-d_6) δ 161.80, 155.06, 145.85, 145.55, 130.55, 125.66, 126.70, 118.30, 60.95, 14.13. Anal. $C_{11}H_8N_3O_2SCl$ (C, H, N).

(5-Chloropyrazolo[1,5-a]thieno[2,3-e]pyrimidin-3-yl)(thiophen-2-yl)methanone (9d). From **6**. Recrystallized by *i*-propanol, yield 95%, cream crystals, mp 186–188 °C. TLC: toluene/ethyl acetate/methanol 8:2:1.5 $v/v/v$ (Rf: 0.5); ^{1}H-NMR (400 MHz, DMSO-d_6) δ 8.80 (s, 1H, H-2); 8.65 (d, 1H, H-7, J = 5.2 Hz); 8.14 (s, 1H, H-5′); 8.07 (m, 2H, H-8 and H-3′); 7.30 (m, 1H, H-4′). ^{13}C-NMR (100 MHz, DMSO-d_6) δ 173.70, 155.06, 145.88, 145.53, 144.35, 135.81, 133.72, 130.50, 127.06, 126.74, 125.66, 126.70, 118.30. Anal. $C_{13}H_6N_3OS_2Cl$ (C, H, N).

General procedure for the synthesis of compounds 10c,d. Compounds **9c,d** (0.4 mmol) was dissolved in a mixture of CH$_2$Cl$_2$/EtOH (7.5 mL/15 mL) and NaBH$_4$ (3.6 mmol, 0.136 g) was added in small portions. The reaction was maintained at room temperature for 40 min, and then the evaporation to dryness of the solvent gave a residue which was recovered with water, filtered and purified with a suitable solvent, obtaining **10c,d**, respectively.

Ethyl 4,5-dihydropyrazolo[1,5-a]thieno[2,3-e]pyrimidine-3-carboxylate (10c). From **9c**. Recrystallized by water, yield 45%, cream crystals, mp 126–128 °C. TLC: toluene/ethyl acetate/methanol 8:2:1.5 $v/v/v$ (Rf: 0.5); ^{1}H-NMR (400 MHz, DMSO-d_6) δ 7.57 (m, 2H, H-2 and H-7); 7.18 (d, 1H, H-8, J = 4.4 Hz); 7.05 (s, 1H, NH, exch.); 4.7 (s, 2H, NCH$_2$); 4.17 (q, 2H, CH$_2$, J = 7.2 Hz); 1.25 (t, 3H, CH$_3$, J = 7.2 Hz). ^{1}H-NMR (400 MHz, CDCl$_3$) δ 8.55 (s, 1H, H-2); 8.08 (d, 1H, H-7, J = 5.6 Hz); 7.95 (d, 1H, H-8, J = 5.6 Hz); 4.70 (s, 2H, NCH$_2$); 4.18 (q, 2H, CH$_2$, J = 6.8 Hz); 1.25 (t, 3H, CH$_3$, J = 6.8 Hz). ^{13}C-NMR (100 MHz, DMSO-d_6) δ 163.29, 140.97, 125.82, 116.47, 59.18, 41.96, 15.02. Anal. $C_{11}H_{11}N_3O_2S$ (C, H, N).

4,5-Dihydropyrazolo[1,5-a]thieno[2,3-e]pyrimidin-3-yl(thiophen-2-yl)methanone (10d). From **9d**. Recrystallized by ethanol, yield 40%, cream crystals, mp 186–188 °C. TLC: toluene/ethyl acetate/methanol 8:2:1.5 $v/v/v$ (Rf: 0.3); ^{1}H-NMR (400 MHz, DMSO-d_6) δ 8.11 (s, 1H, H-2); 7.97 (s, 1H, H-5′); 7.91 (d, 1H, H-7, J = 4.8 Hz); 7.78 (s, 1H, NH, exch.); 7.61 (d, 1H, H-8 J = 4.8 Hz); 7.22 (m, 2H, H-3′ and H-4′); 4.79 (s, 2H, NCH$_2$). ^{13}C-NMR (100 MHz, DMSO-d_6) δ 172.98, 158.01, 144.82, 144.35, 137.85, 135.71, 133.57, 129.06, 127.74, 123.66, 119.30, 51.55. Anal. $C_{13}H_9N_3OS_2$ (C, H, N).

General procedure for the synthesis of compounds 11c,d. Compounds **10c,d** (0.4 mmol) was dissolved in toluene (15 mL), and Pd/C as catalyst was added. The reaction was refluxed for 3–5 h and filtered off the catalyst. The evaporation to dryness of the solution yielded a residue recovered with water, filtered and purified with a suitable solvent, obtaining **11c** and **11d**, respectively.

Ethyl pyrazolo[1,5-a]thieno[2,3-e]pyrimidine-3-carboxylate (11c). From **10c**. Recrystallized by ethanol yield 50%, white crystals, mp 159–160 °C. TLC: toluene/ethyl acetate 8:2 v/v (Rf: 0.1); ^{1}H-NMR (400 MHz, DMSO-d_6) δ 9.42 (s, 1H, H-5); 8.60 (s, 1H, H-2); 8.57 (d, 1H, H-7, J = 5.2 Hz); 7.98 (d, 1H, H-8, J = 5.2 Hz); 4.31 (q, 2H, CH$_2$, J = 6.8 Hz); 1.32 (t, 3H,

CH$_3$, J = 6.8 Hz). ^{13}C-NMR (100 MHz, DMSO-d$_6$) δ 162.34, 148.66, 145.99, 145.25, 141.82, 140.02, 123.02, 115.95, 103.38, 60.08, 14.92. Anal. C$_{11}$H$_9$N$_3$O$_2$S (C, H, N).

Pyrazolo[1,5-a]thieno[2,3-e]pyrimidin-3-yl(thiophen-2-yl)methanone (11d). From **10d**. Recrystallized by *i*-propanol, yield 68%, white crystals, mp 73–75 °C. TLC: toluene/ethyl acetate/methanol 8:2:1.5 *v/v/v* (Rf: 0.2); ^{1}H-NMR (400 MHz, DMSO-d$_6$) δ 9.42 (s, 1H, H-5); 8.74 (s, 1H, H-2); 8.59 (d, 1H, H-7, J = 4.8 Hz); 8.23 (d, 1H, H-5′ J = 2.4 Hz); 8.03 (m, 2H, H-3′ and H-8); 7.28 (m, 1H, H-4′). ^{13}C-NMR (100 MHz, DMSO-d$_6$) δ 173.08, 149.01, 145.82, 144.85, 144.35, 135.81, 133.43, 130.57, 129.06, 126.74, 125.60, 119.10. Anal. C$_{13}$H$_7$N$_3$OS$_2$ (C, H, N).

3-(Thiophen-3-yl)pyrazolo[1,5-a]thieno[2,3-e]pyrimidine (11e). Compound **7**, 3-(thiophene-3-yl)pyrazolo[1,5-a]thieno[2,3-e]pyrimidin-5(4H)-one (0.6 mmol), was suspended in a mixture of POCl$_3$ (5.5 mL) and PCl$_5$ (0.91 mmol, 0.190 g) and refluxed for three hours. The evaporation to dryness to eliminate the excess of POCl$_3$ gave the corresponding 5-chloro derivative (**9e**), not isolated but used as such for the next reduction step through a CTH (catalytic transfer hydrogenation). Thus, this intermediate suspended in EtOH (20 mL) was added of ammonium formate (4.08 mmol, 0.275 g) and 10% Pd/C as catalyst. The reaction was maintained at reflux temperature for several hours, during which it was possible to evidence, by TLC, the formation of the 4,5-dihydro derivative in a mixture with the final 4,5-dehydro compound **11e**. When the reaction finished, the catalyst was filtered off, the solution evaporated to dryness, and the residue recovered with water. Recrystallized by ethanol 80%, yield 90%, cream crystals, mp 157–160 °C. TLC: toluene/ethyl acetate 8:2 *v/v* (Rf: 0.5); ^{1}H-NMR (400 MHz, DMSO-d$_6$) δ 9.24 (s, 1H, H-5); 8.65 (s, 1H, H-2); 8.45 (d, 1H, H-7, J = 5.2 Hz); 8.05 (s, 1H, H-2′); 7.95 (d, 1H, H-8, J = 5.2 Hz); 7.86 (d, 1H, H-5′, J = 4.8 Hz); 7.65 (m, 1H, H-4′). ^{13}C-NMR (100 MHz, DMSO-d$_6$) δ 145.56, 141.21, 138.47, 126.77, 119.61, 115.88. Anal. C$_{12}$H$_7$N$_3$S$_2$ (C, H, N).

General procedure for the synthesis of compounds 12a,c–f and 13b,e,f. A solution of DMF abs. (5 mL), compounds **3–8** (0.40 mmol) and K$_2$CO$_3$ anhydrous (0.80 mmol) was maintained for 15 min at room temperature. After this time, methyl iodide (0.80 mmol) was added and enhanced temperature to 80 °C. After one hour and monitoring the reaction by TLC, adding water gave a precipitate, filtered and purified by a suitable solvent. In the case of compounds **3**, **4** and **6**, only 4-N-CH$_3$ derivatives were formed (**12a,c,d**). From compound **4**, only the 5-methoxyderivative **13b** was recovered, while if the reaction is performed at reflux temperature, only the 4-methyl derivative **12b** was obtained. From **7** and **8**, a mixture of two products was recovered at the end of alkylation. The chromatographic separation permits isolating the 4-NCH$_3$ (**12e,f**) and the 5-OCH$_3$ derivatives (**13e,f**).

4-Methyl-5-oxo-4,5-dihydropyrazolo[1,5-a]thieno[2,3-e]pyrimidine-3-carbonitrile (12a). From **3**. Recrystallized by ethanol, yield 89%, cream crystals, mp 224–226 °C. TLC: toluene/ethyl acetate/methanol 8:2:1 *v/v/v* (Rf: 0.6); ^{1}H-NMR (400 MHz, DMSO-d$_6$) δ 8.44 (s, 1H, H-2); 8.36 (d, 1H, H-7, J = 4.0 Hz); 7.71 29/03/2023 (d, 1H, H-8, J = 4.0 Hz); 3.73 (s, 3H, NCH$_3$). ^{13}C-NMR (100 MHz, DMSO-d$_6$) δ 163.76, 156.51, 148.40, 144.37, 144.30, 135.86, 129.07, 115.41, 106.58, 36.50. Anal. C$_{10}$H$_6$N$_4$OS (C, H, N).

4-Methyl-5-oxo-4,5-dihydropyrazolo[1,5-a]thieno[2,3-e]pyrimidine-3-carboxamide (12b). From **5** at reflux temperature. Recrystallized by ethanol, yield 93%, cream crystals, mp > 300 °C. TLC: dichlorometane/methanol 9:1 *v/v* (Rf: 0.3); ^{1}H-NMR (400 MHz, DMSO-d$_6$) δ 8.28 (d, 1H, H-7, J = 4.8 Hz); 8.11 (s, 1H, H-2); 7.86 (bs, 1H, CONH, exch.); 7.65 (d, 1H, H-8, J = 4.8 Hz); 7.32 (bs, 1H, CONH, exch.); 3.73 (s, 3H, NCH$_3$). ^{13}C-NMR (100 MHz, DMSO-d$_6$) δ 165.30, 163.76, 156.51, 150.40, 145.67, 144.31, 135.86, 132.05, 129.08, 36.69. Anal. C$_{10}$H$_8$N$_4$O$_2$S (C, H, N). The treatment of compound **12a** with sulfuric acid at 60 °C and the subsequent addition of ice/water gave the precipitate, **12b** recovered by filtration.

Ethyl 4-methyl-5-oxo-4,5-dihydropyrazolo[1,5-a]thieno[2,3-e]pyrimidine-3-carboxylate (12c). From **4**. Recrystallized by ethanol, yield 75%, cream crystals, mp 173–174 °C. TLC: toluene/ethyl acetate/methanol 8:2:1.5 *v/v/v* (Rf: 0.5); ^{1}H-NMR (400 MHz, CDCl$_3$) δ 8.19 (s, 1H, H-2); 7.86 (d, 1H, H-7, J = 5.2 Hz); 7.64 (d, 1H, H-8, J = 5.2 Hz); 4.34 (q, 2H,

CH$_2$, J = 7.2 Hz); 4.04 (s, 3H, NCH$_3$); 1.40 (t, 3H, CH$_3$, J = 7.2 Hz). ^{13}C-NMR (100 MHz, DMSO-d$_6$) δ 161.78, 156.10, 145.24, 142.00, 141.69, 138.12, 117.25, 116.95, 98.60, 60.77, 33.57, 14.59. Anal. C$_{12}$H$_{11}$N$_3$O$_3$S (C, H, N).

3-(Thiophene-2-carbonyl)-4-methylpyrazolo[1,5-a]thieno[2,3-e]pyrimidin-5(4H)-one (12d). From **6**. Recrystallized by ethanol, yield 95%, cream crystals, mp 183–186 °C. TLC: toluene/ethyl acetate/methanol 8:2:1.5 $v/v/v$ (Rf: 0.5); ^{1}H-NMR (400 MHz, DMSO-d$_6$) δ 8.37 (d, 1H, H-7, J = 5.2 Hz); 8.33 (s, 1H, H-2); 8.10 (d, 1H, H-5' J = 4.8 Hz); 7.82 (d, 1H, H-3', J = 3.6 Hz); 7.74 (d, 1H, H-8, J = 5.2 Hz); 7.30 (m, 1H, H-4'); 3.56 (s, 3H NCH$_3$). ^{13}C-NMR (100 MHz, DMSO-d$_6$) δ 179.37, 156.18, 145.24, 144.91, 141.71, 138.40, 135.92, 135.62, 129.12, 117.39, 106.50, 33.07. Anal. C$_{14}$H$_9$N$_3$O$_2$S$_2$ (C, H, N).

3-(Thiophene-3-yl)-4-methylpyrazolo[1,5-a]thieno[2,3-e]pyrimidin-5(4H)-one (12e). From **7**, after chromatographic separation, second eluting band (toluene/ethyl acetate/methanol 8:2:1.5 $v/v/v$ as eluent, Rf: 0.5), yield 35%, cream crystals, mp 109–111 °C. ^{1}H-NMR (400 MHz, DMSO-d$_6$) δ 8.29 (d, 1H, H-7, J = 4.8 Hz); 7.85 (s, 1H, H-2); 7.68 (d, 1H, H-8, J = 4.8 Hz); 7.63–7.59 (m, 2H, H-2' and H-4'); 7.25 (m, 1H, H-5'); 3.40 (s, 3H, NCH$_3$). ^{1}H-NMR (400 MHz, CDCl$_3$) δ 7.83 (d, 1H, H-7, J = 5.2 Hz); 7.71 (s, 1H, H-2); 7.65 (d, 1H, H-8, J = 5.2 Hz); 7.40 (m, 1H, H-2'); 7.26 (m, 1H, H-4'); 7.11 (d, 1H, H-5', J = 4.4 Hz); 3.41 (s, 3H, NCH$_3$). ^{13}C-NMR (100 MHz, DMSO-d$_6$) δ 157.94, 144.97, 140.00, 138.10, 137.90, 130.86, 125.99, 114.34, 114.23, 106.12, 20.54. Anal. C$_{13}$H$_9$N$_3$OS$_2$ (C, H, N).

3-(Thiophene-2-yl)-4-methylpyrazolo[1,5-a]thieno[2,3-e]pyrimidin-5(4H)-one (12f). From **8** after chromatographic separation, second eluting band (toluene/ethyl acetate/methanol 8:2:1.5 $v/v/v$ as eluent, Rf: 0.5), yield 36%, yellow light crystals, mp 140–141 °C. ^{1}H-NMR (400 MHz, DMSO-d$_6$) δ 8.30 (d, 1H, H-7, J = 5.2 Hz); 7.90 (s, 1H, H-2); 7.68 (d, 1H, H-8, J = 5.2 Hz); 7.62 (d, 1H, H-5', J = 5.2 Hz); 7.18 (m, 1H, H-3'); 7.13 (dd, 1H, H-4', J_1 = 4.8 Hz, J_2 = 3.6 Hz); 3.31 (s, 3H, NCH$_3$). ^{13}C-NMR (100 MHz, DMSO-d$_6$) δ 165.46, 154.88, 145.71, 137.76, 135.14, 131.79, 129.06, 117.48, 114.01, 56.89. Anal. C$_{13}$H$_9$N$_3$OS$_2$ (C, H, N).

5-Methoxypyrazolo[1,5-a]thieno[2,3-e]pyrimidine-3-carboxamide (13b). From **5** at room temperature. Recrystallized by ethanol, yield 75%, cream crystals, mp 269–270 °C. TLC: dichlorometane/methanol 8:2 v/v (Rf: 0.6); ^{1}H-NMR (400 MHz, DMSO-d$_6$) δ 8.42 (d, 1H, H-7, J = 4.8 Hz); 8.36 (s, 1H, H-2); 7.90 (d, 1H, H-8, J = 4.8 Hz); 7.49 (bs, 1H, CONH, exch.); 7.34 (bs, 1H, CONH, exch.); 4.22 (s, 3H, OCH$_3$). ^{13}C-NMR (100 MHz, DMSO-d$_6$) δ 164.40, 162.26, 145.87, 145.50, 130.56, 125.68, 114.45, 107.32, 53.79. Anal. C$_{10}$H$_8$N$_4$O$_2$S (C, H, N).

5-Methoxy-3-(thiophene-3-yl)pyrazolo[1,5-a]thieno[2,3-e]pyrimidine (13e). From **7**, after chromatographic separation, first eluting band (toluene/ethyl acetate/methanol 8:2:1.5 $v/v/v$ as eluent, Rf: 0.8), yield 35%, cream crystals, mp 127–130 °C. ^{1}H-NMR (400 MHz, DMSO-d$_6$) δ 8.49 (s, 1H, H-2); 8.38 (d, 1H, H-7, J = 5.2 Hz); 7.98 (m, 1H, H-2'); 7.85 (d, 1H, H-8, J = 5.2 Hz); 7.81 (d, 1H, H-5', J = 4.8 Hz); 7.62 (m, 1H, H-4'); 4.20 (s, 3H, OCH$_3$). ^{13}C-NMR (100 MHz, DMSO-d$_6$) δ 162.20, 144.86, 139.60, 133.47, 128.37, 128.20, 126.71, 125.60, 124.75, 114.56, 107.90, 53.75. Anal. C$_{13}$H$_9$N$_3$OS$_2$ (C, H, N).

5-Methoxy-3-(thiophene-2-yl)pyrazolo[1,5-a]thieno[2,3-e]pyrimidine (13f). From **8** after chromatographic separation, first eluting band (toluene/ethyl acetate/methanol 8:2:1.5 $v/v/v$ as eluent, Rf: 0.8), yield 24%, yellow light crystals, mp 118–120 °C. ^{1}H-NMR (400 MHz, DMSO-d$_6$) δ 8.43 (s, 1H, H-2); 8.39 (d, 1H, H-7, J = 5.2 Hz); 7.85 (d, 1H, H-8, J = 5.2 Hz); 7.57 (m, 1H, H-5'); 7.42 (d, 1H, H-3', J = 4.8 Hz); 7.11 (m, 1H, H-4'); 4.20 (s, 3H, OCH$_3$). ^{13}C-NMR (100 MHz, DMSO-d$_6$) δ 162.20, 144.86, 138.20, 133.65, 130.57, 128.67, 128.00, 126.74, 125.60, 114.56, 107.90, 53.50. Anal. C$_{13}$H$_9$N$_3$OS$_2$ (C, H, N).

5-Oxo-4,5-dihydropyrazolo[1,5-a]thieno[2,3-e]pyrimidine-3-carboxylic acid (14). To a suspension of **4** (0.40 mmol) in NaOH 10% solution (10 mL), was added 0.5 mL of methoxyethanol to favour the solubilization. The reaction was refluxed for 1.30 h, then ice/water and HCl 6N until pH 1. The precipitate formed was filtered, washed with water and purified by a suitable solvent. Recrystallized by ethanol, yield 78%, cream crystals, mp 282–284 °C. TLC: toluene/ethyl acetate/acetic acid 8:2:2 $v/v/v$ (Rf: 0.5); ^{1}H-NMR

(400 MHz, DMSO-d$_6$) δ 12.70 (bs, 1H, OH, exch.); 11.37 (bs, 1H, NH, exch.); 8.34 (d, 1H, H-7, J = 4.4 Hz); 8.13 (s, 1H, H-2); 7.68 (d, 1H, H-8, J = 4.4 Hz). ^{13}C-NMR (100 MHz, DMSO-d$_6$) δ 165.25, 164.96, 156.50, 144.35, 143.97, 135.87, 133.20, 129.04, 117.65. Anal. C$_9$H$_5$N$_3$O$_3$S (C, H, N).

5-Oxo-4,5-dihydropyrazolo[1,5-a]thieno[2,3-e]pyrimidine (15). The acid **14** (0.50 mmol) was suspended in 10 mL of HCl conc. and maintained at reflux temperature for 5 h. Adding ice/water gave a residue filtered and purified by recrystallization with ethanol. Recrystallized by ethanol, yield 65%, cream crystals, mp 290–291 °C. TLC: toluene/ethyl acetate/methanol 8:2:1.5 $v/v/v$ (Rf: 0.7); ^{1}H-NMR (400 MHz, DMSO-d$_6$) δ 12.30 (bs, 1H, NH, exch.); 8.26 (s, 1H, H-7); 7.74 (s, 1H, H-2); 7.64 (s, 1H, H-8); 5.93 (s, 1H, H-3). ^{13}C-NMR (100 MHz, DMSO-d$_6$) δ 164.76, 156.26, 148.45, 144.37, 135.85, 133.21, 129.09, 114.15. Anal. C$_8$H$_5$N$_3$OS (C, H, N).

4-Methyl-5-oxo-4,5-dihydropyrazolo[1,5-a]thieno[2,3-e]pyrimidine (16). The ester **12c** (0.50 mmol) was suspended in 10 mL of NaOH 10% solution and maintained at 80 °C until the starting material disappeared. The extraction with ethyl acetate and the next usual work up gave a residue which was filtered and purified by recrystallization with ethanol. Yield 71%, cream crystals, mp 157–159 °C. TLC: toluene/ethyl acetate/acetic acid 8:2:1 $v/v/v$ (Rf: 0.8); ^{1}H-NMR (400 MHz, DMSO-d$_6$) δ 8.18 (d, 1H, H-7, J = 5.6 Hz); 7.83 (d, 1H, H-2 J = 2.0 Hz); 7.62 (d, 1H, H-8, J = 5.6 Hz); 6.22 (d, 1H, H-3 J = 2.0 Hz); 3.86 (s, 3H, N-CH$_3$). ^{1}H-NMR (400 MHz, CDCl$_3$) δ 7.82 (d, 1H, H-7, J = 5.6 Hz); 7.87 (d, 1H, H-2 J = 2.0 Hz); 7.64 (d, 1H, H-8, J = 5.6 Hz); 5.96 (d, 1H, H-3 J = 2.0 Hz); 3.63 (s, 3H, N-CH$_3$). ^{13}C-NMR (100 MHz, DMSO-d$_6$) δ 163.71, 156.56, 148.49, 144.30, 135.80, 133.27, 129.09, 114.20, 37.05. Anal. C$_9$H$_7$N$_3$OS (C, H, N).

General procedure for the synthesis of compounds 17 and 19. A suspension of **12b** or **13b** (0.32 mmol) in H$_2$SO$_4$ conc. (8 mL) was stirred until a solution was obtained and then cooled at 0 °C; to this solution, sodium nitrite (0.22g, 3.2 mmol/5 mL of water) was slowly added and the green suspension was maintained for 3 h at 0 °C. The suspension was made alkaline and extracted with ethyl acetate. After the standard work-up, the evaporation of the organic layer gave a green residue that was purified and characterized.

4-Methyl-3-nitrosopyrazolo[1,5-a]thieno[2,3-e]pyrimidin-5(4H)-one (17). From **12b**. Recrystallized by ethanol, yield 50%, green crystals, mp 240–242 °C. TLC: toluene/ethyl acetate/acetic acid 8:2:1 $v/v/v$ (Rf: 0.9); ^{1}H-NMR (400 MHz, DMSO-d$_6$) δ 8.44 (d, 1H, H-7, J = 5.6 Hz); 7.74 (d, 1H, H-8, J = 5.6 Hz); 7.72 (s, 1H, H-2); 4.21 (s, 3H, NCH$_3$). ^{13}C-NMR (100 MHz, DMSO-d$_6$) δ 163.70, 156.59, 148.49, 144.30, 134.99, 133.25, 129.01, 104.05, 36.99. ESI-HRMS (m/z) calculated for [M+H]$^+$ ion species C$_9$H$_6$N$_4$O$_2$S = 235,0295; found: 235,0284. Anal. C$_9$H$_6$N$_4$O$_2$S (C, H, N).

5-Methoxy-3-nitrosopyrazolo[1,5-a]thieno[2,3-e]pyrimidine (19). From **13b**. Recrystallized by ethanol, yield 48%, green crystals, mp 218–220 °C. TLC: toluene/ethyl acetate/acetic acid 8:2:1 $v/v/v$ (Rf: 0.9); ^{1}H-NMR (400 MHz, DMSO-d$_6$) δ 8.86 (s, 1H, H-2); 8.54 (d, 1H, H-7, J = 5.6 Hz); 7.94 (d, 1H, H-8, J = 5.6 Hz); 4.23 (s, 3H, OCH$_3$). ^{13}C-NMR (100 MHz, DMSO-d$_6$) δ 163.65, 156.60, 148.49, 144.30, 134.99, 133.20, 128.99, 104.10, 55.60. ESI-HRMS (m/z) calculated for [M+H]$^+$ ion species C$_9$H$_6$N$_4$O$_2$S = 235,0294; found: 235,0284. Anal. C$_9$H$_6$N$_4$O$_2$S (C, H, N).

5-Methoxy-3-nitrosopyrazolo[1,5-a]thieno[2,3-e]pyrimidin-3-carboxylic acid (18). From **13b**, after acidification of the alkaline solution. The carboxylic acid was recrystallized by ethanol, yield 30%, white crystals, mp 218–220 °C. TLC: toluene/ethyl acetate/acetic acid 8:2:1 $v/v/v$ (Rf: 0.5); ^{1}H-NMR (400 MHz, DMSO-d$_6$) δ 12.24 (bs, 1H, OH, exch.); 8.44 (d, 1H, H-7, J = 4.8 Hz); 8.38 (s, 1H, H-2); 7.88 (d, 1H, H-8, J = 4.8 Hz); 4.17 (s, 3H, OCH$_3$). ^{13}C-NMR (100 MHz, DMSO-d$_6$) δ 169.35, 162.20, 145.80, 145.59, 130.50, 126.70, 125.69, 114.50, 107.10, 53.70. Anal. C$_{10}$H$_7$N$_3$O$_3$S (C, H, N).

4-Methyl-5-oxo-4,5-dihydropyrazolo[1,5-a]thieno[2,3-e]pyrimidin-3-carbaldehyde (20). A suspension of **16** (150 mg, 0.73 mmol) in glacial acetic acid (6 mL) was added of hexamethylenetetramine (HTMA, 0.36 g) and maintained at reflux temperature for 10 h. After disappearing the starting material, evaluated by TLC (CHX/EtOAc 1:5, v/v as eluent,

Rf: 0.8), the addition of ice gave a precipitate that was recovered by filtration. Yield 85%, white crystals, mp 222–224 °C. ^{1}H-NMR (400 MHz, CDCl$_3$) δ 10.02 (s, 1H, CHO); 8.27 (s, 1H, H-2); 7.91 (d, 1H, H-7, J = 5.2 Hz); 7.88 (d, 1H, H-8, J = 5.2 Hz); 4.03 (s, 3H, OCH$_3$). ^{13}C-NMR (100 MHz, DMSO-d$_6$) δ 170.35, 163.20, 156.25, 145.80, 144.59, 135.84, 133.50, 129.50, 107.15, 37.50. Anal. C$_{10}$H$_7$N$_3$O$_2$S (C, H, N).

4-Methyl-5-oxo-4,5-dihydropyrazolo[1,5-a]thieno[2,3-e]pyrimidin-3-carboxylic acid (21). The aldehyde **20** (150 mg, 0.73 mmol) was suspended in acetone and water (5 mL/5 mL) and a solution of potassium permanganate (1.1 mmol) in water was added after the suspension was made alkaline with sodium hydroxide 10%. The reaction was heated for 8 h, and after cooling and elimination of the manganese dioxide by filtration, the alkaline aqueous phase was extracted to eliminate the starting material not reacting. The next acidification of the aqueous phase gave the corresponding carboxylic acid that was recovered by extraction. Yield 60%, white crystals, mp 220–223 °C. TLC: CHX/EtOAc 1:5, v/v (Rf: 0.2); ^{1}H-NMR (400 MHz, DMSO-d$_6$) δ 12.68 (bs, 1H, OH, exch.); 8.32 (d, 1H, H-7, J = 4.0 Hz); 8.21 (s, 1H, H-2); 7.67 (d, 1H, H-8, J = 4.0 Hz); 3.87 (s, 3H, NCH$_3$). ^{13}C-NMR (100 MHz, DMSO-d$_6$) δ 169.10, 162.40, 145.85, 145.59, 130.50, 126.55, 125.80, 115.50, 107.10, 37.70. Anal. C$_{10}$H$_7$N$_3$O$_3$S (C, H, N).

General procedure for the synthesis of compounds 22a,b. The carboxylic acid **21** (0.5 mmol) was transformed into the corresponding 3-carbonyl chloride by reaction with excess SOCl$_2$ in anhydrous conditions. After the standard work-up, the residue was suspended in dichloromethane (6 mL), and the suitable alcohol (excess 0.15 mL) was added; TLC monitored the reaction until the disappearance of the starting material. Then, the final solution was evaporated to dryness, and the residue recuperated with isopropyl ether and recrystallized.

***tert*-Butyl 4-methyl-5-oxo-4,5-dihydropyrazolo[1,5-a]thieno[2,3-e]pyrimidine-3-carboxylate (22a).** From **21** and *tert*-butanol, white crystals recrystallized by 80% ethanol, yield 25%; mp > 300 °C. TLC: toluene/ethyl acetate/methanol 8:2:1.5 $v/v/v$ (Rf: 0.5); ^{1}H-NMR (400 MHz, CDCl$_3$) δ 8.10 (s, 1H, H-2); 7.85 (d, 1H, H-7, J = 4.8 Hz); 7.63 (d, 1H, H-8, J = 4.8 Hz); 4.01 (s, 3H, NCH$_3$); 1.50 (s, 9H, (CH$_3$)$_3$). ^{13}C-NMR (100 MHz, DMSO-d$_6$) δ 161.78, 156.10, 145.24, 142.00, 141.69, 138.12, 117.25, 116.95, 98.60, 33.57, 14.59. Anal. C$_{14}$H$_{15}$N$_3$O$_3$S (C, H, N).

Thiophen-2-yl-methyl 4-methyl-5-oxo-4,5-dihydropyrazolo[1,5-a]thieno[2,3-e]pyrimidine-3-carboxylate (22b). From **21** and 2-thiophenmethanol, white crystals recrystallized by 80% ethanol, yield 30%; mp 190–195 °C. TLC: toluene/ethyl acetate/methanol 8:2:1.5 $v/v/v$ (Rf: 0.6); ^{1}H-NMR (400 MHz, CDCl$_3$) δ 8.19 (s, 1H, H-2); 7.85 (d, 1H, H-7, J = 5.2 Hz); 7.63 (d, 1H, H-8, J = 5.2 Hz); 7.35 (d, 1H, H-5 Thiophene, J = 5.2 Hz); 7.18 (s, 1H, H-3 Thiophene); 7.03 (d, 1H, H-4 Thiophene, J_1 = 4.8 Hz); 5.47 (s, 2H, OCH$_2$); 4.03 (s, 3H, NCH$_3$). ^{13}C-NMR (100 MHz, CDCl$_3$) δ 145.60, 135.92, 128.25, 126.89, 117.02, 60.81, 33.87. Anal. C$_{15}$H$_{11}$N$_3$O$_3$S$_2$ (C, H, N).

Pyrazolo[1,5-a]thieno[2,3-e]pyrimidine-3-carboxylic acid (23). Compound **11c** (0.40 mmol) was treated with 10% NaOH solution and maintained at 80 °C until the starting material disappeared in TLC (toluene/ethyl acetate/acetic acid 8:2:2 $v/v/v$ as eluent). However, the reaction gave many compounds, and the fast-eluted band's fluorescent spot was recovered by purification with a chromatography column (toluene/ethyl acetate/acetic acid 8:2:2 $v/v/v$ as eluent, Rf: 0.5). Yield 15%, white crystals, mp 246–247 °C. ^{1}H-NMR (400 MHz, DMSO-d$_6$) δ 12.05 (bs, 1H, OH, exch.); 9.34 (s, 1H, H-5); 8.55 (s, 1H, H-2); 8.50 (d, 1H, H-7, J = 5.2 Hz); 7.95 (d, 1H, H-8, J = 5.2 Hz). ^{13}C-NMR (100 MHz, CDCl$_3$) δ 145.60, 135.93, 128.25, 126.89, 117.02, 60.81, 33.87. Anal. C$_9$H$_5$N$_3$O$_2$S (C, H, N).

Pyrazolo[1,5-a]thieno[2,3-e]pyrimidine (24). Compound **11c** (0.40 mmol) were treated with HCl/CH$_3$COOH solution and maintained at 70 °C until the starting material disappeared in TLC (toluene/ethyl acetate/acetic acid 8:2:2 $v/v/v$ as eluent, Rf: 0.6). The reaction from **11c**, monitored by TLC evidenced the formation of 3-carboxylic acid (**23**) that quickly evolved in the decarboxylated compound **24**. The final solution was extracted with ethyl acetate, and after the normal work-up, the residue was purified by recrystallization from ethanol. Yield 25%, white crystals, mp 246–247 °C. ^{1}H-NMR (400 MHz, DMSO-d$_6$) δ 9.14 (s, 1H, H-5); 8.43 (d, 1H, H-7, J = 4.8 Hz); 8.17 (s, 1H, H-2); 7.91 (d, 1H, H-8, J = 4.8 Hz); 6.83 (s, 1H, H-3). ^{13}C-NMR (100 MHz, DMSO-d$_6$) δ 149.90, 146.24, 144.70, 130.52, 126.70, 125.25, 119.15, 101.30. Anal. C$_8$H$_5$N$_3$S (C, H, N).

3-Methylpyrazolo[1,5-a]thieno[2,3-e]pyrimidine (25). Compound 14 (0.34 mmol) was suspended in THF anhydrous (5 mL) and 1.50 mmol of LiAlH$_4$ was added, using a 1M solution of LiAlH$_4$ in THF. After 1 h, the starting material disappeared and adding ice/water quenched the reaction. The extraction with ethyl acetate gave the intermediate 4,5-dihydro derivative not isolated but identified since not fluorescent in TLC, which was treated with toluene and 10% Pd/C at reflux temperature until the dehydrogenation was complete. The final suspension was filtered, toluene was evaporated and the residue was recrystallized by ethanol; yield 35%, white crystals, mp 127–130 °C. TLC: toluene/ethyl acetate/methanol 8:2:1.5 $v/v/v$ (Rf: 0.7); ^{1}H-NMR (400 MHz, DMSO-d$_6$) δ 9.06 (s, 1H, H-5); 8.39 (d, 1H, H-7, J = 5.2 Hz); 8.04 (s, 1H, H-2); 7.87 (d, 1H, H-8, J = 5.2 Hz); 2.33 (s, 1H, CH$_3$). ^{13}C-NMR (100 MHz, DMSO-d$_6$) δ 149.90, 132.74, 132.90, 130.56, 126.70, 125.65, 119.10, 115.63, 14.30. Anal. C$_9$H$_7$N$_3$S (C, H, N).

3.2. Molecular Docking and Molecular Dynamic Simulation

The structure of the binding site was obtained from the Human α1β2γ2-GABAA receptor subtype in complex with GABA and flumazenil, conformation B (PDB ID 6D6T) [21], considering all the amino acids within a distance of about 2 nm from the structure of the Flumazenil. The ligands were placed at the binding site through AUTODOCK 4.2 [20]. The molecular dynamics simulations of ligand binding-site complexes were performed on a minimum number of conformations (maximum 2) such as to cover at least 90% of the poses found by AUTODOCK. A 60 ns MD simulation were performed for all complexes using GROMACS v5.1 program, and it was conducted in vacuum [22]. The DS ViewerPro 6.0 program [19] was used to build the initial conformations of ligands. The partial atomic charge of the ligand structures was calculated with CHIMERA [23] using AM1-BCC method, and the topology was created with ACPYPE [24] based on the routine Antechamber [25]. The OPLS-AA/L all-atom force field [26] parameters were applied to all the structures. To remove bad contacts, the energy minimization was performed using the steepest descent algorithm until convergence is achieved or for 50,000 maximum steps. The next equilibration of the system was conducted in two phases:

(1) Canonical NVT ensemble, a 100 ps position-restrained of molecules at 300 K was carried out using a temperature coupling thermostat (velocity rescaling with a stochastic term) to ensure the proper stabilization of the temperature [27].
(2) Isothermal isobaric NPT ensemble, a 100 ps position-restrained of molecules at 300 K and 1 bar was carried out without using barostat pressure coupling to stabilize the system. These were then followed by a 60 ns MD run at 300 K with position restraints for all protein atoms. The Lincs algorithm [28] was used for bond constraints to maintain rigid bond lengths.

The initial velocity was randomly assigned taken from Maxwell–Boltzman distribution at 300 K and computed with a time step of 2 fs, and the coordinates were recorded every 0.6 ns for MD simulation of 60 ns. During the simulated trajectory, 100 conformations were collected. The 'Proximity Frequencies' (PFs) [1,2] with which the 100 conformations of each binding-site ligand complex intercepts two or more amino acid during the dynamic simulation have calculated. The 'Proximity Frequency' (PF) is the frequency with which the ligand was, during the molecular dynamic simulation, at a distance of less than 0.25 nm from an amino acid of the binding-site and also, simultaneously, from 2, 3 and 4 amino acids of the binding site.

4. Conclusions

The synthesis and the study on the reactivity of the new scaffold 5-oxo-4,5-dihydropyrazolo [1,5-a]thieno [2,3-e]pyrimidine was performed with the aim of isosteric replacement of the benzene ring in our 5-oxo-4,5-dihydro pyrazolo[1,5-a]quinazoline (PQ), already identified as $\alpha1\beta2\gamma2$-GABA$_A$R ligands.

The introduction in this new scaffold of those fragments responsible for the activity in our PQ [6,12] gave six final designed compounds (**12c–f** and **22a,b**), which were, in turn, studied in the 'Proximity Frequency' model [13] to predict their potential profile on the $\alpha1\beta2\gamma2$-GABA$_A$R.

The results indicate for all six products an agonist profile, highlighting the suitability of the nucleus and confirming the importance of the carbonyl group of the ester moiety to engage strong hydrogen bond interaction with the receptor protein. In particular, the esters derivatives (**12c** and **22b**) are able, through the carbonyl group, to interact with the amino acid residue γThr142 with a strong hydrogen bond (2.07 Å). Additionally, **22a** has an ester group in position 3, but the steric hindrance of the *t*-butyloxycarbonyl fragment makes the hydrogen bond interaction of the carbonyl group with the γThr142 residue less probable. Compounds missing the ester group (**12d–f**) show a weak interaction with γThr142 in agreement with the low prediction percentage.

In conclusion, the synthesis of the 5-oxo-4,5-dihydropyrazolo [1,5-a]thieno [2,3-e]pyrimidine scaffold allowed us to identify a new chemical class of compounds potentially active on GABA$_A$ receptor subtype, and the in silico results should be completed and confirmed with biological assays.

Supplementary Materials: The following supporting information can be downloaded at: https://www.mdpi.com/article/10.3390/molecules28073054/s1, Chemistry. Materials and methods; ^{1}H-NMR spectra of all compounds; ^{13}C-NMR spectra of some representative compounds (**6, 8, 10c, 11c, 11e, 12c, 12d, 12e, 12f, 22a, 22b, 23**); Elemental analysis (Table S1); General chemical structure with atoms numeration.

Author Contributions: Conceptualization, G.G. and M.P.G.; methodology, G.G.; software, G.G. and F.M.; validation, G.G. and F.M.; formal analysis, L.C.; investigation, L.C. and G.G.; resources, L.C.; data curation, L.C., G.G. and F.M.; writing—original draft preparation, G.G.; writing—review and editing, G.G., M.P.G., L.C. and C.V.; visualization, L.C.; supervision, G.G. and M.P.G.; project administration, M.P.G. All authors have read and agreed to the published version of the manuscript.

Funding: This research received no external funding.

Institutional Review Board Statement: Not applicable.

Informed Consent Statement: Not applicable.

Data Availability Statement: Not applicable.

Conflicts of Interest: The authors declare no conflict of interest.

References

1. Tolba, M.S.; Ahmed, M.; Kamal El-Dean, A.M.; Hassanien, R.; Farouk, M. Synthesis of New Fused Thienopyrimidines Derivatives as Anti-inflammatory Agents. *J. Heterocycl. Chem.* **2018**, *55*, 408–418. [CrossRef]
2. Kotaiah, Y.; Harikrishna, N.; Nagaraju, K.; Venkata Rao, C. Synthesis and antioxidant activity of 1,3,4-oxadiazole tagged thieno [2,3-d]pyrimidine derivatives. *Eur. J. Med. Chem.* **2012**, *58*, 340–345. [CrossRef] [PubMed]
3. Vikram, V.; Amperayani, K.R.; Umadevi, P. 3-(Methoxycarbonyl)thiophene Thiourea Derivatives as Potential Potent Bacterial Acetyl-CoA Carboxylase Inhibitors. *Russ. J. Org. Chem.* **2021**, *57*, 1336–1345. [CrossRef]
4. Archna; Pathania, S.; Chawla, P.A. Thiophene-based derivatives as anticancer agents: An overview on decade's work. *Bioorg. Chem.* **2020**, *101*, 104026. [CrossRef] [PubMed]
5. Guerrini, G.; Ciciani, G.; Crocetti, L.; Daniele, S.; Ghelardini, C.; Giovannoni, M.P.; Mannelli, L.D.C.; Martini, C.; Vergelli, C. Synthesis and Pharmacological Evaluation of Novel GABAA Subtype Receptor Ligands with Potential Anxiolytic-like and Anti-hyperalgesic Effect. *J. Heterocycl. Chem.* **2017**, *54*, 2788–2799. [CrossRef]
6. Guerrini, G.; Ciciani, G.; Crocetti, L.; Daniele, S.; Ghelardini, C.; Giovannoni, M.P.; Iacovone, A.; Di Cesare Mannelli, L.; Martini, C.; Vergelli, C. Identification of a New Pyrazolo[1,5-a]quinazoline Ligand Highly Affine to γ-Aminobutyric Type A (GABAA) Receptor Subtype with Anxiolytic-Like and Antihyperalgesic Activity. *J. Med. Chem.* **2017**, *60*, 9691–9702. [CrossRef]
7. Guerrini, G.; Ciciani, G.; Daniele, S.; Di Cesare Mannelli, L.; Ghelardini, C.; Martini, C.; Selleri, S. Synthesis and pharmacological evaluation of pyrazolo[1,5-a]pyrimidin-7(4H)-one derivatives as potential GABAA-R ligands. *Bioorg. Med. Chem.* **2017**, *25*, 1901–1906. [CrossRef]

8. Crocetti, L.; Guerrini, G.; Melani, F.; Vergelli, C.; Mascia, M.P.; Giovannoni, M.P. GABAA Receptor Modulators with a Pyrazolo[1,5-a]quinazoline Core: Synthesis, Molecular Modelling Studies and Electrophysiological Assays. *Int. J. Mol. Sci.* **2022**, *23*, 13032. [CrossRef]
9. Balakleevsky, A.; Colombo, G.; Fadda, F.; Gessa, G.L. Ro 19-4603, a benzodiazepine receptor inverse agonist, attenuates voluntary ethanol consumption in rats selectively bred for high ethanol preference. *Alcohol Alcohol.* **1990**, *25*, 449–452.
10. Chambers, M.S.; Atack, J.R.; Broughton, H.B.; Collinson, N.; Cook, S.; Dawson, G.R.; Hobbs, S.C.; Marshall, G.; Maubach, K.A.; Pillai, G.V.; et al. Identification of a Novel, Selective GABAA α5 Receptor Inverse Agonist Which Enhances Cognition. *J. Med. Chem.* **2003**, *46*, 2227–2240. [CrossRef]
11. Selleri, S.; Bruni, F.; Costagli, C.; Costanzo, A.; Guerrini, G.; Ciciani, G.; Costa, B.; Martini, C. Synthesis and BZR affinity of pyrazolo[1,5-a]pyrimidine derivatives. Part 1: Study of the structural features for BZR recognition. *Bioorg. Med. Chem.* **1999**, *7*, 2705–2711. [CrossRef] [PubMed]
12. Guerrini, G.; Vergelli, C.; Cantini, N.; Giovannoni, M.P.; Daniele, S.; Mascia, M.P.; Martini, C.; Crocetti, L. Synthesis of new GABAA receptor modulator with pyrazolo[1,5-a]quinazoline (PQ) scaffold. *Int. J. Mol. Sci.* **2019**, *20*, 1438. [CrossRef] [PubMed]
13. Crocetti, L.; Guerrini, G.; Cantini, N.; Vergelli, C.; Melani, F.; Mascia, M.P.; Giovannoni, M.P. 'Proximity frequencies' a new parameter to evaluate the profile of GABAAR modulators. *Bioorg. Med. Chem. Lett.* **2021**, *34*, 127755. [CrossRef] [PubMed]
14. Huddleston, P.R.; Barker, J.M.; Adamczewska, Y. Preparation and reactions of some 3-hydrazino-2-methoxycarbonylthiophene. *J. Chem. Res.* **1980**, *11*, 238–239.
15. Guerrini, G.; Ciciani, G.; Cambi, G.; Bruni, F.; Selleri, S.; Melani, F.; Montali, M.; Martini, C.; Ghelardini, C.; Norcini, M.; et al. Novel 3-aroylpyrazolo[5,1-c][1,2,4]benzotriazine 5-oxides 8-substituted, ligands at GABAA/benzodiazepine receptor complex: Synthesis, pharmacological and molecular modeling studies. *Bioorg. Med. Chem.* **2008**, *16*, 4471–4489. [CrossRef]
16. Bruni, F.; Selleri, S.; Costanzo, A.; Guerrini, G.; Casilli, M.L.; Giusti, L. Reactivity of 7-(2-dimethylaminovinyl)pyrazolo[1,5-a]pyrimidines: Synthesis of pyrazolo[1,5-a]pyrido[3,4-e]pyrimidine derivatives as potential benzodiazepine receptor ligands. *J. Heterocycl. Chem.* **1995**, *32*, 291–298. [CrossRef]
17. Anwer, M.K.; Sherman, D.B.; Roney, J.G.; Spatola, A.F. Applications of Ammonium Formate Catalytic Transfer Hydrogenation. 6. Analysis of Catalyst, Donor Quantity, and Solvent Effects upon the Efficacy of Dechlorination. *J. Org. Chem.* **1989**, *54*, 1284–1289. [CrossRef]
18. Costanzo, A.; Guerrini, G.; Bruni, F.; Selleri, S. Reactivity of 1-(2-nitrophenyl)-5-aminopyrazoles under basic conditions and synthesis of new 3-, 7-, and 8-substituted pyrazolo[5,1-c][1,2,4]benzotriazine 5-oxides, as benzodiazepine receptor ligands. *J. Heterocycl. Chem.* **1994**, *31*, 1369–1376. [CrossRef]
19. *DS ViewerPro 6.0*, Version 6.0; Accelrys Software Inc.: San Diego, CA, USA, 2023.
20. Morris, G.M.; Ruth, H.; Lindstrom, W.; Sanner, M.F.; Belew, R.K.; Goodsell, D.S.; Olson, A.J. Software news and updates AutoDock4 and AutoDockTools4: Automated docking with selective receptor flexibility. *J. Comput. Chem.* **2009**, *30*, 2785–2791. [CrossRef]
21. Zhu, S.; Noviello, C.M.; Teng, J.; Walsh, R.M.; Kim, J.J.; Hibbs, R.E. Structure of a human synaptic GABAA receptor. *Nature* **2018**, *559*, 67–88. [CrossRef]
22. Pronk, S.; Páll, S.; Schulz, R.; Larsson, P.; Bjelkmar, P.; Apostolov, R.; Shirts, M.R.; Smith, J.C.; Kasson, P.M.; Van Der Spoel, D.; et al. GROMACS 4.5: A high-throughput and highly parallel open source molecular simulation toolkit. *Bioinformatics* **2013**, *29*, 845–854. [CrossRef] [PubMed]
23. Pettersen, E.F.; Goddard, T.D.; Huang, C.C.; Couch, G.S.; Greenblatt, D.M.; Meng, E.C.; Ferrin, T.E. UCSF Chimera—A visualization system for exploratory research and analysis. *J. Comput. Chem.* **2004**, *25*, 1605–1612. [CrossRef] [PubMed]
24. Sousa Da Silva, A.W.; Vranken, W.F. ACPYPE—AnteChamber PYthon Parser interfacE. *BMC Res. Notes* **2012**, *5*, 367. [CrossRef] [PubMed]
25. Wang, J.; Wang, W.; Kollman, P.A.; Case, D.A. Automatic atom type and bond type perception in molecular mechanical calculations. *J. Mol. Graph. Model.* **2006**, *25*, 247–260. [CrossRef]
26. Jorgensen, W.L.; Maxwell, D.S.; Tirado-Rives, J. Development and Testing of the OPLS All-Atom Force Field on Conformational Energetics and Properties of Organic Liquids. *JACS* **1996**, *118*, 11236–11255. [CrossRef]
27. Bussi, G.; Donadio, D.; Parrinello, M. Canonical sampling through velocity rescaling. *J. Chem. Physics.* **2007**, *126*, 014101. [CrossRef]
28. Hess, B.; Bekker, H.; Berendsen, H.J.C.; Fraaije, J.G.E.M. LINCS: A Linear Constraint Solver for molecular simulations. *J. Comput. Chem.* **1997**, *18*, 1463–1472. [CrossRef]

Disclaimer/Publisher's Note: The statements, opinions and data contained in all publications are solely those of the individual author(s) and contributor(s) and not of MDPI and/or the editor(s). MDPI and/or the editor(s) disclaim responsibility for any injury to people or property resulting from any ideas, methods, instructions or products referred to in the content.

Review

The Synthesis and Biological Applications of the 1,2,3-Dithiazole Scaffold

Andreas S. Kalogirou [1], Hans J. Oh [2] and Christopher R. M. Asquith [2,3,*]

1. Department of Life Sciences, School of Sciences, European University Cyprus, 6 Diogenis Str., Engomi, P.O. Box 22006, Nicosia 1516, Cyprus; a.kalogirou@euc.ac.cy
2. Department of Pharmacology, School of Medicine, University of North Carolina, Chapel Hill, NC 27599, USA; hjo0729@live.unc.edu
3. School of Pharmacy, Faculty of Health Sciences, University of Eastern Finland, 70211 Kuopio, Finland
* Correspondence: christopher.asquith@uef.fi; Tel.: +358-(0)50-400-3138; Fax: +358-(0)82-944-4091

Abstract: The 1,2,3-dithiazole is an underappreciated scaffold in medicinal chemistry despite possessing a wide variety of nascent pharmacological activities. The scaffold has a potential wealth of opportunities within these activities and further afield. The 1,2,3-dithiazole scaffold has already been reported as an antifungal, herbicide, antibacterial, anticancer agent, antiviral, antifibrotic, and is a melanin and Arabidopsis gibberellin 2-oxidase inhibitor. These structure activity relationships are discussed in detail, along with insights and future directions. The review also highlights selected synthetic strategies developed towards the 1,2,3-dithiazole scaffold, how these are integrated to accessibility of chemical space, and to the prism of current and future biological activities.

Keywords: antibacterial; anticancer; antifibrotic; antifungal; antimicrobial; antiviral; appel salt; 1,2,3-dithiazole; disulfide bridge; herbicidal

1. Introduction

The 1,2,3-dithiazole core is a five membered heterocycle containing two sulfur atoms and one nitrogen atom. Despite the fact that the 1,2,3-dithiazole is not present in nature, similar to many other heterocycles, it does have a broad range of interesting biological activities. The 1,2,3-dithiazole moiety was first synthesized in 1957 by G. Schindler et al. [1]. This was followed two decades later by a report by J. E. Moore on behalf of Chevron Research Co. (San Ramon, CA, USA) where it showcased antifungal and herbicidal activity [2,3]. In 1985, Appel et al. reported the synthesis of 4,5-dichloro-1,2,3-dithiazolium chloride **1** (Appel's salt), a precursor which allowed access to the 1,2,3-dithiazole core within a single step [4,5].

The synthesis of Appel salt **1** acted as a catalyst to the field and granted access to many 1,2,3-dithiazole derivatives, and to other heterocycles incorporating sulfur and nitrogen atoms [6–11]. The subsequent synthetic reports focused on transformations on the C5 position [6–11]. However, one of the key synthetic interests beyond expanding the scope of 5-substituted 1,2,3-dithiazoles was the limited reactivity of the C4 position. Several different approaches were used to address this C4 reactively issue, including intramolecular cyclization [6] using a multi-step oxime pathway [12,13], or more recently, direct reactions [14], all of which expanded the chemical space around the 1,2,3-dithiazole. Some of these approaches have been covered in past reviews around the chemistry of 1,2,3-dithiazoles [6–11] (Figure 1).

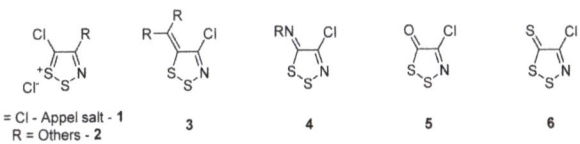

Figure 1. Appel salt (**1**) and other general 1,2,3-dithazoles structures **2**–**6**.

Despite the remaining synthetic challenges, the 1,2,3-dithiazole scaffold has already been reported as an antifungal [2], herbicide [2], antibacterial [15], anticancer agent [16], antiviral [17], antifibrotic [18], and as a melanin [19] and Arabidopsis gibberellin 2-oxidase [20] inhibitor. While there is a wide range of existing biology, there are a wealth of opportunities for expansion, including broader application toward cystine reactive sites [21–25]. In this review, we are primarily focused on the impact of: (1) The chemistry limiting the chemical space, and hence, limiting the biology; (2) The chemistry impacting the biology observed; and (3) How chemistry could be applied to new biology. The chemistry, biology, structure activity relationships, and future directions of research in 1,2,3-dithiazoles are all outlined below.

2. 1,2,3-Dithiazoles Synthesis Overview

2.1. Early Years before Appel Salt

Early work on the synthesis of 1,2,3-dithiazoles used cyanothioformamides as starting materials. Treatment of a variety of arylcyanothioformamides **7** with sulfur dichloride at 0–25 °C gave a number of *N*-aryl-5*H*-1,2,3-dithiazol-5-imines **4** (Scheme 1) [2]. The initial reaction yielded the corresponding hydrochloride salts, which could be converted to the free base by refluxing in a toluene solution.

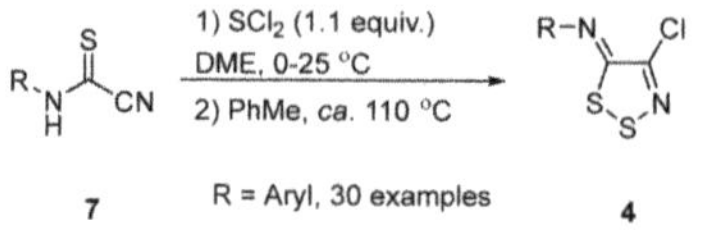
1) SCl2 (1.1 equiv.)
DME, 0-25 °C
2) PhMe, ca. 110 °C
R = Aryl, 30 examples
7
4

Scheme 1. Synthesis of *N*-aryl-5*H*-1,2,3-dithiazol-5-imines **4** from arylcyanothioformamides **7**.

Interestingly, the *N*-aryl-5*H*-1,2,3-dithiazol-5-imines **4** can be degraded to the respective cyanothioformamides **7** by thiophilic ring cleavage after reaction with triphenylphosphine or sodium hydroxide [4,26], oxidative ring cleavage after reaction with *m*-CPBA [27], or by reductive ring cleavage after reaction with sodium cyanoborohydride [28] (Scheme 2).

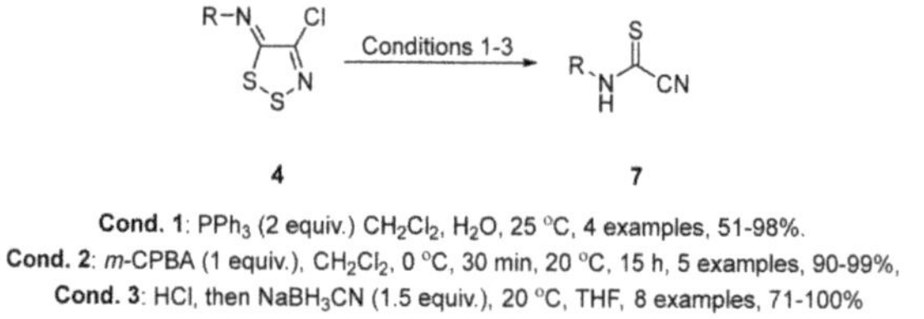
Conditions 1-3
4
7
Cond. 1: PPh3 (2 equiv.) CH2Cl2, H2O, 25 °C, 4 examples, 51-98%.
Cond. 2: m-CPBA (1 equiv.), CH2Cl2, 0 °C, 30 min, 20 °C, 15 h, 5 examples, 90-99%,
Cond. 3: HCl, then NaBH3CN (1.5 equiv.), 20 °C, THF, 8 examples, 71-100%

Scheme 2. Degradation of *N*-aryl-5*H*-1,2,3-dithiazol-5-imines **4** to cyanothioformamides **7**.

2.2. Discovery of Appel Salt and Applications

A significant discovery in the chemistry of 1,2,3-dithiazoles was the synthesis of 4,5-dichloro-1,2,3-dithiazolium chloride **1** (Appel salt) by Appel et al. in 1985, which was readily prepared from chloroacetonitrile and disulfur dichloride [4,29] (Scheme 3). Appel salt **1** was subsequently used as an important reagent for the preparation of other 4-chloro-5*H*-1,2,3-dithiazoles with the most reactive site being the electrophilic C-5 position [4,9,27].

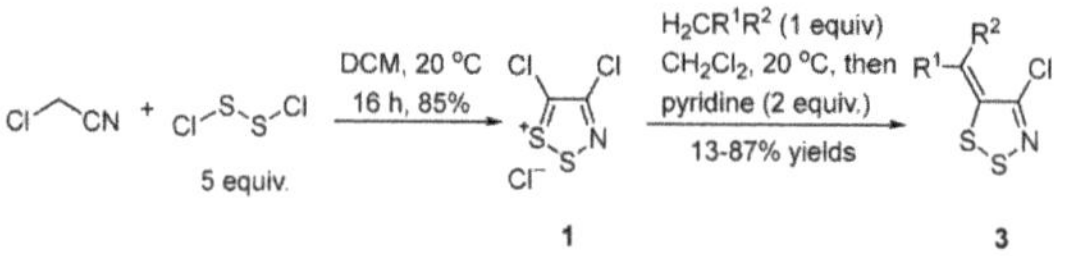
DCM, 20 °C
16 h, 85%
5 equiv.
H2CR1R2 (1 equiv)
CH2Cl2, 20 °C, then
pyridine (2 equiv.)
13-87% yields
1
3

Scheme 3. Synthesis of Appel salt and transformation to 4-chloro-5*H*-1,2,3-dithiazolylidenes **3**.

Appel salt **1** can condense with active methylenes, such as acetonitrile derivatives [4,30,31], diketones, ketoesters, and others [32], to give 4-chloro-5*H*-1,2,3-dithiazolylidenes **3** (Scheme 3).

The condensation of Appel salt with hydrogen sulfide [4] afforded dithiazole-5-thione **6** in 69% yield (Scheme 4). The reaction with oxygen nucleophiles are also common with NaNO$_3$ [4], sulfoxides [33], or formic acid [34] all acting as the source of oxygen to give 4-chloro-5H-1,2,3-dithiazol-5-one **5** in good yields (Scheme 4). Furthermore, the reaction with other carboxylic acids [35] at −78 °C and subsequent treatment with alcohols gave esters **8** in medium to good yields (Scheme 4).

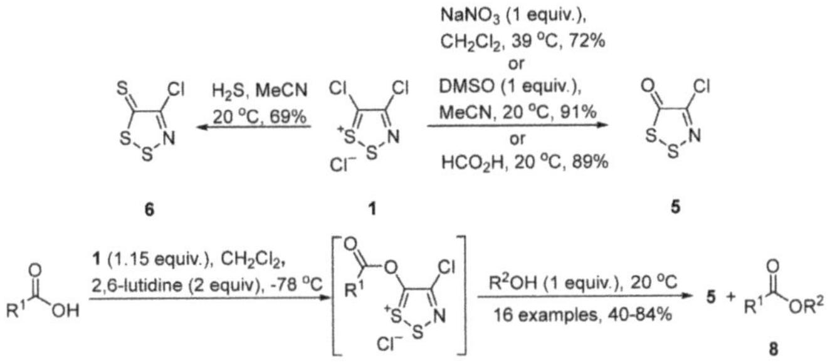

Scheme 4. Reactions of Appel salt **1** with oxygen and sulfur nucleophiles.

The condensation of Appel salt **1** with primary anilines is well studied [4,5,15,36] and typically occurs by treatment with 1 equiv. of the aniline in the presence of pyridine (2 equiv.) as the base to give, in most cases, good yields of N-aryl-5H-1,2,3-dithiazol-5-imines **4** (Scheme 5).

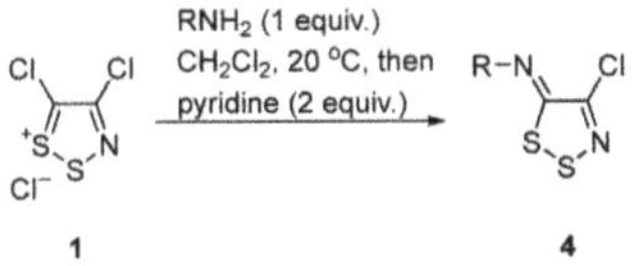

Scheme 5. Synthesis of N-aryl-5H-1,2,3-dithiazol-5-imines **4** from Appel salt **1**.

Some limitations of this chemistry appear when using heterocyclic arylamines, such as aminopyridines. A recent study by Koutentis et al. highlighted that the reactions of the three isomeric aminopyridines with Appel salt **1** gave very different yields based on the position of the amino group. The 2-, 3- and 4-aminopyridines gave 69%, 24%, and 1% yields of the desired 1,2,3-dithiazole, respectively [37] (Scheme 6). Koutentis et al. suggested the low yield of 4-aminopyridine is likely attributed to the reduced nucleophilicity of the primary amine due to a contribution of its zwitterionic resonance form. The low reactivity of the amine leads to complex reaction mixtures due to side reactions.

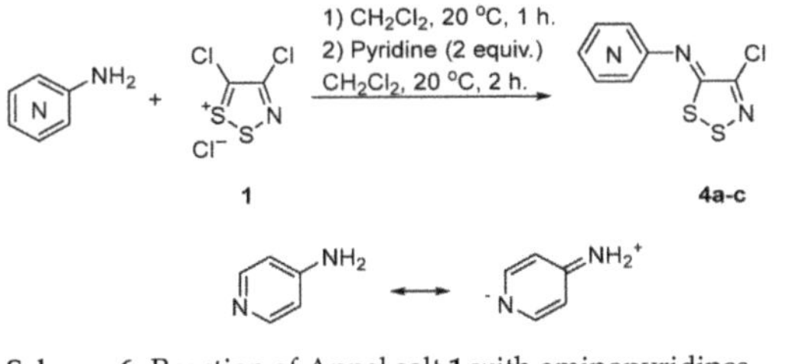

Scheme 6. Reaction of Appel salt **1** with aminopyridines.

2.3. Reactivity of C-4 and the Displacement of the Chloride

The less reactive C4 chlorine of neutral 5H-1,2,3-dithiazoles cannot be directly substituted by nucleophiles. However, utilizing an ANRORC-(Addition of the Nucleophile, Ring

Opening, and Ring Closure)-style mechanism, nucleophilic substitution can occur on the C4 chlorine of the 1,2,3-dithazole. An example of this is where the *N*-Aryl-5*H*-1,2,3-dithiazol-5-imines **4** react with an excess of dialkylamines to give 4-aminodithiazoles **9** in variable yields (Scheme 7). The reaction was found to proceed *via* an ANRORC-style mechanism [38,39] involving ring opening by nucleophilic attack on the S2 position to yield disulfides **10** and subsequent recyclization after amine addition on the cyano group [40]. In another report by Koutentis et al. [14], DABCO was reacted with neutral 5*H*-1,2,3-dithiazoles **4–6** to give *N*-(2-chloroethyl)piperazines **11** in good yields (Scheme 7). The chloroethyl group originating from chloride attack on the intermediate quaternary ammonium salt formed by the displacement of the C4 chloride by DABCO.

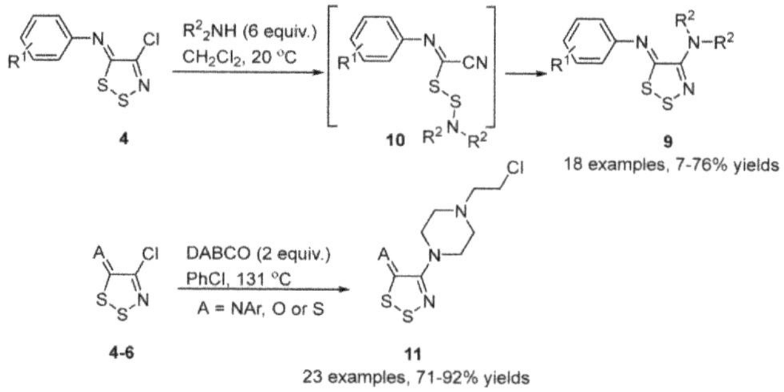

Scheme 7. Displacement of the C4 chlorine of neutral 5*H*-1,2,3-dithiazoles.

2.4. Alternatives beyond Appel Salt Chemistry

A different way to access both monocyclic and ring fused 1,2,3-dithiazoles is by the reaction of oximes with disulfur dichloride. An example of the synthesis of a ring fused dithiazole is the reaction of benzoindenone oxime **12** to give dithiazole **13** in 81% yield [41,42] (Scheme 8). Acetophenone oximes **14** were reacted with disulfur dichloride to yield dithiazolium chlorides **2**, which were subsequently converted to either imines **15**, thiones **16**, or ketone **17** [13] (Scheme 8). Insights in the mechanism of the oxime to dithiazole transformation were given by Hafner et al. [12], who isolated the dithiazole *N*-oxide, which is the intermediate in this reaction.

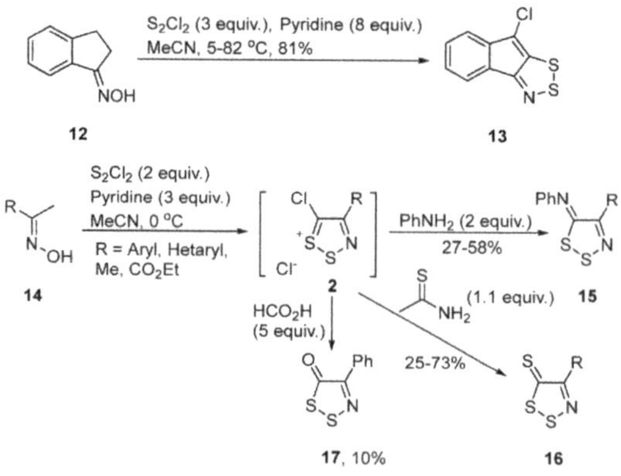

Scheme 8. Synthesis of 1,2,3-dithiazoles from oximes.

2.5. Reactivity of 1,2,3-Dithiazoles

Neutral 1,2,3-dithiazoles can also be transformed to a plethora of other heterocycles, often substituted by a cyano group originating from the imidoyl chloride of the starting material using thermal or reactions with thiophiles. An interesting example of an ANRORC-style mechanism leading to a ring transformation was the reaction of (Z)-N-(4-chloro-5H-1,2,3-dithiazol-5-ylidene)-1H-pyrazol-5-amines **4d** with diethylamine that results in disulfide intermediates **18**. Subsequent treatment with concentrated sulfuric acid gave 1,2,4-dithiazines **19** in good yields [43] (Scheme 9).

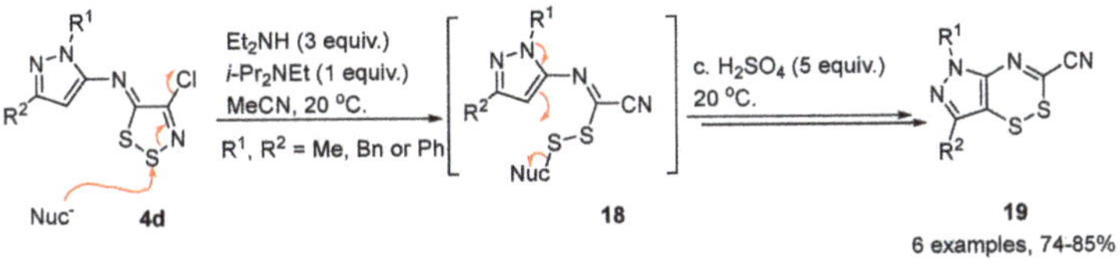

Scheme 9. Synthesis of 1,2,4-dithiazines **19**.

In another example, the pyrazoleimino dithiazoles **20** were converted to 4-methoxy-pyrazolo[3,4-d]pyrimidines **21** in medium to good yields by treatment with sodium methoxide in methanol [16] (Scheme 10). The transformation occurs after addition of the methoxide on the nitrile followed by cyclisation onto the dithiazole C5 position that fragments losing S_2 and chloride to give the final pyrimidine **21**.

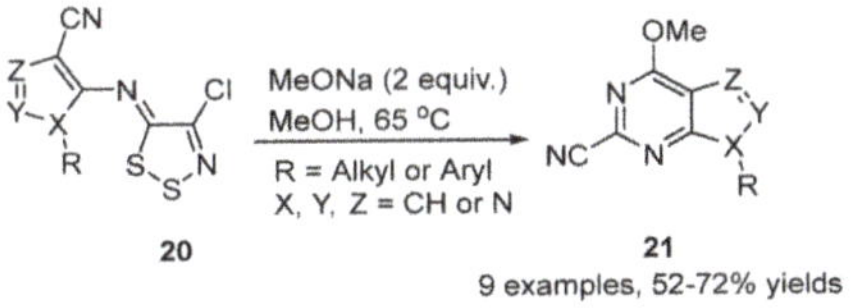

Scheme 10. Synthesis of 4-methoxy-pyrazolo[3,4-d]pyrimidines **21**.

A similar example of ring transformations is that of 2-aminobenzyl alcohol dithiazole-imines **4e** to 1,3-benzoxazines **22** and 1,3-benzothiazines **23** [44]. Treatment of imines **4e** with sodium hydride in THF gave mixtures of benzoxazines **22** and benzothiazines **23**, with the former as the main products (Scheme 11). The formation of the former involves deprotonation of the alcohol and cyclisation of the alkoxide onto the dithiazole C5 position. Subsequent fragmentation with loss of S_2 and chloride gave the final benzoxazine **22**. Alternatively, treatment of imines **4e** with Ph_3P gave exclusively benzothiazines **23** in good yields (Scheme 11). Thiophilic attack on S1 ring opens the dithiazole ring and a second attack by Ph_3P gives the intermediate alkene **24** that cyclizes to benzothiazine **23**.

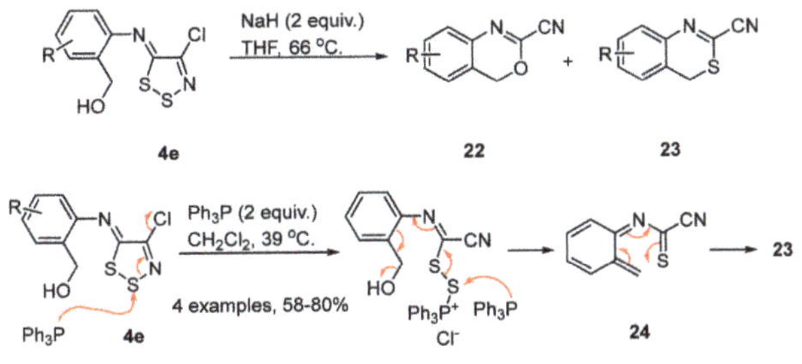

Scheme 11. Synthesis of 1,3-benzoxazines **22** and 1,3-benzothiazines **23**.

1,2,3-Dithiazole derivatives can also be converted to mercaptoacetonitriles by the removal of the S1 atom. One example of this are the 3-(1,2,3-dithiazolylidene)indolin-2-ones **25** reacting with sodium hydride (2 equiv.) to yield the mercaptoacetonitrile products **26** in medium to good yields [45] (Scheme 12).

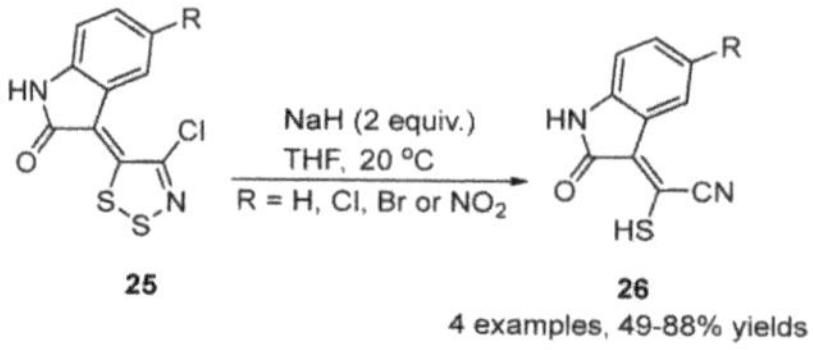

Scheme 12. Conversion of dithiazoles 25 to mercaptoacetonitriles 26.

Perhaps the most unstable 1,2,3-dithiazole is Appel salt itself, which, while relatively stable at ca. 20 °C under a desiccant, in its absence, Appel salt has a tendency to react with moisture. One study by Koutentis et al. revealed that simple stirring in wet MeCN gave elemental sulfur, dithiazole-5-thione 6, dithiazol-5-one 5, and thiazol-5-one 27 [46] (Scheme 13), assisting other scientists working with Appel salt, to identify these products. Interestingly, other dithiazolium salts have also been prepared with increased stability and lower sensitivity to moisture. A series of perchlorate salts of 1,2,3-dithiazoles were prepared by the anion exchange with perchloric acid allowing for more detailed characterization and study of the 1,2,3-dithiazole [29].

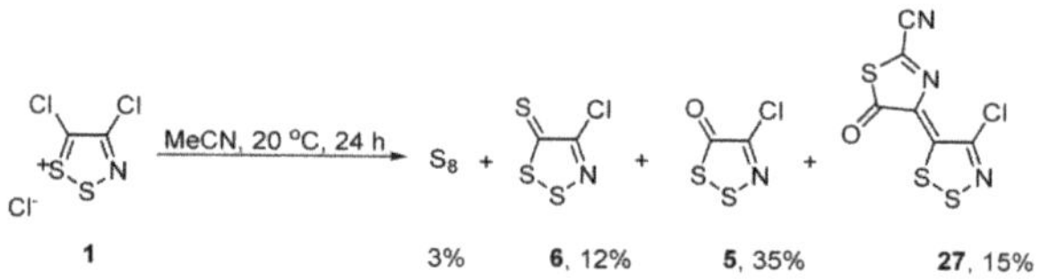

Scheme 13. Degradation of Appel salt 1 in wet MeCN.

In another study by Rakitin et al., 4-substituted 5H-1,2,3-dithiazoles 16 and 17 were converted to 1,2,5-thiadiazoles 28 and 29 by treatment with primary amines [47] (Scheme 14). Mechanistically, the reaction occurs by addition of the amine to the C5 position followed by ring opening of the C-S bond and subsequent ring closing by loss of hydrogen sulfide.

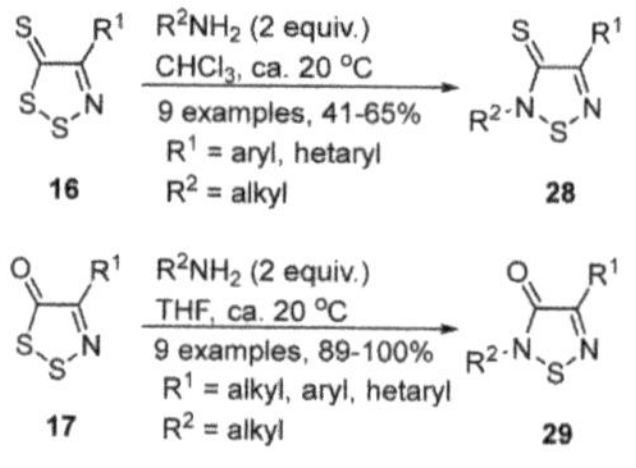

Scheme 14. Transformation of dithiazoles 16–17 to thiadiazines 28–29.

To summarize, 1,2,3-dithiazoles can be converted to other heterocyclic or ring opened derivatives. The six most common mechanisms involved in the transformations of 1,2,3-dithiazoles to other systems are shown below (Scheme 15). These mechanisms begin *via* a nucleophile assisted ring opening of the dithiazole to disulfide intermediates that then can react either intermolecular or intramolecular with other nucleophiles *via* the six paths presented.

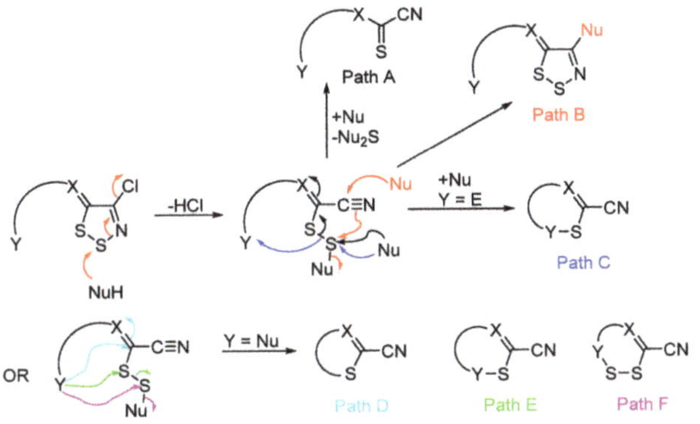

Scheme 15. Overview of the mechanisms of the reactions of 1,2,3-dithiazoles.

3. 1,2,3-Dithiazoles in Medicinal Chemistry

3.1. Antimicrobial Activities of 1,2,3-Dithiazoles, including Antifungal, Herbicidal, and Antibacterial

The first report of biological activity using the 1,2,3-dithiazole scaffold was published in a patent filed by J. E. Moore in 1977 on behalf of Chevron Research Co. [2,3]. The patent disclosed a series of novel 1,2,3-dithiazoles afforded in a 2–3 step sequence from *N*-aryl cyanothioformamide and sulfur dichloride. The main application of these compounds was the controlling of various fungal infections, leaf blights, invasive plant species, and mites.

First, the tomato early blight organism, *Alternaria solani conidia* was tested against 6- to 7-week-old tomato plate seedlings. The tomato plants were sprayed with 250 ppm solutions of a 1,2,3-dithiazole library. This resulted in the identification of (Z)-4-((4-chloro-5H-1,2,3-dithiazol-5-ylidene)amino)benzonitrile (**30**) with a 90% reduction compared with non-treatment. The 2,4-dichloro analogue **31** had weaker activity, with a reduction of just over half of the infection (Figure 1). Next, the tomato late blight organism, *Phytophthora infestans conidia* was tested against seedlings of 5 to 6 weeks old using the same procedure. The 4-cyano analogue **30** was found to afford 97% protection, while the 2-(4-nitrophenoxy) analogue **32** showed an 80% reduction (Figure 2). Then, the celery late blight organism *septoria api* was tested using 11-week-old plants. The 4-cyano analogue **30** afforded less protection at just over 60%, while several other analogues showed improvements, including 2-fluoro **33** and 3-(4-trifluoro, 2-cyanopenoxy) **34** analogues, both reported with 80% protection (Figure 2).

A series of halogenated analogues **35–39** were then identified as active against the powdery mildew pathogen *Erysiphe polygoni* using bean seedlings with well-developed primary leaves. The (Z)-4-chloro-*N*-(4-chloro-2-methylphenyl)-5H-1,2,3-dithiazol-5-imine analogue (**35**) along with the corresponding 3-chloro **36** showed 100% protection at 250 ppm. The corresponding 5-chloro **37** and 4-bromo **38** both showed a small reduction in efficacy, 10% and 1%, respectively, while the 3,5-dichloro **39** was only net 76% effective (Figure 3).

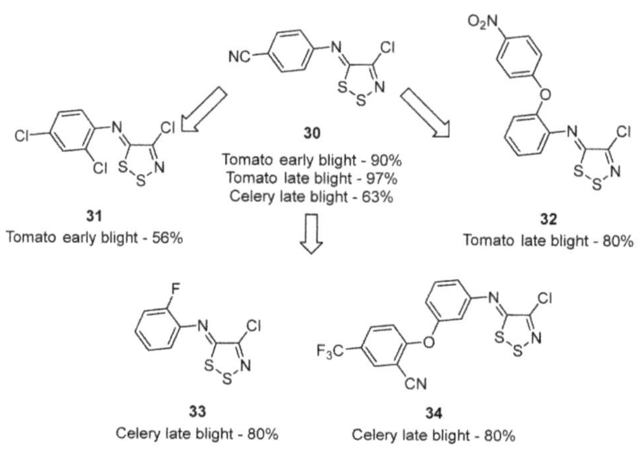

Figure 2. Antifungicidal activities of early 1,2,3-dithiazole derivatives at a concentration of 250 ppm.

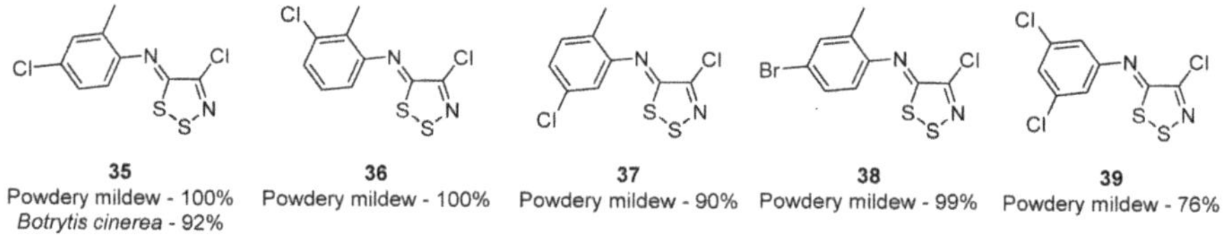

Figure 3. Antifungicidal activities of 1,2,3-dithiazoles against powdery mildew (250 ppm) and *Botriytis cinerea* (40 ppm).

Initial screening was also carried out against necrotrophic fungus *Botrytis cinerea* on the well-developed primary leaves of a 4–6-week-old horsebean plant at a lower concentration (40 ppm). Only 1,2,3-dithiazole **35** was demonstrated to be effective with 92% inhibition (Figure 3). However, after this initial result, screening was carried out on a broader panel of fungal (Figure 4) and herbicidal strains (Figure 5). The fungal panel included *Botrytis cinerea*, *Rhizoctonia solani*, *Fusarium moniloforma*, *Phythium ultimum*, and *Aspergillus niger*. The compounds **30**, **33**, and **39–48** were tested at 500 ppm and fungicidal activities were measured by the zone of inhibited mycelia growth (Figure 4). Interestingly, the unsubstituted phenyl analogue (Z)-4-chloro-N-phenyl-5H-1,2,3-dithiazol-5-imine (**40**) was active on *Botrytis cinerea* at 0.33 µg/cm^2. The addition of a 4-position methyl in analogue **41** reduced the activity against *Botrytis cinerea* by over 2-fold, but increased the activity against *Rhizoctonia solani* and *Fusarium moniloforma*. The 2-position methyl analogue **42** showed a profile switch showing activity only against *Aspergillus niger* (0.98 µg/cm^2). The 2,4,6-trimethyl analogue **43** also only retained activity against on strain *Rhizoctonia solani* (0.98 µg/cm^2). The original 4-chloro, 2-methyl analogue **35** showed activity against *Rhizoctonia solani* (0.63 µg/cm^2), but the dose dependent *Botrytis cinerea* data was not reported. The removal of the methyl group to afford the 4-chloro analogue **44** increased the activity against *Rhizoctonia solani* by 2-fold and showed commensurate activity against *Phythium ultimum* and 3-fold weaker activity against *Aspergillus niger*. The addition of a second chloro in the 3-position in analogue **45** was unfavored with only activity against *Aspergillus niger* retained. When the 4-chloro is removed to afford **46**, the activity is switched again with potency only demonstrated for *Rhizoctonia solani* at the same level as **43**. Addition of a second choro at the 5-position to afford **47** has same activity profile as **46**. The 4-cyano analogue **30** showed good activity against *Rhizoctonia solani* (0.60 µg/cm^2). The 2-fluoro analogue **33**, while having a slightly weaker potency, did show activity against 4 out of 5 of the fungal panel, only excluding *Botrytis cinerea*. The final two analogues identified in this series, 3-(4-nitrophenoxy) **47** and

4-(4-nitrophenoxy) **48**, both showed activity against only *Rhizoctonia solani* with analogue **48** having a 2-fold improvement over **47** at (0.45 μg/cm^2).

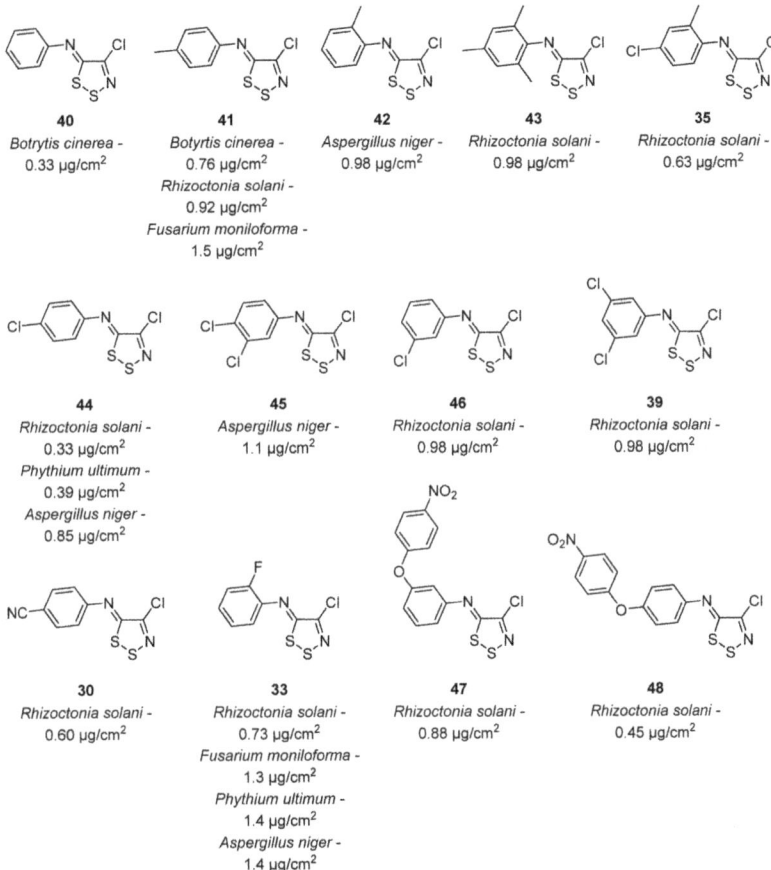

Figure 4. Antifungicidal activities of 1,2,3-dithiazoles, values are amounts required for mycelia inhibition, micrograms/cm^2 for 99% control of fungal growth.

The 1,2,3-dithiazoles were then screened at 33 ppm on a herbal panel that included wild oats (*Awena fatua*), watergrass (*Echinochloa crusgall*), crabgrass (*Digitaria sanguinalis*), mustard (*Brassica arversis*), pigweed (*Amaranthus retroflexus*), and lambsquarter (*Cheropodium album*) (Figure 5). The first analogue (Z)-4-chloro-N-(p-tolyl)-5H-1,2,3-dithiazol-5-imine (**41**), showed good efficacy against *Amaranthus retroflexus* (90%) and total control of *Brassica arversis*. Switching to the 4-fluoro analogue **49** increased coverage across all strains tested, including *Avena fatua* (40%), which was only weakly inhibited across the series and total control of *Amaranthus retroflexus*. The 4-chloro analogue **44** was 3-fold less effective against *Digitaria sanguinalis* and *Avena fatua*. The addition of a 2-position chloro **50** decreased strain coverage, but did mean total control of *Brassica arversis* in addition to *Amaranthus retroflexus*, with additional high efficacy against *Chenopodium Album* (95%). The original 2-methyl 4-chloro analogue **35**, while still showing efficacy across several strains, did not offer total or near total control for any of the strains tested. The 2-chloro analogue **50** showed total control for *Chenopodium album* and *Brassica arversis* and near total for *Amaranthus retroflexus* (93%). However, 2-chloro **50** had a limited effect on *Digitaria sanguinalis* and *Echinochloa crusgalli*, with no impact on *Avena fatua*. The 3,5-dichloro analogue **39** demonstrated good efficacy against most strains, including total control of *Amaranthus retroflexus*, *Chenopdium album*, *Brassica arvensis*, and some activity against *Avena fatua* (35%). The 2-methyl, 5-chloro analogue **51** offered the highest efficacy across the series on *Avena fatua* (45%), total control *of Amaranthus retroflexus* and *Brassica arvensis*, with near total control of *Chenopodium album* (95%). The 3,4-dichloro analogue **45** had a potent but narrower band of activity with total control of *Amaranthus retroflexus*, *Chenopodium album* and *Brassica arvensis*, but weaker activity on the other three strains (30–55%). The 3-bromo analogue **38** has a similar profile to the 4-methyl **41**, while the 2-naphthyl analogue **52** was the most potent in the screening for *Echinochloa crusgalli* (90%) and offered good control over *Amaranthus retroflexus* (85%) and total control over *Brassica arvensis*.

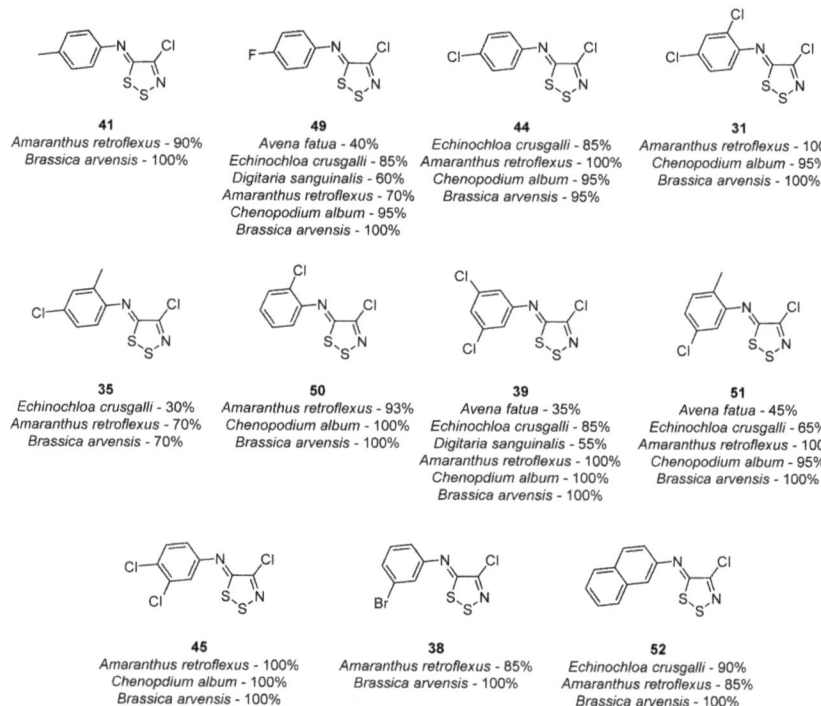

Figure 5. Herbicidal activities of early 1,2,3-dithiazole derivatives tested at 33 ppm.

In order to test for other pests, pinto bean leaves were treated with two spotted mites (*Tetramuchus urticae*). The mites were then allowed to lay eggs on the leaves, and after 48 h, the leaves were treated with 40 ppm of the test compound (Figure 6). A series of halogenated phenyl-5H-1,2,3-dithiazol-5-imines were identified with activity against both *Tetramuchus urticae* and their eggs. The 3,5-dichloro analogue **39** showed a high degree of control with 90% of mites and 85% of eggs suppressed. This increased to almost total control with the 3,4-dichloro **45**. Interestingly, the 2,4-dichloro analogue **31** demonstrated total mite control but had no effect on the eggs. The mono-substituted 2-chloro analogue had a similar profile with no effect on the mite eggs, but only 70% effective control of the mite. The 4-chloro, 2-methyl analogue **35** showed complete egg control and almost complete mite control (94%). The switch to the bromo **38** showed a similar profile, but with 70% mite control. The 2-methyl substituted match pair analogues 3-chloro **56** and 5-chloro **52** both demonstrated a high level of mite and egg control with the 3-position preferred.

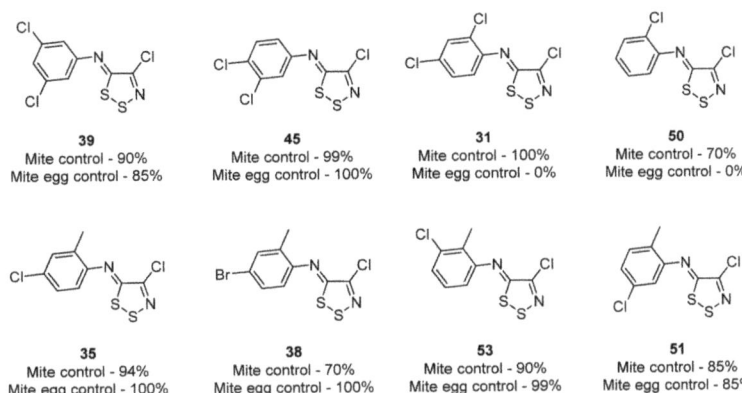

Figure 6. Mite (*Tetranychus urticae*) control activities of early 1,2,3-dithiazole derivatives at 40 ppm.

Subsequent to the work reported by Chevron Research Co., in 1980, a brief patent was filed by Appel, R. et al. on behalf of Bayer AG on the use of 1,2,3-dithiazoles as antifungals specifically against *Trichophyton Mentagrophytes* [48]. This was followed up by another brief patent in 1984 by Mayer R. et al. on behalf of Dresden University of Technology (Technische Universität Dresden) on the use of *N*-arylcyanothioformamides derived from 1,2,3-dithiazoles as herbicides and crop protection agents [49].

The 1,2,3-dithiazoles chemical space and synthesis progressed as outlined in Section 2.2 during the late 1980s and early 1990s. However, it was not until 1996 when Pons et al. disclosed a focused series of *N*-arylimino-1,2,3-dithiazoles and related *N*-arylcyanothioformamides before further biology was elucidated [15]. The unsubstituted aromatic compound **40** and the 2-methoxy analogue **54** were shown to have potent activity on several bacteria strains (Figure 7). Compound **40** had an MIC of 16 µg/mL on *S. aureus*, *E. faecalis*, and *L. monocyotogenes*, while 2-methoxy **54** had the same level of potency, but only on *E. faecalis* and *L. monocyotogenes*. Interestingly, all the *N*-arylcyanothioformamides analogues tested were ineffective, highlighting the need for the 1,2,3-dithiazole ring.

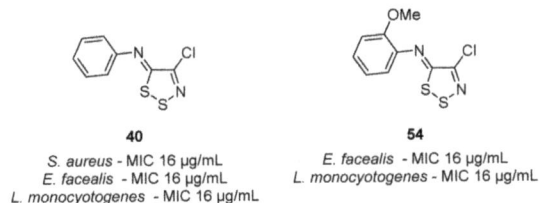

Figure 7. Report on a small panel of 1,2,3-dithiazoles highlighted some nascent antibacterial activity on the dithiazole scaffold.

This work was extended in a subsequent report by Pons et al. [50], where a focused library of 1,2,3-dithiazoles and related analogues were screened on a series of fungal targets. The 1,2,3-dithiazoles were the only compounds that showed antifungal activity, with most potent analogues identified as unsubstituted aromatic **40**, the 2-methoxy **54**, and 4-methoxy analogue **55** (Figure 8). These three most potent analogues all had an MIC of 16 µg/mL on *C. albicans*, *C. glabrata*, *C. tropicalis*, *L. orientalis*, and an MIC of 8 µg/mL on *C. neoformans*.

40 54 55

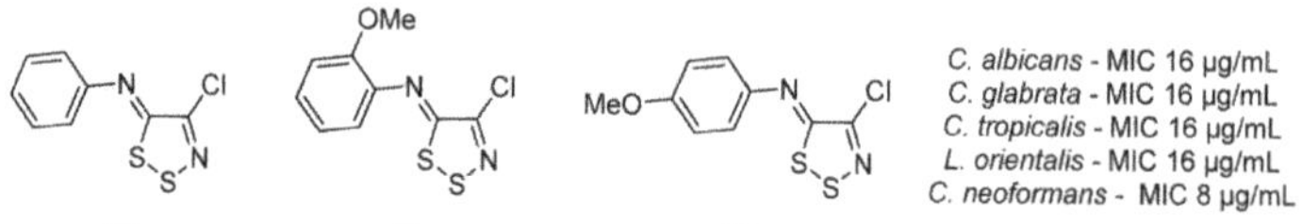

Figure 8. Report on a small panel of 1,2,3-dithiazoles highlighted some nascent antifungal activity on the dithiazole scaffold.

This was followed by a patent filed in 1997 by Joseph, R. W. et al. on behalf of Rohm & Haas Co. [51], a company specializing in the manufacture of coatings. The disclosed innovation involved the use of 1,2,3-dithiazoles to rapidly inhibit microbial and algae growth for industrial applications. These included paints, coatings, treatments, and textiles, among others. The effective amount applied was between 0.1 to 300 ppm, with three main exemplar 1,2,3-dithiazoles highlighted (Figure 9). This included 4-chloro-5*H*-1,2,3-dithiazol-5-one (**5**) with potent antibacterial properties against *R. Rubra TSB* (MIC = 7.5 ppm) and *E. Coli M9G* (MIC = 19 ppm), with potent algae inhibition of *Chlorella, Scenedesmus,* and *Anabaena* (all MIC = 3.9 ppm) and *Phormidium* (MIC = 7.8 ppm). In addition to **5,** the 2-chloro analogue **52** was reported to have potent activity against *R. Rubra TSB* (MIC = 7.5 ppm) and good activity against *E. Coli M9G* (MIC = 32 ppm) and *A. Niger TSB* (MIC = 50 ppm). The 4-nitro analogue **56** also performed well with both *E. Coli M9G* and *A. Niger TSB* having an MIC or 50 ppm. The activity reported between **5** and **52** on *E. Coli M9G* is the first evidence of activity against a Gram-negative bacterium. The company also provided data with time of addition experiments showing that 5 and 10 ppm of **5** are effective at 1 h, whereas 10 ppm of methylene *bis*thiocyanate (MBT), a known commercial antimicrobial compound, is not effective until 24 h.

Subsequently in 1998, more detailed screening and structure activity relationships (SAR) were published from Pons et al. related to the antimicrobial properties of the 1,2,3-dithiazole scaffold [52,53]. These two studies tested activity against bacteria: *S. aureus, E. faecalis, S. pyogenes,* and *L. monocytogenes,* and fungi: *C. albicans, C. glabrata, C. tropicalis,* and *I. orientalis.* This screening supported earlier work on the 1,2,3-dithiazole scaffold, and broadened the scope of this inhibition to several new fungal and bacteria strains (Figure 10). The compounds showed antibacterial activity against Gram-positive bacteria, but as previously described [15], there was no activity against Gram-negative bacteria.

5
Bacteria:
R. Rubra TSB - MIC 7.5 ppm
E. Coli M9G - MIC 19 ppm
Algae:
Chlorella, Scenedesmus, Anabaena - all MIC 3.9 ppm
Phormidium - MIC 7.8 ppm

52
R. Rubra TSB - MIC 7.5 ppm
E. Coli M9G - MIC 32 ppm
A. Niger TSB - MIC 50 ppm

56
R. Rubra TSB - MIC 50 ppm
A. Niger TSB - MIC 50 ppm

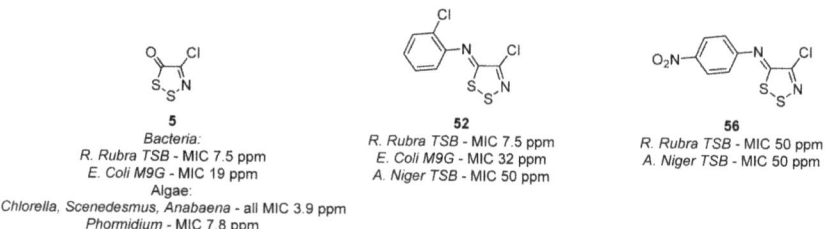

Figure 9. Rohm & Haas Co. filed a patent for industrial applications around three 1,2,3-dithiazoles for antibacterial and antialgae properties.

The unsubstituted analogue **40** was a direct repeat of all activities previously demonstrated with antibacterial *S. aureus; E. faecalis;* and *L. monocyotogenes* (all MIC = 16 μg/mL); and antifungal *C. albicans, C. glabrata; C. tropicalis;* and *L. orientalis* (all MIC = 16 μg/mL). All of the highlighted compounds (**40, 54, 57–66**) had *C. albicans* activity at MIC = 16 μg/mL. The 2-cyano analogue **57** had activity (MIC = 16 μg/mL) across all fungal strains tested but had limited antibacterial effects. Switching to the 2-methylester **58** narrowed the antifungal activity. However, the 2-methoxy **54** had good broad spectrum antimicrobial activity hitting 7 out of the 8 strains tested. The introduction of a second methoxy group in the 5-position to afford (*Z*)-4-chloro-*N*-(2,5-dimethoxyphenyl)-5*H*-1,2,3-dithiazol-5-imine (**59**) increased the potency (MIC = 4 µg/mL) on *C. glabrata*, while maintaining antifungal coverage. Moving the 2-position methoxy to the 4-position in analogue **60** maintained the antifungal coverage but lost the 4-fold boost seen against *C. glabrata* with **59**. The (Z)-(4-chloro-2-((4-chloro-5H-1,2,3-dithiazol-5-ylidene)amino)phenyl)methanol (**61**) analogue showed potency against *E. faecalis*, *C. glabrata*, *C. albicans*, and *L. orientalis* (all MIC = 16 µg/mL); in addition to demonstrating a tolerability for more diverse substitution patterns.

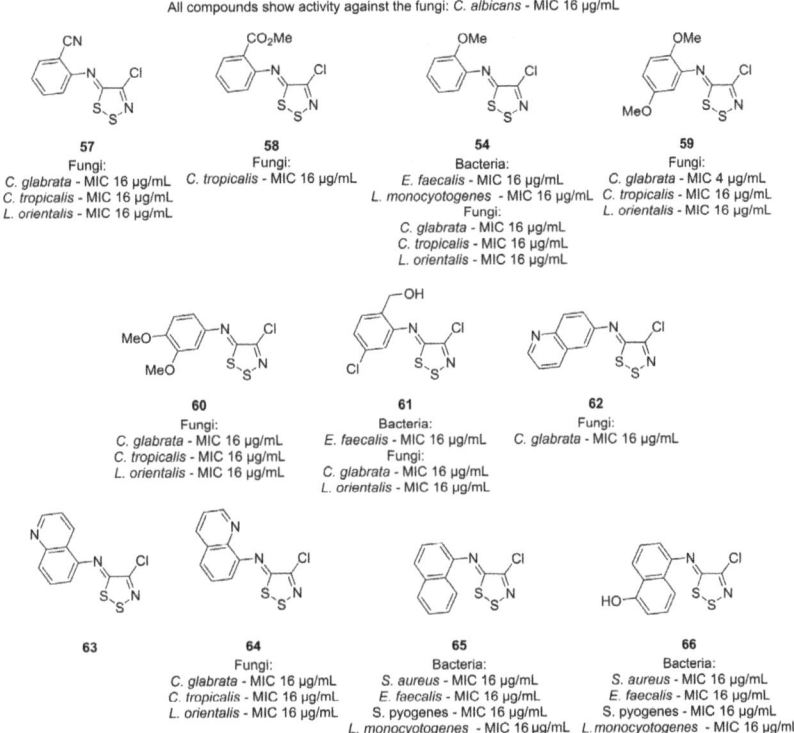

Figure 10. Results of a focused investigation of antibacterial and antifungal activities of selected 1,2,3-dithiazoles [52,53].

A switch to fused heterocycles including quinolines and naphthalene was maintained rather than increased overall potency and coverage. The quinolin-6-yl substituted analogue **62** showed potency against *C. glabrata* and *C. albicans* (both MIC = 16 µg/mL), while the quinolin-5-yl **63** was only active against the *C. albicans*. The naphthalen-1-yl **65** and hydroxy substituted naphthalen-1-yl **66** had the same profile with coverage against all four bacteria tested (*S. aureus*, *E. faecalis*, *S. pyogenes*, and *L. monocyotogenes* all MIC = 16 µg/mL) and *C. albicans*. The hydroxy substitution of **65** to afford **66** did not provide any potency advantage, but did demonstrate there was an ability to alter physicochemical properties without affecting potency.

Access to a series of new substituted 5-phenylimino, 5-thieno, or 5-oxo-1,2,3-dithiazoles was reported in 2009 by Rakitin et al. [13] (synthesis discussed in Section 2.4). A series of sixteen compounds were screened against four fungi strains: *C. albicans*, *C. glabrata*, *C. tropicalis*, and *I. orientalis*; four Gram-negative bacteria strains: *E. coli*, *P. aeruginosa*, *K. pneumoniae*, and *S. Typhimurium*; and four Gram-positive bacteria strains: *E. faecalis*, *S. aureus*, *B. cereus*, and *L. inocua*. The result of this screening were some compounds with limited activity and 4-(pyridin-2-yl)-5H-1,2,3-dithiazole-5-thione (**67**), which was active against bacteria *L. inocua*

(MIC = 16 μg/mL) and fungi *C. galbrata* (MIC = 8 μg/mL) (Figure 11). These results opened up additional chemical space to potentially further investigate the 1,2,3-dithiazole SAR.

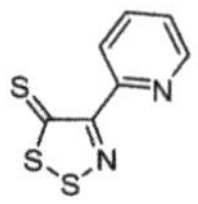

67
Bacteria:
L. inocua - MIC 16 μg/mL
Fungi:
C. galbrata - MIC 8 μg/mL

Figure 11. The 4-(pyridin-2-yl)-5*H*-1,2,3-dithiazole-5-thione (**67**) was the only compound with potent antimicrobial activity from the C4 substituted analogue library.

In 2012, a patent was filed by Benting et al. on behalf of Bayer Cropscience AG focusing on phytopathogenic antifungal crop protection aspects of heteroaromatic substituted 1,2,3-dithiazole analogues [54]. Heteroaromatic substitution patterns had until the point been largely neglected, due in part to electron deficient amines affording lower yields (as in the case of [37]). A library of heteroaromatic 1,2,3-dithiazole derivative were screened against a series of different fungi strains. These included tomato (*Phytophtora*), cucumber (*Sphaerotheca*), apples (*Venturiatest*), tomato (*Alternia*), beans (*Botyrtis*), wheat (*Leptosphaeria nodorum*), wheat (*Septoria tritici*), and rice (*Pyricularia*). Only compounds that showed inhibition above 70% at the respective concentration tested were reported. The results were broadly clustered into three groups, small 5-membered heterocyles, pyrimidine, and 3-pyridyl substituted heterocycles and 2-pyridyl substituted heterocycles. There was broad SAR tolerability for *Phytophtora* (Figure 12), *Sphaerotheca* (Figure 13), *Venturiatest* (Figure 14), *Alternia* (Figure 15), and *Botyrtis* (Figure 16). While with *Leptosphaeria nodorum* (Figure 17), *Septoria tritici* (Figure 18), and *Pyricularia* (Figure 19) the SAR narrowed considerably.

The chemical space around *Phytophtora* inhibition included a broad array of 1,2,3-dithiazol-5-imines **4a** and **68–90** (Figure 12). While most analogues reported were potent; only the 4-methyl 2-pyridyl **87** and 3-methoxy 2-pyridyl **90** achieved total control of the fungi. In the case of *Sphaerotheca*, many compounds demonstrated very high degrees of antifungal control, including **4a**, **4b**, **68–70**, **73**, **77**, **79–81**, **86–87**, and **91–104** (Figure 13), while ten compounds showed complete control including the unsubstituted analogues 2-pyridyl **4a** and 4-methyl 2-pyridyl **91**.

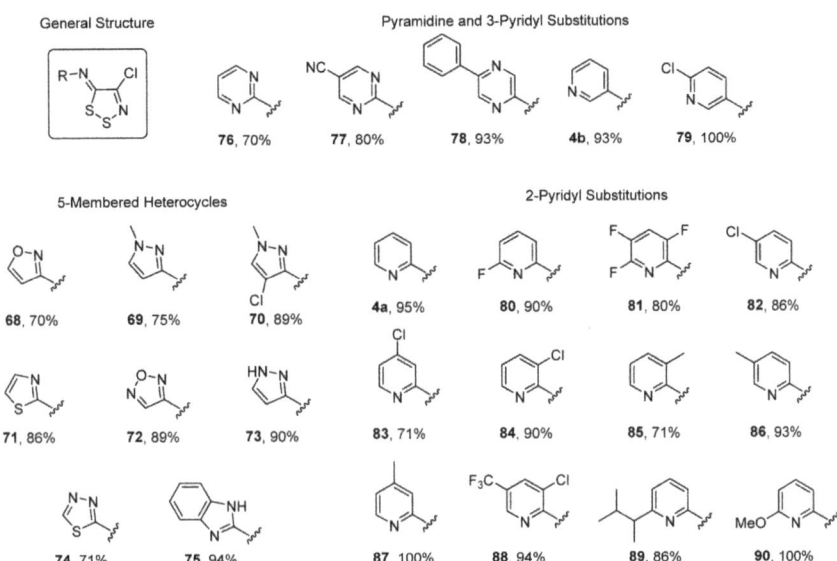

Figure 12. *Phytophtora* (tomato) preventive; ≥70% efficacy at concentration of 1500 ppm.

The *Venturiatest* fungi appears to be easier to target as the reported dose is 6-fold less (250 ppm vs. 1500 ppm), and most reported compounds **4a, 68–69, 75–76, 79–81, 84, 86–87, 90–91, 95–97,** and **105–111** having high potency with a mixture of substitution patterns offering total control (Figure 14). These included isoxazoles **68** and **105**, 1,2,5-oxadiazole **91** and a series of seven substituted 2-pyridyl analogues including **4a, 80–81, 84, 96–97,** and **101** (Figure 14). The *Alternia* fungi appears to be more difficult to effectively target as, while the compounds **4a, 68–69, 75–76, 79–81, 84, 87, 90–91, 95–97, 105,** and **107–111** were similar to the inhibitors identified for *Venturiatest*, none reached total control of *Alternia*. The most potent compounds, pyrazole **69**, unsubstituted 2-pyridyl **4a**, and 3-fluoro 2-pyridyl **80** all had 96% control at 250 ppm (Figure 15). The *Botyrtis* fungi was also not completely controlled by the active compounds **4a, 68–69, 75–76, 81, 84, 91, 95–97, 105, 107,** and **110**, despite a high level of potency. The SAR around *Botyrtis* was considerably narrower with roughly half the number of earlier analogues reported (Figure 16), despite the higher 500ppm concentration tested. The pyrazole analogue **69** was able to potently inhibit *Botyrtis* infection to 99%, while several other analogues also had potent inhibition (>95%).

Only five 1,2,3-dithiazoles were reported to be active against *Leptosphaeria nodorum* (Figure 17) and the need for an increased concentration of test compound to 1000 ppm, potentially highlighting that *Leptosphaeria nodorum* is more difficult to target. The five compounds reported were all 2-pyridyl substituted **80, 84, 87, 90,** and **96**, but only (Z)-4-chloro-N-(6-methoxypyridin-2-yl)-5H-1,2,3-dithiazol-5-imine (**90**) had total control of the *Leptosphaeria nodorum* infection.

The fungi *Septoria tritici* had a similar profile to *Leptosphaeria nodorum*, with only five potent compounds reported: **84, 87, 98, 109,** and **112** (Figure 18). The most potent four of the five compounds reported were 2-pyridyl substituted **84, 87, 98,** and **112**. The 4-methyl 2-pyridyl **87** and 3-methoxy 2-pyridyl **112** were the most potent, with 100% control of *Septoria tritici* infection.

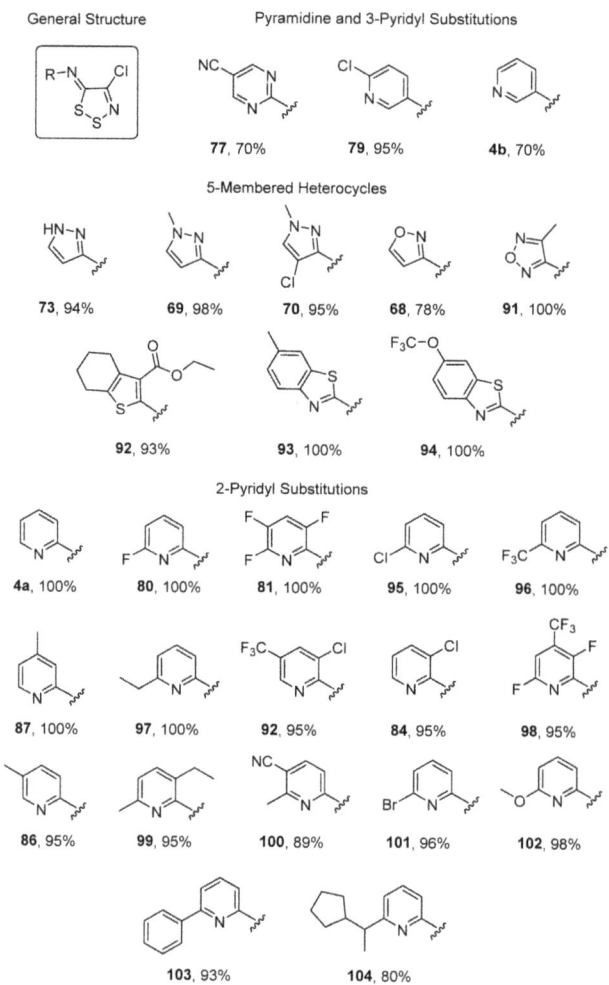

Figure 13. *Sphaerotheca* (cucumber) preventive; ≥70% efficacy at concentration of 1500 ppm.

Interestingly, the final set of results of inhibitors against *Pyricularia* revealed only two highly active compounds. These two compounds, (Z)-4-chloro-N-(6-methoxypyridin-2-yl)-5H-1,2,3-dithiazol-5-imine (**90**) and (Z)-4-chloro-N-(isoxazol-3-yl)-5H-1,2,3-dithiazol-5-imine (**68**), were both able to control 100% of the *Pyricularia* infection even at the lower concentration of 250 ppm. The lack of further SAR may (or may not) indicate that, while two highly active compounds are reported this infection was the most difficult to treat.

More recently, a 2020 study by our group reported a set of 1,2,3-dithiazoles and matched pair 1,2,3-thiaselenazoles as antimicrobials [55]. The rare 1,2,3-thiaselenazoles were synthesized by sulfur extrusion and selenium insertion into 1,2,3-dithiazoles [55,56]. This work was part of the Community for Antimicrobial Drug Discovery (CO-ADD) project to develop new lead compounds for priority targets with an unmet clinical need [57]. The compounds were screened against *S. aureus*, *A. baumannii*, *C. albicans*, and *C. neoformans var. grubii*. with a toxicity counter screen in HEK293 cells and an additional hemolysis assay (Hc10) (Figure 20). These strains are considered by the World Health Organization (WHO) to be the highest priority to develop novel antibiotics for control of these bacteria and fungi [58].

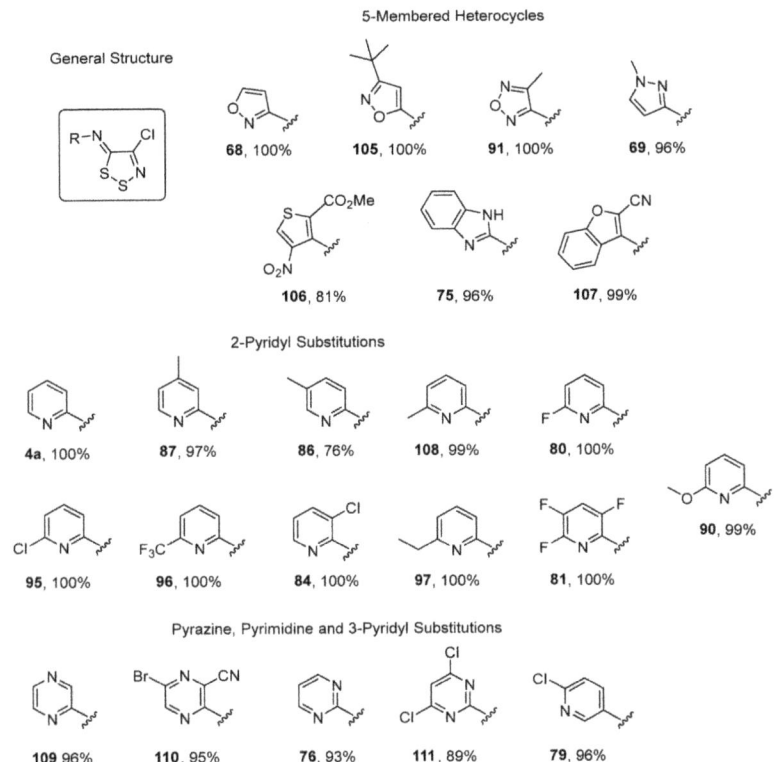

Figure 14. *Venturia test* (apples) preventive; ≥70% efficacy at concentration of 250 ppm.

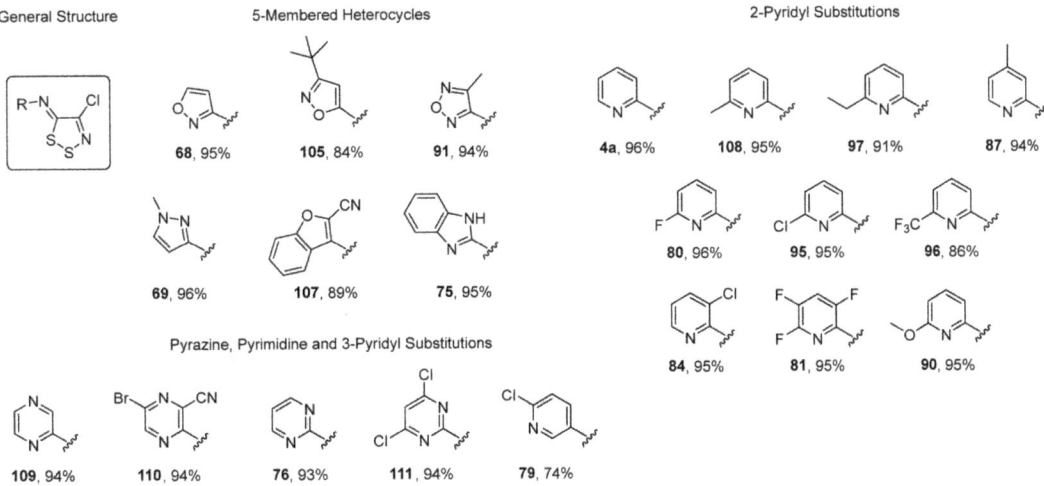

Figure 15. *Alterniα* (tomatoes) preventive ≥70% efficacy at concentration of 250 ppm.

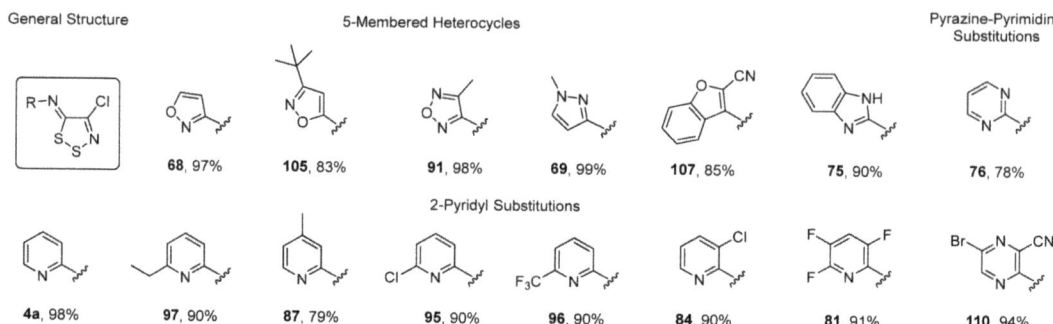

Figure 16. *Botyrtis* (beans) preventive ≥70% efficacy at concentration of 500 ppm.

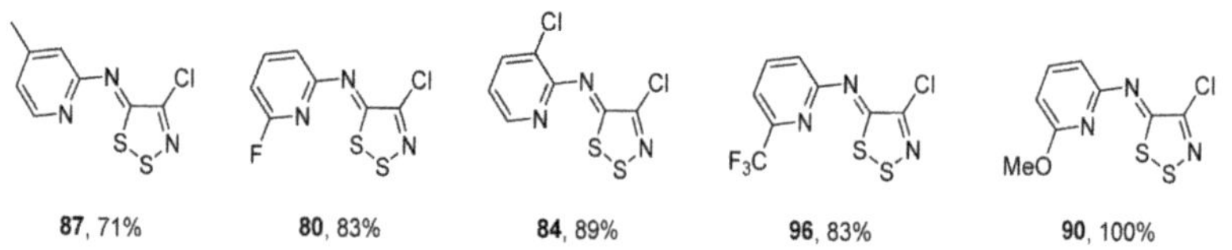

Figure 17. *Leptosphaeria nodorum* (wheat) preventive ≥70% efficacy at concentration of 1000 ppm.

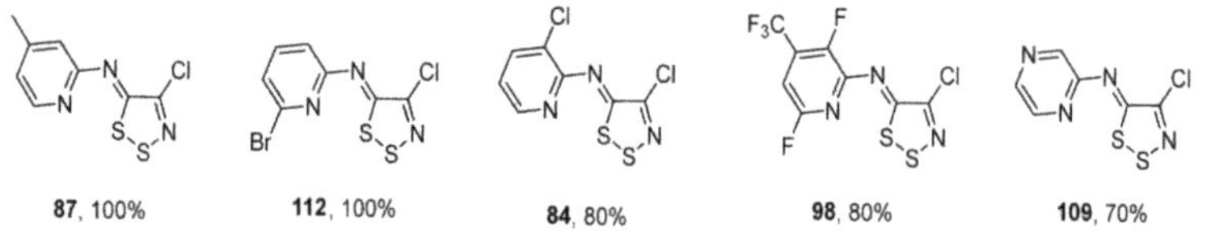

Figure 18. *Septoria tritici* (wheat) preventive ≥70% efficacy at concentration of 1000 ppm.

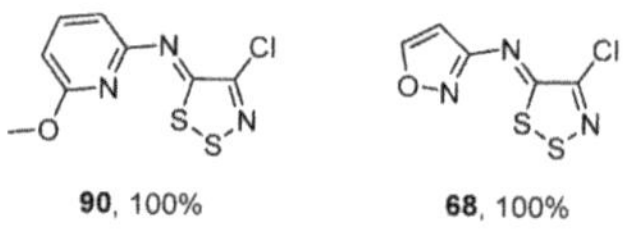

Figure 19. *Pyricularia* (rice) preventive ≥80% efficacy at concentration of 250 ppm.

The compounds **113–120** demonstrated potency against several of the strains tested, with the 1,2,3-thiaselenazoles tending to be more active (Figure 20). The 4,5,6-trichlorocyclopenta[*d*][1,2,3]thiaselenazole (**113**) demonstrated potent activity against Gram-positive bacteria *S. aureus* (MIC = ≤0.25 µg/mL), Gram-negative bacteria *A. baumannii* (MIC = ≤0.25 µg/mL) along with antifungal activity against *C. albicans* and *C. neofromans* (both MIC ≤0.25 µg/mL). The trichoro analogue **113** had some toxicity (CC$_{50}$ = 0.52 µM), whereas both the 4-cyano **114/115** and 4-ethylester **116/117** 1,2,3-dithiazole/1,2,3-thiaselenazole matched pair analogues showed limited to no toxicity (all CC$_{50}$ = >32 µM, apart from **117** = CC$_{50}$ = 7 µM). The 4-cyano analogues **114/115** were both active against *C. albicans* and *C. neofromans* (both MIC ≤0.25 µg/mL); however, the 1,2,3-thiaselenazole also had activity against *S. aureus* (MIC = ≤0.25 µg/mL). This activity trend was matched exactly by the 4-ethylester analogues **116/117**. The 4,5,6-trichlorobenzo[6,7]cyclohepta [1,2-*d*][1,2,3]thiaselenazole (**118**) analogue matched the profile of **115** and **117** albeit with some toxicity (CC$_{50}$ = 0.48 µM). Interestingly, the activity profiles of 8-chloroindeno[1,2-*d*][1,2,3]thiaselenazole (**119**) and benzo[*b*][1,2,3]thiaselenazolo[5,4-*e*][1,4]oxazine (**120**) were similar with antifungal activity against *C. albicans* and *C. neofromans* (all MIC = ≤0.25 µg/mL, apart from **120**, *C. neofromans* = 2 µg/mL). Taken together these results demonstrate an ability for the 1,2,3-dithiazole/

1,2,3-thiaselenazole to inhibit a broad range of challenging and clinically relevant bacteria and fungi [55,58].

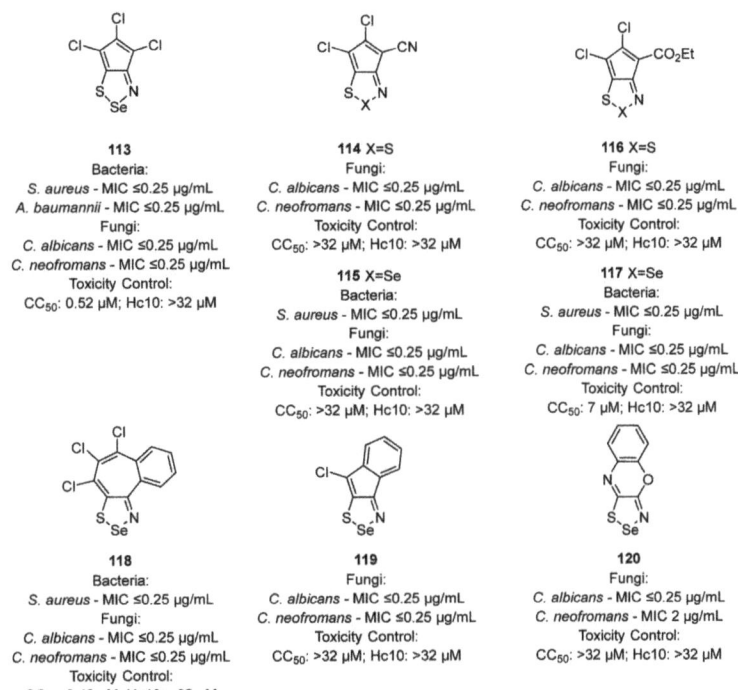

Figure 20. Summary of the most active antifungal 1,2,3-dithiazoles and 1,2,3-thiaselenazoles from the 2020 study by our group [55].

3.2. Antiviral Activities of 1,2,3-Dithiazoles

The first antiviral activities on the 1,2,3-dithazole scaffold were reported in 2016 by Hilton et al. [17]. A series of 5-thien HIV infection. The rationale of using the 1,2,3-dithiazoles to target the nucleocapsid protein was that it could potentially act as ao-, 5-oxo-, and 5-imino-1,2,3-dithiazole derivatives were screened against Feline Immunodeficiency Virus (FIV) as a model for zinc ejector by utilizing the disulfide bridge [59–62]. The compounds were tested for antiviral effects in a feline lymphoid cell line (FL-4) and tested for toxicity using Crandell-Rees feline kidney (CrFK) cells (Figure 21). The four highlighted compounds, **121–124**, were the most potent antivirals with the largest toxicity window (ratio of FL-4/CrFK). The 4-phenyl-5H-1,2,3-dithiazole-5-thione (**121**) had an excellent ratio of activity vs. toxicity (>4000) and potency of EC_{50} = 23 nM. The 4-(4-fluorophenyl)-5H-1,2,3-dithiazol-5-one (**122**) was equipotent to **121** with a small amount of toxicity at higher concentrations (CC_{50} = 64 μM). The 4-methoxy analogue **123** had a drop of almost 8-fold in potency, with the ethyl (Z)-5-(phenylimino)-5H-1,2,3-dithiazole-4-carboxylate (**124**) analogue had an almost 3-fold drop, both with a comparable toxicity profile.

The proposed mechanism of action was modelled on previously experimental reports (Figure 22) [59–62]. Zinc ejection from nucleocapsid protein starts with Zn^{2+} coordinated to cysteine thiol(ate)s reacting with the disulfide of the 1,2,3-dithiazole core to generate a transient intermediate disulfide. This complex then rearranges to form an intramolecular protein disulfide, which has a consequent reduction in zinc ion affinity. This results in the zinc being ejected from the protein in a similar mechanism as previously reported for the HIF1alpha/P300 interaction triple zinc finger [59]. To indirectly prove the mechanism in addition to computational modelling, a disrupted disulfide bridge of analogue **121**, compound **125** was synthesized (Figure 23) [47], demonstrating that the disulfide was required for activity.

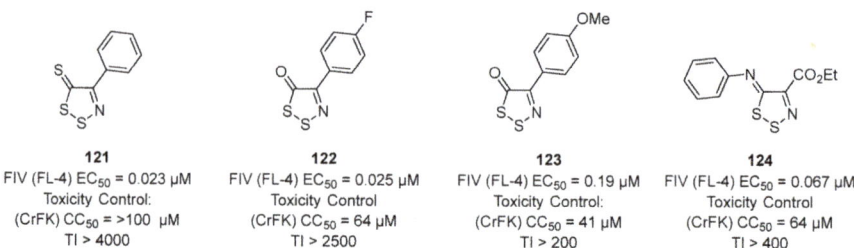

121
FIV (FL-4) EC$_{50}$ = 0.023 µM
Toxicity Control:
(CrFK) CC$_{50}$ = >100 µM
TI > 4000

122
FIV (FL-4) EC$_{50}$ = 0.025 µM
Toxicity Control:
(CrFK) CC$_{50}$ = 64 µM
TI > 2500

123
FIV (FL-4) EC$_{50}$ = 0.19 µM
Toxicity Control:
(CrFK) CC$_{50}$ = 41 µM
TI > 200

124
FIV (FL-4) EC$_{50}$ = 0.067 µM
Toxicity Control:
(CrFK) CC$_{50}$ = 64 µM
TI > 400

Figure 21. Summary of the initial report of antiviral activity of the 1,2,3-dithiazole scaffold.

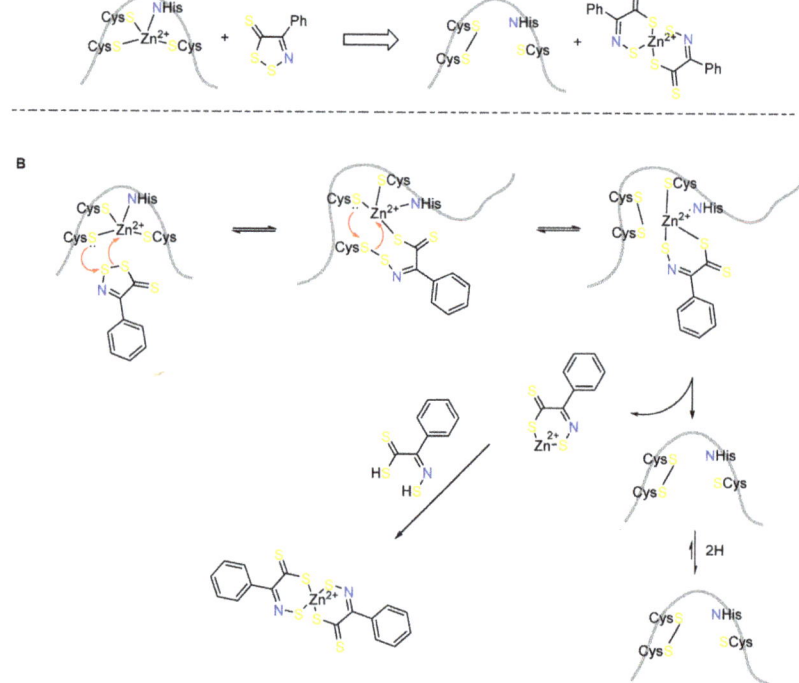

Figure 22. Proposed redox mechanism for 1,2,3-dithiazoles mediated zinc ejection of the FIV nucleocapsid protein. (**A**) Summary of reaction; (**B**) Detailed reaction pathway analysis.

121
FIV (FL-4) EC$_{50}$ = 0.023 µM
Toxicity Control:
(CrFK) CC$_{50}$ = >100 µM
TI = > 4000

125
FIV (FL-4) EC$_{50}$ = 0.39 µM
Toxicity Control:
(CrFK) CC$_{50}$ = 0.68 µM
TI = 1.7

Figure 23. Direct comparison of **121** against **125** with the disrupted 1,2,5-thiadiazole-3(2*H*)-thione ring system.

This idea was followed up in 2019 by our group [63], investigating the same inhibitors later reported as antimicrobials [55]. The key rationale behind this subsequent work was the further investigation of the disulfide bridge involvement on antiviral efficacy with a matched pair side by side comparison between the 1,2,3-dithiazoles and the 1,2,3-thiaselenazole scaffold. Where the weaker S-Se vs. S-S bond should assist in increasing the antiviral efficacy. This followed on from a previous report of a successful selenide isosteric replacement to several literature nucleocapsid protein inhibitors, including DIBA-4 to DISeBA-4 HIV inhibitors, resulting in good potency and only a very limited associated toxicity [64]. The antiviral efficacy of the 1,2,3-dithiazole scaffold was tested using FL-4 cells, but in this study an additional toxicity assay was preformed directly on the FL-4 cells (Figure 24). The 8-phenylindeno[1,2-*d*][1,2,3]dithiazole (**126**) was only weakly active, while the selenium analogue **127** demonstrated a 10-fold boost in potency to EC_{50} = 0.26 µM with only limited toxicity. The ethyl 5,6-dichlorocyclopenta[*d*][1,2,3]dithiazole-4-carboxylate (**116**) had a similar profile with an EC_{50} = 0.26 µM, while the selenium analogue **117** was almost 4-fold more potent. The difference between the benzo[*b*][1,2,3]dithiazolo[1,4]oxazine (**128**) and selenium analogue **120** was even more pronounced with an almost 17-fold increase in potency. These results highlight the advantages of including selenium in the 1,2,3-dithiazole scaffold.

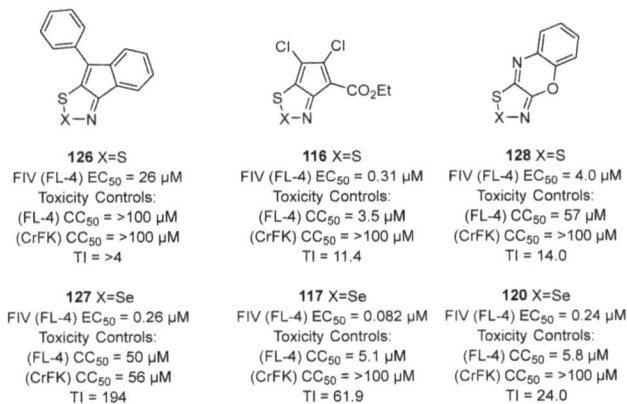

Figure 24. Summary active antiviral matched pair 1,2,3-dithiazoles and 1,2,3-thiaselenazoles.

More recently, a further extension of investigation of the 1,2,3-dithiazoles in 2022 by our group evaluated a further series of 1,2,3-dithiazoles against FIV as a model for HIV infection [36]. The rationale of this investigation was to find a tractable series of 1,2,3-dithiazoles with consistently high potency and lower toxicity to further advance the scaffold. The antiviral screening was performed using FL-4 cells, with a direct toxicity assay on FL-4 cells in addition to CrFK and feline embryo cell line (FEA) cells (Figures 25–27).

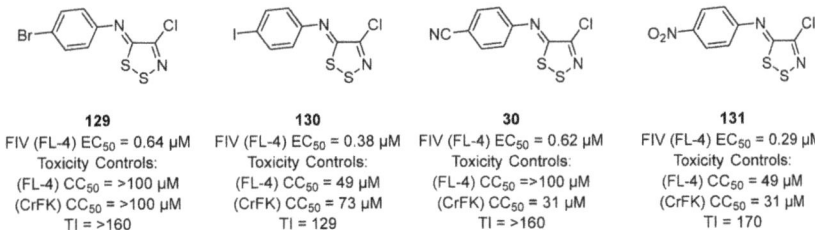

Figure 25. Initial hit compounds from the 1,2,3-dithiazole library.

4a	**132**	**112**	**90**
FIV (FL-4) EC_{50} = 0.38 μM	FIV (FL-4) EC_{50} = 0.42 μM	FIV (FL-4) EC_{50} = 0.48 μM	FIV (FL-4) EC_{50} = 0.26 μM
Toxicity Controls:	Toxicity Controls:	Toxicity Controls:	Toxicity Controls:
(FL-4) CC_{50} = 40 μM	(FL-4) CC_{50} = 51 μM	(FL-4) CC_{50} = 44 μM	(FL-4) CC_{50} = 47 μM
(CrFK) CC_{50} = 29 μM	(CrFK) CC_{50} = 3.1 μM	(CrFK) CC_{50} = 38 μM	(CrFK) CC_{50} = 34 μM
TI = 105	TI = 122	TI = 92	TI = 178

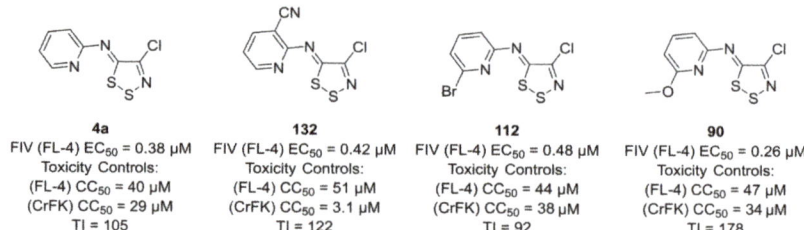

Figure 26. 2-Pyridyl substituted 1,2,3-dithiazoles active against FIV.

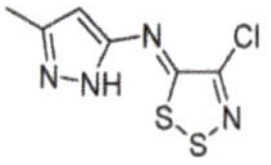

133
FIV (FL-4) EC_{50} = 0.083 μM
Toxicity Controls:
(FL-4) CC_{50} = 50 μM
(FEA) CC_{50} = 55 μM
(CrFK) CC_{50} = >100 μM
TI = 600

Figure 27. Most potent compound **133** active agiants FIV in the 2022 study by our group [36].

The initial hit compounds from the 1,2,3-dithiazole library yielded a series of 4-position substituted phenyl analogues **30** and **129–131** with a range of activities EC_{50} = 0.26–0.48 μM and limited toxicity (Figure 25). Another trend observed within the series was activity across a number of 2-pyridyl substituted analogues **4a**, **90**, **112** and **132**, at a similar level to the earlier analogues but with a divergent SAR profile (Figure 26). The most promising compound identified in this work was the pyrazole (*Z*)-4-chloro-*N*-(3-methyl-1*H*-pyrazol-5-yl)-5*H*-1,2,3-dithiazol-5-imine (**133**) that showed good antiviral potency EC_{50} = 0.083 μM with very limited toxicity (Figure 27).

The proposed mechanism of action on the nucleocapsid protein of this 4-chloro-1,2,3-dithiazol-5-imine series is different to the C4 substituted version previously reported (Figure 22). The DFT calculations and previous ANRORC-style rearrangement reported on this scaffold suggest that there will be a ring opening and chloride elimination. An outline mechanism would be a Zn^{2+}-coordinating cysteine thiol(ate) reacts with 2-S of the 1,2,3-dithiazole core mediated by water to generate a transient trisulfide. This is then followed by a rearrangement to a more thermodynamically stable cyano functionality, resulting in the loss of HCl and water from the system. The, disulfide then rearranges to form an intramolecular protein disulfide with consequent reduction in zinc ion affinity. The zinc ion is then ejected to form a stable complex, with or without adducts (Figure 28). In addition to the literature rearrangement examples [38–40], we also provided extensive computational modelling to support the mechanistic rationale provided.

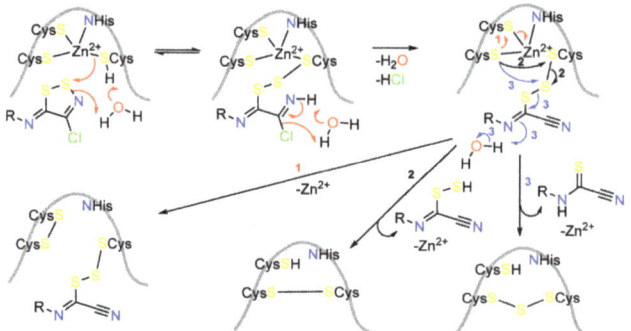

Figure 28. Proposed mechanism of action of the 4-choro-1,2,3-dithiazole series.

3.3. Anticancer Activities of 1,2,3-Dithiazoles

Initial reports of anticancer activity with the 1,2,3-dithiazole scaffold were reported in 2002 by Baraldi et al. [16]. A set of ten 1,2,3-dithiazoles were prepared and screened across multiple antimicrobial and anticancer therapeutic targets. Several of the compounds **134–136** showed low signal digit micromolar potency against the leukemia cell lines L1210 and K562 (Figure 29). While the overall SAR within the series was flat, this first phenotypic report showed tractable activity across both cell lines. The antibacterial screen showed limited activity, but the antifungal screening identified **135** and **136** as having some activity against *Aspergillus niger* at MIC_{50} = 10 μM. This further supports the overall tractability of this scaffold as an antifungal.

134
Cancer:
Leukemia (L1210) - IC_{50} = 5 μM
Leukemia (K562) - IC_{50} = 3 μM

135
Cancer:
Leukemia (L1210) - IC_{50} = 3 μM
Leukemia (K562) - IC_{50} = 5 μM
Fungi:
Aspergillus niger - MIC_{50} 10 μM

136
Cancer:
Leukemia (L1210) - IC_{50} = 3 μM
Leukemia (K562) - IC_{50} = 4 μM
Fungi:
Aspergillus niger - MIC_{50} 10 μM

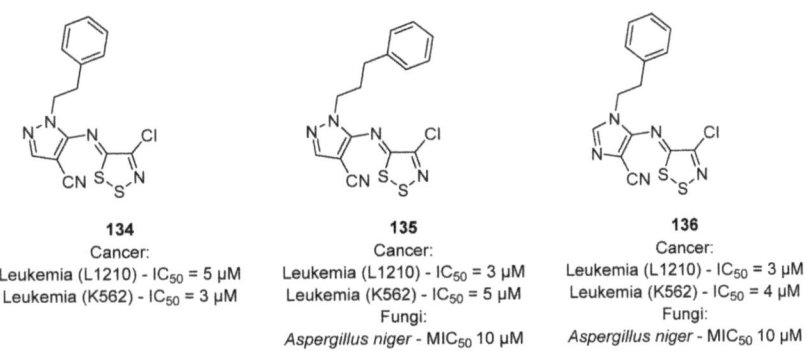

Figure 29. Initial 1,2,3-dithiazoles reported with anticancer activity in 2002 by Baraldi et al. [16].

The earlier 2009 study reported in Section 3.1 by Rakitin et al. also screened the series of C-4 substituted dithiazoles against two breast cancer cell lines, MCF7 and MDA-MB-231 [13]. Limited activity was observed on the MDA-MB-231 cell line across the series. However, (*Z*)-4-(4-nitrophenyl)-*N*-phenyl-5*H*-1,2,3-dithiazol-5-imine (**137**) and the benzofuran-2-yl analogue **138** showed 50% growth inhibition after 72 h at approximately 10 μM (Figure 30). These results lay the groundwork to expand the chemical space to further investigate the anticancer 1,2,3-dithiazole SAR.

137 **138**

Breast cancer (MCF7) - GI_{50} = Approx. 10 μM

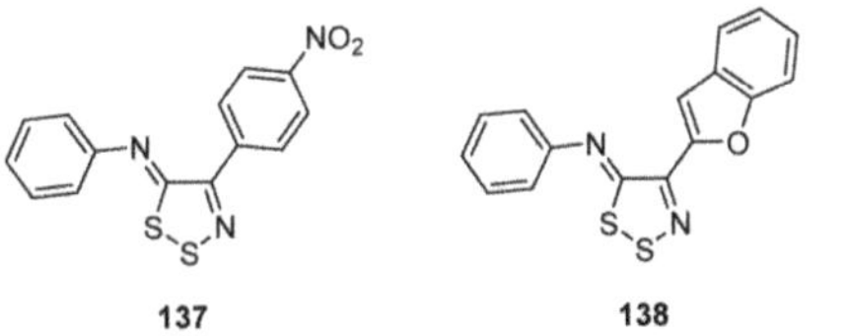

Figure 30. Substituted C4 1,2,3-dithiazoles with activity against breast cancer.

Subsequent to these phenotypic reports of 1,2,3-dithiazoles as anticancer compounds in various cell lines, a just over forty compound 1,2,3-dithiazole library was screened by Indiveri et al. against a transporter target over-expressed in various cancers, the glutamine-amino acid transporter ASCT2 in 2012 [65]. Interactions with scaffold proteins and post-translational modifications regulate the stability, trafficking, and transport activity of ASCT2 [66]. The expression of ASCT2 has been shown to increase in cells with rapid proliferation, including stem cells and inflammation, this enables delivery of the increased glutamine requirements [67]. This same mechanism can be hijacked by cancer promoting pathways to fulfill glutamine demand and facilitate rapid growth by over-expression of ASCT2 [68]. In addition to being described as an anticancer target, ASCT2 also has the ability to traffic virions to infect human cells [69]. A series of 1,2,3-dithiazoles were synthesized and evaluated as transporter inhibitors. While many compounds were in-active at 30 μM, six compounds showed activity at IC$_{50}$ = ~10 μM or below (Figure 31). These compounds potently inhibited the glutamine/glutamine transport catalyzed by ASCT2.

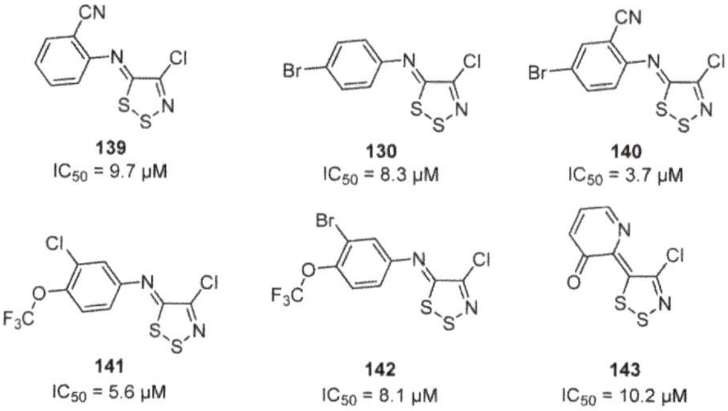

139
IC$_{50}$ = 9.7 μM

130
IC$_{50}$ = 8.3 μM

140
IC$_{50}$ = 3.7 μM

141
IC$_{50}$ = 5.6 μM

142
IC$_{50}$ = 8.1 μM

143
IC$_{50}$ = 10.2 μM

Figure 31. Anticancer 1,2,3-dithiazoles ASCT2 transport inhibitors.

The inhibition was shown to be non-competitive. The inhibition was also reversed by addition of dithiothreitol (DTE), indicating the reaction with protein Cys formed adducts, indicating that the reaction was likely going via an ANRORC-style rearrangement. Modelling, including molecular and quantum mechanical studies (MM and QM, respectively) and Frontier Orbital Theory (FOT) on 1,2,3-dithiazole models showed pathway (ii) was more likely, which is also supported by previous reports on the 1,2,3-dithiazole (Figure 32) [38–40].

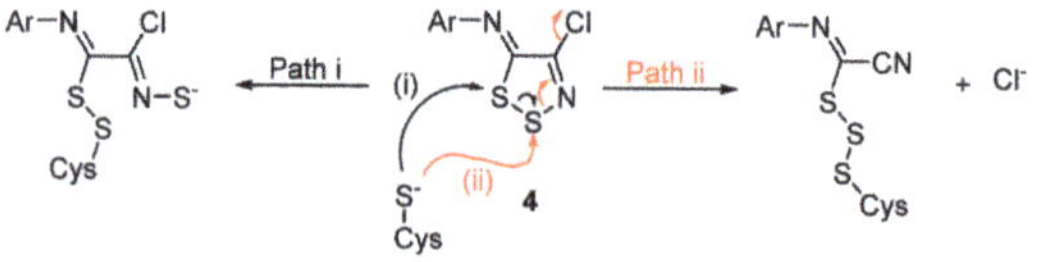

Figure 32. Two mechanisms are proposed, with nucleophilic attack at S2 to likely be preferred.

The ASCT2 report was followed up by a screening of just over fifty 1,2,3-dithiazoles by Indiveri et al. against the LAT1 transporter in 2017 [70]. ASCT2 and LAT1 are both amino acid transporters that are overexpressed in cancer [71]. Subsequently, a number of inhibitors have been reported against both ASCT2 and LAT1, with one LAT1 inhibitor JPH203 used in a recent phase 1 clinical trial [72]. The results of the library screen were eight compounds with inhibition of >90% at 100 μM. The two most potent compounds were **144** and **145** with an IC$_{50}$ = <1 μM (Figure 33).

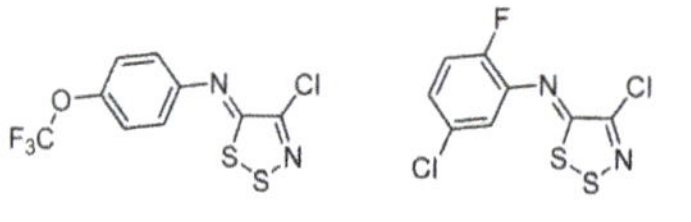

144
IC$_{50}$ = 0.98 μM

145
IC$_{50}$ = 0.89 μM

Figure 33. Anticancer 1,2,3-dithiazoles targeting transporter protein LAT1 (SLC7A5).

The inhibition kinetics, performed on the two best inhibitors (**144** and **145**), indicated a mixed type of inhibition with respect to the substrate. The inhibition of LAT1 was still present after removal of the compounds from the reaction mixture, indicating irreversible binding. However, this effect could be reversed by the addition of dithioerythritol, a S-S reducing agent, which supports the rationale of the formation of disulfide(s) bonds between the compounds and LAT1. Molecular modelling of **144** and **145** on a homology model of LAT1, highlighted the interaction with the substrate binding site and the formation of a covalent bond with the residue C407. This was further supported by a more detailed study reported in 2021 by Marino et al., which also highlighted the need for a molecule of water in the reactive pathway [73].

More recently, an extension of the phenotypic reports of 1,2,3-dithiazoles as anticancer agents was published in 2021 by our group [74]. A library of just under forty 1,2,3-dithiazole analogues were screened on a series of cancer cell lines including breast, bladder, prostate, pancreatic, chordoma and lung; with a skin fibroblast cell line as a non-specific toxicity control (Figures 34–36).

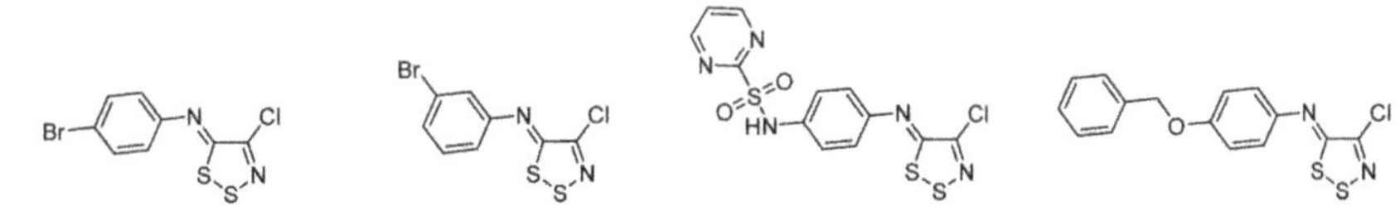

130
Cancer:
Breast (MCF7) - IC_{50} = 11 µM
Toxicity Control:
WS1 - IC_{50} = > 100 µM

54
Cancer:
Breast (MCF7) - IC_{50} = 6.7 µM
Toxicity Control:
WS1 - IC_{50} = 94 µM

146
Cancer:
Breast (MCF7) - IC_{50} = 3.0 µM
Toxicity Control:
WS1 - IC_{50} = > 100 µM

147
Cancer:
Bladder (5637) IC_{50} = 13 µM
Toxicity Control:
WS1 - IC_{50} = > 100 µM

Figure 34. Initial screening highlighted results from the 2021 cancer panel by our group [74].

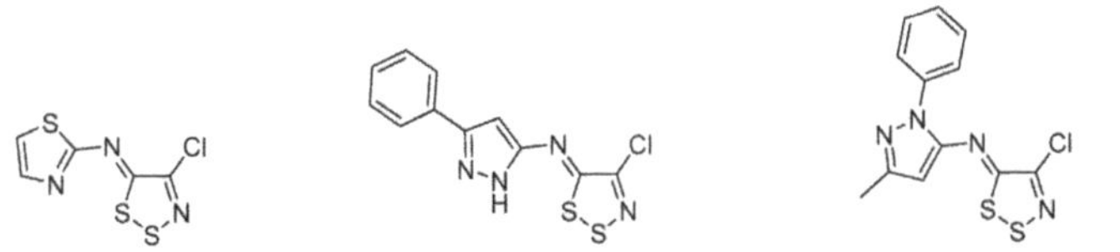

75
Cancer:
Prostate (DU145) - IC_{50} = 8.1 µM
Chordoma (U-CH1) - IC_{50} = 10 µM
Toxicity Control:
WS1 - IC_{50} = 24 µM

148
Cancer:
Prostate (DU145) - IC_{50} = 11 µM
Toxicity Control:
WS1 - IC_{50} = >100 µM

149
Cancer:
Prostate (DU145) - IC_{50} = 10 µM
Bladder (5637) - IC_{50} = 2.1 µM
Toxicity Control:
WS1 - IC_{50} = 15 µM

Figure 35. 5-membered heteroatomic analogues highlighted results from the 2021 cancer panel by our group [74].

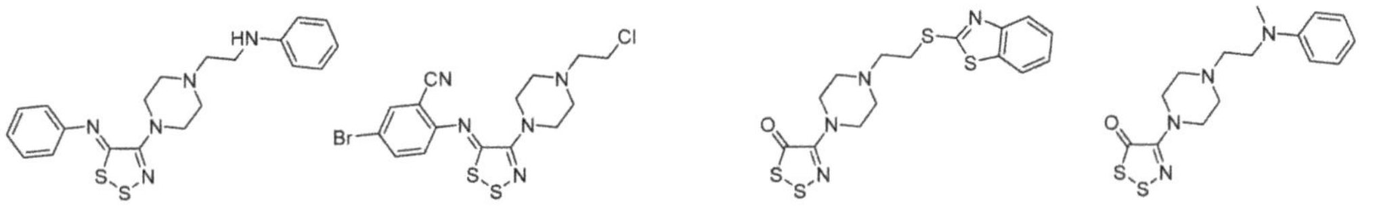

150
Cancer:
Breast (MCF7) - IC_{50} = 10 µM
Toxicity Control:
WS1 - IC_{50} = 55 µM

151
Cancer:
Breast (MCF7) - IC_{50} = 4.1 µM
Bladder (5637) - IC_{50} = 8.0 µM
Toxicity Control:
WS1 - IC_{50} = >100 µM

152
Cancer:
Breast (MCF7) - IC_{50} = 9.9 µM
Toxicity Control:
WS1 - IC_{50} = 92 µM

153
Cancer:
Breast (MCF7) - IC_{50} = 2.2 µM
Prostate (DU145) - IC_{50} = 4.4 µM
Toxicity Control:
WS1 - IC_{50} = 20 µM

Figure 36. C4 substituted 1,2,3-dithazole analogues highlighted results from the 2021 cancer panel by our group [74].

Initial results were encouraging (Figure 34) with (Z)-4-(4-bromophenyl)-N-phenyl-5H-1,2,3-dithiazol-5-imine (**130**) and the corresponding 3-position bromo analogue **54** demonstrated potency against breast cancer cell line MCF7 (IC_{50} = 11 and 6.7 µM, respectfully). This was followed by the identification of (Z)-N-(4-((4-chloro-5H-1,2,3-dithiazol-5-ylidene)amino)phenyl)pyrimidine-2-sulfonamide (**146**) with good activity against breast cancer (MCF7—IC_{50} = 3.0 µM) and no observed toxicity (WS1—IC_{50} = >100 µM). Interestingly, the (Z)-N-(4-(benzyloxy)phenyl)-4-chloro-5H-1,2,3-dithiazol-5-imine (**147**) analogue showed a preference for bladder cancer inhibition (5637—IC$_{50}$ = 13 µM) with no observed toxicity (WS1—IC$_{50}$ = >100 µM).

This was followed by screening a small focused 5-membered heteroatomic compounds, which identified a trend of potency against prostate cancer (Figure 35). (Z)-4-chloro-N-(thiazol-2-yl)-5H-1,2,3-dithiazol-5-imine (**75**) and the two pyrazoles (**148** and **149**) all showed activity against prostate cancer (DU145—IC$_{50}$ = 8–11 µM). The thiazole analogue **75** also showed activity against the chordoma cell line (U-CH1—IC$_{50}$ = 10 µM), albeit with some limited toxicity (WS1—IC$_{50}$ = 24 µM). The (Z)-4-chloro-N-(3-methyl-1-phenyl-1H-pyrazol-5-yl)-5H-1,2,3-dithiazol-5-imine (**149**) also showed low single digit micromolar activity against bladder cancer (5637—IC$_{50}$ = 2.1 µM); unfortunately, this was coupled with some associated toxicity (WS1—IC$_{50}$ = 15 µM).

Subsequent to this screening, a series of C-4 substituted analogues were evaluated resulting in a trend of activity against breast cancer (MCF7) (Figure 36) [14]. The most potent compounds **150–153** (IC$_{50}$ = 2–10 µM) did not show any defining SAR characteristics. In addition to this activity, (Z)-5-bromo-2-((4-(4-(2-chloroethyl)piperazin-1-yl)-5H-1,2,3-dithiazol-5-ylidene)amino)benzonitrile (**151**) demonstrated good potency against bladder cancer (5637—IC$_{50}$ = 8.0 µM), while 4-(4-(2-(methyl(phenyl)amino)ethyl)piperazin-1-yl)-5H-1,2,3-dithiazol-5-one (**153**) was potent against prostate cancer (DU145—IC$_{50}$ = 4.4 µM), albeit with some observed toxicity (WS1—IC$_{50}$ = 20 µM) in the case of **153**. Interestingly, the 4-chloro-1,2,3-dithiazole of the earlier analogues was not required for activity suggesting there may be multiple mechanisms of action.

3.4. Other Biological Applications

3.4.1. Melanin Synthesis Inhibitors

In 2015, a phenotypic screen was carried out using *Xenopus laevis* embryos by Skourides et al. [19]. This led to the identification of a series of 1,2,3-dithiazoles, which caused loss of pigmentation in melanophores and the retinal pigment epithelium (RPE) of developing embryos (Figure 37). This effect was independent of the developmental stage of initial exposure and was reversible. While the target was not elucidated, SAR of the series indicated that the presence of the mesmerically electron-donating methoxy group was important for pigment loss. Compounds with inductive and/or mesmerically electron-withdrawing groups had no effect on pigment loss.

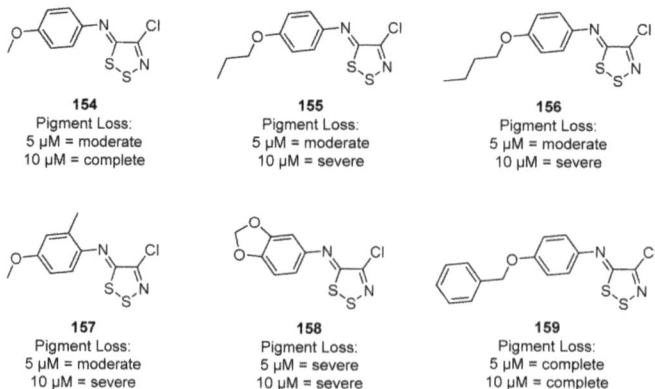

Figure 37. 1,2,3-dithiazoles demonstrating In vivo pigment loss in *Xenopus laevis* embryos.

The (Z)-4-chloro-N-(4-methoxyphenyl)-5H-1,2,3-dithiazol-5-imine (**154**) analogue demonstrated complete pigment loss at 10 μM and moderate at 5 μM. Extension of the methoxy to propyloxy **155** or butyloxy **156** reduced the potency, with the addition of a methyl group in the 2-position had the same effect. The formation of a 3,4-fused methyl catacol **157** increased potency, but did not match the activity of the 4-position methoxy **154**. The extension to form that benzyloxy analogue **158** did boost the potency of **154**, resulting in **158** having complete pigment loss at 5 μM.

Skourides et al. extensively investigated the structural features driving the phenotypic effects observed with **159**. An analogue of **159** was synthesized in two steps via the oxime route (Scheme 8) [12,13], where the 4-chlorine substituent was replaced with a phenyl group to give compound **160** (Figure 38). A second analogue of **159** was furnished where the nitrogen of the 1,2,3-dithiazole was replaced with a chlorocarbon in one step from Boberg salt [75,76] to give compound **161** (Figure 38) [75,77].

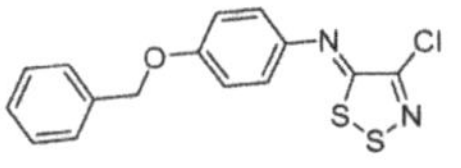

159
Pigment Loss:
5 μM = complete
10 μM = complete

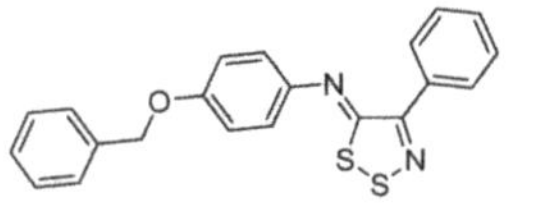

160
Pigment Loss:
5 μM = no effect
10 μM = no effect

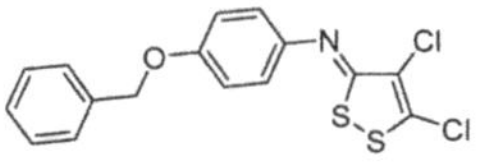

161
Pigment Loss:
5 μM = mild effect
10 μM = no effect

Figure 38. Direct comparison of 1,2,3-dithiazole **159** with disrupted analogues **160** and **161** in In vivo pigment loss in *Xenopus laevis* embryos.

The replacement of the 4-chlorine substituent yielded compound **160**, which showed no phenotypic affect. This supports the idea of an ANRORC-style ring opening mechanism, as the chlorine is a good nucleofuge that facilities the ring opening mechanism [40]. The second analogue **161**, showed some mild activity at 5 μM, while at 10 μM, mild toxicity and developmental defects were observed. This again pointed towards a ring opening ARONOC style mechanism, but more work needs to be done to establish the exact mechanism of action [19].

3.4.2. Antifibrotic Collagen Specific Chaperone hsp47 Inhibitor

Other activities of 1,2,3-dithiazoles include hit compound methyl 6-chloro-3H-benzo[d]dithiazole-4-carboxylate 2-oxide (**162**), which was reported twice, once in 2005 [18] and the second in 2010 [19]. These reports were both high-throughput screens of the compound library, one from Maybridge Chemical Co., Cornwall, U.K. and the other unspecified.

In 2005, Ananthanarayanan et al. screened a Maybridge compound library against Heat shock protein 47 (Hsp47), which, at the time, had no known inhibitors. Hsp47 is a collagen-specific molecular chaperone whose activity has been implicated in the pathogenesis of fibrotic diseases. The regulation of both Hsp47 and collagen expression has been implicated in several different disease indications where changes in the collagen expression are found. These diseases include fibrotic diseases of the liver [78], kidney [79], lung [80], and skin [81], in addition to atherosclerosis [82] and cancer [83]. The screen resulted in a primary hit rate of 0.2%, with 4 out of 2080 compounds being shown to be inhibitors of Hsp47. Secondary screening confirmed **162** (Figure 39), as the most potent compound (IC_{50} = 3.1 μM).

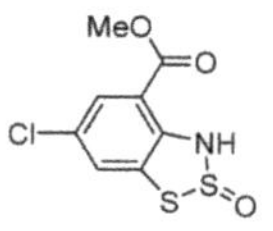

Figure 39. Hsp47 inhibitor methyl 6-chloro-3H-benzo[d][1,2,3]dithiazole-4-carboxylate 2-oxide (**162**).

3.4.3. Arabidopsis Gibberellin 2-Oxidase Inhibitors

In 2010, screening a commercial library of starting points against to Arabidopsis gibberellin 2-oxidases identified compound **162** (Figure 40) [20]. The screening aimed to identify an inhibitor that could both promote Arabidopsis seed germination and seedling growth. Compound **162** was able to do both, without having broad spectrum activity similar to Prohexadione (PHX), which is a broad-spectrum inhibitor of all three 2-oxoglutarate dependent dioxygenase's (2ODD) that were involved in Gibberellin (GA) production (GA 2-oxidase (GA2oxs), GA 3-oxidase (GA3oxs), and GA20-oxidase (GA20oxs)) [84,85]. The 1,2,3-dithiazole **162** was shown to have inhibition GA2oxs with a high degree of specificity, but not on other 2ODDs. The selective inhibition of GA2oxs activity could potentially lead to the delay of GA catabolism in plants, and hence, extend the life of endogenous GA.

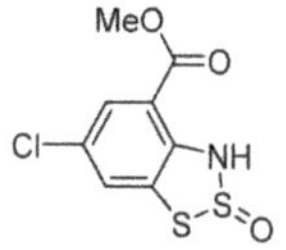

Figure 40. GA2oxs inhibitor methyl 6-chloro-3H-benzo[d][1,2,3]dithiazole-4-carboxylate 2-oxide (**162**).

4. Summary and Overview

The initial observation of the 1,2,3-dithiazole salt 4,5-dichloro-1,2,3-dithiazolium chloride in 1957 [1], was followed by detailed characterization in 1985 [4], and came to be known as Appel salt (**1**) post-1990s [5]. Appel salt (**1**) allowed for one-step access to a range of different chemistries to furnish a wide scope of 5-substituted-1,2,3-dithiazole derivatives [4,5,9,15,27,30–37]. While several additional methods also exist to access C4 substituted derivatives, the main screening has been done on 4-chloro derivatives until more recently [6,12–14,40]. However, synthetic challenges remain, including expanding the chemical space including effective synthesis of N-alkyl-5H-1,2,3-dithiazol-5-imine analogues, and effective access to 4-pyridyl analogues in good yields [37].

The first screening was carried out by Chevron Research Co. in 1977 [2,3], this relatively detailed study has been the foundation of the phenotypic biology observed on this scaffold. It described detailed work on a series of herbicidal effects and anti-mite efficacy, in addition to antifungal activities. This work was followed up in the late 1990s and 2000s by a series of groups extending the understanding of the antifungal and antibacterial SAR scope of the 1,2,3-dithiazole scaffold [15,50–53]. Interestingly, a patent in 1997 by Rohm & Haas Co. [51] highlighted a potential coating application for the 1,2,3-dithiazole with the discovery of potent antialgae and Gram-negative bacteria inhibition. In 2012, a patent filed by Bayer Cropscience AG presented a much broader library of heteroaromatic derivatives [54], highlighting a wider range of antifungal activities, with high degrees of control of commercially important fungi for crop protection. More recently, the antifungal and antibacterial screening has focused on clinically relevant hospital derived infections with good efficacy [86], in part aided by a series of matched pair 1,2,3-thiaselenazoles [55].

More recently, several other phenotypic observations have been reported. These include antiviral efficacy against FIV as a model for HIV, where modelling and mechanistic rationale point to cystine containing nucleocapsid protein (NCp) as the target for the 1,2,3-

dithiazole [17,36,63]. Anticancer effects against a broad range of cancer cell lines have also been reported with limited off-target toxicity [13,16,74]. These were also supported by modelling and mechanistic rationale highlighting ASCT2 [65] and LAT1 [70] as potential targets responsible. This rationale has been further supported by a series of mechanism of action experiments [65,70,73].

In addition to these reports, a series of other studies also highlighted other activities of the 1,2,3-dithiazole scaffold. These included an anti-melanin phenotype in *Xenopus laevis* embryos, where active potent (>5 μM) non-toxic compounds were identified in an in vivo model [19]. An ANRONC style mechanism of action was proposed supported by a series of chemical modifications to the scaffold [38–40]. Finally, two reports of high-throughput screens identified hit compounds against antifibrotic collagen specific chaperone hsp47 [18] and Arabidopsis Gibberellin 2-Oxidase [20].

The full potential of the 1,2,3-dithiazole scaffold has yet to be realized. Key areas of biological activities have been identified with preliminary work in the literature showing encouraging results. These included activities as antifungal [2], herbicidal [2], antibacterial [15], anticancer [16], antiviral [17], antifibrotic [18], and being a melanin [19] and Arabidopsis gibberellin 2-oxidases [20] inhibitors. These results provide a prospective to the versatility as to what is possible with this scaffold. In addition to these interesting reported biology applications, there are potentially significant untapped chemical biology opportunities towards targeting cystine reactive sites [21–25]; using the ANRORC-style 1,2,3-dithiazole chemistry as a latent functionality (Figure 41).

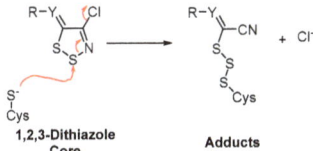

Figure 41. The 1,2,3-dithiazole as a latent cystine reactive functionality.

5. Conclusions

Taken together, the chemistry and biology of the 1,2,3-dithiazoles chemotype has shown a lot of exciting potential. The ANRORC-style rearrangements potentially affording a new route for potential chemical tools and relative cystine within proteins pockets, while the sub-micro molar phenotypic potencies against a series of diverse targets demonstrate potential for further development. Many of these diseases and pathogens have limited treatment options and need new therapies with novel mechanisms of action. The identification of starting points and defined SAR provides the foundation to define a medicinal chemistry trajectory towards optimized inhibitors and potential new treatments for a broad range of diseases.

Author Contributions: Conceptualization, C.R.M.A. and A.S.K.; formal analysis, C.R.M.A. and A.S.K.; investigation, C.R.M.A., H.J.O. and A.S.K.; writing—original draft preparation, C.R.M.A. and A.S.K.; writing—review and editing, C.R.M.A. and A.S.K. All authors have read and agreed to the published version of the manuscript.

Funding: This research received no external funding.

Institutional Review Board Statement: Not applicable.

Informed Consent Statement: Not applicable.

Data Availability Statement: Not applicable.

Conflicts of Interest: The authors declare no conflict of interest.

Sample Availability: Not applicable.

References

1. Wannagat, U.; Schindler, G. Reaktionen des Schwefeldichlorids mit Pyridin und verwandten Verbindungen. *Angew. Chem.* **1957**, *69*, 784-784. [CrossRef]
2. Moore, J.E. (Chevron Research Co.) Certain 4-Halo-5-aryl-1,2,3-dithiazole Compounds and their Preparation. U.S. Patent Application No. US4059590, 22 November 1977.
3. Moore, J.E. (Chevron Research Co.) Method for control of fungi using 4-halo-5-aryl-1,2,3,-dithiazoles. U.S. Patent Application No. US4119722 A, 10 October 1978.
4. Appel, R.; Janssen, H.; Siray, M.; Knoch, F. Synthese und Reaktionen des 4,5-Dichlor-1,2,3-dithiazolium-chlorids. *Chem. Ber.* **1985**, *118*, 1632–1643. [CrossRef]
5. Cuadro, A.M.; Alvarez-Buila, J. 4,5-Dichloro-1,2,3-dithiazolium chloride (Appel's Salt): Reactions with N-nucleophiles. *Tetrahedron* **1994**, *50*, 10037–10046. [CrossRef]
6. Rees, C.W. Polysulfur-nitrogen Heterocyclic Chemistry. *J. Heterocycl. Chem.* **1992**, *29*, 639–651. [CrossRef]
7. Rakitin, O.A. 1,2-Oxa/thia-3-azoles. In *Comprehensive Heterocyclic Chemistry III*; Katritzky, A.R., Ramsden, C.A., Scriven, E.F.V., Taylor, R.J.K., Zhdankin, V.V., Eds.; Elsevier: Oxford, UK, 2008; Volume 6, Chapter 6.01; pp. 1–36.
8. Navo, C.D.; Peccati, F.; Mazo, N.; Núñez-Franco, R.; Jiménez-Osés, G. 1,2-Oxa/thia-3-azoles. In *Comprehensive Heterocyclic Chemistry IV*; Black, D.S., Cossy, J., Stevens, C.V., Eds.; Elsevier: Oxford, UK, 2022; Volume 6, Chapter 6.01; pp. 1–55.
9. Konstantinova, L.S.; Rakitin, O.A. Synthesis and properties of 1,2,3-dithiazoles. *Russ. Chem. Rev.* **2008**, *77*, 521–546. [CrossRef]
10. Kim, K. Recent Advances in 1,2,3-Dithiazole Chemistry. *Phosphorus Sulfur Silicon Relat. Elem.* **1997**, *120*, 229–244. [CrossRef]
11. Kim, K. Synthesis and Reactions of 1,2,3-Dithiazoles. *Sulfur Rep.* **1998**, *21*, 147–207. [CrossRef]
12. Hafner, K.; Stowasser, B.; Sturm, V. Synthesis and properties of 4,6-Di-t-Butyl-Cyclopenta-1,2-Dithiole and its 3-aza-derivative. *Tetrahedron Lett.* **1985**, *26*, 189–192. [CrossRef]
13. Konstantinova, L.S.; Bol'shakov, O.I.; Obruchnikova, N.V.; Laborie, H.; Tanga, A.; Sopéna, V.; Lanneluc, I.; Picot, L.; Sablé, S.; Thiéry, V.; et al. One-pot synthesis of 5-phenylimino, 5-thieno or 5-oxo-1,2,3-dithiazoles and evaluation of their antimicrobial and antitumor activity. *Bioorg. Med. Chem. Lett.* **2009**, *19*, 136–141. [CrossRef]
14. Koyioni, M.; Manoli, M.; Koutentis, P.A. The Reaction of DABCO with 4-Chloro-5H-1,2,3-dithiazoles: Synthesis and Chemistry of 4-[N-(2-Chloroethyl)piperazin-1-yl]-5H-1,2,3-dithiazoles. *J. Org. Chem.* **2016**, *81*, 615–631. [CrossRef]
15. Cottenceau, G.; Besson, T.; Gautier, V.; Rees, C.W.; Pons, A.-M. Antibacterial Evaluation of Novel N-Arylimino-1,2,3-dithiazoles and N-Arylcyanothioformamides. *Bioorg. Med. Chem. Lett.* **1996**, *6*, 529–532. [CrossRef]
16. Baraldi, P.G.; Pavani, M.G.; Nuñez, M.C.; Brigidi, P.; Vitali, B.; Gambari, R.; Romagnoli, R. Antimicrobial and antitumor activity of N-heteroimmine-1,2,3-dithiazoles and their transformation in triazolo-, imidazo-, and pyrazolopirimidines. *Bioorg. Med. Chem.* **2002**, *10*, 449–456. [CrossRef] [PubMed]
17. Asquith, C.R.M.; Konstantinova, L.S.; Meli, M.L.; Laitinen, T.; Poso, A.; Rakitin, O.A.; Hofmann-Lehmann, R.; Hilton, S.T. Evaluation of Substituted 1,2,3-Dithiazoles as Inhibitors of the Feline Immunodeficiency Virus (FIV) Nucleocapsid Protein via a Proposed Zinc Ejection Mechanism. *ChemMedChem* **2016**, *11*, 2119–2126. [CrossRef]
18. Thomson, C.A.; Atkinson, H.M.; Ananthanarayanan, V.S. Identification of Small Molecule Chemical Inhibitors of the Collagen-Specific Chaperone Hsp47. *J. Med. Chem.* **2005**, *48*, 1680–1684. [CrossRef]
19. Charalambous, A.; Koyioni, M.; Antoniades, I.; Pegeioti, D.; Eleftheriou, I.; Michaelidou, S.S.; Amelichev, S.A.; Konstantinova, L.S.; Rakitin, O.A.; Koutentis, P.A.; et al. 1,2,3-Dithiazoles—New reversible melanin synthesis inhibitors: A chemical genomics study. *Med. Chem. Comm.* **2015**, *6*, 935–946. [CrossRef]
20. Otani, M.; Yoon, J.M.; Park, S.H.; Asami, T.; Nakajima, M. Screening and characterization of an inhibitory chemical specific to Arabidopsis gibberellin 2-oxidases. *Bioorg. Med. Chem. Lett.* **2010**, *20*, 4259–4262. [CrossRef]
21. Maurais, A.J.; Weerapana, E. Reactive-cysteine profiling for drug discovery. *Curr. Opin. Chem. Biol.* **2019**, *50*, 29–36. [CrossRef] [PubMed]
22. Bak, D.W.; Bechtel, T.J.; Falco, J.A.; Weerapana, E. Cysteine Reactivity Across the Sub-Cellular Universe. *Curr. Opin. Chem. Biol.* **2019**, *48*, 96–105. [CrossRef]
23. Yang, F.; Chen, N.; Wang, F.; Jia, G.; Wang, C. Comparative reactivity profiling of cysteine-specific probes by chemoproteomics. *Curr. Res. Chem. Biol.* **2022**, *2*, 100024. [CrossRef]
24. Hoch, D.G.; Abegg, D.; Adibekian, A. Cysteine-reactive probes and their use in chemical proteomics. *Chem. Commun.* **2018**, *54*, 4501–4512. [CrossRef]
25. Chaikuad, A.; Koch, P.; Laufer, S.A.; Knapp, S. The Cysteinome of Protein Kinases as a Target in Drug Development. *Angew. Chem. Int. Ed.* **2018**, *57*, 4372–4385. [CrossRef]
26. Besson, T.; Emayan, K.; Rees, C.W. 1,2,3-Dithiazoles and new routes to 3,1-benzoxazin-4-ones, 3,1 -benzothiazin-4-ones and N-arylcyanothioformamides. *J. Chem. Soc. Perkin Trans.* **1995**, *1*, 2097–2102. [CrossRef]
27. Besson, T.; Rees, C.W. Some chemistry of 4,5-dichloro-l,2,3-dithiazolium chloride and its derivatives. *J. Chem. Soc. Perkin Trans.* **1995**, *1*, 1659–1662. [CrossRef]
28. Lee, H.; Kim, K. A new procedure to N-Arylcyanothioformamides from 5-arylimino-4-chloro-5H-1,2,3-dithiazoles. *Bull. Korean Chem. Soc.* **1992**, *13*, 107–108. [CrossRef]
29. Koutentis, P.A. The Preparation and Characterization of 5-Substituted-4-chloro-1,2,3-dithiazolium Salts and their Conversion into 4-Substituted-3-chloro-1,2,5-thiadiazoles. *Molecules* **2005**, *10*, 346–359. [CrossRef]

30. Clarke, D.; Emayan, K.; Rees, C.W. New synthesis of isothiazoles from primary enamines. *J. Chem. Soc. Perkin Trans. 1* **1998**, *1*, 77–82. [CrossRef]
31. Kalogirou, A.S.; Christoforou, I.C.; Ioannidou, H.A.; Manos, M.J.; Koutentis, P.A. Ring transformation of (4-chloro-5H-1,2,3-dithiazol-5-ylidene)acetonitriles to 3-haloisothiazole-5-carbonitriles. *RSC Adv.* **2014**, *4*, 7735–7748. [CrossRef]
32. Moon-Kook, J.; Kim, K. Synthesis of new 5-alkylidene-4-chloro-5H-1,2,3-dithiazoles and their stereochemistry. *Tetrahedron* **1999**, *55*, 9651–9667. [CrossRef]
33. Kalogirou, A.S.; Koutentis, P.A. The reaction of 4,5-dichloro-1,2,3-dithiazolium chloride with DMSO: An improved synthesis of 4-chloro-1,2,3-dithiazol-5H-one. *Tetrahedron* **2009**, *65*, 6855–6858. [CrossRef]
34. Kalogirou, A.S.; Koutentis, P.A. A qualitative comparison of the reactivities of 3,4,4,5-tetrachloro-4H-1,2,6-thiadiazine and 4,5-dichloro-1,2,3-dithiazolium chloride. *Molecules* **2015**, *20*, 14576–14594. [CrossRef]
35. Folmer, J.J.; Weinreb, S.M. Generation of esters from carboxylic acids using Appel's salt (4,5-dichloro-1,2,3-dithiazolium chloride). *Tetrahedron Lett.* **1993**, *34*, 2737–2740. [CrossRef]
36. Laitinen, T.; Meili, T.; Koyioni, M.; Koutentis, P.A.; Poso, A.; Hofmann-Lehmann, R.; Asquith, C.R.M. Synthesis and evaluation of 1,2,3-dithiazole inhibitors of the nucleocapsid protein of feline immunodeficiency virus (FIV) as a model for HIV infection. *Bioorg. Med. Chem.* **2022**, *68*, 116834. [CrossRef] [PubMed]
37. Koutentis, P.A.; Koyioni, M.; Michaelidou, S.S. Synthesis of [(4-Chloro-5H-1,2,3-dithiazol-5-ylidene)amino]azines. *Molecules* **2011**, *16*, 8992–9002. [CrossRef] [PubMed]
38. Van der Plas, H.C. Chapter II SN(ANRORC) Reactions in Azines, Containing an "Outside" Leaving Group. *Adv. Heterocycl. Chem.* **1999**, *74*, 9–86. [CrossRef]
39. Van der Plas, H.C. Chapter III SN(ANRORC) Reactions in Azaheterocycles Containing an "Inside" Leaving Group. *Adv. Heterocycl. Chem.* **1999**, *74*, 87–151. [CrossRef]
40. Lee, H.; Kim, K.; Whang, D.; Kim, K. Novel Synthesis of 5-(arylimino)-4-(dialkylamino)-5H-1,2,3-dithiazoles and the mechanism of their formation. *J. Org. Chem.* **1994**, *59*, 6179–6183. [CrossRef]
41. Konstantinova, L.S.; Baranovsky, I.V.; Irtegova, I.G.; Bagryanskaya, I.Y.; Shundrin, L.A.; Zibarev, A.V.; Rakitin, O.A. Fused 1,2,3-dithiazoles: Convenient synthesis, structural characterization, and electrochemical properties. *Molecules* **2016**, *21*, 596. [CrossRef]
42. Plater, M.J.; Rees, C.W.; Roe, D.G.; Torroba, T. Cyclopenta-1,2,3-dithiazoles and related compounds. *J. Chem. Soc. Chem. Commun.* **1993**, *7*, 293–294. [CrossRef]
43. Koyioni, M.; Manoli, M.; Manos, M.J.; Koutentis, P.A. Reinvestigating the Reaction of 1H-Pyrazol-5-amines with 4,5-Dichloro-1,2,3-dithiazolium Chloride: A Route to Pyrazolo [3,4-c]isothiazoles and Pyrazolo[3,4-d]thiazoles. *J. Org. Chem.* **2014**, *79*, 4025–4037. [CrossRef]
44. Besson, T.; Guillaumet, G.; Lamazzi, C.; Rees, C.W. Synthesis of 3,1-Benzoxazines, 3,1-Benzothiazines and 3,1-Benzoxazepines via N-Arylimino-1,2,3-dithiazoles. *Synlett* **1997**, *6*, 704–706. [CrossRef]
45. Letribot, B.; Delatouche, R.; Rouillard, H.; Bonnet, A.; Chérouvrier, J.-R.; Domon, L.; Besson, T.; Thiéry, V. Synthesis of 2-Mercapto-(2-Oxoindolin-3-Ylidene)Acetonitriles from 3-(4-Chloro-5H-1,2,3-Dithiazol-5-Ylidene)Indolin-2-ones. *Molecules* **2018**, *23*, 1390. [CrossRef] [PubMed]
46. Kalogirou, A.S.; Koutentis, P.A. The degradation of 4,5-dichloro-1,2,3-dithiazolium chloride in wet solvents. *Tetrahedron* **2009**, *65*, 6859–6862. [CrossRef]
47. Konstantinova, L.S.; Bol'shakov, O.I.; Obruchnikova, N.V.; Golova, S.P.; Nelyubina, Y.V.; Lyssenko, K.A.; Rakitin, O.A. Reactions of 4-substituted 5H-1,2,3-dithiazoles with primary and secondary amines: Fast and convenient synthesis of 1,2,5-thiadiazoles, 2-iminothioacetamides and 2-oxoacetamides. *Tetrahedron* **2010**, *66*, 4330–4338. [CrossRef]
48. Appel, R.; Janssen, H.; Haller, I.; Plempel, M. (Bayer AG) 1,2,3-Dithiazolderivate, Verfahren zu ihrer Herstellung Sowie ihre Verwendung als Arzneimittel. Ger. Patent Application No. DE2848221 A1, 7 November 1980.
49. Mayer, R.; Fçrster, E.; Matauschek, B.D. Verfahren zur Herstellung von Aromatisch oder Heteroaromatisch Substituierten Cyanthioformamiden. Ger. Patent Application No. DD212387, 8 August 1984.
50. Besson, T.; Rees, C.W.; Cottenceau, G.; Pons, A.-M. Antimicrobial evaluation of 3,1-benzoxazin-4-ones, 3,1-benzothiazin-4-ones, 4-alkoxyquinazolin-2-carbonitriles and N-arylimino-1,2,3-dithiazoles. *Bioorg. Med. Chem. Lett.* **1996**, *6*, 2343–2348. [CrossRef]
51. Joseph, R.W.; Antes, D.L.; Osei-Gyimah, P. (Rohm & Haas Co.) Antimicrobial Compounds with Quick Speed of Kill. U.S. Patent Application No. US5688744 A, 18 November 1997.
52. Thiéry, V.; Rees, C.W.; Besson, T.; Cottenceau, G.; Pons, A.-M. Antimicrobial activity of novel N-quinolinyl and N-naphthylimino-1,2,3-dithiazoles. *Eur. J. Med. Chem.* **1998**, *33*, 149–153. [CrossRef]
53. Thiéry, V.; Bébéteau, V.; Guillard, J.; Lamazzi, C.; Besson, T.; Cottenceau, G.; Pons, A.M. Antimicrobial activity of novel N-arylimino-1,2,3-dithiazoles. *Pharm. Pharmacol. Commun.* **1998**, *4*, 39–42. Available online: https://onlinelibrary.wiley.com/doi/epdf/10.1111/j.2042-7158.1998.tb00315.x (accessed on 21 March 2023).
54. Benting, J.; Dahmen, P.; Wachendorff-Neumann, U.; Hadano, H.; Vors, J.-P. (Bayer Cropscience AG) Int. 5-heteroarylimino-1,2,3-dithiazoles PCT Pub. No. WO2012045726 A2, 12 April 2012.
55. Laitinen, T.; Baranovsky, I.V.; Konstantinova, L.S.; Poso, A.; Rakitin, O.A.; Asquith, C.R.M. Antimicrobial and antifungal activity of rare substituted 1,2,3-Thiaselenazoles and Corresponding Match Pair 1,2,3-Dithiazoles. *Antibiotics* **2020**, *9*, 369. [CrossRef]

56. Konstantinova, L.S.; Baranovsky, I.V.; Pritchina, E.A.; Mikhailov, M.S.; Bagryanskaya, I.Y.; Semenov, N.A.; Irtegova, I.G.; Salnikov, G.E.; Lyssenko, K.A.; Gritsan, N.P.; et al. Fused 1,2,3-Thiaselenazoles Synthesized from 1,2,3-Dithiazoles through Selective Chalcogen Exchange. *Chem. Eur. J.* **2017**, *23*, 17037–17047. [CrossRef]
57. Blaskovich, M.A.; Zuegg, J.; Elliott, A.G.; Cooper, M.A. Helping chemists discover new antibiotics. *ACS Infect. Dis.* **2015**, *1*, 285–287. [CrossRef]
58. WHO. Global Priority List of Antibiotic-Resistant Bacteria to Guide Research, Discovery, and Development of New Antibiotics. 2017. Available online: http://www.who.int/mediacentre/news/releases/2017/bacteria-antibiotics-needed/en/ (accessed on 28 February 2023).
59. Cook, K.M.; Hilton, S.T.; Mecinović, J.; Motherwell, W.B.; Figg, W.D.; Schofield, C.J. Epidithiodiketopiperazines block the interaction between hypoxia-inducible factor-1α (HIF-1α) and p300 by a zinc ejection mechanism. *J. Biol. Chem.* **2009**, *284*, 26831–26838. [CrossRef]
60. Sekirnik, R.; Rose, N.R.; Thalhammer, A.; Seden, P.T.; Mecinović, J.; Schofield, C.J. Inhibition of the histone lysine demethylase JMJD2A by ejection of structural Zn(II). *Chem. Commun.* **2009**, *42*, 6376–6378. [CrossRef]
61. Woodcock, J.C.; Henderson, W.; Miles, C.O. Metal complexes of the mycotoxins sporidesmin A and gliotoxin, investigated by electrospray ionisation mass spectrometry. *J. Inorg. Biochem.* **2001**, *85*, 187–199. [CrossRef]
62. Woodcock, J.C.; Henderson, W.; Miles, C.O.; Nicholson, B.K. Metal complexes of sporidesmin D and dimethylgliotoxin, investigated by electrospray ionisation mass spectrometry. *J. Inorg. Biochem.* **2001**, *84*, 225–232. [CrossRef] [PubMed]
63. Asquith, C.R.M.; Meili, T.; Laitinen, T.; Baranovsky, I.V.; Konstantinova, L.S.; Poso, A.; Rakitin, O.A.; Hofmann-Lehmann, R. Synthesis and comparison of substituted 1,2,3-dithiazole and 1,2,3-thiaselenazole as inhibitors of the feline immunodeficiency virus (FIV) nucleocapsid protein as a model for HIV infection. *Bioorg. Med. Chem. Lett.* **2019**, *29*, 1765–1768. [CrossRef]
64. Sancineto, L.; Mariotti, A.; Bagnoli, L.; Marini, F.; Desantis, J.; Iraci, N.; Santi, C.; Pannecouque, C.; Tabarrini, O. Design and Synthesis of DiselenoBisBenzamides (DISeBAs) as Nucleocapsid Protein 7 (NCp7) Inhibitors with anti-HIV Activity. *J. Med. Chem.* **2015**, *58*, 9601–9614. [CrossRef] [PubMed]
65. Oppedisano, F.; Catto, M.; Koutentis, P.A.; Nicolotti, O.; Pochini, L.; Koyioni, M.; Introcaso, A.; Michaelidou, S.S.; Carotti, A.; Indiveri, C. Inactivation of the Glutamine/Amino Acid Transporter ASCT2 by 1,2,3-dithiazoles: Proteoliposomes as a Tool to Gain Insights in the Molecular Mechanism of Action and of Antitumor Activity. *Toxicol. Appl. Pharmacol.* **2012**, *265*, 93–102. [CrossRef]
66. Scalise, M.; Pochini, L.; Console, L.; Losso, M.A.; Indiveri, C. The Human SLC1A5 (ASCT2) Amino Acid Transporter: From Function to Structure and Role in Cell Biology. *Front. Cell Dev. Biol.* **2018**, *6*, 96. [CrossRef]
67. Bröer, S. Amino Acid Transporters as Disease Modifiers and Drug Targets. *SLAS Discov.* **2018**, *23*, 303–320. [CrossRef] [PubMed]
68. Liu, Y.; Zhao, T.; Li, Z.; Wang, L.; Yuan, S.; Sun, L. The role of ASCT2 in cancer: A review. *Eur. J. Pharmacol.* **2018**, *837*, 81–87. [CrossRef]
69. Tailor, C.S.; Marin, M.; Nouri, A.; Kavanaugh, M.P.; Kabat, D. Truncated forms of the dual function human ASCT2 neutral amino acid transporter/retroviral receptor are translationally initiated at multiple alternative CUG and GUG codons. *J. Biol. Chem.* **2001**, *276*, 27221–27230. [CrossRef]
70. Napolitano, L.; Scalise, M.; Koyioni, M.; Koutentis, P.; Catto, M.; Eberini, I.; Parravicini, C.; Palazzolo, L.; Pisani, L.; Galluccio, M.; et al. Potent Inhibitors of Human LAT1 (SLC7A5) Transporter Based on Dithiazole and Dithiazine Compounds for Development of Anticancer Drugs. *Biochem. Pharmacol.* **2017**, *143*, 39–52. [CrossRef]
71. Lopes, C.; Pereira, C.; Medeiros, R. ASCT2 and LAT1 Contribution to the Hallmarks of Cancer: From a Molecular Perspective to Clinical Translation. *Cancers* **2021**, *13*, 203. [CrossRef]
72. Okano, N.; Naruge, D.; Kawai, K.; Kobayashi, T.; Nagashima, F.; Endou, H.; Furuse, J. First-in-human phase I study of JPH203, an L-type amino acid transporter 1 inhibitor, in patients with advanced solid tumors. *Investig. New Drugs* **2020**, *38*, 1495–1506. [CrossRef] [PubMed]
73. Prejanò, M.; Romeo, I.; La Serra, M.A.; Russo, N.; Marino, T. Computational Study Reveals the Role of Water Molecules in the Inhibition Mechanism of LAT1 by 1,2,3-Dithiazoles. *J. Chem. Inf. Model.* **2021**, *61*, 5883–5892. [CrossRef]
74. Maffuid, K.A.; Koyioni, M.; Torrice, C.D.; Murphy, W.A.; Mewada, H.K.; Koutentis, P.A.; Crona, D.J.; Asquith, C.R.M. Design and evaluation of 1,2,3-dithiazoles and fused 1,2,4-dithiazines as anti-cancer agents. *Bioorg. Med. Chem. Lett.* **2021**, *43*, 128078. [CrossRef]
75. Wentrup, G.-J.; Koepke, M.; Boberg, F. Über 1,2-Dithiacyclopentene; XXIX. 3-Thioxo-3H-1,2-Dithiole aus 3-Chloro-1,2-dithio-liumchloriden. *Synthesis* **1975**, 525–526. [CrossRef]
76. Lowe, P.A. *The Chemistry of the Sulphonium Group*; Stirling, C.J.M., Ed.; Wiley: Chichester, UK, 1981; Volume 1, Chapter 11, pp. 267–312. [CrossRef]
77. Ogurtsov, V.A.; Rakitin, O.A.; Rees, C.W.; Smolentsev, A.A.; Lyssenko, K.A. New routes to 1,2-dithiole-3-thiones and 3-imines. *Mendeleev Commun.* **2005**, *15*, 20–21. [CrossRef]
78. Masuda, H.; Fukumoto, M.; Hirayoshi, K.; Nagata, K. Coexpression of the collagen-binding stress protein Hsp47 gene and the alpha 1(I) and alpha 1(III) collagen genes in carbon tetrachloride induced rat liver fibrosis. *J. Clin. Investig.* **1994**, *94*, 2481–2488. [CrossRef]
79. Razzaque, M.S.; Taguchi, T. Collagen-binding heat shock protein (Hsp) 47 expression in anti-thymocyte serum (ATS)-induced glomerulonephritis. *J. Pathol.* **1997**, *183*, 24–29. [CrossRef]
80. Razzaque, M.S.; Nazneen, A.; Taguchi, T. Immunolocalization of collagen and collagen-binding heat shock protein 47 in fibrotic lung diseases. *Mod. Pathol.* **1998**, *11*, 1183–1188.

81. Razzaque, M.S.; Ahmed, A.R. Collagens, collagen-binding heat shock protein 47 and transforming growth factor-beta 1 are induced in cicatricial pemphigoid: Possible role(s) in dermal fibrosis. *Cytokine* **2002**, *17*, 311–316. [CrossRef]
82. Rocnik, E.; Chow, L.H.; Pickering, J.G. Heat shock protein 47 is expressed in fibrous regions of human atheroma and is regulated by growth factors and oxidized low-density lipoprotein. *Circulation* **2000**, *101*, 1229–1233. [CrossRef] [PubMed]
83. Nagata, K.; Yamada, K.M. Phosphorylation and transformation sensitivity of a major collagen-binding protein of fibroblasts. *J. Biol. Chem.* **1986**, *261*, 7531–7536. [CrossRef]
84. Nakayama, I.; Miyazawa, T.; Kobayashi, M.; Kamiya, Y.; Abe, H.; Sakurai, A. Effects of a New Plant Growth Regulator Prohexadione Calcium (BX-112) on Shoot Elongation Caused by Exogenously Applied Gibberellins in Rice (*Oryza sativa* L.) Seedlings. *Plant Cell Physiol.* **1990**, *31*, 195–200. [CrossRef]
85. Rademacher, W. GROWTH RETARDANTS: Effects on Gibberellin Biosynthesis and Other Metabolic Pathways. *Annu. Rev. Plant Physiol. Plant Mol. Biol.* **2000**, *51*, 501–531. [CrossRef]
86. Mazum, T.K.; Bricker, B.A.; Flores-Rozas, H.; Ablordeppey, S.Y. The Mechanistic Targets of Antifungal Agents: An Overview. *Mini. Rev. Med. Chem.* **2016**, *16*, 555–578. [CrossRef] [PubMed]

Disclaimer/Publisher's Note: The statements, opinions and data contained in all publications are solely those of the individual author(s) and contributor(s) and not of MDPI and/or the editor(s). MDPI and/or the editor(s) disclaim responsibility for any injury to people or property resulting from any ideas, methods, instructions or products referred to in the content.

Review

Advances in the Synthesis of Heteroaromatic Hybrid Chalcones

Ajay Mallia * and Joseph Sloop

Department of Chemistry, School of Science & Technology, Georgia Gwinnett College, 1000 University Center Lane, Lawrenceville, GA 30043, USA
* Correspondence: amallia@ggc.edu

Abstract: Chalcones continue to occupy a venerated status as scaffolds for the construction of a variety of heterocyclic molecules with medicinal and industrial properties. Syntheses of hybrid chalcones featuring heteroaromatic components, especially those methods utilizing green chemistry principles, are important additions to the preparative methodologies for this valuable class of molecules. This review outlines the advances made in the last few decades toward the incorporation of heteroaromatic components in the construction of hybrid chalcones and highlights examples of environmentally responsible processes employed in their preparation.

Keywords: chalcone; heteroaromatic; hybrid chalcone

1. Introduction

The chalcone class of enones has been a privileged scaffold in organic synthesis for more than a century. Kostanecki and Tambor are credited with the first reported preparation of E-1,3-diphenylprop-2-en-1-one and coined the term "chalcone" in 1899 [1]. Figure 1 shows the structure of E-chalcone, the most energetically favorable stereoisomer, as well as the sterically encumbered and less common Z-chalcone, both of which contain benzene rings at C_1 and C_3 joined by a three-carbon α,β-unsaturated ketone unit. The absolute configuration of solid chalcone stereochemistry obtained during synthesis can often be determined with X-ray crystallography [2,3].

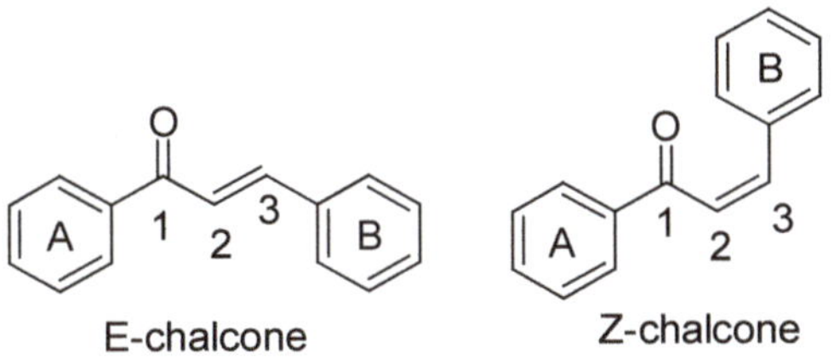

Figure 1. Chalcone structure and stereochemistry.

By convention, the aromatic ring attached to C_1 is designated as ring A while the aromatic ring attached to C_3 is designated as ring B. For the purposes of this review, we will adhere to the conventional ring designations in describing preparations of heteroaromatic hybrid chalcones.

The utility of chalcones both as a pharmacophore and as a scaffold in the synthesis of a wide variety of heterocycles ranging from pyrazoles, isoxazoles, triazoles, barbituric acid derivatives, etc. has been investigated thoroughly over the years, with numerous research articles as well as several reviews appearing in the last decade describing the current chalcone synthetic strategies, the heterocycles derived from them, and the bioactivity and pharmaceutical uses of these compounds [4–13]. Within that context, the preparation of more highly functionalized chalcones that contain heteroaromatic components has been an area of intense research over the last decade [10,14–30].

Research has established that heteroaromatic hybrid chalcones themselves possess broad medicinal value as anticancer [16,19,23], antimicrobial [11,20,23,28], antifungal [16], anti-tuberculosis [25] and anti-inflammatory agents [22] as well as having other important pharmacological functions [9,10], agrochemical utility as photosynthesis inhibitors [18] and industrial use as photoinitiators in 3D printing [17]. Figure 2 shows a representative selection of heteroaromatic hybrid chalcone pharmacophores and industrially important compounds.

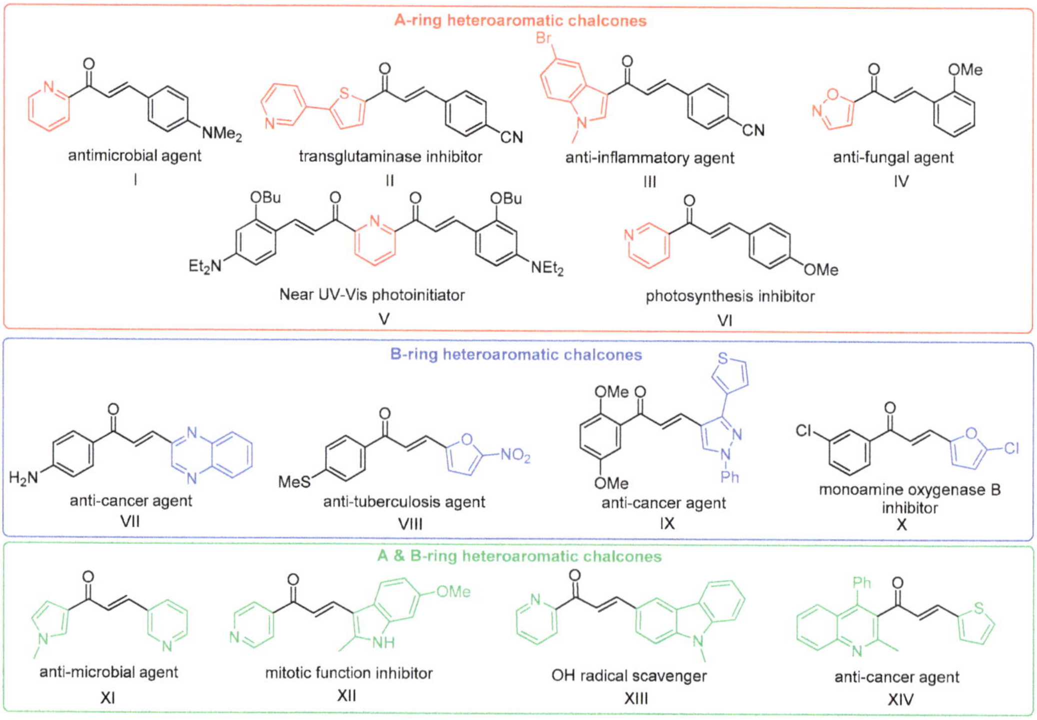

Figure 2. Medicinally and industrially important heteroaromatic hybrid chalcones.

Synthetic methodologies to prepare hybrid chalcones have developed rapidly over the last two decades. To the best of our knowledge, no reviews have been found that focus on heteroaromatic chalcone synthesis and the green synthesis methods employed to prepare them. This review will focus on the construction of heteroaromatic hybrid chalcones with the Claisen–Schmidt condensation, 1,3-dipolar additions, ring-opening reactions, 3+2 annulations and Wittig reactions. The review will discuss four different heteroaromatic hybrid chalcone types: A-ring and B-ring-substituted mono-heteroaromatic hybrid chalcones, hybrid chalcones possessing heteroaromatic moieties on both the A and B rings, and the synthesis strategies used to prepare heteroaromatic bis chalcone hybrids. Herein, we also detail the green methods that have been employed to prepare these hybrid chalcones including microwave irradiation, sonication, ball milling, continuous flow reactions, the use of benign solvents, solvent-free/solid-state processes and nanocatalysis. See Figure 3.

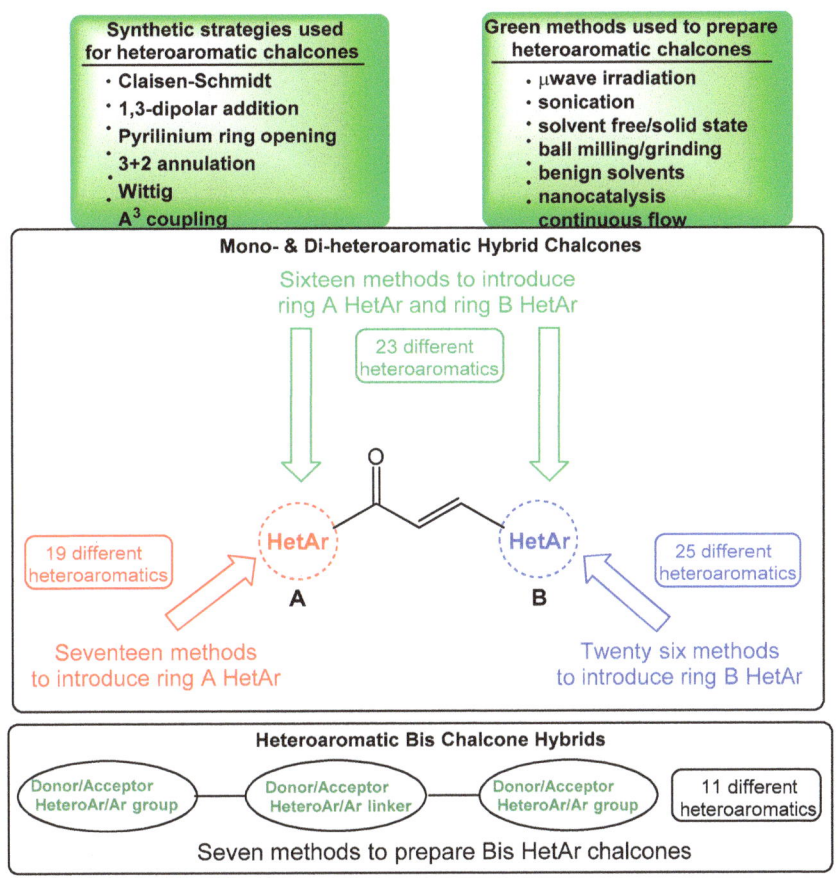

Figure 3. Heteroaromatic hybrid chalcone construction.

2. A-Ring Heteroaromatic Hybrid Chalcone Synthesis

This section catalogues several representative conventional and green processes by which hybrid chalcones bearing a heteroaromatic species at ring A may be prepared. Heteroaromatic components of the chalcone products include a variety of single-ring (furan, pyrrole, thiazole, thiophene, pyridine, pyrimidine) and fused-ring (indole, benzimidazole, benzothiazole, benzofuran, pyrazolopyridine, quinoline) systems.

2.1. Claisen–Schmidt Condensations

The Claisen–Schmidt (C-S) condensation has been widely used to prepare chalcones for many years. This reaction, which can be catalyzed by acids or bases, offers mild conditions that tolerate a wide scope of functionality in both the ketone donors and aldehyde acceptors.

2.1.1. Base-Catalyzed C-S Condensations

The hydroxide bases KOH, NaOH and to a lesser extent $Ba(OH)_2$ are the bases used to promote the condensations depicted below in Schemes 1–12. These bases may be introduced to the reaction medium as dilute or concentrated aqueous solutions or as solids. Ethanol or methanol are the solvents of choice in most reactions depicted herein. The reaction temperatures vary from 0 °C to those obtained by refluxing the alcoholic solvents. The reaction times range from less than a minute in the case of selected microwave-mediated reactions and can extend to 72 h for the conventional condensations.

Scheme 1. Synthesis of pyrrolyl chalcone.

a: 60% KOH (aq), EtOH, 30 min, centrifuge (22%)
b: 10% NaOH (aq), EtOH, RT, overnight (49%)
c: 20 mol% NaOH (aq), EtOH, RT, overnight (91%)

Scheme 2. Synthesis of furyl chalcone.

a: 0.1 mol% KOH (aq), MeOH, RT, 4h, (87%)
b: 20 mol% NaOH (aq), EtOH, RT, 6 h (88%)

Scheme 3. Synthesis of furyl chalcone derivatives.

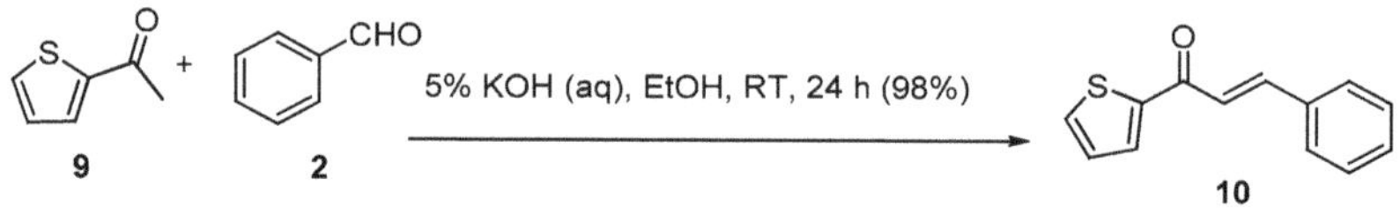

40% NaOH in EtOH, RT, 2–3 h (12–59%)

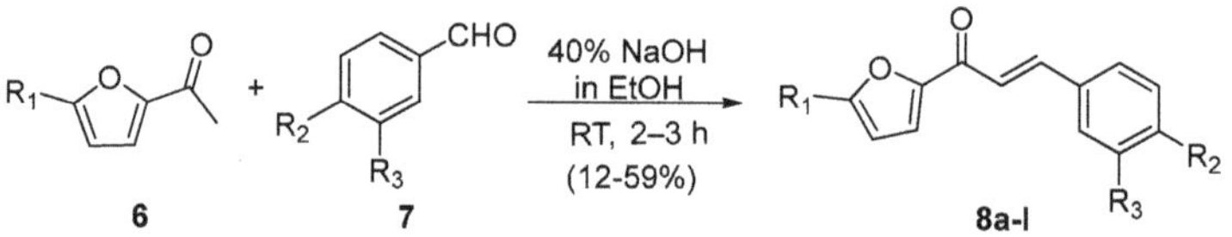

a: R_1 = H, R_2 = Cl, R_3 = H, 17%
b: R_1 = H, R_2 = OMe, R_3 = H, 22%
c: R_1 = Me, R_2 = Cl, R_3 = H, 12%
d: R_1 = Me, R_2 = H, R_3 = Cl, 41%
e: R_1 = Me, R_2 = Br, R_3 = H, 22%
f: R_1 = Me, R_2 = H, R_3 = Br, 48%
g: R_1 = Me, R_2 = F, R_3 = H, 21%
h: R_1 = Me, R_2 = H, R_3 = F, 44%
i: R_1 = Me, R_2 = Br, R_3 = F, 59%
j: R_1 = Me, R_2 = CF_3, R_3 = H, 58%
k: R_1 = Me, R_2 = H, R_3 = CF_3, 23%
l: R_1 = Me, R_2 = H, R_3 = OMe, 46%

Scheme 4. Synthesis of thienyl chalcone.

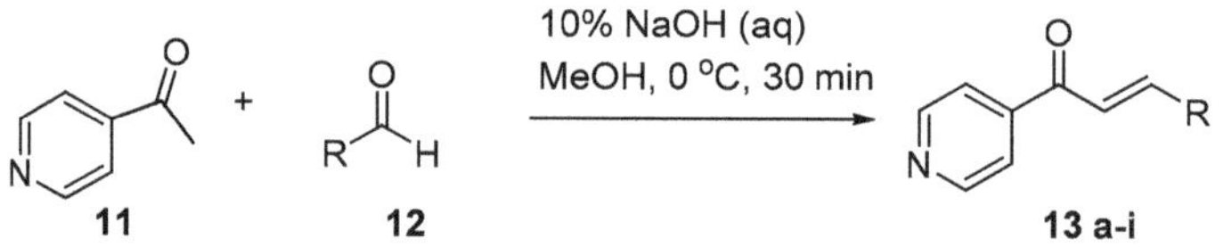

5% KOH (aq), EtOH, RT, 24 h (98%)

Scheme 5. Synthesis of pyridyl chalcone.

10% NaOH (aq), MeOH, 0 °C, 30 min

a: R = phenyl, 75%
b: R = 4-methoxyphenyl, 67%
c: R = 2,5-Dimethoxyphenyl, 67%
d: R = 2,4,5-trimethoxyphenyl, 67%
e: R = 3,4,5-trimethoxyphenyl, 67%
f: R = 4-tolyl, 76%
g: R = 4-Methylsulfanylphenyl, 70%
h: R = 3,4-Dimethylphenyl, 67%
i: R = 4-Chlorophenyl, 68%

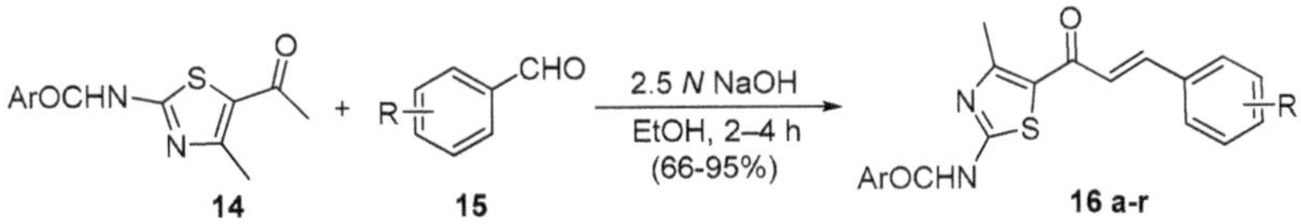

Scheme 6. Synthesis of thiazolyl chalcones.

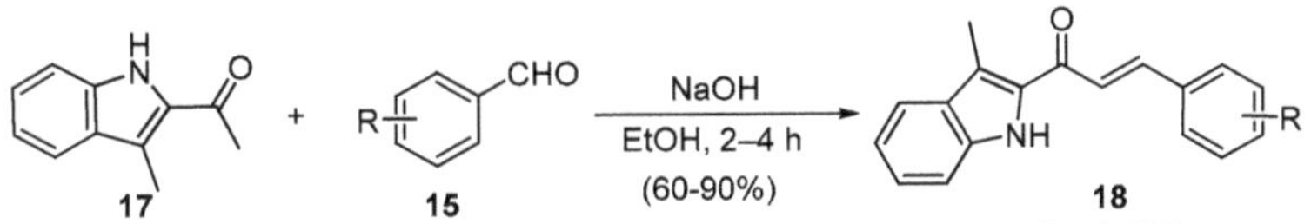

a: Ar = Ph, R = H, 83%
b: Ar = Ph, R = 4-F, 85%
c: Ar = Ph, R = 4-Cl, 79%
d: Ar = Ph, R = 4-Br, 89%
e: Ar = Ph, R = 3-Br, 66%
f: Ar = 4-MeOC$_6$H$_4$, R = 4-F, 88%
g: Ar = 4-O$_2$NC$_6$H$_4$, R = 4-Cl, 92%
h: Ar = Ph, R = 3-O$_2$NC$_6$H$_4$, 92%
i: Ar = Ph, R = 4-MeO, 85%
j: Ar = Ph, R = 3,4-diMeO, 77%
k: Ar = Ph, R = 3,4,5-triMeO, 81%
l: Ar = Ph, R = 4-Me, 95%
m: Ar = 4-FC$_6$H$_4$, R = 4-Me, 92%
n: Ar = 4-MeOC$_6$H$_4$, R = 4-Me, 87%
o: Ar = Ph, R = 4-Me$_2$N, 82%
p: Ar = 4-FC$_6$H$_4$, R = 4-Me$_2$N, 86%
q: Ar = Ph, R = 6-Br-1,3-dioxo-5-yl, 95%
r: Ar = 4-FC$_6$H$_4$, R = 6-Br-1,3-dioxo-5-yl, 94%

Scheme 7. Synthesis of indolyl chalcones.

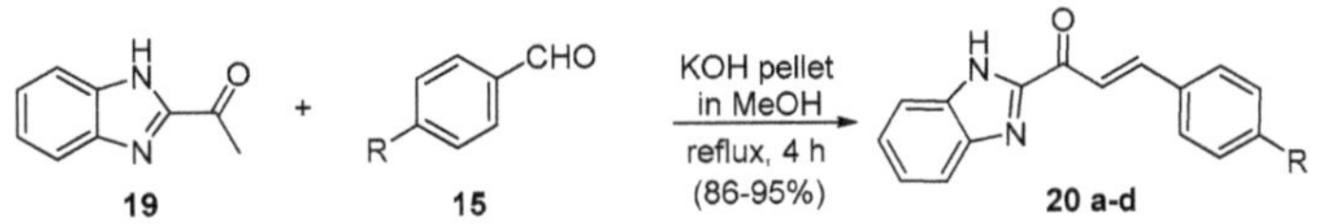

a: R = 2-OMe
b: R = 4-Cl
c: R = 2,3-diCl

Scheme 8. Synthesis of benzimidazolyl chalcones.

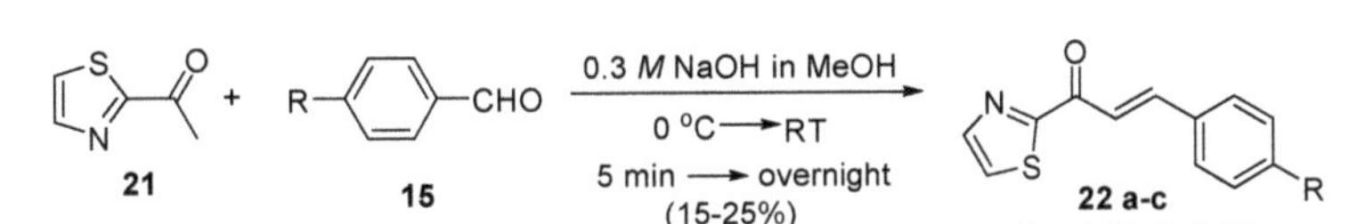

a: R = H, 88%
b: R = CH$_3$, 90%
c: R = Cl, 86%
d: R = OCH$_3$, 95%

22 a-c
a: R = 4-MeO, 24%
b: R = 4-Me$_2$N, 25%
c: R = -O$_2$NC$_6$H$_4$, 15%

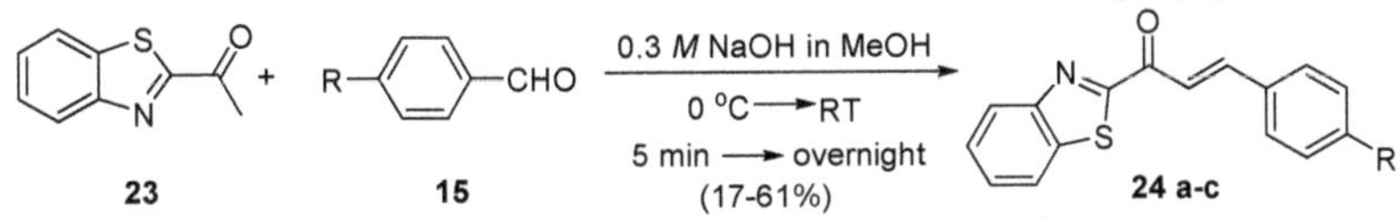

24 a-c
a: R = 4-MeO, 61%
b: R = 4-Me$_2$N, 17%
c: R = 4-O$_2$NC$_6$H$_4$, 42%

Scheme 9. Synthesis of thiazolyl and benzothiazolyl chalcones.

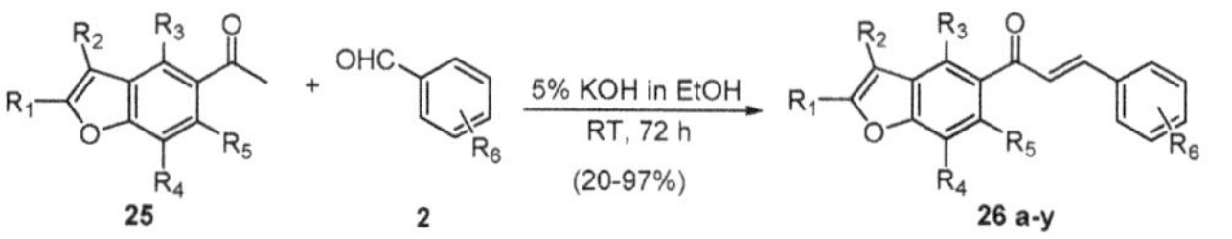
5% KOH in EtOH
RT, 72 h
(20-97%)
25
2
26 a-y

a: $R_1 = R_2 = R_3 = R_4 = R_6 = H$, $R_5 = OH$, 63%
b: $R_1 = R_3 = R_4 = R_6 = H$, $R_2 = Me$, $R_5 = OH$, 27%
c: $R_1 = R_2 = Me$, $R_3 = R_4 = R_6 = H$, $R_5 = OH$, 81%
d: $R_1 = R_3 = R_4 = R_6 = H$, $R_2 =$ t-Bu, $R_5 = OH$, 20%
e: $R_1 = R_3 = R_4 = R_6 = H$, $R_2 = Ph$, $R_5 = OH$, 27%
f: $R_1 = R_2 = R_3 = R_6 = H$, $R_4 = Me$, $R_5 = OH$, 72%
g: $R_1 = R_3 = R_6 = H$, $R_2 = R_4 = Me$, $R_5 = OH$, 60%
h: $R_1 = R_2 = R_4 = Me$, $R_3 = R_6 = H$, $R_5 = OH$, 81%
i: $R_1 = R_3 = R_6 = H$, $R_2 =$ t-Bu, $R_4 = Me$, $R_5 = OH$, 30%
j: $R_1 = R_3 = R_6 = H$, $R_2 = Ph$, $R_4 = Me$, $R_5 = OH$, 71%
k: $R_1 = R_3 = R_6 = H$, $R_2 =$ 4-HOPh, $R_4 = Me$, $R_5 = OH$, 44%
l: $R_1 = R_2 = R_6 = H$, $R_3 = R_4 = OMe$, $R_5 = OH$, 35%
m: $R_1 = R_2 = R_3 = R_4 = R_6 = H$, $R_5 = OMe$, 34%
n: $R_1 = R_2 = Me$, $R_3 = R_4 = H$, $R_5 = OH$, $R_6 =$ 2-OMe, 70%
o: $R_1 = R_2 = Me$, $R_3 = R_4 = H$, $R_5 = OH$, $R_6 =$ 3-OMe, 61%
p: $R_1 = R_2 = Me$, $R_3 = R_4 = H$, $R_5 = OH$, $R_6 =$ 4-OMe, 68%
q: $R_1 = R_2 = Me$, $R_3 = R_4 = H$, $R_5 = OH$, $R_6 =$ 2,4-diOMe, 76%
r: $R_1 = R_2 = Me$, $R_3 = R_4 = H$, $R_5 = OH$, $R_6 =$ 3,4-diOMe, 71%
s: $R_1 = R_2 = R_4 = R_5 = R_6 = H$, $R_3 = OH$, 87%
t: $R_1 = R_4 = R_5 = R_6 = H$, $R_2 = Me$, $R_3 = OH$, 97%
u: $R_1 = R_2 = Me$, $R_3 = OH$, $R_4 = R_5 = R_6 = H$, 80%
v: $R_1 = R_2 = Me$, $R_3 = R_4 = H$, $R_5 = OH$, $R_6 =$ 4-OMe, 68%
w: $R_1 = R_4 = R_5 = R_6 = H$, $R_2 = Ph$, $R_3 = OH$, 38%
x: $R_1 = R_2 = R_4 = R_5 = R_6 = H$, $R_3 = OMe$, 71%
y: $R_1 = R_4 = R_5 = R_6 = H$, $R_2 =$ 4-HOPh, $R_3 = OH$, 48%

Scheme 10. Synthesis of benzofuryl chalcones.

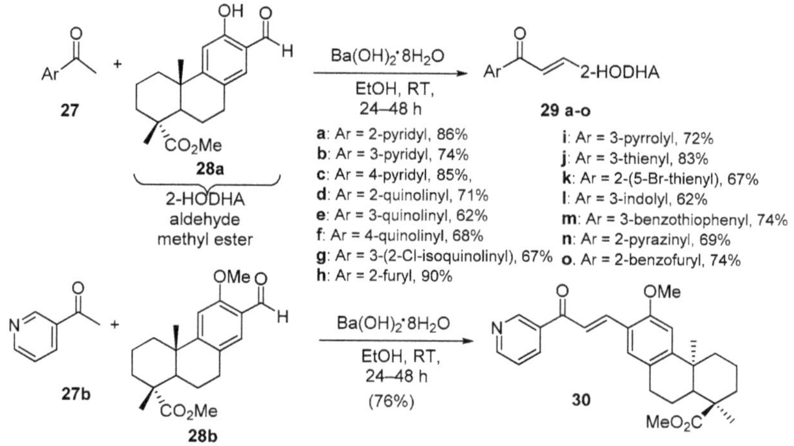
Ba(OH)2·8H2O
EtOH, RT,
24–48 h
27
28a
2-HODHA
aldehyde
methyl ester
29 a-o
a: Ar = 2-pyridyl, 86%
b: Ar = 3-pyridyl, 74%
c: Ar = 4-pyridyl, 85%,
d: Ar = 2-quinolinyl, 71%
e: Ar = 3-quinolinyl, 62%
f: Ar = 4-quinolinyl, 68%
g: Ar = 3-(2-Cl-isoquinolinyl), 67%
h: Ar = 2-furyl, 90%
i: Ar = 3-pyrrolyl, 72%
j: Ar = 3-thienyl, 83%
k: Ar = 2-(5-Br-thienyl), 67%
l: Ar = 3-indolyl, 62%
m: Ar = 3-benzothiophenyl, 74%
n: Ar = 2-pyrazinyl, 69%
o. Ar = 2-benzofuryl, 74%
Ba(OH)2·8H2O
EtOH, RT,
24–48 h
(76%)
27b
28b
30

Scheme 11. Synthesis of heteroaromatic dehydroabietic acid-chalcone hybrids.

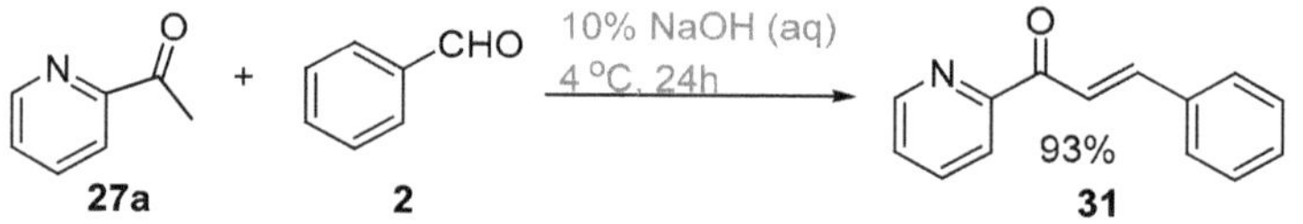
10% NaOH (aq)
4 °C, 24h
93%
27a
2
31

Scheme 12. Green synthesis of pyridyl chalcone.

In our first entry, three room-temperature C-S preparations of pyrrolyl chalcone **3** are presented that have differing reaction times and different base concentrations. Sweeting et al. (Scheme 1a) used strongly basic conditions (60% aqueous KOH) and centrifugation mixing to prepare the pyrrolyl chalcone **3** in a modest yield. The low yield is likely attributed to the short reaction time. Ref. [31] Robinson et al. reported that increasing the reaction time ([32], Scheme 1b) using NaOH (aq) in ethanol increased the yield of the pyrrolyl chalcone. Using 20 mol % NaOH (aq) in ethanol, Song et al. obtained a 91% yield in the preparation of the chalcone (Scheme 1c). Ref. [33] Lokeshwari's team (Scheme 2a) and Liu's group prepared furyl chalone **5** in an 87% yield using 0.1 mol % KOH (aq) in 4 h, while Liu's group (Scheme 2b) obtained equally high yields with 20 mol% NaOH (aq) in 6 h [34,35]. Robinson et al. (Scheme 3) condensed 2-acetylfuran and 2-acetyl-5-methylfuran with assorted benzaldehydes at room temperature en route to the twelve furyl chalcones **8** in modest to medium yields [36].

Parveen et al. reported a nearly quantitative conversion for the room-temperature C-S condensation of 2-acetylthiophene and benzaldehyde using aqueous KOH (Scheme 4) in ethanol to the thienyl chalcone **10** [37].

Sunduru et al. reported the preparation of pyridyl chalcone derivatives **13** by condensing 4-acetylpyridine with the respective aromatic aldehyde (Scheme 5) [38]. In this reaction, one equivalent of 4-acetylpyridine was added dropwise to a cooled methanolic solution containing 10% aqueous NaOH. Then, one equivalent of aldehyde was added slowly at 0 °C. After workup and recrystallization, the pyridyl chalcones were obtained in yields ranging from 67 to 76% (Scheme 5).

Sinha and coworkers (Scheme 6) used similar conditions to synthesize eighteen 1,3-thiazolylchalcones **16** in very good overall yields [39].

Zhao et al. (Scheme 7) used reflux conditions to achieve yields in excess of 60% for the small series of fused-ring indolyl chalcones **18** [40]. In two separate publications, Hsieh and coworkers used base-catalyzed C-S condensations to prepare indolyl (Scheme 8, [41]), thiazolyl and benzothiazolyl hybrid chalcones (Scheme 9, [42]).

Saito's team used 5% KOH in ethanol at room temperature to prepare a series of functionalized benzofuran hybrid chalcones in yields as high as 97% (Scheme 10) [43].

Grigoropoulou's team found barium hydroxide octahydrate effective in promoting the condensation of both single- and fused-ring heteroaromatic ketones with dehydroabietic acid methyl ester en route to sixteen hybrid chalcones in good overall yields (Scheme 11) [44].

Base-catalyzed C-S condensations have also been demonstrated using green principles. These processes include the use of benign solvents including water and microwave irradiation. Mubofu and Engberts reported a C-S condensation reaction of 2-acetylpyridine and benzaldehyde using 10% NaOH (Scheme 12) [45]. The reagents were finely dispersed in water at 4 °C and after workup the pyridyl chalcone **31** was obtained in a good yield (Scheme 12).

Jianga et al. showed that the condensation of 2-acetylfuran or 2-acetylthiophene and benzaldehyde using 2 mol% NaOH (aq) gave (E)-1-(Furan-2-yl)-3-phenylprop-2-en-1-one or (E)-1-(thiophen-2-yl)-3-phenylprop-2-en-1-one at room temperature in nearly quantitative yields (Scheme 13) [46].

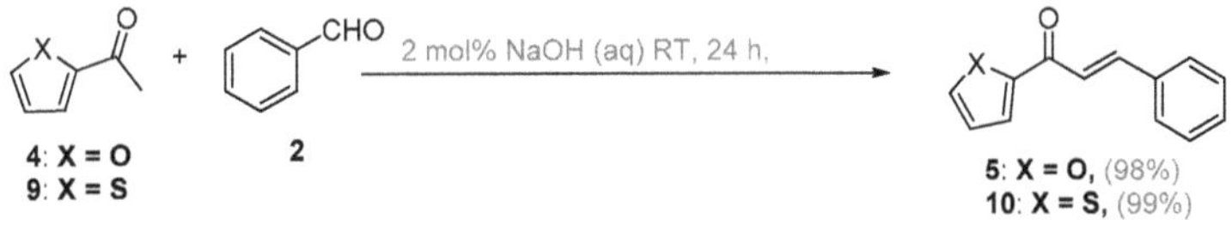

Scheme 13. Green synthesis of furyl and thienyl chalcones.

Ritter et al. (Scheme 14) used 2-acetylthiophene **9** and assorted benzaldehydes in glycerin solvent to prepare seven 2-thienochalcones **10** and **32a–f** in very good yields [47].

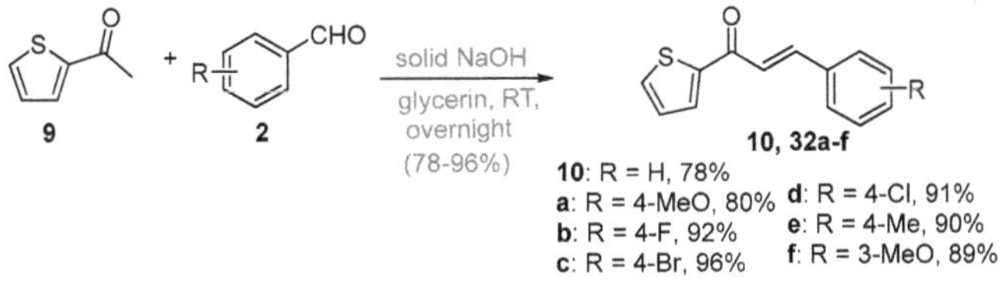

Scheme 14. Green synthesis of 2-thienyl chalcones.

Khan and Asiri (Scheme 15) showed that 3-acetylthiophene **33** underwent a microwave-mediated C-S condensation with several benzaldehydes in less than a minute to give thienyl chalcones **34a–f** in yields exceeding 82% [48].

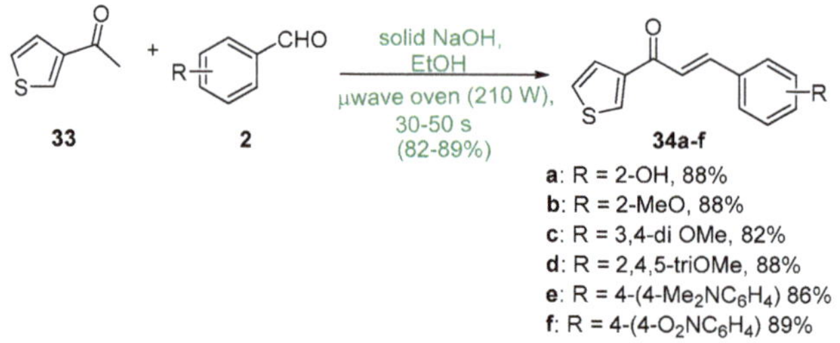

Scheme 15. Microwave synthesis of 3-thienyl chalcones.

Sarveswari and Vijayakumar (Scheme 16) conducted a comparative study of conventional and microwave processes in which four examples of highly substituted quinolinyl hybrid chalcones **36a–d** were prepared [49]. Both processes gave the desired chalcones in yields greater than 75%. Particularly noteworthy is the fact that the microwave reaction time is 1/144 of the conventional reaction time.

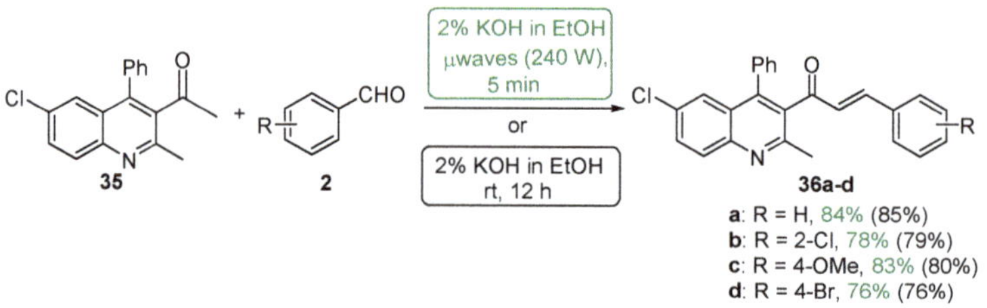

Scheme 16. Synthesis of quinolinyl chalcones.

Polo et al. demonstrated that sonochemical mediation was very effective in preparing a series of pyrazolopyridyl hybrid chalcones **38a–e** (Scheme 17) in high yields that compare favorably with conventional base-catalyzed C-S condensations [50].

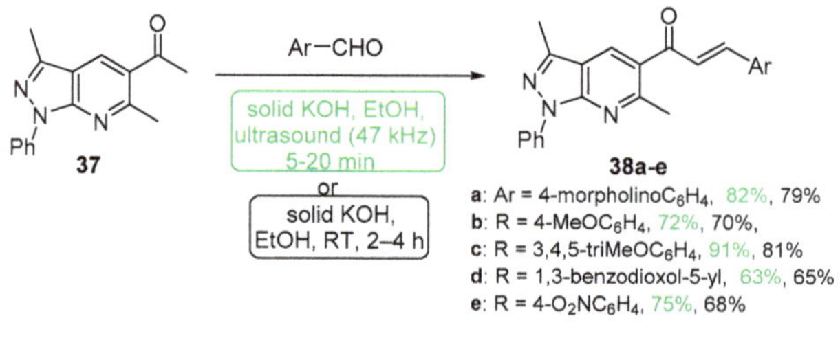

Scheme 17. Sonochemical synthesis of pyrazolopyridyl chalcones.

2.1.2. Acid-Catalyzed C-S Condensations

In the recent literature, Adnan et al. showed that *p*-toluenesulfonic acid (PTSA) effectively catalyzed the condensation of 2-acetylthiophene (**9**) and *p*-tolualdehyde (**2**) in a green solventless process in which the reactants were ground in a warm mortar and pestle for 4 min to give the thienyl chalcone **32e** in a very good yield [13]. See Scheme 18.

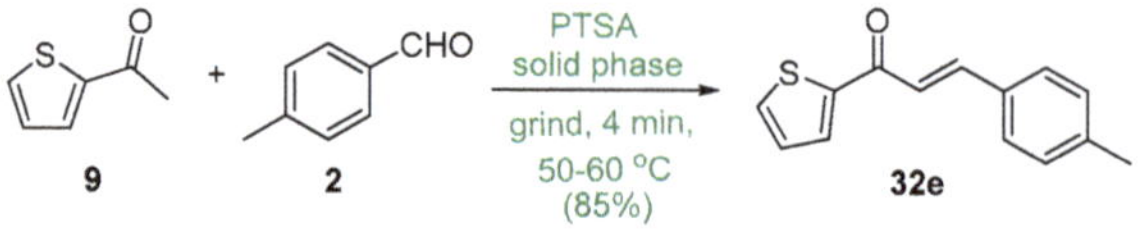

Scheme 18. PTSA-catalyzed synthesis of thienyl chalcone.

Shaik et al. reported an acid-catalyzed condensation reaction of 2,4-dimethyl-5-acetylthiazole with 2,4-difluorobenzaldehyde to prepare (E)-1-(2′,4′-dimethyl)-(5-acetylthiazole)-(2,4″-difluorophenyl)-prop-2-en-1-one (Scheme 19) [23].

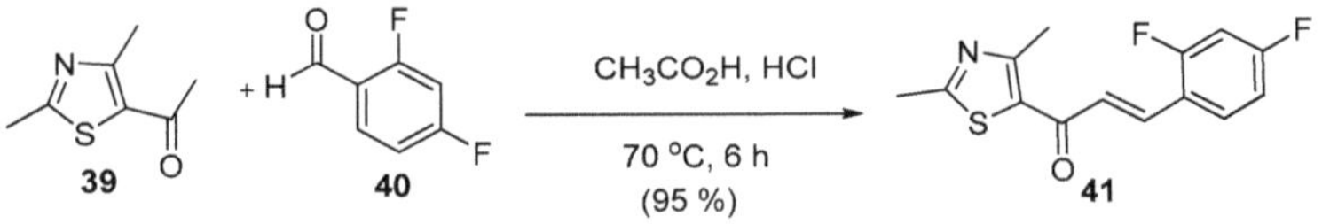

Scheme 19. Acid-catalyzed synthesis of thiazolyl chalcone.

2.2. Non C-S Condensations

Our final installment of A-ring hybrid chalcone synthesis is an interesting green coupling reaction between a series of arylacetylene derivatives (**42a–j**) and various pyridine and benzopyridine carboxaldehydes (Scheme 20). Yadav's group showed that a copper-based silica-coated magnetic nanocatalyst (Cu@DBM@ASMNPs) used in conjunction with a piperidine base was very effective in preparing ten hybrid chalcones in yields ranging from 49 to 94% [51]. A noteworthy feature of this reaction was the ability to recover the catalyst via a magnet. The catalyst was reported to be efficient for up to seven reaction cycles.

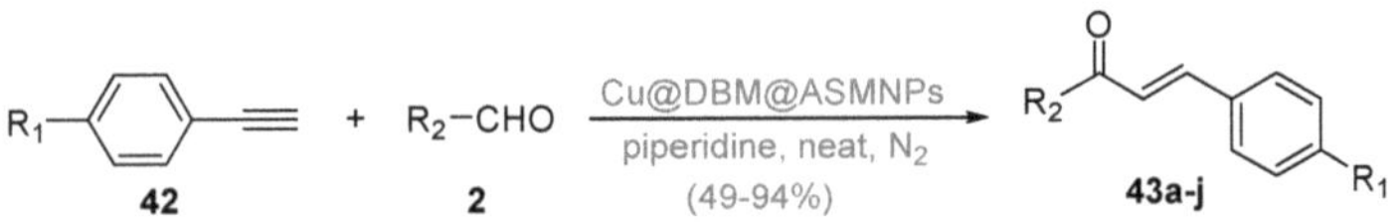

a: R_1 = H, R_2 = 2-pyridyl, 94%
b: R_1 = t-Bu, R_2 = 2-pyridyl, 90%
c: R_1 = Me, R_2 = 2-pyridyl, 76%
d: R_1 = F, R_2 = 2-pyridyl, 55%
e: R_1 = H, R_2 = 2-(4-Clpyridyl), 88%
f: R_1 = t-Bu, R_2 = 2-(4-Clpyridyl), 56%
g: R_1 = Me, R_2 = 2-(4-Clpyridyl), 49%
h: R_1 = F, R_2 = 2-(4-Clpyridyl), 71%
i: R_1 = H, R_2 = 2-(5,6-benzopyridyl), 75%
j: R_1 = MeO, R_2 = 2-(4-Clpyridyl), 90%

Scheme 20. Cu-based nanocatalyzed A^3 synthesis of pyridyl- and benzopyridyl chalcones.

3. B-Ring Heteroaromatic Hybrid Chalcone Synthesis

This section catalogues selected conventional and green processes by which hybrid chalcones containing a heteroaromatic component at ring B may be prepared. In addition, examples of tandem ring-opening dipolar additions to obtain ring B heteroaromatic substituted chalcones are presented. The heteroaromatic components of the chalcone products highlighted in this section include a variety of single-ring (furan, pyrrole, pyrazole, thiazole, thiophene, pyridine) and fused-ring (indole, benzimidazole, benzothiazole, benzofuran, quinoline, imidazo [1,2-a]pyrimidine or imidazo [1,2-a]pyridine, quinoxaline, carbazole) systems.

3.1. Claisen–Schmidt Condensations

As in the preceding section, Claisen–Schmidt (C-S) condensation has been widely used to prepare B-ring heteroaromatic chalcones. This reaction, which can be catalyzed by bases or acids, offers mild conditions that tolerate a wide scope of functionality in both the ketone donors and aldehyde acceptors.

Base-Catalyzed C-S Condensations

In the preparations shown below, NaOH and KOH are the bases of choice. Shown in Scheme 21, Li et al. used dilute aqueous KOH to prepare pyrrolyl chalcone (**46**) in a very good yield. Using mild conditions, Robinson et al. (Scheme 22) condensed acetophenones **47** and furfural derivatives **48** to prepare five furyl chalcones (**49a–e**) that show promise as monoamine oxidase inhibitors in low to medium yields [36].

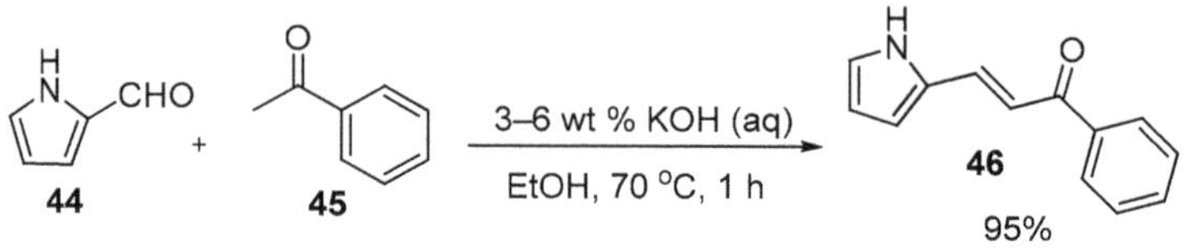
CHO
44
45
3–6 wt % KOH (aq)
EtOH, 70 °C, 1 h
46
95%

Scheme 21. Pyrrolyl chalcone synthesis.

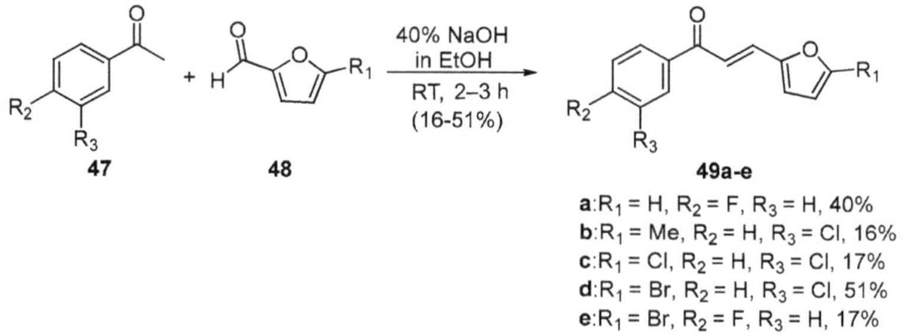
40% NaOH
in EtOH
RT, 2–3 h
(16-51%)
47
48
49a-e
a:R1 = H, R2 = F, R3 = H, 40%
b:R1 = Me, R2 = H, R3 = Cl, 16%
c:R1 = Cl, R2 = H, R3 = Cl, 17%
d:R1 = Br, R2 = H, R3 = Cl, 51%
e:R1 = Br, R2 = F, R3 = H, 17%

Scheme 22. Furyl chalcone synthesis.

In Scheme 23, Fu and coworkers reacted 1,2,3-triazole-substituted acetophenones **50** with furfural **48** and thiophene-2-carbaldehyde **51** in ethanolic KOH for 3 h to prepare hybrid chalcones **52a** and **52b** in satisfactory yields. Condensation of **50** and pyridine carbaldehydes **53a–b** under the same conditions provided eight additional pyridyl hybrid chalcone examples **54a–h** in yields ranging from 50 to 79% [52].

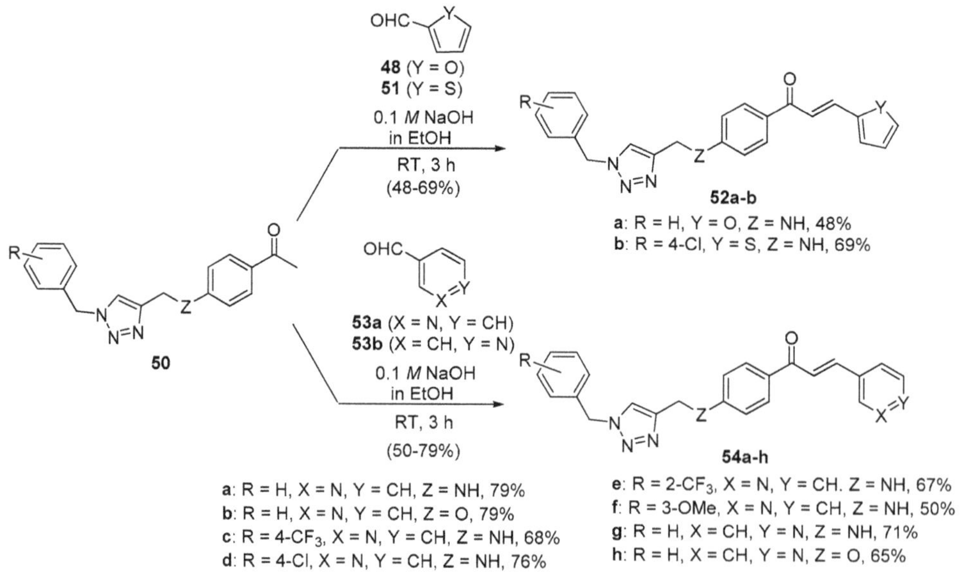
48 (Y = O)
51 (Y = S)
0.1 M NaOH
in EtOH
RT, 3 h
(48-69%)
52a-b
a: R = H, Y = O, Z = NH, 48%
b: R = 4-Cl, Y = S, Z = NH, 69%
50
53a (X = N, Y = CH)
53b (X = CH, Y = N)
0.1 M NaOH
in EtOH
RT, 3 h
(50-79%)
54a-h
a: R = H, X = N, Y = CH, Z = NH, 79%
b: R = H, X = N, Y = CH, Z = O, 79%
c: R = 4-CF3, X = N, Y = CH, Z = NH, 68%
d: R = 4-Cl, X = N, Y = CH, Z = NH, 76%
e: R = 2-CF3, X = N, Y = CH. Z = NH, 67%
f: R = 3-OMe, X = N, Y = CH, Z = NH, 50%
g: R = H, X = CH, Y = N, Z = NH, 71%
h: R = H, X = CH, Y = N, Z = O, 65%

Scheme 23. Furyl, thienyl and pyridyl chalcone synthesis.

Gadhave and Uphade demonstrated the satisfactory condensation of 4-morpholinoacetophenone **55** with 4-pyrazolocarbaldehydes **56** conducted at room temperature, which provided five examples of 4-pyrazolylchalcones **57** [53]. See Scheme 24.

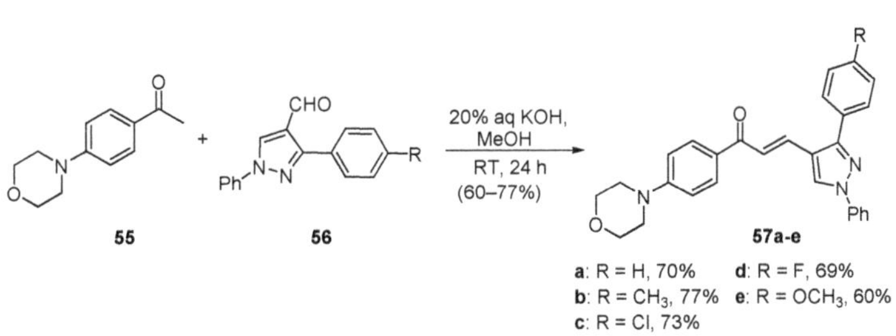

Scheme 24. Pyrazolyl chalcone synthesis.

An interesting study conducted by Mallik and associates involves the preparation of pyrrole-substituted hybrid chalcones from the C-S condensation of several acetophenones **58** and 2-formylpyrrole **44** under different molar ratios of **58**:**44** [54]. As Scheme 25 shows, the desired product **59** predominated when the reactant molar ratios were 1:1, but when the ratio was lowered to 1:2, a nearly equal proportion of the product mixture was found to be the heteroaromatic ketone **60**. Upon increasing the molar proportion of **58** to four times that of **44**, ketone **60** was the major product. The authors propose an interesting mechanism by which **60** is formed—a twin aldol addition—intramolecular cyclization-dehydration.

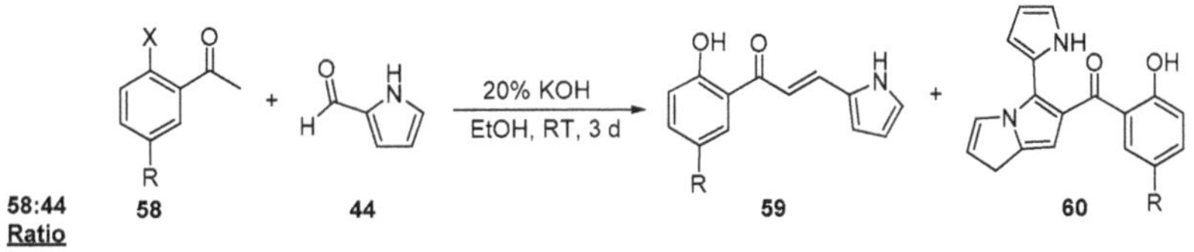

58:44 Ratio	58	59	60
1:1	**a**: X = OH, R = H	**a**: 46%	**a**: 8%
1:2	**a**: X =OH, R = H	**a**: 34%	**a**: 31%
	b: X= OH, R = Cl	**b**: 27%	**b**: 24%
	c: X = OH, R = Me	**c**: 32%	**c**: 26%
	d: X = H, R = H	**d**: 29%	**d**: 29%
1:4	**a**: X = OH, R = H	**a**: 13%	**a**: 41%
	b: X = OH, R = Cl	**b**: 9%	**b**: 44%
	c: X = OH, R = Me	**c**: 11%	**c**: 46%
	d: X = H, R = H	**d**: 14%	**d**: 39%

Scheme 25. Pyrrolyl chalcone synthesis.

Fused-ring heteroaromatic aldehydes have also been successfully condensed with various acetophenones to prepare B-ring hybrid chalcones under typical C-S reaction conditions. Zhao et al. prepared indole hybrid chalcones **63a–e** (Scheme 26) from assorted acetophenones and N-methylindolycarbaldehydes **62** in yields ranging from 60 to 90% [40].

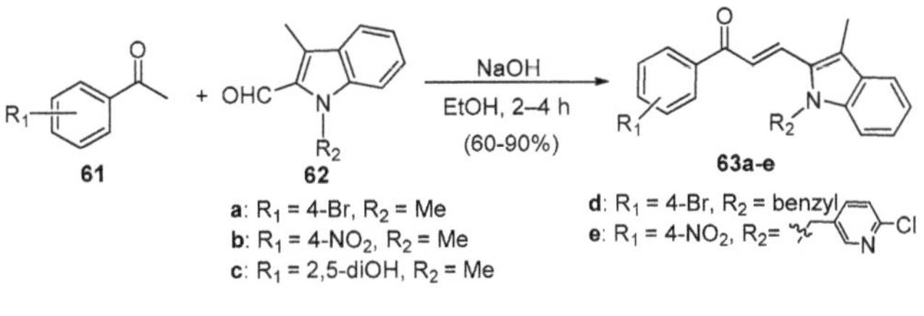

Scheme 26. Indolyl chalcone synthesis.

Bandgar and coworkers (Scheme 27) synthesized a diverse library of carbazole hybrid chalcones **66** [30], while Bindu's team condensed acetophenone derivatives with quinoline carboxaldehdes **68** under mild C-S conditions (Scheme 28) to prepare eight examples of

B-ring-substituted quinolinoid hybrid chalcones **68a–h** [55]. Abonia et al. prepared the chromen-4-one—quinoline hybrid chalcone **71** under similar conditions [56]. See Scheme 29.

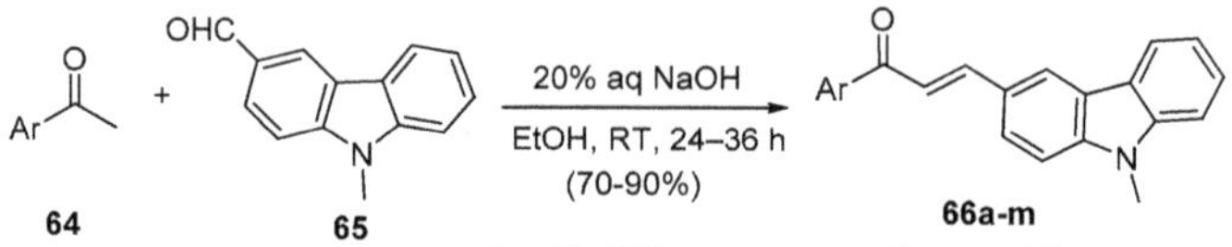

64 **65** **66a-m**

a: Ar = Ph, 83%
b: Ar = 4-MeOC_6H_4, 70%
c: Ar = 3,4-diMeOC_6H_3, 77%
d: Ar = 2,4,6-triMeOC_6H_2, 78%
e: Ar = 4-FC_6H_4, 85%
f: Ar = 2,4-diFC_6H_3, 88%
g: Ar = 3,4-diFC_6H_3, 87%
h: Ar = 4-ClC_6H_4, 86%
i: Ar = 2,4-diClC_6H_3, 88%
j: Ar = 3,4-diClC_6H_3, 88%
k: Ar = 2,5-diClC_6H_3, 90%
l:Ar = 4-BrC_6H_4, 90%
m: Ar = 4-$O_2NC_6H_4$, 87%

Scheme 27. Carbazolyl hybrid chalcone synthesis.

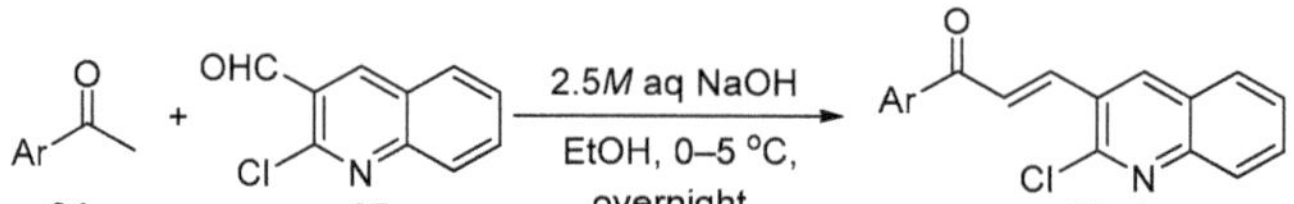

64 **67** **68a-h**

a: Ar = Ph, 85%
b: Ar = 4-MeC_6H_4, 87%
c: Ar = 4-MeOC_6H_4, 80%
d: Ar = 4-HOC_6H_2, 82%
e: Ar = 4-ClC_6H_4, 85%
f: Ar = 4-BrC_6H_4, 88%
g: Ar = 4-$O_2NC_6H_4$, 80%
h: Ar = 4-MeHNC_6H_4, 73%

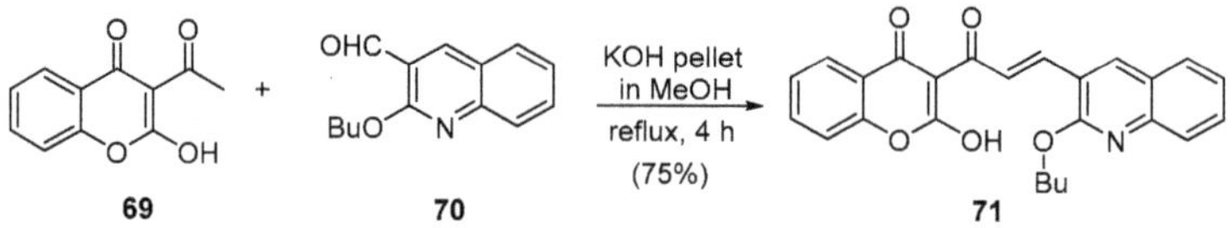

69 **70** **71**

Scheme 28. Quinolinyl hybrid chalcone synthesis.

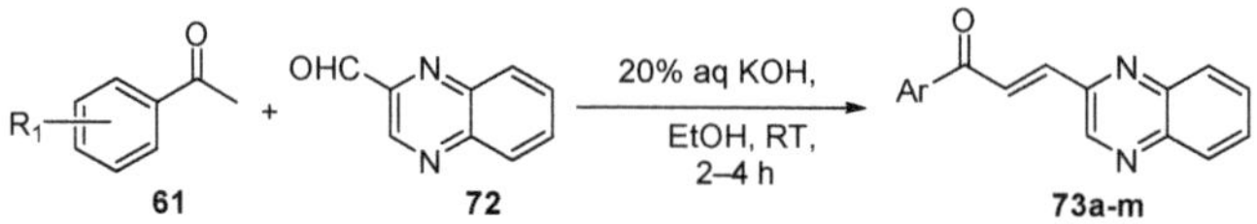

61 **72** **73a-m**

a: Ar = Ph, 70%
b: Ar = 4-HOC_6H_4, 78%
c: Ar = 4-$H_2NC_6H_4$, 75%
d: Ar = 4-BrC_6H_2, 80%
e: Ar = 4-MeOC_6H_4, 85%
f: Ar = 4-ClC_6H_4, 75%
g: Ar = 4-FC_6H_4, 80%
h: Ar = 3-HOC_6H_4, 80%
i: Ar = 3-BrC_6H_4, 95%
j: Ar = 3-$H_2NC_6H_4$, 75%
k: Ar = 4-$O_2NC_6H_4$, 60%
l: Ar = 3-$O_2NC_6H_4$, 60%
m: Ar = naphthyl, 90%

Scheme 29. Quinoxalinyl hybrid chalcone synthesis.

Desai and coworkers used mild C-S reaction conditions to prepare a series of thirteen quinoxalinyl hybrid chalcones **73a–m** in yields ranging from 60 to 95%, as shown in Scheme 29 [24].

In a study of microtubule polymerization inhibition, Sun et al. synthesized a library of fused-ring heteroaromatic chalcones featuring indoles, benzofurans, dibenzofurans, benzothiophenes, dibenzothiophenes, and benzimidazoles [57]. See Figure 4. Of particular note were the numerous methods used in the preparation of these hybrid chalcones, which included both base-promoted processes (piperidine, NaOH, KOH, NaOMe, Cs_2CO_3 and NaH) in methanolic and ethanolic solvents, Lewis acid catalysis (BF_3•etherate) in dioxane solvent and Brønsted (glacial acetic acid) acid catalysis in toluene. Scheme 30 depicts the scope of this work.

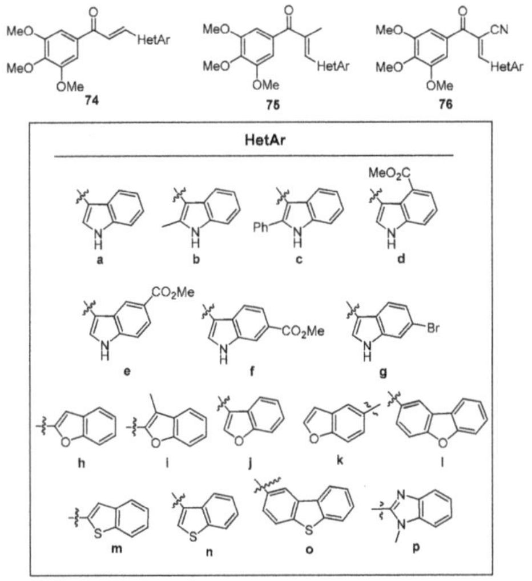

Figure 4. Hybrid chalcone heteroaromatic components prepared by Sun et al.

Base-catalyzed C-S condensations that employ green chemistry principles to produce B-ring-substituted hybrid chalcones have also been successfully conducted. See Scheme 31. These processes include the use of benign solvents, solvent-free reactions, microwave irradiation, ultrasound and ball milling. For example, Ashok's group compared a typical base-catalyzed C-S condensation of **83** and **84** with a solvent-free, microwave-mediated process to prepare a series of carbazolyl hybrid chalcones **85** [58]. The yields for the short-duration microwave-mediated reactions exceeded those of the lengthy conventional C-S reactions in every case. Bhatt et al. prepared the furyl chalcone **87** using both conventional C-S and ultrasound processes to condense furfural **48** and 2,4-dihydroxyacetophenone **86** [59]. The effectiveness of sonication is evident—a 10% increase in yield in 1/20 the reaction time. Jadhava's team used PEG-400 as a benign solvent to mediate the condensation of 4-fluoroacetophenone **84** and a series of pyrazole carbaldehydes **85** en route to eight fluorinated pyrazolyl hybrid chalcones **86** [60]. Kudlickova and coworkers employed a mechanochemical ball-milling process to prepare a series of indoylchalcones **92** in yields ranging from 28 to 79% in only 30 min [61]. Nimmala's group used a solventless process to condense various acetophenones and imidazo [1,2-a]pyrimidine **93** or imidazo [1,2-a]pyridine **95** en route to hybrid chalcones **94a–f** and **96a–f**, respectively, in very good yields [62]. Joshi and Saglani employed ultrasound to assist in the condensation of the fused-ring ketone **97** and a series of quinoline carbaldehydes **98** to prepare the quinolinyl hybrid chalcones **99** [63].

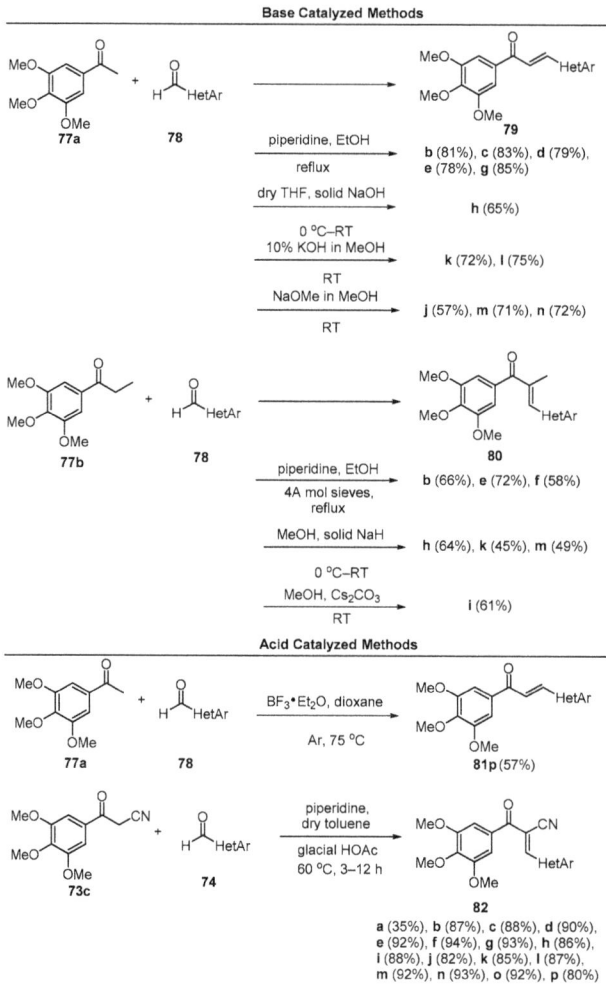

Scheme 30. N, O, S Fused-ring heteroaromatic hybrid chalcone synthesis.

3.2. Non C-S Condensations

The final entries describing ring-B-substituted heteroaromatic hybrid chalcones feature unique tandem reactions involving pyrylium tetrafluoroborate derivatives. Devi and colleagues conducted a very interesting examination of a single-pot, base-mediated, tandem-ring-opening, 1,3-dipolar addition reaction between several electron withdrawing group (EWG)-substituted diazo compounds **101** with tri-substituted pyrylium salts **100**, producing an extensive array of pyrazole hybrid chalcones **102** in moderate to high yields, as shown in Scheme 32 [64].

Tan and Wang leveraged a similar pyrilium ring-opening strategy in a single-pot 3+2 reductive annulation with benzil derivatives **103** to prepare a comprehensive library of tetra-substituted Furano chalcones **105a–ii** in yields as high as 70% [65]. See Scheme 33. A noteworthy observation in both works was the finding that Z-chalcone derivatives were the major or sole product in all instances.

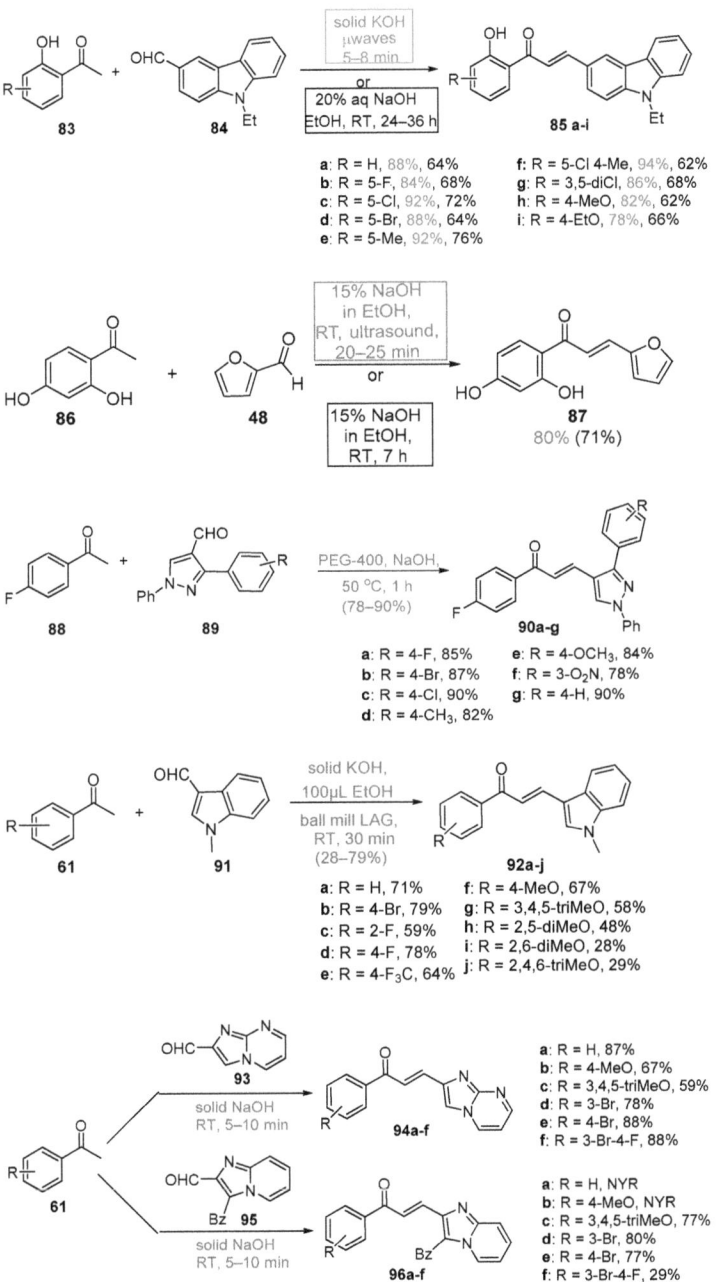

Scheme 31. Cont.

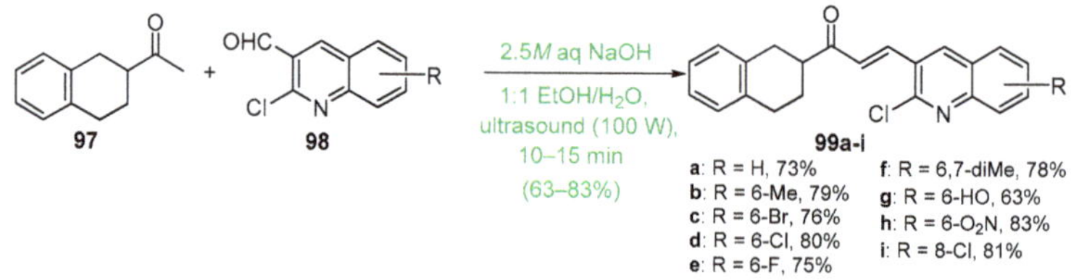

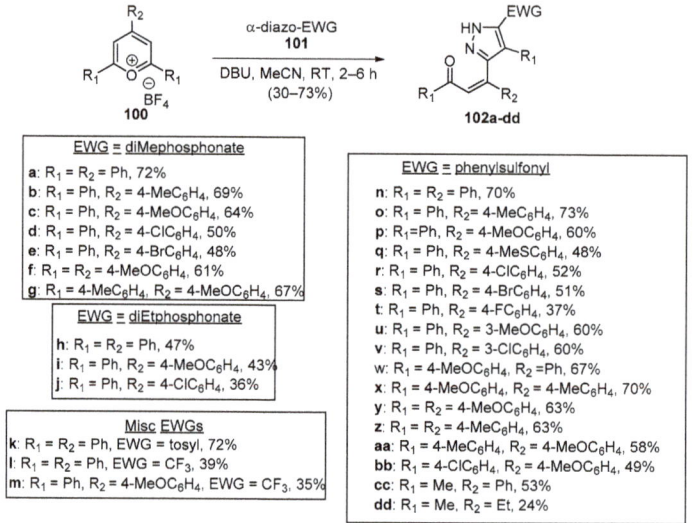

Scheme 31. Green syntheses of B-ring heteroaromatic hybrid chalcones.

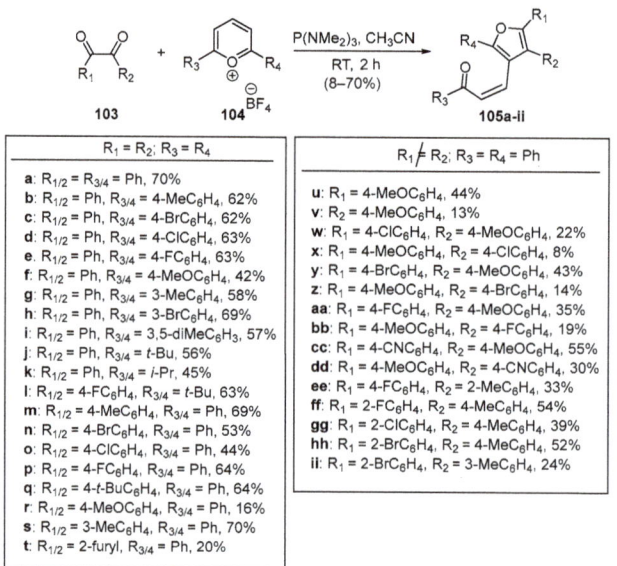

Scheme 32. Synthesis of pyrazole hybrid Z-chalcones via a pyrilium.ring-opening dipolar addition.

Scheme 33. Synthesis of furanyl hybrid Z-chalcones via pyrilium ring-opening benzil-derivative reductive 3+2 annulation.

4. A–B Ring Dual Heteroaromatic Hybrid Chalcone Synthesis

This section catalogues selected processes by which hybrid chalcones bearing a heteroaromatic species at both rings A and B may be prepared. Of particular note is the incredibly diverse array of chalcones produced that feature 21 different heteroaromatic A–B ring-substituted groups on the hybrid chalcones shown in Schemes 34–46.

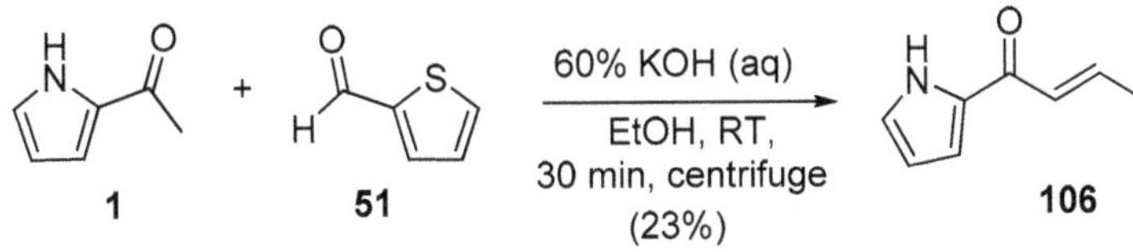

Scheme 34. Synthesis of pyrrolyl–thienyl hybrid chalcones.

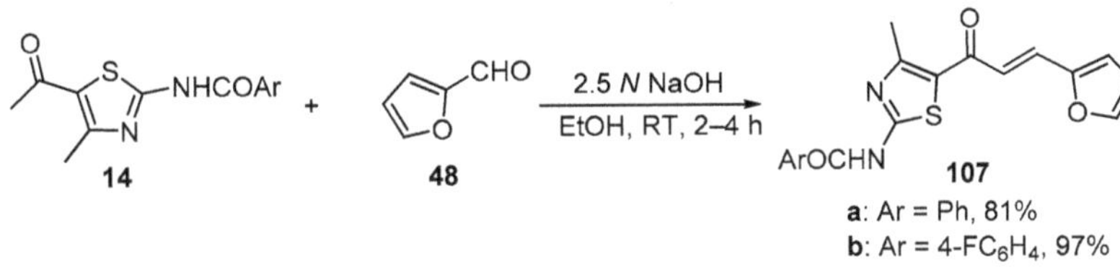

Scheme 35. Synthesis of thiazolyl–furyl hybrid chalcones.

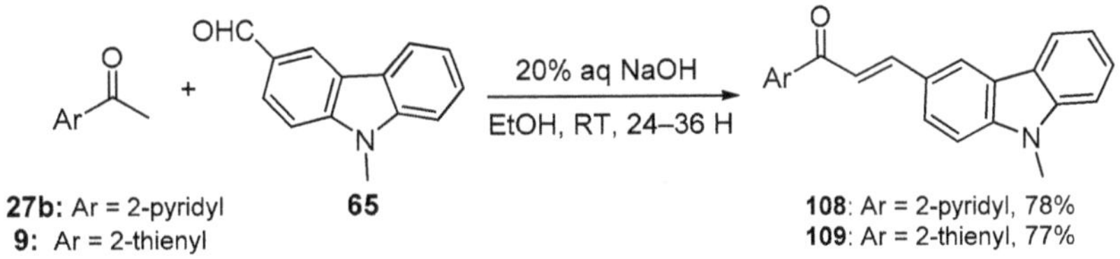

Scheme 36. Synthesis of pyridyl– and thienyl–carbazole hybrid chalcones.

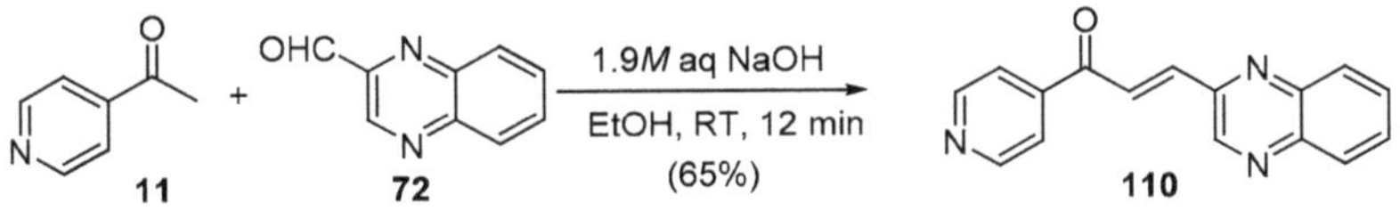

Scheme 37. Synthesis of pyridyl–quinoxazolyl hybrid chalcone.

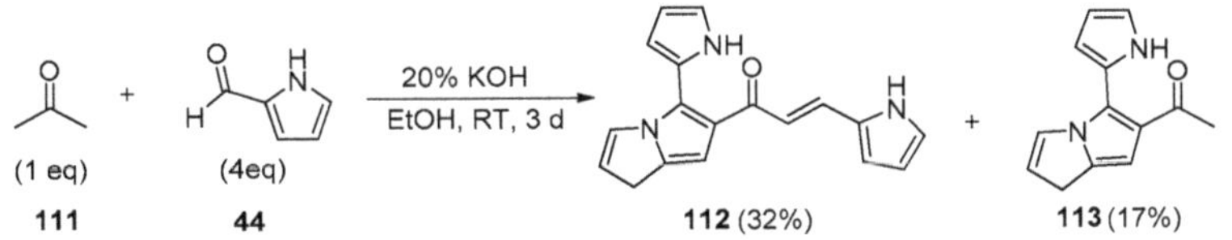

Scheme 38. Synthesis of pyrrole–[(2-pyrrolyl)-3H-pyrrolizinyl] hybrid chalcone.

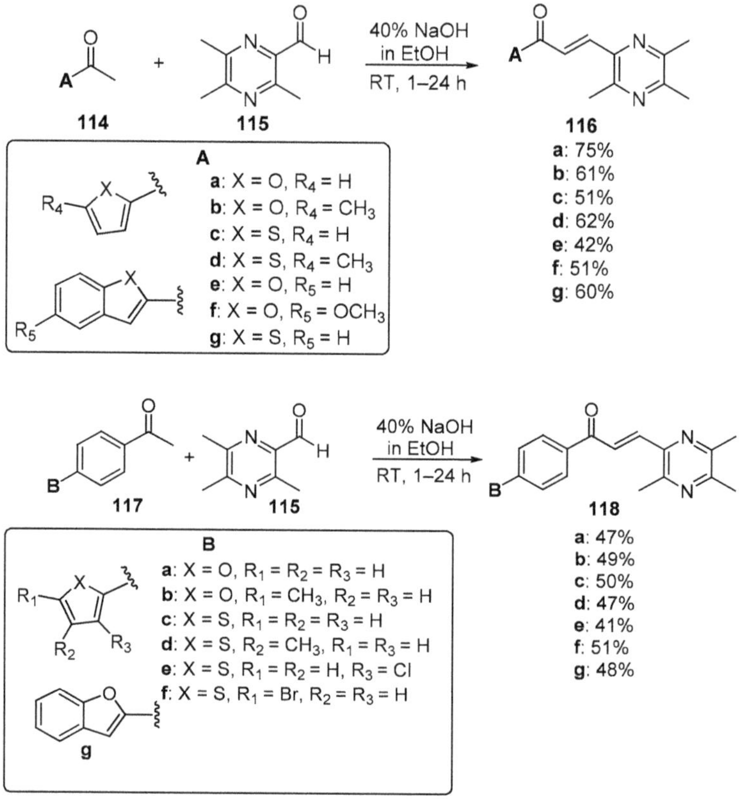

Scheme 39. Synthesis of pyrazinyl hybrid chalcones.

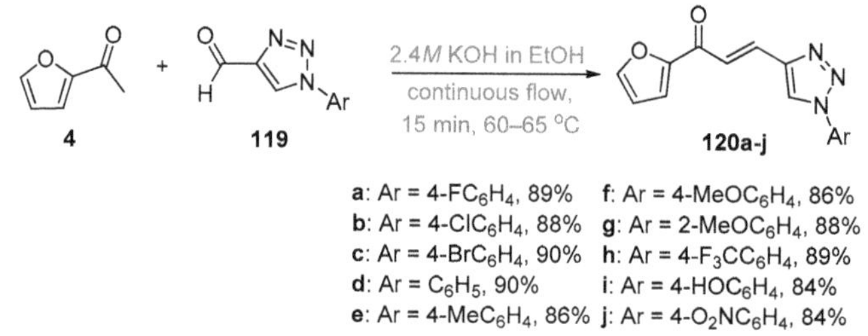

Compound 116:
- a: 75%
- b: 61%
- c: 51%
- d: 62%
- e: 42%
- f: 51%
- g: 60%

A:
- a: X = O, R_4 = H
- b: X = O, R_4 = CH_3
- c: X = S, R_4 = H
- d: X = S, R_4 = CH_3
- e: X = O, R_5 = H
- f: X = O, R_5 = OCH_3
- g: X = S, R_5 = H

Compound 118:
- a: 47%
- b: 49%
- c: 50%
- d: 47%
- e: 41%
- f: 51%
- g: 48%

B:
- a: X = O, R_1 = R_2 = R_3 = H
- b: X = O, R_1 = CH_3, R_2 = R_3 = H
- c: X = S, R_1 = R_2 = R_3 = H
- d: X = S, R_2 = CH_3, R_1 = R_3 = H
- e: X = S, R_1 = R_2 = H, R_3 = Cl
- f: X = S, R_1 = Br, R_2 = R_3 = H
- g: (benzofuran)

Scheme 40. Synthesis of furyl–triazolyl hybrid chalcones.

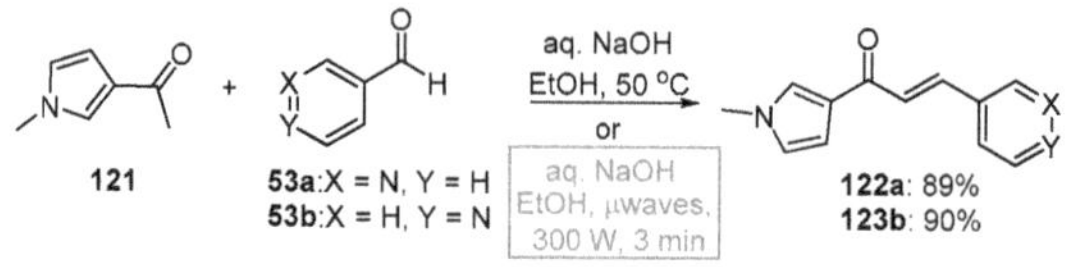

Compound 120a–j:
- a: Ar = 4-FC_6H_4, 89%
- b: Ar = 4-ClC_6H_4, 88%
- c: Ar = 4-BrC_6H_4, 90%
- d: Ar = C_6H_5, 90%
- e: Ar = 4-MeC_6H_4, 86%
- f: Ar = 4-$MeOC_6H_4$, 86%
- g: Ar = 2-$MeOC_6H_4$, 88%
- h: Ar = 4-$F_3CC_6H_4$, 89%
- i: Ar = 4-HOC_6H_4, 84%
- j: Ar = 4-$O_2NC_6H_4$, 84%

Conditions: 2.4M KOH in EtOH, continuous flow, 15 min, 60–65 °C

53a: X = N, Y = H
53b: X = H, Y = N

Conditions: aq. NaOH, EtOH, 50 °C or aq. NaOH, EtOH, μwaves, 300 W, 3 min

122a: 89%
123b: 90%

Scheme 41. Synthesis of pyrrolyl–pyridyl hybrid chalcones.

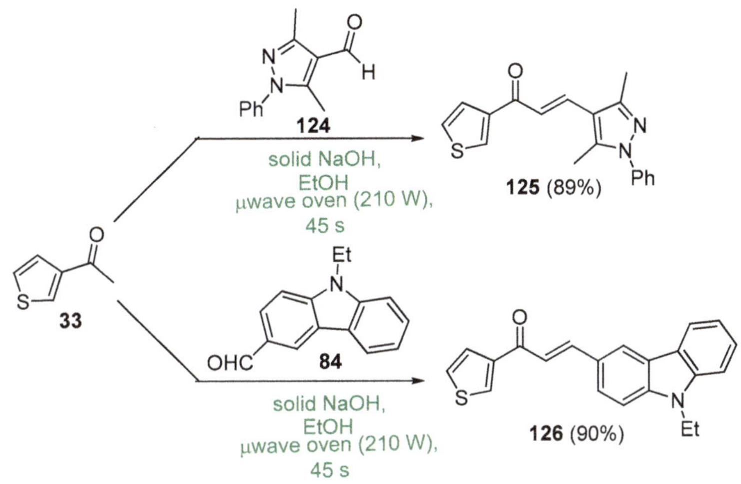

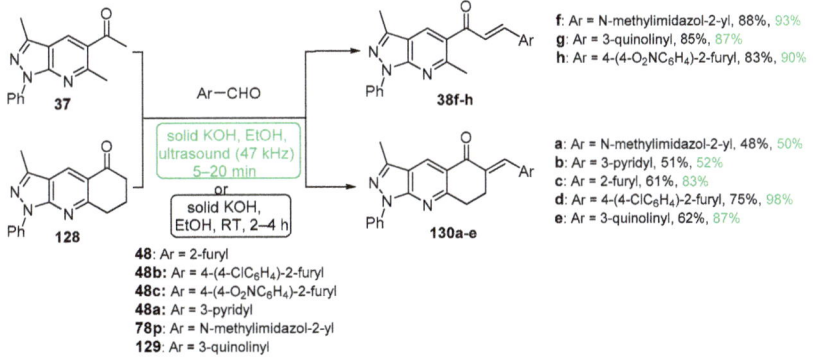

Scheme 42. Synthesis of thienyl–pyrazolyl/carbazolyl hybrid chalcones.

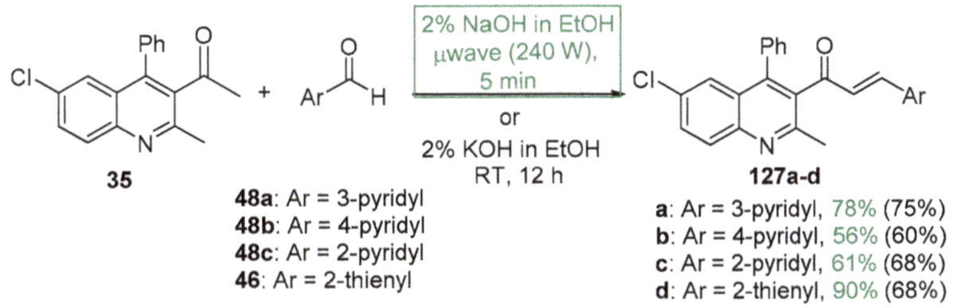

Scheme 43. Synthesis of quinolinyl–pyridyl/thienyl hybrid chalcones.

Scheme 44. Synthesis of pyrazolopyridyl–heteroaryl hybrid chalcones.

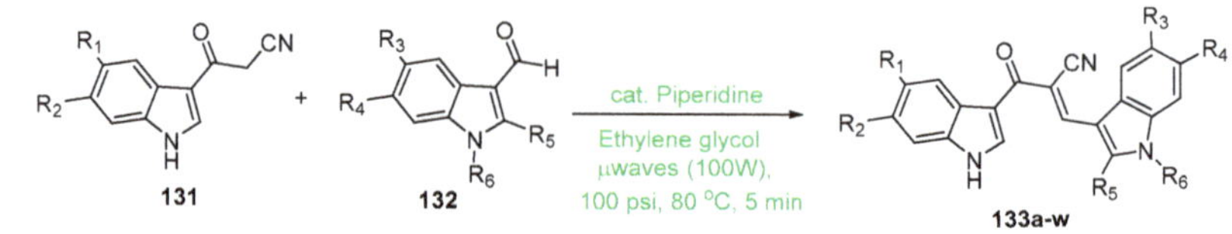

a: $R_1 = R_2 = R_3 = R_4 = R_5 = R_6$ = H, 92%
b: $R_1 = R_2 = R_4 = R_5 = R_6$ = H, R_3 = OMe, 85%
c: $R_1 = R_2 = R_3 = R_5 = R_6$ = H, R_4 = F, 85%
d: $R_1 = R_2 = R_3 = R_5 = R_6$ = H, R_4 = OMe, 82%
e: $R_1 = R_3 = R_4 = R_5 = R_6$ = H, R_2 = OMe, 85%
f: $R_2 = R_4 = R_5 = R_6$ = H, $R_1 = R_3$ = Br, 78%
g: $R_2 = R_4 = R_5 = R_6$ = H, $R_1 = R_3$ = OMe, 85%
h: $R_2 = R_3 = R_5 = R_6$ = H, R_1 = OMe, R_4 = F, 85%
i: $R_1 = R_3 = R_5 = R_6$ = H, $R_2 = R_4$ = F, 80%
j: $R_1 = R_4 = R_5 = R_6$ = H, R_2 = F, R_3 = OMe, 85%
k: $R_1 = R_3 = R_5 = R_6$ = H, R_2 = F, R_4 = OMe, 90%
l: $R_1 = R_4 = R_5 = R_6$ = H, $R_2 = R_3$ = OMe, 85%
m: $R_1 = R_3 = R_5 = R_6$ = H, $R_2 = R_4$ = OMe, 84%
n: $R_1 = R_3 = R_5 = R_6$ = H, R_2 = OMe, R_4 = F, 75%
o: $R_1 = R_2 = R_3 = R_4 = R_6$ = H, R_5 = 4-OMeC_6H_4, 72%
p: $R_1 = R_2 = R_3 = R_4 = R_5$ = H, R_6 = Me, 82%
q: $R_1 = R_2 = R_3 = R_4 = R_5$ = H, R_6 = 4-Cl$C_6H_4CH_2$, 85%
r: $R_1 = R_2 = R_3 = R_4 = R_5$ =H, R_6 = 4-OMe$C_6H_4CH_2$, 82%
s: $R_1 = R_2 = R_4 = R_5$ = H, R_3 = Br, R_6 = 4-Cl$C_6H_4CH_2$, 80%
t: $R_1 = R_3 = R_4 = R_5$ = H, R_2 = OMe, R_6 = 4-Cl$C_6H_4CH_2$, 85%
u: $R_1 = R_3 = R_4 = R_5$ = H, R_2 = OMe, R_6 = 4-OMe$C_6H_4CH_2$, 85%
v: $R_1 = R_3 = R_4 = R_5$ = H, R_2 = F, R_6 = 4-Cl$C_6H_4CH_2$, 85%
w: $R_1 = R_3 = R_4 = R_5$ = H, R_2 = F, R_6 = 4-OMe$C_6H_4CH_2$, 82%

Scheme 45. Synthesis of twin indolyl hybrid chalcones.

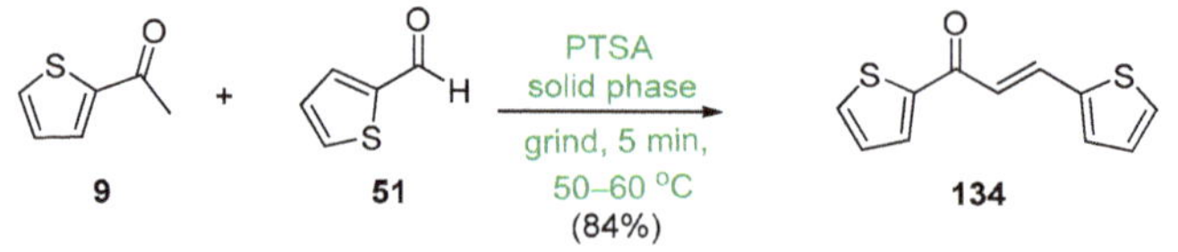

Scheme 46. Synthesis of twin thienyl hybrid chalcone.

4.1. *Claisen–Schmidt Condensations*

As noted in the preceding sections, the Claisen–Schmidt (C-S) condensation is the most common method used to prepare A–B ring heteroaromatic chalcones. This reaction, which can be catalyzed by bases or acids, offers mild conditions that tolerate a wide scope of functionality in both the ketone donors and aldehyde acceptors.

4.1.1. Base-Catalyzed C-S Condensations

In most instances, NaOH and KOH are the most widely used bases. Sweeting's group synthesized and obtained an X-ray crystal structure for the pyrrolyl–thienyl hybrid chalcone **106** as part of a chalcone solubility and stability study [30]. See Scheme 34. While the use of centrifuging to mix the reagents is of interest, the low yield is likely attributable to the limited reaction time of 30 min. Sinha and coworkers prepared two thiazolyl–furyl hybrid chalcones in high yields (Scheme 35) while investigating potential ant-lipoxygenase agents [37].

Fused-ring A–B hybrid chalcone examples have also been successfully prepared under very mild, base-catalyzed C-S conditions. Bandgar's team prepared the pyridyl and thienyl–carbazolyl heteroaromatic hybrid chalcones **108–109** in very good yields (Scheme 36) [29]. While investigating ACP reductase inhibition, Desai's group prepared the pyridyl/quinoxazolyl chalcone **110** in a good yield as shown in Scheme 37 [23]. Mallik et al. found that when one equivalent of acetone and four equivalents of 2-pyrrole carbaldehyde were condensed in 20% KOH, the unusual pyrrolizinyl–pyrrolyl chalcone **112** was formed in modest yield (32%), accompanied by the acetylpyrrolizine **113** (17%) [53]. See Scheme 38. This finding is complementary to the work shown in Scheme 25 in which similar pyrrolizine products were formed. In an examination of chalcones with potential anticancer properties, Bukhari prepared a diverse set of furyl-, thienyl-, benzofuryl, and benzothienyl-1,4-pyrazinyl chalcones **116** in yields ranging from 42 to 75%. Extending that work to include condensations of 4-heteroaromatic acetophenones **117** with pyrazine carbaldehyde **115** gave rise to an array of hybrid chalcones **118** in moderate yields [18]. See Scheme 39.

4.1.2. Green C-S Condensations

The recent literature reports a number of green, base-promoted C-S condensations used to prepare A–B ring heteroaromatic hybrid chalcones. While studying potential antimicrobial agents, Kumar et al. synthesized ten furyl-triazolyl chalcones **120a–j** via a continuous-flow reactor [66]. Of note are the exceptional yields (84–90%) obtained in only 15 min. See Scheme 40. Moreover, in pursuit of suitable chalcones that have antimicrobial properties, Usta's team prepared two pyrrole–pyridyl chalcones using both conventional and microwave processes [27]. The yields reported were as high as 90% after only 3 min of irradiation. See Scheme 41.

Several syntheses of A–B ring heteroaromatic chalcones having fused-ring systems have also been reported. Khan and Asiri prepared two hybrid chalcones and tested them for antibacterial activity, a thienyl–pyrazole chalcone as well as a thienyl–carbazolyl chalcone using a microwave oven [46]. See Scheme 42. The base-catalyzed process, completed in only 45 s, provided the chalcones in 89–90%. Quinolinyl chalcones, such as those prepared by Sarveswari and Vijayakumar in Scheme 43, have also shown promise as antibacterial and antifungal agents [47]. Again, yields for the short-duration, microwave-mediated process was on par with or exceeded those obtained by the conventional C-S reactions conducted in their comparative study.

Acetylated pyrazolo pyridines **37** and **128** were condensed with five heteroaryl aldehydes by Polo et al. under both ultrasonic and conventional conditions to prepare interesting A–B ring hybrid chalcones substituted with furyl, pyridyl, imidazolyl and quinolinyl groups [48]. See Scheme 44. Chalcone series **38** was part of a larger study discussed earlier in the review (Scheme 17). Yields for the short-duration ultrasound-assisted condensation met or exceeded those obtained by the conventional, base-promoted C-S condensations performed by the group.

In Scheme 45, Kumar et al. employed piperidine base to catalyze the microwave-mediated condensation of indoles **131** and **132** en route to a large array of highly differentially functionalized twin indolyl hybrid chalcones **133** [67]. The yields reported were excellent, ranging from 72 to 92%, especially given the reaction time of 5 min.

Our final entry in this section is a green, solid-state, acid-catalyzed condensation of 2-acetylthiophene **9** and the thienyl carboxaldehyde **51** conducted by Adnan and associates, which produced the twin thienyl chalcone **134** in an excellent yield [13]. See Scheme 46.

5. Heteroaromatic Bis Chalcone Hybrid Synthesis

This section catalogues several processes by which heteroaromatic bis chalcone hybrids bearing two or more heteroaromatic species have been prepared. The reactions feature both heteroaromatic donors and acceptors as the linker unit in the bis hybrid chalcone systems. Conventional and green condensations as well as a unique Wittig preparation are discussed.

5.1. Claisen–Schmidt Condensations

The Claisen–Schmidt (C-S) condensation is the most widely used method to prepare heteroaromatic bis chalcone hybrids. In this section, we present base-promoted condensations that tolerate a wide scope of functionality in both the bis-ketone donors and bis-aldehyde acceptors.

5.1.1. Base-Catalyzed C-S Condensations

As seen in the previous sections, NaOH and KOH are the most widely used bases. Methanol and ethanol are the solvents of choice in these condensations. In the first entry of bis hybrid chalcone preparation (Scheme 47), Alidmat et al. prepared three examples of mono- and dichlorinated bis-thienyl chalcones with potential as anticancer agents [68]. Of note is the one-pot preparation of the non-symmetric bis hybrid chalcone **138** from the condensation of 4-formylbenzaldehyde **135** (1 mole) and equimolar quantities of acetylthiophenes **136** and **137**. In contrast, the condensation of **135** (1 mole) with two moles of **136** or **137** resulted in the symmetric bis hybrid chalcones **139** or **141**, respectively.

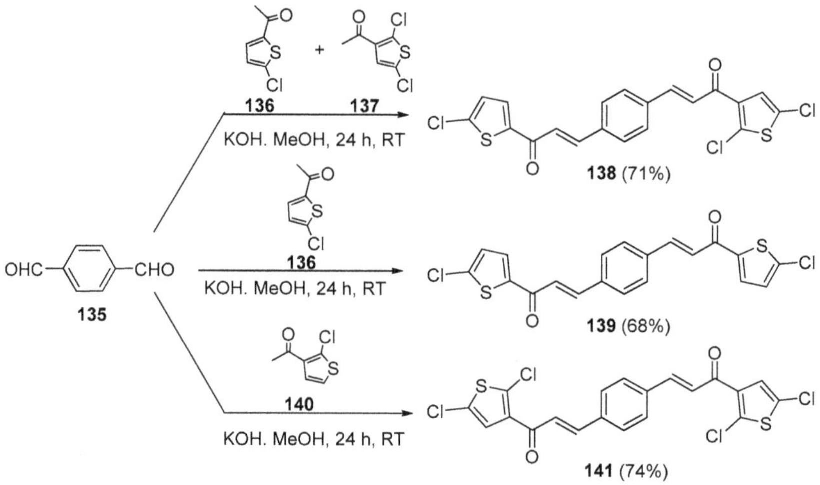

Scheme 47. Synthesis of bis thienyl hybrid chalcones.

While investigating photoinitiators with applications in 3D/4D printing, Chen's group prepared several bis hybrid chalcones that show promise as light-sensitive photoinitiators. See Scheme 48. 4,4′-diacetylbiphenyl **142** was condensed with 2-formylthiophene under mild, base-promoted conditions to synthesize the bis thienyl biphenyl chalcone **143** in a good yield [17]. Under the same reaction conditions, 2,6-diacetylpyridine **144** was condensed with several substituted benzaldehydes **145** en route to three pyridyl bis aryl hybrid chalcones **146a–c** in yields ranging from 58 to 86%.

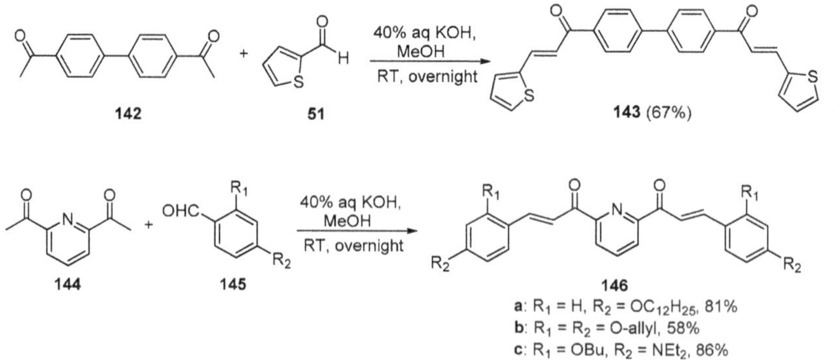

Scheme 48. Synthesis of biphenyl bis thienyl and pyridyl bis aryl hybrid chalcones.

While investigating lung cancer cell growth inhibitors, Zhao et al. prepared the indole bis phenyl chalcone **148** by condensing 1,2-diacetyl-3-methylindole **147** with benzaldehyde in 60% yield [54]. See Scheme 49.

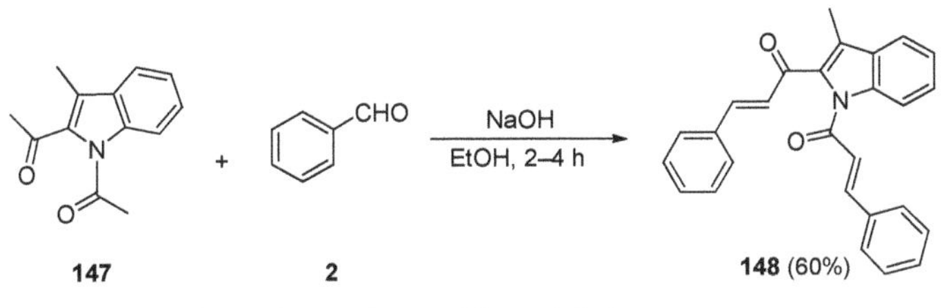
CHO
NaOH
EtOH, 2–4 h
147
2
148 (60%)

Scheme 49. Synthesis of indolyl bis aryl hybrid chalcones.

Presented in Schemes 50 and 51 are green methods used to prepare bis heteroaromatic chalcones. Asir and coworkers used sonochemical mediation to prepare examples of bis thienyl and bis furyl hybrid chalcones **150a–b**. The reaction time of 5 min was sufficient to give product yields in excess of 70%. [69] In a study of the anti-inflammatory activity of 3,4-bis-chalcone-N-arylpyrazoles, Abdel-Aziz et al. prepared eight examples of assorted aryl- and heteroaryl-substituted chalcone pyrazoles **152** using an aqueous KOH/EtOH medium at 60 °C and microwave irradiation [70]. The total reaction time reported was only four minutes to achieve yields ranging from 70 to 93%. Analogous conventional C-S condensations were also carried out over a 12 h period; the yields obtained were about 75–85% of those obtained with μwave mediation.

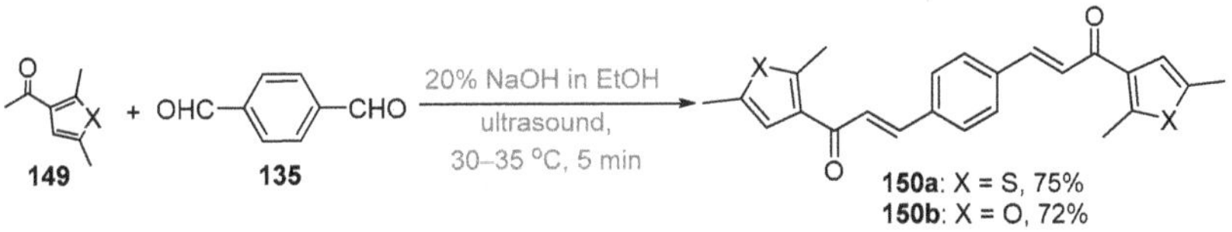
OHC
CHO
20% NaOH in EtOH
ultrasound,
30–35 °C, 5 min
149
135
150a: X = S, 75%
150b: X = O, 72%

Scheme 50. Sonochemical synthesis of bis thienyl and bis furyl hybrid chalcones.

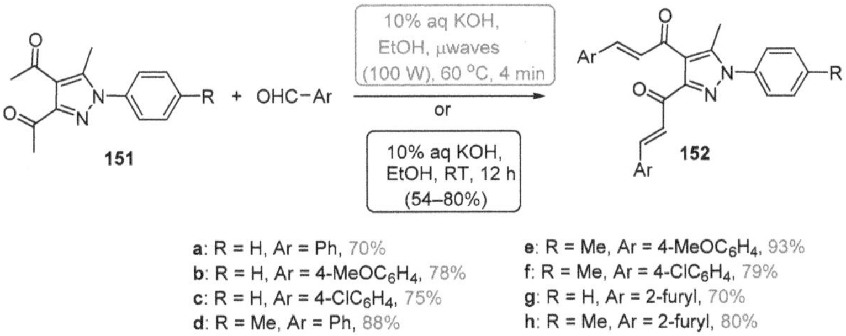
OHC–Ar
10% aq KOH,
EtOH, μwaves
(100 W), 60 °C, 4 min
or
10% aq KOH,
EtOH, RT, 12 h
(54–80%)
151
152
a: R = H, Ar = Ph, 70%
b: R = H, Ar = 4-MeOC6H4, 78%
c: R = H, Ar = 4-ClC6H4, 75%
d: R = Me, Ar = Ph, 88%
e: R = Me, Ar = 4-MeOC6H4, 93%
f: R = Me, Ar = 4-ClC6H4, 79%
g: R = H, Ar = 2-furyl, 70%
h: R = Me, Ar = 2-furyl, 80%

Scheme 51. Microwave-mediated synthesis of bis aryl/heteroaryl chalcone pyrazoles.

5.1.2. Non C-S Condensations

Our final installment for the bis hybrid chalcone section is an early example published by Saikachi and Muto in 1971 [71]. Their work, shown in Scheme 52, which focused on the preparation and utility of bisphosphoranes in oligimerization studies, exemplified how the bis-Wittig reagents **153, 155** and **157** could be successfully coupled with furan or thienylcarbaldehydes to provide a series of bis heteroaromatic chalcones **154, 156** and **158** in yields ranging from 45 to 99%. This work was unique in providing the bis hybrid chalcone system with benzene, biphenyl, diphenyl ether, diphenylmethylene, and diphenylethylene linker units.

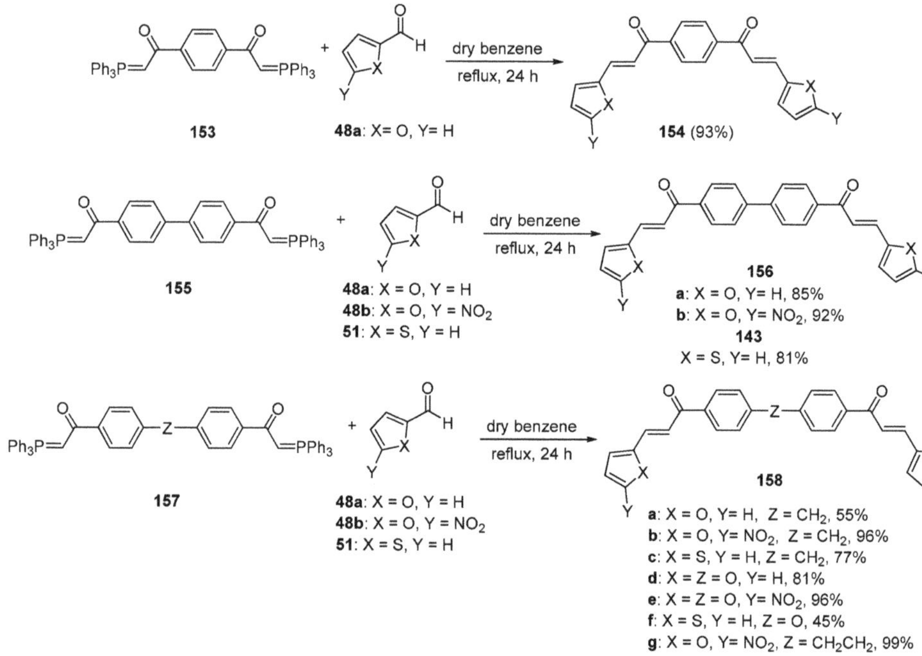

Scheme 52. Wittig synthesis of bis thienyl and bis furyl hybrid chalcones.

6. Conclusions and Future Directions

This review of the preparation of heteroaromatic hybrid chalcones gives a robust accounting of more than 50 historic and current synthetic processes leading to more than 430 different hybrid chalcone examples that include single-ring and multi-ring heteroaromatic moieties. We have shown that the venerable Claisen–Schmidt reaction, by far the most common condensation method discussed herein, has been successfully used in either base-promoted or acid-catalyzed processes en route to heteroaromatic hybrid chalcones. We note that variations in the base or acid identity, solution concentration and physical state often make direct comparisons of the yields challenging. Also discussed has been the wide array of reaction conditions, such as the temperature and reaction time, which likewise impact the overall yield. Finally, the topology and electronic reactivity of the ketone donors and aldehyde acceptors likely modulate the product stereochemistry and yields as well.

Additionally, this review has provided the reader with an appreciation of alternative methods used to prepare these hybrid chalcones. Presented in our review are metal-catalyzed coupling reactions, cycloadditions, ring-opening processes and Wittig reactions that enable the formation of more than 75 hybrid chalcone examples.

A key thrust of this review has been to highlight the application of green chemistry methods in heteroaromatic hybrid chalcone synthesis. From the use of benign/renewable solvents and solvent-free and solid-state processes, researchers have demonstrated the ability to minimize waste streams. Through the use of sonochemical, mechanochemical, microwave irradiation, continuous-flow reactions and nanocatalytic methods, scientists minimize the reagent costs, reaction times and energy expenditure while optimizing the yields. Taken together, the important advances in green method uses noted herein portend well for future investigations of heteroaromatic chalcone synthesis.

Author Contributions: Conceptualization, A.M. and J.S.; writing—original draft preparation, A.M. and J.S.; writing—review and editing, A.M. and J.S.; visualization, A.M. and J.S. All authors have read and agreed to the published version of the manuscript.

Funding: This research received no external funding.

Institutional Review Board Statement: Not applicable.

Informed Consent Statement: Not applicable.

Data Availability Statement: Not applicable.

Conflicts of Interest: The authors declare no conflict of interest.

References

1. Kostanecki, S.; Tambor, J. Ueber die sechs isomeren Monooxybenzalacetophenone (Monooxychalkone). *J. Chem. Ber.* **1899**, *32*, 1921–1926. [CrossRef]
2. Faass, M.; Mallia, A.; Sommer, R.; Sloop, J. (Z)-2-(3, 5-Dimethoxybenzylidene) Indan-1-one. CCDC 2081969: Experimental Crystal Structure Determination. *CSD Commun.* **2021**. [CrossRef]
3. Encarnacion-Thomas, E.; Sommer, R.D.; Mallia, V.A.; Sloop, J. (E)-2-(3,5-Dimethoxy-benzylidene)indan-1-one. *IUCrData* **2020**, *5*, x200759. [CrossRef]
4. Makarov, A.; Sorotskaja, L.; Uchuskin, M.; Trushkov, I. Synthesis of quinolines via acid-catalyzed cyclodehydration of 2-(tosylamino)chalcones. *Chem. Heterocycl. Comp.* **2016**, *52*, 1087–1091. [CrossRef]
5. Jin, H.; Jiang, X.; Yoo, H.; Wang, T.; Sung, C.; Choi, U.; Lee, C.-R.; Yu, H.; Koo, S. Synthesis of Chalcone-Derived Heteroaromatics with Antibacterial Activities. *ChemistrySelect* **2020**, *5*, 12421–12424. [CrossRef]
6. Kalirajan, R.; Sivakumar, S.; Jubie, S.; Gowramma, B.; Suresh, B. Synthesis and Biological evaluation of some heterocyclic derivatives of Chalcones. *Int. J. ChemTech Res.* **2009**, *1*, 27–34. [CrossRef]
7. El-Gohary, N. Arylidene Derivatives as Synthons in Heterocyclic Synthesis. *Open Access Lib. J.* **2014**, *1*, e367. [CrossRef]
8. Albuquerque, H.; Santos, C.; Cavaleiro, J.; Silva, A. Chalcones as Versatile Synthons for the synthesis of 5- and 6-membered Nitrogen Heterocycles. *Curr. Org. Chem.* **2014**, *18*, 2750–2775. [CrossRef]
9. Zhuang, C.; Zhang, W.; Sheng, C.; Zhang, W.; Xing, C.; Miao, Z. Chalcone: A Privileged Structure in Medicinal Chemistry. *Chem. Rev.* **2017**, *117*, 7762–7810. [CrossRef] [PubMed]
10. Ardiansah, B. Chalcones bearing N, O, and S-heterocycles: Recent notes on their biological significances. *J. Appl. Pharm. Sci.* **2019**, *9*, 117–129. [CrossRef]
11. Sharshira, E.; Hamada, N. Synthesis and Antimicrobial Evaluation of Some Pyrazole Derivatives. *Molecules* **2012**, *17*, 4962–4971. [CrossRef] [PubMed]
12. Borge, V.V.; Patil, R.M. Comparative Study on Synthesis and Biological, Pharmaceutical Applications of Aromatic Substituted Chalcones. *Mini. Rev. Org. Chem.* **2023**, *20*, 260–269. [CrossRef]
13. Mhaibes, R.M. Antimicrobial and Antioxidant Activity of Heterocyclic Compounds Derived from New Chalcones. *J. Med. Chem. Sci.* **2023**, *6*, 931–937. [CrossRef]
14. Adnan, D.; Singh, B.; Mehta, S.; Kumar, V.; Kataria, R. Simple and solvent free practical procedure for chalcones: An expeditious, mild and greener approach. *Curr. Res. Green Sustain. Chem.* **2020**, *3*, 100041. [CrossRef]
15. Urbonavîcius, A.; Fortunato, G.; Ambrazaitytė, E.; Plytninkienė, E.; Bieliauskas, A.; Milišiunaitė, V.; Luisi, R.; Arbâciauskienė, E.; Krikštolaitytė, S.; Šâckus, A. Synthesis and Characterization of Novel Heterocyclic Chalcones from 1-Phenyl-1H-pyrazol-3-ol. *Molecules* **2022**, *27*, 3752. [CrossRef]
16. Kitawata, B.; Singha, M.; Kale, R. Solvent Free Synthesis, Characterization, Anticancer, Antibacterial, Antifungal, Antioxidant and SAR Studies of Novel (E)-3-aryl-1-(3-alkyl-2-pyrazinyl)-2-propenone. *New J. Chem.* **2013**, *37*, 2541–2550. [CrossRef]
17. Chen, H.; Noirbent, G.; Liu, S.; Brunel, D.; Graff, B.; Gigmes, D.; Zhang, Y.; Sun, K.; Morlet-Savary, F.; Xiao, P.; et al. Bis-Chalcone Derivatives Derived from Natural Products as Near-UV/Visible Light Sensitive Photoinitiators for 3D/4D Printing. *Mater. Chem. Front.* **2021**, *5*, 901–916. [CrossRef]
18. Mara Silva de Padua, G.; Maria De Souza, J.; Celia Moura Sales, M.; Gomes de Vasconcelos, L.; Luiz Dall'Oglio, E.; Faraggi, T.M.; Moreira Sampaio, O.; Campos Curcino Vieira, L. Evaluation of Chalcone Derivatives as Photosynthesis and Plant Growth Inhibitors. *Chem. Biodivers.* **2021**, *18*, e2100226. [CrossRef]
19. Bukhari, S. Synthesis and evaluation of new chalcones and oximes as anticancer agents. *RSC Adv.* **2022**, *12*, 10307–10320. [CrossRef]
20. Gaber, M.; El-Ghamry, H.A.; Mansour, M.A. Pd(II) and Pt(II) chalcone complexes. Synthesis, spectral characterization, molecular modeling, biomolecular docking, antimicrobial and antitumor activities. *J. Photochem. Photobiol. A Chem.* **2018**, *354*, 163–174. [CrossRef]
21. Baretto, M.; Fuchi, N. Tissue Transglutaminase Inhibitor, Chalcone Derivative, and Pharmaceutical Application Thereof. Japan Patent 2013-180955A, 12 September 2013.
22. Özdemir, A.; Altıntop, M.D.; Turan-Zitouni, G.; Çiftçi, G.A.; Ertorun, I.; Alataş, Ö.; Kaplancıklı, Z.A. Synthesis and evaluation of new indole-based chalcones as potential antiinflammatory agents. *Eur. J. Med. Chem.* **2015**, *89*, 304–309. [CrossRef] [PubMed]
23. Shaik, A.; Bhandare, R.; Palleapati, K.; Nissankararao, S.; Kancharlapalli, V.; Shaik, S. Antimicrobial, Antioxidant, and Anticancer Activities of Some Novel Isoxazole Ring Containing Chalcone and Dihydropyrazole Derivatives. *Molecules* **2020**, *25*, 1047. [CrossRef] [PubMed]

24. Desai, V.; Desai, S.; Gaonkar, S.N.; Palyekar, U.; Joshi, S.D.; Dixit, S.K. Novel quinoxalinyl chalcone hybrid scaffolds as enoyl ACP reductase inhibitors: Synthesis, molecular docking and biological evaluation. *Bioorg. Med. Chem. Lett.* **2017**, *27*, 2174–12180. [CrossRef] [PubMed]
25. Gomes, M.N.; Braga, R.C.; Grzelak, E.M.; Neves, B.J.; Muratov, E.N.; Ma, R.; Klein, L.K.; Cho, S.; Oliveira, G.R.; Franzblau, S.G.; et al. QSAR-driven design, synthesis and discovery of potent and selective chalcone derivatives with antitubercular activity. *Eur. J. Med. Chem.* **2017**, *137*, 126–138. [CrossRef]
26. Hawash, M.; Kahraman, D.C.; Eren, F.; Atalay, R.C.; Baytas, S.N. Synthesis and biological evaluation of novel pyrazolic chalcone derivatives as novel hepatocellular carcinoma therapeutics. *Eur. J. Med. Chem.* **2017**, *129*, 12–26. [CrossRef]
27. Minders, C.; Petzer, J.; Petzer, A.; Lourens, A. Monoamine oxidase inhibitory activities of heterocyclic chalcones. *Bioorg. Med. Chem. Lett.* **2015**, *25*, 5270–5276. [CrossRef]
28. Usta, A.; Öztürk, E.; Beriş, F. Microwave-assisted preparation of azachalcones and their N-alkyl derivatives with antimicrobial activities. *Nat. Prod. Res.* **2014**, *28*, 483–487. [CrossRef]
29. Li, P.; Jiang, H.; Zhang, W.; Li, Y.; Zhao, M.; Zhou, W. Synthesis of carbazole derivatives containing chalcone analogs as non-intercalative topoisomerase II catalytic inhibitors and apoptosis inducers. *Eur. J. Med. Chem.* **2018**, *145*, 498–510. [CrossRef]
30. Bandgar, B.; Adsul, L.; Lonikar, S.; Chavan, H.; Shringare, S.; Patil, S.; Jalde, S.; Koti, B.; Dhole, N.; Gacche, R.; et al. Synthesis of novel carbazole chalcones as radical scavenger, antimicrobial and cancer chemopreventive agents. *J. Enzym. Inhib. Med. Chem.* **2013**, *28*, 593–600. [CrossRef]
31. Sweeting, S.; Hall, C.; Potticary, J.; Pridmore, N.; Warren, S.; Cremeens, M.; D'Ambruoso, G.; Matsumoto, M.; Hall, S. The solubility and stability of heterocyclic chalcones compared with transchalcone. *Acta Cryst. B* **2020**, *B76*, 13–17. [CrossRef]
32. Robinson, T.P.; Hubbard, R.B.; Ehlers, T.J.; Arbiser, J.L.; Goldsmith, D.J.; Bowen, J.P. Synthesis and biological evaluation of aromatic enones related to curcumin. *Bioorg. Med. Chem.* **2005**, *13*, 4007–4013. [CrossRef] [PubMed]
33. Song, T.; Duan, Y.; Yang, Y. Chemoselective transfer hydrogenation of α,β-unsaturated carbonyls catalyzed by a reusable supported Pd nanoparticles on biomass-derived carbon. *Catal. Commun.* **2019**, *120*, 80–85. [CrossRef]
34. Lokeshwari, D.M.; Rekha, N.D.; Srinivasan, B.; Vivek, H.K.; Kariyappa, A.K. Design, synthesis of novel furan appended benzothiazepine derivatives and in vitro biological evaluation as potent VRV-PL-8a and H+/K+ ATPase Inhibitors. *Bioorg. Med. Chem. Lett.* **2017**, *27*, 3048–3054. [CrossRef]
35. Liu, W.; Shi, H.-M.; Jin, H.; Zhao, H.-Y.; Zhou, G.-P.; Wen, F.; Yu, Z.-Y.; Hou, T.-P. Design, Synthesis and Antifungal Activity of a Series of Novel Analogs Based on Diphenyl Ketones. *Chem. Biol. Drug. Des.* **2009**, *73*, 661–667. [CrossRef] [PubMed]
36. Robinson, S.J.; Petzer, J.P.; Petzer, A.; Bergh, J.J.; Lourens, A.C.U. Selected furanochalcones as inhibitors of monoamine oxidase. *Bioorg. Med. Chem. Lett.* **2013**, *23*, 4985–4989. [CrossRef]
37. Parveen, H.; Iqbal, P.F.; Azam, A. Synthesis and Characterization of a New Series of Hydroxy Pyrazolines. *Synth. Commun.* **2008**, *38*, 3973–3983. [CrossRef]
38. Sunduru, N.; Agarwal, A.; Katiyar, S.B.; Goyal, N.; Gupta, S.; Chauhana, P.M.S. Synthesis of 2,4,6-trisubstituted pyrimidine and triazine heterocycles as antileishmanial agents. *Bioorg. Med. Chem.* **2006**, *14*, 7706–7715. [CrossRef]
39. Sinha, S.; Manju, S.; Doble, M. Chalcone-Thiazole Hybrids: Rational Design, Synthesis, and Lead Identification against 5-Lipoxygenase. *Med. Chem. Lett.* **2019**, *10*, 1415–1422. [CrossRef]
40. Zhao, X.; Dong, W.; Gao, Y.; Shin, D.-S.; Ye, Q.; Su, L.; Jiang, F.; Zhao, B.; Miao, J. Novel indolyl-chalcone derivatives inhibit A549 lung cancer cell growth through activating Nrf-2/HO-1 and inducing apoptosis in vitro and in vivo. *Sci. Rep.* **2017**, *7*, 3919. [CrossRef]
41. Hsieh, C.-Y.; Ko, P.-W.; Chang, Y.-J.; Kapoor, M.; Liang, Y.-C.; Chu, H.-L.; Lin, H.-H.; Horng, J.-C.; Hsu, M.-H. Design and Synthesis of Benzimidazole-Chalcone Derivatives as Potential Anticancer Agents. *Molecules* **2019**, *24*, 3259. [CrossRef]
42. Hsieh, C.; Kuiying Xu, K.; Lee, I.; Graham, T.; Tu, Z.; Dhavale, D.; Kotzbauer, P.; Mach, R. Chalcones and Five-Membered Heterocyclic Isosteres Bind to Alpha Synuclein Fibrils in Vitro. *ACS Omega* **2018**, *3*, 4486–4493. [CrossRef] [PubMed]
43. Saito, Y.; Kishimoto, M.; Yoshikawa, Y.; Kawaii, S. Synthesis and Structure–Activity Relationship Studies of Furan-ring Fused Chalcones as Antiproliferative Agents. *Anticancer Res.* **2015**, *35*, 811–818. [PubMed]
44. Grigoropoulou, S.; Manou, D.; Antoniou, A.I.; Tsirogianni, A.; Siciliano, C.; Theocharis, A.D.; Athanassopoulos, C.M. Synthesis and Antiproliferative Activity of Novel Dehydroabietic Acid-Chalcone Hybrids. *Molecules* **2022**, *27*, 3623. [CrossRef] [PubMed]
45. Mubofu, E.B.; Engberts, J.B.F.N. Specific acid catalysis and Lewis acid catalysis of Diels–Alder reactions in aqueous media. *J. Phys. Org. Chem.* **2004**, *17*, 180–186. [CrossRef]
46. Jianga, X.; Jina, H.; Wanga, T.; Yoob, H.; Koo, S. Synthesis of Phenyl-2,2-bichalcophenes and Their Aza-Analogues by Catalytic Oxidative Deacetylation. *Synthesis* **2019**, *51*, 3259–3268. [CrossRef]
47. Ritter, M.; Martins, R.; Rosa, S.; Malavolta, J.; Lund, R.; Flores, A.; Pereira, C. Green Synthesis of Chalcones and Microbiological Evaluation. *J. Braz. Chem. Soc.* **2015**, *26*, 1201–1210. [CrossRef]
48. Khan, S.; Asiri, A. Green Synthesis, Characterization and biological evaluation of novel chalcones.as anti bacterial agents. *Arab. J. Chem.* **2013**, *10*, S2890–S2895. [CrossRef]
49. Sarveswari, S.; Vijayakumar, V. A rapid microwave assisted synthesis of 1-(6-chloro-2-methyl-4-phenylquinolin-3-yl)-3-(aryl)prop-2-en-1-ones and their anti bacterial and anti fungal evaluation. *Arab. J. Chem.* **2016**, *9*, S35–S40. [CrossRef]

50. Polo, E.; Ferrer-Pertuz, K.; Trilleras, J.; Quiroga, J.; Guti'errez, M. Microwave-assisted one-pot synthesis in water of carbonylpyrazolo [3,4-b]pyridine derivatives catalyzed by InCl3 and sonochemical assisted condensation with aldehydes to obtain new chalcone derivatives containing the pyrazolopyridinic moiety. *RSC Adv.* **2017**, *7*, 50044–50050. [CrossRef]
51. Yadav, P.; Yadav, M.; Gaur, R.; Gupta, R.; Arora, G.; Rana, P.; Srivastava, A.; Sharma, R.K. Fabrication of copper-based silica-coated magnetic nanocatalyst for efficient one-pot synthesis of chalcones via A3 coupling of aldehydes-alkynes-amines. *Chem. Cat. Chem.* **2020**, *12*, 2488–2496. [CrossRef]
52. Fu, D.-J.; Zhang, S.-Y.; Liu, Y.-C.; Yue, X.-X.; Liu, J.-J.; Song, J.; Zhao, R.-H.; Li, F.; Sun, H.-H.; Zhang, Y.-B.; et al. Design, synthesis and antiproliferative activity studies of 1,2,3-triazole–chalcones. *Med. Chem. Commun.* **2016**, *7*, 1664–1671. [CrossRef]
53. Gadhave, A.G.; Uphade, B.K. Synthesis of Some Pyrazole Containing Chalcones and Pyridine-3-Carbonitriles and Study of their Anti-inflammatory Activity. *Orient. J. Chem.* **2017**, *33*, 219–225. [CrossRef]
54. Mallik, A.; Dey, S.; Chattopadhyay, F.; Patra, A. Novel formation of 6-acyl-5-(2-pyrrolyl)-3H-pyrrolizines bym base-catalysed condensation of pyrrole-2-aldehyde with methyl ketones. *Tetrahedron Lett.* **2000**, *43*, 1295–1297. [CrossRef]
55. Bindu, P.; Mahadevan, K.; Naik, T.; Harish, B. Synthesis, DNA binding, docking and photocleavage studies of quinolinyl chalcones. *Med. Chem. Commun.* **2014**, *5*, 1708–1717. [CrossRef]
56. Abonia, R.; Gutiérrez, L.; Quiroga, J.; Insuasty, B. (E)-3-[3-(2-Butoxyquinolin-3-yl)acryloyl]-2-hydroxy-4H-chromen-4-one. *Molbank* **2018**, *2018*, M1001. [CrossRef]
57. Sun, M.; Yuyang, W.; Minghua, Y.; Qing, Z.; Yixin, Z.; Yongfang, Y.; Yongtao, D. Angiogenesis, Anti-Tumor, and Anti-Metastatic Activity of Novel α-Substituted Hetero-Aromatic Chalcone Hybrids as Inhibitors of Microtubule Polymerization. *Front. Chem.* **2021**, *9*, 766201. [CrossRef]
58. Ashok, D.; Ravi, S.; Ganesh, A.; Lakshmi, B.; Adam, S.; Murthy, S. Microwave-assisted synthesis and biological evaluation of carbazole-based chalcones, aurones and flavones. *Med. Chem. Res.* **2016**, *25*, 909–922. [CrossRef]
59. Bhatt, K.; Vishal Rana, V.; Patel, N.; Parikh, J.; Pillai, S. Novel Oxygen Fused Bicyclic Derivatives and Antioxidant Labelling: Bioactive Chalcone Based Green Synthesis. *Biointerface Res. Appl. Chem.* **2023**, *13*, 130. [CrossRef]
60. Jadhava, S.; Peerzadeb, N.; Gawalia, R.; Bhosaleb, R.; Kulkarnic, A.; Varpe, B. Green synthesis and biological screening of some fluorinated pyrazole chalcones in search of potent anti-inflammatory and analgesic agents. *Egypt Pharmaceut. J.* **2020**, *19*, 172–181. [CrossRef]
61. Kudlickova, Z.; Stahorsky, M.; Michalkova, R.; Vilkova, M.; Balaz, M. Mechanochemical synthesis of indolyl chalcones with antiproliferative activity. *Green Chem. Lett. Rev.* **2022**, *15*, 2089061. [CrossRef]
62. Nimmala, S.; Chepyala, K.; Balram, B.; Ram, B. Synthesis and antibacterial activity of novel imidazo [1,2-a]pyrimidine and imidazo [1,2-a]pyridine chalcone derivatives. *Der Pharma Chem.* **2012**, *4*, 2408–2415.
63. Joshi, H.; Saglani, M. Rapid and greener ultrasound assisted synthesis of series (2-((substituted 2-Chloroquinolin-3-yl) methylene)-3,4-dihydronaphthalen-1(2h)-one) derivatives and their biological activity. *J. Adv. Sci. Res.* **2020**, *11*, 147–152.
64. Devi, L.; Sharma, G.; Kant, R.; Shukla, S.; Rastogi, N. Regioselective Synthesis of Functionalized Pyrazole-Chalcones via Base Mediated Reaction of Diazo Compounds with Pyrylium Salts. *Org. Biomol. Chem.* **2021**, *19*, 4132–4136. [CrossRef] [PubMed]
65. Tan, P.; Wang, S.R. Reductive (3 + 2) Annulation of Benzils with Pyrylium Salts: Stereoselective Access to Furyl Analogues of Cis-Chalcones. *Org. Lett.* **2019**, *21*, 6029–6033. [CrossRef] [PubMed]
66. Kumar, K.; Siddaiah, V.; Lilakar, J.; Ganesh, A. An efficient continuous-flow synthesis and evaluation of antimicrobial activity of novel 1,2,3-Triazole-Furan hybrid chalcone derivatives. *Chem. Data Collect.* **2020**, *28*, 100457. [CrossRef]
67. Kuamr, D.; Kumar, N.M.; Tantak, M.P.; Ogura, M.; Eriko Kusaka, E.; Ito, T. Synthesis and identification of α-cyano bis(indolyl)chalcones as novel anticancer agents. *Bioorg. Med. Chem. Lett.* **2014**, *24*, 5170–5174. [CrossRef]
68. Alidmat, M.; Khairuddean, M.; Salhimi, S.; Al-Amin, M. Docking studies, synthesis, characterization, and cytotoxicity activity of new bis-chalcone derivatives. *Biomed. Res. Ther.* **2021**, *8*, 4294–4306. [CrossRef]
69. Asiri, A.; Marwani, H.; Alamry, K.; Al-Amoudi, M.; Khan, S.; El-Daly, S. Green Synthesis, Characterization, Photophysical and Electrochemical Properties of Bis-chalcones. *Int. J. Electrochem. Sci.* **2014**, *9*, 799–809.
70. Abdel-Aziz, H.A.; Al-Rashood, K.A.; ElTahir, K.E.H.; Ibrahim, H.S. Microwave-assisted Synthesis of Novel 3,4-Bis-chalcone-N-arylpyrazoles and Their Anti-inflammatory Activity. *J. Chin. Chem. Soc.* **2011**, *58*, 863–868. [CrossRef]
71. Saikachi, H.; Muto, H. Preparation of Resonance-Stabilized Bisphophoranes. *Chem. Pharm. Bull.* **1971**, *19*, 2262–2270. [CrossRef]

Disclaimer/Publisher's Note: The statements, opinions and data contained in all publications are solely those of the individual author(s) and contributor(s) and not of MDPI and/or the editor(s). MDPI and/or the editor(s) disclaim responsibility for any injury to people or property resulting from any ideas, methods, instructions or products referred to in the content.

MDPI
St. Alban-Anlage 66
4052 Basel
Switzerland
Tel. +41 61 683 77 34
Fax +41 61 302 89 18
www.mdpi.com

Molecules Editorial Office
E-mail: molecules@mdpi.com
www.mdpi.com/journal/molecules